# Molecular Embryology

# METHODS IN MOLECULAR BIOLOGY™

## John M. Walker, SERIES EDITOR

METHODS IN MOLECULAR BIOLOGY™

# Molecular Embryology

## *Methods and Protocols*

Edited by

## Paul T. Sharpe

and

## Ivor Mason

*Dental and Medical Schools of Guy's, King's,
and St. Thomas's Hospitals, King's College, London, UK*

**Humana Press** ✳ Totowa, New Jersey

© 1999 Humana Press Inc.
999 Riverview Drive, Suite 208
Totowa, New Jersey 07512

This publication is printed on acid-free paper. ∞
ANSI Z39.48-1984 (American Standards Institute)
Permanence of Paper for Printed Library Materials.

Cover illustration: Figure 7A from Chapter 2, "Culture of Postimplantation Mouse Embryos," by Paul Martin and David L. Cockroft.

Cover design by Jill Nogrady.

For additional copies, pricing for bulk purchases, and/or information about other Humana titles, contact Humana at the above address or at any of the following numbers: Tel.: 973-256-1699; Fax: 973-256-8341; E-mail: humana@humanapr.com; or visit our Website: http://humanapress.com

Printed in the United States of America. 10 9 8 7 6 5 4 3 2 1

Library of Congress Cataloging in Publication Data

Main entry under title:

Methods in molecular biology™.

Molecular embryology : methods and protocols / edited by Paul T. Sharpe and Ivor Mason.
       p.  cm. -- (Methods in molecular biology™ ; 97)
   Includes index.
   ISBN 0-89603-387-2 (alk. paper)
   1. Embryology--Vertebrates--Methodology. 2. Chemical embryology--Methodology I. Sharpe, Paul T. II. Mason, I. III. Series: Methods in molecular biology (Totowa, NJ) ; 97.
   QL959.M65    1998
   571.8'616--dc21
                                     98-23234
                                     CIP

# Preface

Most people have some interest in embryos; this probably results, in part, from their interest in understanding the biological origins of themselves and their offspring and, increasingly, concerns about how environmental change such as pollution might affect human development. Obviously, ethical considerations preclude experimental studies of human embryos and, consequently, the developmental biologist has turned to other species to examine this process. Fortunately, the most significant conclusion to be drawn from the experimental embryology of the last two decades is the manner in which orthologous or closely related molecules are deployed to mediate similar developmental processes in both vertebrates and invertebrates. The molecular mechanisms regulating processes fundamental to most animals, such as axial patterning or axon guidance, are frequently conserved during evolution. (It is now widely believed that the differences between phyla and classes are the result of new genes, arising mostly by duplication and divergence of extant sequences, regulating the appearance of derived characters.)

Other vertebrates are obviously most likely to use the same developmental mechanisms as humans and, within the vertebrate subphylum, the apparent degree of conservation of developmental mechanism is considerable. It has long been recognized that particular vertebrate species offer either distinct advantages in investigating particular stages of development or are especially amenable to particular manipulations. No single animal can provide all the answers because not all types of experiments can be carried out on a single species. Traditionally, developmental biologists have worked on their particular experimental favorite, working, for example, solely on *Drosophila,* or *Xenopus,* or the mouse. In the last few years, this has started to change and there are now increasing numbers of laboratories that have acquired the expertise to work on several different animals and are thus able to harness the experimental advantages of different developmental systems to address specific developmental questions. Alternatively, Developmental Biology Departments are becoming organized such that they have expertise in several model organisms. It is the increasing necessity to be able to move between embryos of different vertebrate classes as a project progresses that prompted us to as-

*v*

semble *Molecular Embryology: Methods and Protocols.* We hope that it will allow researchers to familiarize themselves with the various commonly studied vertebrate embryos, to make informed choices about which might be best suited to their investigations, and to understand the techniques by which they might be manipulated.

Sadly, while this book was going to press, Nigel Holder, one of its contributors, died. Nigel was an excellent developmental biologist, a founder of the Developmental Biology Research Group at King's College, and had recently been appointed to the Chair of Anatomy and Human Biology at University College London. He was both a colleague and friend to ourselves and to many of the other contributors to this volume. He is greatly missed.

*Paul T. Sharpe*
*Ivor Mason*

# List of Color Plates

Color plates 1–12 appear as an insert following p. 368.

# Contents

# Contributors

ENRIQUE AMAYA • *Wellcome, CRC Institute, Cambridge, UK*

JUAN CARLOS IZPÍSUA BELMONTE • *Gene Expression Laboratory, The Salk Institute for Biological Sciences, La Jolla, CA*

BRUCE BLUMBERG • *Gene Expression Laboratory, The Salk Institute for Biological Sciences, La Jolla, CA*

VICTORIA BOSTOCK • *School of Biological Sciences, University of Manchester, UK*

GERARD BRADY • *School of Biological Sciences, University of Manchester, UK*

JANE BRENNAN • *BBSRC Centre for Genome Research, University of Edinburgh, Scotland, UK*

CLARE BRUNET • *School of Biological Sciences, University of Manchester, UK*

JONATHAN D. W. CLARKE • *Department of Anatomy and Developmental Biology, University College, London*

DAVID L. COCKROFT • *Imperial Cancer Research Fund, Department of Zoology, University of Oxford, UK*

JONATHAN COOKE • *National Institute for Medical Research, London, UK*

PETER D. CURRIE • *Molecular Embryology Unit, Imperial Cancer Research Fund, London, UK*

KIM DALE • *Developmental Genetics Programme, The Krebs Institute, University of Sheffield, UK*

CLIVE DICKSON • *Imperial Cancer Research Fund, London, UK*

SUSANNE DIETRICH • *Department of Craniofacial Development, Guy's Campus, King's College, London, UK*

TAMIRA ELUL • *Graduate Group in Biophysics, University of California, Berkeley, CA*

PHILIPPA FRANCIS-WEST • *Department of Craniofacial Development, Guy's Campus, King's College, London, UK*

JONATHAN D. GILTHORPE • *MRC Brain Development Programme, Department of Developmental Neurobiology, Guy's Campus, King's College, London, UK*

MONICA GORDON • *Embryology Unit, Children's Medical Research Institute, Wentworthville, Australia*

ANTHONY GRAHAM • *Department of Experimental Pathology, Guy's Campus, King's College, London, UK*

SARAH GUTHRIE • *MRC Brain Development Programme, Department of Developmental Neurobiology, Guy's Campus, King's College, London, UK*

ANNA-KATERINA HADJANTONAKIS • *Samuel Lunenfeld Research Institute, Mount Sinai Hospital, Toronto, Canada*

BRIGID L. M. HOGAN • *Department of Cell Biology, Howard Hughes Medical Institute, Vanderbilt University Medical School, Nashville, TN*

NIGEL HOLDER • *Department of Anatomy and Developmental Biology, University College London, UK (deceased)*

PETER W. H. HOLLAND • *School of Animal and Microbial Sciences, The University of Reading, Whitenights, UK*

AMATA HORNBRUCH • *MRC Brain Development Programme, Department of Developmental Neurobiology, Guy's Campus, King's College, London, UK*

PHILIP W. INGHAM • *Developmental Genetics Programme, The Krebs Institute, University of Sheffield, UK*

ALISON ISAAC • *National Institute for Medical Research, London, UK*

DAVID I. JACKSON • *Imperial Cancer Research Fund, London, UK*

C. MICHAEL JONES • *Chester Beatty Laboratories, London, UK*

TREVOR JOWETT • *Department of Biochemistry and Genetics, University of Newcastle Upon Tyne Medical School, Newcastle Upon Tyne, UK*

RAY KELLER • *Department of Biology, University of Virginia, Charlottsville, VA*

ANDREW KENT • *Division of Anatomy, Cell, and Human Biology, Guy's Campus, King's College, London, UK*

KRISTEN L. KROLL • *Wellcome, CRC Institute, Cambridge, UK*

PATRICIA A. LABOSKY • *Department of Cell Biology, Howard Hughes Medical Institute, Vanderbilt University Medical School, Nashville, TN*

NICOLE M. LE DOUARIN • *Centre National de la Recherche Scientifique, Instituit d'Embryologie Cellulaire et Moleculaire, College de France, Nogent-sur-Marne, France*

CAIRINE LOGAN • *Neuroscience Research Group, Department of Anatomy and Cell Biology, University of Calgary, Canada*

ANDREW LUMSDEN • *MRC Brain Development Programme, Department of Developmental Neurobiology, Guy's Campus, King's College, London, UK*

MARLA B. LUSKIN • *Department of Anatomy and Cell Biology, Emory University School of Medicine, Atlanta, GA*

MALCOLM MADEN • *Developmental Biology Research Centre, King's College London, UK*

RADMA MAHMOOD • *MRC Brain Development Programme, Department of Developmental Neurobiology, Guy's Campus, King's College, London, UK*

PAUL MARTIN • *Department of Anatomy and Developmental Biology, University College London, UK*

IVOR MASON • *MRC Brain Development Programme, Department of Developmental Neurobiology, Guy's Campus, King's College, London, UK*

GILLIAN M. MORRISS-KAY • *Department of Human Anatomy, Oxford, UK*

DAVID MURPHY • *Department of Medicine, University of Bristol, UK*

ANDRÁS NAGY • *Samuel Lunenfeld Research Institute, Mount Sinai Hospital, Toronto, Canada*

MARK NOBLE • *Huntsman Cancer Institute, University of Utah Health Sciences Center, Salt Lake City, UT*

CÉSAR NÚÑEZ • *School of Biological Sciences, University of Manchester, UK*

MAALA PARAMESWARAN • *Embryology Unit, Children's Medical Research Institute, Wentworthville, Australia*

GRACE K. PAVLATH • *Department of Anatomy and Cell Biology, Emory University School of Medicine, Atlanta, GA*

MELINDA PIRITY • *Samuel Lunenfeld Research Institute, Mount Sinai Hospital, Toronto, Canada*

MARYSIA PLACZEK • *Developmental Genetics Programme, The Krebs Institute, University of Sheffield, UK*

ANN POZNANSKI • *Department of Biology, University of Virginia, Charlottsville, VA*

GABRIEL A. QUINLAN • *Embryology Unit, Children's Medical Research Institute, Wentworthville, Australia*

MARIA REX • *Department of Genetics, University of Notingham, UK*

PETER W. J. RIGBY • *Chester Beatty Laboratories, London, UK*

CARIN SAHLBERG • *Biocentre, University of Helsinki, Finland*

THOMAS F. SCHILLING • *Department of Anatomy and Developmental Biology, University College, London, UK*

PAUL J. SCOTTING • *Department of Genetics, University of Notingham, UK*

HUMA SHAMIM • *MRC Brain Development Programme, Department of Developmental Neurobiology, Guy's Campus, King's College, London, UK*

PAUL T. SHARPE • *Department of Craniofacial Development, Guy's Campus, King's College, London, UK*

SAN LING SI-HOE • *Department of Medicine, University of Bristol, UK*

WILLIAM C. SKARNES • *BBSRC Centre for Genome Research, University of Edinburgh, Scotland, UK*

JAMES C. SMITH • *Laboratory of Developmental Biology, National Institute for Medical Research, London, UK*

CLAUDIO D. STERN • *Department of Genetics and Development, College of Physicians and Surgeons of Columbia University, New York, NY*

KARIN STURM • *Embryology Unit, Children's Medical Research Institute, Wentworthville, Australia*

PATRICK P. L. TAM • *Embryology Unit, Children's Medical Research Institute, Wentworthville, Australia*

MARIE-AIMÉE TEILLET • *Centre National de la Recherche Scientifique, Instituit d'Embryologie Cellulaire et Moleculaire, College de France, Nogent-sur-Marne, France*

IRMA THESLEFF • *Biocentre, University of Helsinki, Finland*

CHERYL TICKLE • *The Wellcome Trust Building, University of Dundee, UK*

PAUL A. TRAINOR • *National Institute for Medical Research, London, UK*

ANTONIO TUGORES • *Gene Expression Laboratory, The Salk Institute for Biological Sciences, La Jolla, CA*

HIROSHI WADA • *School of Animal and Microbial Sciences, The University of Reading, UK*

DAMIAN L. WEAVER • *School of Biological Sciences, University of Manchester, UK*

RICHARD J. T. WINGATE • *MRC Brain Development Programme, Department of Developmental Neurobiology, Guy's Campus, King's College, London, UK*

QILING XU • *National Institute for Medical Research, London, UK*

SHEILA X. ZHOU • *Embryology Unit, Children's Medical Research Institute, Wentworthville, Australia*

CATHERINE ZILLER • *Centre National de la Recherche Scientifique, Instituit d'Embryologie Cellulaire et Moleculaire, College de France, Nogent-sur-Marne, France*

# I

# THE MOUSE EMBRYO

# 1

## The Mouse as a Developmental Model

### Paul T. Sharpe

The laboratory mouse, *Mus musculus*, is the developmental biologist's mammal of choice for studies of development. Its embryology and genetics have been extensively studied for over 100 years. However, it is the advent of in vivo gene manipulation in the last few years that has established the mouse as probably the single most powerful animal system in vertebrate biology.

Mouse developmental biology, as it exists today, has its origins in genetics and embryology. These days it is hard to separate mouse development from mouse genetics, the two having become intertwined as genetics provides increasingly more powerful tools for studying development.

By the early 1980s, progress in understanding mouse embryology had started to stagnate. Molecular biology had provided the tools to clone genes, but identification of important genes in mouse development from the estimated 80,000–100,000 genes in the genome seemed almost intractable. One breakthrough that was to change the course of developmental studies came in 1984 with the discovery of the homeobox in flies, and the realization that mice and other vertebrates had the same genes *(1–3)*. It seems hard to believe now, when evolutionary conservation of genes and functions between in vertebrates and mammals is the norm, that in the 1980s many developmental biologists believed there would be no such homologies, and even if certain gene (protein) sequences, such as the homeobox, were conserved, it was believed they would have completely different functions in mammals. Evolution of developmental mechanisms has provided developmental biologists with their most powerful tool: conservation of gene function.

The ability to clone potentially important developmental genes by screening mouse libraries with *Drosophila* gene probes, together with the advent of *in situ* hybridization to study their spatial expression in embryos, provided the

From: *Methods in Molecular Biology, Vol. 97: Molecular Embryology: Methods and Protocols*
Edited by: P. T. Sharpe and I. Mason © Humana Press Inc., Totowa, NJ

impetus for the explosion in mouse developmental studies over the last 15 years. Armed with potentially interesting developmental genes, mouse embryologists were able to begin to utilize transgenesis to investigate their functions and regulation. The first transgenic mice were produced using pronuclear injection in 1980/1981 in the laboratories of Frank Ruddle at Yale *(4)* and Frank Costantini and Elizabeth Lacy in Oxford *(5)* and the methodology they described is now widely used. The limitations of studying developmental gene function using transgenic animals produced by pronuclear injection soon became apparent, since it allowed only gain-of-function gene manipulations, which were not always informative. Concurrent with the progress in identifying important regulatory genes in mouse development, groups working on pluripotentiality of mouse embryo cells produced the first embryonic stem cells (ES cells) *(6,7)*. The isolation of ES cells and the subsequent development of gene targeting thus came at the perfect time. Mouse developmental biologists had the genes and their expression patterns, but could only surmise the functions. Gene knockout provided the missing tool in the bag, enabling gene function to be assayed directly in vivo. Although perhaps not fully appreciated at the time, these two strands were to come together in the most dramatic way to provide the basis for the understanding mouse development and to start to approach that of *Drosophila*. ES cells also provided a way of identifying important developmental genes based on function rather than homology by gene trapping, and large-scale gene traps have been undertaken in several laboratories (Chapter 8).

The prospects of using large-scale mutagenesis to identify mouse developmental genes, as used in *Drosophila* and more recently zebrafish, was not considered viable for many years. Although mouse developmental mutants, generated by traditional mutagenesis methods, have over the years proved a valuable resource, the advent of targeted mutation techniques greatly reduced the need for traditionally generated mutants. However, more recently, the possibility of large-scale mutagenesis screens for developmental genes has been revisited by several groups using *N*-ethyl-*N*-nitrosourea (ENU) to generate point mutations. Such screens, in conjunction with the mouse genome mapping and sequencing projects, could provide an important contribution to mouse developmental biology in the future.

Basic development of mouse embryos have been more than adequately described in several landmark texts, such as *The Mouse* by Roberts Rugh *(8)*, *The House Mouse* by Karl Theiler *(9)*, and *The Atlas of Mouse Development* by M. H. Kaufman *(10)*, and therefore we did not consider it necessary for this to be duplicated in this text. The chapters in this section on the mouse as a developmental model provide the background and detail to some of the latest manipulation techniques. We recommend that readers consult *Manipulating the Mouse Embryo* by Hogan et al. *(11)*, which provides considerable detail on the production of transgenic mice.

## References

1. McGinnis, W., Levine, M. S., Hafen, E., Kuroiwa, A., and Gehring, W. J. (1984) A conserved DNA sequence in homeotic genes of the *Drosophila Antennapedia* and bithorax complexes. *Nature* **308,** 428–433.
2. Scott, M. P. and Weiner, A. J. (1984) Structural relationships among genes that control development: sequence homology between *Antennapedia, Ultrabithorax* and *fushi tarazu* loci of *Drosophila. Proc. Natl. Acad. Sci. USA* **81,** 4115–4119.
3. Carrasco, A. E., McGinnis, W., Gehring, W. J., and De Robertis, E. M. (1984) Cloning of a Xenopus leavis gene expressed during early embryogenesis that codes for a peptide region homologous to the Drosophila homeotic genes. *Cell* **37,** 409–414.
4. Gordon, J. W., Scangos, G. A., Plotkin, D. J., Barbosa, J. A., and Ruddle, F. H. (1980) Genetic transformation of mouse embryos by microinjection of purified DNA. *Proc. Natl. Acad. Sci. USA* **77,** 7380–7384.
5. Costantini, F. and Lacy, E. (1981) Introduction of a rabbit β-globin gene into the mouse germ line. *Nature* **294,** 92–94.
6. Evans, M. J. and Kaufman, M. H. (1981) Establishment in culture of pluripotential cells from mouse embryos. *Nature* **292,** 154–156.
7. Martin, G. (1981) Isolation of a pluripotent cell-line from early mouse embryos cultured in medium conditioned by teratocarcinoma stem cells. *Proc. Natl. Acad. Sci. USA* **78,** 7634–7638.
8. Rugh, R. (1990) *The Mouse: Its Reproduction and Development.* Oxford Science Publications, Oxford, UK.
9. Theiler, K. (1989) *The House Mouse: Atlas of Embryonic Development.* Springer-Verlag, New York.
10. Kaufman, M. H. (1992) *The Atlas of Mouse Development.* Academic, London, UK.
11. Hogan, B., Beddington, R., Costantini, F., and Lacy, E. (1994) *Manipulating the Mouse Embryo. A Laboratory Manual.* Cold Spring Harbor Laboratory Press, Cold Spring Harbor, NY.

# 2

# Culture of Postimplantation Mouse Embryos

## Paul Martin and David L. Cockroft

## 1. Introduction

A major disadvantage of working with postimplantation mammalian embryos is their relative inaccessibility to experimentation while they develop within the maternal uterus. Two techniques allow us to get around this problem to a large extent. The first, which is the subject of this chapter, can be used for mouse embryos explanted between 7.5 d of gestation (E7) and 12.5 d of gestation (E12), and involves dissecting embryos from the uterus and culturing them in roller bottles *(1)*. In this way, embryos can be surgically or chemically manipulated or labeled and will develop quite normally in culture for periods of 12–60 h, depending on the stage at explantation *(2,3)*. The second technique, which allows experimentation on more advanced stages, is that of *exo utero* or open uterus surgery, in which fetuses are suspended in the fluid-filled abdominal cavity of the female mouse while retaining their placental attachment to the uterine wall *(4)*. This procedure is suitable for fetuses at 12.5 d of gestation (E12) and beyond. This chapter will focus on the techniques for culturing E11 mouse embryos with open yolk sacs at limb-bud stages *(5)* and will also include protocols for culturing earlier stage embryos. We will describe studies in the mouse, but similar manipulations are possible with rat embryos, bearing in mind that they are generally 1 or 2 d behind mouse development; for example, an E11 rat embryo closely resembles a mouse embryo between E9 and E10.

## 2. Materials

1. Microscope: A good-quality dissecting microscope is essential, preferably with both transmitted (bright- and dark-field) and incident (e.g., fiber-optic) illumination—we use Wild microscopes with an overall magnification range of 6–50×.
2. Instruments: **Table 1** shows the tools you will need and what they should be used for.

From: *Methods in Molecular Biology, Vol. 97: Molecular Embryology: Methods and Protocols*
Edited by: P. T. Sharpe and I. Mason © Humana Press Inc., Totowa, NJ

**Table 1**
**Tools for the Job**

| Items | Use | Remarks |
|---|---|---|
| 1 Pair coarse scissors, 1 pair coarse forceps | Opening the abdominal cavity, when explanting embryos or bleeding rats | Can also be used for removing uterus |
| 1 Pair fine scissors | Removing uterus | |
| 2 Pairs fine serrated, or watchmaker forceps | Removing uterus, opening uterus, exposing aorta during bleeding | |
| 2 Pairs watchmaker's forceps (no. 5), carefully ground to a tip diameter of 0.05–0.1 mm (*see* **Note 1**) | Removing decidua of E7–E9 embryos, opening Reichert's membrane | Do not use for opening uterus and so on, since they are too delicate and easily damaged by such heavy work |
| Iridectomy scissors (large) | Removing decidua/Reichert's membrane of E10–E11 embryos | Can use watchmaker's forceps instead |
| Iridectomy scissors (small) | Opening yolk sac of E10–Ell embryos | |
| Hemostat (Spencer-Wells forceps) | Holding colon out of the way during bleeding | |

3. Dishes: 5-cm plastic Petri dishes (Sterilin) are suitable for all the operations of explanting embryos at all ages described here. You will need 3–6/litter (more for older embryos); always transfer embryos to a fresh dish if the medium becomes cloudy with blood, and so forth, and ensure that dishes contain enough medium to submerge embryos completely.

4. Saline recipes—*see* **Table 2**—we use a 1:1 mix of Tyrode's and Earle's salts (=explanting saline) for explanting open yolk sac embryos. To this, we add extra bicarbonate and glucose (=culture saline) when it is used in the culture medium, in order to equilibrate it with the $CO_2$ in the gas mixture used during culture and to increase the energy sources.

Embryos cultured with closed yolk sacs are explanted in PB1, and cultured in pure rat serum. The pH of all media should be about 7.2.

5. Rat serum: This is the basic culture medium for all postimplantation embryo culture, and is used undiluted for embryos cultured with closed yolk sacs (E7–E10 mouse) (*6*), or diluted to 25% for culture with open yolk sacs (E10–E11 mouse) (*5*). Rat serum is available commercially, at a price, but because of the method of preparation, it is inferior, particularly for the culture of the earlier stages (*7,8*), to what you can prepare yourself (*see* **Subheading 3.1.3.**).

**Table 2**
**Media**

|  | Explanting saline, g/L | Culture saline, g/L | PB1 medium, g/L |
|---|---|---|---|
| NaCl | 6.9 | 6.9 | 8.0 |
| KCl | 0.3 | 0.3 | 0.2 |
| MgSO$_4$·7H$_2$O | 0.1 | 0.1 | — |
| MgCl$_2$·6H$_2$O | 0.05 | 0.05 | 0.1 |
| NaH$_2$PO$_4$·2H$_2$O | 0.1 | 0.1 | — |
| Na$_2$HPO$_4$·12H$_2$O | — | — | 2.88 |
| KH$_2$PO$_4$ | — | — | 0.2 |
| CaCl$_2$·2H$_2$O | 0.265 | 0.265 | 0.13 |
| Glucose | 1.5 | 2.0 | 1.0 |
| Na pyruvate | — | — | 0.036 |
| NaHCO$_3$ | 0.5 | 2.0 | — |
| Fetal calf serum | — | — | 10% v/v |

6. Sterility: Provided all dishes, instruments, and media used are sterile, we find that explantations and experimental manipulations can be performed on the open bench without problems of infection.

## 3. Methods

### 3.1. General

#### 3.1.1. Obtaining Timed Pregnant Animals

The best way to obtain embryos at precisely the stage of development needed is to oversee the mouse mating procedure yourself. If only small numbers of litters are required (1–10 litters/wk), then a colony of 5 studs, with virgin females bought in as needed (say 20 females/2 wk) is quite satisfactory. For larger numbers of litters, it is probably more economical to set up your own breeding colony, with weaned newborns grown up to replace the female stock. We usually mate female mice at 8–12 wk old, and retire our breeding studs before they are a year old. With albino varieties of mice, it is relatively easy to detect the one day in the cycle of 4 or 5 d that females are in estrus, because the vagina appears pink and swollen by comparison with nonestrus females. In nonalbino strains it is harder to distinguish estrus, and if this cannot be judged, then three or four times as many breeding pairs should be set up to make allowance for the nonestrus females. We introduce females individually to studs late in the afternoon, and check for vaginal plugs the following morning. The hard whitish plug occluding the vagina is usually very obvious, but occasionally it is located deep, and this can be checked using a round-ended probe. If you have too few studs for paired matings, then it is possible to put up to three females with one stud, but plugging efficiency and litter number will sometimes be compromised. Generally, we expect 30–80% of our females judged to be in estrus to become pregnant. Depending on strain of mouse, the litters will contain from 5–15 embryos.

### 3.1.2. Killing Pregnant Animals

In the United Kingdom there are "Home Office" (HO) regulations stipulating methods of killing animals. The most satisfactory HO Schedule 1 method for killing a pregnant mouse is by cervical dislocation. This is done humanely by placing the fingers of one hand behind the animal's neck and stretching it by pulling the mouse's tail with your other hand. Alternative methods of killing include overdoses of anesthetic, but this may have adverse effects on the later health of the embryos.

### 3.1.3. Making Rat Serum

Rats (preferably fasted overnight) should be anesthetized with halothane or (if permitted) ether. Lay the anesthetized animal on its back, douse it with 70% ethanol, and open the abdominal cavity. Displace the small intestines to the left, pull the colon to the right, and hold it there with a hemostat (Spencer-Wells forceps). Expose the dorsal aorta with fine forceps (it is generally smaller and paler than the adjacent vein), and insert, bevel down, a 19- or 21-gage needle connected to a 10- or 20-mL syringe, with all air excluded. Gently withdraw blood until the rat stops breathing, after which only a further milliliter or two can be obtained. Remove the needle and transfer the blood to a 12- or 15-mL centrifuge tube, running it gently down the side of the tube. Immediately spin the tube for at least 5 min at 2000$g$. This will precipitate the blood cells, and a whitish clot will form in the supernatant serum. When all rats have been bled (20–30 is a reasonable number for a single session), squeeze the clots with long slender forceps to expel serum, spin again as above, and decant the serum into 50-ml centrifuge tubes with a pipet. Spin again to bring down any persisting blood cells, and then pool the day's serum in a suitable container (e.g., a 250-mL tissue-culture flask). Add antibiotics if desired (e.g., 100 µg/mL Streptomycin, 100 IU/mL penicillin), and aliquot the serum in quantities of 5–20 mL before freezing. It will keep for several months at –20°C, or years at –70°C (freeze first at –20°C, and then transfer to –70°C; otherwise tubes may crack). *See* **refs. 3** and **9** for further information on rat bleeding.

### 3.1.4. Preparing Rat Serum for Culture

Thaw the serum at room temperature or in a 37°C water bath, then heat-inactivate it for 30 min at 56°C. Gassing the serum while hot will help drive off any persisting anesthetic dissolved in it. A residual clot often appears at this stage, and can be removed by centrifugation or, more effectively, by filtration through a 0.45-μm syringe filter (Sigma). Serum is then ready for use, either pure or diluted, according to embryonic age.

## 3.2. Explanting E11 Embryos (Cultured with Opened Yolk Sacs)

1. Although the procedure for explanting with open yolk sacs described and illustrated here is for E11 embryos, it is equally applicable to E10 embryos. The freshly killed pregnant animal is laid on its back and the abdomen doused with 70% alcohol. The abdominal skin can be pinched between thumb and forefinger of both hands, and the skin torn back towards head and tail to reveal a clean peritoneal surface, which can then be opened with scissors and forceps. This method avoids contaminating the abdominal cavity with hair. Alternatively, all abdominal layers can be opened at once with a large U-shaped scissor incision, with the two ends of the U at the hindlimbs.

2. The uterus with embryos is then lifted clear of the abdominal cavity. It can be held with forceps midway along one horn at a site between two embryos and needs to be severed with scissors at the tip of that horn (near the ovary), then where it communicates with the cervix (without separating the two horns), and finally at the tip of the second horn. As this is being done, the uterus can also be trimmed free of mesentery and fat.

3. The whole uterus should then be rinsed in PBS and remaining mesentery cleared away before transfer to a fresh dish of PBS. The uterus can now be opened to expose the embryos. This is done by carefully tearing along the antimesometrial wall of the uterus with fine forceps (**Fig. 1**). It is easy to damage embryos at this stage, and the best way to avoid this is to keep the two pairs of forceps close to one another, so that as they tug apart, all the effort goes into tearing the uterus and not squashing the embryos. Once the uterus is open, the embryos (conceptuses) appear like peas in a pod attached to the uterus only in their placental regions. The easiest way to free this attachment is to hold the uterus tightly with one pair of forceps, slip the other pair of forceps on either side of the uterine sheet, and then drag them across each placental attachment in turn, gently teasing the embryos free (**Fig. 2**).

4. The individual conceptuses can now be transferred by pipet to a fresh dish of explanting saline, and their decidua removed. The decidua of an E11 embryo is thin and is best peeled off, beginning on the side opposite the placenta, by shallow pinching with two pairs of fine forceps, which allows a gentle tearing action. Often the very thin and transparent Reichert's membrane, which sits beneath the decidua and clings to the yolk sac, will rupture during this operation, in which case it is easy to remove it with the decidua. When the decidua has been torn back to about the level of the placenta, it can be tidied up by trimming off with a pair of iridectomy scissors (**Fig. 3**). If Reichert's membrane is still intact, it must be torn open and trimmed back to the placenta also.

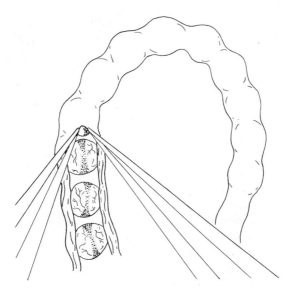

Fig. 1. The uterus (Ell) is gently torn open along its antimesometrial wall using fine forceps to expose the conceptuses.

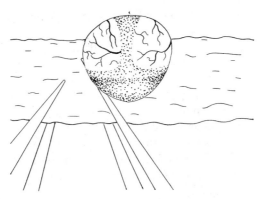

Fig. 2. Conceptuses are scraped free of their placental attachment to the uterus by drawing a fine pair of forceps between conceptus and uterus.

5. Next, the yolk sac must be cut with fine iridectomy scissors, close to where it abuts the placenta, taking care to avoid damaging any of the larger yolk sac vessels (**Fig. 4**). First use two pairs of watchmaker's forceps to make a small hole in the yolk sac adjacent to the placenta, to allow access for the iridectomy scissors. The yolk sac should not be cut completely free of the placenta—rather cut about 4/5 of the way around, creating enough of an opening for the embryo to be pulled out of the yolk sac head first. The region

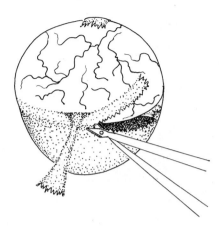

Fig. 3. The decidua and thin Reichert's membrane overlying the E11 yolk sac are trimmed back level with the placenta.

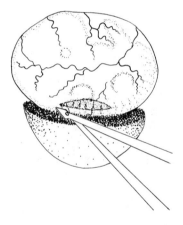

Fig. 4. A small hole is made in the yolk sac close to where it abuts the placenta and adjacent to the forelimb bud. Iridectomy scissors are then used to cut around 4/5 of the base of the yolk sac.

of yolk sac left uncut should be that adjacent to the tail of the embryo—otherwise delivery is made significantly more difficult. The embryo is drawn out of the yolk sac by pulling on the amniotic membrane overlying the embryo's head with one pair of forceps, while holding the mouth of the yolk-sac incision with a second pair of forceps (**Fig. 5**). After the head is outside the yolk sac, it is necessary to rupture the amniotic membrane in order to exteriorize the rest of the embryo. Last of all, the yolk sac and amniotic membranes are flipped under the tail, and the embryo is now available for manipulations or immediate culture.

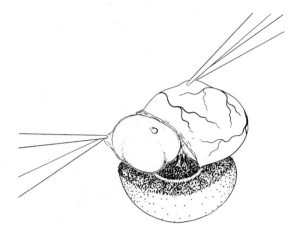

Fig. 5. The E11 embryo is delivered from the yolk sac by tugging on the amniotic membrane overlying the embryo's head.

6. If the developmental stage of the embryos is critical for your experiment, you should stage them now (*see* **Note 2**).

7. All of the above procedures and any planned subsequent manipulations (*see* **Note 5**) should take no more than about 2 h, or subsequent development in culture will be compromised. This generally imposes a limit of no more than 2 litters of embryos/culture session. When embryos are ready for culture, they are individually transferred into 50-mL Falcon tubes containing 5 mL of 25% (v/v) rat serum in culture saline (*see* **Table 2** for recipes). To ensure a gaseous seal, we apply a thin coat of silicone vacuum grease (Dow-Corning) to the rim of the tube. The tube is then gassed with 95% $O_2$, 5% $CO_2$ (for 1 min at a gas flow rate sufficient to ruffle the surface of the medium gently—**Fig. 6**). The Falcon lid is then screwed tight and the tube is placed in a 37°C incubator containing rollers-rotating at 30 rpm. We find that the custom-made BTC roller-incubators (*see* **Note 9**) are excellent for this purpose. By carefully stacking Falcon tubes in the incubator, it is possible to culture up to 18 tubes (embryos) at one time.

8. The health of embryos can be determined at any stage during culture by brief removal of the Falcon tube from the incubator and viewing under a dissecting microscope with transmitted light. Because the walls of the tube are translucent, it is easy to check how well the heart is beating, and usually whether there is good circulation in the larger vessels of the yolk sac and over the brain. Also, check whether there is any swelling of the pericardium or blistering of distal limb ectoderm (usually signs that the embryo is faring badly). In any case, embryos should be regassed every 12 h or so. Only badly damaged embryos will fail to culture successfully for 12 h and, of these, 90% will successfully make it through to 24 h: From this stage on, survival rates get steadily worse with only 1 in 3 embryos

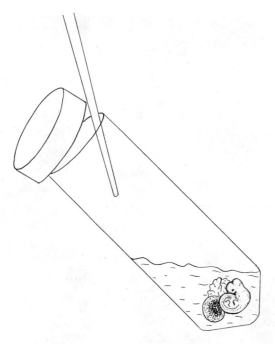

Fig. 6. Embryos are transferred to a roller bottle containing culture medium and gassed with 95% oxygen before culture begins.

healthy at 36 h and 1 in 6 or 7 making it through to 48 h. **Figure 7A** shows the appearance of E11 embryos before and after culture for 24 h.

## 3.3. Explanting E9 Embryos (Cultured with Closed Yolk Sacs)

1. Embryos younger than E10 at explantation are cultured with the visceral yolk sac intact. The initial stages of killing the mouse, opening the uterus, and separating the conceptuses are the same as for the E11 embryos, though of course the conceptuses are smaller.
2. Removal of the decidua of E9 embryos is similar to the E11 procedure, since it already forms a relatively thin layer over the conceptus. It is advisable to remove all of the decidua, starting at the equator and tearing toward and over the placenta, but being careful not to damage the latter, since it has an extensive blood circulation at this stage.
3. Next, open Reichert's membrane; although this is thin and transparent, it is mostly overlain with a layer of trophoblast and blood cells, and it may be necessary to pick through this layer before rupturing Reichert's membrane, after which removal is straightforward, trimming it up to, but not beyond, the placental border (**Fig. 8**).

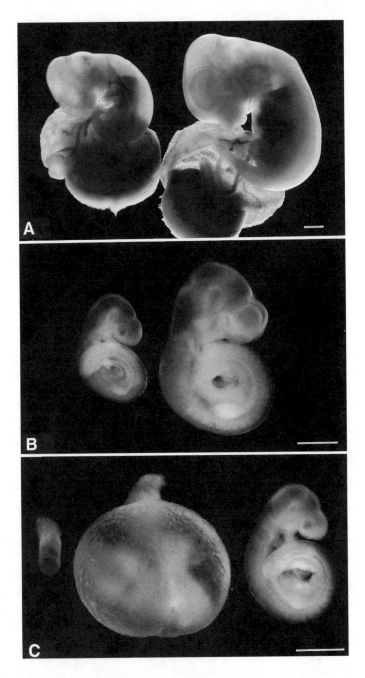

Fig. 7. (**A**) Two mouse embryos taken from the same uterus at E11. The one on the left was refrigerated at explantation. The one on the right was cultured with open yolk sac as described for 24 h before photography. (**B**) E9 embryos before and after culture.

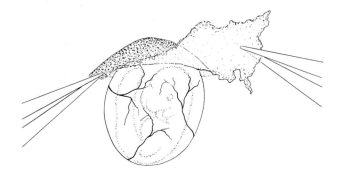

Fig. 8. The Reichert's membrane of the E9 embryo is opened using watchmaker's forceps, and the surplus is trimmed to the border of the placenta.

4.  If the visceral yolk sac or placenta is damaged, causing deflation or bleeding, the embryo should be discarded; otherwise, it is ready for culture.
5.  E9 embryos may be cultured in 50-mL Falcon tubes as above or 30-mL Universal containers (Nunc, with a smear of silicone grease round the internal rim of the lid to provide a gas-tight seal) with 1–1.5 mL pure serum/embryo, and 2–5 embryos/ tube, depending on size. Initially, the embryos should be gassed with 5% $CO_2$ in air (i.e., 20% $O_2$, 75% $N_2$, 5% $CO_2$), which is replaced with 40% $O_2$, 55% $N_2$, 5% $CO_2$ after 18–24 h of culture. **Figure 7B** shows the extent of development of E9 embryos after 48 h in culture, though 18–24 h are a more realistic period if you need the majority of embryos to be healthy at termination.

### 3.4. Explanting E7 Embryos (Cultured with Closed Yolk Sacs)

1.  The initial stages are as above, but with E7 embryos, removal of the decidua requires a rather different strategy, since the conceptus is embedded in a relatively thick mass of decidua. Start by impaling the decidual mass parallel with its long axis, but off-center, with one arm of a sharpened slender pair of forceps. Then squeeze the forceps together to form an incision in the decidua.
2.  Repeat on the other side of the decidua, then unite the two incisions around the base of the conceptus (usually the thicker end).
3.  Now grasp the two flaps of decidua with two pairs of watchmaker forceps, and pull apart (**Fig. 9**), when the conceptus should be exposed, and usually it will

---

Fig. 7. *(continued)* The embryo on the left was refrigerated at explantation; on the right is a littermate cultured for 48 h. The embryos are shown dissected free of the extraembryonic membranes with which they are cultured. (**C**) E7 embryos before and after culture. An E7 conceptus (including the visceral yolk sac and ectoplacental cone with, which it would be cultured) is shown on the left. This was refrigerated while littermates (center—with membranes as cultured; right—dissected free of membranes) were cultured for 48 h. Scale bars for A, B, and C are all 1 mm.

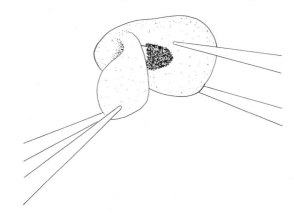

Fig. 9. As the decidua of the E7 conceptus is pulled apart, the ectoplacental cone of the embryo within can be seen.

remain on one of the decidual halves as they are separated. If the embryo sticks to both decidual halves, so that separating them further might damage it, either hold the flaps apart with one pair of forceps, whilst teasing the embryo free on one side with the apposed tips of the other pair of forceps, or make a further incision in one of the decidual halves, so that only a quarter is removed initially, followed by the second quarter.

4. Once this has been accomplished, further divide the base of the decidua containing the embryo along the long axis, and peel apart along the length of the conceptus (**Fig. 10**). Repeat if necessary until the embryo is attached only to a thin sliver of decidua, like a segment of an orange.

5. Impale the sliver of decidua on either side of the embryo with the points of one pair of forceps, stretching it slightly, and tease the embryo free with the apposed tips of the other pair of forceps (**Fig. 11**).

6. All that remains is to open Reichert's membrane (**Fig. 12**); as with the E9 embryos, this may be overlaid with blood and trophoblast, though the membrane itself is thin and transparent. Sometimes Reichert's membrane stands clear over the embryo (at the end opposite the reddish ectoplacental cone), where it can be grasped and pulled apart. Otherwise, it will be necessary to use two pairs of watchmaker forceps to rupture the membrane midway along the length of the conceptus, in a region overlying the visceral yolk sac, where unseen damage to the underlying tissue will be less serious than damage to the embryo itself. Once opened, Reichert's membrane is trimmed off to the border of the ectoplacental cone, which is left intact.

7. The E7 embryo is now ready for culture. These embryos are cultured with 1 mL pure serum/embryo, with three embryos per 30-mL Nunc tube, or up to five embryos per 50-mL Falcon tube. They are gassed initially with 5% $O_2$, 90% $N_2$, 5% $CO_2$, and then with 5% $CO_2$ in air (i.e., 20% $O_2$) after 24–36 h of culture (i.e.,

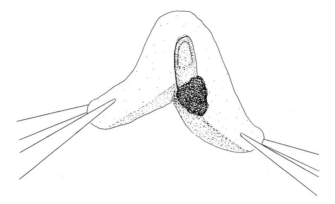

Fig. 10. The decidua is removed by successively peeling off strips along the long axis of the conceptus.

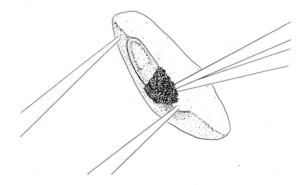

Fig. 11. The conceptus is teased free from a narrow sliver of decidua.

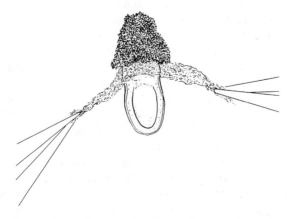

Fig. 12. Reichert's membrane is opened with finely ground watchmaker's forceps.

when the heart beat and visceral yolk sac circulation are established). **Figure 7C** shows the extent of development of E7 embryos after 48 h in culture.

## 4. Notes

1. Sharpening forceps: A combination Japanese waterstone (1000/6000 grit, e.g., King brand) is ideal. Wet it with distilled water and grind the forceps along their length until a tip diameter of 0.05–0.1 mm is achieved. It is very important that the tips meet precisely, without crossing or one protruding beyond the other. Then round off all but the innermost (mating) edges, so that when apposed, the tips together form a blunt-ended probe.

2. Within and between litters of embryos of the same age, the range of developmental stages can vary dramatically. There are a number of established staging guides: **ref. *10*** covers all stages from fertilization to birth, **ref. *11*** from fertilization to 4 wk postpartum, **ref. *12*** from 8–16 d, **ref. *13*** from 7.5–10.5 d, **ref. *14*** from 6.75–8.0 d, and **ref. *15*** from 9 d to post natal stages. Between E9 and E14 a useful method of accurately staging embryos is to use the shape of the fore- and hindlimbs *(16)*. These publications will also be useful in assessing the stage of development reached after culture and how it compares with growth in vivo. **References *10*** and ***12*** also provide a wealth of additional information, including sectioned embryos, which can help determine the success of your experiment.

3. An alternative method of transferring embryos from dish to dish while they are surrounded by their decidua is to pick them up with watchmaker forceps where the decidua is thickest, thus minimizing carryover of medium from one dish to another.

4. When the time comes to harvest your cultured embryos, they should be gently slipped into ice-cold PBS, and trimmed free of their yolk sac and placenta by cutting the umbilical vessels. Embryos can then be transferred to the fixative of your choice. We favor half-strength Karnovsky's fixative *(17)* if embryos are to be processed for scanning electron microscopy or resin histology, or 4% paraformaldehyde if they are for whole-mount or sectioned immunocytochemistry or *in situ* hybridization studies. For general wax histology, good results are obtained with Bouin's fixative followed by staining with Hematoxylin and Eosin after sectioning *(10)*.

5. Manipulations: We have successfully performed a number of manipulations on the E11 embryo after it has been delivered from its yolk sac and prior to culture. It is possible to inject reagents into various epithelial lumens—for example, we have injected the marker dye Monastral Blue into the otic vesicle *(18)*. Similarly, it is possible to perform simple surgical operations *(19–21)* and to label groups of cells with the lipophilic dye, DiI *(21)*. Even if the embryos bleed a little after surgical manipulations, they generally survive and can be successfully cultured. The two biggest problems with any such manipulations are being able to see well enough what you are doing and keeping the embryo from moving around the dish as you operate on it. The first of these problems is easily resolved with good lighting (both incident and transmitted) and the second requires gentle holding or supporting of the embryo with forceps.

6. Transgenic mice: It is possible to culture transgenic mice, but these will generally be derived from heterozygote crosses, so only a fraction of the embryos cultured from each litter will be of the required genotype. It is sometimes possible to take a small piece of tissue (tail tip or a small piece of yolk sac) at the time of explant, so that embryos can be genotyped by PCR during the culture period. Otherwise, this can be done with spare embryonic tissue taken after the culture period, but before fixation.

7. Synthetic medium: We have recently begun to culture E11 embryos with open yolk sacs in a serum-free medium made by Gibco BRL. This medium is excellent for E11 cultures up to 24 h, but we have not tried it for longer culture periods, or its efficacy for the more sensitive younger embryo stages. Good results can also be obtained over a range of stages with partially defined medium, made by extensively dialyzing rat serum, and then supplementing it with glucose, vitamins, and amino acids *(22)*.

8. Adding glucose to serum: If you have fasted your rats before bleeding, they will have lowered blood glucose levels. Addition of 8 µL/mL of a 50 mg/mL stock solution of glucose to the serum will restore the glucose levels to that in serum obtained from fed rats.

9. Address for BTC: BTC Engineering, 12 Shirley Close, Milton, Cambridge CB4 4BG, UK.

## Acknowledgments

We are particularly indebted to Catherine Haddon for the lucid diagrams illustrating the dissections. We also thank The Medical Research Council and the Wellcome Trust (P. M.), and the Imperial Cancer Research Fund (D. L. C.) for financial support.

## References

1. New, D. A. T., Coppola, P. T., and Terry, S. (1973) Culture of explanted rat embryos in rotating tubes. *J. Reprod. Fert.* **35,** 135–138.
2. New, D. A. T. (1978) Whole-embryo culture and the study of mammalian embryos during organogenesis. *Biol. Rev.* **53,** 81–122.
3. Cockroft, D. L. (1990) Dissection and culture of postimplantation embryos, in *Postimplantation Mammalian Embryos—A Practical Approach* (Copp, A. J. and Cockroft, D. L., eds.), IRL, Oxford, pp. 15–40.
4. Muneoka, K., Wanek, N., Trevino, C., and Bryant, S. V. (1990) Exo utero surgery, in *Postimplantation Mammalian Embryos—A Practical Approach* (Copp, A. J. and Cockroft, D. L., eds.), IRL, Oxford, pp. 41–59.
5. Cockroft, D. L. (1973) Development in culture of rat foetuses explanted at 12.5 and 13.5 days of gestation. *J. Embryol. Exp. Morph.* **29,** 473–483.
6. Sadler, T. W. and New, D. A. T. (1981) Culture of mouse embryos during neurulation. *J. Embryol. Exp. Morph.* **66,** 109–116.
7. Steele, C. E. and New, D. A. T. (1974) Serum variants causing the formation of double hearts and other abnormalities in explanted rat embryos. *J. Embryol. Exp. Morph.* **31,** 707–719.

8. New, D. A. T., Coppola, P. T., and Cockroft, D. L. (1976) Improved development of head-fold rat embryos in culture resulting from low oxygen and modifications of the culture serum. *J. Reprod. Fert.* **48,** 219–222.

9. Freeman, S. J., Coakley, M. E., and Brown, N. A. (1987) Post-implantation embryo culture for studies of teratogenesis, in *Biochemical Toxicology—A Practical Approach* (Snell, K. and Mullock, B., eds.), IRL, Oxford, UK, pp. 83–107.

10. Kaufman, M. H. (1992) *The Atlas of Mouse Development.* Academic Press, London.

11. Theiler, K. (1989) *The House Mouse. Atlas of Embryonic Development.* Springer-Verlag, New York.

12. Rugh, R. (1990) *The Mouse. Its Reproduction and Development.* Oxford University Press, Oxford.

13. Brown, N. A. and Fabro, S. (1981) Quantitation of rat embryonic development in vitro: a morphological scoring system. *Teratology* **24,** 65–78.

14. Downs, K. and Davies, T. (1993) Staging of gastrulating mouse embryos by morphological landmarks in the dissecting microscope. *Development* **118,** 1255–1266.

15. Wanek, N., Muneoka, K., Holler-Dinsmore, G., Burton, R., and Bryant, S. V. (1989) A staging system for mouse limb development. *J. Exp. Zool.* **249,** 41–49.

16. Martin, P. (1990) Tissue patterning in the developing mouse limb. *Int. J. Dev. Biol.* **34,** 323–336.

17. Karnovsky, M. J. (1965) A formaldehyde-gluteraldehyde fixative of high osmolarity for use in electron microscopy. *J. Cell Biol.* **27,** 137,138.

18. Martin, P. and Swanson, G. (1993) Descriptive and experimental analysis of the epithelial remodellings that control semicircular canal formation in the developing mouse inner ear. *Dev. Biol.* **159,** 549–558.

19. Martin, P. Dickson, M. C., Millan, F. A., and Akhurst, R. J. (1993) Rapid induction and clearance of TGF$\beta$1 is an early response to wounding in the mouse embryo. *Dev. Genet.* **14,** 225–238.

20. Hopkinson-Woolley, J., Hughes, D. A., Gordon, S., and Martin, P. (1994) Macrophage recruitment during limb development and wound healing in the embryonic and foetal mouse. *J. Cell Sci.* **107,** 1159–1167.

21. McCluskey, J. and Martin, P. (1995) Analysis of the tissue movements of embryonic wound healing—DiI studies in the wounded E11.5 mouse embryo. *Dev. Biol.* **170,** 102–114.

22. Cockroft, D. L. (1988) Changes with gestational age in the nutritional requirements of postimplantation rat embryos in culture. *Teratology* **38,** 281–290.

# 3

# Organ Culture in the Analysis of Tissue Interactions

## Irma Thesleff and Carin Sahlberg

## 1. Introduction

Interactions between epithelial and mesenchymal tissues constitute a central mechanism regulating the development of most embryonic organs. Studies on the nature of such interactions require the separation of the interacting tissues from each other and the follow-up of their advancing development in various types of recombined explants. The tissues can be either transplanted and their development followed in vivo, or they can be cultured as explants in vitro. Although the transplantation methods offer certain advantages, including physiological environment and the possibility for long-term follow-up, organ culture techniques are superior in many other aspects. The cultured tissues can be manipulated in multiple ways, and their development can be continuously monitored. The culture conditions are reproducible, and the composition of the medium is known exactly and it can be modified. Furthermore, the in vitro culture conditions allow analyses of the nature of the inductive signals.

Many types of organ culture systems have been used over the years for studies on embryonic organ development. The Trowell method *(1)* has been widely applied, and it has proven to be suitable for the analysis of the morphogenesis of many different organs *(2–6)*. In this system, the explants are cultured in vitro at the medium/gas interface on thin membrane filters that are supported by a metal grid. We have used the Trowell technique as modified by Saxén *(7)*.

The embryonic tooth is a typical example of an organ in which reciprocal epithelial–mesenchymal interactions regulate morphogenesis and cell differentiation *(8)*. We have used the Trowell-type organ culture method for the analysis of the mechanisms of tissue interactions at various stages of tooth development *(9–11)*. In the following, we describe the protocols for separation

From: *Methods in Molecular Biology, Vol. 97: Molecular Embryology: Methods and Protocols*
Edited by: P. T. Sharpe and I. Mason © Humana Press Inc., Totowa, NJ

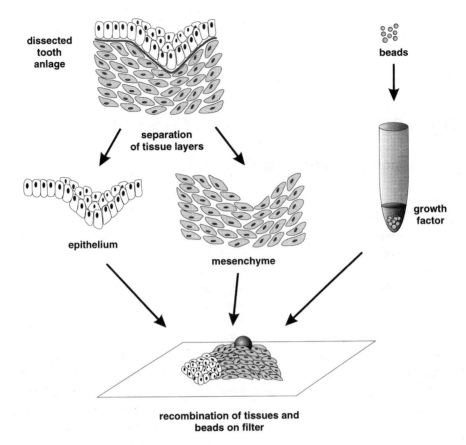

Fig. 1. Schematic representation of the method.

and culture of dental epithelial and mesenchymal tissues, and for bead experiments in which the local effects of growth factors and other diffusible molecules can be analyzed (**Fig. 1**) *(12–15)*.

## 2. Materials

All solutions and equipment should be sterile. The glassware and metal instruments should be autoclaved and solutions filtered or autoclaved. All work should be done in a laminar flow hood; also the dissection microscope should be placed in hood.

### 2.1. Salt Solutions, Enzymes, and Culture Media

1. Phosphate-buffered saline (PBS), pH 7.4: Dulbecco's phosphate-buffered saline (D-PBS, Gibco-BRL Laboratories, Detroit, MI, cat. no. 14080-048) supplemented with penicillin-streptomycin (PS), 20 IU/mL.

2. PS (Gibco-BRL, Life Technologies, Paisely, UK, cat. no. 15148-186), 10,000 IU/mL.
3. Enzyme solution for tissue separation: trypsin (Difco, cat. no. 0152-15); stock solution of pancreatin (Gibco BRL, cat. no. 15725-013), 1.25 g/mL; Tyrode's solution: NaCl (8.0 g), KCl (0.2 g), $NaH_2PO_4$ (0.05 g), glucose (1.0 g), $NaHCO_3$ (1.0 g). Adjust pH to 7.2, make up to 1000 ml with distilled $H_2O$, and sterile filter. Store at +4°C. Dissolve 0.225 g trypsin in 6 mL Tyrode's solution on ice, using a magnetic stirrer. Add 1 mL pancreatin and 20 µL PS. Adjust pH to 7.4 with NaOH. Make up to the final volume 10 mL with Tyrode's solution and sterile filter. Aliquot 1 mL in Eppendorf tubes and store at –20°C. The enzyme solution can be stored at –20°C for 1 wk.
4  Medium for tissue dissection and culture: Dulbecco's Modified Eagle Medium with glutaMAX-1 (DMEM, Gibco-BRL, cat. no. 61965-026), supplemented with 10% heat-inactivated fetal calf serum (FCS) and PS (20 IU/ml). Store at +4°C (*see* **Note 6**).

## 2.2. Dissection and Culture

1. For tissue dissection: 10-cm diameter plastic bacteriological Petri dishes and 4–10 cm diameter glass Petri dishes, small scissors and forceps, disposable 20- and 26-gage needles attached to 1-mL plastic syringes (*see* **Note 1**).
2. Culture dishes: 3.5-cm diameter plastic Petri dishes (bacteriological or cell culture dishes).
3. Metal grids: Prepared from stainless-steel mesh (corrosion-resistant, size of mesh 0.7 mm) by cutting approx 3-cm diameter disks and bending the edges on a cutting form to give a 3-mm height (the height of the metal grids can be altered to allow the use of more or less culture media). Holes in the grid are produced either by nails in the cutting form or by punching. The holes facilitate the examination and photography of the explants (and they are handy when transfilter cultures are prepared) (**Note 5**) (**Fig. 2**). There are commercially available organ culture dishes featuring a central well in which a metal grid (even without bended edges) can be placed (Falcon 3037, Becton Dickinson Ltd., Oxford, UK).
4. Filters: Nuclepore polycarbonate filters (Nuclepore, Pleasanton, CA). The pore size routinely used is 0.1 µm (*see* **Note 4**). The filters are cut in halves, washed in detergent, rinsed under running water for 2 h and 10× in distilled water, and stored in 70% ethanol.
5. Glass Pasteur pipets are used for transferring tissues. They are siliconized (to prevent sticking of tissues), stuffed with cotton wool, and autoclaved. Before use, they are drawn by heating to adjust to the size of the tissues. Ideally, the diameter should be the minimal to allow free passage of the tissue.
6. Beads for growth factors: Affi-Gel Blue agarose beads (Bio-Rad Laboratories, Hercules, CA) or heparin-coated acrylic beads (Sigma, St. Louis, MO) are divided into aliquots and stored at +4°C.

# 3. Methods
## 3.1. Treatment of Beads

Pipet agarose beads or heparin-coated acrylic beads to PBS in a Petri dish. Count 100–200 beads under the microscope, and transfer to Eppendorf tube. Spin down the

Fig. 2. The Trowell-type organ culture dish. The metal grid supports six pieces of filters. The cultured explants (like the one in the bottom of **Fig. 1**) are seen on the filters over the holes of the grid.

beads, and remove PBS. Add growth factors in a small volume (10–50 µL) of 0.1% bovine serum albumin (BSA) in PBS. (In general, high concentrations of proteins are used. We use FGF-4 at 20 ng/µL and TGFβ-1 at 1 ng/µL). An equal amount of 0.1% BSA in PBS is pipetted to control beads. Incubate for 30 min at +37°C and store at +4°C. The beads can be used at least for 14 d (depending on the stability of the protein).

## 3.2. Preparation of Culture Dishes

1. Take a sheet of Nuclepore filter from ethanol and rinse in PBS in a plastic 10-cm diameter Petri dish. Cut the filter, using small scissors and watchmaker forceps, in approx 3 × 3 mm pieces, and leave in PBS.
2. Place metal grids in 3.5-cm plastic culture dishes. Add approx 2 mL culture medium (D-MEM +10% FCS, *see* **Note 6**) by pipeting through the grid. The surface of the medium should contact the plane of the grid, but not cover it (excess medium results in floating of the filters and tissues). No air bubbles should remain under the grid (if present, they can be sucked empty with a thin Pasteur pipet). Using forceps, transfer the Nuclepore filter pieces on the grids placing them over the holes.

## 3.3. Dissection of Tissues

1. Place the mouse uterus (E 12) in a 10-cm plastic Petri dish containing D-PBS, and cut open the uterine wall using small scissors and forceps. Continue the work

under the stereomicroscope with transmitted light. Remove the embryos from fetal membranes, and transfer them to a fresh dish of D-PBS. Cut off the heads using disposable needles as "knives." The needles are used during all subsequent steps of dissection. Transfer the heads to a glass Petri dish containing D-PBS, and dissect out the lower jaw. Dissect out the tooth germs of the first mandibular molar with some surrounding tissue left in place (*see* **Notes 1** and **2**).

2. With a drawn Pasteur pipet (preferably mouth-controlled), transfer the tooth germs to an Eppendorf tube. Remove most of the liquid. Melt an aliquot of pancreatin/trypsin, and spin immediately for 30 s at 8000*g*. Add cold supernatant on the tooth germs and incubate on ice for 2 min. Remove most of the liquid, add culture medium, mix, and transfer the tooth germs to a glass Petri dish containing culture medium. Leave the tissues for 30 min at room temperature.

3. Gently separate the epithelia from the mesenchymes using needles and remove excess surrounding nondental tissue (*see* **Note 3**). Transfer the tissues on the Nuclepore filters in culture dishes that have been prepared in advance. Avoid air bubbles in the pipet, and avoid sucking the tissue beyond the capillary part of the pipet. Ideally, the tissues should be placed directly in their final position, but if needed, they can be gently pushed with needles.

4. Wash the beads quickly in culture medium in a glass Petri dish (they tend to stick on plastic dishes). Under the microscope, transfer the beads one at a time to the tissues.

## 3.4. Culture and Fixation

1. Culture the tissues in a standard incubator at 37°C, in an atmosphere of 5% $CO_2$ in air and 100% humidity for 24 h.

2. Photograph the explants before fixation (the translucent zone cannot be seen after fixation). To avoid detachment of tissues from the filters, prefix the explants in ice-cold methanol on the grids as follows: Remove the culture medium by sucking, and pipet methanol gently on the tissues. Leave for 5 min, and transfer filters by watchmaker forceps to Eppendorf tubes for subsequent treatments (*see* **Notes 8** and **9**). Typical explants are shown in **Fig. 3**.

## 4. Notes

1. For tissue dissection, the disposable needles are superior to other instruments, such as scalpels or iris knives, because they need no sharpening or sterilization. The size of the needles can be chosen according to the sizes of tissues. The syringes need not be absolutely sterile and can be used many times. For best preservation of tissue vitality, dissecting should be done by determined cuts avoiding tearing. Glass Petri dishes are preferable to plastic ones during dissection of the tissues, because cutting with the needles tends to scrape the bottom of the plastic dish and loosen pieces of it.

2. The preparation and dissection of tissues should be done as quickly as possible to promote survival of the tissues. One uterus at a time should be prepared and the rest stored in D-PBS at +4°C. The dissected tissues should not be stored for long times (2–3 h max) before transfer to the culture dishes and incubator.

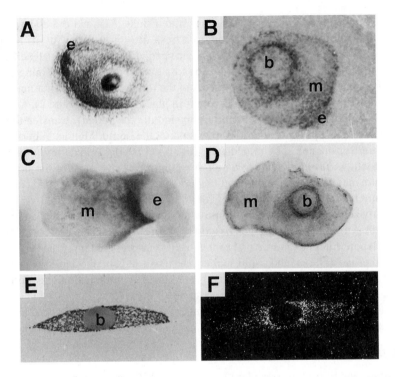

Fig. 3. Examples of the analysis of cultured explants. (**A**) Appearance of an explant in the stereomicroscope. The epithelium as well as the TGFβ-1-releasing bead have induced a translucent zone in dental mesenchyme. (**B**) Localization of cell proliferation with BrdU incorporation under the epithelium as well as around an FGF-4-releasing bead. (**C**) Whole-mount *in situ* hybridization analysis of Msx-1 gene expression indicating induction by the epithelium in the mesenchyme. (**D**) Whole-mount immunohistochemical staining showing stimulation of tenascin expression in the mesenchyme around an FGF-4-releasing bead. (**E**) Section of an explant of dental mesenchyme and a TGFß-1-releasing bead (the filter has been detached during processing). (**F**) Dark-field illumination of the explant in E showing the induction of tenascin-C transcripts by *in situ* hybridization analysis. e, dental epithelium; m, dental mesenchyme; b, bead.

3. The dissection and culture techniques are basically similar when different organs or different developmental stages of tooth germs are studied. Separation of the epithelium and mesenchyme can be accomplished in young tissues, after enzyme treatment, even without dissection by briefly vortexing the tissues. On the other hand, more advanced tissues require a longer incubation in the enzyme solution (up to 10 min). The time needed for best separation depends also on the batches of enzymes, and therefore the optimal time must always be checked for new batches of enzyme and for different tissues.

4. Different supporting materials can be used for the cultured explants. Lens paper may be used for large tissue pieces. The supporting material must allow good diffusion of the medium to the tissue, and therefore Millipore filters of 100-μm thickness are not suitable. (The thickness of Nuclepore filters is approx 10 μm.) Different pore size Nuclepore filter may be used. Small pores (0.05–0.2 μm) allow better examination of the explants in the stereomicroscope using transmitted light, but the tissues tend to detach from these filters more readily during fixation and other treatments after culture. Therefore, larger pore sizes (up to 1 μm) may be preferable, depending on the experiment.

5. The Trowell-type organ culture can be used for a variety of other organ culture designs. One example is the transfilter culture, where the interacting tissues are cultured on opposite sides of the filter *(4,9)*. The tissue to be grown below the filter is glued by heated 1% agar, after which the filter is turned upside down, and the other tissue is placed on top of the filter.

6. The composition of the optimal culture medium depends on the tissues. The medium in this protocol is good for a number of different organs at early stages of development, but during more advanced stages, different organs may have special requirements. Chemically defined media with varying compositions have been designed. For cultures of whole tooth germs, we routinely use chemically defined medium composed of D-MEM and F12 (Ham's Nutrient Mixture, Gibco-BRL) 1:1, supplemented with 50 μg/mL transferrin (Sigma, T-2252,10 mg/mL, 20-μL aliquots, stored at –20°C). For more advanced stages of tooth development, ascorbic acid is added at 150 μg/mL to allow deposition of dentin collagen *(16)*. During prolonged culture, the medium should be changed at 2–3 d intervals.

7. Isolated epithelial tissue does not survive as well as the mesenchymal tissue, when cultured alone. The growth of the dental epithelium (as well as epithelium from other organs) is significantly improved by culture on extracellular matrix material. Collagen has been used, but by far the best results are obtained with the basement membrane matrix, Matrigel (Collaborative Biomedical Products, Bedford, MA, cat. no. 40234), which also promotes epithelial morphogenesis *(17)*. Matrigel is kept on ice and pipeted on the filters. Dishes are transferred to the incubator for 30 min, allowing gelling of Matrigel. At room temperature, the tissue is placed on Matrigel. Covering of the tissue with a drop of Matrigel further improves epithelial growth.

8. *Bromodeoxyuridine* (BrdU) incorporation is commonly used for the analysis of cell proliferation (**Fig. 3B**). The explants are labeled by adding BrdU 0.5–3 h before fixation (we use cell proliferation kits from Amersham International, Little Chalfont, UK or Boehringer-Mannheim, Mannheim, Germany). After fixation in ice-cold methanol, the explants are washed in PBS and immunostained as whole mounts using antibodies against BrdU *(12)*.

9. Usually, the tissues are analyzed after culture either as whole mounts (**Fig. 3B–D**), or they are paraffin-embedded and serially sectioned (**Fig. 3E,F**). For most purposes, they are fixed for 1 h in 4% paraformaldehyde in PBS (PFA) (after 5 min prefixation in ice-cold methanol). PFA should be fairly freshly made (not more

than 7-d-old). The procedure used for whole-mount immunostaining is described in (**ref. *11***), and that for *in situ* hybridization in (**ref. *12***).

## References

1. Trowell, O. A. (1959) The culture of mature organs in a synthetic medium. *Exp. Cell Res.* **16,** 118–147.
2. Grobstein, C. (1953) Inductive epithelio-mesenchymal interaction in cultured organ rudiments of the mouse. *Science* 118, 52–55.
3. Saxén, I. (1973) Effects of hydrocortisone on the development in vitro of the secondary palate in two inbred strains of mice. *Arch. Oral Biol.* **18,** 1469–1479.
4. Saxén, L., Lehtonen, E., Karkinen-Jääskeläinen, M., Nordling, S., and Wartiovaara, J. (1976) Morphogenetic tissue interactions: Mediation by transmissible signal substances or through cell contacts? *Nature* **259,** 662,663.
5. Nogawa, H. and Takahashi, Y. (1991) Substitution for mesenchyme by basement-membrane-like substratum and epidermal growth factor in inducing branching morphogenesis of mouse salivary epithelium. *Development* **112,** 855–861.
6. Nogawa, H. and Ito, T. (1995) Branching morphogenesis of embryonic mouse lung epithelium in mesenchyme-free culture. *Development* **121,** 1015–1022.
7. Saxén, L. (1966) The effect of tetracycline on osteogenesis in vitro. *J. Exp. Zool.* 162, 269–294.
8. Thesleff, I., Vaahtokari, A., and Partanen, A. M. (1995) Regulation of organogenesis. Common molecular mechanisms regulating the development of teeth and other organs. *Int. J. Dev. Biol.* **39,** 35–50.
9. Thesleff, I., Lehtonen, E., Wartiovaara, J., and Saxén, L. (1977) Interference of tooth differentiation with interposed filters. *Dev. Biol.* **58,** 197–203.
10. Partanen, A. M., Ekblom, P., and Thesleff, I. (1985) Epidermal growth factor inhibits tooth morphogenesis and differentiation. *Dev. Biol.* **111,** 84–94.
11. Vainio, S. and Thesleff, I. (1992) Coordinated induction of cell proliferation and syndecan expression in dental mesenchyme by epithelium: evidence for diffusible signals. *Dev. Dyn.* **194,** 105–117.
12. Vainio, S., Karavanova, I., Jowett, A., and Thesleff, I. (1993) Identification of BMP- 4 as a signal mediating secondary induction between epithelial and mesenchymal tissues during early tooth development. *Cell* **75,** 45–58.
13. Jernvall, J., Kettunen, P., Karavanova, I., Martin, L.B., and Thesleff, I. (1994) Evidence for the role of the enamel knot as a control center in mammalian tooth cusp formation: non-dividing cells express growth stimulating Fgf-4 gene. *Int. J. Dev. Biol.* **38,** 463–469.
14. Vaahtokari, A., Åberg, T., and Thesleff, I. (1996) Apoptosis in the developing tooth: Association with an embryonic signalling center and suppression by EGF and FGF-4. *Development* **122,** 121–129.
15. Mitsiadias, T., Muramatsu, T., Muramatsu, H., and Thesleff, I. (1995) Midkine (MK), a heparing-binding growth/differentiation factor, is regulated by retinoic acid and epithelial-mesenchymal interactions in the develping mouse tooth, and affects cell proliferation and morphogenesis. *J. Cell Biol.* **129,** 267– 281.

16. Laine, M. and Thesleff, I. (1986) Development of mouse embryonic molars in vitro: An attempt to design defined culture conditions allowing mineralization. *J. Biol. Buccale* **14,** 15–23.
17. Gittes, G., Galante, P. E., Hanahan, D., Rutter, W. J., and Debas, H. T. (1996) Lineage-specific moprhogenesis in the developing pancreas: role of mesecnhymal factors. *Development* **122**, 439–447.

# 4

## Treatment of Mice with Retinoids In Vivo and In Vitro
## Skeletal Staining

### Gillian M. Morriss-Kay

### 1. Introduction

The retinoids comprise a large group of natural and synthetic compounds related to vitamin A (retinol). The family name is derived from the early observation of the necessity of vitamin A for normal vision, and the association of vitamin A deficiency with night blindness *(1)*. With the exception of the visual cycle in the rod photoreceptor cells of the retina, in which protein-bound 11-*cis*-retinal is reversibly isomerized to free all-trans-retinal, retinoic acid is the active retinoid for biological processes. However, the visual cycle illustrates two characteristics of retinoids that are relevant to the use of this family of molecules as tools in experimental embryology and for the interpretation of results. (1) retinoids are light-sensitive, and will undergo isomerisation or degradation if exposed to light; (2) in nature, their activity and their metabolism are associated with binding to specific proteins.

The most important natural retinoids are retinyl esters, retinol, and retinoic acid (RA). In the all-*trans* form, RA is the ligand for the nuclear RA receptors RAR-$\alpha$, RAR-$\beta$, and RAR-$\gamma$; in the 9-*cis* form, it is the ligand for the retinoid "X" receptors (RXRs), which form heterodimers with RARs to form the transcriptionally active complex that binds to RA response elements of target genes, as well as with the RARs (*see* **ref. 2** for details and references). In the past, retinyl esters and retinol have been used for treating embryos *(3,4)*; their effects on developmental pattern probably depend on conversion to RA, although some other natural metabolites, such as 4-oxo-RA, may also be directly active *(5)*. Most recent studies have used RA (mainly in the all-*trans* form) for experimental purposes.

From: *Methods in Molecular Biology, Vol. 97: Molecular Embryology: Methods and Protocols*
Edited by: P. T. Sharpe and I. Mason © Humana Press Inc., Totowa, NJ

The disadvantage of using RA, both in vivo and in vitro, is that it is highly toxic. Although it is the principal biologically active retinoid, it is not normally transported long distances from its source of synthesis. Retinoids are stored in the liver as esters and transported in the bloodstream as retinol bound to retinol binding protein (RBP). In early pregnancy in the mouse (i.e., at stages appropriate for studies on hindbrain and limb bud development), the maternal blood is in direct contact with the parietal yolk sac (Reichert's membrane), a thick basement membrane-like structure with dispersed attached cells. Immediately internal to this, the visceral yolk sac endoderm is rich in transcripts of cellular retinol binding protein (CRBP I) *(6)*. The mechanism of transfer involves receptor-mediated uptake of retinol from maternal RBP-retinol by the visceral yolk sac endoderm, where it binds to CRBPI and interacts with the enzymes mediating RA synthesis *(7)*. The assumption that retinol is retained specifically in CRBP I-expressing embryonic tissues has been verified by using $^{14}C$-labeled retinyl acetate (delivered intravenously to the pregnant dam) as a source of retinol *(8)*.

RA administered to pregnant rats and mice is rapidly transferred to the embryos, where it remains at high levels for only a few hours. In the mouse, embryonic all-*trans*-RA levels are raised within 30 min of maternal oral administration, reaching a peak at 2 h and then falling rapidly to low levels by 8 h *(9)*. In the rat, the embryonic peak for RA is also at 2 h, falling even more rapidly than in the mouse to a low level at 4 h, and then more gradually to undetectable levels *(10)*. These pharmacokinetic profiles indicate that administration of RA to pregnant rats or mice results in exposure of the embryos to RA for relatively short periods of time; if it is desired to allow development to continue for 24 h or more, we can be confident that they are not exposed to raised RA levels for more than the first few hours after the time of administration, with a peak at 1–2 h. It should also be noted that RA administered in vivo or in vitro undergoes isomerisation to other active retinoids (e.g., 13-*cis*-RA and 9-*cis*-RA) both within the maternal tissues and in the embryo *(9–11)*.

The following protocols cover only administration with RA. Retinol can be used in exactly the same way, but higher concentrations are required to induce an equivalent effect *(6)*. It may be more appropriate to use retinol than RA in vitamin A deficiency studies, in which much lower (physiological) concentrations should be used than in studies using RA to alter gene expression. Retinyl esters must be made up as an emulsion; they are only useful in in vivo studies, being inactive in vitro, where the necessary conversion enzymes are not available. In addition to the protocols described below (*see* **Subheading 3.**), RA has also been applied on a bead directly to the mouse limb bud using the *ex utero* · technique; the effects on limb development were reduction defects resembling the effects of maternal administration *(12)*.

## 2. Materials

1. Retinoic acid is available in crystalline form from Sigma (UK). Once open, it must be protected from light and oxygen. Protection from oxygen can be achieved by replacing the air with nitrogen or argon, as available, immediately before sealing the container. All solutions and suspensions must be made up fresh, protected from light, and used within 24 h.
2. For oral administration (gavage) of pregnant mice, use a 1–5 in. stainless-steel 18-gage dosing cannula with a luer fitting and a bulbous tip (available from Harvard Apparatus Ltd., Kent, UK), fitted to a standard 1 mL disposable syringe. For intraperitoneal (ip) injection, use a 25-gage × 5/8-in. needle.

## 3. Methods

The following examples of retinoic acid treatments are from our own studies. The timing is critical. Our timings are included to illustrate what has worked for us in the past. For each new study and for repeats of previous studies, even in the same laboratory, it is essential to invest some time in establishing the time period during which the desired developmental process occurs. Timed matings, in which the male is removed after 2–4 h, minimize but do not abolish the spread of developmental stages between litters; there is also a range of stages within each litter. It is likely that implantation time is more significant than the time of mating for determining the precise developmental stage later on. Time of year, even in an air-conditioned, light- and temperature-controlled animal house, can affect developmental timing in the mouse.

### 3.1. Production of Hindbrain Abnormalities

Embryos exposed to RA prior to the onset of somitic segmentation fail to develop rhombomeres, and *Hoxb-1* is expressed throughout the preotic hindbrain instead of being specific to rhombomere 4; embryos exposed to RA after the onset of somitic segmentation form rhombomeres of variable regularity, with *Hoxb-1* expression in rhombomere 2 as well as rhombomere 4, and rhombomeres 1,2 and 5 showing a normal pattern of gene expression with respect to *HoxB* genes and *Krox-20 (13)*. These abnormalities were induced by oral dosing with RA on d 7.75 and 8.25, respectively (dosing on d 8.0 gave mixed litters), giving 10 or 12 mg RA/kg (a mouse weighs 20–25 g).

Make up crystalline RA as a 1 mg/mL suspension in peanut oil. Mix very thoroughly, using a magnetic stirrer, and use immediately. If kept overnight for using again, the suspension should be gassed, sealed, stored in the dark at 4°C, and mixed thoroughly again immediately before reuse. For 10 mg/kg, give 0.25 mL to a 25 g mouse. For gavage (right-handed person), proceed as follows: place the mouse on the cage laid, holding its tail with your right hand; it will grip the bars of the lid. Pick up the mouse with your left hand so that the skin of its neck and back is held firmly between the thumb and first finger; head movement is restricted, and the snout tilted upward. Crook your little finger around the tail if necessary. The mouse should be comfortable, but immobile; if it struggles the hold is wrong, and you should start again. Introduce the bulb of the cannula between the jaws at the diastema (the gap between the incisors and molars), place it on the tongue, and then use it to tilt the head into a position so that the cannula is in line with the esophagus. Slide the cannula gently down the esophagus so that the tip enters the stomach. This should be easy. If at any stage it is not easy, stop. Deliver the measured amount of RA suspension, and remove the cannula. The mouse should be lively immediately on replacing into the cage, and there should be no sign of oil around the mouth. This is a skilled procedure and should be learned from an experienced user. *See* **Subheading 4.**

### 3.2. Production of Limb Abnormalities

Limb abnormalities can be induced by RA treatment on d 11 or early on d 12. Like all RA-induced effects, the resulting abnormalities of skeletal pattern are stage-related. In our experiments, administration on d 11.0 resulted in partial or complete fusion of digits 4 and 5, administration on d 11.5 to loss of digits 1 and 5, and administration on d 12.0 to partial or complete loss of digit 1; skeletal differentiation was delayed by all of these timed exposures to RA *(14)*.

Make up a 10 mg/mL suspension in peanut oil, and give 100–120 mg/kg (0.25–0.3 mL for a 25-g mouse) by gavage. Limb abnormalities can be analyzed at late fetal stages (d 17 or 18) by double skeletal staining for cartilage and bone (*see* **Subheading 3.5.**).

### 3.3. Intraperitoneal administration

We have used ip injection for low-dose RA supplementation to *curly tail* mutant mice *(15)*. This treatment results in a decrease in the incidence of neural tube defects and is a good example of the use of RA in studying genotype–phenotype interactions. A thick, oily suspension is unsuitable for delivery through a small-gage needle, so the RA is first dissolved in 5–10% ethanol (this takes 15–20 min), then a small amount of oil, and then mixed with the remainder of the oil. Thorough mixing at each stage is essential, e.g., for administration of 5 mg/kg to a 25 g mouse, dissolve 5 mg RA in 0.8 mL Analar ethanol, add 1.2 mL peanut oil, and mix until dissolved (at least 10 min); add 8 mL peanut oil, and mix again.

### 3.4. Treatment of Embryos and Tissues In Vitro

Exposure of embryos to RA in vitro has the advantage that the precise developmental stage at the time of exposure is known, the concentration in the medium can be precisely calculated, and the duration of the exposure controlled. An exposure time of 2 h is sufficient to induce an effect on morphogenesis and gene expression, and avoids the complication that deleterious effects on the yolk sac placenta may affect development if RA is present throughout the whole culture period. The required concentration and exposure time need to be worked out by using a range of concentrations and times for each new experiment. Hindbrain defects can be induced by a concentration of 0.25 µg/mL medium. For this, dissolve 1 mg RA in 0.8 mL Analar ethanol and add 1 µL of this solution to 5 mL culture medium. One microliter ethanol should be added to control cultures. This amount of ethanol is insufficient to cause abnormalities, and probably evaporates rapidly at 38°C. Some laboratories have used DMSO as a solvent for RA *(11)*. RA should always be added to the medium before the embryos are added. After exposure, embryos should be washed in Tyrode saline (do not use PBS, because they require calcium for normal morphogenesis) and replaced in fresh culture medium.

Both RA and retinol can similarly be introduced to the tissue-culture medium of embryonic cell, tissue, and organ cultures.

## 3.5. Staining of Fetal and Neonatal Skeletons

The effects of retinoids in vivo include important alterations of skeletal pattern. The following procedures are also used for analysis of skeletal defects of genetic origin. Alcian blue staining alone is appropriate for early stages of skeletogenesis; double staining to show both cartilage and bone is appropriate to late fetal stages, and also works well for early postnatal mice and rats. Alcian blue stains the proteoglycans of cartilage matrix; Alizarin red S stains the mineralized matrix of bone.

### 3.5.1. Cartilage Staining of Whole Embryos

1. Fix from fresh in Bouin's fluid for 4 h.
2. Drain and wash off fixative with distilled water.
3. Immerse for 1 h in each of four changes of 1% ammonia solution in 70% alcohol (with frequent agitation).
4. Stain in 95 mL 5% acetic acid plus 5 mL of 1% Alcian blue 8GX in 5% acetic acid (use 40–50 mL/embryo) for 1–4 h or overnight (depending on age of embryo).
5. Wash for 1 h in two changes of 5% acetic acid.
6. Transfer to a third change of 5% acetic acid for 16 h.
7. Dehydrate in graded ethanols (70, 95, 100% × 3).
8. Clear and store in three parts methyl salicylate to one part benzyl benzoate.

Result: Cartilage matrix: intense blue/green; other tissues: only very slight staining (of glycosaminoglycans).

### 3.5.2. Double Staining for Cartilage and Bone

1. Fix in 95% alcohol. Remove skin, including tail skin.
2. Stain in 80 mL 95% alcohol, 20 mL glacial acetic acid, 15 mg Alcian blue 8GX (or 8GN) for 20–48 h.
3. Differentiate in 95% alcohol, three changes over 7 d.
4. Macerate in weak (0.2–2%) KOH until bones are visible (we use 1%).
5. Wash for 12 h in running tap water.
6. Stain in freshly made 0.1% aqueous Alizarin red S + 10–20 drops of 1% KOH 3–12 h (check color—add acid if necessary to get red color).
7. Wash for 30 min in running tap water.
8. Decolorize 1–2 wk in 20% glycerine in 1% KOH.
9. Dehydrate through mixture of 70% alcohol:glycerine:water sequentially in the following proportions: 1:2:7; 2:2:6; 3:3:4; 4:4:2; 5:5:0, allowing approx 3 d in each mixture. The soft tissues are clear before the end of this procedure, but it is important to complete the process for storage. Skeletal preparations can then be stored for years.

### 3.5.3 Alternative Double Staining for Cartilage and Bone in Adults

This procedure has been found to work well for adult tissues and can also be used for tissues that have been previously fixed in formalin.

1. Fix in ethanol for 4 d (room temperature).
2. Keep in acetone for 3 d (room temperature).
3. Rinse in distilled autoclaved water.
4. Stain for a minimum of 10 d in a staining solution consisting of:
   1 vol of 0.3% alcian blue in 70% ethanol;
   1 vol of 0.1% alizarin red in 95% ethanol;
   1 vol of 100% acetic acid;
   17 vol of 100% distilled autoclaved water.
5. Rinse in distilled autoclaved water.
6. Add 20% glycerol:2% potassium hydroxide solution and keep at 37°C overnight.
7. Change solution every 2–3 d, keeping at 37°C during the day and at room temperature overnight, for approx 2–3 wk, until the specimens have completely cleared.
8. For storage, transfer specimens to 50, 80, and 100% glycerol in succession.

## 4. Notes

In the Untied Kingdom, all administration procedures to pregnant mice require Home Office personal and project licenses. Embryo culture beyond midgestation (d 9 in the mouse), and removal of tissues from embryos and fetuses, also require Home Office permission. Licenses will only be granted after attendance at a course on laboratory animal management and welfare (Home Office Training Modules 1–3). Readers from other countries should seek advice from the appropriate authorities.

# References

1. Moore, T. (1957) *Vitamin A.* Elsevier, Amsterdam.
2. Mangelsdorf, D. J., Umesono, K., and Evans, R.M. (1994) The retinoid receptors in *The Retinoids: Biology, Chemistry and Medicine,* 2nd ed. (Sporn, M. B., Roberts, A. B., and Goodman, D. S., eds.), Raven, New York, pp. 319–349.
3. Morriss, G. M. (1972) Morphogenesis of the malformations induced in rat embryos by maternal hypervitaminosis A. *J. Anat.* **113,** 241–250.
4. Morriss, G. M. and Steele, C. E. (1977) Comparison of the effects of retinol and retinoic acid on postimplantation rat embryos in vitro. *Teratology* **15,** 109–120.
5. Nau, H. (1994) Retinoid teratogenesis: toxicokinetics and structure specificity, in *Use of Mechanistic Information in Risk Assessment* (Bolt, H. M., Hellman, B., and Denker, L., eds.) (*Arch. Toxicol.* **Suppl. 16**), Springer-Verlag, Berlin, pp. 118–127.
6. Ruberte, E., Doll, P., Chambon, P., and Morriss-Kay, G. (1991) Retinoic-acid receptors and cellular binding proteins: II. Their differential pattern of transcription during early morphogenesis in mouse embryos. *Development* **111,** 45–60.
7. Båvik, C., Ward, S. J., and Ong, D. E. (1997) Identification of a mechanism to localize generation of retinoic acid in rat embryos. *Mech. Dev.* **69,** 155–167.
8. Gustafson, A.-L., Dencker, L., and Eriksson, U. (1993) Non-overlapping expression of CRBP I and CRABP I during pattern formation of limbs and craniofacial structures in the early mouse embryo. *Development* **117,** 451–460.
9. Creech-Kraft, J., Löfberg, B., Chahoud, I., Bochert, G., and Nau, H. (1989) Teratogenicity and placental transfer of all-*trans*-, 13-*cis*-, 4-oxo-all-*trans*-, and 4-oxo-13-*cis*-retinoic acid after administration of a low oral dose during organogenesis in mice. *Toxicol. Appl. Pharmacol.* **100,** 162–176.
10. Ward, S. J. and Morriss-Kay, G. M. (1995) Distribution of all-trans-, 13-cis and 9-cis-retinoic acid to whole rat embryos and maternal serum following oral administration of a teratogenic dose of all-trans-retinoic acid. *Pharmacol. Toxicol.* **76,** 196–201.
11. Klug, S., Creech-Kraft, J., Wildi, E., Merker, H.-J., Persaud, T. V. N., Nau, H., and Neubert, D. (1989) Influence of 13-cis and all-trans retinoic acid on rat embryonic development in vitro: correlation with isomerisation and drug transfer to the embryo. *Arch. Toxicol.* **63,** 185–192.
12. Bryant, S. V. and Gardiner, D. M. (1992) Retinoic acid, local cell-cell interactions, and pattern formation in vertebrate limbs. *Dev. Biol.* **152,** 1–25.
13. Wood, H. B., Pall, G. S., and Morriss-Kay, G. M. (1994) Exposure to retinoic acid before or after the onset of somitogenesis reveals separate effects on rhombomeric segmentation and 3' *HoxB* gene expression domains. *Development* **120,** 2279–2285.
14. Wood, H. B., Ward, S. J., and Morriss-Kay, G. M. (1996) Effects of retinoic acid on skeletal pattern, 5'*HoxD* gene expression, and RAR-$\beta 2/\beta 4$ promoter activity in embryonic mouse limbs. *Dev. Genet.* **18,** 74–84.
15. Chen, W.-H., Morriss-Kay, G. M., and Copp, A. (1995) Genesis and prevention of spinal neural tube defects in the *curly tail* mutant mouse: involvement of retinoic acid and its nuclear receptors RAR-$\alpha$ and RAR-$\gamma$. *Development* **121,** 681–691.

# 5

## Cell Grafting and Labeling in Postimplantation Mouse Embryos

**Gabriel A. Quinlan, Paul A. Trainor, and Patrick P. L. Tam**

### 1. Introduction

Fate mapping experiments provide direct information on the differentiation pathways normally taken by cells or tissues during embryogenesis. Systematic analyses of the developmental fate of cell populations localized in different parts of the embryo enables the construction of fate maps. A comparison of the expression pattern of lineage-specific genes and the fate map allows the identification of precursor tissue for cell lineages well before definitive histogenesis takes place. The ability to trace the early lineage history of cells greatly facilitates the elucidation of the forces and processes which lead to the specification of cell lineages and the determination (or commitment) of cell fate. The knowledge of cell fate may also assist the interpretation of the phenotype of mutant embryos produced either by spontaneous mutation or by gene knockout experiments.

This chapter describes the technical aspect of fate mapping the mouse embryo during gastrulation (6.5 d post-coitum [p. c.]) *(1–3)* and organogenesis (8.5 d p.c.) *(4–8)*. Two experimental strategies are used to study the developmental fate of cells. First, a specific population of cells can be marked by labeling with vital carbocyanine dyes *in situ*, and second, the same population of cells can be isolated from a transgenic embryo followed by trasnplantation (grafting) to a host embryo. The pattern of tissue colonization and differentiation of the descendants of these marked or transplanted cells is then analyzed after a period of in vitro development to assay their development fate.

Cell labeling and grafting procedures have their special advantages. When grafting a genetically identifiable population of cells, there is no dilution of the label owing to cell proliferation, so the contribution of transgenic cells to every

From: *Methods in Molecular Biology, Vol. 97: Molecular Embryology: Methods and Protocols*
Edited by: P. T. Sharpe and I. Mason © Humana Press Inc., Totowa, NJ

available lineages can be assessed. Cell transplantation techniques can also be applied to the study of the developmental potency of a population of cells, by confronting the cells with novel tissue environment or inductive signals *(9)*. The usefulness of the cell grafting approach depends critically on the ability to isolate a defined cell population for transplantation and to place these cells at the appropriate site in the host. By contrast to cell transplantations, *in situ* labeling experiments do not require tedious dissection of tissue fragments. Fate mapping studies can be carried out directly after *in situ* labeling with minimal disruption of existing tissue architecture. However, the label must be noncytotoxic and should remain only among the descendants of the labeled cells.

## 2. Materials

### 2.1. Culturing 6.5–8.5 D Embryos

1. Roller/rotator bottle culture apparatus (B.T.C. Engineering, Milton, Cambridge, UK).
2. Water-jacketed $CO_2$ incubator (Forma Scientific, Marietta, OH, Model 3336).
3. Four-well chamber slides (NUNC, Naperville, IL).
4. Glass culture bottles, thin-walled,15-mL, 30-mL capacity (B.T.C. Engineering).
5. Refrigerated bench top centrifuge, (CENTRA-7R, International Equipment Company, Needingham Heights, MA).
6. 15-mL sterile centrifuge tubes (Corning, Cambridge, MA, cat. no. 25310-15).
7. Penicillin/streptomycin, 5000 µg/mL (Trace Biosciences, Sydney, NSW, Australia).
8. Glutamine 200 µ*M* (Trace Bioscience).
9. Dulbecco's Modified Eagle's Medium (DMEM) (Gibco-BRL, Grand Island, NY, cat. no. 12100-103, glucose 4 g/L).
10. Rat serum (RS) (*see* **Subheading 3., step 1**).
11. Human cord serum (HCS) (*see* **Subheading 3., step 1**).

### 2.2. Isolation of Tissue Fragments for Grafting

Alloy metal needles are made by electrolytically sharpening orthodontic wire (Rocky Mountain Orthodontics, Denver, CO) with the a wire polishing unit (Dental Corporation of America, Hagerstown, MD).

Glass needles are made from thick-walled glass capillaries (Leitz, Rockleigh, NJ, #520-119). The capillaries are heated over a small flame to fuse a segment of the glass. This fused segment is then pulled with a vertical pipet puller (David Kopf Instruments, Tujunga, CA, Model 720) to give two needles. These needles are coated with Repelcote (BDH Chemicals, Poole, UK) to prevent the tissue from adhering.

Dissecting microscope (Wild M3Z or MDG 17).

Tissue culture dishes (60-mm, Corning).

Fetal calf serum (FCS) (Trace Biosciences): The FCS is thawed and inactivated by heating at 56°C for 30 min immediately before use.

Polyvinylpyrrolidone (PVP, Sigma, St. Louis, MO). Dialyze a 0.5% aqueous PVP solution against water at 4°C overnight followed by lyophilization.

Enzymatic solution contains 0.5% trypsin (Trace Biosciences), 2.5% pancreatin (Boehringer Mannheim, Indianapolis, IN), 0.2% glucose (Sigma), and 0.1% PVP dissolved in calcium-magnesium-free phosphate-buffered saline (PBS) (Flow Laboratory, Costa Mesa, CA).

NaCl (BDH Chemicals).

KCl (Fisons, Beverly, MA).

$Na_2HPO_4$ (Ajax Chemicals, Sydney, NSW, Australia).

$KH_2PO_4$ (BDH Chemicals).

$CaCl_2 \cdot 2H_2O$ (BDH Chemicals).

$MgCl_2 \cdot 6H_2O$ (Ajax Chemicals).

Sodium pyruvate (Sigma): Dissolve 85 mg of sodium pyruvate in 10 mL of 0.9% NaCl. Dilute 1:50 in 0.9% (w/v) NaCl before use. It can keep for 2 wk at 4°C.

Phenol red (Sigma): Add 13 mg of phenol red and 129 mg of $NaHCO_3$ (BDH Chemicals) to 10 mL of $dH_2O$. It can be stored for 2 wk at 4°C.

Penicillin (Sigma): Add 599 mg to 1 mL of 0.9% (w/v) NaCl. Dilute 1:100 before use. Store at –20°C.

PB1 is prepared according to the formulation in **Table 1**. After preparation, the solution is equilibrated with 5% $CO_2$ in air for 5–10 min, 130 mg of glucose and 520 mg of bovine serum albumin (Fraction V, Sigma) are added, and the solution is sterilized with a 0.22-μm Millipore filter. The solution is stored in 10- or 40-mL aliquot at 4°C.

## 2.3. Labeling and Grafting of Cell in 6.5-Day Embryos

Embryo culture requirement as in **Subheading 1**.

Vertical pipet puller (David Kopf Instruments, Model 720).

Microforge (Narishige Scientific Instrument Laboratory, Greenvale, NY, MF79).

Holding, injection, and grafting pipets (**Fig. 1A–D**). Holding pipets are made from thick-walled glass capillaries (Leica, cat. no. 520-119). Injection and grafting pipets are made from glass capillaries (outer diameter: 1 mm, inner diameter: 0.75 mm, Thomas, Swedesboro, NJ).

Transfer pipets made from Pasteur pipets.

Diamond glass cutter (Thomas).

Micromanipulation apparatus: base plate with fixing elements for both manipulators (Leitz, cat. no. 335 520 139); manipulators (Leitz, cat. no. 335 520 137 and cat. no. 335 520 138); instrument holders (Leitz, cat. no. 335 520 142 and cat. no. 335 520 143), and instrument sleeves.

Laborlux S microscope with fixed mechanical stage (Leitz).

Dissecting microscope (Wild, MDG 17).

de Fonbrune syringe (Alcatel, Malakoff, France).

Micrometer syringe (Wellcome, London, UK).

Manipulation chamber (**Fig. 5**).

Tissue-culture dishes (60-mm, Corning).

**Table 1**
**The Compostition of PB1 Medium for Handling Embryos and Tissue Fragments**

| Stock solution (g/500 mL) | Volume, mL, to add to make 100 mL solution |
|---|---|
| NaCl (4.5) | 65.8 |
| KCl (5.75) | 1.8 |
| $Na_2HPO_4$ (10.93) | 5.2 |
| $KH_2PO_4$ (10.5) | 0.9 |
| $CaCL_2 \cdot 2H_2O$ (8.1) | 0.8 |
| $MgCl_2 \cdot 6H_2O$ (16.55) | 0.3 |
| Na pyruvate[a] | 21.4 |
| Phenol red[a] | 0.8 |
| Penicillin[a] | 0.3 |
| Distilled $H_2O$ | 0.27 |

[a]*See text* for method of preparation.

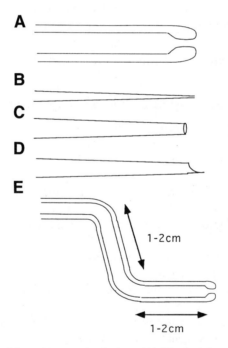

Fig. 1. Pipets used for micromanipulation. (**A**) The holding pipet, (**B**) the injection pipet used for labeling embryos, (**C**) the pipet used for grafting cell clumps, and (**D**) a beveled pipet that can also be used for grafting tissue. The pipets in (A–D) are used for manipulating 6.5-d embryos in hanging drops. (**E**) An angled holding pipet that is used for manipulating 7.5–8.5-d embryos in drops on a Petri dish. Angled grafting or labeling pipets can be made by introducing similar bends in pipets shown in (B–D).

Coverslips (24 × 50-mm, Mediglass, Sydney, NSW, Australia).

Light and heavy paraffin oil (BDH Chemicals).

1,1-Dioctadecyl-3,3,3'3'-tetramethylindocarbocyanine percholate (DiI) (Molec–ular Probes, Eugene, OR) and 3,3'-dioctadecyloxacarbobyanide percholate (DiO) (Molecular probes): Stock dye solutions (0.5% w/v) were prepared by dissolving the crystals in 100% ethanol. Dilute 1:10 (by volume) for DiI and 1:5 (by volume) for DiO in 0.3 $M$ sucrose (BDH Chemicals) immediately before use for cell labeling.

## 2.4. Labeling and Grafting Cells in 7.5-D Embryos

The materials are the same as outline in **Subheadings 2.1.** and **2.3.**

Fluovert FS (fixed-stage) inverted microscope (Leitz).

Holding, injection, and grafting pipets (**Fig. 1**).

Tissue-culture dishes (60 mm, Corning).

## 2.5. Transplanting and Labeling Cells in the Cranial Region of 8.5-D Embryos

The materials are the same as outlined in **Subheadings 2.1.** and **2.4.**

## 3. Methods

### 3.1. Culturing 6.5–8.5 D Embryos

### 3.1.1. Preparation of Culture Media

#### 3.1.1.1. PREPARATION OF DMEM

1. Dissolve the contents of one packet of powdered formula in 4.75 L (less 5% of the final volume required).
2. Stir gently at room temperature until the contents dissolve.
3. Add 11 g of $NaHCO_3$.
4. Bring the volume up to 5 L by adding reverse osmosis (RO) water.
5. Adjust the pH to 7.2–7.4 using 1 $N$ NaOH or 1 $N$ HCl.
6. Sterilize immediately by membrane filtration (pH value usually rises 0.1–0.3 on filtration).
7. Test sterility of a 50-µL aliquot from each bottle by incubating at 37°C for 3–5 d.
8. Before DMEM is used for culture, add 10 mL each of glutamine and penicillin/streptomycin solution to 1 L of DMEM. Fresh glutamine and penicillin/strepto-mycin should be added to the DMEM working solution after 2 wk. The working solution is good for 4 wk after preparation when kept at 4°C.

#### 3.1.1.2. COLLECTION OF RAT SERUM (RS)

1. Anesthetize the rat using 5% Halothane in 1–1.5 L of $O_2$.
2. Blood is collected from the anesthetized rat by drawing blood from the aorta into a nonheparinized syringe using a G20 hypodermic needle.
3. Dispense the freshly collected blood in 15-mL centrifuge tubes and centrifuge at 3000 rpm for 10 min (*see* **Notes 3–5**).

4. Grasp the fibrin clot with a flame-sterilized forceps, and spool it around the shaft of the forceps to squeeze out the serum trapped in the clot.
5. Transfer the serum to a new centrifuge tube using an autoclaved Pasteur pipet, and spin again at 3000 rpm for 10 min.
6. Collect the serum from the second spin aseptically and store at –20°C.

### 3.1.1.3. COLLECTION OF HUMAN CORD SERUM (HCS)

1. Collect cord blood from the placenta obtained following Caesarean delivery, and keep the blood on ice.
2. Centrifuge the blood as soon as possible in 15-mL centrifuge tubes at 3000 rpm for 10 min at 4°C.
3. Removed the serum and store at –20°C (*see* **Note 3–5**).

### *3.1.2. Static Culture in 4-Well Chamber Slides at 37°C in 5% $CO_2$ in Air*

1. Thaw the required volume of HCS and RS, and inactivate the sera by heating for 30 min in a water bath at 56°C, 3.2 mL of culture medium are required for one chamber slide, and this is made up of 0.8 mL HCS, 1.6 mL RS, and 0.8 mL DMEM (1:2:1 by volume).
2. Put 0.5 mL in the first well (**Note 3**) and 0.9 mL in the remaining three wells. This volume is sufficient for culturing groups of 8–10 6.5-d embryos/well (**Note 4**) for up to 48 h (**Note 5**).
3. 7.5-D embryos can be cultured for up to 24 h by the static culture method.

### *3.1.3. Roller Bottle Culture (with Continuous Gassing) for 7.5–8.5-D Embryos*

1. Embryos are cultured for 24 h in a medium of DMEM:RS (1:1 by vol). Use 1 mL of culture medium/embryo. Four to five embryos can be cultured in one 30-mL bottle. Bottles are maintained in a BTC embryo culture chamber at 37°C in 5% $O_2$, 5% $CO_2$, and 90% N, and are rotated on a roller/rotator at 30 rpm.
2. Embryos are transferred to fresh medium of DMEM:RS (1:3 by vol) if culturing for another 24 h is required. The gas phase is changed to 20% $O_2$, 5% $CO_2$, and 75% $N_2$ *(10)*.

## *3.2. Isolation of Tissue Fragments for Grafting*

### *3.2.1. Isolation of Epiblast Fragment from 6.5-D Embryos*

1. Position the embryo for easiest access to the epiblast cells required for transplantation. For example, if anterior or posterior epiblast cells are to be isolated, the embryo could be positioned with the sagittal plane in view (**Note 6**). However, in order to dissect cells from the lateral epiblast, the embryo is best oriented with the frontal view in sight.
2. Pin the embryo to the Petri dish using a sharp metal needle. Hold it at the site immediately adjacent to the tissue fragment required for grafting. Another tungsten needle is used to slice through the epiblast in a scissor-like action against the first needle. Another cut is then made at an angle to the first, so that the tissue fragment is released from the epiblast. The dissection is shown in **Fig. 2**.

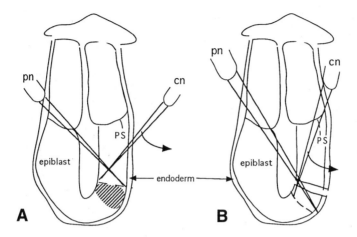

Fig. 2. Schematic representation of the step taken to isolate epiblast fragments from sites adjacent to the distal cap region on the posterior side of a 6.5-d embryo. **(A)** Position of the pinning needle (pn) and the line of cut that will be made by the cutting needle (cn) just proximal to the required tissues (shaded area). **(B)** Second cut that is made to isolate the tissue fragment. Similar cutting actions are employed to isolate tissue fragments from other regions of the embryo. Curved arrows indicate the direction of slicing made by the cutting needle. Abbreviation: ps, primitive streak.

3. The endoderm usually remains attached to the epiblast. Make several scratch marks on the bottom of the Petri dish with a metal needle. To remove the endoderm, place the tissue fragment endoderm side down on the grid, and nudge the fragment onto the scratched surface. When the endoderm sticks to the surface, the epiblast layer can be torn away with metal needles or glass needles.
4. Cut the epiblast fragments into clumps of 5–10 cells for grafting.

### 3.2.2. Separation of the Germ Layers of 7.0–7.5-D Embryos

1. Cut the embryo at the junction between the embryonic and extraembryonic tissue (**Fig. 3A**).
2. Incubate the embryonic tissue in a trypsin-pancreatin solution (*see* **Subheading 2., item 2**) for 5–10 min or until the endoderm is loosened from the mesoderm.
3. Transfer the embryonic fragment to three changes of PB1 + 10% FCS to stop the enzyme digestion.
3. Separate the germ layers as shown in **Fig. 3B–D**.
4. Cut the isolated germ layers into clumps of 5–10 cells using glass needles.

### 3.2.3. Isolation of Mesoderm and Premigratory Neural Crest Cells of 8.5-D Embryos

1. Remove the yolk sac and the amnion using fine watchmaker forceps.
2. Bisect the embryo using tungsten needles along its midline.

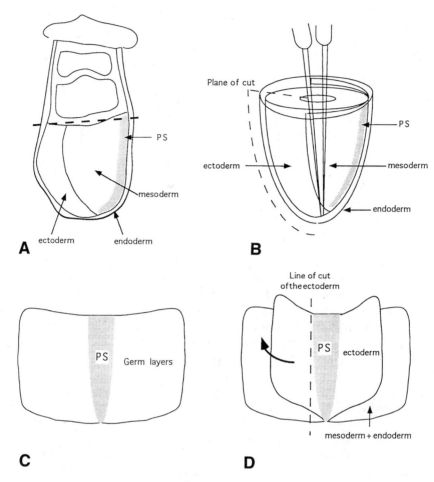

Fig. 3. Schematic representation of steps of germ layer separation of 7.0- to 7.5-d embryos. (**A**) Tissue organization of the 7.5-d embryo. The dashed line marks the position of the first cut at the junction between the embryonic and extraembryonic parts of the egg cylinder. (**B**) Embryonic portion of the embryo that reveals the anatomical relationship of the three germ layers. A metal needle is pushed into the amniotic cavity and pins the egg cylinder in the upright position. The second metal needle is then brought inside the amniotic cavity, and the embryo is cut along the anterior side (indicated by the dashed line) by slicing the second needle through all germ layers. (**C**) Embryo opened out flat and ready for enzyme digestion. (**D**) Ectoderm layer that has been loosened by enzymatic digestion. The ectoderm can be lifted up and torn away from the mesoderm using needles. It is not possible to separate the germ layers at the site of the primitive streak. The ectoderm is therefore cut along the dashed line where no further separation from the underlying primitive streak could be made. The mesoderm layer can be separated from the endoderm in the same manner, followed again by cutting them close to the primitive streak. The remaining tissue is the primitive streak (PS).

3. Make transverse cuts (using metal needles) along the neuromeric junctions to isolate wedge-shaped fragments containing the tissue to be transplanted by *(7,8)*. For example, to isolate somitomere IV and the middle hindbrain neural crest cells, transverse cuts should be made at the preotic sulcus and the otic sulcus.
4. Incubate the embryonic fragments in the trypsin-pancreatin solution for 20 min at 37°C. When the mesoderm and ectoderm layers are loosened, separate the tissue layers using glass needles.
5. Dissect the isolated mesoderm or neuroectoderm into small clumps of to 10 cells (*see* **Note 7**).

## *3.3. Labeling and Grafting of Cell in 6.5-D Embryos*

### *3.3.1. Making Pipets*

#### 3.3.1.1. ANGLED PIPETS

1. Make holding, grafting, and injection pipets as described in **Subheadings 3.3.1.2.** and **3.3.1.3.**
2. Heat the pipet, over a flame, 1–2 cm from the tip of the injection pipet. It is heated until it is bent to an angle of approx 100°.
3. Turn the pipet 180°, and heat the pipet at a position 1–2 cm further proximal of the first bend. Again the pipet is bent at an angle of approx 100° (**Fig. 1E**).

#### 3.3.1.2. MAKING HOLDING PIPETS

The holding pipets are made using Leitz thick-wall glass capillaries. The internal diameter of holding pipets should be 10–50 μm depending on the size of the embryo used for the experiment.

1. Hold the middle portion of the capillary over a small flame until the glass begins to melt.
2. Take the capillary away from the flame, and pull from both ends to produce a fine segment of capillary.
3. Break the fine capillary with a diamond pencil.
4. Polish the holding pipets on a microforge. A small glass bead is melted onto the platinum wire of the microforge. Heat the glass bead up by increasing the electric current until the platinum wire glows red hot. Bring the tip of the pipet as close as possible to the glowing glass melts under the heat of the glass bead. Retract the pipet from the glass bead when a polished tip is produced (**Fig. 1A,** *see* **Note 8**).

#### 3.3.1.3. MAKING INJECTION AND GRAFTING PIPETS

Injection pipets with an inner diameter of about 1–2 μm are made from thin-wall glass capillaries. The internal diameter of grafting pipets is larger and varies according to the size of the clumps of cells to be grafted.

1. Pull glass capillaries on a vertical pipet puller to produce pipets with a fine tip, a long shaft, and a short shoulder.

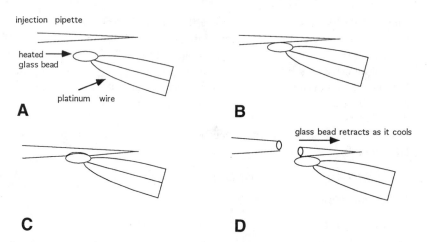

Fig. 4. Making grafting pipets. A glass bead is attached to the tip of the heating filament of the microforge. The bead can be heated and cooled instantly by switching on and off the current of the heating filament. The current is adjusted to yield optimal heating that is appropriate to the thickness and the thermal property of the glass capillary. (**A,B**) The heated glass bead contacts the shaft of the pipet at the position that will give the desired internal diameter. (**C**) Melting of the capillary wall into the glass bead. (**D**) Resultant break in the pipet after the bead cools down and retracts when the current is switched off. This produces a pipet of the desired internal diameter and an even tip.

2. Break off the tip of the injection pipet by touching it against the cold glass bead on the filament of a microforge. This will produce a pipet for dye labeling (**Fig. 1B**).
3. In order to make grafting pipets, pull the glass capillary as in **step 1**. Determine a position on the shaft of the pipet by measuring with a ocular micrometer on the eyepiece of the microscope of the microforge, where the glass capillary has an inner diameter appropriate for the size of the cell clumps . Break the shaft of the pipet (**Fig. 4**) by (1) bringing the pipet to the heated glass bead on the microforge filament so that the site of intended break just touches the bead, (2) as the capillary begins to fuse with the glass bead, turn off the power supply to the filament to cool down the filament instantly. The retraction of the filament as it cools down will snap the pipet precisely at where the capillary fuses to the bead.
4. Beveled pipet (**Fig. 3D**) are made from the grafting pipet made in **step 3** (**Fig. 3C**) (*see* **Notes 9** and **10**). Touch one side of the pipet with the heated glass bead. When the pipet tip fuses with the bead, slowly withdraw the pipet to pull a sharp bevel of about 5 µm in length (*see* **Notes 11–15**).

### 3.3.1.4. Making Transfer Pipets for 6.5–7.5 D Embryos

1. Heat the shaft of the Pasteur pipet over a flame.
2. When the glass begins to melt, take the pipet away from the flame, and hand-pull the pipet to produce a segment of thin capillary.

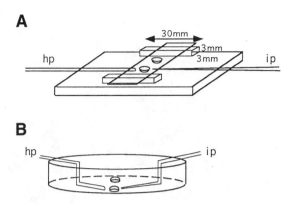

Fig. 5. **(A)** Schematic representation of the manipulation chamber used for manipulating 6.5-d embryos in hanging drops. The base of the manipulation chamber is made of glass of good optical quality supplied by the optometrist. The dimension of the glass strips supporting the coverslip is 3 × 3 × 30 mm. They are set between 4 and 6 mm from the edge. Two drops of solution are placed on the coverslip approx 1 cm apart, and then the coverslip is inverted over the paraffin oil-filled chamber. hp, Holding pipet; ip, injection pipet. **(B)** Schematic representation of the Petri dish set-up used for manipulating 7.5- and 8.5-d embryos. Two drops are placed approx 1 cm apart in the center of the dish and are then covered with paraffin oil. Angled pipets are used: hp, holding pipet; ip, injection pipet.

3. Break the capillary by bending the pipet. Check the size of the pipet tip under the dissecting miscroscope (0.5 mm for 6.5-d embryos and 1–1.5 mm for 7.5-d embryos).
4. Connected the pipet with a flexible plastic tubing to a mouthpiece. The movement of fluid and embryo in the pipet is controlled by suction or blowing into the mouthpiece.

### 3.3.2. Labeling the Epiblast Cells of 6.5-D Embryos with Carbocyanine Dye

1. Embryos are manipulated in a hanging drop under a Laborlux S microscope. Set up the micromanipulation chamber as in **Fig. 5A**.
2. Place a 5-μL drop of media (10% FCS in PB1) for holding the embryos during manipulation on a coverslip using a transfer pipet. A second 5-μL drop of dye is placed approx 1 cm from the first drop (*see* **Note 11**; **Fig. 5A**).
3. Invert the coverslip, and place it across the two glass strips of the chamber.
4. Fill the chamber with light paraffin oil.
5. To transfer embryos into the manipulation chamber, place the chamber under a dissecting microscope with the drops in focus. Pick up the embryos using a mouth-controlled transfer pipet. Pass the pipet through the oil to reach the drop, and gently expel the embryo with a small volume of medium.
6. Set up the manipulation apparatus as shown in **Fig. 6**. Attach the holding pipet to

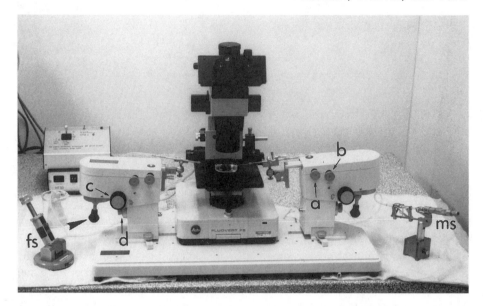

Fig. 6. The micromanipulation assembly. The manipulators are clamped to the base plate. Each manipulator can hold up to two instrument holders. Movement of the instruments is controlled by screw-knobs (**a,b**) that allow positioning of the manipulator in the *X-Y* horizontal plane. There is a third control (**c**) by which the angle of inclination of the manipulator can be adjusted. Coarse and fine adjustment of the position of the manipulator in the *Z*-vertical plane is controlled by the fourth knob (**d**). Fine *X-Y* movement of the instrument under the microscope is controlled by joysticks (arrowhead). The holding instrument is on the lefthand side of the assembly and the suction and expulsion of fluid are controlled by the micrometer syringe (ms) on the right hand-side. Conversely, the manipulating (labeling or grafting) instruments are on the right-hand side and are controlled by the de Fonbrune syringe (fs) on the left-hand side. This set-up enables the positioning of the instruments and the control of syringe to be accomplished simultaneously.

   the instrument holder on the left manipulator. Attach injection pipet to the instrument holder on the right manipulator. The holding pipet is controlled by the micrometer syringe on the right, whereas the injection pipet is controlled by the de Fonbrune syringe on the left. With this configuration, the positioning of the manipulator and the control of the syringe can be performed simultaneously.

7. Back-fill the holding and injection pipets with heavy paraffin oil by adjusting the syringes.
8. Position the manipulation chamber, and adjust the microscope to focus on the medium drop containing the embryos.
9. Bring the holding pipet into the field of view by pushing it through the oil into the drop. Draw a small amount of medium into the holding pipet to create an oil-medium meniscus.

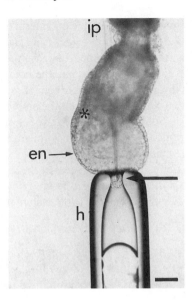

Fig. 7. Labeling of the epiblast by microinjecting DiI into the distal cap of the embryo. To inject in the midline, the embryo is held at the distal tip of the egg cylinder by gentle suction with the holding pipet (h). The injection pipet (ip) passed through the extraembryonic tissues of the egg cylinder. The injection pipet is brought to the site of labeling from within the pro-amniotic cavity to avoid inadvertent labeling of other embryonic germ layers. The arrow points to the tip of the labeling pipet in the epiblast layer. en, primitive endoderm; *primitive streak. Bar = 20 μm.

10. Bring the injection pipet into view. Make sure that no medium is taken into the pipet, because the contact with an aqueous solution causes precipitation of the carbocyanine dye.
11. Retract the holding and injection pipets from the medium drop (**Note 12**). Bring the dye drop into the field of view by moving the stage of the microscope (**Note 13**).
12. Dip the tip of the injection pipet into the dye drop, and draw a small amount of dye into the pipet using the de Fonbrune syringe. Always keep the dye-oil meniscus in view, since this offers the only means for monitoring the flow of dye during labeling. Take the injection pipet out of the dye when the dye-oil meniscus has stopped moving.
13. Bring the medium drop containing the embryos back into the field of view.
14. Position the embryo by pushing it with the pipets and rotating it by lifting the pipet from beneath the embryo.
15. After the embryo is oriented correctly, bring the embryo into focus using the focusing control of the microscope, and then bring the injection and holding pipets into focus using the control knob on the manipulators (**Fig. 6**). Touch the injection and holding pipets lightly against the embryo to confirm that they are in the same plane.

16. Bring the holding pipet into contact with the endoderm of the embryo next to the epiblast to be labeled. Apply a suction force using the micrometer syringe to draw the embryo against the holding pipet. Increase the suction so that a small area of the endoderm layer is partly drawn into the holding pipet (**Fig. 7**).

17. Push the injection pipet through the extraembryonic tissue into the amniotic cavity (*see* **Note 14**). Then push the pipet tip into the epiblast layer (**Fig. 7**).

18. Apply pressure via the de Fonbrune syringe to expel a small volume of dye into the epiblast. It is important to monitor the movement of the dye front to avoid injecting any oil into the embryo.

19. Once the dye front has stopped advancing, retract the injection pipet (**Note 15**).

20. Release the embryo from the holding pipet by applying a positive pressure via the micrometer syringe. Return the embryo to the culture medium (**Subheading 1.**).

### 3.3.3. Grafting Epiblast Fragments to 6.5-D Embryos

1. Place two drops of PB1 (+10% FCS) medium approx 1 cm apart on the coverslip, and set up the manipulation chamber as **Subheading 3.3.2., steps 1–4**.

2. Set up the manipulation apparatus, and position the pipets as described in **Subheading 3.3.2., steps 5–9**.

3. Position and hold the embryo as in **Subheading 3.3.2., steps 14–16**, except that the embryos is held on the side opposite to the site of grafting (**Fig. 8**).

4. Pick up the cell clumps into the grafting pipet with a small amount of medium. Draw the cell clumps about 20 μm into the grafting pipet.

5. Insert the grafting pipette by a sharp jabbing action through all tissue layers into the pro-amniotic cavity at the site of transplantation. If necessary, a beveled pipet may be used.

6. Apply a positive pressure to the de Fonbrune syringe to expel the graft from the pipet as the pipet is slowly withdrawn from the embryo (**Note 16**). A coordinated movement of the pipet and the expulsion of the graft is critical to the precise positioning of the graft in the epiblast (**Note 17**).

7. Return the embryo to the culture medium (**Note 18**) (**Subheading 1.**).

### 3.3.4. Labeling and Grafting of Cells

1. Place a 50-μL drop of medium (the manipulation drop) slightly off-center in the lid of a Petri dish (**Note 19**). Place another drop of medium (for holding cells for grafting) or a drop of carbocyanine dye (for labeling of embryos) next to the first drop (**Fig. 4B**).

2. Fill the Petri dish with light paraffin oil to cover the drops.

3. Transfer embryos from the culture to the manipulation drop in the Petri dish by mouth pipeting.

4. Angled pipets are used for holding and labeling the embryos. Place the pipets in the instrument holders of the manipulator. Tilt the manipulators so that the tips of the pipets are immersed in the paraffin oil and are pointing to the bottom of the Petri dish. **Figure 6** shows the setup for manipulating 7.5-d embryos.

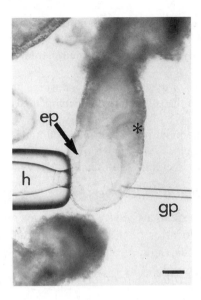

Fig. 8. Grafting of cells to a 6.5-d embryo. The embryo is held on the anterior side opposite the intended site of grafting by the holding pipet (h). The cells are grafted into the epiblast (ep) on the posterior side of the embryo (*marks the position of the primitive streak) immediately proximal to the margin of the distal cap. Bar = 20 μm.

5. Focus the microscope on the embryos in the manipulation drop. Lower the pipets into the drop. Manipulation is carried out according to **Subheading 3.3.2., steps 7–20** and **Subheading 3.3.3., steps 3–8** and **Note 20**.

## 3.4. Transplanting and Labeling Cells in the Cranial Region of 8.5-D Embryos

### 3.4.1. Making Transfer Pipets

Embryos that are 8.5 d old are transferred using Pasteur pipets that are cut at the shaft to give an internal diameter of about 3–4 mm. This allows the embryos to be transferred without any damage to the yolk sac.

### 3.4.2. Manipulation

1. Embryos that are 8.5 d old are manipulated in a Petri dish in the same way as described in **Subheading 3.3.4., steps 1–4**.
2. Orient the embryo using fine watchmaker forceps.
3. Hold the embryo by the yolk sac with the holding pipet by applying negative pressure to the micrometer syringe. The embryo is positioned so that a clear silhouette of the cranial neural fold is visible.

4. Insert the injection/grafting pipet through the yolk sac, and then the amnion into the amniotic cavity.
5. Push the pipet gently into the cranial neural fold or further into the cranial mesoderm.
6. Expel the cell clumps or dye by applying a positive pressure to the de Fonbrune syringe. Expulsion of the cell clumps is accompanied by the simultaneous withdrawal of the pipet.
7. For dye delivery, the pipet remains in position until the dye front has stopped moving toward the tip.
8. Returned the embryo to culture (**Subheading 1.**).

### 3.5. Analysis of Labeled Embryos

Dye labeled embryos are analyzed by fluorescence microscopy. Fluorescent detection indicates the location of the labeled cells within the embryos. More detailed analysis can be carried out by confocal microscopy.

### 3.5.1. Preparation of Embryos for Confocal Microscopy

1. For confocal microscopy embryos are analyzed in whole mount. Embryos that are 8.5 d old and older need to have the extraembryonic tissues dissected away. Embryos that are 6.5 and 7.5 d old are analyzed intact.
2. Embryos are mounted flat under a coverslip (coverglass number $1^1/_2$). To allow sufficient space for the embryos between the slide and the coverslip, small blue-tac feet are placed at the four corners of the coverslip (*see* **Note 21**).
3. The slide is filled with PBS and sealed with nail polish.

### 3.5.2. Confocal Analysis

A confocal microscope allows you to obtain an image of the fluorescent label and a transmission image of the embryo. The two images can be overlaid to identify where along the anterior posterior axis of the embryo the labeled cells are located. The depth of the sections of the fluorescent label gives some indication of which tissue layer contains the labeled cells.

### 3.5.3. Analysis of Grafted Embryos

The distribution of graft-derived cells may be analyzed in whole mount or on serial sections of the embryo if the cells are marked by *lacZ* transgene. For protocols used to visualize the transgenic cells, *see* Chapter 7.

## 4. Notes

### 4.1. Preparation of Culture Media

1. It is important that the rat and cord blood are spun as soon as possible after collection, preferably before clotting commences. The serum obtained by squeezing the fibrin clot is known as the immediately centrifuged serum.

2. Serum used as medium supplement should have a low lipid content (clear instead of cloudy). Hemolyzed serum does not support normal culture of mouse embryos.

3. The medium in the first well of the NUNC slide is used to wash the embryo once before transferring to the culture wells.

4. Growth rates decline if the embryos are overcrowded and, likewise, if they are cultured alone.

5. Best development for 48 h is achieved with embryos explanted shortly after the appearance of the primitive streak.

6. The appearance of the primitive streak provides an unequivocal indication of the anterior–posterior axis and helps with the orientation during dissection. Early gastrulation embryos are therefore used in experiments that demand knowledge of the source of cells in the anterior–posterior and medial–lateral axis. However, if only a distinction between proximal and distal epiblast is required, both pre- (without the primitive streak) and early gastrulation embryos can be used.

7. The choice of developmental stage is important in order to avoid the contamination of the cranial mesoderm by migrating neural crest cells. Previous studies have established that the first population of neural crest cells to leave the neural plate are those of the mesencephalon at the 5–6 somite stage *(11–13)*. Therefore, in experiments involving the cranial mesoderm, only embryos having ≤5 somites are used so that the mesodermal explants are free of any migrating neural crest cells. In order to obtain an enriched neural crest cell population for cell fate analysis, premigratory neural crest cells are isolated from the lateral region of the neural plate of these embryos for transplantation experiments.

8. It is important that the holding pipette has a smooth surface so that the embryo is not damaged when it is held against the tip by suction.

9. If the internal diameter of the grafting pipet is large relative to the size of the embryo, it may be difficult to puncture the tissue layers with the pipet during grafting. Beveled pipets can be used in this situation. The sharp point of the beveled pipet slices the tissue layers to create a passage for the grafting pipet.

10. Do not to overheat the glass bead, since this may cause distortion and contriction of the pipet tip. It is useful to mark the side of the pipet with the bevel for later orientation.

11. The drops should be placed far apart enough to avoid mixing of contents as the pipets are moved around in the chamber.

12. When moving between the dye drop and the media drop, the holding pipet should be retracted, so that its tip does not become contaminated with the dye.

13. When drawing up the dye solution, it is best to keep the edge of the dye drop in sight, so that only the tip of the pipet is dipped into the dye drop. Pushing the pipet too far into the drop will result in excessive amount of dye adhering to the outer surface of the pipet.

14. Labeling of the embryonic tissue is best accomplished through the extraembryonic route, because it reduces the chance of accidentally labeling other germ layers. Even if dye is leaking from the pipet or coming off the surface of the pipet during the passage of the pipet, only the extraembryonic tissues will be inadvertently labeled.

15. The pipets may clog with precipitate of the dye after repeated uses. The clogged pipet can be recovered by breaking its tip by brushing against the holding pipet while in the chamber or by wiping it gently with a lint-free tissue when it has been taken out of the chamber.

16. If the cell clumps become sticky, they will follow the grafting pipet out of the embryo. Leave the pipet with the graft partly out of the tip at the transplantation site for 10–30 s to allow the graft to adhere to the surrounding tissue. Following that, a snappy retraction of the grafting pipet or tapping gently on the base plate of the manipulator may dislodge the graft from the pipet tip. Siliconizing the injection glassware also stops the cell clumps from adhering to the injection pipets.

17. A small graft is readily incorporated into the host embryo. Larger grafts may be squeezed out of the embryo as the wound closes. Pushing the tip of the grafting pipet against the graft that has been lodged in the germ layer for approx 10–30 s helps the retention of the graft.

18. Occasionally, the host embryos will collapse because of fluid leakage through the wound made to accommodate the graft. Inflating the embryonic cavity by injecting a small amount of medium to replenish the loss of fluid during transplantation may improve the development of the embryo.

19. The lids of Petri dishes are used because the rim has a lower clearance, which gives better accessibility of the angled pipets to the embryos and tissues in the dish.

20. The principle for manipulating 7.5-d embryos is the same as for 6.5-d embryos. However, manipulations are carried out in a Petri dish and this therefore requires the use of the fluovert FS inverted microscope.

21. Despite the blue-tac feet, the embryos are squashed to some extent. With 6.5- and 7.5-d embryos, it can be difficult to orient the embryos once it has been squashed. For ease of analysis, mount the embryos in the sagittal plane, and mark the direction of the anterior–posterior axis on the slide.

## References

1. Lawson, K. A., Meneses, J. J., and Pedersen, R. A. (1991).Clonal analysis of epiblast fate during germ layer formation in the mouse embryo. *Development* **113,** 891–911.

2. Quinlan, G. A., Williams, E. A., Tan, S.-S., and Tam, P. P. L. (1995).Neuroectodermal fate of epiblast cells in the distal cap region of the mouse egg cylinder: implication for body plan organisation during early embryogenesis. *Development* **121,** 87–984.

3. Parameswaran, M. and Tam, P. P. L. (1995) Regionalisation of cell fate and morphogenetic movement of the mesoderm during mouse gastrulation. *Develop. Genet.* **17,** 16–28.

4. Tam, P. P. L. and Beddington, R. S. P. (1987) The formation of mesodermal tissues in the mouse embryo during gastrulation and early organogenesis. *Development* **99,** 109–126.

5. Tam, P. P. L. and Beddington, R. S. P. (1992) Establishing and organisation of germ layers in the gastrulating mouse embryo, in *Postimplantation Development in the Mouse*, Ciba Foundation Symp. 165, Wiley, Chichester, pp. 27–49.

6. Tam, P. P. L. (1989) Regionalisation of mouse embryonic ectoderm: allocation of prospective ectodermal tissue during gastrulation. *Development* **107,** 55–67.
7. Trainor, P. A., Tan, S.-S., and Tam, P. P. L. (1994) Cranial paraxial mesoderm: regionalisation of cell fate and impact upon craniofacial development in mouse embryos. *Development* **120,** 2397–2408.
8. Trainor, P. A. and Tam, P. P. L. (1995) Cranial paraxial mesoderm and neural crest cells of the mouse embryo: co-distribution in the craniofacial mesenchyme but distinct segregation in branchial arches. *Development* **121,** 2569–2582.
9. Beddington, R. S. P. (1994) Induction of a second neural axis. *Development* **120,** 613–620.
10. Sturm, K. and Tam, P. P. L. (1992) Isolation and culture of whole postimplantation embryos and their germ layer derivatives, in *Methods in Enzymology. Guide to Techniques in Mouse Development* (Wassarman, P. and DePamphilis, M., eds.), Academic, New York, pp. 164–190.
11. Jacobson, A. G. and Tam, P. P. L. (1982) Cephalic neurulation in the mouse embryo analysed by SEM and morphometry. *Anat. Rec.* **203,** 375–395
12. Chan, W. Y. and Tam, P. P. L. (1988) A morphological and experimental study of the mesencephalic neural crest cells in the mouse embryo using wheat germ agglutinin-gold conjugates as the cell marker. *Development* **102,** 427–442.
13. Serbedzija, G. N., Bronner-Fraser, M., and Fraser, S. E. (1989) A vital dye analysis of timing and pathways of avian neural crest cell migration. *Development* **106,** 809–816.

# 6

## Production of Transgenic Rodents by the Microinjection of Cloned DNA into Fertilized One-Celled Eggs

### San Ling Si-Hoe and David Murphy

## 1. Introduction

A pioneering experiment in the early 1980s demonstrated that microinjection of recombinant growth hormone into the pronuclei of fertilized one-celled mouse embryos resulted in inheritable changes in the growth of these mice *(1)*. Mammalian transgenic experiments have since contributed tremendously to our understanding of numerous complex biological processes. The power of the technique lies in that it allows the function, and developmental and physiological regulation of almost any protein to be studied within the context of the normal processes occurring in the whole animal.

It is unknown how DNA integrates into the host chromosome, but some information can be inferred from a study made on the state and organization of the inserts found in transgenic mice *(2)*. Approximately 70% of the mice carry exogenous DNA in all their somatic and germ cells, implying that integration took place prior to the first round of DNA replication. The remaining 30% of the mice showed some degree of mosaicism, indicating that the DNA must have integrated after this first round of replication. Integration events have been observed on many different autosomes *(3)*, on the X-chromosome *(4)*, and on the Y-chromosome (Murphy, unpublished observations). Transgene copy number varies (one to several hundred), but within a transgenic animal, there is usually only one integration site where if in multiple copies, the transgene is arranged in a head-to-tail tandem array.

Once incorporated in the DNA, the transgene transmits in a classic Mendelian fashion. Appropriate tissue-specific and physiologically regulated expres-

From: *Methods in Molecular Biology, Vol. 97: Molecular Embryology: Methods and Protocols*
Edited by: P. T. Sharpe and I. Mason © Humana Press Inc., Totowa, NJ

sion of the transgene is a prerequisite in many transgenic experiments, and can only be obtained by including the appropriate genetic elements in the structure. Once appropriate expression is obtained, gain-of-function (overexpressing the gene of interest) or loss-of-function (cell-specific expression of cytotoxic proteins) approaches can be employed to derive information about almost any gene or cell type. Studies of the nervous and immune systems and oncogenesis *(5–7)* are some examples where transgenic animal studies have contributed significantly to our knowledge.

This chapter details the procedures necessary for the generation of transgenic mice and rats by the injection of cloned DNA into the pronuclei of fertilized one-celled eggs. The technique demands precise technical skill and expensive equipment, but its reliability and speed make it currently the most efficient way of generating transgenic mice. It is still the only route of making transgenic rats with inheritable genetic alterations. There are other methodologies, such as the infection of preimplantation embryos with retroviruses or homologous recombination in embryonic stem cells.

### 1.1. Summary of the Microinjection Method

1. Immature donor females are superovulated and mated with stud males (**Subheading 5.**).
2. Approximately 12 h postcoitus (pc), the oviducts from donor females are collected, and eggs are harvested and placed into culture (**Subheading 8.**).
3. The eggs are microinjected with purified cloned DNA (**Subheading 11.**).
4. Eggs that survive microinjection are returned to the natural environment by implantation into pseudopregnant surrogate mothers produced by mating sexually mature females with vasectomized males (**Subheading 13.**).
5. A percentage of the transplanted eggs will survive to full term, and will be delivered either naturally or by Caesarean section (**Subheading 14.**).
6. Transgenic animals are identified by analysis of genomic DNA isolate from the tail tissue.
7. Transgenic animals are bred to produce a line and analyzed for transgene expression.

The efficiency of producing transgenic mice or rats varies considerably between experiments. Mice have been more often used in transgenic experiments, and the efficiencies of each stage are well recorded. Under optimal conditions, 60–80% of the eggs survive injection, and 10–30% implant in the pseudopregnant recipient, proceed through normal development, and are born. Ten to 30% of the pups are transgenic. In rats, the available data are that 50–75% survive injection, 10–30% implant, and are born as pups of which 10–33% are transgenic *(8)*.

## 2. Animal Welfare

The conduct of scientific experiments involving animals is strictly regulated by most governments and institutions. Scientists must consult with the relevant regulatory bodies before commencing on any experiment involving genetic engineering and/or live animals. Apart from the legalities, the scientist must always abide by the basic principle that the animals under his or her care should not suffer any avoidable stress. Guidelines for such care are published by the National Institutes of Health *(9)*.

The techniques using live animals detailed in this chapter are only a guide for the novice. All manipulations involving animals must be taught to a novice by a skilled and experienced operator. For the surgical procedures (vasectomy and oviduct transfer), it is advised that the inexperienced novice first practices on cadavers until confidence is gained to operate on live animals. If an animal at any point of the experiment appears to be suffering from undue stress, kill it immediately by decapitation or cervical dislocation.

## *2.1. Killing Mice and Rats*

The recommended way of killing mice is by breaking the neck (cervical dislocation). This is quick, and causes the animal the minimum of distress. Pick the mouse up by the base of the tail, and place it on the top of the cage. Allow it to run away such that the animal is stretched out with its hindlegs almost in the air and its forelimbs gripping the cage bars firmly. Using a blunt instrument, press down firmly on the base of the skull and pull on its tail. The stretching action breaks the neck, the pressure at the base of the skull defining the point of dislocation.

Kill rats either by cervical dislocation or by decapitation with a rodent guillotine. The rats' greater size and strength require that they are stunned before either procedure. Hold the animal firmly by the base of its tail, and swing it to hit the back of the rats' head firmly on a hard surface (e.g., a wooden tabletop). While the animal is stunned, dislocate its neck using a large pair of scissors. Alternatively, a guillotine (e.g., Harvard Apparatus Ltd., Kent, UK) can be used to decapitate the animal after it is stunned.

## *2.2. Anesthesia*

### *2.2.1. Materials*

#### 2.2.1.1. MICE

1. Avertin: A 100% stock is made by mixing 10 g of tribromoethylalcohol with 10 mL of tertiary amyl alcohol. The 2.5% working solution is diluted from stock in sterile water. Both are stored at 4°C, protected from light.
2. Sterile 1-mL disposable syringe.
3. Sterile $0.5 \times 16$ mm disposable needles.
4. Animal balance

#### 2.2.1.2. RAT

1. CRC: Mix one part Dormicum™ (5 mg/mL midazolam, Hoffmann-La Roche, Basel, Switzerland), one part Hypnorm™ (10 mg/mL fluanisone and 0.2 mg fen-

tanyl, Janssen Pharmaceuticals, Beerse, Belgium), and two parts sterile water. The madeup solution is stable at room temperature for 1 wk.
2. Sterile 1-mL disposable syringe.
3. Sterile 0.5 × 16 mm disposable needles.
4. Animal balance.

## 2.2.2. Methods

### 2.2.2.1. MICE

1. Weigh the mouse to determine the dose of Avertin to administer (15–17 µL 100% Avertin/g body wt).
2. Draw up the appropriate volume into a sterile 1-mL disposable syringe. Exclude any air bubbles. (Have the needle of the syringe facing up, flick gently to dislodge any bubbles, and push the barrel of the piston until the bubbles are expelled.)
3. Restrain the animal, and introduce the needle into the abdomen of the mouse, avoiding both the diaphragm and the bladder. Withdraw the barrel of the syringe slightly—a small air bubble is observed if the needle has been inserted correctly into the intraperitoneal space. Inject the anesthetic and wait awhile before withdrawing the needle. Accidental subcutaneous injection is revealed by backleakage of drops of Avertin through the skin. The mouse will remain fully anesthetized for 0.5–1 h.
4. An adequate depth of anesthesia is indicated by an absence of a blinking reflex while blowing on the eyes and the maintenance of rapid breathing. Following surgery, the animal is observed to have regained mobility before being left to recover in a quiet, warm place.

### 2.2.2.2. RATS

1. Weigh the rat to determine the dose of CRC to administer (0.275 µL CRC/g body wt). Inject the rat intraperitoneally as described for the mouse (above). If the rat is too big to be restrained with one hand, use one of the many commercially available plastic animal restrainers or ask another person to assist in restraining the rat.
2. The rat will remain fully anesthetized for 20–30 min. An adequate depth of anesthesia is indicated by an absence of a jerk reflex on pinching the animal's paw and the maintenance of rapid breathing.
3. Following surgery, the animal is observed to have regained some mobility before being left to recover in a quiet, warm place.

## 2.2.3. Notes

1. If an adequate level of anesthesia does not develop, inject an additional 25–50% of the initial volume of anesthesia.
2. Two other anesthetics that combine a suitably deep level of anesthesia with rapid reliable recovery in rats are:
   a. 1:1 Mixture of Vetalar (100 mg/mL ketamine hydrochloride; Parke Davis Veterinary, Morris Plains, NJ) and Rompun (2% solution equivalent to 20

mg/mL xylazine; Bayer, Leverkusen, Germany): 0.6 mL of the 1:1 Vetalar/ Rompun mixture being administered to rats weighing between 250–300 g.

b. Sagittal (60 mg/mL sodium pentobarbitone; Rhone-Merieux, Rhone-Merieux, Lyon, France): the recommended dose being 26 mg/kg.

## 3. Animal Stocks and Their Maintenance
### 3.1. Introduction

In order to generate and maintain transgenic animals, a scientist must have access to and be responsible for hundreds if not thousands of rodents. Thus, only those scientists working in institutions equipped with the necessary facilities are able to undertake transgenic experiments. Practically, a suitable animal facility should be able to provide:

1. Spacious caging and a regular change of clean, comfortable bedding for the animals: Cage-washing facilities should also be available.
2. A food and water supply.
3. Environmental control to regulate the lighting, ventilation, temperature, and humidity.
4. Access to veterinary care.

A variety of animals are necessary to produce a regular supply of fertilized one-celled eggs for microinjection. In addition, many more animals are produced in the process of generating and maintaining transgenic lines. The investigator and his or her staff must be prepared to devote much time to the maintenance and care of the animals if they wish to engage in transgenic experiments.

### 3.1.1. Mice Stocks Required for the Production of Transgenics

Fertilized one-celled eggs for microinjection are produced by mating a donor female and a stud male. Choosing the right strain of mouse for egg production is pivotal, since the ease and efficacy of generating transgenic mice is highly strain-dependent. Brinster et al. *(10)* compared C57BL/6J inbred eggs and C57BL/6J × CBA/J hybrid eggs in terms of parameters, such as egg yield and survival after injection. Overall, the experiments on hybrid eggs were eightfold more efficient than inbred eggs. Inbred zygotes should only be used when the genetic background of the host animal needs to be carefully controlled. We use F1 animals generated from matings between CBA/J and C57BL/6J.

#### 3.1.1.1. Production of Fertilized One-Celled Eggs

Immature (12–14 g, 4–5 wk old) F1 hybrid females are superovulated and mated with F1 hybrid males. Ten immature superovulated females will yield at least 250 viable eggs (F2 zygotes). Fifty breeding pairs (CBA/J male and C57BL/6J female) can provide 30 F1 females/wk. Alternatively, immature F1 hybrid females can be purchased from a reputable commercial and institutional supplier. Twenty F1 hybrid studs caged individually are required to mate with the immature females and are replaced every 8 mo to a year.

### 3.1.1.2. RECIPIENTS OF THE MICROINJECTED EGGS

Estrous females (0.5 d pc) are used as surrogate mothers to nurture surviving eggs to birth. Females can be of any strain with good maternal characteristics, but need to be sexually mature and >19 g in weight. We use C57BL/6J x CBA/J F1 females. Estrous females are made pseudopregnant by mating with vasectomized males.

### 3.1.1.3. VASECTOMIZED MALE MICE

Sexually mature males are sterilized by vasectomy and used to engender pseudopregnancy in sexually mature females. Any strain of mouse can be used, but Parkes and Swiss males are particularly suitable because they perform well. Twenty to 30 vasectomized males, mated on alternate days with females analyzed for their stage in the estrous cycle, should be able to provide at least five pseudopregnant females a day. Sexually mature male mice (about 2 mo old) are used. The vasectomy procedure is described in **Subheading 4.** of this chapter.

### 3.1.1.4. FOSTER MOTHERS

If only a very low number of fetuses develop in the pseudopregnant recipient, the resultant fetuses may be too large for natural delivery and a Caesarean section must be performed. The pups are fostered to mothers with natural pups of a similar age.

## 3.1.2. Rat Stocks Required for the Production of Transgenics

The generation of transgenic rats is a more recent development. As such, less research has been done on the efficiency of the technique using different rat strains. In our laboratory, we have used only the Sprague-Dawley (SD) strain of rats, but Mullins et al. *(11)* have successfully used fertilized eggs derived from mating SD females and WKR males.

### 3.1.2.1. PRODUCTION OF FERTILIZED ONE-CELLED EGGS

Fertilized eggs are obtained from immature (4–5 wk old) superovulated females mated with mature stud males. These females will yield 100–200 injectable eggs. To supply 20 immature females/week, 30 breeding pairs should be more than sufficient. At least 20 stud males caged individually are required for mating with the immature females and should be replaced every 8 mo to a year. An individual male should be presented with a female only every alternate day.

### 3.1.2.2. RECIPIENTS OF THE MICROINJECTED EGGS

Estrous exbreeder females (females that have successfully pupped at least one litter) of between 8 and 16 wk are mated with vasectomized males the night before they are implanted with the microinjected eggs. Although virgins tend to produce larger litters than exbreeders, they often eat the resulting pups.

### 3.1.2.3. VASECTOMIZED MALE RATS

Vasectomized males are needed to engender pseudopregnancy in the exbreeder females. Experienced males are preferred, and once vasectomized, have to be replaced every 6–8 mo as they grow quickly to a large size.

### 3.1.2.4. FOSTER MOTHERS

Foster mothers (with pups of about the same age) are occasionally required for the rare litter that is delivered by Caesarean section.

## *3.2. Note*

1. Following the production of transgenic founders, mate the animals as soon as possible to establish a transgenic line. Analysis of the founder phenotype should not begin until the transgene has passed successfully to a subsequent generation. In vitro fertilization may be necessary if the founders do not mate.
2. Unless the animals are kept in an SPF unit, sporadic microbial infections will occur. A six-monthly check of the colony's health status by a veterinarian is advisable.

## 4. Vasectomy of Mice and Rats

## *4.1. Materials*

1. Sexually mature male mice (2 mo old) or rats (5 wk old).
2. Dissection instruments (e.g., from Arnold Horwell, West Hampstead, London, UK).
3. Dissection scissors (large and small).
4. Fine blunt forceps (one pair curved, one pair straight).
5. Sharp curved forceps.
6. Watchmaker's forceps (size 5).
7. Curved surgical needle (size 10, triangular, pointed).
8. Surgical silk sutures (size 5).
9. Autoclips and applicator.
10. Anaesthetics: Avertin for mice, CRC for rats.
11. Bunsen burner.
12. 70% Ethanol in squeeze bottle.

## *4.2. Methods*

### *4.2.1. Mice*

1. Anesthetize the mouse as in **Subheading 2.2**. Place the animal ventral (abdomen) side up on the lid of a 9-cm glass or plastic Petri dish.
2. Spray the lower abdomen with 70%, (v/v) ethanol. Comb the hair away from the proposed site of incision (*see* **Fig. 1A**) with a pair of fine forceps.

3. Lift the skin away from the body wall with a pair of fine, blunt forceps. Make a 1-cm transverse cut—level with the top of the hindlimbs (**Fig. 1A**)—with a pair of fine dissecting scissors. To reduce bleeding, stretch the incision to about 1.5 cm with the outer edges of the opened scissors.
4. Cut the body wall parallel to the skin cut. Again, stretch the incision.
5. Introduce a single stitch into one side of the body wall wound and leave in place.
6. Pull out the fat pad on one side of the animal and with it the testis, epididymis, and vas deferens (**Fig. 1B**) using blunt forceps.
7. Identify the vas deferens (**Fig. 1B**; held by left forceps). The vas deferens links the epididymis to the penis. It is rigid with a blood vessel running alongside it. Free the vas deferens from the surrounding membranes using fine forceps and scissors (**Fig. 1C,D**). The vas deferens is clearly seen against the black background (**Fig. 1E**).
8. Hold the vas deferens in a loop with a pair of forceps. Heat a pair of blunt forceps until glowing red. Grip the vas deferens loop with the hot forceps to burn away the tube and cauterize the ends (**Fig. 2**).
9. Separate the cauterized ends.
10. Return the organs into the body cavity with a pair of blunt forceps.
11. Repeat with the other side of the reproductive tract.
12. Sew up the body wall with two or more stitches.
13. Clip the skin with two wound clips.
14. Allow the animal to recover in a warm quiet place.
15. Cage the vasectomized males individually. Allow a few weeks for full recovery before using the animal to produce pseudopregnant recipient females.

### 4.2.2. Rats

1. Anesthetize the rat as in **Subheading 2.2.**
2. Place the rat abdomen side up.
3. Swab the scrotal area with 70% ethanol.
4. Lift the scrotal skin of one testis with a pair of fine forceps, and make a 1-cm longitudinal cut on the more anterior outer portion of the scrotum.
5. Lift the body wall (which is extremely thin) with the fine forceps, and make a parallel cut.
6. Place a single stitch into the body wall wound and leave in position.
7. Pull out the fat pad, and with it the testis, epididymis, and vas deferens.
8. Identify the vas deferens located underneath the testis. Tie off each end of the tube with cotton thread, and then cut the tube between the two ties with a pair of fine scissors.
9. Return the organs into the body cavity using a pair of blunt forceps. It is often hard to relocate the body wall—use the single stitch placed in the body wall at the start of the surgery as a marker.
10. Stitch up the body wall with two or more sutures.
11. Stitch up the scrotal skin.
12. Repeat with the other side of the rat.
13. Make sure the animal is regaining consciousness and some semblance of mobility before leaving it to recover in a warm quiet place.

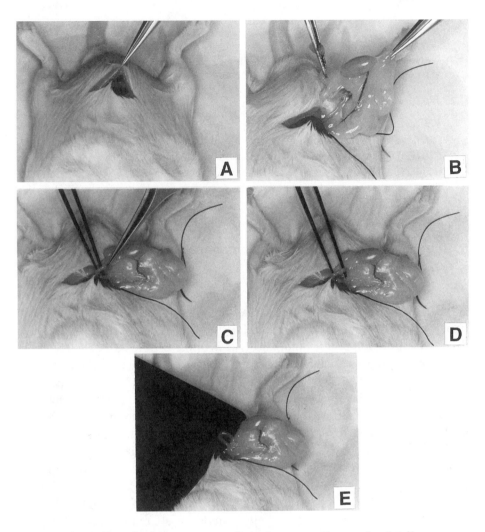

Fig. 1. Steps in the vasectomy of a male mouse. *See text* for details.

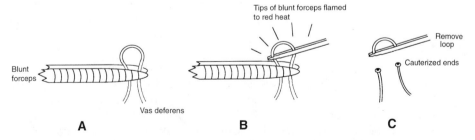

Fig. 2. Cauterization of the vas deferens.

14. Cage the vasectomized males individually. Allow a few weeks for full recovery before using the animal to produce pseudopregnant recipient females.

### 4.3. Notes

1. Following surgery, the mice or rats are caged individually and left to recuperate for at least 3 wk prior to the presentation of the first female.
2. Check that the males are vasectomized, but capable of mating by presenting each male with a different female for 2 wk running. Successful mating is confirmed by the presence of a copulatory plug in the female and male sterility by the females not becoming pregnant. Any male not meeting both criteria should be replaced.
3. Vasectomized males should be presented with females only once every 2 d at the most.

## 5. Superovulation of Immature Rats and Mice

### 5.1. Introduction

For maintaining stocks of normal and transgenic strains, mice and rats are allowed to mate and breed with minimal intervention. This is simply a matter of regulating the transfer of males and females between cages, the provision of nesting material to pregnant mothers, and weaning off the pups.

Natural matings between mature female mice (6–7 wk) and stud males (over 7–8 wk) can be used to supply the eggs for microinjection, but yield around 10 F2 hybrid eggs/mouse. Administering gonadotrophins to immature females to induce an increased release of ova (superovulation) and mating these with male studs can yield 30–40 eggs/mouse depending on the strain used. In superovulation, the timing of ovulation, copulation, and fertilization is controlled to optimize egg yield. Eighty to 100% of treated mice, and 60–100% of similarly treated rats are successfully impregnated.

It is essential both for natural and timed matings that the animals are maintained on a constant light–dark cycle. Simple on/off regimens are adequate (e.g., 1700–0600 dark; 0600–1700 light), but sophisticated regimens that simulate natural conditions are preferable (e.g., 1830–0630 0% light; 0630–0830 25% light; 0830–1130 75% light; 1130–1300 100% light; 1300–1600 75% light; 1600–1830 25% light).

### 5.2. Materials

1. Follicle-stimulating hormone (FSH): Either Folligon (Intervet Laboratories Ltd, Cambridge UK) or Pregnant Mares Serum Gonadotrophins (PMSG; Sigma, St. Louis, MO; cat. no. GH879) can be used. Working solution: 50 IU/mL made up with sterile water or 0.9% (w/v) NaCl. Store frozen at 4°C in 1-mL aliquots.
2. Human chorionic gonadotrophin (hCG): Either Chorulon (Intervet Laboratories Ltd., Cambridge, UK) or hCG (Sigma CG-5) can be used. Working solution: 50 IU/mL made up with sterile water or 0.9% (w/v) NaCl. Store frozen at 4°C in 1-mL aliquots.
3. Sterile 1-mL disposable syringes and 25-gage needles.

## 5.3. Methods

### 5.3.1. Mice

1. To generate approx 200 viable eggs for microinjection, use 10 F1 hybrid (e.g., CBA/J X C57BL/6J) females (12–14 g, 4–5 wk old).
2. Inject each mouse ip with 100 µL of 50 IU/mL FSH (5 IU/mouse) between 1000 and 1200 h.
3. Forty-six to 48 h later, inject the same mice with 100 µL of 50 IU/mL hCG (5 IU/ mouse). Immediately place each female with a male stud.
4. On the following morning, check each female for the presence of a copulatory plug. The plug is a creamy white mass of coagulated sperm and protein blocking the vagina. It is normally obvious, but is sometimes deep within the vagina— examine the mouse carefully with a smooth blunt probe.

### 5.3.2. Rats

1. To generate between 100 and 200 viable eggs for microinjection, use 10 females (4- and 5-wk old).
2. Inject each rat intraperitoneally with 400 µL of 50 IU/mL FSH (20 IU/rat) at 0900 h.
3. Two days later, ip inject the same rat with 200 µL of 50 IU/mL hCG (10 IU/rat) at 1400 h. Immediately place each female with a male stud.
4. On the following morning, check each female for the presence of a copulatory plug. In rats, the plug has often fallen out by the morning. A vaginal smear should be taken from any animal in which it is not detected (*see* **Subheading 12.**) and the contents examined under 400× magnification for sperm.

## 5.4. Notes

1. Other labs have reported using a continual "minipump" delivery of porcine FSH *(11)* to rats instead of IP injection of the hormone to yield 60–100 eggs/ rat, but Waller et al. *(8)* found that only a small percentage of treated rats ovulate and mate.
2. In the light–dark cycle described, males will copulate with females at around 0030 h. Fertilization occurs 30 min to 2 h later. Harvesting the eggs between 1000 and 1130 h of the same day gives a subsequent window of 12 h during which the eggs can be injected. After this time, the fertilized eggs will start to undergo the first cleavage division to give two-cell embryos—injections near this time-point may result in chimeric animals.
3. Pseudopregnant females 0.5 d pc can be implanted with microinjected eggs anytime during the day following an infertile mating.
4. For superovulating mice, we routinely inject at 1030 h for both FSH and hCG.
5. Occasionally, a high proportion of eggs from superovulated rats are unhealthy or unfertilized. In this instance, consider switching FSH and hCG injection times to 1100 and 1200 h, respectively.

## 6. Culture of Fertilized One-Celled Eggs

### 6.1. Introduction

For the purpose of making transgenic rats and mice, fertilized one-celled embryos have to be maintained outside of the natural environment for up to 36 h. Fertilized eggs are collected at approx 12 h pc (coitus will normally occur at about 0030 h) and are injected at some point during the following 12 h. The eggs are then returned to the natural environment of the pseudopregnant surrogate mother. This procedure occurs either immediately after microinjection while the eggs are still in the one-celled stage or after overnight culture when the eggs are transferred in the two- celled stage.

### 6.1.1. Preparation of Culture Media for Fertilized One-Celled Eggs

Two types of media are required for the in vitro manipulations of the eggs *(12)*.

1. M16: for maintaining the eggs in microdrop cultures in a 37°C incubator gassed with 5% $CO_2$. M16 is buffered with bicarbonate alone and unsuitable for maintaining the eggs outside of the incubator, since the eggs are very susceptible to pH changes.
2. M2: essentially similar to M16, except that the bicarbonate is partially replaced by HEPES buffer to facilitate the survival of the eggs outside the $CO_2$ incubator. Eggs should not be in M2 longer than 30 min.

Both M2 and M16 contain bovine serum albumin (BSA), which reduces the stickiness of the eggs and absorbs low-level poisons. Some batches of BSA are toxic to the eggs, so it is crucial that each stock is tested for its ability to sustain mouse embryo development through to the blastocyst stage (e.g., BSA; Sigma cat. no. 4161: Ho et al. [unpublished observations] in Waller et al. *[8]*)

Media preparation:

1. Use Sigma tissue-grade chemicals throughout.
2. Make up all stocks using sterile disposable plastic containers and pipets. Washed glass items can be contaminated with detergents that are toxic to the eggs. Filter all concentrated stocks through 0.45-μm Millipore filters into sterile plastic tubes. Store frozen at 20°C.
3. Make up the M2 and M16 culture media as follows:
   a. Concentrated stocks:
      i. 100X A: Weigh out the following reagents in a 50-mL sterile tube: 2.767 g NaCl (Sigma S5886), 0.178 g KCl (Sigma P5405), 0.081 g $KH_2PO_4$ (Sigma P5655), 0.1465 g $MgSO_4 \cdot 7H_2O$ (Sigma M2643), 0.5 g glucose (Sigma G6138), 0.03 g penicillin (Sigma P3032), 0.025 g streptomycin (Sigma S9137). Weigh out 1.305 g sodium lactate (Sigma L4263) into a microcentrifuge tube and add this to the 50-mL tube. Rinse the microcentrifuge tube with double-distilled water and use the rinsings to make 10X A to 50 mL final volume

    ii. 10X B: Dissolve 1.0505 g NaHCO$_3$ (Life Sciences, cat. no. 895-1810 1 *N*) and 0.005 g phenol red (Sigm P5530) in water and make up to 50 mL final volume

    iii. 100X C: 0.18 g Na pyruvate (Sigma P5280) in 50 mL water

    iv. 100X D: 1.26 g CaCl$_2$·2H$_2$O (Sigma C7902) in 50 mL water

    v. 100X E: 2.979 g HEPES (Sigma H9136), 0.005 g phenol red (Sigma P5530) into a sterile 50-mL tube and dissolve in 25-mL double-distilled water. Adjust to pH 7.4 with 0.2 M NaOH, then make up to 50 mL final volume.

b. Preparation of M2 nad M16 media from concentrated stock: To prepare M2 and M16, mix the stock solutions in the appropriate volumes as detailed below. Measure the double-distilled water in a sterile 50 plastic tube. Aliquot the concentrated stocks into the water, then carefully rinse the pipet by sucking the liquid up and down. Add the BSA and mix gently until dissolved. Pass the mixed solutions through a 0.45-μm Millipore filter using a large sterile 60 mL disposable syringe, aliquoting into sterile containers. Store at 4°C. Prepare fresh every 2 wk.

| Stock | M2 | M16 |
|---|---|---|
| 10X A | 5.0 mL | 5.0 mL |
| 10X B | 0.8 mL | 5.0 mL |
| 100X C | 0.5 mL | 0.5 mL |
| 100X D | 0.5 mL | 0.5 mL |
| 10X E | 4.2 mL | — |
| Double-distilled water | 39.0 mL | 39.0 mL |
| BSA (Sigma A4161) | 0.2 g | 0.2 g |
| Total volume | 50.0 mL | 50.0 mL |

## 6.2. Notes

1. Osmolarity of the medium does not have to be routinely checked but for reference, the values should be: M2 (285–287 mosM/L) and M16 (288–292 mosM/L).
2. Modified M2 and M16 media are commercially available (e.g., Specialty Media, Lavallette, NJ) in a powdered form that needs only to be made up with sterile water.

# 7. Preparation of Egg-Transfer Pipets

## 7.1. Introduction

Two types of egg-transfer pipets need to be prepared in advance; general-transfer pipets and oviduct-transfer pipets. These are made from hard glass capillaries (BDH Ltd., Poole, UK) and are assembled into a mouth-operated system made up of a mouthpiece, rubber tube (approx 60 cm), and a pipet holder (components available from Arnold Horwell).

## 7.2. Materials

1. Hard glass capillaries (BDH Ltd.), 1.5 mm od.
2. Diamond pencil.
3. Microforge (e.g., Narishige MF-9, Tokyo, Japan).
4. Gas flame supplier.

## 7.3. Method

### 7.3.1. General Transfer Pipets

1. Using a small gas flame, soften the middle of a BDH hard glass capillary tube by rotating it slowly over the flame. As soon as the glass softens, withdraw it from the heat, and simultaneously pull the ends apart sharply, but not until the two halves snap.
2. Score the capillary with a diamond pen, and snap it into two. If the pulled portion is too long (>5 cm) or too narrow (<200 μm), score and snap again. The internal diameter of the pipet should be about 300 μm, and the end should be flush with no jagged ends.

### 7.3.2. Oviduct Transfer Pipets

1. Follow the above procedure for making general-transfer pipets, but aim for pipets with internal diameters of around 150 mm.
2. Flame polish the end in a microforge. Ensure that the aperture of the pipet is large enough to permit easy passage of a mouse or rat egg (~120 mm).

## 8. Collection of Fertilized One-Celled Eggs

### 8.1. Materials

1. M2 and M16 media.
2. 35-mm Sterile tissue-culture dishes.
3. Hyaluronidase (Sigma H1136): 10 mg/mL in M2 medium. Store in 1-mL aliquots at –20°C (stable for several months).
4. Light paraffin oil (Fluka, Buchs, Switzerland, cat. no. 76235).
5. Egg transfer pipet and mouthpiece (**Subheading 7.3.1.**).
6. Dissecting scissors, one fine and one regular.
7. Watchmaker's forceps. 2X #5.
8. Stereomicroscope with understage illumination (e.g., Nikon [Tokyo] SMZ10TD).
9. Fiberoptic light source (e.g., Nikon).
10. 37°C Incubator gassed with 5% $C_2O_2$.

### 8.2. Methods

1. At least 1 h before collecting the eggs, set up four 35-mm tissue-culture dishes containing 2–3 mL M16 medium and two 35-mm culture dishes containing M16 microdrops (40 μL) covered with liquid paraffin. Allow these to equilibrate in a 37°C incubator gassed with 5% $CO_2$. At the same time, prepare five 35-mm tissue-culture dishes containing 2–3 mL M2 medium and leave at room temperature.

2. Kill the plugged donor females as described in **Subheading 5.**
3. Lay the animal on its back, and soak the abdomen with 70% ethanol.
4. Pinch up the skin with fingers, cut into the midline, and skin the animal. Cut the body wall, and enter the abdominal tract. Push aside the coils of the gut to reveal one arm of the reproductive tract, which is associated with a fat pad. Identify the coiled oviduct lying between the uterus and the ovary.
5. Gripping the uterus with a watchmaker's forceps, lift up the reproductive tract. Use a pair of fine forceps to puncture the membrane (mesometrium) that joins the reproductive tract to the body wall. Trim the membrane away from the oviduct.
6. Still gripping the uterus, cut between the ovary and the oviduct (cut A in **Fig. 3**).
7. Transfer the grip to the oviduct, and cut between the uterus and the oviduct (cut B in **Fig. 3**). The cuts should be made as close to the oviduct as possible.
8. Place the oviduct into one of the dishes of M2 prepared earlier.
9. Dissect the oviduct from the other horn of the reproductive tract, and then proceed with the rest of the female donors. All the oviducts should be placed in the same dish of M2 medium.
10. View the oviducts under the 10–20× magnification of the dissecting stereomicroscope. The oviduct should appear as mass of opaque coils with a single transparent swollen region, the ampulla (**Fig. 4**). The ampulla is the target of the egg collection, since, at this stage, it contains the cumulus mass (numerous eggs surrounded by cumulus cells). Eggs may be visible through the walls of the ampulla.
11. Use one pair of sharp watchmaker's forceps to hold the oviduct down, and another to tear the ampulla. The cumulus mass should spill out of the hole. Sometimes, it is necessary to tease the eggs out of the ampulla with forceps. Discard the empty oviduct, and repeat the procedure with the rest of the oviducts.
12. Mix the cumulus mass with 50–100 µL of 10 mg/mL hyaluronidase. Enzymatic digestion is required to separate the eggs from the cumulus cells—a few minutes of treatment is sufficient—the eggs should not be in contact with the enzyme for a longer period. Gently pipeting the cells up and down with a general transfer pipet facilitates the process.
13. Using the general transfer pipet, transfer the eggs to a fresh dish of M2 medium to wash away traces of the hyaluronidase. Repeat in another dish of M2. In each wash, try to leave behind the cumulus cells.
14. Wash the eggs twice in two of the prewarmed dishes of M16 medium and finally, transfer the eggs to the microdrop culture (20–30 eggs/drop).
15. Incubate the eggs at 37°C with 5% $CO_2$ until required for microinjection. It is best to leave the eggs for at least an hour before microinjection.

### *8.3. Note*

Paraffin oil used to prevent evaporation of the media is often a source of toxins that may kill the eggs. Paraffin oil from Fluka (cat. no. 76235) allows more than 50% of the eggs to develop to the blastocyst stage in culture.

## 9. Transgene Design

Generating a transgenic animal is a long process—great weight must be placed on the design of the transgene. After months of generating sufficient

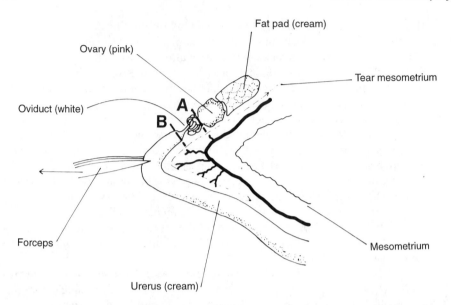

Fig. 3. Dissection of a female reproductive tract of a mouse illustrating the procedure for removal of the oviduct prior to egg collection.

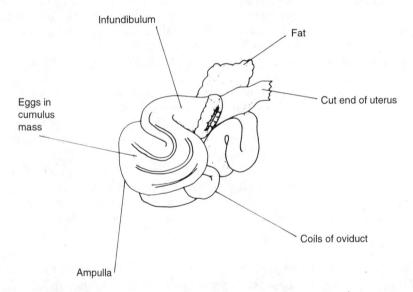

Fig. 4. The mouse oviduct showing the position of eggs in the swollen ampulla.

transgenic animals for analysis, one does not like to discover that the transgene needs to be redesigned. However, it cannot always be avoided, since little is known about the vast majority of genes or the regulatory elements that govern their expression. A dose of serendipity is sometimes required to achieve the desired pattern of expression of the transgene.

Some principles have been derived that can aid in planning the construct of the transgene:

1. The presence of contiguous vector-derived prokaryotic DNA sequences in a fragment of microinjected DNA can severely hinder the expression of some eukaryotic transgenes *(13)*. Bacterial coding sequences, such as chloramphenicol acetyltransferase (CAT) and β-galactosidase, are often incorporated into transgenes as reporters of transgene expression driven by the flanking eukaryotic promoter sequences. Unlike some vector sequences, these do not inhibit the expression of the eukaryotic genes. It may be that the inhibitory effect of some prokaryotic-derived DNA is specific to sequences contained within the commonly used λ- and phage-derived vectors. As a rule, remove all vector sequences before injecting the cloned DNA. Linear DNA integrates fivefold more efficiently than supercoiled DNA. Also, the structure of the fragment ends created by different restriction enzymes has little effect *(4)*.

2. For most transgenic experiments, appropriate tissue-specific and physiologically regulated expression is desired. Since the DNA elements conferring these specificities are often unknown, it is best to use as much of the gene as possible. It is advisable to include introns, 3'-untranslated regions, and upstream sequences, because regulatory elements can be located in these regions *(14,15)*. Adjacent genes can contain sequences that determine the correct expression of their neighbors as in the case of the neuropeptides oxytocin and vasopressin *(16)*. Locus control regions (LCRs) allow position-independent and copy number-dependent gene expression, but have been identified for only a very small percentage of genes and are usually a very long distance away from the coding region. The technology for manipulating fragments of DNA up to one megabase has recently been developed in the form of yeast artificial chromosomes (YACs). Schedl et al. *(17)* have described the first transgenic mice produced by the microinjection of a YAC. Their 250-kb transgene contained the tyrosinase gene, was inserted without major rearrangements, and was able to rescue completely the albino phenotype of the recipient mice.

3. Reporter genes should ideally be located in the first or last exon. Insertion into an internal exon can disrupt splicing efficiency.

4. The use of a chimeric transgene, combining pieces of DNA from different genes, may result in unpredictable ectopic expression of the transgene *(18)*. It is best to keep the transgene simple by using regulatory elements from just one gene.

## 10. Purification of Microinjection DNA from Agarose Gels

### *10.1. Introduction*

DNA for microinjection can be prepared from the plasmid or the cosmid clones by any of the standard techniques of lysozyme/Triton X-100 lysis or lysozyme/alkaline lysis followed by banding of supercoiled molecules in ethidium bromide/CsCl gradients. The many commercially available plasmid preparation kits, using alkaline lysis followed by anion-exchange column purification of DNA, provide DNA of a suitable quality with the minimum of time and effort.

Vector-free DNA is isolated by restriction enzyme digestion followed by preparative agarose-gel electrophoresis. This latter step is very important. DNA fragments for microinjection must be free from all contaminants that may be toxic to the eggs and all particulate matter that could clog up the injection pipet. Isolation of DNA from agarose gels by binding to glass beads *(19)* followed by passage through a Sephadex G- 50 column and filtration though a 0.45-μm filter provides such DNA. The method is described in the following protocol. However, glass bead DNA isolation kits are also commercially available (e.g., Geneclean from Bio101, Vista, CA).

## 10.2. Materials

1. Glass beads: These can be purchased or prepared in the laboratory as described below. The preparation involves boiling nitric acid so take full safety precautions. Mix 250 mL of powdered glass flint available from glass supply companies (e.g., Eagle Ceramics Inc., 12267 Wilkins Avenue, Rockville, MD 20852) with 500 mL sterile water. Allow the heavier particles to settle over 1 h. Decant the fines into centrifugation tubes, and spin for 5 min at 1000$g$. Resuspend the pelleted fines in 200 mL sterile water. Add 200 mL of concentrated nitric acid, and bring to boil. Carry out this step in an efficient fume hood. Allow to cool and then centrifuge (1000$g$, 5 min) to recover the glass beads. Discard the nitric acid carefully. Wash several times in distilled water (i.e., cycles of resuspension and spinning) until the pH of the suspension is neutral. Store the glass bead slurry as a 50% (w/v) slurry at –20°C, working stocks at 4°C.
2. Gel electrophoresis equipment.
3. 50X TAE buffer stock: 2.0 $M$ Tris base, 1.0 $M$ sodium acetate, 0.05 $M$ EDTA.
4. Ethidium bromide (10 mg/mL) in sterile water.
5. Long-wave (365 n$M$) UV light transilluminator.
6. 6 $M$ NaI: Dissolve 90.8 g of NaI and 1.5 g $Na_2SO_3$ in water to a final volume of 100 mL. Filter through a 0.45-μm Nalgene filter or Whatman no. 1 paper. Add 0.5 g $Na_2SO_3$ to saturate. The $Na_2SO_3$ crystals do not dissolve, but serve to prevent the oxidation of the NaI. Store at 4°C protected from light.
7. Ethanol wash solution: 50% (v/v) ethanol, 0.1 $M$ NaCl, 10 m$M$ Tris-HCl, pH 7.5, 1 m$M$ EDTA. Store at –20°C.
8. Sterile (autoclaved) filtered (0.2 μm) Microinjection TE (MITE): 10 m$M$ Tris-HCl, pH 7.4, and 0.2 m$M$ EDTA.
9. Sephadex G50 slurry: Prepared by swelling G-50 powder in water followed by autoclaving.
10. 0.45-μm Millipore filter attached to a 1-mL disposable syringes.
11. Baked glass wool: Bake at 250°C for at least 3 h to destroy all contaminating nucleases.

## 10.3. Method

1. Digest the DNA with the appropriate restriction enzymes according to the recommendations of the manufacturer.
2. Fractionate the digested DNA in an agarose gel prepared in 1X TAE buffer containing 0.5 μg/mL ethidium bromide.
3. Visualize the DNA on the long-wave UV light transilluminator. Use a fresh scalpel blade to excise the fragment of interest carefully in as small a volume as possible. Transfer the excised fragment to a preweighed Eppendorf tube.
4. Add as much NaI as possible, the minimum ratio being 3:1 (v:w).
5. Incubate at 55–65°C with occasional agitation until the agarose has completely dissolving.
6. Add 1-μL of glass beads (fully dissolved and mixed) for every 2 μg of DNA. Mix well, and incubate on ice for 45 min inverting every 10–15 min.
7. Pellet the glass milk by centrifugation (pulse to 13,600g over and release). Discard the supernatant.
8. Wash with 500 μL of NaI twice more, spinning briefly (1–2 s) each time to recover the glass milk.
9. Wash with 500 μL of ethanol wash twice.
10. After discarding the supernatant, vortex the glass milk in the small volume of ethanol that remains and respin. Carefully remove as much of the supernatant as possible, but do not let the pellet dry out.
11. Add more than 50 μL of MITE immediately, and incubate at 55–65°C for 15 min to elute the DNA.
12. Pellet the glass beads by spinning at full speed for 2 min. Transfer the supernatant containing the DNA to a fresh tube. **After this point, all tubes and tips must be prerinsed with filtered water before use**.
13. Prepare a spun Sephadex G-50 column. Block the end of a 1-mL disposable syringe with glass wool. Fill with Sephadex G-50 until all the excess fluid drains out and the beads are packed. Position the column in a disposable 15-mL plastic conical centrifuge tube, and centrifuge at 4000g for 3 min. Equilibrate this column with 100 mL MITE and spinning as before. Do this five times.
14. Pass the eluted DNA (maximum volume 100 μL) through the equilibrated Sephadex G-50 column, collecting the purified DNA in a fresh 1.5-mL Eppendorf tube (with the lid cut off) placed at the bottom of a fresh 15-mL centrifuge tube.
15. Filter the elute through a small 0.45-μm Millipore filter attached to a 1-mL syringe.
16. Determine the DNA concentration by spectrophotometry at 260 nm or comparative ethidium staining with DNA standards of known concentration.

## 10.4. Notes

1. Many manipulations of DNA and RNA leave the nucleic acid in a solution containing unwanted salts, nucleotides, or radioactive moieties. In most cases, ethanol precipitation will not entirely remove these or will add further salts. Sephadex gel-exclusion chromatography is often used to purify DNA fragments. Sephadex

is a bead formed, crosslinked dextran. The cross linking is carefully controlled to give beads consisting of a network of holes of uniform size and shape. A molecule passing through a Sephadex column will pass through a volume dependent on the size of the molecule. A molecule too large to enter a pore will pass through the volume of the column not occupied by beads. A molecule small enough to enter the pores will do so, and pass though a much larger volume and take longer to pass through the column. The exclusion limit is the size limit of the molecule that can just enter the pore. Molecules of intermediate size will be fractionated by their molecular weight. Thus, for example, Pharmacia makes G-50 Sephadex gel-filtration matrices that fractionate in the range of 500–10,000 daltons. This matrix can fractionate a mixture of DNA and salts cleanly, the DNA being excluded from the pores and passing straight through and the salts and nucleic acids being retained within the pores. Additionally, the sample does not become appreciably diluted.

2. Glass bead DNA isolation kits are available (e.g., Geneclean II by Bio 101).
3. The recovery of DNA fragments between 500 and 800 bp by glass bead isolation is relatively low. Also there is a risk that large DNA fragments (>15 kb) may be broken during the wash cycles. If a large DNA fragment binds to two or more beads, then on washing and separation of the beads the DNA strand may be broken.

## 11. Microinjection

### 11.1. Introduction

Central to the process of making transgenic mice and rats is the physical introduction of the cloned DNA fragments into one-celled eggs. First described 10 years ago by a number of investigators (1), microinjection remains the most popular method of generating transgenic animals because the advantages of speed and reliability outweigh the demands placed on the investigator for precise technical skill and expensive technical equipment.

### 11.2. Materials

1. Inverted microscope (e.g., Nikon Diaphot, Tokyo, Japan) with the following features:
   a. Image-erected optics.
   b. A condenser with a long working distance.
   c. A fixed stage (the objective lens rather than the stage moves when focusing).
   d. 10× Magnification eyepieces.
   e. 4× Objective for low-magnification work.
   f. 40× Objective for microinjection.
   g. Suitable objectives: Nomarski differential interference contrast (DIC) optics are best for visualizing the internal structures of the egg. DIC optics are expensive, and glass-injection systems are required. Hoffman modulation contrast optics are compatible with plastic injection chambers, but give inferior resolution to DIC. If both are not available, eggs can be viewed under bright field. Phase-contrast microscopy is not compatible with any microinjection system.
   The Nikon Diaphot-TMD equipped with the diascopic DIC Normarski attachment TMD-NT2 is excellent, robust, affordable, and widely used for microinjection.

2. Micromanipulators: Two are required—for the holding pipet, which holds the egg in place, and for the microinjection pipet. The Leitz M (Wetzlor, Germany) type, with joystick control of horizontal movement in two planes, is most commonly used. The micromanipulators and the microscope must be positioned on a purpose-built baseplate (**Fig. 5**), which must be custom-engineered. Narishige produces economically priced micromanipulation systems compatible with its Diaphot-TMD microscope. These are very flexible and do not require a custom-built base plate.

3. Micropipet holders: Leitz single-instrument holders fitted with Leitz single-instrument tubes if Leitz Type M micromanipulators are used.

4. Holding pipet (**Fig. 6**)
   a. Draw a Leitz hard-glass capillary (cat. no. 520119) in a gas flame. Grip the ends of the capillary, and turn the middle of the capillary in the hottest part of the flame until the glass softens. Quickly withdraw the pipet from the flame, and simultaneously pull on both ends.
   b. Score the drawn section 2 cm from the shoulder with a diamond pen, and break the capillary at this point.
   c. Mount the capillary vertically in a microforge (e.g., Narishige MF-9). Focus on the tip of the capillary with the 4× objective. The capillary must be straight and its tip perfectly flush with no jagged edges and an external diameter of 100 mm (±20 mm). Bring the tip close to, but not touching the microforge filament. Heat the filament, and melt the tip until its internal diameter reaches 10–15 mm. The size of the hole is extremely important; too small a hole will make controlling the holding pipet difficult; too large a hole may allow eggs to be sucked into the pipet.
   d. Move the capillary to a horizontal position. Position the filament 1–2 mm below the capillary tip. Move the filament close to the pipet, and heat to bend the capillary by about 15° to the horizontal. One holding pipet will last one microinjection session and is not reused. Large numbers can be made in advance and stored in a sterile plastic tube.

5. Microinjection pipets for physically introducing the DNA fragment into the nucleus of the one-celled egg: Microinjection pipets are made from thin-walled glass capillaries with an external diameter of 1 mm. Capillaries with an internal filament (Clark Electromedical Instruments, Reading, UK; cat. no. GC 100 TF-15) are useful, since they can be backfilled by capillary action from the distal end to the injection tip. Pipets are drawn on a commercially available pipet puller (e.g., David Kopf Instruments, Tujunga, CA; Vertical Pipet Puller Model 720 or the Narishige PB-7). Injection tips should have a 1-mm opening. Larger tips burst eggs, and smaller ones become blocked easily. Microinjection pipets are pulled when needed.

6. The injection chamber: Glass depression slides are compatible with inverted microscopes fitted with any optic system. Siliconize slides by rinsing in a 3% (v/v) solution of dichloromethyl silane in chloroform. The slides are rinsed thoroughly with water and a standard household detergent. Prior to use, rinse the slide in ethanol, and dry with a paper tissue. It must be devoid of dust particles.

7. Agla micrometer syringe (Wellcome, Kent, UK, or equivalent).

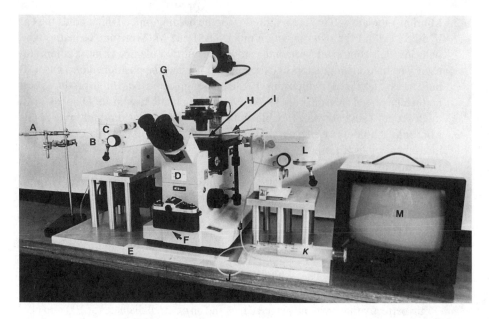

Fig. 5. A typical arrangement of the equipment needed for the microinjection of fertilized one-cell eggs. **A.** Agla micrometer syringe. **B.** Liquid-paraffin-filled tube. **C.** Left-hand micromanipulator. **D.** Inverted microscope. **E.** Base plate. **F.** Camera (optional). **G.** Left- hand instrument tube for holding pipet. **H.** Microinjection chamber (depression slide) sitting on fixed stage. **I.** Right-hand instrument tube for injection pipet. **J.** Air-filled tube. **K.** Glass 50-mL syringe. **L.** Right-hand micromanipulator. **M.** Video system (optional).

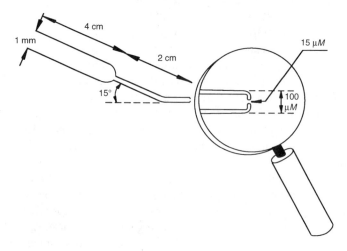

Fig. 6. Construction of a typical holding pipet.

8. There are two possible microinjection systems:
   a. Manual injections involve the use of a 50-mL syringe with a ground-glass plunger connected by an air-filled tube to the microinjection needle. Manual squeezing of the syringe squeezes DNA from the pipet into the egg.
   b. An automatic injection system that uses compressed air to expel the DNA, triggered by a foot-operated pedal (e.g., Narishige: Picoinjector PL1-188 with compressed air supplied from the Hitachi Package Oilfree Bebicon PO-O 75PSB compressor). Although more expensive than manual systems, the advantages are:
      i. The foot-operated injection trigger leaves both hands free to control the micromanipulators.
      ii. A constant low positive pressure applied to the microinjection pipet produces a continual outflow of the DNA, preventing the backflow of M2 medium and clogging of the pipets.

      The assembly of an automatic injection system should be done by a trained professional. The injection pressure must be determined empirically.
9. M2 and M16 culture media.
10. $CO_2$ tissue-culture incubator (37°C, 5% $CO_2$).
11. Light liquid paraffin (Fluka 76235).
12. 26-Gage needles.
13. Tygon (Akron, OH) tubing (3/32-in. id; 5/32-in. od).
14. One clamp stand.
15. Diamond pen.
16. Disposable 1-mL syringes.
17. Fluorinert electronic liquid (3M Company, St. Paul, MN; cat. no. FC77).
18. Dissecting microscope.

## 11.3. Methods

### 11.3.1. The Microinjection Chamber

1. Place 100 mL of M2 medium in the well of a ethanol-washed, dried depression slide.
2. Cover the M2 drop with light paraffin oil (Fluka) to prevent evaporation.
3. Place the slide on the stage of the microscope, and using a 4× objective, focus on the bottom of the M2 drop.

### 11.3.2. The Holding Pipet

1. Connect a Leitz instrument tube to an Agla micrometer syringe with 1 $M$ Tygon tubing. Fill the system with light paraffin oil making sure that all the air bubbles have been excluded. Position the Agla micrometer syringe in a clamp stand in a convenient position close to the left-hand manipulator.
2. Fill the holding pipet with Flourinert using a needle (26-gage, 5 cm) attached to a 1-mL disposable syringe. Insert the holding pipet into the oil-filled Leitz instrument tube, and tighten the ring to hold it in place.

3. Clamp the instrument tube into the instrument tube holder of the left-hand micromanipulator (if one is right-handed). Adjust the Agla micrometer until fluid stops flowing out of the holding pipet. Do not allow air to flow back into the holding pipet.
4. Position the holding pipet 2 cm above the center of the microinjection chamber making the necessary adjustments, such that the tip of the holding pipet is horizontal and straight. Monitoring under the 4× magnification, use the fine controls of the micromanipulator to lower the holding pipet into the injection chamber. Adjust the manipulator such that the tip of the holding pipet is just above the floor of the chamber. The holding pipet should move freely in the horizontal plane throughout the field of vision and not catch on the floor of the chamber.
5. Using the Agla micrometer, draw M2 medium into the holding pipet until the meniscus between the M2 and Fluorinert is just at the shoulder of the holding pipet.

## 11.3.3. The Microinjection Pipet

1. Connect a Leitz instrument tube holder via 1 m of Tygon tubing to either a 50-mL syringe with ground-glass plunger (lubricated with liquid paraffin) or an automatic injection system. The system is air-filled.
2. Backfill a freshly drawn microinjection pipet with DNA. Place the end distal to the tip into the DNA solution, allow the liquid to ride into the tip by capillary action and withdraw when liquid can be seen in the tip. Avoid contaminating the DNA stock solution, e.g., with glove powder or enzymes from the exposed hand.
3. Assemble the microinjection pipet into the instrument tube. Clamp the instrument tube onto the instrument tube holder of the micromanipulator.
4. Position the microinjection pipet until the tip is 2 cm above the center of the microinjection chamber and 15–20° to the horizontal. Using the fine vertical control of the micromanipulator, lower the tip into the injection chamber until it is just above the chamber floor. Monitor the position of the tip throughout using the 4X objective.
5. Using the horizontal controls of the micromanipulators, bring both the injection pipet and the holding pipet to the center of the field of view.
6. Switch to the 40× Normarski objective, and focus on the HP. Using the fine controls of the micromanipulator, bring the tip of the microinjection pipet into the same focal plane as the holding pipet. The microinjection pipet should move freely in the horizontal plane when operated by the joystick control.

## 11.3.4. Microinjection of Fertilized One-Celled Mouse and Rat Eggs

1. Remove about 20 one-celled eggs from storage in M16 medium at 37°C using a general-transfer pipet.
2. Wash the eggs twice in M2 medium. Load them in as small a volume as possible in the transfer pipet, and discharge them into the injection chamber. Observe the entry of the eggs into the chamber using a 4× objective. Try to keep the eggs in a group positioned below the holding and injection pipets. Avoid releasing any air bubbles into the chamber—eggs become obscured or even lost, necessitating the reassembly of the injection chamber.

3. Readjust the vertical positions of the holding and injection pipets so they are in the same plane as the eggs.
4. Bring the tip of the HP close to an egg, and by adjusting the Agla syringe, apply light pressure such that the egg is held onto the tip of the pipet.
5. Bring the egg to the center of the field of vision. Switch to the ×40 DIC objective and focus on the egg. Focus up and down to locate the two egg pronuclei. The larger (male) pronucleus is the target and must be positioned for convenient injection. The egg can be moved by gentle expulsion and resuction and rolling it with the holding pipet.
6. Ensure that the egg is tightly held by applying slightly more suction. The zona pellucida can be mildly distorted without harming the egg.
7. Focus on the target pronucleus. Position the microinjection pipet close beneath the egg so that its tip is vertically below the pronucleus. Use the fine vertical micromanipulator control to bring the tip into the same focal plane as the pronucleus. Without changing its vertical plane, bring the tip up to the zona pellucida at a level horizontal with the pronucleus.
8. Squeeze hard on the 50-mL syringe or if using the automated system, clear the microinjection pipet. It is possible to see DNA solution being ejected by its mixing with the M2 medium, a slight movement of the egg, or the presence of contaminating particulate matter in the medium.
9. Inject the egg. The zone pellucida is easily pierced. When the tip appears to be within the pronucleus, squeeze on the injection syringe. Three things may be observed:
    a. The pronucleus swells. This is a successful injection. Apply pressure until the pronucleus is roughly twice its original volume, and then withdraw the pipet in a single smooth rapid movement (**Fig. 7**).
    b. A small clear bubble appears at the tip of the microinjection pipet, and the perivitelline space may swell. The extremely elastic egg membrane has not been pierced. To penetrate the membrane, continue to push the microinjection pipet as far as the holding pipet, and then pull the tip back into the pronucleus before injecting again. Experienced operators can "feel" the egg membrane give way.
    c. Nothing. The injection pipet is probably blocked and should be changed. Alternatively, the pipet puller may be producing microinjection pipet with sealed tips or tips with excessively small openings—adjust the pipet puller. If particulate matter in the DNA stock is suspected, change the DNA stock.
10. Following injection, cytoplasmic granules may flow out into the perivitelline space—indicating egg lysis. If eggs lyse on two or three successive occasions, change the microinjection pipet.
11. Switch back to the 4× objective, and place the injected egg above the holding and microinjection pipet. Injected eggs should be divided into two groups: eggs that have survived injection and those that have not.
12. Eggs should be maintained in M2 medium for a maximum of 15 min. A skilled operator can inject 15–40 eggs in this period. Once all the eggs in a batch have been injected, return the survivors to M16 microdrop culture at 37°C in a 5% $CO_2$ incubator through two washes of M16 equilibrated at 37°C, 5% $CO_2$.

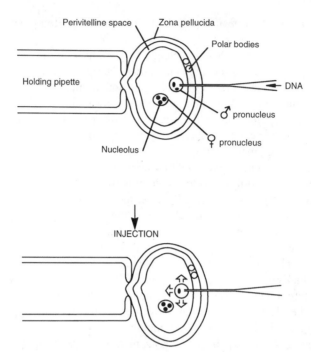

Fig. 7. Injection of fertilized one-cell eggs. High-magnification (400×) microscopic view.

13. Eggs that have survived injection are returned by oviduct transfer to the natural environment afforded by a recipient pseudopregnant female. Eggs can be transferred on the same day as microinjection or transferred following overnight culture.

## 11.4. Notes

1. The quality of the DNA is crucial to the survival of the eggs. DNA should be purified as described in **Subheading 10.3., step 13**. The composition of the solvent in which the DNA is dissolved is very important. The presence of minute amounts of $Mg^{2+}$ will kill an injected egg as will an overly high concentration of EDTA. The optimal buffer is 10 m$M$ Tris-HCl, pH 7.4, containing 0.1–0.25 m$M$ EDTA (Microinjection TE or MITE; **10**).
2. There is no correlation between the concentration of DNA and the resultant copy number of the transgene. However, excessively high concentrations of DNA are toxic to the egg (**10**).
3. An estimated 1–2 pL of DNA solution enter the egg pronucleus in each microinjection. Most investigators, use concentrations of 1–5 mg/mL and are thus injecting about 500 copies of the DNA fragment depending on the size of the fragment (**10**). For different fragments, the DNA concentration combining the optimum efficiency of integration and egg survival will vary. One can make up different dilutions of the DNA (1, 2, and 5 mg/mL) and rotate between these dilutions during a microinjection session.

4. The physical state of the DNA has little bearing on the success of the experiment. Cosmid clones of 50 kb can be introduced as easily as DNA of smaller sizes (*20*) Linearized DNA integrates with a fivefold greater efficiency than circularized DNA (*10*). The structure of the DNA ends created by different restriction enzymes has no effect on the efficiency of integration or the organization of the resulting transgene (*10*).

5. Avoid the sticky nucleolus on microinjecting. If this attaches to the injection pipet, it may be drawn out of the egg, thus killing it.

6. A technique has been developed in which DNA complexed with polysine is microinjected into the egg cytoplasm rather than the pronuclei (thus affording a much easier and larger target), 12.5% of pups born from zygotes injected in this manner were transgenic compared with 21.7% for pronuclear injection. No transgenic pups were born from microinjection DNA alone into the cytoplasm (*21*).

## 12. Identification of Ovulating Mice and Rats

### 12.1. Introduction

Pseudopregnant recipient females are used as surrogate mothers to nurture surviving microinjected eggs to birth. Females are made pseudopregnant by mating them with vasectomized males. To increase the chances of a successful mating, only females in estrous are coupled with the males. Female rats and mice maintained on the light–dark cycle described here will ovulate roughly every 4 d. The stage of the estrous cycle for mice can be determined by the appearance of the vagina as detailed in **Table 1**. For rats, a vaginal smear is required. On the afternoon before the day of implantation, estrous females are paired with a vasectomized male. On the following morning, female mice are examined for the presence of a copulatory plug, evidence that a successful mating has occurred (50–80% of females that are visually determined to be in estrous will be impregnated). In rats, the copulatory plug will often have fallen out by the morning, and there will not be sperm in the vagina to indicate that a successful mating has taken place. However, vaginal stimulation can induce pseudopregnancy, so as a backup procedure, female rats are gently prodded with a smooth blunt probe for 1 min before being placed with the vasectomized males.

### 12.2. Materials

1. Phosphate-buffered saline (PBS): 20 m$M$ sodium phosphate buffer, pH 7.4, containing 0.8% NaCl or 0.9% NaCl alone.
2. 1-mL disposable syringe (without needle).
3. Petri dish.
4. Light microscope with 40× magnification (and 400× magnification if checking for the presence of sperm).
5. Female rats and mice (preferably experienced mothers).

**Table 1**
**Identification of Ovulating Mice**

| Stage of the oesterous cycle | Vaginal characteristics |
|---|---|
| Dioestrous | Opening small, tissues small and moist |
| Prooestrous | Opening gaping, tissues red-pink and moist Dorsal and ventral folds |
| Oestrous (ovulating) | Opening gaping, tissues pink and moist Pronounced folds |
| Metoestrous | Tissues pale and dry, white cell debris |

## 12.3. Method

1. To screen for potential surrogate mothers, use female rats that have given birth to at least one litter and weaned for more than a week.
2. Mark rats on tail for identification.
3. Fill a 1-mL syringe with 0.2–0.3 mL PBS or 0.9% saline.
4. Expose the rat's vagina by lifting its tail.
5. Gently insert the syringe into the vaginal opening until resistance is felt. Expel the PBS or 0.9% NaCl into the vagina, and then draw the fluid back into the syringe to collect vaginal cells into the fluid.
6. Transfer the contents onto a suitably labeled Petri dish. Repeat with several other rats.
7. Examine the cells under a light microscope with 40× magnification, and identify ovulating rats from the characteristics detailed below.

| Stage of estrous cycle | Cell characteristic |
|---|---|
| Diestrous | Few epithelial cells and leukocytes |
| Proestrous (pair rats at this stage) | Many nucleated epithelial cells, yellow appearance |
| Estrous | Many large cornified epithelial cells, brownish appearance |
| Metestrous | Fewer cornified epithelial cells, many leukocytes |

8. If checking for the presence of sperm, examine the vaginal smear under 400× magnification.

## 12.4. Notes

1. Although virgin female rats can make suitable surrogate mothers, they tend to consume a part of their litters. Experienced mothers tolerate the traumas associated with the experimental procedures better to care for the unusually large or small litters that may eventually be produced.
2. Some female rats are anestrous, remaining always in a particular phase of the estrous cycle. These are not suitable for mating.

## 13. Delivery of Microinjected Eggs to Surrogate Mothers by Oviduct Transfer

### 13.1. Introduction

In the oviduct transfer, microinjected embryos are implanted into the ampulla of recipient pseudopregnant females. It is preferably performed on the same day as the microinjections to minimize the time the injected eggs are in culture. Alternatively, it can be performed on the following day when it is possible to observe if the injected eggs have developed to the two cell stage. Fifteen to 20 eggs delivered to each oviduct (30–40 eggs overall) should give an ideal litter size of 5–10 pups. In practice, the number of eggs needed to achieve this number depends on the competence of the experimenter. Oviduct transfer is not an easy technique, and novice experimenters are advised to practice with cadavers until they are happy with the procedure. A convenient dye can be used in place of eggs to visualize correct delivery of the pipet contents into the ampulla.

### 13.2. Materials

1. Microinjected eggs in microdrop culture.
2. M2 medium.
3. 5-mm Sterile tissue-culture dishes.
4. Pseudopregnant recipient mouse or rat.
5. CRC (for rats) or Avertin (for mice) anesthetic.
6. Dissection instruments (as in vasectomies).
7. Artery clip (1.5 in available from Arnold Horwell).
8. Fiberoptic illumination (e.g., Nikon).
9. Surgical microscope with optional assistant's viewing head (e.g., Carl Zeiss, Jena, Germany, OPM 212T with head and model 050).
10. Oviduct transfer pipets and mouthpiece (**Subheading 7.3.2.**).
11. Epinephrine: prepare 0.5% (W/V) epinephrine in 0.1 $M$ HCl, and dilute to 0.1% (v/v) in PBS for a working solution.
12. Ampicillin (Binotal from Bayer).
13. Paraffin oil (Fluka cat. no. 76235).
14. 70% ethanol in squeeze bottle.

### 13.3. Methods

Preliminary procedure: Loading the oviduct transfer pipet (**Fig. 8**):

Fill the pipet with liquid paraffin oil to the shoulder. The viscosity of the oil affords a greater degree of control over the movement of the eggs. Then, take up a small amount of air, then a small amount of M2, and then another bubble. The eggs are then collected, preferably in a stacked rank with a minimum of medium. A third air bubble is taken up followed by a final column of M2. The total length of the eggs/bubbles/medium should not exceed 2 cm. The loaded pipet can be conveniently stuck on a piece of plasticene on the surgical microscope.

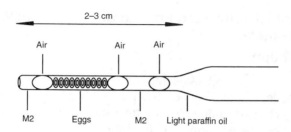

Fig. 8. The tip of an oviduct transfer pipet enlarged to show the arrangement of eggs, air bubbles, and media.

1.  Anesthetize a 0.5-d pc pseudopregnant recipient female as in **Subheading 2.2.**
2.  Place the animal ventral side down on the lid of a 9-cm Petri dish if operating on a mouse, but directly onto the stage of the dissecting stereomicroscope if using a rat. Spray the back with 70% ethanol.
3.  Comb the hair away from the incision site using a pair of fine forceps. Make a 1-cm transverse cut in the skin about 1 cm left of the spinal cord, at the level of the last rib, using large sharp scissors.
4.  For mice, locate the orange-colored ovary beneath the body wall. A 3–5 mm cut should be made through the body wall at a point a few millimeters away from the ovary, using fine scissors. Stretch the cut to prevent bleeding.
5.  For rats, the ovary is not normally visible through the body wall. Make a parallel 1-cm cut into the body wall using a pair of fine scissors and stretch to prevent bleeding.
6.  Introduce a single stitch into the body wall on one side of the incision, and leave the silk suture in place.
7.  Pull out the fat pad joined to the ovary using a pair of fine, blunt forceps. The ovary, uterus, and oviduct will be pulled out as well.
8.  Attach an artery clip to the fat pad (avoid the ovary), and position the reproductive tract over the back of the animal such that the coils of the oviduct are exposed and the ovary is toward the left.
9.  The mouse should be moved (while on the Petri dish) to the stage of the dissecting stereomicroscope. Illuminate the oviduct with a fiberoptic light source and view under the 10–20× magnification.
10. Orient the oviduct coils to reveal a cavity (**Fig. 9**) that lies below the ovary and behind the coils of the oviduct. The opening of the oviduct (the infundibulum— target of the transfer process) is located within the cavity behind a transparent membrane—the bursa—which covers the cavity, oviduct, and ovary. Find an area of the membrane, preferably above the infundibulum, that is free of capillaries. Tear it gently with watchmaker's forceps. If the infundibulum cannot be seen through the bursa, just rip the membrane at a convenient point, and continue the search. The infundibulum may be lifted out of the cavity when gently gripped with watchmaker's forceps. Excessive bleeding can be halted with an application of epinephrine (*see* **Subheading 13.4.**).

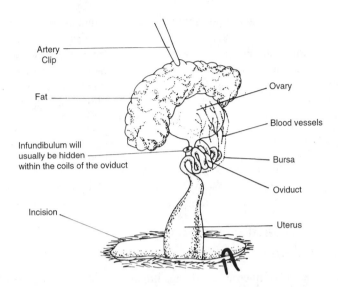

Fig. 9. Schematic diagram of the mouse ovary and oviduct prepared for oviduct transfer. The infundibulum is located within the coils of the oviduct and may be accessed by penetrating the transparent membrane that covers the oviduct.

11. Prepare the eggs for transfer. Remove a maximum of 20 microinjected eggs from the microdrop culture, and wash them in M2 medium. Load the eggs into an oviduct transfer pipet (**Fig. 8**) as described in the preliminary procedure.
12. Return to the animal. Use small screws of tissue paper held in the watchmaker's forceps to mop up any excess blood.
13. Grip the tip of the infundibulum with a sharp pair of watchmaker's forceps such that the opening can be accessed by the oviduct transfer pipets. Push the tip of the OTP into the mouth of the infundibulum (the opening is not visible until penetrated and may be located by gentle prodding with the tip of the pipet). Push the pipet into the infundibulum until it has entered the ampulla. The pipet tip must be far enough into the infundibulum so it does not fall out when the eggs are expelled, but not so far in that the opening is against the wall of the ampulla, so restricting the escape of the eggs.
14. Expel the contents of the OTP into the ampulla, and monitor delivery of the eggs by the appearance of the bubbles. When three bubbles have appeared, one can be certain that the eggs have been deposited into the ampulla.
15. Withdraw the pipet. Remove the artery clip, grip the fat pad with a pair of blunt forceps, and return the reproductive tract into the body cavity. Sew up the body wall with one or two stitches, and then clip the skin together with an autoclip.
16. Repeat with the other side of the reproductive tract if the availability of the microinjected eggs allow and the animal is sufficiently anesthesized.
17. If the recipient is a rat, inject it with 50 mg ampicillin (Binotal) intraperitoneally.

This prevents low-grade postsurgery infections and can optimize the number of pups subsequently delivered. This step is not necessary for mice.

18. After confirming that the animal has regained some degree of consciousness and mobility, leave it to recover in a warm, quiet place, and then transfer to an individual cage.

19. Examine the animals regularly during the days after the operation to ensure that they do not pick up any infections. Mouse pups should be delivered 19–20 d after the operation, and rats 21–22 d.

### 13.4. Note

Excessive bleeding can be prevented by a prior application of epinephrine (Sigma; 20–50 µL of 0.1% solution diluted in 0.01 *M* HCl) made through the bursa at the top of the oviduct using a fine needle and a 1-mL disposable syringe.

## 14. Caesarean Section and Fostering
### 14.1. Introduction

The survival rate of microinjected eggs implanted into the pseudopregnant recipient female varies tremendously from 0 to 50%, resulting in abnormally small or large litters. In the former case, the few eggs that do develop tend to be overnourished. Fetuses grow too large to be delivered normally and risk dying *in utero*. If pregnancy proceeds for 2–3 d beyond the normal gestation period of 19–21 d, the pups can be rescued by Caesarean section and fostering. It may also be necessary to foster pups born normally to a mother that dies unexpectedly or pups in a large litter that are not fed sufficiently.

### 14.2. Materials

1. 70% Ethanol in a squeeze bottle.
2. Sharp, fine dissecting scissors.
3. Sharp watchmaker's forceps (size 5).
4. Blunt forceps.

### 14.3. Methods

#### 14.3.1. Caesarean Section

1. Kill the mother by cervical dislocation.
2. Soak the abdomen with 79% ethanol.
3. Skin the lower half of the animal. Cut open the body wall to reveal the pregnant uterus. Dissect the uterus carefully by cutting at the oviduct and the cervix, and tearing away at the attached membranes.
4. Cut the uterus into sections, each containing a single pup. Gently squeeze each pup out with a blunt forceps.
5. Dissect away the membranes that surround the pup, and cut the umbilical cord. Wipe away any fluid from the mouth and nose. Rhythmically and gently squeeze the chest with blunt forceps to simulate breathing.
6. Place the pups on warm tissue, and keep warm until fostered.

### 14.3.2. Fostering

1. Remove the foster mother from her pups.
2. Mix the foster pups with the natural pups.
3. Try to make the foster mother urinate on the mixed litter. The animal will often urinate on being picked up and restrained with one hand. SD rats will often foster pups without this stimulus.
4. Leave the mother undisturbed with the mixed litter for at least 3 h.
5. Remove some of the natural pups to leave a litter of 12. The mother will be unable to care for a larger litter.

## 14.4. Notes

1. Foster mothers should have natural litters born within 1–2 d of the pups they are fostering.
2. The foster mother should be of a strain with good maternal characteristics (e.g., Swiss or Parkes mice, SD rats). Preferably, the coat color of the natural pups can be used to differentiate them from the fostered pups. Transgenic mice derived from (CBA/J)/(C57B1) F2 eggs will be black or agouti, readily distinguishable from the albino pups of the Swiss 3T3 strain.

## 15. In Vitro Fertilization

### 15.1. Introduction

Some transgenic animals may be infertile owing to an inability to mate or to rear a litter. With mice, in vitro fertilization (IVF) can be used to continue the line. the technique involves superovulating females and fertilizing the eggs with sperm taken from the epididymis of males. IVF eggs are then transferred to a 0.5-d pc pseudopregnant recipient female. The strain of mouse used should be the same as those used for the original transgenic mouse production (e.g., C57BL/6J X CBA/JF1). Males should be more than 6 wk of age and preferably studs rested the previous night. Females should be at least 4 wk of age. The described protocol employs a single male and 10 females.

### 15.2. Materials

1. Fertilization medium (FM): All chemicals should be tissue-culture grade. Prepare the following concentrated stocks, and filter each through a 0.4-$\mu$m Millipore filter into a sterile plastic tube. Store frozen at –20°C.
   a. 10X A stock: Dissolve 7.013 g NaCl (Sigma S5886), 1.0 g glucose (Sigma G6138), 0.201 g KCl (Sigma P5405), 21.3 mg $Na_2HPO_4$ (Sigma S5136), 0.102 g $MgCl_2 \cdot 6H_2O$ (Sigma M2393) in double-distilled or Milli-Q water. Bring to 100 mL final volume.
   b. 10X B stock: Dissolve 2.106 g $NaHO_3$ (Life Technologies 895-1810P), 1.0 g phenol red (Sigma P5530) in water. Bring to 100 mL final volume.
   c. 100X C stock: Dissolve 55 mg sodium pyruvate (Sigma P5280) in 10 mL water.

    d. 100X D stock: Dissolve 0.264 g $CaCl_2 \cdot 2H_2O$ (Sigma C7902) in 10 mL water. To prepare fertilization medium, mix 10 mL of 10X A, 10 mL of 10X B, 1 mL of 100X C, and 1 mL of 100X D. Make up to 100 mL by adding 78 mL double-distilled water. Add 3.0 g BSA (Sigma A4161), and mix gently until dissolved. Pass through a 0.45-µm Millipore filter using a large disposable syringe. Store in sterile plastic containers at –20°C.

2. M16 medium.
3. 70% Ethanol in a squeeze bottle.
4. Light paraffin oil (Fluka).
5. 35-mm Sterile tissue-culture dishes.
6. Micropipet and sterile tips (e.g., Gilson, Villiers-le-Bel, France, Pipetman, P200).
7. Egg transfer and mouth-operated pipet system.
8. Dissecting scissors: 1 regular, 1 fine.
9. Watchmaker's forceps: 2X #5.
10. Stereomicroscope with understage illumination (e.g., Nikon; SMZ-1OTD).
11. Fiberoptic light source.
12. 37°C incubator gassed with 5% $CO_2$.

## 15.3. Methods

1. Two days before the IVF procedure, start the superovulation of the donor females as described in **Subheading 4**.
2. On the day before the IVF or at least 3 h before, preincubate one 35-mm sterile tissue-culture dish containing 1 mL FM without paraffin and six dishes containing 0.5 mL FM under paraffin oil at 37°C with 5% $CO_2$.
3. Early on the morning of the IVF, prepare three tissue-culture dishes containing 2–3 mL M16 medium and one tissue-culture dish containing M16 microdrop cultures. Incubate at 37°C with 5% $CO_2$.
4. At 0645 h, kill the male donor by cervical dislocation. Place it abdomen side up, and soak with 70% ethanol. Open up the body cavity. Pull out the testes with a pair of watchmaker's forceps. Dissect out the epididymis (a white mass of coils at the base of the testes) with a pair of fine scissors, and immediately transfer to the pregassed tissue culture dish containing 0.5 mL FM under paraffin. Repeat for the other epididymis.
5. Rapidly tease away the fine membrane with a pair of fine forceps, and squeeze the sperm out of the epididymis. The sperm should exit in a continuous stream. Do this fast since the sperm should be at room temperature for the minimum possible time.
6. Incubate the sperm at 37°C with 5% $CO_2$ for 30 min.
7. To each preincubated tissue-culture dish containing 0.5 mL FM under paraffin, add 50 µL of sperm. The final concentration of sperm should be approx $1–2 \times 10^6$ sperm/mL.
8. Incubate the diluted sperm mixture for 2 h at 37°C with 5% $CO_2$.
9. At 0940–1000 h, kill the superovulated females by cervical dislocation, and dissect out the oviducts from both sides (**Subheading 13.3.**). Collect all the oviducts into a pregassed dish containing 1 mL of FM without oil.

10. Tease away the ampulla to release the cumulus mass from the oviducts. Transfer the cumulus masses from four oviducts to each pregassed dish containing 0.5 mL FM plus sperm.
11. Incubate the sperm–egg mixture for 4 h (min 3 h) at 37°C with 5% $CO_2$.
12. At 1500 h, wash the eggs three times in pregassed M16 medium to remove excess sperm. Transfer the eggs to the M16 microdrops and culture overnight.
13. On the next day, surgically transfer the two-celled embryos to the oviducts of 0.5-d pc pseudopregnant recipient females.

### 15.4. Note

Should sperm prove difficult to obtain from the epididymis of one male, it is advisable to kill another male quickly and repeat the dissection.

## 16. Cryopreservation of Transgenic Rodent Lines

### 16.1. Introduction

Valuable transgenic lines can be frozen and preserved indefinitely by cryopreservation. Cryopreservation protects the transgene line from environmental catastrophes, but is also an economical and labor-saving method of preserving lines for future detailed analysis. Mouse blastocysts can be generated by culturing fertilized one-celled embryos in M16 medium (12,22). Rat eggs do not survive very well in culture, so rat blastocysts are preferably obtained directly from the animal at 5 d post-hCG.

There are two general methods for cryopreservation of mouse embryos (23, 24). One is an equilibrium method (slow cooling), which uses a programmable cooling machine; embryos are exposed to moderate cryoprotectant concentrations and are cooled slowly at 0.3–2°C/min. Embryos are dehydrated during this slow-cooling process.

The second is a fast-cooling method that requires only a –70°C freezer: embryos are exposed to high molar concentrations of cryoprotectants and cooled rapidly. Exposure time to cryoprotectants is reduced and vitrification occurs soon after the cooling begins. This method is more sensitive to minor variations in protocol, especially time.

Taking into account the survival rate of the embryos as they undergo the different manipulations (e.g., from 100 frozen embryos, 88 are recovered from straws of which 77 survive freeze–thawing, and 7–8 pups are born after oviduct transfers of which 4 are transgenic), at least 400 embryos must be frozen in two different sessions to be sure of obtaining 15–16 transgenic animals.

### 16.2. Slow-Cooling (Equilibration) Method of Freezing Mouse Embryos

#### 16.2.1. Materials

1. PBS with $Ca^{2+}$ and $Mg^{2+}$. Prepare PBS-A by dissolving 10 g NaCl (Sigma S5885), 0.25 g KCl (Sigma P5405), 1.44 g $Na_2HPO_4$ (Sigma S5136), 0.25 g $KH_2PO_4$

(Sigma P5655) in 1 L of water. Autoclave. To 160 mL of PBS-A, add 20 mL of autoclaved 6.8 m$M$ CaCl$_2$·2H$_2$O (Sigma C7902) and 20 mL of autoclaved MgCl$_2$·6H$_2$O (Sigma M2393).

2. Fetal calf serum (FCS) (Sigma, Ghent, Belgium).
3. Glycerol (Life Technologies, cat. no. 5514VA).
4. Freezing solution 1: PBS with Ca$^{2+}$ and Mg$^{2+}$ containing 10% (v/v) FCS.
5. Freezing solution 2: PBS with Ca$^{2+}$ and Mg$^{2+}$ containing 10% (v/v) FCS and 5% (v/v) glycerol.
6. Freezing solution 3: PBS with Ca$^{2+}$ and Mg$^{2+}$ containing 10%, (v/v) FCS and 10% (v/v) glycerol.
7. Freezing solution 4: PBS with Ca$^{2+}$ and Mg$^{2+}$ containing 10% (v/v) FCS 1, % (v/v) glycerol, and 0.6 $M$ sucrose.
8. Sucrose (Sigma S1888).
9. M16 culture medium.
10. Freezing straws and plugs (Biotechnologies international, cat. no. ZA475 and ZA511, respectively).
11. Forceps.
12. Programmable freezing machine (e.g., Kryo 10 series II controlled-rate freezer from Planer Products Ltd., Sunbury, UK).
13. Liquid nitrogen storage tank.
14. 37°C incubator gassed with 5% CO$_2$.
15. Stereomicrosope.
16. 35-mm Tissue-culture dishes.
17. Egg-transfer pipets and mouthpiece.

### 16.2.2. Method

#### 16.2.2.1. FREEZING THE EMBRYO

1. Prepare two 35-mm tissue-culture dishes containing freezing solution 1, one dish containing freezing solution 2, and one containing freezing solution 3. Incubate at room temperature.
2. Transfer the mouse embryos (eight cell to blastocyst stage) from M16 to freezing solution 1. Rinse briefly.
3. Transfer the embryos to the second dish of freezing solution 1 and rinse.
4. Place the embryos in a dish containing freezing solution 2, and incubate at room temperature for 10 min.
5. Transfer the embryos to freezing solution 3, and incubate at room temperature for 10 min. During this period, label the straws, load them with embryos, and seal the straws.
6. Begin cooling within 10–30 min from when the embryos are first exposed to freezing solution 3. Cool the straws in the programmable freezing machine from room temperature to –6°C at –2°C/min and hold at –6°C for 5 min.
7. Remove the straws from the freezing machine. Hold the straws in a vertical position, and vertically seed them by grasping the straws at a point above the embryos with a pair of forceps dipped in liquid nitrogen. Hold for a few seconds until the liquid at the point of contact freezes.

8. Return the straws to the freezing machine and maintain at $-6°C$ for 10 min.
9. Cool the straws at $0.5°C/min$ to $-30°C$.
10. Finally plunge the straws into liquid nitrogen and store.

### 16.2.2.2. THAWING THE FROZEN EMBRYOS

1. Prepare one 35-mm tissue-culture dish with freezing solution 4, two other dishes containing freezing solution 1, and two containing M16 medium. Incubate at room temperature.
2. Remove the straws quickly from the liquid nitrogen, and hold at room temperature for 30 s before placing in a 37°C water bath until the ice melts.
3. Expel the embryos into the dish of freezing solution 4, and incubate at room temperature for **exactly** 10 min.
4. Rinse the embryos twice in two dishes of freezing solution 1.
5. Transfer the embryos into the M16 medium and rinse twice.
6. Culture the embryos, or transfer them into recipients.

## 16.3. Fast-Cooling (nonequilibration) Method of Freezing Mouse Embryos

### 16.3.1. Materials

1. The same as for **Subheading 16.2.1.**
2. Freezing solution 5: PBS with $Ca^{2+}$ and $Mg^{2+}$ containing 10% (v/v) FCS, 3.25 $M$ glycerol and 0.5 $M$ sucrose.
3. Freezing solution 6: PBS with $Ca^{2+}$ and $Mg^{2+}$ containing 10% (v/v) FCS and 0.5 $M$ sucrose.
4. $-70°C$ Freezer.

### 16.3.2. Methods

#### 16.3.2.1. FREEZING EMBRYOS

1. Prepare two 35-mm tissue-culture dish with freezing solution 1 and one with freezing solution 5. Incubate at room temperature.
2. Transfer the mouse embryos (eight cells to the morula stage or $2\frac{1}{2}$ d post-hCG) from M16 culture medium to one of the dishes containing freezing solution 1.
3. Rinse the embryos twice in the second dish of freezing solution 1.
4. Place the embryos in the dish of freezing solution 5, and incubate at room temperature for 20 min.
5. Load the embryos into prelabeled straws, and seal.
6. Keep the embryos at room temperature, so that the total exposure to freezing solution 5 is 20 min.
7. Place the straws horizontally onto the bottom of the $-60$ to $-80°C$ freezer.
8. Cool the straws for 5–15 min.
9. Plunge the straws into liquid nitrogen using cooled forceps as before.

16.3.2.2. Thawing Embryos

1. Prepare two 35-mm tissue-culture dish containing M16 medium, and warm in a 37°C incubator pregassed at 5% $CO_2$. Also, prepare two dishes containing freezing solution 1 and one containing freezing solution 6. Incubate at room temperature.
2. Remove the straws quickly from the liquid nitrogen, and place in a 25°C water bath until the ice melts.
3. Expel the embryos into the dish of freezing solution 6, and incubate at room temperature for **exactly** 12 min.
4. Transfer the embryos into a dish of freezing solution 1, and rinse twice in the second dish of freezing solution 1.
5. Transfer the embryos into the prewarmed M16 medium, and rinse twice.
6. Culture the embryos or transfer them into recipients.

## *16.4. Cryopreservation of Rat Embryos*

Rat cryopreservation is less well researched. This protocol describes a method for freezing rat embryos by direct plunging into liquid nitrogen *(25)*.

### *16.4.1. Materials*

1. The same as for **Subheading 7.2.** (although with different freezing medium).
2. Freezing medium VS1: PBS with $Ca^{2+}$ and $Mg^{2+}$ containing 0.4% BSA (Sigma A4161), 20% (v/v) DMSO (Sigma D2650), 15.5% (w/v) acetamide (Sigma A0500), 10% (v/v) propylene glycol (Sigma 1009), 6% (w/v) polyethylene glycol 8000 (Sigma P2139).
3. Freezing medium VS2: PBS with $Ca^{2+}$ and $Mg^{2+}$ containing 0.4% BSA and 12.5% (v/v) freezing medium VS1.
4. Freezing medium VS3: PBS with $Ca^{2+}$ and $Mg^{2+}$ containing 0.4°/O BSA and 25% (v/v) freezing medium VS1.
5. Freezing medium VS4: PBS with $Ca^{2+}$ and $Mg^{2+}$ containing 0.4% BSA and 50% (v/v) freezing medium VS1.
6. 0.4% BSA in PBS.

### *16.4.2. Methods*

16.4.2.1. Freezing Embryos

1. Prepare one 35-mm tissue-culture dish containing VS1 and chill at 4°C. Prepare another 35-mm tissue-culture dish containing VS2 and with VS3. Incubate at room temperature.
2. Place the rat blastocysts (5 d post-hCG) in the dish containing VS2, and incubate for 5 min at room temperature.
3. Transfer the embryos to a dish containing VS3, and incubate at room temperature for 5 min.
4. Wash the embryos in the chilled (4°C) dish of VS1. Transfer the embryos in a volume of 40 μL to a precooled 4°C polypropylene tube or freezing straw. Chill at 4°C or 15 min.
5. Plunge the tube into liquid nitrogen and store.

### 16.4.2.2. THAWING EMBRYOS

1. Chill freezing media VS4 and VS3 to 4°C.
2. Prepare one 35-mm tissue-culture dish containing VS2 and two dishes containing 0.4% BSA in PBS with $Ca^{2+}$ and $Mg^{2+}$. Incubate all three dishes at room temperature. Prepare two 35-mm dishes containing M16, and incubate in a 37°C incubator pregassed at 5% $CO_2$.
3. Thaw the tubes on ice.
4. Add 200 µL of chilled VS4 to the thawed tube, and incubate at 4°C for 10 min.
5. Add 400 µL of chilled VS3, and leave at 4°C for 10 min.
6. Transfer the embryos to the dish containing VS2, and maintain at room temperature for 5 min.
7. Wash the embryo twice in the dishes of 0.4% BSA in PBS with $Ca^{2+}$ and $Mg^{2+}$.
8. Transfer the embryos into pseudopregnant mothers.

## References

1. Palmiter, R. D., Brinster, R. L., Hammer, R. E., Trumbauer, M. E., Rosenfeld, M. G., Birberg, N. C., et al. (1982) Dramatic growth of mice that develop from eggs microinjected with metallothioneine-growth hormone fusion genes. *Nature* **300,** 611–615.
2. Palmiter, R. D. and Brinster, R. L. (1986) Germ line transformation of mice. *Ann. Rev. Genet.* **20,** 465–499.
3. Lacy, E., Roberts, S., Evans, E. P., Burtenshaw, M. D., and Constantini, F. (1983) A foreign beta globin gene in transgenic mice—integration at abnormal chromosomal positions and expression in inappropriate tissues. *Cell* **34,** 343–358.
4. Krumlauf, R., Chapman, V. M., Hammer, R. E., Brinster, R. L., and Tilghman, S. M. (1985) Differential expression of a-fetoprotein genes on the inactive X chromosome in extra-embryonic and somatic tissues of a transgenic rodent line. *Nature* **319,** 224–226.
5. Carter, D. A. (1993) Transgenic rodents and the study of the central nervous system, in *Methods in Molecular Biology, vol. 18: Production of Transgenic Rodents by Microinjection* (Murphy, D. and Carter, D. A., eds.), Humana, Totowa, NJ, pp. 7–22.
6. Hui, K. M. (1993) Transgenic animals and the study of the immune system, in *Methods in Molecular Biology, vol. 18: Production of Transgenic Rodents by Microinjection* (Murphy, D. and Carter, D. A., eds.), Humana, Totowa, NJ, pp. 37–52.
7. Murphy, D. (1993) Transgenic animals and the study of cancer, in *Methods in Molecular Biology, vol. 18:Production of Transgenic Rodents by Microinjection* (Murphy, D. and Carter, D. A., eds.), Humana, Totowa, NJ, pp. 23–36.
8. Waller, S. J., Ho, M., and Murphy, D. (1995) in *DNA Cloning 4: A Practical Approach* (Glover, D. M., and Hames, B. D., eds.), IRL, New York, pp. 185–229.
9. (1985) NIH Guide for the Care and Use of Laboratory Animals National Institutes of Health publication No 85-23, Washington, DC.
10. Brinster, R. L., Chen, H. Y., Trunbauer, M. E., Yagle, M. K., and Palmiter, R. D. (1985) Factors affecting the efficiency of introducing foreign DNA into mice by microinjecting eggs. *Proc. Natl. Acad. Sci. USA* **82,** 4438–4442.
11. Mullins, J. J., Peters, J., and Ganten, D. (1990) Fulminant hypertension in transgenic rats harbouring the mouse Ren-2 gene. *Nature* **344,** 541–544.

12. Hogan, B., Constantini, F., and Lacy, E. (ed.) (1986) *Manipulating the Mouse Embryo—A Laboratory Manual,* Cold Spring Harbor Laboratory Press, Cold Spring Harbor, NY.

13. Chada, K., Magram, J., Raphael, K., Radice, G., Lacy, E., and Constantini, F. (1985) Specific expression of a foreign B-globin gene in erythroid cells in transgenic mice. *Nature* **314,** 377–380.

14. Oberdick, J., Smeyne, R. J., Mann, J. R., Zackson, S., and Morgan, J. I. (1990) The promoter that drives transgene expression in cerebellar Purkinje and retinal bipolar neurons. *Science* **248,** 223–236.

15. Brinster, R. L., Allen, J. M., Behringer, R. R., Gelinas, R. E., and Palmiter, R. D. (1988) Introns increase transcriptional efficiency in transgenic mice. *Proc. Natl. Acad. Sci. USA* **85,** 836–840.

16. Young, W. S. III, Reynolds, K., Shepard, E. A., Gainer, H., and Castel, M. (1990) Cell specific expression of the rat oxytocin gene in transgenic mice. *J. Neuroendocrinol.* **2,** 917.

17. Schedl, A., Montoliu, L., Kelsey, G., and Schutz, G. (1993) A yeast artificial chromosome covering the tyrosinase gene confers copy number-dependent expression in transgenic mice. *Nature* **362,** 258–261.

18. Swanson, L. W., Simmons, D. M., Azzira, J., Hammer, R., and Brinster, R. (1985) Novel developmental specificity in the nervous system of transgenic animal expressing growth hormone fusion gene. *Nature* **317,** 363–366.

19. Vogelstein, B. and Gillespie, D. (1979) Preparative and analytical purification of DNA from agarose. *Proc. Natl. Acad. Sci. USA* **82,** 615–619.

20. Constantini, F. and Lacy, E. (1981) Introduction of a rabbit B-globin gene into the mouse germ line. *Nature* **294,** 92–94.

21. Page, R. L., Butler, S. P., Subramaniam, A., Gwazdauskas, F. C., Johnson, J. L., and Velander, W. H., (1995) Transgenesis in mice by cytoplasmic injection of polylysinetDNA mixtures. *Transgenic Res.* **4,** 353–360.

22. Whittingham, D. G. (1971) Culture of mouse ova. *J. Reprod. Fertil. Suppl.* **14,** 7–21.

23. Pomeroy, K. O. (1991) Cryopreservation of transgenic mice. *GATA* **8,** 95–101.

24. Kasai, M., Komi, J. H., Takakomo, A., Tsudura, H., Sakurai, T., and Machada, T. (1990) A simple method for mouse embryo cryopreservation in a low toxicity vitrification solution, without appreciable loss of viabilty. *J. Reprod. Fertil.* **89,** 91–97.

25. Kono, T., Suzuki, O., and Tsunoda, Y. (1988) Cryopreservation of rat blastocysts by vitrification. *Cryobiology* **25,** 170–173.

# 7

# Cre Recombinase Mediated Alterations of the Mouse Genome Using Embryonic Stem Cells

## Anna-Katerina Hadjantonakis, Melinda Pirity, and András Nagy

## 1. Introduction

The introduction and establishment of transgenic, and in particular embryonic stem (ES) cell-based gene "knockout" technologies have made the mouse a key player in studying embryonic development and disease *(1,2)*. In recent years, methods for the production of more complex genomic alterations have become increasingly widespread, hinting at an ability to manipulate and study a mammalian genome to an extent never previously thought possible. Such methodologies often partner homologous recombination-mediated gene targeting or random integration with site-specific recombination events.

This chapter is concerned with the utilization of the bacteriophage P1 derived site-specific recombinase protein Cre *(3–5)*, and its employment as a means to catalyze modifications in homologously recombined and randomly integrated target sites within the mouse genome.

Cre is a 38-kDa protein that recombines DNA between two loxP target sites. loxP sequences are 34 basepairs (bp) long comprising two 13-bp inverted repeats flanking an asymmetric 8bp core sequence. The recombination between two loxP sites with same orientation on the same DNA leaves two products each containing a single loxP site *(6)* (**Fig. 1**). This type of site-specific recombination, of which there are several other well-characterized systems in addition to the Cre/loxP, generates precise rearrangements of DNA but dispenses with the requirement for extensive homology between DNA partaking in the recombination. Recombination occurs through the recognition of the target sites by the recombinase, which then catalyzes strand exchange between them by precise breakage and rejoining events that are restricted to an internal region of identical sequence contained within the specific sites *(6)*.

From: *Methods in Molecular Biology, Vol. 97: Molecular Embryology: Methods and Protocols*
Edited by: P. T. Sharpe and I. Mason © Humana Press Inc., Totowa, NJ

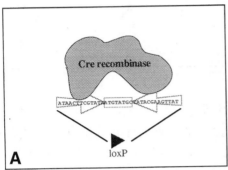 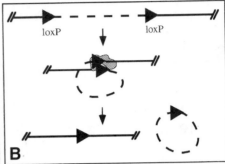

Fig. 1. The Cre recombinase has a 34-bp recognition site comprising two inverted repeats and a core sequence (**A**). It can catalyze a site-specific recombination event leading to the deletion of the intervening DNA (**B**).

In addition to the Cre/loxP system, another one of the many recombinase that does not require *cis*-elements, but utilizes short recognition sites for recombination is the yeast FLP/FRT system. This system has also been widely used and applied to genome alterations *(7–10)*, though to date the Cre protein has been shown to be more amenable to use in mammalian cells, and is therefore currently favored by most laboratories for use in ES cells and transgenic mice *(11)*. As a consequence, we will solely refer to the Cre/loxP system in the strategies we present, though it should be noted that if the FLP/FRT were to be as efficient as the Cre/loxP system it could be substituted in all methodologies. Additionally, it might also be anticipated that future experiments may require a multistep, site-specific recombination strategy, thereby requiring the use of two separate recombination systems.

This chapter will illustrate and provide the methodologies for some of the applications of such a site-specific recombination system to experiments aimed at analyzing mouse embryonic development and/or disease conditions, from single-gene alterations, lineage-restricted and/or conditional gene ablations or ectopic expression through to chromosome engineering, and finally the use of such a system for lineage analysis.

## 1.1. Combining Homologous and Site-Specific Recombination

Homologous recombination in ES cells allows the precise disruption (knock-out) of a target gene. Many new approaches require a defined alteration of a gene or the genome. By combining the homologous and site-specific recombinations, we are now in the position of creating most desired alterations in the mouse genome *(12,13)*. In the following section, we will briefly introduce some of the most important current applications. The list will not be complete, since novel applications for the use of this system are continuously being reported.

### 1.1.1. Eliminiating Any Regional Effect of a Knockout: Removal of a Selectable Marker

To identify targeted events, an introduction of a positive selectable marker, usually neomycin, into the targeted locus is required. Recently, there has been an increasing concern regarding the repressor effect of the selectable marker cassette on the genes in the vicinity of its insertion. Therefore, removal of the marker from all targeted genes is advisable. This can most easily be performed by flanking (floxing[flanking with loxP]) the selectable marker cassette by loxP sites, which on introduction of the Cre recombinase will result in the removal of neomycin.

### 1.1.2. Introducing Subtle Changes in a Gene of Interest

Cre/loxP type approaches can also be used to introduce subtle changes into any gene, including point mutations *(13a)* or small deletions into genes of interest, particularly if domain deletions *(14)* or domain swaps within the protein coding regions of a particular gene are desired. Such an approach is illustrated in **Fig. 2**.

### 1.1.3. Introducing Specific Chromosomal Changes

Over the past few years, strategies have been developed for chromosome engineering in ES cells. Such approaches have been used to design novel chromosomal variants, or to mimic altered chromosomes associated with human disease or metastasis *(15)*. Such approaches rely on the sequential targeting of two loxP sites, either in *trans* (i.e., to different chromosomes) or in *cis* (some distance apart on the same chromosome), followed by the transient expression of the Cre protein in order to mediate the site-specific recombination event between the loxP sites, leading to the formation of the new chromosomal variant *(15,16)* (**Fig. 3**). In this case, usually three ES cell electroporations and resulting screening for the alteration are required: the two end points are targeted separately, and are followed by the introduction of the recombinase, which will mediate the recombination event between them.

This type of strategy can be used to create almost all cataloged forms of chromosomal aberration. Additionally, application of this technology could allow the creation of multiple large-scale chromosomal alterations, for example, a set of nested hemizygous deletions (also known as deficiencies) covering an entire chromosome. These could then be used to reveal novel tumor suppressor genes or functional haploinsufficiencies mapping within the deleted DNA. If a panel of deficiencies is available, screens for interesting phenotypes can be carried out either in culture, or in mice, the latter being particularly amenable to ES cell ↔ tetraploid embryo aggregation *(17,18)* as a means of creating completely ES cell-derived embryos, therefore bypassing the germline for accessing embryonic phenotypes *(19)*.

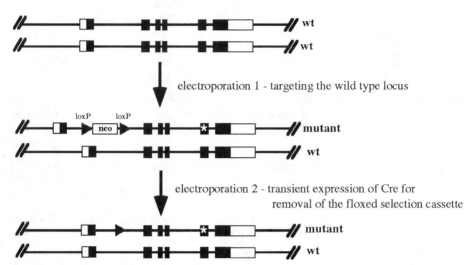

Fig. 2. Simple gene alterations—introducing a subtle change into a gene of interest. In this example, a point mutation, small deletion, or domain swap (*) is introduced into an exon of a gene using a relacement type vector, containing a floxed neomycin resistance cassette. Transient expression of Cre results in removal of the region between the loxP sites, and as a result the neo selection cassette is removed, thereby rendering the genomic structure identical to wt with the exception of the introduced change and the single loxP site remaining in the first intron.

Since a recombination event between two loxP sites some distance apart or positioned on different chromosomes, is relatively rare, presumably because of physical constraints, a reflection of chromosome architecture and decreased proximity, such strategies are designed incorporating a binary positive selection system that is only activated after a successful recombination recreates the cassette. Thus, the desired recombination event will reconstruct the selectable genetic marker, from two silent portions placed adjacent to each of the loxP sites. The most commonly used selection systems include the reconstitution of a human hypoxanthine phosphoribosyltransferase (HPRT) minigene *(16–20)*, or the juxtaposition of a strong promoter upstream of a selectable marker.

## 1.1.4. Creating Lineage and/or Inducible Gene Alteration

### 1.1.4.1. Lineage-Specific Gene Knockouts

The combination of a lineage-restricted promoter and the Cre/loxP system can be used to create a modified locus that is restricted to a certain spatiotemporal domain within the mouse. This has recently been demonstrated using a keratin 5 promoter-driven Cre to ablate the X-linked pig-a gene in skin *(21)* and αCamkinase II promoter-driven Cre to ablate the NMDAR1 gene in a sub-

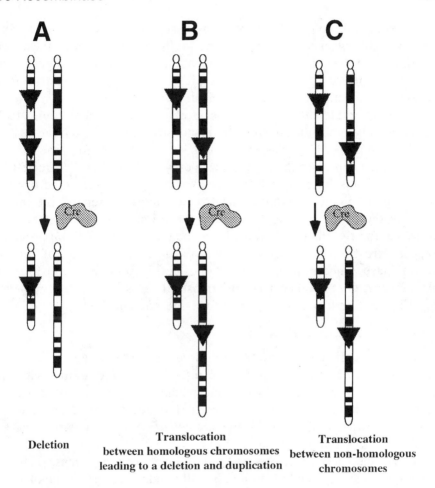

**Deletion**

**Translocation between homologous chromosomes leading to a deletion and duplication**

**Translocation between non-homologous chromosomes**

Fig. 3. Chromosomal modifications. Targeting of the two loxP sites some distance apart, either in *cis* (**A**) or in *trans*, in the same (**B**), or different chromosomes (**C**) can give rise to various chromosomal modifications.

set of postnatal cells of the CNS *(22)*. Combinations of conditional and inducible Cre/loxP gene targeting regimes can be utilized for a more sophisticated assessment of gene function in the developing embryo and adult animal.

### 1.1.4.2. LINEAGE-SPECIFIC GENE REPAIR

This approach allows one to study specific cellular phenotypes over restricted time-points or spatial locations during development or adult life. Here, the targeted allele should contain the original gene structure designed to be silent or compromised *(13a)* owing to the interruption by the loxP-flanked "stop" sequence (resulting in no or comprimised gene expression). The Cre recombinase will be expressed from a transgene in a restricted set of cells, thereby resulting in the excision of the loxP-flanked region in these cells (resulting in normal gene expression). Consequently, the original gene structure will be restored solely in cells expressing the Cre transgene, but the remaining population will still be deficient.

If the reparable allele phenotype is characterized, and found to be embryonically lethal, then the primary responsible lineage/organ can be identified, and a proper Cre transgenic line made (or selected from the existing lines), which expresses the recombinase only in the primarily affected lineage. When this lineage-specific Cre-expressing line is crossed over the homozygous mutant genotype, the Cre recombinase repairs the mutant allele in the primary lineage, rescuing the primary deficiency, therefore allowing for the manifestation of secondary defects. This approach is expected to be less sensitive to a possible mosaic action of the Cre recombinase, since in many cases, a mosaic repair is sufficient for complete rescue. On the other hand, in almost all cases, high-fidelity lineage-specific deletion is necessary for lineage-specific knockout.

### 1.1.4.3. INDUCIBLE GENE KNOCKOUT

Conditional lineage-restricted ablations can be obtained through the incorporation of an inducible system into a transgenic regime. Here the Cre protein can be induced where and when appropriate. This can either be achieved by placing the Cre gene under the control of an inducible promoter (either ubiquitous or lineage-specific) or to construct the Cre cassette as an inducible fusion protein.

Several approaches have been utilized for inducible gene expression in both experimental animals and in culture. Initially, inducible systems involved the use of heat shock, isopropylithio-β-D-galactoside (IPTG), and heavy metals as inducing agents *(23,24)*, but owing to their lack of specificity and toxic side effects, these systems are primarily restricted to use in prokaryotes, yeast, and *Drosophila*. Unfortunately, at present there is no totally satisfactory inducible system available for use in transgenic mice, though recently several laboratories have reported the successful use of drug- and hormone-inducible systems in mammalian cell culture *(25,26)*. A common aspect of these various approaches is that the majority comprise binary systems involving the use of chimeric transcription factors that can reversibly bind target gene sequences in response to the administered drug or hormone. Modifications of the bacterial tetracycline system *(27)*, the *Drosophila* ecdysone receptor system *(26)*, and molecular dimerizer systems based on FK506 or its analog rapamycin *(28)* have been shown to work in cells in culture and are presently being developed for use in transgenic mice.

1.1.4.4. Requirement for Lineage-Specific or Inducible Cre Transgenic
Lines—A Cre Transgenic Mouse Database

All of the above technologies rely on the availability of properly working lineage-specific or inducible Cre transgenic lines. To this end, we are coordinating the assimilation of information concerning the available and planned Cre transgenic lines, and have compiled them into a continually updated database, which can be accessed through the World Wide Web at www.mshri.on.ca/develop/nagy/nagy.htm.

## 1.2. Future Directions

### 1.2.1. Fate Mapping

A great deal of interest has centered around the fate of individual cells within the developing embryo *(29,30)*. In lower organisms, following the fate and genesis of cells is less complex than in a mammal such as the mouse embryo, where marking a single cell by injection or transplantation and following its descendants during the course of development is technically demanding, requiring expensive equipment and expertise.

ES cell-mediated transgenic technologies utilizing the Cre/loxP system may be able to facilitate fate mapping the embryo greatly. Here a lineage-restricted promoter is used to drive the Cre recombinase. A second transgene containing a reporter gene flanked by the loxP sites is also required. The second transgene should contain a "stop" sequence between the ubiquitous promoter and the marker gene to keep the gene silent. Double transgenic animals will neither express Cre nor the marker gene until the specific developmental stage permitting Cre expression and subsequent recombination, resulting in marker gene activation. This will result in activation of the recombinase, resulting in the excision of the "stop" sequence, and expression of the marker in all the progenitors cells, regardless of the later expression status of the Cre-driving promoter. If conditions are optimized, then the Cre can be used for noncomplete excision and, thus, result in excision in a very limited number of cells or even in single cell. Another possible improvement would be the utilization of an inducible Cre recombinase, allowing the precise regulation of excision frequency and timing, thereby making it relatively straightforward to follow the fate of individual cells and their descendants based on the expression pattern of the marker.

### 1.2.2. Gene Trap

It is feasible that gene insertion strategies utilizing loxP sites may gain popularity in certain gene-trap experiments, the goal of such experiments being to identify novel ubiquitous and/or lineage-restricted promoter/enhancer elements or genes.

Using a vector carrying a splice acceptor sequence placed upstream of a reporter gene (such as *lacZ* or GFP [green fluorescent protein]), different types

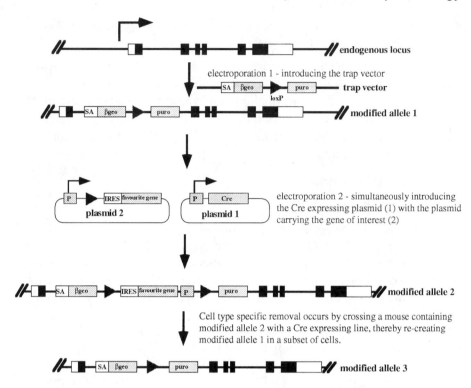

Fig. 4. Trapping an endogenous locus with the option of introducing a new gene. Here two electroporation steps are required. The first is used to identify a locus of interest, and the second is used to introduce the new gene into that locus. Two alternative alleles are created on the second electroporation, one doubly drug-resistant, the other singly. Thus, the drug selection will determine which clones can grow. P, promoter driving gene expression in ES cells; SA, splice acceptor; βgeo, βgal-neo gene fusion; puro, puromycin resistance cassettes; IRES, internal ribosome entry site.

of regulatory or gene sequences can be trapped *(31–33)*. ES cell-chimeric embryos are stained for the histochemical marker to reveal expression domains of the trapped elements *(31–33)*. On the basis of the expression pattern information gained on the trap cell lines, a subset is chosen for further study. If a specially designed trapping vector is used, such as that illustrated in **Fig. 4**, the trapped locus can be retargeted via loxP sites, and different transgenes can be knocked-in leading to their spatiotemporal expression being governed by the trapped element. Later, the expression can also be abolished by introducing the Cre recombinase into the system.

## 2. Materials

For all procedures, solutions should be made to the standard required for molecular biology using molecular biology-grade and/or "tissue-culture-tested" reagents. All solutions should be made using sterile double-distilled or MilliQ water, and where appropriate, autoclaved or filter-sterilized.

## 2.1. ES Cell-Culture Media and Solutions

1. DMEM+: For all procedures, we use Dulbecco's Modified Eagle's Medium (DMEM—Flow Labs [Finland], powder, cat. no. 430-1600) supplemented with the following:
   a. 0.1 m$M$ nonessential amino acids (100X stock, Gibco, Grand Island, NY, cat. no. 320-1140AG; *see* **Note 1**).
   b. 1 m$M$ sodium pyruvate (100X stock, Gibco cat. no. 320-1360).
   c. 100 m$M$ β-mercaptoethanol (100X stock stored as aliquots at –20°C, Sigma, St. Louis, MO, cat. no. 600564AG).
   d. 2 m$M$ L-glutamine (100X stock, stored as aliquots at –20°C, Gibco cat. no. 320-5030AG).
   e. 15% Fetal calf serum (FCS) (*see* **Note 2**).
   f. Penicillin and streptomycin (final concentration 50 µg/mL each, Gibco cat. no. 600-564AG).
   g. Leukemia inhibitory factor (different sources, for example, Gibco) 1000 U/mL. The supplemented DMEM used for propagation of ES cells is referred to as DMEM+ (*see* **Notes 1–3**).
2. 0.1% Gelatin: 1 g (w/v) gelatin is 0.1% gelatin (Sigma or BDH) made up in 1 L water, autoclaved and stored at 4°C (*see* **Note 4**).
3. 2X ES cell-freezing medium: 2X ES cell-freezing medium should be made up fresh each time it is to be used, and should comprise freshly prepared 60% DMEM+, 20% FCS, and 20% DMSO (Sigma, cat. no. D-5879).
4. Phosphate-buffered saline (PBS): For all tissue-culture work, we use PBS without calcium and magnesium. This is made from 10 g NaCl, 0.25 g KCl, 1.5 g Na$_2$HPO$_4$, 0.25 g KH$_2$PO$_4$, pH 7.2. The solution is autoclaved and stored at 4°C.
5. Trypsin (0.1%): Dissolve 0.5 g trypsin powder (Gibco, cat. no. 0153-61-1) in 500 mL saline/EDTA solution. Adjust the pH to 7.6, sterilize through a 0.22-µm filter, and store at –20°C. This constitutes a 5% stock, which needs to be diluted to 0.1% on defrosting.
6. Saline/EDTA solution: 1L of saline/EDTA solution comprises 0.2 g EDTA, 8.0 g NaCl, 0.2 g KCl, 1.15 g Na$_2$HPO$_4$, 0.2 g KH$_2$PO$_4$, 0.01 g phenol red, 0.2 g glucose, pH 7.2. The solution should be sterilized through a 0.22-µm filter, and stored at room temperature.
7. Tissue-culture-treated plasticware: We routinely use NUNC, Corning, and Falcon plasticware.
8. Humidified incubator: This is maintained at 37°C and 5% CO$_2.$
9. Electroporation apparatus: Use apparatus, such as a Bio-Rad GenePulser, and appropriate cuvets (Bio-Rad, Hercules, CA, cat. no. 165-2090).
10. Selection reagents: There are drugs, such as G418 (Gibco, cat. no. 11811-031), gancyclovir (Syntex, cat. no. 00865516), puromycin (Sigma, cat. no. P8833), and 6-thioguanine (Sigma, cat. no. A4660) or HAT (Gibco, cat. no. 31062-037) for HPRT-negative or positive selection, and mitomycin C (Sigma, cat. no. M-0563).

11. Cre recombinase-expression vectors: Cre recombinase-expressing vectors are described in detail in a number of publications *(34–36)*.

## 3. Methods

### *3.1. Preparation of Tissue Culture Plates*

Tissue-culture plates require special treatment prior to the plating of ES cells. This can either be by coating them with gelatin or mitotically inactivated fibroblast cells.

#### *3.1.1. Gelatinized Plates*

In our experience, R1 ES cells *(17)* can be propagated on both types of plates without losing their totipotency or their ability to contribute to the germline (*see* **Note 7**).

1. To prepare gelatinized plates rinse the surface of tissue-culture dishes with a 0.1% gelatin solution (approx 100 µL/well of 96-well plates, or 3 mL for 6-cm plates, 5 mL for 10-cm plates).
2. Aspirate off the excess gelatin, and then allow the surface of the plates to dry a little (2–4 min).
3. Add fresh medium to the plates, and place them into the incubator until required.

#### *3.1.2. Plates with Feeder Cells*

Primary mouse embryonic fibroblast (EMFI) cells *(37)* or the STO fibroblast cell line *(38)* is the most commonly used feeder layers. Details for the preparation of a stock of EMFI or STO cells are described elsewhere *(39)*. A brief protocol for the preparation of EMFI feeder layers will be given here.

1. Quickly defrost a vial of EMFI or STO cells, and then transfer the cell suspension to a sterile 15-mL tube containing 10 mL prewarmed feeder cell media (DMEM supplemented with 10% FCS). Then spin at 1000*g* for 5 min, at room temperature.
2. Aspirate the supernatant, and then gently resuspend the cell pellet in 10 mL media.
3. Plate the cell suspension onto five 15-cm plates each containing 25 mL media, and place in an incubator.
4. When the cells form a confluent monolayer (usually takes 3 d) they are ready to be treated with mitomicin C.
5. Briefly, medium is aspirated from the confluent plates and replaced with 10 mL media containing 100 µL of 1 mg/mL mitomycin C, and then placed in an incubator for 2–2.5 h.
6. The medium is aspirated, and the plates washed twice each with 10 mL PBS, followed by the addition of 10 mL of trypsin/EDTA/dish.
7. Plates are placed in an incubator until the cells begin to detach.
8. Ten milliliters of media are added to each plate, and the cell suspension broken up by gentle pipeting.

9. The cell density is determined (hemocytometer) and adjusted to $2 \times 10^5$ cells/mL. Then the cells are plated directly onto dishes suitable for ES cell culture. We routinely plate approx $1 \times 10^6$ R1 cells per 10-cm dish (*see* **Notes 5–7**).

## 3.2. Passaging ES Cells

Optimally ES cells should be fed every day and split every second day (by which time they should be 70–80% confluent). It is important not to let them overgrow, since this may induce them to differentiate.

1. If the cells are split into gelatinized plate, the plates should be prepared (as detailed previously) before starting the trypsinization.
2. Trypsinize the cells by first aspirating the medium off the dishes and then rinsing twice with PBS.
3. Aspirate off any remaining PBS, and add trypsin (0.1%) to the cells. For 10-cm plates, we use 2.5 mL trypsin. This volume should be scaled according to the size of plate used.
4. Place the dish containing the cells in trypsin in a 37°C incubator for 3–6 min.
5. Check under an inverted microscope to see if the cells have detached. When they have, add 5 mL medium to the dish.
6. Resuspend the cells and transfer them to a 12-mL tube.
7. Spin down at 1000*g* for 5 min at room temperature.
8. Aspirate the medium, and then add 1 drop of PBS to the pellet.
9. Flick the tube hard in order to resuspend the cells.
10. Add 5 mL of medium to the tube, pipet again to mix, and split the contents at 1:5 or 1:7 ratio in new plates containing sufficient volume of medium.

## 3.3. Electroporation of ES Cells

Cells are routinely passaged two days prior to electroporating. Cells are ready for electroporating when their density is optimal. Usually one 10-cm plate at approx 80% confluency will provide enough cells for 1–2 electroporations. Our standard electroporation protocol is given below.

1. Gelatinize 10-cm plates, and then add 10 mL medium to each.
2. Place them in a 37°C incubator until they are required.
3. Switch on the electroporation apparatus.
4. Harvest the cells by trypsinization.
5. Resuspend the cell pellet in ice-cold PBS (1 mL for each 10-cm plate).
6. Determine the cell density (hemocytometer), and dilute with PBS to the required density for electroporation. We regularly electroporate at a relatively high cell density: $7 \times 10^6$ cells/mL (this number varies between different labs).
7. For each electroporation, mix together 20–40 μg vector DNA (for an approx 10-kb vector; *see* **Notes 8** and **9**) and 0.8 mL of the ES cell suspension in an electroporation cuvet (Bio-Rad, cat. no. 165-2088) (*see* **Note 8**).
8. Set up the electroporation conditions prior to placing the cuvet into the electroporation chamber. We routinely use 250 V, 500 μF for the Bio-Rad GenePulser (*see* **Note 9**).

9. Zap the cuvet, then place it on ice for 20 min to 1 h.
10. Transfer the cells from the cuvet into the prewarmed medium containing dishes. (The contents of one cuvette are routinely seeded into two 10-cm dishes).
11. Change medium daily.
12. If drug selection is required, start this on the second day after electroporation (*see* **Note 10**).
13. Continue the selection until colonies become apparent, and grow to a size that is amenable to picking (usually takes 7–10 d) (*see* **Note 10**).

### 3.4. Picking Colonies After Selection

This can either be done with the naked eye or by placing a dissecting microscope (such as a Lietz M3B) into the laminar flow hood.

1. Colonies are ready for picking if they are large and well separated (usually 7–9 d after electroporation). For picking start by gently washing the plate twice with PBS.
2. After the second rinsing, leave a little of the PBS behind in the dish (about 1 mL) in order to keep the surface of the plate wet, therefore preventing the colonies from drying out.
3. Choose a colony to pick. The colony is optimal if it is neither too small nor too big, and contains nondifferentiated ES cells with characteristic morphology. Fill a P200 pipeter with approx 20 µL PBS, and gently pour this over the colony thereby rinsing it. Retain about 5–10 µmL of the PBS in the pipet, and with this, try to "suck up" the colony from the bottom of the plate. Hold the pipet perpendicular (if picking with the naked eye) or at 45° (if picking under the dissecting microscope) to the surface of the plate, since this facilitates the lifting of the colony from the plastic.
4. Using the P200, transfer each individual colony into separate wells of a 96-well plate containing 50 µL trypsin in each well. In doing so, pipet the colony up and down several times in order to dissociate the cells. Be careful to avoid creating too many air bubbles.
5. When all the colonies from a plate have been picked and transferred to the trypsin, the 96-well plate is placed in a 37°C incubator for 5–10 min.
6. During this time a new gelatinized 96-well flat-bottom plate with medium (200 µL/well, containing the selection agent) is prepared.
7. Working row by row with a multichannel pipeter the cell-trypsin solution is transferred to the gelatinized plate.
8. Pipet thoroughly, without creating too many air bubbles, so as to promote the formation of a single cell suspension.
9. Return the plate to the 37°C incubator.
10. Change the media daily until the cells are ready to passage (80% confluency).
11. When passaging the cells, split them into two or three new plates. These can each be used for the preparation of DNA for genotype screening and for creating frozen stocks, which can then be used for thawing the required clones.

### 3.5. Passaging Cells in 96-Well Plates

Optimally 3 or 4 d after picking colonies into the 96-well plates, the cells are at a density required for splitting. Since cells in different wells generally exhibit different growth potential, they will not grow at a synchronous rate. Therefore the optimal time for splitting the whole plate needs to be determined. It is best to choose a time when the majority of the cells have reached 80–90% confluency. Another more laborious alternative is to passage the clones at different stages (pooling them into groups depending on their growth rate) and replating them into different 96-well plates.

1. To passage cells in 96-well plates, first prepare several gelatinized 96-well plates.
2. Add 200 μL medium/well, and place plate in a 37°C incubator.
3. Aspirate the medium from the plate to be split, and then wash with PBS by multipipeting 200 μL of PBS into each well followed by aspiration.
4. Remove all traces of PBS, and then add 50 μL trypsin per well.
5. Incubate at 37°C for 5–10 min. The cells should detached with gentle tapping on the plate.
6. Multipipet 50 μL medium/well into each of the wells. Pipet up and down about five times so as to resuspend completely (*see* **Note 11**). Then split them (working row by row) into two or three newly gelatinized plates.
7. Return these plates to the 37°C incubator (*see* **Notes 12** and **13**).

### 3.6. ES Cell Freezing

#### 3.6.1. In Cryovials

The general protocol for freezing cells grown in a standard 10-cm dish at 70% confluency is given below (*see* **Notes 14** and **15**):

1. Change media 2–3 h before freezing the cells.
2. Freshly prepare 2X freezing media.
3. Harvest the cells in a 15-mL tube containing DMEM+ after trypsinization (*see* **Note 15**).
4. Spin down at 1000*g* for 5 min at room temperature.
5. Remove the supernatant, and then add one or two drops of DMEM+ to the tube. Shake gently, but thoroughly to disperse the cells.
6. Add an additional DMEM+ medium to a total volume of 1.5 mL, and disperse the cells carefully so that they comprise a single-cell suspension.
7. Add an equal volume (1.5 mL) of 2X freezing medium, and mix by pipeting several times.
8. Quickly aliquot the cell suspension into three vials, and immediately place them in a styrofoam box (this will allow them to cool down gradually). Alternatively, special boxes dedicated to this task can be purchased from a number of manufacturers (for example, Stratagene) (*see* **Note 16**).
9. Place the box in a –70°C freezer for 1–2 d, and then transfer the individual cryovials into a liquid nitrogen container for long-term storage (*see* **Note 16**).

#### 3.6.2. In 96-Well Plates

1. Working one row at a time using a multichannel pipetter, change the medium 2–3 h prior to freezing.

2. Freshly prepare 2X cell-freezing media.
3. Aspirate the medium from each well, and wash the cells with PBS (approx 200 μL).
4. Add 50 μL trypsin to each well, and then place plate in an incubator for 5–10 min.
5. Working on ice, preferably in a wide, flat container, aliquot 50 μL of DMEM+ into each well. Pipet the cells several times in order to get them into a homogenous suspension.
6. Then add 100 μL 2X cell freezing media to the wells, and again pipet to mix.
7. Finally add 80–100 μL sterile mineral oil (Sigma, cat. no. M-8410) to cover the cell/freezing medium mixture.
8. Wrap the plates in parafilm, place in a styrofoam box, and store in a –70°C freezer until such time as the desired clones have been identified and need to be recovered (*see* **Note 17**).

## 3.7. Thawing ES Cells

1. Prepare the appropriate-sized feeder layer containing plates for the cells that are to be recovered (24-well plates for 96-well frozen plates or 6 cm plates for cryovials containing approx $5$–$10 \times 10^6$ cells).
2. Add medium to the plates, and prewarm them in the incubator at 37°C.
3. The following steps vary with the type of tissue-culture plates used (*see* **Note 18**).

### 3.7.1. From 96-Well Plates

1. Remove the sample from the freezer, and as it begins to melt, carefully aspirate the overlying oil from the cell/medium mixture.
2. Using a multichannel pipetter and working one row at a time, quickly, but gently multipipet the cells twice in order to resuspend them thoroughly.
3. Change the pipet to a P200 (set to 200 μL), and then quickly transfer the well contents one at a time to the individual wells of a 24-well plate. Pipet quickly to resuspend the cells in the 96-well plate. Then transfer. Once the sample is transferred to the 24-well plate, pipet again to resuspend the cells and distribute them evenly in the prewarmed fresh media.
4. Aliquot another 200 μL into the now empty well of the 96-well plate to rinse it, and remove any remaining cells.
5. Transfer this additional 200 μL to the equivalent well of the 24-well plate.
6. Repeat the above for each required well of the 96-well plate.
7. Place the 24-well plate containing the newly transferred cells in a humidified $CO_2$ incubator at 37°C.
8. Change the media after 8 h or the following morning (if defrosting is carried out late in the evening), and then daily until the cells are ready to passage.

### 3.7.2. From Cryovials

1. Defrost the cells quickly, then transfer the cell suspension to a sterile 15-mL tube containing prewarmed media (approx 10 mL), and then spin at 1000*g* for 5 min at room temperature.
2. Aspirate the supernatant, and then add 1 drop of PBS.
3. Resuspend the cells by either flicking the bottom of the tube or by gently pipeting up and down.

4. Add few milliliters of media to the tube, and again gently pipet up and down to dissociate the cells.
5. Plate the cell suspension onto a 6-cm plate, and place in an incubator.
6. Change the media after 8 h or the following morning (if defrosting is carried out late in the evening), and then daily until passaging is required.
7. When the cells reach approx 70% confluence (usually taking 2 d), they are ready to passage.

### 3.8. Removal of loxP-Flanked Short Genomic Segment in ES Cells by Transient Expression of Cre Recombinase

1. Electroporate the correctly targeted ES cell line with 50 µg/mL of circular plasmid containing the Cre recombinase gene driven by an ES cell transcriptionally active promoter, for example, the pBS185 plasmid, which contains Cre under the control of the hCMV promoter *(35)*.
2. After electroporation, plate the cells very sparsely (approx 1000 cells/10-cm dish) onto gelatinized 10 cm plates each containing 10 mL of DMEM+.
3. Change media daily until the colonies have attained a size that is ready to pick (usually takes 7 d).
4. Pick colonies into 96-well plates, and expand into two or more plates. One plate is frozen, and the others are used for PCR or Southern screening to detect the required recombination event. A subset of the colonies will be mosaic for the Cre-mediated excision, therefore requiring further subcloning for the derivation of pure lines (*see* **Notes 19** and **20**).

### 3.9. Deletion of loxP-Flanked Large Genomic Segment or Selection for Site-Specific Chromosomal Alteration in ES Cells by Transient Expression of Cre Recombinase

This is carried out in much the same manner as in **Subheading 3.8.**, except that cells are usually subjected to selection, since a bipartite cassette that will be reconstituted after the desired recombination is routinely used (for example, HAT for the reconstitution of the HPRT minigene). Routine selection is applied 2 d after the electroporation. As a consequence, the cells need to be plated at regular density (one cuvet into one or two 10-cm plates) after the electroporation.

### 3.10. Introducing ES Cells Into Mice Preparation of Cells for Aggregation with and Injection into Embryos

When genetically altered ES cell lines are identified, thaw the 96-well plates, or cells from vials, onto feeders, and grow them up. Avoid taking them through too many passages, since with increasing passage number, the possibility of cells loosing their germline-transmitting ability may increase. Our standard protocol for preparing cells for aggregation is given below. Preparation of cells for blastocyst injection is performed in a similar manner, except that for this procedure, the cells need to be seeded at the usual density on d 3, trypsinized slightly longer, spun down and washed in PBS to obtain single-cell suspension on d 5 *(below)*.

Day 1:   Thaw cells 4 d prior to aggregation on a feeder cell layer containing plate.
Day 2:   Change the medium.
Day 3:   Split cells onto gelatinized plates, but instead of the usual 1:5 ratio, pass them 1:50 or even more dilute (*see* **Notes 21** and **22**).
Day 4:   Change the medium.
Day 5:   Trypsinize the cells briefly, until the colonies lift up as a loosely connected clump of cells. Stop trypsin by adding DMEM+ to the plate. Select clumps directly form the plate for aggregation (*see* **Note 23**).

Protocols and descriptions detailing the introduction of cells into mice by aggregation of ES cells with preimplantation embryos are provided elsewhere *(40)*. Our own lab protocols can be obtained through the World Wide Web at http://www.mshri.on.ca/develop/nagy/nagy.htm.

### 3.11. Removal of loxP-Flanked Short Genomic Segment In Situ by Crossing Germline-Transmitter Chimeras with a Stable Cre Transgenic Line

This approach requires the availability of a Cre-expressing transgenic line. In our laboratory, we have established a Cre recombinase-expressing transgenic mouse line (tgCre-1) by pronuclear injection of the hCMV Cre gene. Briefly, the insert purified from pBS185 plasmid *(34)* was directly injected at a concentration of 5 ng/mL into the pronucleus of zygotes fertilized by germline-transmitting males. This line is now routinely used to remove the loxP-flanked genomic piece efficiently from a targeted locus, by crossing this line with the targeted mouse *(13a)*. This and other lines with similar properties *(41)* can be found in the Cre transgenic database mentioned previously (**Subheading 1.1.4.4.**).

1. Males homozygous or heterozygous for the loxP-flanked DNA sequence are crossed with Cre transgenic females.
2. Offspring are screened for the required recombination event by PCR (from ear punch-derived sample DNA) or Southern analysis (from tail biopsy sample-derived DNA).

### 3.12. Removal of loxP-Flanked Genomic Segment In Situ by Transiently Expressing the Cre Recombinase in F1 Preimplantation Stage

An alternative method for excising loxP-flanked DNA is to express the Cre recombinase in an early embryo transiently *(42)*.

1. Briefly, germline transmission of the primary loxP-flanked DNA sequence is achieved.

2. Then a Cre-expressing plasmid is injected in circular form into the pronucleus of zygotes produced by a cross between males carrying the loxP-flanked sequence and wild-type females.
3. Injected embryos are transferred into pseudopregnant recipients.
4. Offspring are screened for the required excision event.

## 4. Notes

### 4.1. Cell-Culture Media and Solutions

1. Because glutamine and LIF are unstable, DMEM+ that is kept at 4°C for a period exceeding 2 wk needs to be supplemented with a new aliquot of 100X glutamine stock and LIF.
2. The quality of the FCS is critical to the propagation and maintenance of ES cells. We recommend that several batches be tested for plating efficiency and toxicity from different suppliers for their ability to support growth of pluripotent ES cells, and that then a bulk order of the best (to last approx 1 yr) be purchased, and the bottles stored at −20°C for up to 2 yr.
3. LIF helps maintain ES cells in an undifferentiated state especially when they are growing on gelatinized plates in the absence of feeder cell layers. However, it should be noted that too high a concentration of LIF can be deleterious for the cells. We suggest using twice the lowest concentration in which the cells stay undifferentiated. For our cell line, R1 it is 1000 U/mL.

### 4.2. Preparation of Tissue-Culture Plates

4. Gelatinized Plates: Using R1 ES cells, it is possible to carry out all ES cell manipulations on gelatinized plates from the initial passaging through to the introduction into mice. After replating cells from feeder containing plates onto gelatin plates, their characteristic morphology may be seen to change. This is usually only transient, and in time, they will revert to their usual appearance.
5. After mitomycin C treatment, the feeders are plated onto dishes for ES cell culture, the cells usually take an overnight incubation to attach they are then ready for use.
6. The medium should be changed from feeder cell medium (DMEM+ 10% FCS) to ES cell medium (DMEM+) before the addition of the ES cells.
7. An alternative to mitomycin C treatment is to treat the cells with 6000–10,000 rad of γ-irradiation.

### 4.3. Electroporation of ES Cells

8. The targeting vector needs to be linearized for electroporation. The DNA is pre-pared by standard "maxiprep" procedures, for example, cesium chloride gradient centrifugation, or popular kits (such as Qiaex, Promega Wizard, Geneclean). Ethanol-precipitate the digested DNA before the electroporation, and dissolve the pellet in sterile TE. The concentration of the DNA should be around 1 µg/µL.
9. For the transient Cre electroporation, plasmid DNA can be used straight after "maxiprep" purification. In this case, the electroporation procedure is carried out

according to the standard protocol provided. The Cre expressing vector can be coelectroporated along with a second selectable marker containing vector, or can be introduced alone *(35)*. If a selection-based coelectroporating strategy for identifying cells taking up the Cre plasmid is used, then the cells should be plated at normal density (one cuvet into one or two 10-cm plates). Transient expression of the Cre recombinase in ES cell culture results in almost 100% excision between loxP sites placed a few kilobases apart *(14)*.

10. When selecting for loss or gain of HPRT function, the cells should be maintained in the positive or negative selection prior to the selection switch that will assay the altered HPRT activity. The transiently expressed Cre-mediated excision is often mosaic. Therefore, a PCR screen should be carefully designed to detect such situations. Additionally, Southern blot analyses should be performed as a final check on candidate clones identified by PCR.

## 4.4. Passaging ES Cells

11. The cell pellet should be carefully resuspended. If this is not carried out, the ES cells will grow as large clumps, containing necrotic centers and differentiated cells at the periphery.

12. ES cells are growing optimally if they are ready for passing on each alternate day, and if their morphology does not change during this time. The typical ES cell morphology is when single cells are not visible, and they grow as characteristically shaped small clumps.

13. The number of cells to be plated at each passage and length of growth between two passages is critical. Cells should not be left to overgrow, and care should be taken not to split cells too diluted or too dense.

## 4.5. ES Cell Freezing

14. We usually freeze ES cells in cryovials at a density of $5–10 \times 10^6$ cells/mL of freezing medium. A single 10-cm dish usually gives approx three cryovials each containing 1 mL of cells.

15. To freeze cells from different-sized dishes, proportionally altered volumes of cell medium are used (for example, a 6-cm dish requires 1 mL, and a 3.5-cm dish requires 0.5 mL).

16. It is important to note that the cells should be frozen down gradually (in a styrofoam box or isopropanol container), and not to be kept at $-70°C$ for too long a period of time. Cryovials should be transferred to liquid Nitrogen for long-term storage.

17. Unfortunately, 96-well plates can only be kept at $-70°C$. We do not recommend keeping the plates for more than 2 mo at this temperature. Therefore, all screening for the required alleles needs to be performed within this time period.

## 4.6. Thawing ES Cells

18. It is important to thaw cells as rapidly as possible in order to avoid long crystal formation as the frozen vial passing through critical temperatures. Immediately

after removal from frozen storage, the cell-containing vials (cryovials or 96-well plates) are placed at 37°C in a water bath until they are almost completely defrosted (1–3 min).

## 4.7. Removal of loxP-Flanked Short Genomic Segment in ES Cells by Transient Expression of Cre Recombinase

19. No selectable marker is needed for this Cre-mediated excision, since 2–5% of cells are picking up the Cre expressing DNA. Practically in almost all these cells excision occurs between loxP sites placed a few kilobases apart. As a consequence, transient expression of the Cre recombinase in ES cells results in a 1 in 30 average frequency of excision.

20. If the region of DNA to be deleted contains a drug selection marker, then the newly acquired sensitivity to the marker can be tested when replica 96-well plates containing cells are available. Here, however, one should be aware of the frequent mosaic type of excision. Therefore this step does not replace the PCR or Southern blot screening.

## 4.8. Introducing ES Cells into Mice

21. In general cells should be replated on the appropriate size plate (96- or 24-well plate when passaging from 96-well plate, 6-cm dish when passaging from a cryovial), and grown up so that there are enough for freezing into cryovials and introducing into mice.

22. A separate plate should be prepared for aggregation as was described in the protocol. The reason for highly diluted single-cell plating is to produce clumps of 10–25 loosely connected cells prior to aggregation. Then the required-size colonies are predominantly found on the plate. Care should also be taken in order not to disaggregate the cell clumps (by pipeting the cells too vigorously or overtrypsinization).

23. For blastocyst injection purposes, cells do not need to be maintained as clumps. Therefore they should be completely dissaggregated.

## 4.9. Removal of loxP-Flanked Short Genomic Segment In Situ by Crossing Germline-Transmitter Chimeras with a Stable Cre Transgenic Line

In our experience, not all offspring that inherit the Cre transgene will undergo excision as shown in **Fig. 4**, indicating that although the efficacy of such a transgene is very high (complete excision in 90% of animals), it is not active in all embryos that inherit it.

## References

1. Capecchi, M. R. (1989) The new mouse genetics: altering the genome by gene targeting. *Trends Genet.* **5,** 70–76.
2. Thomas, K. R., Deng, C., and Capecchi, M. R. (1992) High-fidelity gene targeting in embryonic stem cells by using sequence replacement vectors. *Mol. Cell. Biol.* **12,** 2919–2923.

3. Sternberg, N. (1994) The P1 cloning system: past and future. *Mamm. Genome* **5**, 397–404.

4. Argos, P., Landy, A., Abremsky, K., Egan, J. B., Haggard-Ljungquist, E., Hoess, R. H., et al. (1986) The integrase family of site-specific recombinases: regional similarities and global diversity. *EMBO J.* **5**, 433–440.

5. Sauer, B. and Henderson, N. (1988) Site-specific DNA recombination in mamalian cells by the Cre recombinase of bacteriophage P1. *Proc. Natl. Acad. Sci. USA* **85**, 5166–5170.

6. Hoess, R. H. snd Abremski, K. (1985) Mechanism of strand cleavage and exchange in the Cre-lox site-specific recombination system. *J. Mol. Biol.* **181**, 351–362.

7. Babineau, D., Vetter, D., Andrews, B. J., Gronostajski, R. M., Proteau, G. A., Beatty, L. G., et al. (1985) The FLP protein of the 2-micron plasmid of yeast. Purification of the protein from *Escherichia coli* cells expressing the cloned *FLP* gene. *J. Biol. Chem.* **260**, 12,313–12,319.

8. Cox, M. M. (1988) in *Genetic Recombination* (Kucherlapati, R. and Smith, G. R., eds.), American Society for Microbiology, Washington DC, pp. 429–443.

9. O'Gorman, S., Fox, D. T., and Wahl, G. M. (1991) Recombinase-mediated gene activation and site-specific integration in mammalian cells. *Science* **251**, 1351–1355.

10. Fiering, S., Kim, C. G., Epner, E. M., and Groudine, M. (1993) An "in-out" strategy using gene targeting and FLP recombinase for the functional dissection of complex DNA regulatory elements: analysis of the b-globin locus control region. *Proc. Natl. Acad. Sci. USA* **90**, 8469–8473.

11. Buchholz, F., Ringrose, L., Angrand, P.O., Rossi, F. and Stewart, A.F. (1996) Different thrermostabilities of FLP and Cre recombinases: implications for applied site-specific recombination. *Nucleic Acids Res.* **24**, 4256–4262.

12. Rossant, J. and Nagy, A. (1995) Genome engineering: the new mouse genetics. *Nat. Med.* **1**, 592–594.

13. Nagy, A. (1996) in *Mammalian Development* (Lonai, P., ed.), Harwood Academic Publishers, Amsterdam, pp. 339–371.

13a. Nagy, A., Moens, C., Ivanyi, E., Pawling, J. Gertsenstein, M,. Hadjantonakis, K., Pirity, M., and Rossant, J. (!998) Dissecting the role of N-myc in development using a single targeting vector to generate a series of alleles. *Curr. Biol.* **82**, 661–664.

14. Gu, H., Zou, Y. R., and Rajewsky, K. (1993) Independent control of immunoglobulin switch recombination at individual switch regions evidenced through Cre-*loxP*-mediated gene targeting. *Cell* **73**, 1155–1164.

15. Smith, A. J. H., Sousa, M. A. D., Kwabi-Addo, B., Heppell-Parton, A., Impey, H., and Rabbitts, P. (1995) A site-directed chromosomal translocation induced in embryonic stem cells by Cre-*lox*P recombination. *Nature Genetics* **9**, 376–385.

16. Ramirez-Solis, R., Liu, P., and Bradley, A. (1995) Chromosome engineering in mice. *Nature* **378**, 720–724.

17. Nagy, A., Rossant, J., Nagy, R., Abramow-Newerly, W., and Roder, J. (1993) Derivation of completely cell culture-derived mice from early-passage embryonic stem cells. *Proc. Natl. Acad. Sci. USA* **90**, 8424–8428.

18. Nagy, A. and Rossant, J. (1996) Targeted mutagenesis: Analysis of phenotype without germ line transmission. *J. Clin. Invest.* **97**, 1360–1365.
19. Carmeliet, P., Ferreira, V., Breier, G., Pollefeyt, S., Kieckens, L., Gertsenstein, M., et al. (1996) Abnormal blood vessel development and lethality in embryos lacking a single VEGF allele. *Nature* **380**, 435–439.
20. Zhang, H. & Bradley, A. (1996) Mice deficient for BMP2 are nonviable and have defects in amnion/chorion and cardiac development. *Development* **122**, 2977–2986.
21. Takahama, Y., Ohishi, K., Tokoro, K., Sugawara, T., Yoshimura, Y., Okabe, M., et al. (1997) T cell-specific disruption of Pig-a gene involved in glycosylphosphatidylinositol (GPI) biosynthesis: Generation and function of T cells deficient in GPI-anchored proteins. *Proc. Natl. Acad. Sci. USA,* in press.
22. Tsien, J. Z., Chen, D. F., Gerber, D., Tom, C., Mercer, E. H., Anderson, D. J., et al. (1996) Subregion- and cell type-restricted gene knockout in mouse brain. *Cell* **87**, 1317–1326.
23. Bonner, J. J., Heyward, S., and Fackenthal, D. L. (1992) Temperature-dependent regulation of heterologous transcriptional activation domain fused to yeast heat shock transcription factor. *Mol. Cell. Biol.* **12**, 1021–1030.
24. Palmiter, R. D., Norstedt, G., Gelinas, R. E., Hammer, R. E., and Brinster, R. L. (1983) Metallothionein-human GH fusion genes stimulate growth of mice. *Science* **222**, 809–814.
25. No, D., Yao, T. P., and Evans, R. M. (1996) Ecdysone-inducible gene expression in mammalian cells and transgenic mouse. *Proc. Natl. Acad. Sci. USA* **93**, 3346–3351.
26. Christopherson, K. S., Mark, M. R., Bajaj, V., and Godowski, P. J. (1992) Ecdysteroid-dependent regulation of genes in mammalian cells by a *Drosophila* ecdysone receptor and chimeric transactivators. *Proc. Natl. Acad. Sci. USA* **89**, 6314–6318.
27. Gossen, M. and Bujard, H. (1992) Tight control of gene expression in mammalian cells by tetracycline-responsive promoters. *Proc. Natl. Acad. Sci. USA* **89**, 5547–5551.
28. Rivera, V. M., Clackson, T., Natesan, S., Pollock, R., Amara, J., Keenan, T., et al. (1996) A humanized system for pharmacologic control of gene expression. *Nat. Med.* **2**, 1028–1032.
29. Lawson, K. A. and Pedersen, R. A. (1987) Cell fate, morphogenetic movement and population kinetics of embryonic endoderm at the time of germ layer formation in the mouse. *Development* **101**, 627–652.
30. Smith, J. L., Gesteland, K. M., and Schoenwolf, G. C. (1994) Prospective fate map of the mouse primitive streak at 7.5 days of gestation. *Dev. Dyn.* **201**, 279–289.
31. Gossler, A., Joyner, A. L., Rossant, J., and Skarnes, W. C. (1989) Mouse embryonic stem cells and reporter constructs to detect developmentally regulated genes. *Science* **244**, 463–465.
32. Joyner, A. L., Auerbach, A., and Scarnes.W.C. (1992) The gene trap approach in embryonic stem cells: the potential for genetic screens in mice. *Ciba Foundation Symp.* **165**, 277–288.

33. Forrester, L. M., Nagy, A., Sam, M., Watt, A., Stevenson, L., Bernstein, A., et al. (1996) An induction gene trap screen in embryonic stem cells: Identification of genes that respond to retinoic acid *in vitro*. *Proc. Natl. Acad. Sci. USA* **93,** 1677–1682.

34. Sauer, B. and Henderson, N. (1989) Cre-stimulated recombination at *loxP*-containing DNA equences placed into the mammalian genome. *Nucleic Acids Res.* **17,** 147–161.

35. Sauer, B. and Henderson, N. (1990) Targeted insertion of exogenous DNA into the eukaryotic genome by the Cre recombinase. *New Biologist* **2,** 441–449.

36. Sauer, B. (1993) in *Guide to Techniques in Mouse Development, Methods in Embryology,* vol 225 (Wasserman, P. M. and DePamphilis, M. L., eds.), Academic, San Diego, pp. 890–900.

37. Doetschman, T. C., Eistetter, H., Katz, M., Schmidt, W., and Kemler, R. (1985) The *in vitro* development of blastocyst-derived embryonic stem cell lines: formation of visceral yolk sac, blood islands and myocardium. *J. Embryol. Exp. Morph.* **87,** 27–45.

38. Williams, R. L., Hilton, D. J., Pease, S., Willson, T. A., Stewart, C. L., Gearing, D. P., et al. (1988) Myeloid leukaemia inhibitory factor maintains the developmental potential of embryonic stem cells. *Nature* **336,** 684–687.

39. Wurst, W. and Joyner, A. L. (1993) *Gene Targeting: A Practical Approach.* (Joyner, A. L., ed.), Oxford University Press, Oxford, p. 33.

40. Nagy, A., Merentes, E., Gocza, E., Ivanyi, E., and Rossant, J. (1991) Developmetal potential of mouse embryonic stem cells and inner cell mass: a comparison. [Abstract] *Symposium of Mouse Molecular Genetics,* Heidelberg, August 21–25, 1991.

41. Schwenk, F., Baron, U., and Rajewsky, K. (1995) A *cre*-transgenic mouse strain for the ubiquitous deletion of *loxP*-flanked gene segments including deletion in germ cells. *Nucleic Acids Res.* **23,** 5080,5081.

42. Araki, K., Araki, M., Miyazaki, J., and Vassalli, P. (1995) Site-specific recombination of a transgene in fertilized eggs by transient expression of Cre recombinase. *Proc. Natl. Acad. Sci. USA* **92,** 160–164.

# 8

## Gene Trapping in Mouse Embryonic Stem Cells

### Jane Brennan and William C. Skarnes

## 1. Introduction

Gene trapping in mouse embryonic stem (ES) cells offers a method to create random developmental mutants with a direct route to cloning and defining the expression pattern of the disrupted gene *(1)*. Gene trapping involves the use of reporter gene constructs that are activated following insertion into endogenous transcription units. A number of plasmid- and retroviral-based vectors have been developed, which differ in their requirements for reporter gene activation (reviewed in **refs.** *2* and *3*). "Promoter trap" vectors simply consist of a promoterless reporter gene that is activated following insertions in exons of genes. In contrast, "gene trap" vectors contain a splice acceptor sequence upstream of a reporter and are activated following insertions into introns of genes. Both promoter and gene trap insertions create a fusion transcript from which a portion of the endogenous gene may be readily cloned *(4,5)*. The pattern of reporter gene activity can be monitored in ES-cell derived chimeric embryos *(6)* or in transgenic embryos following germline transmission *(7)*. With two plasmid-based vectors, reporter gene expression has been shown to reflect accurately that of the endogenous gene *(5,8)*. Ultimately, the function of the trapped gene can be tested following germline transmission. Using this approach, a number of embryonic lethal mutations and visible adult phenotypes have been isolated *(4,5,7,9–11)*.

Before initiating a screen, it is important to consider carefully the design of individual vectors, their efficiency in detecting gene trap events, and their potential biases (discussed in **refs.** *2* and *3*). For example, although gene trap vectors are more efficient at detecting insertions in genes than promoter trap

From: *Methods in Molecular Biology, Vol. 97: Molecular Embryology: Methods and Protocols*
Edited by: P. T. Sharpe and I. Mason © Humana Press Inc., Totowa, NJ

vectors, they will favor the detection of genes composed of large intronic regions, and are likely to miss genes possessing few or no introns. Gene trap vectors are also constrained by the reading frame imposed by the splice acceptor sequence. One solution has been to incorporate the splice acceptor derived from Moloney murine leukemia virus (MoMuLv) *env* gene that is capable of splicing in all three reading frames simultaneously *(12)*. However, splice acceptors with this property may be weak, so they may fail to mutate the gene at the site of insertion effectively. Alternatively, an internal ribosome entry site (IRES) may be used to initiate translation of the reporter gene independent of the upstream open reading frame *(13)*. Our own answer to this problem has been to construct separate vectors in each of the three reading frames. Given that individual vector designs each have their own inherent biases, we recommend using a combination of vectors to ensure the most representative sampling of the genome.

Perhaps the most useful vectors are those based on the βgeo reporter system *(7)*. βgeo encodes a polypeptide fusion possessing both β-galactosidase and neomycin phosphotransferase *(neo)* activities, thereby providing direct drug selection for gene trap events and obviating the need to screen through a large background of nongene trap events. However, there is some debate over whether *neo* driven by its own promoter may be better suited to trap genes activated at later stages of development. In this regard, the sensitivity of the drug selection marker becomes an important consideration. It has been shown, for example, that the original βgeo reporter contained a mutation in *neo* and thus tended to preselect for genes expressed at high levels *(8)*. Correction of this mutation has enabled the isolation of genes expressed at low levels in ES cells that are activated on differentiation.

The isolation of so-called white colonies (cells expressing less than detectable levels of β-galactosidase [βgal] activity) does not necessarily indicate that the trapped gene is expressed at a low level, but rather may reflect inactivation of the βgal enzyme in the resulting fusion. One important class of genes that produce inactive βgal fusion products are those that encode N-terminal signal sequences *(8)*. To capture specifically this class of genes, which include secreted and membrane-spanning proteins, we modified the conventional gene trap design by adding a type II transmembrane domain upstream of βgeo. With the secretory trap vector pGT1.8TM, only fusions that acquire an N-terminal signal sequence will produce an active βgal fusion. One surprising result to emerge from this study was the isolation of two independent insertions in the same gene in a sample of six cell lines, suggesting that gene trapping may be far less random than originally anticipated. This finding further emphasizes the need to recognize and ultimately overcome inherent biases imposed by individual vector designs.

This chapter will focus on the use of βgeo-based plasmid vectors in mouse ES cells to screen for developmentally regulated genes in the mouse. Methods for the maintenance and electroporation of a feeder-independent ES cell line are described. To identify βgal-positive clones, we employ a screening protocol that simply involves staining one set of duplicate wells with X-gal. This procedure can also be used to screen for genes induced or repressed in response to specific growth factors or other inducers of ES cell differentiation. Several methods have been used to clone a portion of the endogenous gene associated with gene trap and promoter trap insertions. These include the construction of cDNA libraries *(10)*, inverse PCR *(4)*, ligation-mediated PCR *(11)*, and 5' rapid amplification of cDNA ends (RACE) *(5)*. This chapter, provides a detailed method for 5' RACE cloning, used routinely in the authors' laboratory, in which a number of improvements on previously published protocols have been added *(3,14)*. The generation of ES cell-derived chimeras by blastocyst injection or morulae aggregation is used to monitor expression patterns in embryos and to transmit the insertions to the germline of mice. These methods are not covered in this chapter, since they have been described extensively elsewhere *(15–18)*. Finally, we have included a rapid dot-blot method for genotyping transgenic mice carrying gene trap insertions. This method can be used to analyze as many as 400 tail biopsies in a single day, and can reliably distinguish between heterozygous and homozygous animals based on signal intensity.

## 2. Materials
### 2.1. Maintenance of Mouse ES Cells

1. Cell culture medium (1X): store up to 1 mo at 4°C:
   a. 10X Glasgow MEM/BHK12 (Gibco, Paisley, UK): store at 4°C     40 mL
   b. 7.5% sodium bicarbonate (Gibco): store at 4°C     13.2 mL
   c. 1X MEM nonessential amino acids (Gibco): store at 4°C     4 mL
   d. 200 m*M* glutamine, 100 m*M* sodium pyruvate (Gibco):
      store at –20°C     8 mL
   e. 0.1 *M* 2-mercaptoethanol (Sigma, Dorset, UK):     0.4 mL
      store at 4°C for 1 mo
   f. Fetal calf serum (Globepharm, Surrey, UK), batch-tested
      (*see* **Note 1**), store at –20°C     40 mL
   g. Sterile, deionized water (dH$_2$O)     340 mL
   h. Differentiation inhibiting activity (DIA) (*see* **Note 1**)
2. Phosphate-buffered saline (PBS), filter-sterilized.
3. Trypsin solution: Dissolve 250 mg of trypsin (Difco, Surrey, UK) and 372 mg EDTA disodium salt (Sigma) in 1 L of PBS. Add 10 mL of chicken serum (Flow Labs, Herts, UK and filter sterilize. Store in 20-mL aliquots at –20°C.
4. 1% (w/v) Gelatin: add 1 g gelatin (Type A from bovine skin, Sigma) to 100 mL dH$_2$O, autoclave, and store in 20-mL aliquots at 40°C. For a working solution of 0.1%, add 10 mL of 1% gelatin to 90 mL of PBS.

5. 0.1 $M$ 2-mercaptoethanol: Add 100 mL 2-mercaptoethanol (Sigma) to 14.1 mL PBS. Store up to 1 mo at 4°C.
6. Geneticin (G 418; Boehringer-Mannheim, Sussex, UK): 200 mg/mL dissolved in dH$_2$O.
7. Freezing solution: 1 mL of DMSO to 9 mL of cell-culture medium (made fresh).

## 2.2. Vector Preparation

1. Restriction endonuclease (*Hin*dIII for pGT1.8βgeo and pGT1.8TM) to linearize the gene trapping vector and a 10X concentration of the appropriate digestion buffer.
2. Absolute ethanol.
3. 70% Ethanol.
4. Sterile PBS.

## 2.3. X-Gal Staining Colonies

1. Phosphate buffer (0.1 $M$) made by adding 21 parts of 0.1 $M$ Na$_2$HPO$_4$ (Sigma) to 4 parts 0.1 $M$ NaH$_2$PO$_4$ (Sigma) to give a final pH of 7.5.
2. 5-Bromo-4-chloro-3-indolyl-β-D-galactopyranoside (X-gal, Boerhinger-Mannheim) stock solution, 50 mg/mL, dissolved in dimethyl formamide (Sigma). Store in the dark at –20°C.
3. Fix buffer: 0.1 $M$ Phosphate buffer, pH 7.3, 2 m$M$ MgCl$_2$, 0.2% Gluteraldehyde (Sigma), 0.5 m$M$ EGTA (pH 8.0), Store at 4°C for up to 1 mo.
4. Wash buffer: 0.1 $M$ phosphate buffer, 2 m$M$ MgCl$_2$, 0.1% sodium deoxycholate (Sigma), 0.02% Nonidet P-40 (Sigma). Store at 4°C.
5. X-gal staining solution:
      0.106 g Potassium ferrocyanide (Sigma).
      0.082 g Potassium ferricyanide (Sigma).
      1 mL X-gal (50 mg/mL).
   Filter to remove crystals, and store in the dark at 4°C. Add to 50 mL of X-gal wash buffer.

## 2.4. RNA Preparation from ES Cells

1. Guanidinium lysis buffer: 4 $M$ guanidinium thiocyanate dissolved in 0.1 $M$ Tris-HCl (pH 7.5), filtered through Whatman 3MM paper. Just prior to use, add 2-mercaptoethanol to a final concentration of 1% and sodium lauryl sarcosinate to 0.5% (w/v).
2. 5.7 $M$ cesium chloride (CsCl) dissolved in 10 m$M$ EDTA, pH 7.5, filtered, DEPC-treated, and autoclaved.
3. 8 $M$ urea dissolved in 10 m$M$ HEPES, pH 7.0.
4. DEPC H$_2$O prepared in a fume hood by adding 1 mL of diethyl pyrocarbonate (Sigma) to 1 L of dH$_2$O, shake thoroughly and leave overnight before autoclaving.
5. PBS, sterile.
6. 70% Ethanol.
7. 0.4 $M$ NaCl, DEPC-treated.
8. 1:1 Phenol/chloroform.

## 2.5. RNA Dot Blot

1. Formamide (Gibco-BRL).
2. Formaldehyde, 37% solution (BDH, Leicester, UK).
3. 1 $M$ Sodium phosphate buffer: Dissolve 70.5 g of anhydrous $Na_2HPO_4$ in $dH_2O$, add 4 mL of phosphoric acid and make up to 1 L with $H_2O$. Filter sterilize.
4. Hybond N membrane (Amersham, Buckhinghamshire, UK).
5. Prehybridization/hybridization buffer (40 mL, made fresh): weigh out 0.4 g of bovine serum albumin (Sigma), and dissolve in 14 mL of 1 $M$ phosphate buffer. Add 12 mL of formamide and 14 mL 20% SDS.

## 2.6. Rapid Amplification of cDNA Ends (RACE)

1. Primers used for 5' RACE cloning are shown in **Table 1**.
2. DEPC-treated $dH_2O$ (*see* **Subheading 2.4.**, item 4).
3. Superscript II reverse transcriptase (200 U/mL), 0.1 $M$ dithiothreitol (DTT) and 5X buffer: 250 $M$ Tris-HCl (pH 8.3), 0.375 $M$ KCl, 15 m$M$ $MgCl_2$ (Gibco-BRL 18064-014).
4. 1 $M$ Sodium hydroxide.
5. 1 $M$ Hydrochloric acid.
6. Microdialysis filters: 0.025 and 0.1 μm pore size (Millipore, Watford, UK VSWP02500 and VCWP02500).
7. TE buffer: 10 m$M$ Tris-HCl, pH 8.0 (20°C), 1 m$M$ EDTA, pH 8.0.
8. Terminal deoxynucleotidyl transferase (TdT): 15 U/μL and 5X reaction buffer: 0.5 $M$ potassium cacodylate (pH 7.2), 10 m$M$ cobalt chloride, 1 m$M$ DTT (Gibco-BRL 18008-011).
9. dATP for tailing reaction: 2 m$M$ dATP.
10. Restriction buffer M: 10 m$M$ Tris-HCl, pH 7.5, 50 m$M$ NaCl, 1 m$M$ DTT (Boehringer-Mannheim 1417 983).
11. dNTP mix: dNTPs each at a final concentration of 10 m$M$ (Pharmacia, St. Albans, UK 27203501).
12. Klenow enzyme, 2 U/μL (Boehringer-Mannheim)
13. Amplitaq (5 U/μL) and 10X buffer: 100 m$M$ Tris-HCl (pH 8.3), 500 m$M$ KCl (Applied Biosystems N8080161).
14. 25 m$M$ Magnesium chloride.
15. Restriction endonucleases *Xba*I and *Kpn*I for cloning PCR products.
16. Glycogen, 5 mg/mL (Boehringer-Mannheim).
17. 10 $M$ Ammonium acetate.
18. Absolute ethanol.
19. 70% Ethanol.
20. pBluescript SKII$^+$ plasmid DNA (Stratagene, Cambridge, UK).
21. T4 DNA ligase (1 U/μL) and 10X ligation buffer (Boehringer-Mannheim).
22. Electrocompetent bacteria (e.g., DH5α).
23. LB agar: 1% Bacto tryptone (w/v) (Difco, Detroit, MI), 0.5% Bacto-yeast extract (w/v), 1% (w/v) sodium chloride (w/v), and 50 mg/mL ampicillin.

## *2.7. Tail DNA Dot Blot*

1. Tail buffer (made up fresh): to 10 m$M$ Tris-HCl, pH 8.0, 100 m$M$ NaCl, 50 m$M$ EDTA, 0.5% SDS. Add proteinase K (Sigma) to a final concentration of 0.5 mg/mL.
2. 5 $M$ NaCl.
3. Chloroform.
4. 0.53 $M$ NaOH.
5. Hybond N$^+$ membrane.
6. Prehybridization/hybridization buffer (40 mL, made fresh): Weigh out 0.2 g of Marvel milk powder, and dissolve in 6 mLwater. Add 20 mL of phosphate buffer and 14 mL 20% SDS.

# 3. Methods

## *3.1. Maintenance of ES Cells*

Careful maintenance of ES cells is crucial for successful germline transmission of gene trap cell lines. We generally change the medium every day, and do not allow the cells to grow to confluence. Several previously published protocols describe procedures for maintaining ES cells on fibroblast feeder layers *(15–18)*. The conditions outlined below apply to the maintenance of a feeder-independent line of ES cells, CGR8, which rely on an exogenous source of DIA (also known as leukocyte inhibitory factor [LIF]). These cells are karyotypically male and were derived from the 129/Ola strain of mice as described *(19)*. Feeder-independent cell lines are more convenient to work with and contribute as efficiently to the germline of mice as feeder-dependent cell lines. We find that 80% of our gene trap cell lines contribute to the germline at an average rate of 1 germline male/10 C57Bl/6 blastocysts injected (Skarnes, unpublished results).

All ES cell manipulations should be carried out in a laminar flow hood. All solutions should be warmed in a 37°C waterbath prior to use.

### *3.1.1. Thawing ES Cells*

1. Coat a 25-cm$^2$ tissue-culture flask with 0.1% gelatin and aspirate off.
2. Quickly thaw ES cells (one-half of a confluent 25-cm$^2$ flask or approx 5 × 10$^6$ cells) in a 37°C water bath, and transfer them to a disposable centrifuge tube containing 10 mL of prewarmed medium.
3. Spin the cells down in a bench-top centrifuge at 260$g$ for 3 min.
4. Aspirate medium off and gently resuspend cells in 10 mL of pre-warmed medium.
5. Transfer to 25-cm$^2$ tissue-culture flask and grow in a humidified 37°C/6% CO$_2$ incubator.
6. Change medium after about 8 h of growth to remove dead cells and any remaining DMSO (an inducer of ES cell differentiation).
7. Medium should be changed every day.

### 3.1.2. Passage and Expansion of ES Cells

1. ES cells are passed once the have nearly reached confluence. For a 25-cm$^2$ flask, aspirate medium off, and add 5–10 mL of PBS down the opposite side of the flask to where the cells are growing. Rock the flask gently and aspirate off. Repeat.
2. Cover cells with 1 mL of trypsin, and incubate at 37°C for 1–2 min. Incubate for longer if the cells are not in a uniform suspension.
3. Add 9 mL of medium to stop trypsinization.
4. Count cells, and add 10$^6$ (approx 1/10 of a 25-cm$^2$ flask) to a freshly gelatinized flask.
5. If expanding ES cells for an electroporation, where a total of 10$^8$ cells are needed, plate 3 × 10$^6$ in a 80-cm$^2$ flask containing 30 mL of medium. Feed cells the following day with an additional 20 mL of medium. Once the cells reach confluence, trypsinize and plate 5 × 10$^6$ into three 175-cm$^2$ flasks containing 50 mL of medium each. On the next day, add an additional 40 mL of medium. Each confluent flask should yield about 8 × 10$^7$ cells.

### 3.1.3. Freezing ES Cells

1. Trypsinize 25-cm$^2$ flask as in **Subheading 3.1.2., steps 2** and **3**.
2. Collect trypsinized cells in 9 mL of medium and spin down at 260$g$ for 3 min.
3. Resuspend cell pellet in 1 mL of freezing solution, and dispense 0.5 mL of suspension into two 1-mL cryotubes.
4. Freeze cells at –80°C overnight and transfer to liquid nitrogen for long-term storage.

## 3.2. Electorporation of Gene Trap Vector into ES Cells and Selection of Colonies

### 3.2.1. Vector Preparation

1. In a volume of 0.3 mL, linearize 150 µg of vector DNA with the appropriate restriction enzyme. (Plasmid DNA for electroporation is prepared by alkaline hydrolysis and banded on a cesium chloride gradient as described in **ref. 20**).
2. Precipitate the digested DNA sample in 2 vol of absolute ethanol on ice for 5 min. Spin in microfuge, wash pellet several times with 70% ethanol, and drain off as much of the ethanol as possible.
3. Evaporate off the remaining ethanol in the hood by keeping the lid of the tube open (approx 1 h). Resuspend DNA pellet in 100 µL of sterile PBS, and vortex sample occasionally over a period of at least 4 h to ensure that the DNA is completely dissolved.

### 3.2.2. Electroporation of ES Cells and Picking G418-Resistant Colonies

1. Trypsinize three 175-cm$^2$ flasks using 5 mL trypsin/flask as described above. Add 8 mL of medium to each flask, and combine cells in a 50-mL disposable centrifuge tube. Spin for 5 min at 260$g$, and resuspend in 20 mL of PBS.
2. Count cells, and resuspend at a concentration of 10$^8$ cells in 0.7 mL of PBS.

3. Add cells to the tube containing the 150 µg of linearized plasmid, and transfer immediately to a 0.4-cm electroporation cuvet. Electroporate in Bio-Rad Gene Pulsar unit set at 3 mF/800 V (time constant = 0.1 ms).

4. Leave cells to recover in the cuvet for 20 min, and then transfer them to 200 mL of medium. Plate 10 mL ($5 \times 10^6$ cells) of the cell suspension onto 20 10-cm diameter gelatinized tissue-culture dishes.

5. After 24 h, aspirate medium, and replace with medium containing 200 µg/mL of G418. For the first 5 d, change medium daily. Once G418-resistant colonies appear, the medium may be changed every other day (*see* **Note 2**).

6. After about 10–12 d of growth, colonies should be about 1 mm in diameter. Circle colonies with a marker pen on the bottom of the dish.

7. Gelatinize the required number of 24-well plates.

8. Aspirate medium, and add 10 mL of PBS to each dish. Ideally, colonies should be picked in a tissue-culture hood. However, it is possible to pick colonies on the bench, resulting in minimal or no contamination as long as procedures are performed swiftly. Using a P200 Pipetman set at 100 µL and sterile tips, break up the colony, pipet the cells up in a volume of 100 µL PBS, and transfer to a 24-well dish. Once 24 colonies have been picked, add 100 µL of trypsin to each well, and incubate at 37°C for 10 min. Tap the dish several times to disperse the cells, and then add 2 mL of medium to each well.

9. To identify βgal-positive colonies, we normally split four-fifths of a nearly confluent well into a new 24-well plate (Experimental) for staining with X-gal, and split the remaining one-fifth into a second 24-well plate (Master) for maintenance and subsequent expansion of selected cell lines (*see* **Note 3**). Since the colonies grow at different rates, the cells are split in batches over a period of several days. Cells in the Experimental plates are stained with X-gal the following day, and βgal activity is scored after an overnight incubation. By this time, the cells on the Master plate should be ready to expand for further analysis.

### 3.2.4. Staining Cells for β-Galactosidase Activity

1. Aspirate medium and wash cells once with PBS.
2. Fix cells for 10 min at room temp.
3. Wash twice for 5 min.
4. Stain with X-gal overnight in a humidified chamber at 37°C. Colonies expressing low levels of β-gal may appear slightly discolored compared to β-gal negative cell lines. βgal staining is difficult to see using phase contrast. We recommend using either bright field illumination or Nomarski optics.

### 3.3. RACE Cloning

### 3.3.1. RNA Preparation from ES Cells (see Note 4)

1. Wash cells grown to confluence in a 25-cm² flask twice with PBS.
2. Add 1.4 mL of guanidinium lysis buffer/$10^7$ cells (approx a confluent 25-cm² flask) swirl flask for about 30 s and collect lysate.

3. Rinse the required number of SW50.1 centrifuge tubes with DEPC-treated water and dry. Add 3.6 mL of CsCl solution to each tube. Draw lysate through a 23-gage needle several times to shear genomic DNA, and carefully layer on top of the CsCl cushion.

4. Spin at 15,000$g$ in a SW50.1 rotor for 16 h at 20°C.

5. Remove the top 4 mL of the gradient, and wash the sides of the tube twice with guanidinium lysis buffer. Invert tube, and drain off remainder of gradient. Cut the bottom off the tube with a scalpel, and rinse RNA pellet twice with 70% ethanol. RNA pellet should become visible in 70% ethanol.

6. Resuspend pellet in 0.3 mL 8 $M$ urea buffer, transfer to new tube, and vortex occasionally over a period of 1 h.

7. Add 0.1 mL 0.4 $M$ NaCl. Extract sample twice with an equal volume of phenol/chloroform followed by one chloroform extraction.

8. Precipitate RNA with 2 vol of absolute ethanol and store RNA as a precipitate at −20°C.

9. To determine the concentration of RNA in the sample, vortex to produce a uniform suspension, and spin down 0.1 mL in a microfuge. Wash RNA pellet in 70% ethanol, and resuspend in 0.1 mL water. The OD$_{260nm}$ of this sample gives a reasonably accurate measure of the concentration of RNA in the precipitated sample.

### 3.3.2. RNA Dot Blot to Eliminate Intron-Containing Lines

We have observed that a significant proportion of gene trap insertions fail to utilize properly the splice acceptor of the vector and produce fusion transcripts that hybridize to intron sequences of the vector. Inefficient splicing is predicted to occur if the vector inserts either into exons of genes or in non-pol II transcription units. We have characterized two such insertions and found that both occurred within rRNA genes transcribed by RNA pol I (Sleeman and Skarnes, unpublished results). To eliminate these nongene trap events prior to RACE cloning, we use an RNA dot-blot method to detect cell lines producing intron-containing transcripts.

1. Spin down and resuspend 10 µg of RNA in 20 mL of DEPC-treated water. Add 40 mL of formamide, 14 mL of formaldehyde, and 88 mL of NaP buffer.

2. Heat samples to 55°C for 15 min, and apply one-half of each sample to duplicate sets of wells in the dot-blot apparatus. Leave for 30 min before applying a vacuum.

3. To assemble the dot-blot apparatus, and cut a piece of Hybond N membrane and a piece of Whatman paper to fit the dot-blot apparatus. Prewet the membrane in water, and then soak it in 0.5 $M$ NaP buffer for 10 min. Prewet the Whatman in 0.5 $M$ NaP buffer, and place it underneath the membrane.

4. Draw samples through with a gentle vacuum, and wash wells once with 0.5 $M$ NaP buffer. Disassemble apparatus, remove membrane with paper attached, and crosslink RNA to membrane using Stratalinker (Stratagene) set on autocrosslink. Probe duplicate sets of membrane with intron and reporter gene probes. Results from a typical dot-blot experiment are shown in **Fig. 1**.

**lacZ probe**                          **intron probe**

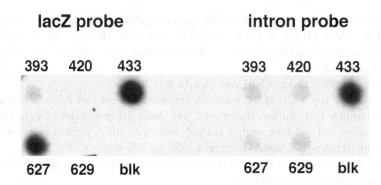

393     420     433                     393     420     433

627     629     blk                     627     629     blk

Fig. 1. RNA dot blots used to eliminate cell lines that produce intron-containing fusion transcripts. Duplicate RNA dot blots containing 10 µg total RNA were hybridized with *lacZ* and intron probes of the gene trap vector. 433 is a typical example of an intron-containing cell line.

### 3.3.3. RACE Cloning

The following protocol has been streamlined and contains a number of modifications to the original method described by Frohman, et al. (**ref. *14***). The most important of these are:

1. Alkaline hydrolysis of the RNA (required for efficient T-tailing with terminal deoxytransferase);
2. Synthesis of second-strand cDNA with Klenow instead of *Taq* polymerase; and
3. The use of microdialysis filters, which serves a dual role of removing primers/ buffers between steps and size-selecting informative cDNA fragments above 300 bp in length.

By way of example, **Table 1** lists the oligos used for RACE cloning from cell lines obtained with the pGT1.8βgeo and pGT1.8TM vectors.

1. Spin down 5–10 µg of total RNA in an Eppendorf tube, wash RNA pellet twice with 70% ethanol, and partially dry pellet in Speed Vac. Repeat this step, since it is important to remove any residual urea. Resuspend RNA sample on ice in 10 µL of DEPC-treated water. Add 1 µL of 10 ng/µL primer 1, and heat for 5 min at 70°C. Cool on ice and spin briefly.
2. Set up first-strand reaction by adding:
     4 µL of first-strand buffer (5X).
     2 µL of 0.1 *M* DTT.
     1 µL of Superscript II RT.
     Incubate first-strand reaction at 37°C for 1 h.
3. To improve the efficiency of the tailing reaction, hydrolyze RNA by adding 2.2 µL of 1 *M* sodium hydroxide for 20 min at 65°C. Neutralize with 2.2 µL of 1 *M* hydrochloric acid.

**Table 1**
**Primer Combinations Used in 5' RACE[a]**

| Primer | Gene trap vector, pGT1.8geo | Secretory trap vector, pGT1.8tm |
|---|---|---|
| 1 | 5' TAATGGGATAGGTTACG | 5' CCAGAACCAGCAAACTGAAGGG |
| 2 | 5' GGTTGTGAGCTCTTCTAGATGG(T)$_{17}$ | 5' GGTTGTGAGCTCTTCTAGATGG(T)$_{17}$ |
| 3 | 5' GGTTGTGAGCTCTTCTAGATGG | 5' GGTTGTGAGCTCTTCTAGATGG |
| 4 | 5' ATTCAGGCTGCGCAACTGTTGG | 5' AGTAGACTTCTGCACAGACACC |
| 5 | 5' TGCTCTGTCAGGTACCTGTTGG | 5' TGCTCTGTCAGGTACCTGTTGG |

[a]*Xba*I and *Kpn*I sites used in cloning the RACE products are underlined.

4. Microdialyze sample on a 0.025-µm filter floating in a Petri dish of TE for 4 h. Transfer the remainder of the sample to an Eppendorf tube, and wash the filter with water to bring the final volume to 20 µL.
5. Set up tailing reaction by adding 6 µL of TdT buffer (5X) and 2 µL of 2 m*M* dATP. Incubate for 2 min at 37°C. Add 2 µL of TdT enzyme, and incubate for a further 5 min. Stop the reaction by heating to 70°C for 2 min.
6. Carry out second-strand synthesis by adding the following to 15 µL of tailed cDNA:
   2 µL restriction buffer M (10X).
   1 µL dNTPs (10 m*M*).
   1 µL primer 2 (10 ng/mL).
   1 µL Klenow enzyme.
   Incubate for 30 min at room temperature, 30 min at 37°C, and 5 min at 70°C.
7. Microdialyze for 4 h on a 0.1-µm filter. Recover cDNA from filter into a final volume of 37 µL of H$_2$O.
8. For the first-round PCR reaction, add the following to 37 µL of sample:
   5 µL 10X AmpliTaq buffer
   4 µL 25 m*M* MgCl$_2$
   1 µL dNTP mix
   1 µL primer 3 (100 ng)
   1 µL primer 4 (100 ng)
   1 µL AmpliTaq
   Carry out 30 cycles using the following parameters:
   | Denature | 94°C for 1.5 min |
   | Anneal | 60°C for 1.5 min |
   | Extend | 72°C for 3.0 min |
9. Microdialyze for 4 h on a 0.1-µm filter to remove smaller, uninformative PCR products and excess primers. Recover sample from filter.
10. Perform second-round PCR using 5 µL of the first-round PCR reaction and:
   5 µL PCR buffer
   4 µL 25 m*M* MgCl$_2$
   1 µL dNTP mix

1 µL primer 3 (100 ng/mL)
1 µL primer 5 (100 ng/mL)
1 µL AmpliTaq
37 µL $H_2O$

Use same cycle parameters as first-round PCR, but use a hot start. After the final cycle, polish the second-round PCR products by adding 50 ng of each primer, 0.5 µL of dNTPs, and 0.5 µL of Amplitaq. Perform one cycle using the following conditions:

| | |
|---|---|
| Denature | 94°C for 1.5 min |
| Anneal | 60°C for 1.5 min |
| Extend | 72°C for 20.0 min |

11. Microdialyze for 4 h on a 0.1-µm filter, and recover sample from filter. Analyze 5 µL of sample by gel electrophoresis and Southern blot hybridization (*see* **Note 5**). Digest the remainder of the sample with *Xba*I and *Kpn*I.

12. Following digestion, extract the sample twice with phenol/chloroform, extract once with chloroform, and precipitate the RACE products on ice for 10 min by adding 5 µg of glycogen, 25 µL of 10 *M* ammonium acetate, and 300 µL of ethanol. Microfuge for 10 min, wash pellet with 70% ethanol, partially dry pellet in Speed Vac and resuspend in 20 µL of TE.

13. Ligate 100–200 ng of *Xba*I/*Kpn*I-digested RACE products with 50–100 ng *Xba*I/*Kpn*I-digested plasmid DNA in a final volume of 15 mL at room temperature for at least 2 h:

    4–8 µL Digested PCR products (100–200 ng)
    1 µL Digested plasmid DNA (100 ng)
    1.5 µL Ligation buffer (10X)
    1 µL T4 DNA ligase
    3.5–7.5 µL $H_2O$

14. Transform into competent bacteria, and screen colonies for desired inserts (*see* **Note 6**).

### 3.4. Genotyping Animals

The following is a one-tube method for preparing genomic DNA from tail biopsies. To obtain uniform signals, it is important to apply enough DNA to saturate the membrane. This was determined empirically using a Bio-Rad 96-well dot-blot apparatus and Amersham Hybond $N^+$ membrane. **Figure 2** shows the results of a typical dot blot from intercrosses of a line of mice carrying a single-copy insertion. Homozygous mice, which in this case carry two copies of the vector, are easily distinguishable from their heterozygous litter mates.

1. Tail biopsies (1.5 cm in length) are taken at weaning age and digested overnight at 55–65°C in 0.4 mL of tail buffer.

2. While tubes are still warm, add 0.1 mL 5 *M* NaCl and vortex at high speed for 5–10 s. Add 0.5 mL of chloroform, and vortex again for 5–10 s. Spin in microfuge for 5 min.

3. Transfer 50 µL of the top aqueous phase to a 96-well plate. Denature DNA by adding 150 µL of 0.53 *M* NaOH and incubating at 37°C for 30 min.

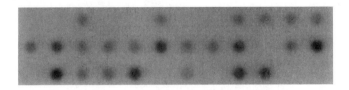

Fig. 2. Genotyping mice by dot-blot analysis. Tail DNA from intercross litters carrying a single-copy gene trap insertion were hybridized to a *lacZ* probe. Dots showing weak signals represent heterozygous animals, whereas those showing signals of twice the intensity represent homozygotes.

4. Apply samples to dot-blot apparatus, and leave for 30 min before applying a vacuum.
5. Prepare the dot-blot apparatus by cutting a piece of Hybond N$^+$ membrane and a piece of Whatman paper to fit the apparatus. Prewet the membrane in H$_2$O, and then soak in 0.4 $M$ NaOH for 10 min. Prewet the Whatman in 0.4 $M$ NaOH, and place it underneath the membrane on the apparatus.
6. Draw samples through with gentle vacuum. Disassemble apparatus and wash membrane with 30 m$M$ NaP/0.1% SDS buffer. Hybridize membrane with reporter gene probe.

## 4. Notes

1. DIA is produced by transient expression in COS cells using the method described in **ref. *21***. Serial dilutions of the medium are tested on ES cells plated in 24-well plates. A 100-fold higher dilution than the minimal dilution required to keep ES cells undifferentiated is typically used. Serum batches are tested for their ability to sustain the growth, differentiation, and viability of ES cells grown at clonal density in the presence and absence of DIA.
2. Electroporation of 10$^8$ ES cells normally generates between 200 and 400 G418$^R$ colonies. Using the gene trap vector pGT1.8geo *(8)*, we find 50% of colonies stain for βgal activity. A proportion of the white colonies represents genes expressed at low levels in ES cells, which can be induced on differentiation. The remainder probably represents inactive βgal fusion proteins, such as secretory molecules, and events where sequences from the 5'-end of βgal have been removed by endogenous exonucleases prior to insertion into the genome. Using the secretory trap vector pGT1.8TM, we find approx 25% of colonies stain positive for βgal. Almost all show the secretory pattern of staining in the peri-nuclear compartment of ES cells and in multiple cytoplasmic inclusions, and represent fusions to genes encoding a 5'-signal sequence.
3. To screen for genes induced or repressed under various growth conditions, more than one Experimental plate can be seeded. If the assays are to be conducted over a period of days, the Master plate can be frozen at –80°C *(22)*, allowing the recovery of selected colonies at a later date.

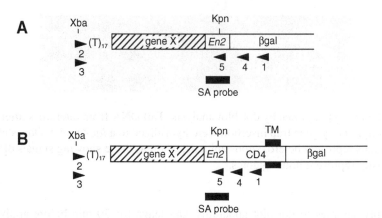

Fig. 3. Arrangement of oligos used to RACE clone genes disrupted by gene trap vector pGT1.8geo (**A**) and the secretory trap vector pGT1.8TM (**B**). Both vectors contain the mouse *Engrailed-2 (En2)* splice acceptor (SA) linked to a βgeo reporter containing βgal and neomycin phosphotransferase activities. The secretory trap vector contains a fragment of CD4 that encodes the transmembrane domain (TM). Oligo 1, primer for first-strand cDNA synthesis; oligo 2, primer for second-strand cDNA synthesis; oligos 3 and 4, first-round PCR primers and oligos 3 and 5, second-round PCR primers (*see* **Table 1**).

4. Other methods of RNA preparation have recently been shown to work well for RACE cloning. These include acid phenol (RNazol B, Biogenesis Ltd.) and poly dT-based bead selection (Dynabeads, Dynal Ltd.) protocols.

5. Before digesting second-round products, it is advisable to check the fidelity of the preceding reactions. This is done by running 5 μL of second-round PCR products on a 1.5% agarose gel and carrying out Southern blot analysis using a vector probe that includes sequences contained in the RACE products (*see* **Fig. 3**). If the reactions have worked, the autoradiograph will require less than a 15-min exposure. In most cases, we see a heterogeneous smear of amplified products. However, if the fusion transcripts are small, a band may be seen.

6. Colonies are selected by digesting plasmid minipreps with the enzymes used for cloning the RACE products and carrying out Southern analysis using the SA probe (*see* **Fig. 3**). Colonies can also be screened using a PCR strategy employing primer 5 and a primer in the pBluescript polylinker. Only clones containing gene trap vector sequences should be amplified. We normally recover products that contain between 100 and 700 bp of sequence from the disrupted gene. To confirm that the RACE clone truly represents the disrupted gene, Northern blot analysis should be performed using the RACE clone as a probe. Alternatively, if the RACE clones are small, RNase protection experiments can be used.

## References

1. Gossler, A., Joyner, A. L., Rossant, J., and Skarnes, W.C. (1989) Mouse embryonic stem cells and reporter constructs to detect developmentally regulated genes. *Science* **244,** 463–465.
2. Skarnes, W. C. (1993) The identification of new genes: gene trapping in transgenic mice. *Curr. Op. Biotech.* **4,** 684–689.
3. Gossler, A. and Zachgo, J. (1993) Gene and enhancer trap screens in ES cell chimaeras, in *Gene Targeting: A Practical Approach* (Joyner, A., ed.), Oxford University Press, Oxford, pp. 181–227.
4. von Melchner, H., DeGregori, J. V., Rayburn, H., Reddy, S., Friedel, C., and Ruley, H. E. (1992) Selective disruption of genes expressed in totipotent embryonal stem cells. *Genes Dev.* **6,** 919–927.
5. Skarnes, W. C., Auerbach, B. A., and Joyner, A. L. (1992) A gene trap approach in mouse embryonic stem cells: the lacZ reporter is activated by splicing, reflects endogenous gene expression, and is mutagenic in mice. *Genes Dev.* **6,** 903–918.
6. Wurst, W., Rossant, J., Prideaux, V., Kownacka, M., Joyner, A., Hill, D. P., Guillemot, F., Gasca, S., Cado, D., Auerbach, A., and Ang, S.-L. (1995). A large-scale gene-trap screen for insertional mutations in developmentally regulated genes in mice. *Genetics* **139,** 889–899.
7. Friedrich, G. and Soriano, P. (1991) Promoter traps in embryonic stem cells: a genetic screen to identify and mutate developmental genes in mice. *Genes Dev.* **5,** 1513–1523.
8. Skarnes, W. C., Moss, J. E., Hurtley, S. M., and Beddington, R. S. P. (1995) Capturing genes encoding membrane and secreted proteins important for mouse development. *Proc. Natl. Acad. Sci. USA* **92,** 6592–6596.
9. DeGregori, J., Russ, A., von Melchner, H., Rayburn, H., Priyaranjan, P., Jenkins, N. A., Copeland, N. G., and Ruley, H. E. (1994) A murine homolog of the yeast RNA1 gene is required for postimplantation development. *Genes Dev.* **8,** 265–276.
10. Chen, Z., Friedrich, G. A., and Soriano, P. (1994) Transcriptional enhancer factor 1 disruption by a retroviral gene trap leads to heart defects and embryonic lethality in mice. *Genes Dev.* **8,** 2993–2301.
11. Takeuchi, T, Yamazaki, Y., Katoh-Fuki, Y., Tsuchiya, R., Kondo, S., Motoyama, J., and Higashinakagawa, T. (1995) Gene trap capture of a novel mouse gene *jumonji*, required for neural tube development. *Genes Dev.* **9,** 1211–1222.
12. Kerr, W. G., Nolan, G. P., Serafini, A. T., and Herzenberg, L. A. (1989) Transcriptionally defective retroviruses containing *lacZ* for the in situ detection of endogenous genes and developmentally regulated chromatin. *Cold Spring Harbor Symp. Quant. Biol.* **54,** 767–776.
13. Mountford, P. S. and Smith, A. G. (1995) Internal ribosome entry sites and dicistronic RNAs in mammalian transgenesis. *Trends Genet.* **11(5),** 179–184.
14. Frohman, M. A., Dush, M. K., and Martin, G. R. (1988) Rapid production of full-length cDNAs from rare transcripts: Amplification using a single gene specific oligonucleotide primer. *Proc. Natl. Acad. Sci. USA* **85,** 8998–9002.

15. Robertson, E. (ed.) (1987) *Teratocarcinomas and Embryonic Stem Cells: A Practical Approach.* IRL, Oxford.

16. Joyner, A. (ed.) (1993) *Gene Targeting: A Practical Approach.* Oxford University Press, Oxford.

17. Wassarman, P. M. and DePamphilis, M. L. (eds.) (1993) *Methods in Enzymology, vol. 225: Guide to Techniques in Mouse Development.* Academic, San Diego, CA.

18. Hogan, B., Beddington, R., Costanini, F., and Lacy, E. (eds.) (1994) *Manipulating the Mouse Embryo. A Laboratory Manual,* 2nd ed. Cold Spring Harbor Laboratory, Cold Spring Harbor, NY.

19. Nichols, J., Evans, E. P., and Smith, A. G. (1990) Establishment of germline-competent embryonic stem (ES) cells using differentiation inhibiting activity. *Development* **110,** 1341–1348.

20. Maniatis, T., Fritsch, E. F., and Sambrook, J. (eds.) (1982) *Molecular Cloning, A Laboratory Manual.* Cold Spring Harbor Laboratory, Cold Spring Harbor, NY, pp 1.21–1.52.

21. Smith, A. G. (1991) Culture and differentiation of embryonic stem cells. *J. Tissue Culture Meth.* **13,** 89–94.

22. Ure, J. M., Fiering, S., and Smith, A. G. (1992) A rapid and efficient method for freezing and recovering clones of embryonic stem cells. *Trends Genetics* **8(1),** 6.

# 9

## Production and Growth of Conditionally Immortal Cell Lines from the H-2K$^b$tsA58 Transgenic Mouse

### Mark Noble

### 1. Introduction

The purpose of this chapter is to provide some of the rationale leading up to the development of the H-2K$^b$tsA58 transgenic mouse, a prototype of animals that can be used to create conditionally immortal cell lines readily from any tissue of interest. In addition, examples are presented of some of the cell lines thus far derived from the animals and of general principles that can be derived from these successes. The reader is also referred to **refs. 1–3**, as well as the indicated primary references.

### 2. Generation of Conditionally Immortal Cell Lines by Genetic Modification In Vitro

There are two major approaches that can be used to generate cell lines of interest. The first of these, and still the most frequently used, is to transduce an immortalizing gene construct into cultures of dividing cells. This can be accomplished by transfection or by use of retroviruses to insert the immortalizing gene randomly into the cellular genome.

Despite the undoubted importance of in vitro gene insertion in the creation of cell lines, there remain certain drawbacks intrinsic to this approach. For example, transfection requires a large number of target cells in order to ensure that some cells of interest stably integrate the chosen DNA in a position suitable for expression. Viral-mediated gene transfer can be carried out with smaller numbers of cells by cocultivation of target cells with virus-producing feeder layers, but still requires the division of target cells of the desired cell type to achieve the necessary goal of integration of the selected DNA into the genome. Moreover, both of these technologies require the growth of cells for

From: *Methods in Molecular Biology, Vol. 97: Molecular Embryology: Methods and Protocols*
Edited by: P. T. Sharpe and I. Mason © Humana Press Inc., Totowa, NJ

extended periods of time in culture under selective pressure to obtain sufficient numbers of immortalized cells for experimentation.

An additional problem associated with the use of immortalized cell lines is that the introduction of immortalizing genes into cells can alter normal cellular physiology in two different ways. First, because the functional definition of an immortalizing gene is that it prevents cells from terminally differentiating, expression of characteristics of differentiated cells may be prevented in immortalized cell lines. In addition, expression of immortalizing genes can also alter the response of cells to exogenous signals, such as mitogens or regulators of differentiation. Such alterations can be so dramatic that they actually cause cells to respond to particular signals in a manner fundamentally different from that which occurs in the normal counterpart of the immortalized cell. One example of this problem is offered by studies on cooperative interactions between SV40 large T-antigen (TAg, an immortalizing oncogene) and a mutationally activated *ras* gene (a transforming oncogene, the product of which is thought to be involved in the control of cell division). When expressed in Schwann cells harboring TAg, *ras* converts slowly dividing cells dependent on exogenous mitogens to rapidly dividing cells capable of growing in mitogen-free conditions, an effect characteristic of a transforming oncogene. In contrast, when *ras* is expressed in Schwann cells that are not also expressing an immortalizing oncogene, then the constitutively activated *ras* protein rapidly induces the diametrically opposite effect of proliferation arrest *(4)*. Since it is thought that the *ras* protein normally functions in nontransformed cells as a principal component of signal transduction systems (for review, *see* **ref.** *5*), the ability of an immortalizing oncogene to alter the effects of *ras* activation raises the possibility that expression of immortalizing oncogenes can also alter the effects of other components of the signal transduction apparatus.

The potential alterations in cellular physiology associated with expression of immortalizing genes theoretically can be overcome through the use of conditional immortalizing genes, which allow the generation of cell lines in which the activity of the product of the experimentally introduced gene can be turned off by manipulation of the cellular environment. For example, the *ts*A58 mutant of TAg *(6)* encodes a thermolabile protein capable of immortalizing cells only at the permissive temperature of 33°C, and has been successfully used in the generation of a variety of conditionally immortal cell lines, as discussed below. In addition, it has turned out that cell lines rendered immortal by overexpression of the c-*myc* gene are, in at least some cases, able to differentiate normally in vivo.

The construction of cell lines able to differentiate has proven to be a powerful approach for studying the developmental biology of various neural cell types, including both neurons and glia *(7–11)*. In these studies, it has been possible to create cell lines that are able to participate in normal development after injection into the developing brain. The range of genes used to generate cell lines is large, and includes the SV40 large T-gene *(8,12–16)*, polyoma virus large T and adenovirus 5 E1A *(17)*, *src* *(18,19)* and c-and v-*myc* *(10,20)*. Recently, Louis and coworkers also have described a spontaneously generated cell line called CG-4 *(21)*; since this cell line arose spontaneously, the immortalizing mutation is unknown.

The capacity of conditionally immortalized neural cells to regain normal patterns of behavior in certain conditions (e.g., after *ts*TAg is inactivated at nonpermissive temperatures) is indicated by both in vitro and in vivo experimentation. In vitro, in the studies on Schwann cells referred to earlier, the effects of expression of activated *ras* protein were identical in normal Schwann cells and in Schwann cells harboring tsTAg and switched to nonpermissive temperatures *(4)*. More dramatically, studies by McKay and colleagues *(8,9,22)* have demonstrated that a hippocampal cell line immortalized with *ts*TAg in vitro can undergo neuronal differentiation and apparent integration into normal tissue when injected into the rodent central nervous system (CNS) (where the normal body temperature is sufficiently high to cause *ts*TAg to be degraded rapidly and thus inactivated). The capacity of c-*myc*-expressing cells to integrate normally into CNS tissues is elegantly demonstrated by the recent studies of Snyder et al. *(10)*. In addition, oligodendrocyte-type-2 astrocyte (O-2A) progenitor cell lines made with either *ts*A58 TAg or c-*myc* are useful tools for the in vitro and in vivo study of this lineage, because these cells are able to differentiate in a similar manner to O-2A progenitor cells from primary cultures *(11,20)*. The in vitro differentiation potential of the O-2A/TAg cell line correlates well with its differentiation potential in vivo. Cell lines transplated into white matter lesions depleted of host glial cells are able to remyelinate axons and also to generate astrocytes in vivo *(11,20)*.

Despite the merits of using conditionally active oncogenes to immortalize cells, these approachs still share many of the problems associated with the introduction of constitutively active immortalizing cells by in vitro means of gene insertion. Thus, the inability to target rare cells, the need to promote cell division to achieve integration and continued function of the immortalizing gene, and the need for extended growth of immortalized cells in vitro before enough cells exist for experimental use are all problems that are not affected by whether oncogene activity is conditional or constitutive.

One final problem associated with the generation of cell lines by in vitro gene insertion is that every cell line produced by these procedures has the immortalizing oncogene integrated into a different site in the genome. For reasons presumably associated with these differing sites of integration, putatively identical cell lines can express markedly different levels of oncogene product and markedly different behaviors. This problem can even confound the use of conditional immortalizing genes. For example, introduction of *ts*TAg into rat embryo fibroblasts leads to the ready isolation of cell lines that can be grown indefinitely when maintained at 33°C (the permissive temperature for function of this gene), but that rapidly senesce when switched to 39.5°C (the nonpermissive temperature). However, the degree of conditionality expressed by different fibroblast lines varies over several orders of magnitude, with some lines showing only a modest reduction in growth at 39.5°C, whereas other cell lines derived from the same infected plate of fibroblasts exhibited an almost complete cessation of growth when switched to nonpermissive temperatures *(23)*. As a consequence of this variability, it is obviously difficult to compare the properties of cell lines made in different laboratories, or even of lines made within a single laboratory.

## 2.1. Cell Lines from Transgenic Animals

### 2.1.1. Targeted Cell Lines

One way to make certain that all cell lines produced in an experiment contain the same site of integration of the immortalizing gene is to create cell lines from transgenic animals. Since all cells derived from such an animal will share the same insertion site, at least this source of variability in the biology of cell lines will be brought under more stringent control. The possibility of making cell lines from such transgenic animals was recognized from the earliest studies in which immortalizing genes were inserted into the animal genome. For example, the first studies indicating that expression of oncogenes would disrupt normal development *(24)* showed that mice expressing SV40 TAg under the control of the metallothionein promotor developed tumors of the choroid plexus and showed that cell lines could be isolated from the transformed tissue. Studies in which the 5' regulatory sequences of the insulin gene were used to control the in vivo expression of SV40 TAg showed that it was possible to use tissue-specific promoters to cause an immortalizing gene to be expressed precisely in the cell type of interest *(25)*. Further studies demonstrated that targeted oncogene expression allowed isolation of cell lines from a variety of tissues in which the oncogene caused neoplastic development in vivo (e.g., *26,27*). It has also been found that it was not necessary for full transformation to occur in order to allow generation of cell lines from affected tissue. For example, nonmalignant hepatocyte cell lines (which become increasingly transformed with further growth in culture) can also be isolated from mice in which SV40 TAg expression was driven by the mouse metallothionein promoter sequences *(28,29)*. The use of targeted expression of immortalizing genes to create cell lines of particular interest to neurobiologists has been recently demonstrated by studies of Hammang et al. *(30)* and Mellon et al. *(31)*.

Despite the success of targeted transgene expression in the generation of neural cell lines, there are two major drawbacks to the use of this technique.

First, in all cases reported thus far, cell lines have been isolated from tumor tissue. Since tumor formation is associated with the acquisition of multiple genetic aberrations (e.g., *26,32–36*), it is likely that the resultant cell lines differ from their normal counterparts in a variety of important ways. Indeed, given that the acquisition of cooperating mutations is necessary to allow transformed growth to occur *(37–39)*, it is almost inconceiveable that the resultant cell lines do not have multiple mutations. Furthermore, since expression of a single activated immortalizing gene seems to be sufficient to generate the phenotype of a benign tumor in vivo *(35)*, it is likely that these cells will have undergone a protracted period of abnormal growth in vivo prior to the point of having generated a tumor. The extent to which such cells can be expected to mimic their untransformed ancestors is unknown.

The second drawback associated with the approach of specifically targeting oncogene expression to individual cell types is the necessity of identifying cell-type-specific promoter elements for every cell type of interest. At present, there are relatively few cell types for which appropriate 5'-regulatory elements have been identified. In addition, since many of the cell-type-specific proteins identified thus far are expressed at relatively late stages of differentiation, it may be difficult to target oncogene expression to precursor populations by utilizing the promoter regions for such proteins. In specific regard to the generation of precursor cell lines, it may be more useful to use promoter regions associated with such developmentally regulated genes as the homeobox proteins.

## 2.1.2. The H-2K^b^tsA58 Mouse, a Multipotential Source of Conditionally Immortal Cell Lines

An approach to cell line production that would overcome many of the difficulties described thus far would be to create a strain of transgenic animals in which expression of a conditionally active immortalizing gene was regulated in such a way that the gene was not functionally active in vivo (and thus did not perturb normal development), but could be turned on in cells isolated from any tissue of the body by simple manipulation of tissue-culture conditions. Ideally, the oncogene utilized should be capable of immortalizing as wide a range of cell types as possible, and the regulatory regions used to control expression of this gene should also be capable of functioning in the widest possible range of cell types. Theoretically, transgenic animals of this generic class would allow the direct derivation of cell lines from a wide variety of tissues simply by dissection and growth of cells in an appropriate in vitro milieu. In addition, each cell line generated would possess an identical integration site of the immortalizing gene. Such animals would also have two further distinct advantages. First, the presence of the immortalizing gene in the genome of the transgenic animal would not cause neoplastic development *in situ*, because functionally active protein would not be expressed in vivo at levels sufficient to perturb normal development. Second, expression of the functionally active immortalizing gene could be turned off again after desired cell lines were obtained in tissue culture, thus allowing study of normal processes of differentiation either in vitro or following cell transplantation in vivo.

The first transgenic animals expressing the ideal characteristics described above are the H-2K$^b$*ts*A58 transgenic mice *(40)*. These mice harbor the *ts*A58 TAg gene under the control of the H-2K$^b$ Class I antigen promoter *(41–43)*. The combination of existing studies on the effects of TAg in tissue culture and in transgenic animals indicates clearly that this gene expresses its immortalizing function in a very wide variety of cell types (e.g. *9,27,44–48*). In addition, expression from the H-2K$^b$ promoter can be induced or enhanced in almost all cell types by exposure of cells to interferons *(48–50)*. Thus, in cells or tissues (such as brain) that normally express little or no Class I antigens, exposure to interferons activates transcription from this promoter. Moreover, cells that constitutively express high levels of Class I antigens can still be induced to express even higher levels of expression by exposure to interferons.

In initial experiments on the H-2K$^b$*ts*A58 transgenic mice, 34 different transgenic founder animals were created *(40)*. Skin fibroblasts from normal and transgenic animals were placed in culture at 33°C, the permissive temperature for *ts*A58 TAg, in the presence of γ-interferon (IFN-γ), which is known to increase expression from the H-2K$^b$ promoter. Skin fibroblasts derived from the transgenic mice all readily yielded proliferating cultures that could be continuously passaged when grown in permissive conditions. In contrast, fibroblasts stopped dividing within a limited number of passages, with a significant decline in even this limited passage number seen in cultures established from older animals.

The conditionality of growth observed in the fibroblasts derived from transgenic animals was correlated with the levels of *ts*A58 TAg. In all cultures, the level of *ts*A58 TAg was reduced by increasing temperature and/or by removal of IFN-γ. It is interesting that when the most conditional cultures were grown at 33°C in the absence of IFN-γ, a condition where these cells did not grow, low levels of TAg could still be detected. This suggests that the level of TAg produced in these conditions was below the threshold needed to support continued rapid cell growth or to allow single-cell cloning.

In general, H-2K$^b$*ts*A58 transgenic mice appear to undergo normal development. However, these animals do routinely exhibit hyperplastic development of the thymus. This enlargement appears to be owing to hyperplasia rather than to malignant transformation, since thymic histology, T-cell repertoire, and T-cell clonality were all normal in the transgenic animals, and cells derived from enlarged thymuses did not generate tumors in syngeneic recipients. Despite their enlargement in vivo, thymuses of transgenic mice yielded conditionally immortal cultures containing cells of both epithelial and fibroblastic morphologies. Both morphological cell types could be readily cloned, exhibited optimal growth in fully permissive conditions, and did not grow in nonpermissive conditions. Clones that exhibited epithelial-like morphologies expressed cytokeratin and had the ability to rosette T-lymphocytes. Thus, in our initial experiments, we were able to derive conditionally immortal lines of epithelial cells, as well as of fibroblasts, readily from these mice *(40)*.

One founder animal survived to the age of 6 mo and fathered multiple offspring, which harbor a functional transgene. Sibling matings of transgenic offspring generated homozygous animals that have been bred successfully through many generations. As discussed below, this transgenic mouse strain has been a ready source of novel cell lines.

### 2.1.3. Astrocyte Clones Derived from H-2KᵇtsA58 Transgenic Mice Express Properties of Glial Scar Tissue

As part of our investigation of the use of H-2Kᵇ*ts*A58 transgenic for neuroscience research, we generated clonal astrocyte cell lines that exhibit the properties of glial scar tissue. Such cell lines are of potential importance, since it is believed that glial scarring within the CNS may play an important role in inhibiting regrowth of axons *(51–62)* and perhaps also in inhibiting repair of extensive breakdown of myelin in the CNS (as occurs in multiple sclerosis; *63*).

To generate clonal lines of conditionally immortal astrocytes, cortical astrocytes were purified from neonatal H-2Kᵇ*ts*A58 transgenic mice by simple and well-established methods *(64)*, with the only modificiation being that cultures were grown at 33°C in the presence of 25 U/mL of IFN-γ. To generate clonal cell lines, cultures of purified astrocytes were infected with the BAG retrovirus, which expresses bacterial β-galactosidase and the neomycin resistance gene *(65)*. Subsequent selection in medium containing G418 allowed the ready generation of clonal colonies. Four of these clones were chosen for detailed analysis

All four astrocyte lines chosen for further analysis expressed glial fibrillary acidic protein (GFAP, an astrocyte-specific cytoskeletal protein), although at lower levels than that seen in primary cultures of mouse cortical astrocytes. Characterization of expression of membrane and extracellular matrix molecules indicated that all of these clones expressed a phenotype much like that associated with glial scar tissue *(66)*. All four lines expressed tenascin, laminin, and chondroitin sulfate proteoglycans, all of which are present in some glial populations during development and are particularly expressed in CNS lesions *(67–70)*.

The extent of neurite outgrowth promoted by our astrocyte cell lines was consistent with the possibility that these cells expressed a phenotype functionally similar to glial scar tissue, in that monolayers of the four transgenic astrocyte lines were less effective than monolayers of primary cortical astrocytes at promoting outgrowth of cerebellar neurons. The mean total neurite length on all four astrocyte lines was less than 50% of that seen on primary astrocyte monolayers. Despite this dramatic failure to promote growth, the astrocyte cell lines did not cause the cerebellar neurons to clump together or fasciculate, as has been reported for neurons of the CNS growing on fibroblasts or meningeal cells derived from the CNS *(64–71)*. Thus, although the extent of neurite outgrowth was markedly reduced on monolayers of our H-2K$^b$*ts*A58 astrocyte cell lines, the organization of neuronal cell bodies and processes on the surfaces of these astrocytes suggested that they still expressed glial, rather than nonglial, surface properties.

The four transgenic astrocyte cell lines were also less effective than primary astrocytes at supporting outgrowth of neurites from dorsal root ganglion neurons derived from 7-d postnatal rats. However, there was no inhibition of the growth of dorsal root ganglion neurons derived from ganglia of 18-d-old embryos. In this respect also, the astrocyte cell lines appeared to behave like scar tissue, which is thought to be markedly less inhibitory for the growth of immature neurons as compared with mature neurons *(58)*.

### 2.1.3.1. INHIBITION OF O-2A PROGENITOR MIGRATION

As previously observed for purified cortical astrocytes *(72,73)*, all of the clonal astrocyte cell lines derived from H-2K$^b$*ts*A58 transgenic mice produced platelet-derived growth factor (PDGF) and promoted the division of O-2A *(74)* progenitors in vitro. However, the astrocyte cell lines *(66)* differed markedly from primary astrocytes in their support of O-2A progenitor migration. O-2A progenitors formed small, tight colonies on monolayers of all four astrocyte lines after 7 d of growth, whereas on primary cortical astrocytes, the O-2A progenitor cells were distributed much more evenly over the entire monolayer. The possibility that this failure to migrate represented a functional inhibition of migration was tested by preparing confrontation experiments. O-2A progenitors were either plated onto primary astrocytes and allowed to migrate onto the astrocyte cell lines, or were plated onto the astrocyte cell lines and allowed to migrate onto primary astrocytes. The interface between primary and transgenic astrocyte monolayers could be clearly seen owing to the lower level of GFAP expression in the transgenic astrocytes.

In experiments where O-2A progenitor cells growing on primary astrocytes were challenged with a monolayer of transgenic astrocyte lines, very few progenitors crossed the interface between the primary and transgenic astrocytes. Instead, the progenitor cells appeared to migrate to the astrocyte interface, but no further, frequently aligning their processes along the interface. The apparent failure of O-2A progenitors to cross the interface from primary to transgenic astrocytes was not the result of a failure to cross an astrocyte interface *per se*. In the majority of cases where O-2A progenitors were plated onto transgenic astrocyte monolayers and challenged with monolayers of primary astrocytes, the progenitors were able to cross the astrocyte interface and migrate considerable distances over the primary astrocyte monolayer.

The availability of clonal cell lines that exhibit properties much like that of glial scar tissue offers us a simple in vitro system for analyzing the biochemical cues that hinder neurite outgrowth from mature neurons and for identifying factors that might inhibit migration of O-2A progenitors in demyelinated lesions.

## 2.2. Direct Derivation from H-2K$^b$tsA58 Transgenic Mice of Conditionally Immortal Myoblasts Able to Differentiate into Myotubes In Vivo

Transplantation of myogenic cells has been discussed as a treatment for genetic disease, in muscle and other tissues *(75,76)*, and also holds promise as a tool for investigating the regulation of expression of muscle genes in vivo. However, there are two major obstacles to full exploitation of these techniques: the finite mitotic capacity of primary myoblasts limits the cell numbers available from clones, whereas the tumorigenic tendency of established myogenic lines makes them difficult to use in vivo *(77)*.

To generate cell lines with the above characteristics, J. Morgan and T. Partridge and their colleagues isolated conditionally immortal myoblast cell lines from the limb muscles of 19-d-old embryonic, and 10-d-old and 4-wk-old H-2K$^b$*ts*A58 transgenic mice *(78)*. Cell growth was sufficiently robust to allow the generation of clones directly by growth at low density. Clones exhibiting a typical myogenic morphology grew readily in the permissive conditions, but not in semipermissive conditions (of 33°C, IFN-γ-negative). Of the clones chosen, 14 of 15 readily formed myotubes when grown at high density in nonpermissive conditions. Confluent cultures switched from permissive to nonpermissive conditions showed greatly reduced DNA synthesis and began to form myotubes within 24 h. The number of myotubes continued to increase for the next several days. Some fusion also occurred in permissive conditions if cultures were allowed to become very dense. In both cases, the resultant myotubes expressed dystrophin and muscle-specific myosin. Nonfused cells did not express muscle-specific myosin or dystrophin.

Cells from eight different clones were injected into the leg muscles of dystrophin-negative MDX nu/nu mice of the Gpi Is$^a$ isotype in which either the right or both legs had been subjected to 18G X-rays at 15–17 d of age. Irradiation inhibits the regeneration of mdx muscle, which eventually atrophies, thus presenting a good environment for the formation of new muscle from implanted myogenic cells *(79,80)*.

Seven of eight clones injected in vivo formed new muscle histologically indistinguishable from that formed by injection of primary myoblasts *(81)*, but did not form tumors in both irradiated and nonirradiated mdx muscles. No systematic difference in the ability to form muscle was seen between different clones, or between irradiated vs. nonirradiated legs.

Even after 19 passages (approx 60 doublings) in culture, cells remained diploid and formed normal muscle in the absence of tumor formation on injection into irrradiated mdx nu/nu muscle, in contrast to the neoplastic behavior of established muscle cell lines *(77,82,83)*, no tumors were found up to 120 d postimplantation *(78)*.

One striking aspect of the above experiments was that rare clones of H-2K$^b$*ts*A58-derived cells could be reisolated from injected muscle after several weeks of in vivo growth, injected into a second generation of hosts (where the injected cells formed new muscle), reisolated again, and injected to form normal muscle in still a third generation of hosts. Clones were reisolated successfully from both irradiated and nonirradiated muscle. Only rare clones were obtained in these reisolation experiments, suggesting that myoblast cell lines derived from H-2K$^b$*ts*A58 mice are able both to form myotubes and also to enter the pool of slowly dividing, or quiescent, myogenic cells known to exist in normal muscle.

In contrast with the problems of senesence and neoplasia associated with the use of primary cells or spontaneously arising cell lines, respectively, clonal myoblast lines derived from H-2K$^b$*ts*A58 mice can be grown for extended periods in vitro without exhibiting any loss of conditionality or capability of undergoing normal differentiation in vitro or in vivo. Moreover, the ease of isolation of cell lines from H-2K$^b$*ts*A58 mice of different ages raises the possibility of comparing the properties of myoblast lines representing different developmental stages.

### 2.3. H-2K$^b$tsA58 Mice and the Generation of Mutant Cell Lines

One particularly exciting application of the H-2K$^b$*ts*A58 transgenic mice will be in the generation of cell lines from other strains of mice that are genetically aberrant, either as a result of being bred as a mutant strain of animals or as a result of a mutation introduced by choice (such as mice in which the function of a particular gene of interest has been disrupted, for example). Since cell lines can be generated from heterozygous H-2K$^b$*ts*A58 transgenic mice, it thus follows that mating of homozygote breeding stock with a homozygous mutant will yield animals expressing both properties of interest. If the mutation of interest is dominantly active, then direct generation of cell lines from F1 litters would be appropriate, whereas recessive mutations would have to be rebred to homozygosity in order for the cell lines of interest to be derived. These strategies have already been applied to the study of myoblasts from *mdx* mice, which have a mutation in the dystrophin gene. H-2K$^b$*ts*A58 × *mdx* F1 pups have proven to be a ready source of *mdx*-deficient skeletal myoblast lines, thus providing an important new tool for the study of dystrophin function *(78)*.

## 3. General Principles Guiding Cell Line Production from H-2K$^b$*ts*A58 Transgenic Mice

A number of general principles emerge from our experiments to date. One point of central importance is that we think it most appropriate to utilize heterozygotes for the generation of cell lines in most (and possibly all) circumstances. The successful derivation of cell lines from heterozygotes offers a number of practical advantages, not least of which is the fact that it is not necessary to maintain a homozygous breeding colony if one's intention is to use only animals from a limited series of dissections. Instead, male stud homozygotes can be mated with normal females, with cells of interest being derived from the heterozygous F1 animals. It is easy to obtain a steady stream of heterozygous animals with a few homozygous stud males and a harem of normal females.

It does seem reasonable to consider whether cell types that have proven difficult to grow from heterozygous animals may be more readily obtained from homozygotes, owing to the higher level of TAg expression achieved per unit of interferon applied. Thus far, however, our tendency has been to expend more effort on determining appropriate growth conditions for each cell type of interest. Part of the reason for choosing this approach is that we are primarily interested in using cells derived from H-2K$^b$*ts*A58 transgenic animals as a tool for exploring the cellular biology of normal cells, and we prefer to express the minimal amount of TAg needed to achieve immortalization.

A second useful observation we have made is that one can often (although not invariably) turn off *ts*TAg function by shifting temperature only up to 37°C. This reduces the need for extra incubators from 2 (33°C and 39.5°C) to 1 (33°C), with regular 37°C incubators being used to turn off TAg activity. For a large number of laboratories, this second advantage is particularly useful. More importantly, the lack of an absolute requirement for shift to 39.5°C means that transplantation in vivo has a higher chance of turning off functional expression of *ts*TAg, thus allowing cells to differentiate normally *in situ*.

A further point to note is that we have not seen any great differences in effectiveness of IFN-γ from different manufacturers. Moreover, it should be possible to substitute any other inducer of Class I antigen expression (e.g., other interferons) to promote expression from the H-2K$^b$ promoter, in the event that IFN-γ is specifically detrimental to the growth of the cells of interest (a problem we have not encountered).

Although it appears likely that the H-2K$^b$*ts*A58 transgenic mice and their generic relatives will greatly facilitate cell line production, it is important to note that the probability of successfully generating a cell line of interest is still enhanced by knowledge about the biological properties of the cell of interest. Such a result is expected from previous studies on cell lines produced by retroviral infection of cultured cells. For example, even though expression of TAg in Schwann cells simplifies the growth factor requirements of these cells, promotion of cell division in the immortalized Schwann cells still requires the presence of two out of three of the mitogenic stimuli used in the growth of primary Schwann cells *(4)*. To date, our studies with cells derived from the H-2K$^b$*ts*A58 transgenic mice have indicated that generation of novel cell lines is indeed a straightforward matter for many tissues. However, when attempting to grow cells for which little is known about the control of division in the cell type of interest, it has also become clear that experimentation is required to determine suitable conditions for optimizing cell line production.

In addition to the cells discussed thus far, it also has been possible to isolate cell lines from an increasing variety of tissue. Primitive kidney cells have been isolated *(84)* and characterized for their chloride conductance properties *(85)*. Hippocampal progenitor cells have also been described with the ability to generate neurons *(86,87)*. Isolation of colonic and intestinal epithelial cells *(88)* has been followed by the derivation of cell lines from a cross-breed of H-2K$^b$*ts*A58 transgenics with Min mice, yielding colonic epithelial cells with mutations in the APC gene *(89)*. Vascular smooth muscle lines *(90)* and biliary epithelial cell lines *(91)* also have been recently described. In addition, it is possible to target expression of *ts*A58 to specific lineages, thus yielding very specific cells lines (e.g., *92*). Thus, the H-2K$^b$*ts*A58 transgenic mice appear to be useful in the manner we originally envisaged. Moreover, the success of our transplantation studies with myoblast lines derived from H-2K$^b$*ts*A58 mice suggests that these animals might provide a generally useful source of cell lines suitable for application in studies in which transplantation and incorporation into normal tissue are a desired goal.

## 4. Availability of H-2K$^b$*ts*A58 Mice

Charles River Laboratories has agreed to supply H-2K$^b$tsA58 transgenic mice to the international scientific community. Requests to purchase animals should be directed to Charles River Laboratories, 251 Ballardvale St., Wilmington, MA 01887.

## 5. Sample Protocol: Generation of Astrocyte Cell Lines from H-2K$^b$*ts*A58 Transgenic Mice

In general, the methodology used to establish cultures of interest from H-2K$^b$*ts*A58 transgenic mice is to grow cells in whatever conditions favor their division, in the presence of IFN-γ at 33°C. Some cell types require a period of growth of at least 24 h in tissue culture before they are cloned, and others can be cloned directly. Cloning can be carried out by limited dilution cloning (as with studies on myoblasts; *78*) or can be cloned by infection with a retrovirus encoding a antibiotic resistance gene (as with studies on astrocytes; *66*). In the latter case, clonality can be confirmed by Southern blot analysis to ensure that putative clones possess only a single integration site for the antibiotic resistance gene.

Enriched astrocyte cultures were prepared by a modification of the methods of Noble et al. *(64)*. Cortices from perinatal mice were removed and dissected free of meninges, chopped finely, and incubated at 37°C for 30 min in EDTA solution (200 mg/mL solution of EDTA [FDS-grade, Sigma] in $Ca^{2+}/Mg^{2+}$-free DMEM) containing 8500 U/mL trypsin (Sigma, St. Louis, MO). SBTI-DNase (a solution of 0.52 mg/mL soybean trypsin inhibitor, 0.04 mg/mL bovine pancreatic DNase, and 3 mg/mL BSA fraction V [Sigma]) was added at a ratio of 4 mL for every 7 mL cortical cell suspension and incubated for a further 5 min before being centrifuged at 200$g$ for 5 min. The tissue was resuspended in DMEM containing 10% fetal calf serum (Imperial), 2 m*M* glutamine (Gibco-BRL), and 25 mg/mL gentamycin (DMEM-FCS) with 0.4% BSA and dissociated by repeated trituration through 21- and 25-gage needles. The dissociated cells were centrifuged through a cushion of DMEM-FCS containing 4% BSA at 200$g$ for 5 min. The pellet was resuspended in DMEM-FCS and the cells were seeded into flasks coated with 5 mg/mL poly-*L*-lysine at a density of $10^7$ cells/75-cm$^2$ flask. The cultures were grown for 6 d, and cells on top of the monolayer were removed by shaking overnight at 37°C on a rotary platform (100 rpm). After 24 h, the cells were pulsed with $2 \times 10^{-5}$ *M* cytosine arabinoside (AraC) for 4 d. This procedure routinely produced astrocyte cultures of >95% purity as assessed by staining with a polyclonal antiserum against GFAP.

## 5.1. Generation of Astrocyte Cell Lines

Cortical astrocytes were purified from neonatal H-2K^b *ts*A58 transgenic mice as described above, except that the cultures were prepared at 33°C in the presence of 25 U/mL of IFN-γ. The cultures were passaged at a ratio of 1:4 and refed. The following day, cultures were incubated with 1 mL of BAG retroviral supernatant (Price et al., 1989), 1 mL of DMEM-FCS and 2 mL of Polybrene (10 mg/mL; Sigma) for 2 h at 37°C. The cultures were then washed with DMEM-FCS, fed with DMEM-FCS containing 25 U/mL of IFN-γ, and shifted back down to 33°C. The cells were passaged at a ratio of 1:4 once more the following day and fed with the same medium. After 4 d, the culture medium was supplemented with 250 mg/mL G418 (Geneticin; Gibco-BRL) and the cultures fed every 3 d for 2 wk, after which time drug-resistant colonies were clearly visible. Fifty colonies were passaged using autoclaved cloning rings and cultured in 96-well trays. The cell lines were expanded progressively to 24- and 6-well trays, and then to T25 and T75 tissue-culture flasks. The passage at which the lines were plated into a T25 flask was designated as Passage 1. Ten lines were chosen for Southern blot analysis using a 1.1-kb probe for the *neo* gene to determine whether each line contained a single retroviral integration site.

## References

1. Noble, M., Groves, A. K., Ataliotis, P., Ikram, Z., and Jat, P. S. (1995) The H-2K$^b$*ts*A58 transgenic mouse: a new tool for the rapid generation of novel cell lines. *Transgenic Res.* **4,** 215–225.
2. Noble, M., Groves, A. K., Ataliotis, P., Ikram, Z., and Jat, P. S. (1995) The H-2K$^b$*ts*A58 transgenic mouse: a single source of a wide range of conditionally immortal cell lines, in *Strategies in Transgenic Animal Science* (Monastersky, G. and Robl, J.M., eds.), American Society for Microbiology Press, Washington, DC, pp. 325–346.
3. Noble, M. and Barnett, S. C. (1996) Production and growth of conditionally immortal primary glial cell cultures and cell lines, in *Culture of Immortalized Cells* (Freshney, I. and Freshney, M.G., eds.), Wiley-Liss, New York, pp. 331–366.
4. Ridley, A., Paterson, H., Noble, M., and Land, H. (1988) *Ras*-mediated cell cycle arrest is altered by nuclear oncogenes to induce Schwann cell transformation. *EMBO J.* **7,** 1635–1645.
5. Marshall, C. J. (1991) How does p21*ras* transform cells? *Trends. Genet.* **7(3),** 91–95.
6. Tegtmeyer, P. (1975) Function of simian virus 40 gene A in transforming infection. *J. Virol.* **15,** 613–618.
7. Cepko, C. L. (1989) Immortalization of neural cells via retrovirus-mediated oncogene transduction. *Ann. Rev. Neurosci.* **12,** 47–65.
8. Frederiksen, K., Jat, P., Valtz, N., Levy, D., and McKay, R. (1988) Immortalization of precursor cells from the mammalian CNS. *Neuron* **1,** 439–448.
9. Renfranz, P. J., Cunningham, M. G., and McKay, R. D. G. (1991) Region-specific differentiation of the hippocampal stem cell line HiB5 upon implantation into the developing brain. *Cell* **66,** 713–729.
10. Snyder, E. Y., Deitcher, D. L., Walsh, C., Arnold-Aldea, S., Hartwieg, E. A., and Cepko, C. (1992) Multipotent neural cell lines can engraft and participate in development of mouse cerebellum. *Cell* **68,** 33–51.
11. Barnett, S. C. and Crouch, D. H. (1995) The effect of oncogenes on the growth, differentiation and transformation of oligodendrocyte-type-2 astrocytes progenitor cells. *Cell Growth Differ.* **6,** 69–84.
12. Geller, H. M. and Dubois-Dalcq, M. (1988) Antigenic and functional characterization of a rat central nervous system-derived cell line immortalized by a retroviral vector. *J. Cell Biol.* **107,** 1977–1986.
13. Peden, K. W. C., Charles, C., Sanders, L., and Tennekoon, G. (1989) Isolation of rat Schwann cell lines: use of SV40 T antigen gene regulated by synthetic metallothionine promoters. *Exp. Cell Res.* **185,** 60–72.

14. Galiana, E., Borde, I., Marin, P., Cuzin, F., Gros, F., Rouget, P., et al. (1990) Establishment of permanent astroglial cell lines, able to differentiate *in vitro*, from transgenic mice carrying the polyoma virus large T antigen: an alternate approach to brain cell immortalization. *J. Neurosci. Res.* **26,** 269–277.

15. Allinquant, B., D'Urso, D., Almazan, G., and Colman, D. R. (1990) Transfection of transformed *shiverer* mouse glial cell lines. *Dev. Neurosci.* **12,** 340–348.

16. Almazan, G. and McKay, R. (1992) An oligodendrocyte precursor cell line from rat optic nerve. *Brain Res.* **579,** 234–245.

17. Evrard C., Galiana, E., and Rouget, P. (1988) Immortalization of bipotential glial progenitors and generation of permanent "blue" cell lines. *J. Neurosci. Res.* **21,** 80–87.

18. Giotta, G. J., Heitzmann, J., and Cohen, M. (1980) Properties of the temperature-sensitive Rous sarcoma virus transformed cerebellar cell lines. *Brain Res.* **202,** 445–458.

19. Trotter, J., Boulter, C. A., Sontheimer, H., Schachner, M., and Wagner, E. F. (1989) Expression of v-src arrests murine glial cell differentiation. *Oncogene* **4,** 457–464.

20. Barnett, S. C., Franklin, R. J. M., and Blakemore, W. F. (1994) In vitro and in vivo analysis of a rat bipotential O-2A progenitor cell line containing the temperature sensitive mutant gene of the SV40 large T antigen. *Eur. J. Neurosci.* **5,** 1247–1260.

21. Louis, J. C., Magal, E., Muir, D., Manthorpe, M., and Varon, S. (1992) CG-4, a new bipotential glial cell line from rat brain, is capable of differentiating *in vitro* into either mature oligodendrocytes or type-2 astrocytes. *J. Neurosci. Res.* **31,** 193–204.

22. Cunningham, M. G., Nikkhah, G., and McKay, R. D. G. (1993) Grafting immortalized hippocampal cells into the brain of the adult and newborn rat. *Neuroprotocols* **3,** 260–272.

23. Jat, P. S. and Sharp, P. A. (1989) Cell lines established by a temperature-sensitive simian virus 40 large-T-antigen are growth restricted at the nonpermissive temperature. *Mol. Cell. Biol.* **9,** 1672–1681.

24. Brinster, R. L., Chen, H. Y., Messing, A., van Dyke, T., Levine, A. J., and Palmiter, R. D. (1984) Transgenic mice harboring SV40 T-antigen genes develop characteristic brain tumors. *Cell* **37,** 367–379.

25. Hanahan, D. (1985) Heritable formation of pancreatic β-cell tumors in transgenic mice expressing recombinant insulin/simian virus 40 oncogenes. *Nature* **315,** 115–122.

26. Efrat, S., Linde, S., Kofod, H., Spector, D., Delannoy, M., Grant, S., et al. (1988) Beta-cell lines derived from transgenic mice eexpressing a hybrid insulin gene-oncogene. *Proc. Natl. Acad. Sci. USA* **85,** 9037–9041.

27. MacKay, K., Striker, L. J., Elliot, S., Pinkert, C. A., Brinster, R. L., and Striker, G.E. (1988) Glomerular epithelial mesangial, and endothelial cell lines from transgenic mice. *Kidney Int.* **33,** 677–684.

28. Paul, D., Höhne, M., and Hoffmann, B. (1988) Immortalized differentiated hepatocyte lines derived from transgenic mice harboring SV40 T-antigen genes. *Exp. Cell Res.* **175,** 354–365.

29. Paul, D., Höhne, M., Pinkert, C., Piasecki, A., Ummelmann, E., and Brinster, R. L. (1988) Immortalization and malignant transformation of hepatocytes by transforming genes of polyoma virus and of SV40 virus in vitro and in vivo. *Klin. Wochenstr.* **66(Suppl. 11),** 134–139.
30. Hammang, J. P., Baetge, E. E., Behringer, R. R., Brinster, R. L., Palmiter, R. D., and Messing, A. (1990) Immortalized rat neurons derived from SV40 TAg induced tumors in transgenic mice. *Neuron* **4,** 775–782.
31. Mellon, P. L., Windle, J. J., Goldsmith, P. C., Padula, C. A., Roberts, J. L., and Weiner, R .I. (1990) Immortalization of hypothalamic GnRH neurons by genetically targeted tumorigenesis. *Neuron* **5,** 1–10.
32. Hamilton, S. R. and Vogelstein, B. (1988) Point mutations in human neoplasia. *J. Pathol.* **154(3),** 205,206.
33. James, C. D., Carlbom, E., Dumanski, J. P., Hansen, M., Nordenskjold, M., Collins, V.P., et al. (1988) Clonal genomic alterations in glioma malignancy stages. *Cancer Res.* **48(19),** 5546–5551.
34. Vogelstein, B., Fearon, E. R., Hamilton, S. R., Kern, S. E., Preisinger, A. C., Leppert, M., et al. (1988) Genetic alterations during colorectal cancer development. *N. Engl. J. Med.* **319(9),** 525–532.
35. Thompson, T. C., Southgate, J., Kitchener, G., and Land, H. (1989) Multistage carcinogenesis induced by *ras* and *myc* oncogenes in a reconstituted organ. *Cell* **56,** 917–930.
36. Compere, S. J. (1989) The *ras* and myc oncogenes cooperate in tumor induction in many tissues when introduced into midgestation mouse embryos by retroviral vectors. *Proc. Natl. Acad. Sci. USA* **86,** 2224–2228.
37. Land, H., Parada, L. F., and Weinberg, R. A. (1983) Cellular oncogenes and multistep carcinogenesis. *Science* **222,** 771–778.
38. Land, H., Parada, L. F., and Weinberg, R. A. (1983) Tumorigenic conversion of primary embryo fibroblasts requires at least two cooperating oncogenes. *Nature* **304,** 596–602.
39. Ruley, H. E. (1983) Adenovirus early region 1A enables viral and cellular transforming genes to transform primary cells in culture. *Nature* **304,** 602–606.
40. Jat, P. S., Noble, M. D., Ataliotis, P., Tanaka, Y., Yannoutsos, N., Larssen, L., et al. (1991) Direct derivation of conditionally immortal cell lines from an H-2K$^b$-*ts*A58 transgenic mouse. *Proc. Natl. Acad. Sci. USA* **88,** 5096–5100.
41. Weiss, E. H., Mellor, A., Golden, L., Fahrner, K., Simpson, E., Hurst, J., et al. (1983) The structure of the mutant *H-2* gene suggests that the generation of polymorphism in *H-2* genes may occur by gene conversion-like events. *Nature* **301,** 671–674.
42. Kimura, A., Israel, A., Le Bail, O., and Kourilsky, P. (1986) Detailed analysis of the mouse H-2K$^b$ promoter: enhancer-like sequences and their role in the regulation of Class I gene expression. *Cell* **44,** 261–272.
43. Baldwin, A. S., Jr. and Sharp, P. A. (1987) Binding of a nuclear factor to a regulatory sequence in the promoter of the mouse H-2K$^b$ class I major histocompatibility gene. *Mol. Cell. Biol.* **7,** 305–313.

44. Santerre, R. F., Cook, R. A., Crisel, R. M., Sharp, J. D., Schmidt, R. J., Williams, D. C., et al. (1981) Insulin synthesis in a clonal cell line of simian virus 40-transformed hamster pancreatic beta cells. *Proc. Natl. Acad. Sci. USA* **78,** 4339–4343.

45. Bayley, S. A., Stones, A. J., and Smith, C. G. (1988) Immortalization of rat keratinocytes by transfection with polyomavirus large T gene. *Exp. Cell Res.* **17,** 232–236.

46. Williams, D. A., Rosenblatt, M. F., Beier, D. R., and Cone, R. D. (1988) Generation of murine stromal cell lines supporting hematopoietic stem cell proliferation by use of recombinant retrovirus vectors encoding simian virus 40 large T antigen. *Mol. Cell. Biol.* **8,** 3864–3871.

47. Burns, J. S., Lemoine, L., Lemoine, N. R., Williams, E. D., and Wynford-Thomas, D. (1989) Thyroid epithelial cell transformation by a retroviral vector expressing SV40 large T. *Br. J. Cancer* **59,** 755–760.

48. Wallach, D., Fellous, M., and Revel, M. (1982) Preferential effect of γ interferon on the synthesis of HLA antigens and their mRNAs in human cells. *Nature* **299,** 833–836.

49. Basham, T. Y. and Merigan, T. C. (1983) Recombinant interferon-gamma increases HLA-DR synthesis and expression. *J. Immunol.* **130,** 1492–1494.

50. Israel, A., Kimura, A., Fournier, A., Fellous, M., and Kourilsky, P. (1986) Interferon response sequence potentiates activity of an enhancer in the promoter of a mouse H-2 gene. *Nature* **322,** 743–746.

51. Reier, P. J., Stensaas, L. J., and Guth, L. (1983) The astrocytic scar as an impediment to regeneration in the central nervous system, in *Spinal Cord Reconstruction* (Kao, C. C., Bunge, R. P., and Reier, P. J., eds.), Raven, New York, pp. 163–195.

52. Barrett, C. P., Donati, E. J., and Guth, L. (1984) Differences between adult and neonatal rats in their astroglial response to spinal injury. *Exp. Neurol.* **84,** 374–385.

53. Carlstedt, T. (1985) Dorsal root innervation of spinal cord neurons after dorsal root implantation into the spinal cord of adult rats. *Neurosci. Lett.* **55,** 343–348.

54. Carlstedt, T., Dalsgaard, C.-J., and Molander, C. (1987) Regrowth of lesioned dorsal root nerve fibres into the spinal cord of neonatal rats. *Neurosci. Lett.* **74,** 14–18.

55. Liuzzi, F. J. and Lasek, R. J. (1987) Astrocytes block axonal regeneration in mammals by activating the physiological stop pathway. *Science* **237,** 642–645.

56. Smith, G. M., Miller, R. H., and Silver, J. (1986) Changing role of forebrain astrocytes during development, regenerative failure and induced regeneration upon transplantation. *J. Comp. Neurol.* **251,** 23–43.

57. Rudge, J. S., Smith, G. M., and Silver, J. (1989) An *in vitro* model of wound healing in the CNS: Analysis of cell reaction and interaction at different ages. *Exp. Neurol.* **103,** 1–16.

58. Fawcett, J. W., Housden, E., Smith-Thomas, L., and Meyer, R. L. (1989) The growth of axons in three-dimensional astrocyte cultures. *Dev. Biol.* **135,** 449–458.

59. Smith, G. M., Rutishauser, U., Silver, J., and Miller, R. H. (1990) Maturation of astrocytes *in vitro* alters the extent and molecular basis of neurite outgrowth. *Dev. Biol.* **138,** 377–390.

60. Rudge, J. S. and Silver, J. (1990) Inhibition of neurite outgrowth on astroglial scars *in vitro*. *J. Neurosci.* **10,** 3594–3603.
61. Geisert, E. E. and Stewart, A. M. (1991) Changing interactions between astrocytes and neurons during CNS maturation. *Dev. Biol.* **143,** 335–345.
62. McKeon, R. J., Schreiber, R. C., Rudge, J. S., and Silver, J. (1991) Reduction of neurite outgrowth in a model of glial scarring following CNS injury is correlated with the expression of inhibitory molecules on reactive astrocytes. *J. Neurosci.* **11,** 3398–3411.
63. ffrench-Constant, C., Miller, R. H., Burne, J. F., and Raff, M. C. (1988) Evidence that that migratory oligodendrocyte-type-2 astrocyte (O-2A) progenitor cells are kept out of the rat retina by a barrier at the eye-end of the optic nerve. *J. Neurocytol.* **17,** 13–25.
64. Noble, M. D., Fok-Seang, J., and Cohen, J. (1984) Glia are a unique substrate for the *in vitro* growth of central nervous system neurons. *J. Neurosci.* **4,** 1892–1903.
65. Price, J., Turner, D., and Cepko, C. L. (1987) Lineage analysis in the vertebrate nervous system by retrovirus-mediated gene transfer. *Proc. Natl. Acad. Sci. USA* **84,** 156–160.
66. Groves, A. K., Entwistle, A., Jat, P. S., and Noble, M. (1993) The characterisation of astrocyte cell lines that display properties of glial scar tissue. *Dev. Biol.* **159,** 87–104.
67. Liesi, P., Dahl, D., and Vaheri, A. (1983) Laminin is produced by early rat astrocytes in primary culture. *J. Cell Biol.* **96,** 920–924.
68. Liesi, P. (1985) Laminin-immunoreactive glia distinguish regenerative adult CNS systems from non-regenerative ones. *EMBO J.* **4,** 2505–2511.
69. Laywell, E. and Steindler, D. (1991) Boundaries and wounds, glia and glycoconjugates: Cellular and molecular analyses of developmental partitions and adult brain lesions. *Ann. NY Acad. Sci.* **633,** 122–141.
70. Laywell, E., Dörries, U., Bartsch, U., Faissner, A., Schachner, M., and Steindler, D. (1992) Enhanced expression of the developmentally regulated extracellular matrix molecule tenascin following adult brain injury. *Proc. Natl. Acad Sci. USA* **89,** 2634–2638.
71. Fallon, J. R. (1985) Preferential outgrowth of central nervous system neurites on astrocytes and Schwann cells as compared with nonglial cells *in vitro*. *J. Cell Biol.* **100,** 198–207.
72. Noble, M. D., Murray, K., Stroobant, P., Waterfield, M. D., and Riddle, P. (1988) Platelet-derived growth factor promotes division and inhibits premature differentiation of the oligodendrocyte/type-2 astrocyte progenitor cell. *Nature* **333,** 560–562.
73. Richardson, W. D., Pringle, N., Mosley, M., Westermark, B., and Dubois-Dalcq, M. (1988) A role for platelet-derived growth factor in normal gliogenesis in the central nervous system. *Cell* **53,** 309–319.
74. Raff, M. C., Miller, R. H., and Noble, M. (1983) A glial progenitor cell that develops *in vitro* into an astrocyte or an oligodendrocyte depending on the culture medium. *Nature* **303,** 390–396.
75. Dhawan, J., Pan, L. C., Pavlath, G. K., Travis, M. A., Lanctot, A. M., and Blau, H. M. (1991) Systemic delivery of human growth hormone by injection of genetically engineered myoblasts. *Science* **254,** 1509–1511.

76. Barr, E. and Leiden, J. M. (1991) Systemic delivery of recombinant proteins by genetically modified myoblasts. *Science* **254,** 1507–1509.

77. Morgan, J. E., Moore, S. E., Walsh, F. S., and Partridge, T. A. (1992) Formation of skeletal muscle in vivo from the mouse C2 cell line. *J. Cell Sci.* **102,** 779–787.

78. Morgan, J., Beauchamp, J. R., Peckham, M., Ataliotis, P., Jat, P., Noble, M., et al. (1994) Myogenic cells derived from H-2K^btsA58 transgenic mice are conditionally immortal in vitro but differentiate normally in vivo. *Dev. Biol.* **162,** 486–498.

79. Partridge, T. A., Morgan, J. E., Coulton, G. R., Hoffman, E. P., and Kunkel, L. M. (1989) Conversion of mdx myofibres from dystrophic-negative to -positive by injection of normal myoblasts. *Nature* **337,** 176–178.

80. Wakeford, S., Watt, D. J., and Partridge, T. A. (1991) X-irradiation improves mdx mouse muscle as a model of muscle fibre loss in DMD. *Muscle Nerve* **14,** 42–50.

81. Morgan, J. E., Hoffman, E. P., and Partridge, T. A. (1990) Normal myogenic cells from newborn mice restore normal histology to degenerating muscles of the *mdx* mouse. *J. Cell Biol.* **111,** 2437–2449.

82. Wernig, A., Irintchev, A., Hartling, A., Stephan, G., Zimmerman, K., and Stariznski-Powitz, A. (1991) Formation of new muscle fibres and tumours after injection of cultured myogenic cells. *J. Neurocytol.* **20,** 982.

83. Partridge, T. A., Morgan, J. E., Moore, S. E., and Walsh, F. S. (1988) Myogenesis in vivo from the mouse C2 muscle cell-line. *J. Cell Biochem.* **Suppl. 12C,** 331.

84. Woolf, A. S., Kolatsi-Joannou, M., Hardman, P., Andermarcher, E., Moorby, C., Fine, L. G., et al. (1995) Roles of hepatocyte growth factor/scatter factor and the met receptor in the early development of the metanephros. *J. Cell Biol.* **128,** 171–184.

85. Barber, R. D., Woolf, A. S., and Henderson, R. M. (1995) A characterization of the chloride conductance in mesangial cells from the H-2Kb-*ts*A58 transgenic mouse. *Biochim. Biophys. Acta.* **1269,** 267–274.

86. Rashid-Doubell, F., Kershaw, T. R., and Sinden, J. D. (1994) Effects of basic fibroblast growth factor and gamma interferon on hippocampal progenitor cells derived from the H-2Kb-*ts*A58 transgenic mouse. *Gene Ther.* **1 (Suppl. 1),** S63.

87. Kershaw, T. R., Rashid-Doubell, F., and Sinden, J. D. (1994) Immunocharacterization of H-2Kb-A58 transgenic mouse hippocampal neuroepithelial cells. *Neuroreport* **5,** 2197–2200.

88. Whitehead, R. H., VanEeden, P. E., Noble, M. D., Ataliotis, P., and Jat, P.S. (1993) The establishment of conditionally immortalised epithelial cell lines from both colon and small intestine of H-2K^btsA58 transgenic mice. *Proc. Natl. Acad. Sci. USA* **90,** 587–591.

89. Whitehead, R. H. and Joseph, J. L. (1994) Derivation of conditionally immortalized cell lines containing the Min mutation from the normal colonic mucosa and other tissues of an "Immortomouse"/Min hybrid. *Epithelial Cell Biol.* **3,** 119–125.

90. Ehler, E., Jat, P. S., Noble, M. D., Citi, S., and Draeger, A. (1995) Vascular smooth muscle cells of H-2Kb-*ts*A58 transgenic mice. Characterization of cell lines with distinct properties. *Circulation* **92,** 3289–3296.

91. Paradis, K., Le, O. N., Russo, P., St. Cyr, M., Fournier, H., and Bu, D. (1995) Characterization and response to interleukin 1 and tumor necrosis factor of immortalized murine biliary epithelial cells. *Gastroenterology* **109,** 1308–1315.

92. Cairns, L. A., Crotta, S., Minuzzo, M., Moroni, E., Granucci, F., Nicolis, S., et al. (1994) Immortalization of multipotent growth-factor dependent hemopoietic progenitors from mice transgenic for GATA-1 driven SV40 tsA58 gene. *EMBO J.* **13,** 4577–4586.
93. Price, J., Turner, D., and Cepko, C. H. (1987) Lineage analysis int he vertebrate nervous system by retrovirus-mediated gene transfer. *Proc. Natl. Acad. Sci. USA* **84,** 156–160.

# 10

## Reporter Genes for the Study of Transcriptional Regulation in Transgenic Mouse Embryos

### Jonathan D. Gilthorpe and Peter W. J. Rigby

## 1. Introduction

The development of transgenic technology during recent years has allowed researchers to probe much more deeply than was previously possible into the molecular mechanisms influencing embryonic development. Transgenic procedures allow the transfer of a cloned gene into a host genome or the mutation of specific genomic sequences via targeting in embryonic stem (ES) cells (*see* Chapter 7). Most commonly, transgenic gene transfer experiments have been used to study the phenotypic effects caused by the misexpression or overexpression of a transgene *(1–3)*, or to investigate the transcriptional mechanisms underlying developmental and tissue-specific gene regulation *(4–6)*. The use of reporter genes in this latter application will be dealt with in this chapter.

Prior to the advent of gene transfer methods, research into the developmental regulation of gene expression was restricted to the use of indirect embryological and in vitro techniques. However, transgenesis employing reporter genes has provided a model assay to identify *cis*-acting DNA sequences that can influence particular aspects of gene expression during development in vivo. Providing that the transgene integrates into the germline of the founder transgenic animal, it will be present in every cell of the $F_1$ generation, and may be passed on to subsequent generations. This enables the expression of a gene to be studied throughout development and in all cell types, which is not possible using cell-culture techniques. The system employs a transgene consisting of genomic sequences linked to a reporter gene, the product of which can be easily detected. The expression of the wild-type construct and of truncated or mutated forms can be monitored by histochemical analysis and compared, thus

From: *Methods in Molecular Biology, Vol. 97: Molecular Embryology: Methods and Protocols*
Edited by: P. T. Sharpe and I. Mason © Humana Press Inc., Totowa, NJ

allowing specific regulatory sequences to be defined. In the final analysis, individual transcription factor binding sites can be identified enabling the characterization of the cognate *trans*-acting factors. The most suitable reporters for this purpose are genes encoding enzymes, the activity of which can be detected with a chromogenic substrate enabling expression patterns to be directly visualized. Two such reporter genes are currently available, the *Escherichia coli lacZ* gene encoding β-galactosidase, and the human gene encoding placental alkaline phosphatase *(PALP-I)*.

Three basic methods are available for the production of transgenic mice carrying reporter genes. The manipulation of ES cells and retroviral infection of embryos are useful for certain types of experiments *(7)*. The most commonly used method, however, is the microinjection of DNA into the pronuclei of single-cell embryos *(8)*, and this is best suited to studies of gene regulation. This process begins with the construction of a suitable transgenic reporter construct, which is then purified and microinjected into fertilized eggs. Microinjection is a time-consuming procedure requiring a high degree of skill. An experienced operator can microinject and transfer between 150 and 300 fertilized eggs/d. Approximately 30–60% of these will develop to give viable embryos, of which 10–20% will usually carry the transgene and a proportion will express it in a construct-dependent manner. Generally this translates to 1–3 d of injection/construct to give a reliable number of data points (ideally six or more expressing mice or embryos). In addition, an expensive microinjection setup is required as is a well-organized animal facility to provide and maintain the mice.

This chapter covers the design, construction, and purification of a reporter gene for microinjection and the subsequent analysis. Staining protocols for β-galactosidase and alkaline phosphatase activities are included. Details of transgenic mouse production have not been incorporated, and readers should refer to earlier chapters in this book and to the "transgenic bible" of Hogan et al. *(7)*.

## 2. Materials

All solutions should be made up using molecular biology-grade reagents and sterile distilled water, unless otherwise indicated. Molecular biological and transgenic techniques are based on those described in **refs. 9** and 7, respectively.

### 2.1. Calcium Phosphate-Mediated Transfection

1. Cell-culture medium: Dulbecco's Modified Eagle's Medium (DMEM) supplemented with 4-m$M$ L-glutamine and 10% (v/v) fetal calf serum (Imperial Laboratories Europe, Winchester Hants, UK). Media may be optionally supplemented with the antibiotics penicillin (Sigma, Poole, Dorset, UK) at a concentration of 100 U/mL, and streptomycin (Gibco-BRL, Paisley, Scotland) at a concentration of 50 µg/mL. Store at 4°C.

2. DNA for transfection: 10 μg at a concentration of 0.5–1 μg/μL in TE (10 m$M$ Tris-HCl, 1 m$M$ EDTA, pH 8.0). Plasmid DNA should be in the closed, circular form purified by CsCl-ethidium bromide density gradient centrifugation *(9)*. Store at 4°C.

3. 2X HEPES-buffered saline (2X HBS): 280 m$M$ NaCl, 10 m$M$ KCl, 1.5 m$M$ Na$_2$HPO$_4$, 12 m$M$ dextrose, 50 m$M$ HEPES, pH 7.05, with NaOH. Sterilize by filtration through a 0.22 μm filter. Store in 5-mL aliquots at –20°C.

4. 2 $M$ CaCl$_2$: Sterilize by filtration through a 0.22-μm filter. Store in 5-mL aliquots at –20°C.

5. Phosphate-buffered saline (PBS): 0.138 $M$ NaCl, 2.7 m$M$ KCl, 10 m$M$ Na$_2$HPO$_4$, 1.4 m$M$ KH$_2$PO$_4$, pH 7.2. Sterilize by autoclaving. Store at 4°C or room temperature.

6. Wash solution: 0.02% (v/v) NP-40 in PBS. Store at 4°C.

7. Cell fixative: 0.2% (w/v) glutaraldehyde in wash solution. Make fresh from a 25% (w/v) solution (Sigma).

8. X-gal stain solution: PBS (pH 7.3) containing 5 m$M$ K$_3$Fe(CN)$_6$, 5 m$M$ K$_4$Fe(CN)$_6$·3H$_2$O, 2 m$M$ MgCl$_2$, 0.02% (v/v) NP-40, 0.5 mg/mL X-gal (5-bromo-4-chloro-3-indolyl-β-D-galactoside, Sigma). Store at 4°C in the dark. X-gal can be made up as a 100X stock solution (50 mg/mL in $N,N$-dimethylformamide—store in dark at –20°C). K$_3$Fe(CN)$_6$ and K$_4$Fe(CN)$_6$·3H$_2$O can be made up as 0.5 $M$ (100X) stock solutions and stored at 4°C in the dark.

9. Alkaline phosphatase (AP) stain solution: 100 m$M$ Tris-HCl (pH 9.5), 100 m$M$ NaCl, 50 m$M$ MgCl$_2$, 1 mg/mL NBT (nitro blue tetrazolium, Boehringer, Lewes, East Sussex, UK), 0.1 mg/mL 5-bromo-4-chloro-3-indolyl-phosphate (BCIP), X-Phos, (Boehringer). Make up freshly. NBT can be made as a 50X stock solution (50 mg/mL in 70% [v/v] $N,N$-dimethylformamide—store in dark at –20°C). BCIP can be made as a 100X stock solution (10 mg/mL in distilled water—store in the dark at –20°C).

## 2.2. Isolation and Purification of DNA Fragments for Microinjection

1. Recombinant DNA containing the reporter construct to be microinjected. Maxiprep DNA purified by CsCl density gradient centrifugation is desirable, although miniprep DNA may also be used.

2. Tris-Borate EDTA buffer (TBE): 89 m$M$ Tris-HCl (pH 8.3), 89 m$M$ boric acid, 2 m$M$ EDTA. Made as a 5X stock solution and diluted to 1X as required with the addition of deionized water (Elga, High Wycombe, Bucks, UK) containing 0.5 μg/mL ethidium bromide (from a 10 mg/mL stock).

3. 0.5–1% (w/v) Agarose (Ultrapure, Bio Rad, Hemel Hempstead, Herts, UK) in 1X TBE/ethidium bromide.

4. 1% (w/v) low-melting-point (LMP) agarose (NuSieve GTG, FMC Bioproducts, Lichfield, Straffs, UK) in 1X TBE/ethidium bromide.

5. β-Agarase I (1000 U/mL) and 10X Agarase buffer (New England Biolabs, Hitchine, Herts, UK).

6. Phenol (Tris-HCl buffered, pH 8.0, Amresco, Luton, Beds, UK). Store at 4°C.

7. Chloroform.

8. Diethyl ether.
9. 3 *M* sodium acetate, pH 5.2.
10. Propan-2-ol (isopropanol).
11. Microinjection buffer: 10 m*M* Tris-HCl, pH 7.4, 0.1 m*M* EDTA. Make up using high-quality, endotoxin-free water that has preferably been embryo-tested (e.g.,W1053, Sigma). Sterilize by passage through a 0.22-μm filter. Store in 10-mL aliquots at –20°C.

## 2.3. Whole-Mount Staining for Reporter Gene Activity

1. PBS: *See* **Subheading 2.1., item 5**.
2. Mirsky's fixative (National Diagnostics, Hull, Humberside, UK): Make from 10X concentrate and 10X buffer in distilled water. Store at 4°C for up to 1 mo.
3. Embryo wash: 0.02% (v/v) NP-40 in PBS.

### 2.3.1. β-Galactosidase Staining

1. X-gal stain solution (**Subheading 2.1., item 8**).
2. Bluo-gal stain solution. PBS (pH 7.3) containing 5 m*M* K$_3$Fe(CN)$_6$, 5 m*M* K$_4$Fe(CN)$_6$·3H$_2$O, 1 m*M* MgCl$_2$, 150 m*M* NaCl and 0.4 mg/mL Bluo-gal (5-bromo-3-indolyl-β-D-galactoside, Sigma). Store at 4°C in the dark. Bluo-gal can be made up as a 100X stock solution (40 mg/mL in *N,N*-dimethylformamide—store in dark at –20°C).

### 2.3.2. AP Staining

For AP stain solution, *see* **Subheading 2.1., item 9**.

## 2.4. Transgenic Diagnosis by the Polymerase Chain Reaction (PCR)

1. PK digestion buffer (10X stock): 100 m*M* Tris-HCl (pH 8.3), 500 m*M* KCl, 1% (v/v) Tween-20. Store at room temperature. Dilute to 1X with distilled water.
2. Proteinase K (Boehringer): 10 mg/mL (100X) stock solution in 50 m*M* Tris-HCl (pH 8.0), 1 m*M* CaCl$_2$. Filter-sterilize, and store in aliquots at 4°C.
3. *Taq* polymerase: Amplitaq (5 U/μL) and 10X PCR buffer II (Perkin Elmer, Warrington, Cheshire, UK).
4. MgCl$_2$ solution: 10X stock for PCR, depending on optimal Mg$^{2+}$ concentration for primers being used (e.g., 0.5–2.5 m*M* final). Store at –20°C in 1-mL aliquots.
5. dNTP mix: 100 m*M* solutions (Ultrapure, Pharmacia, Little Chalfont, Bucks, UK). Combine and dilute to 2 m*M* each with TE (pH 7.5) to give a 10X dNTP stock. Store in aliquots at –20°C.
6. PCR primers: Suitable transgene-specific and internal control primers. Store at 10X concentration (2 μ*M*, 2 pmol/μL) in distilled water at –20°C.
7. Mineral oil (Sigma).
8. 1.5% (w/v) Agarose (Ultrapure, Bio Rad) in 1X TBE/ethidium bromide.

## 3. Methods

### 3.1. Reporter Transgene Design

When designing a transgenic reporter construct, careful planning is of the utmost importance. The production and analysis of transgenic animals is both labor- and time-intensive, and a well-designed construct will give you a good base on which to plan your subsequent experiments. Although the exact layout of a given reporter depends on the aims of the experiment in question, certain general considerations should be addressed. These will be discussed below.

### 3.1.1. General Considerations

1. A reporter transgene requires the capacity to work like any other gene. It must contain all of the structural elements recognized by the transcriptional, posttranscriptional, and translational machinery of the host, leading to the production of a functional reporter gene product.

   The primary goal of an experiment aimed at identifying control sequences responsible for particular aspects of gene expression during development is to reproduce the exact pattern of expression of the endogenous gene. This provides a baseline from which sequences controlling this expression can be defined. If there is no previous knowledge about regulatory elements, it is best to begin working with the largest available fragment of DNA, thus maximizing the chances of encompassing all of the required control sequences. Cloning strategies that employ artificial chromosome vectors derived from yeast (YAC), bacteria (BAC), or bacteriophage P1 (PAC) enable the sequences of very large (>100 kb) genomic regions to be tested *(10–12)*.

2. A well-designed vector will prove to be a great help both for the initial cloning of your reporter and its subsequent analysis. Some plasmid vectors, such as pPolyIII *(13)*, have been especially designed with extensive multiple cloning sites for the cloning and excision of transgenic constructs. It may, however, be advantageous to engineer your own.

3. The reporter gene fragment needs to be purified away from vector sequences prior to microinjection (*see* **Note 1**). For this reason, the construct should be flanked by unique restriction enzyme sites. The most commonly used enzyme for this purpose is *Not*I (8-bp recognition sequence), which is an infrequent cutter within eukaryotic DNA. Other suitable enzymes are now commercially available, such as *Asc*I, *Pac*I, *Pme*I, *Sfi*I, and *Fse*I (8-bp recognition sequences, New England Biolabs) or the extremely rare cutting intron-encoded endonucleases (e.g., I-*Sce*I, 18-bp recognition sequence, Boehringer).

4. Try to utilize a vector that does not contain the *lacZα* region, since we have found that problems with high frequencies of recombination may arise when trying to insert the *lacZ* gene into such a vector (*see* **Note 2**).

5. Certain eukaryotic DNA sequences may cause problems during cloning, giving rise to a high frequency of recombination events. If this is the case, try switching to a lower copy number cloning vector, or switch to a different bacterial host strain, since this may circumvent the problem (*see* **Note 3**).

6. When developing transgenic experiments to define regulatory elements, you will undoubtedly wish to make deletions of your basic construct. It is useful to include suitable restriction sites that will allow the generation of 5'- and 3'-nested deletions using exonuclease III *(14)*. At later stages of the analysis, you may wish to introduce point mutations into specific *cis*-regulatory elements. It is a good idea to utilize a vector that will allow the employment of site-directed mutagenesis without the need to subclone specific fragments *(15)*.

### 3.1.2. Intact Gene Constructs

The usual starting point for investigations into gene regulation requires the structural organization of the transgene to be as close as possible to that of the endogenous gene, thus allowing the context of regulatory elements to remain unchanged. For this reason, an intact gene construct is normally used in which the reporter is inserted, in frame, into the coding sequence of the gene leading to the production of a fusion protein. Alternatively, it is possible to create a bicistronic transgene, whereby translational initiation of the reporter is mediated by a viral internal ribosome entry site (IRES) sequence (*see* **Note 4**). It is desirable to insert the reporter either at, or as close as possible to, the translational initiation codon (ATG). This is particularly important if your gene product has known biological activity, since overexpression of the fusion protein may interfere with the regulation of your transgene or have more widespread biological effects in the transgenic organism. Such constructs should contain:

1. 5'-Flanking sequences, including 5'-untranslated regions (5'-UTR), and should at least encompass the minimal elements capable of promoter activity (usually located proximally to the transcription start site) enabling the production of mRNA with a functional translational initiation codon (AUG). More distal 5'-sequences (sometimes many kilobases upstream) may also contain important regulatory regions.
2. 3'-Sequences containing elements that specify correct transcriptional termination and the addition of poly(A) to the 3'-end of the mRNA (*see* **Note 5**). The 3'-untranslated regions (3'-UTR) of certain genes include sequences that confer rapid degradation of the encoded mRNA and may have to be removed to obtain satisfactory levels of transgene activity (*see* **Note 6**). 3'-Transcriptional regulatory regions may also be present.
3. Intragenic regions: Introns have been shown to increase the overall levels of transgene activity *(16,17)* and, in many cases, contain transcriptional regulatory elements. However, there is no absolute requirement for splicing for the production of a functional reporter gene product. The intronic sequences of some genes are not required for appropriate transcriptional regulation and may, in some cases, be omitted. For example, in the case of the *myogenin* gene (*see* **Fig. 1**), only upstream flanking sequences appear to be necessary for appropriate transcriptional regulation *(6)*.

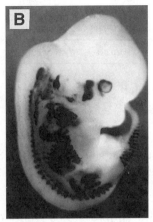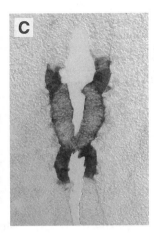

Fig. 1. **(A,B)** Dynamic progression of *myogenin* gene expression during the development of the skeletal musculature. The embryos shown, stained with X-gal for β-galactosidase activity, are from a transgenic line carrying a nuclear localized *lacZ* reporter under the control of 1.1 kb of myogenin 5'-flanking sequence (Ashby and P. W. J. R., unpublished data; *see* **ref. 6**). At 11.5 dpc (A) intense bars of staining can be seen within the myotomal component of the somitic mesoderm. By 13.5 dpc (B), specific muscle blocks are clearly distinct. **(C)** Use of dual reporter transgenes within the same embryo. A coronal section through the hindbrain of a 9.5-dpc embryo is shown (anterior is uppermost). Blue staining resulting from β-galactosidase activity is derived from expression of the *lacZ* gene under the control of a *Hoxb*-2 enhancer, which is active in rhombomeres 3 and 5. A second construct harbors the alkaline phosphatase gene under the control of an enhancer from *Hoxb*-2, which directs expression in rhombomere 4 (revealed as brown staining). Photographs were kindly provided by N. Collins and B. Singh (myogenin) and J. Sharpe (*Hoxb*-2), Division of Developmental Neurobiology, N.I.M.R, Mill Hill. (*See* color plate 9 appearing after p. 368.)

### 3.1.3. Basic Promoter Constructs

1. Once specific regulatory regions have been defined, they may then be tested individually on a simpler, basic promoter construct. Sequences can be tested in both 5'- and 3'-contexts and in both orientations. This will provide information on what subset(s) of the endogenous gene's expression pattern is specifiable in the absence of other elements and test for enhancer function. The basic structure of this type of construct consists of a minimal promoter (*see* **Note 7**) linked to a reporter gene, e.g., *lacZ*/SV-40p(A), to which putative regulatory sequences can be apposed. The most widely used promoters for this purpose are listed in **Table 1**.

2. Negative regulatory regions can be tested on a construct that can direct ubiquitous or widespread expression to see if they are capable of restricting this pattern. Various heterologous promoters have been identified that may be suitable for

**Table 1**
**Some Heterologous Transgene Promoters**

| Name | Source | Size, bp | Notes | Refs. |
|------|--------|----------|-------|-------|
| Minimal promoters | | | | |
| β-Globin[a] | Human β-globin gene (–37 to +21) | 48 | Weak; contains TATA box and initiator element; clean | *6,18* |
| Hsp68[a] | Mouse heat-shock gene (*hsp70.1*) | 330 | Strong; no activity in embryo in absense of enhancer elements[b] | *5,19,20* |
| Hoxb-4 | Mouse *Hoxb-4* gene | 1200 | Contains two independent initiators; expression only in the superior coliculi in the midbrain | *5,21* |
| HSV-TK | Herpes simplex virus thymidine kinase gene (–105 to +51) | 155 | Ectopic expression only | *22–25* |
| Ubiquitous promoters | | | | |
| β-actin | Chicken β-actin gene | 340 | Ubiquitous; larger, intron containing, human promoter gives stronger expression | *1,26,27* |
| HMGCR | Mouse hydroxy methylglutaryl CoA reductase gene | 5500 | Ubiquitous embryonic expression; contains untranslated exon and intron | *28,29* |
| Hoxa-7 | Mouse *Hoxa-7* gene | 1590 | Uniform expression throughout mid-gestation embryo | *30* |

[a]*See* **Note 38**.
[b]*See* **Note 39**.

this purpose (*see* **Table 1**). It is probably more desirable though to test such regions on a construct containing the homologous promoter.

**Table 2**
**Reporter Genes and Substrates**

| Reporter | Cellular localization | Form | Suitability | Substrates | Refs. |
|---|---|---|---|---|---|
| β-Galactosidase *(lacZ)* | Cytoplasm | Native form | All genes | | *31,49,50* |
| | Nucleus | SV-40 T-Ag nuclear localization signal fused to *lacZ* | All genes | X-gal (light blue or Bluo-gal (dark blue) | |
| | Cytoplasmic membrane | Synthetic trans-membrane domain fused to *lacZ* | Transmem-brane or secreted | | |
| | Axonally transported | Tau (microtubule associated protein) fused to *lacZ* | Axonally expressed | | |
| P *(PALP-1)* | Cytoplasmic membrane (extracellular) | Native form | All genes and double staining with *lacZ* | BCIP/NBT (blue-purple), BM purple (deep purple), NATP/fast red RC (red), or NAGP/fast blue BN (green) | *32–34* |

### 3.1.4. Choice of Reporter Gene

1. Various reporter systems are available, the choice of which depends largely on the particular experimental aims and the cell types in which the construct is expressed (*see* **Table 2**). *LacZ* reporters can be targeted to the cytoplasm, the nucleus, and the cytoskeleton, or anchored to the cell membrane *(31)*. Two chromogenic substrates are available for use with *lacZ* reporters, X-gal, and Bluo-gal (*see* **Subheading 3.4.1.**).

2. Although *lacZ* is often the reporter gene of choice, another option is the AP system, which utilizes the human placental alkaline phosphatase gene (*PALP-I: 32–34*). *PALP-I* has a number of attributes that make it useful as a transgenic reporter: The cDNA is small (approx 2 kb as opposed to 3.2 kb for *lacZ*) and is readily expressed in mammalian cells. In addition, the human AP protein is 100-fold more resistant to heat treatment than that of the mouse, and there are several inhibitors of mouse AP activity that do not affect the human enzyme *(33)*. A range of substrates is available for AP that give different colored products (*see* **Table 2**). Its main advan-

tage is that it may be used in conjunction with *lacZ* to produce "double transgenics," where the expression of two reporter genes can be compared in the same embryo (*see* **Note 8** and **Fig. 1**). Another potentially useful application is in the study of the transcriptional regulation of two linked genes. In this case, *lacZ* can then be used to monitor the expression of one gene and *PALP-1* that of the other, allowing the analysis of elements that may act on one or both genes.

3. The green fluorescent protein (GFP) from the jellyfish *Aequorea victoria* has received widespread interest as a reporter owing to its potential for the direct real-time imaging of gene expression and protein localization in vivo *(35–37)*. GFP has been successfully employed as a reporter within a range of organisms *(38,39)*, although its application to mammalian species has been hindered by problems with both levels of protein expression and intensity of fluorescence. These problems may, in part, be due to thermal effects and the requirement for molecular oxygen for the formation of the GFP chromophore *(40,41)*. The continued development of variant GFPs, which have altered spectral properties and improved transcriptional, translational, and posttranslational characteristics, lends great hope for the success of this versatile marker gene in mammalian systems *(41–48)*. A useful forum for the discussion of GFP and related topics is the Fluorescent Proteins Newsgroup, available via USENET at bionet.molbio. proteins.fluorescent or via the World Wide Web at http://www.bio.net/hypermail/ FLUORESCENT-PROTEINS/.

## 3.2. Testing Functionality

Having designed and built a transgenic reporter gene, it is extremely prudent to test whether it is capable of acting as a functional expression unit prior to microinjection. This is easily achieved by transfection of the intact plasmid into cultured cells. Most cell lines are suitable for this purpose, although there may be instances where your particular reporter gene is not expressed in a certain cell line (*see* **Note 9**). A standard, inexpensive, transfection method using calcium phosphate is described *(51)*, although other methods, such as those utilizing liposome formulations, are equally suitable.

1. Harvest cells by trypsinization, and reseed at a density of $1–3 \times 10^5/cm^2$ in 85-mm diameter tissue-culture dishes. Incubate for 4–16 h at 37°C in a $CO_2$ incubator to allow the cells to adhere to the dish.

2. Prepare DNA for each point in a sterile microfuge tube containing; 10 μg of DNA, 95 μL of 2 $M$ $CaCl_2$, and sterile distilled water up to a total volume of 750 μL (*see* **Note 10**).

3. Slowly add each DNA sample dropwise to a tube (e.g., Falcon 2059) containing 750 μL of 2X HBS, with continuous mixing by gentle vortexing. Incubate the samples at room temperature for 30 min to allow a DNA–calcium phosphate coprecipitate to form.

4. Add fresh cell-culture medium, and then overlay the precipitate onto the cells, swirling gently to ensure even distribution. Incubate cells for 20 h at 37°C in a $CO_2$ incubator.

5. Remove the medium, and wash the cells twice with warm (37°C) PBS. Add fresh medium, and reincubate the cells for 24 h.
6. Rinse the cell monolayer twice with wash solution, and incubate at room temperature for 5 min with 8 mL of cell fixative. Wash three times for 5 min at room temperature with wash solution.
7. For *lacZ* reporters, add 8 mL of X-gal stain solution, and incubate at 37°C in the dark for 1 h to overnight (*see* **Note 11**). Wash cells to remove stain solution, and store at 4°C in 50% (v/v) ethanol in PBS.
8. For alkaline phosphatase reporters, heat-treat the cells for 45 min at 65°C in PBS to reduce preferentially the activity of endogenous alkaline phosphatases. Add 8 mL of AP stain solution, and incubate at room temperature in the dark for 1 h to overnight. Wash cells with PBS containing 20 m$M$ EDTA to stop the staining reaction, and store at 4°C in 50% (v/v) ethanol in PBS.

## 3.3. Isolation and Purification of DNA Fragments for Microinjection

The purification of DNA fragments for microinjection is extremely important. Linear DNA fragments integrate much more efficiently than circular forms, and prokaryotic vector sequences can reduce the efficiency of transgene expression. Contamination of the DNA with impurities (e.g., phenol, ethanol, or detergents) can adversely affect the survival of injected eggs, and particulate matter can block needles during microinjection (*see* **Note 12**).

A simple purification method is described here for isolating DNA fragments from agarose gels using β-Agarase I (New England Biolabs), although GELase (Epicenter Technologies, Cambridge, Cambs, UK) is equally suitable.

1. Excise the reporter construct to be microinjected from the vector by digestion with the appropriate restriction enzyme(s) according to the conditions recommended by the supplier.
2. Fractionate 5–20 μg of the digested DNA by agarose-gel electrophoresis using 1X TBE buffer containing 0.5 μg/mL ethidium bromide (*see* **Note 13**).
3. Wearing appropriate protective wear, view the gel on a long-wave transilluminator, and using a scalpel blade, cut a large well in front of the fragment band to be purified. Place the gel at 4°C, and fill the well with molten (45–50°C) LMP agarose. Allow to solidify, and then return the gel to the electrophoresis tank until the band of interest has migrated into the LMP agarose. Using a clean scalpel blade excise the band, and place in a clean microcentrifuge tube (*see* **Note 14**).
4. Incubate the sample at 65°C until molten (approx 10 min) After estimating the sample volume, add 1/10 vol of 10X Agarase buffer, mix, and equilibrate at 40°C.
5. Add 2 U of β-Agarase I/200 μL of 1% (w/v) LMP agarose. Mix well and incubate for 1 h at 40°C (*see* **Notes 15** and **16**).
6. Extract the sample twice with phenol, twice with chloroform, and once with diethyl ether, each time retaining the aqueous phase (*see* **Notes 17** and **18**).
7. Transfer the aqueous phase to a fresh tube, and precipitate with 0.6–1X vol of propan-2-ol. Pellet the DNA by centrifugation at 15,000$g$ for 20 min. Wash the pellet thoroughly with 70% (v/v) ethanol, and allow to dry at room temperature.

8. Dissolve the DNA pellet in microinjection buffer, and pass through a 0.22-μm filter (Spinex, Costar, High Wycombe, Bucks, UK). Store frozen at –20°C.

9. The concentration and integrity of your purified DNA fragment can be accurately judged by agarose gel-ethidium bromide electrophoresis against samples of known amount (*see* **Note 19**).

10. Prior to microinjection, DNA is diluted to a concentration of 1–5 μg/mL with filtered microinjection buffer (*see* **Note 20**).

## 3.4. Whole-Mount Staining for Reporter Gene Activity

Staining protocols for both *lacZ* and AP reporter gene activity are presented below. For embryonic stages up to 13.5 days postcoitum (dpc), embryos may be stained intact. After this age, problems arise due to the developing skin, which acts as a barrier to fixation and staining. Older embryos may be partially dissected or processed as frozen sections to overcome this *(7)*.

1. Dissect embryos out of uteri and decidua and free of extraembryonic membranes in PBS. Retain the placentas for transgenic identification if necessary (*see* **Subheading 3.5.**). Rinse the embryos with PBS to remove serum (*see* **Notes 21** and **22**).

2. Transfer the embryos into fixative (*see* **Notes 23** and **24**). Fixation time depends on the size of the embryo. Ten to 15 min are sufficient for embryos up to 8.5 dpc, 15–30 min for 8.5–11.5 dpc and up to 90 min for 14.5 dpc embryos.

3. Wash the fixed embryos twice in a large volume of embryo wash for 20 min each at room temperature. Proceed with the appropriate staining method as described in **Subheadings 3.4.1.** and **3.4.2.**

### 3.4.1. β-Galactosidase Staining

Two possible substrates are available for β-galactosidase staining (*see* **Table 2**). 5-bromo-4-chloro-3-indolyl-β-D-galactopyranoside (X-gal) was the first chromogenic substrate available for the detection of ß-galactosidase activity and is still the most widely used. 5-bromoindolyl-β-*O*-galactopyranoside (Bluo-gal) is a more recent development and has some merits over X-gal for certain applications. The advantages of X-gal are that it gives a light-blue reaction product that is more photogenic than the dark-blue of Bluo-gal in whole-mount embryos, and it is also cheaper. It does have some drawbacks, however, particularly in the prolonged staining of embryos older than 11.5 dpc, where levels of endogenous β-galactosidase activity increase and may lead to a greenish-blue background staining (*see* **Note 25**). This problem does not seem to arise when staining embryos with Bluo-gal, and it is particularly useful when dealing with weak expression at these stages. The other advantages of Bluo-gal over X-gal are in the histological examination of stained tissue sections where greater sensitivity and morphological resolution can be obtained with the use of hematoxylin and eosin (H&E) counterstaining methods *(49)*.

1. X-gal staining: Incubate the fixed and washed embryos in X-gal stain solution in the dark. Staining conditions depend on the levels of β-galactosidase activity and on the size of the samples (*see* **Note 26**). Maximal staining is usually seen after 16–24 h. Embryos of ages up to 10.5 dpc can be stained at 37°C, whereas older embryos should be stained at lower temperatures (30°C or room temperature).
2. Bluo-gal staining: Incubate in Bluo-gal stain solution. The conditions are as for X-gal staining. However, higher temperatures can be used with older embryos owing to the reduced problem of background staining.

### 3.4.2. AP Staining

Several possible chromogenic substrates producing a range of colors are available for AP detection (**Table 2**). BCIP/NBT (e.g., Boehringer, Sigma) is the most commonly used substrate, producing a blue-purple color, and is suitable for most applications. BM Purple (Boehringer), which gives a deep purple color, is more expensive than BCIP/NBT but has a greater sensitivity and produces less background staining. The purple colors of these substrates may not contrast well with blue when also staining for β-galactosidase activity. In this case, red (Napthol-AS-TR-phosphate (NATP)/Fast Red RC) or green (Napthol-AS-GR-phosphate/Fast Blue BN) color producing substrates may prove useful (Boehringer, Sigma). For double staining proceedures the β-galactosidase step should be performed first before proceeding with the alkaline phosphatase protocol, as heat treatment will inactivate the β-galactosidase enzyme *(20)*.

1. Incubate the fixed, washed embryos at 65°C in PBS for 10–45 min to preferentially reduce endogenous AP activity (*see* **Note 27**).
2. Stain the embryos in AP stain solution at room temperature in the dark.
3. Terminate the staining reaction by immersing the embryos in PBS containing 20 m$M$ EDTA.

### 3.4.3. Poststaining Treatment of Samples

1. Wash the embryos for 10 min in two changes of PBS to remove any stain solution.
2. If samples are to be sectioned, postfix in Mirsky's or 4% (w/v) paraformaldehyde prior to embedding.
3. Store at 4°C in 70% (v/v) ethanol after dehydration through a series of 30% (v/v) and 50% (v/v) ethanol (*see* **Note 28**).

## 3.5. Transgenic Diagnosis by PCR

Presented here is a simple and quick method for the identification of transgenic animals or embryos by PCR from small tissue samples. The starting sample can be a small (0.5 cm) tail biopsy for animals or a portion of extraembryonic tissue (placenta or extraembryonic membranes) for embryos (*see* **Note 29**).

1. Boil the tissue sample to be analyzed for 5 min in 0.5 mL of proteinase K digestion buffer (*see* **Note 30**).
2. Add 5 μL of proteinase K solution (final concentration 0.1 mg/mL), mix well by inversion, and digest the tissue samples overnight at 55°C.
3. Vortex the samples briefly to disperse any clumps of partially digested tissue. Reboil the tissue sample for 5 min (*see* **Note 30**).
4. Spin at 15,000*g* for 20 min (4°C) to pellet any cellular debris. Use 1 μL as a template for subsequent PCR diagnosis.
5. Set up a 20-μL PCR reaction containing; 1X PCR buffer, 1X dNTP mix, 1X MgCl$_2$, 1X transgene-specific primers (0.2 μ*M* each), control primers (0.2 μ*M* each), template, and 1.5 U of *Taq* polymerase. Overlay with mineral oil, and amplify using suitable conditions in a thermal cycler. It is useful to include both positive and negative controls (*see* **Notes 31** and **32**).
6. Run out the reaction products on an agarose-ethidium bromide gel. Positive samples should show the presence of both the transgene-specific and control bands.

## 3.6. Analysis of Reporter Gene Expression

Owing to the large amount of work required for the production and analysis of transgenic mice, it is important to employ careful planning in the design and execution of experiments. The rate-limiting step in the process is the production of transgenics, so any information that can reduce the number of constructs required can greatly speed up the operation (*see* **Notes 33** and **34**).

Once a baseline reporter has been defined, it is extremely important to establish clearly its characteristics of expression throughout various developmental stages. This will enable subsequent comparison with mutant constructs and aid the definition of specific regulatory elements. Because of the enormous variety of results that can be obtained from such work, it is impossible to advise a standard schedule for priorities. It is advisable though to be very thorough with the analysis and to review continually the situation as new data arrive (*see* **Note 35**).

Once regulatory elements have been localized to relatively small regions (e.g., a few hundred basepairs), it is often better to employ in vitro methods (e.g., electrophoretic mobility shift assays or footprinting) and sequence analysis for the identification of specific potential regulatory elements. This may be much more rapid than continued mutational transgenic analysis. When individual transcription factor binding sites have been identified, the effect of mutations in them can be further tested in the transgenic context.

### 3.6.1. Transient vs Line Analysis

Many benefits can be drawn from directly analyzing founder transgenic embryos (transients) rather than breeding them through to the next generation and producing a transgenic line. The disadvantage is that each embryo represents only a snapshot of a particular expression pattern at a particular developmental time-point. However, it does allow the rapid analysis of a large number of transgenic constructs without the need for a time-, space- and labor-intensive breeding program. Transient analysis is therefore recommended for an initial assessment of a large number of constructs and line analysis for more specific, in-depth analyses.

### 3.6.2. Position Effects and Ectopic Expression

The position of integration of a transgene in the host genome is considered to be random, and it can greatly affect both the levels and pattern of its expression (termed position effects). A certain proportion of transgenic animals will not express the reporter gene either as a result of rearrangements during integration or because integration has occurred in a region of inactive chromatin. Position effects are often more evident with smaller or mutated transgenic constructs where the reporter is presumably more exposed to interference from surrounding regulatory elements. It should also be noted that there are reported instances where β-galactosidase activity does not reliably report levels of transcription (*see* **Note 36**).

Two forms of ectopic expression from a transgenic reporter are possible: Inconsistent ectopic expression owing to position effects is discernible by the comparison of a number of transgenic animals carrying the same construct. Consistent ectopic expression owing to regulatory interactions within the reporter gene *per se* is less clear-cut, since they may only be apparent in a proportion of the analyzed samples (if dependent on expression levels, and so forth). Another source of consistent, but apparently ectopic expression is an unavoidable drawback of reporter genes, and is a function of the reporter gene products (mRNA and protein) and their stability within the cellular environment (*see* **Note 37**).

## 4. Notes

1. Linear DNA has been shown to integrate into the germline more efficiently than circular forms *(52)*. Prokaryotic vector sequences have been shown to reduce drastically the expression of eukaryotic transgenes *(53)*. It is therefore necessary to remove as much vector sequence as possible from the transgene prior to microinjection. We have not experienced problems with expression from transgenes containing vector polylinker sequences, but these should be removed as far as possible.
2. Many modern pUC-derived vectors, such as the pBluescript (Stratagene, Cambridge, Cambs, UK) or pGEM (Promega, Southampton, Hants, UK) series, contain the *lacZα* region, allowing the use of α-complementation for blue/white screening during cloning procedures.
3. Problems of recombination may be experienced particularly when working with DNA fragments that contain repetitive sequences or have an irregular structure.

Characteristically, cloning steps will result in a very high proportion of recombinant clones that are truncated or rearranged. Initially, it may be worthwhile simply to alter the bacterial host. Strains that carry the *recA* mutation, for example, DH1, DH5α, or XL1-Blue, are deficient for homologous recombination. Certain strains, such as SURE® cells (Stratagene), carry additional mutations (e.g., *recB*, *recJ*, and *sbcC*) that interrupt specific pathways for the repair of DNA that is capable of forming irregular structures, such as inverted repeats and Z-DNA, or undergoes plasmid-by-plasmid recombination. An alternative approach that we have found to be very successful is to employ a lower copy number vector. Plasmids may vary considerably in their copy number, and this is partly dependent on the origin of replication that they contain. Potentially, higher copy number vectors may be more prone to interplasmid recombination, and lowering the number of copies per cell may effectively stabilize an otherwise unstable plasmid. The majority of modern pUC-derived cloning vectors (e.g., pBluescript or pGEM) carry derivatives of the pMB1/ColE1 origin of replication and are maintained at copy numbers between 500 and 700/cell. pBR322 and its derivatives, for example, pPolyIII *(13)*, also carry the pMB1 replicon, but have a much lower copy number (15–20/cell). Plasmids that are based on the pSC101 origin have only 1–5 copies/cell and may be the most suitable, although DNA yields are necessarily much lower and these plasmids cannot be amplified using inhibitors of protein synthesis *(9)*. Alternatively, BAC or PAC vectors are thought to be extremely reliable for the cloning and propagation of unstable inserts *(11,12)*.

4.  The IRES sequence, derived from the picornaviridae family of viruses, which includes encephalomyocarditis virus (ECMV), foot and mouth disease virus (FMDV), and poliovirus, is able to initiate direct ribosome binding and protein translation *(54)*. This is an alternative mechanism to the classical model of translation, which involves ribosome entry at the 5'-cap structure of a mRNA molecule *(55)*. Potentially, an IRES-linked reporter gene can be introduced at any location within the transcribed region of the gene of interest, leading to the translation of both open reading frames from one mRNA *(56)*.

5.  There are two elements important for polyadenylation: a hexanucleotide poly(A) signal (AAUAAA) located 10–30 nucleotides 5' of the poly(A) site and 3' U-rich sequences. If these sequences are not present in the reporter construct or have been removed during deletional analysis, an artificial one will have to be added. Most commonly used are the transcriptional termination and polyadenylation sequences from the SV-40 early region, SV-40p(A), (*see* **ref. 5**). It must be noted that heterologous poly(A) sequences are not always advantageous. In the case of *Hoxb-4*, stronger transgene expression is seen in the presence of its own transcriptional termination sequences than with those of SV-40 (Morrison and Krumlauf, personal communication).

6.  Certain genes have high rates of mRNA turnover, for example those encoding gene products required rapidly, but for a short time, which may lead to very weak levels of reporter gene activity. The rapid turnover of mRNA is generally conferred by AU-rich elements (AREs) consisting of repeated AUUUA sequences

and located in the 3'-UTR of the gene (*57,58* and references therein). We have found that it is possible to increase the activity levels from constructs containing AREs by the addition of the SV-40p(A) sequence to the 3'-end of the *lacZ* gene (Summerbell and P. W. J. R., personal communication).

7. To simplify analyses using constructs of this type, the promoter used should be either inactive or unable to direct any specific expression in the absence of regulatory regions.

8. In this case the staining protocol will need to be optimized for the two reporters. Since AP is a membrane-associated protein, staining for AP activity will tend to mask any intracellular staining for β-galactosidase. The most suitable *lacZ* reporter in this case is probably the nuclear localized form. Another option is to use immunohistochemical detection methods to distinguish better between the activities of AP and of β-galactosidase, either in whole mount or sectioned samples.

9. We generally use Neuro 2a cells (a mouse neuroblastoma cell line), which we have found to have a high efficiency of transfection and to have worked with a range of reporters.

10. Suitable negative and positive control plasmids can be included, e.g., the basic construct without the reporter gene and pSV-βgal (Promega), respectively.

11. Depending on the efficiency of transfection, between 1 and 15% of the cells should exhibit a high level of reporter gene staining, usually visible within 1 h. Longer periods may be required for weakly expressing constructs.

12. Use clean, dust-free labwear throughout, preferably rinsed with filtered water. Filter solutions through a 0.22-μm filter to prevent the addition of particulate matter to the DNA sample. Wear powder-free gloves or rinse powder-coated gloves with water prior to handling samples.

13. It is important to start with a large amount so you can afford to lose some on the way. The more concentrated the final DNA preparation is, the more any impurities will be diluted at the microinjection stage.

14. By using a slightly higher percentage of LMP agarose, the migrating band can be tightened, reducing the sample volume for subsequent Agarase digestion. Make sure the fragment has migrated well into the LMP Agarose to avoid the carryover of normal agarose, which will not be digested by β-Agarase. Ensure sample is fully molten. Allow sufficient time for temperature equilibration.

15. This is the minimum time required for complete Agarase digestion by this method, although samples may be incubated for extended periods (e.g., overnight) without degradation of the DNA.

16. An optional step may be included to remove any undigested carbohydrates. Adjust the salt concentration of the sample for alcohol precipitation by the addition of sodium acetate to a final concentration of 0.3 *M*. Chill on ice for 10–15 min and centrifuge at 15,000*g* for 15 min to pellet any undigested carbohydrates. This is unnecessary if Agarase digestion is complete.

17. Perform the extractions in a suitable volume, taking care to leave all of the interface and some aqueous phase behind each time. Centrifugation at 15,000*g* for 5 min is sufficient to ensure adequate phase separation. Diethyl ether has a lower

density than that of water and partitions as the upper phase. Residual ether can be evaporated by leaving the uncapped sample in a fume hood for 5–10 min.

18. Phenol is corrosive and can cause severe burns. Chloroform is an irritant and a carcinogen. Work with these chemicals should be performed in a fume hood with appropriate protection. Irrigate areas of the skin that come into contact with phenol with a large volume of water or PEG-300. Wash with soap and water. Seek medical advice if necessary.

19. We routinely run serial dilutions of our samples against known amounts of digested λ DNA (*Hin*dIII, *Bst*EII, or *Eco*RI/*Hin*dIII) run in several control lanes. For example, 485 ng of λ-*Hin*dIII cut DNA will yield markers ranging from 23 to 1.25 ng. It is advisable to compare the fluorescence of the transgene fragment with other DNA fragments of similar size, since the relationship between these two parameters is not strictly linear. For this reason, we also run standards consisting of similar-sized microinjection fragments that have yielded good frequencies of transgenics in previous experiments.

20. Excessively high concentrations of DNA are detrimental to the embryo and reduce survival rates. We routinely use DNA at a concentration of 1.5–2 μg/mL.

21. Glass liquid-scintillation vials are good containers for bulk processing of embryos. When you wish to keep individual embryos separate (e.g., for subsequent transgenic diagnosis), plastic multiwell plates (24-well, Sigma-Aldrich) can be used.

22. Plastic, disposable Pasteur pipets (e.g., Pastettes, Alpha Laboratories Ltd., Eastleigh, Hants, UK) are useful for the handling of embryos and can be cut to the required diameter.

23. Use a large enough volume of fixative to avoid significant dilution by carried over PBS. Alternatively replace the fixative once the embryos have been transferred.

24. The type of fixative used depends on personal choice and subsequent histological procedures. Many researchers use a solution containing 1% (w/v) formaldehyde, 0.2% (w/v) glutaraldehyde *(5)*, or 4% (w/v) paraformaldehyde. We prefer to use Mirsky's Fixative (National Diagnostics), since it is simple to make up (from a 10X concentrate and 10X buffer), has a long shelf life (1 mo at 4°C), and is relatively nontoxic. It is a mild fixative, leaving the embryos soft and easy to section.

25. By performing the β-galactosidase assay at the optimal pH for the bacterial enzyme (pH 7.3), there should be relatively little activity of the endogenous mammalian lysosomal β-galactosidase (pH optimum between 3.0 and 6.0). However, during prolonged incubations, Tris-HCl (pH 7.3) should be added to the stain solution to stabilize the pH of the phosphate buffer. The effects of endogenous β-galactosidase activity can also be reduced by performing the staining reaction at lower temperatures.

26. For preliminary investigations where the expression patterns of transgenes are to be compared, staining conditions (i.e., time and temperature) should be kept as constant as possible. Prolonged staining, especially at 37°C, will lead to a gradual degeneration of sample tissue and may cause a loss in morphological resolution with sectioned samples. In this case, it is worthwhile to keep the staining conditions as conservative as possible.

27. For AP and β-galactosidase/AP double staining, conditions should ideally be optimized for the particular construct and developmental stage. Staining/heat-inactivation times and the addition of AP inhibitors, e.g., 1 m$M$ levamisole (L[–]-2,3,5,6,-tetrahydro-6-phenylimidazo[2,1-b]thiazole, Sigma) may be tested to reduce the level of endogenous alkaline phosphatase activity present in the embryo *(59)*.

28. Use tightly capped storage tubes to prevent ethanol loss by evaporation. During long-term storage, change the solution periodically.

29. Care is required throughout this procedure to avoid sample contamination with plasmid DNA or PCR products. Preferably perform the method in a separate, clean environment. Keep all solutions separate from plasmid manipulations and store in aliquots. Use separate pipets for manipulations and/or use filtertips (e.g., FilterPro, Rainin Instrument Co., Luton, Beds, UK) to reduce the risk of contaminating pipet barrels.

30. We have found that the two boiling steps prior to and after digestion significantly increase both the sensitivity and quality of the PCR reaction. This is most conveniently achieved by placing the samples in a water bath at 60–70°C, which is then turned up to boiling for 5 min. The water bath may then be turned down to 55°C, and the samples allowed to cool prior to the addition of proteinase K.

31. It is useful to make up a large volume of 1X diagnostic PCR mix containing buffer, $MgCl_2$, dNTPs, primers, and water. This can be stored at –20°C in aliquots of 20–100 reaction volumes to which only *Taq* polymerase and template need to be added.

32. For *lacZ*/SV-40 transgenes, we routinely use the primers LZ3 (5'-GCGACTTCCAGTTCAACATC) complementary to the minus strand of *lacZ* and STB (5'-GATGAGTTTGGACAAACCAC) complementary to the plus strand of the SV-40 poly(A). As a qualitative internal control, we use MGP1 (5'-CCAAGTTGGTGTCAAAAGCC) and MGP2 (5'-CTCTCTGCTTTAAGGAGT CAG), specific to the endogenous *myogenin* gene *(6)*. These primer pairs amplify fragments of 590 and 172 bp, respectively. Conditions used for the reaction are 94°C/3 min, 94°C/30 s, 55°C/30 s, 72°C/30 s for 26 cycles at a final $Mg^{2+}$ concentration of 1.5 m$M$.

33. What is known about the transcriptional regulation of the gene of interest? Have in vitro studies identified regulatory elements that should be included or eliminated from the transgene or might be involved in particular aspects of expression? Can sequence comparisons pinpoint any areas of conservation within noncoding regions that may have regulatory functions?

34. The TRANSFAC transcription factor database *(60)* is a useful reference source of transcription factors and their associated DNA target sequences. The database is available with an online search facility via the internet (http://transfac.gbf-braunschweig.de/).

35. The expression patterns obtained from any given construct can be extremely variable, particularly when working with small fragments of DNA out of their normal genomic context. For this reason, it is important to compare a number of transgenic animals (ideally six or more) for a particular construct, so that variabilities and consistencies in the expression pattern can be assimilated.

36. For example, denervation of muscles in adult mice carrying a *lacZ* reporter driven
    by regulatory sequences from the acetylycholine receptor α-subunit leads to the
    transcriptional activation of the transgene, but to a decrease in β–galactosidase
    staining probabaly because of an increase in proteolytic activity *(61)*. This
    discrepency is not seen when the gene encoding chloramphenicol
    acetyltransferase is used in an essentially identical transgene construct.
37. *LacZ* mRNA and protein may persist for 3–4 d in transgenic animals. Therefore, a
    reporter gene may continue to show activity in cell lineages long after the reporter
    and endogenous gene have been switched off in them *(62)*. This can be particularly
    problematic at early stages of development when rapid growth is most evident.
38. We have had the greatest experience with the β-globin and hsp68 promoter sys-
    tems. The advantages of β-globin over hsp68 are that the minimal promoter is
    small, simple, and well defined. Containing both TATA box and initiator ele-
    ments, it is able to work with a wide range of enhancers to generate faithful pat-
    terns of reporter gene expression. One disadvantage of the β-globin promoter is
    that compared to hsp68, it is relatively feeble, and with weak enhancers, it may
    give a low percentage of expressing transgenic offspring.
39. Several researchers, including ourselves, have experienced a pattern of consis-
    tent, but ectopic expression in the ventral neural tube of transgenic embryos with
    hsp68 promoter constructs. This has been attributed to an element in the hsp68
    promoter that is able to interact with enhancer elements juxtaposed to it *(63–65)*,
    and at some frequency, we have seen this pattern of expression with many differ-
    ent enhancer constructs. It is particularly useful as a control for effective
    transgenesis in situations in which essential transcriptional regulatory elements
    in the enhancer under the test have been inactivated by mutation.

## References

1. Balling, R., Mutter, G., Gruss, P., and Kessel, M. (1989) Craniofacial abnormalities induced by ectopic expression of the homeobox gene *Hox-1.1* in transgenic
   mice. *Cell* **58,** 337–347.
2. Wolgemuth, D. J., Behringer, R. R., Mostoller, M. P., Brinster, R. L., and Palmiter,
   R. D. (1989) Transgenic mice overexpressing the mouse homeobox-containing
   gene *Hox-1.4* exhibit abnormal gut development. *Nature* **337,** 464–467.
3. Lufkin, T., Mark, M., Hart, C. P., Dollé, P., LeMeur, M., and Chambon, P. (1992)
   Homeotic transformation of the occipital bones of the skull by ectopic expression
   of a homeobox gene. *Nature* **359,** 835–841.
4. Goring, D. R., Rossant, J., Clapoff, S., Breitman, M. L., and Tsui, L.-C. (1987) *In
   situ* detection of β-galactosidase in lenses of transgenic mice with a γ-Crystallin/
   *lacZ* gene. *Science* **235,** 456–458.
5. Whiting, J., Marshall, H., Cook, M., Krumlauf, R., Rigby, P. W. J., Stott, D., and
   Allemann, R. K. (1991) Multiple spatially specific enhancers are required to
   reconstruct the pattern of *Hox-2.6* gene expression. *Genes Dev.* **5,** 2048–2059.
6. Yee, S.-P. and Rigby, P. W. J. (1993) The regulation of *myogenin* gene expression
   during the embryonic development of the mouse. *Genes Dev.* **7,** 1277–1289.

7. Hogan, B., Beddington, R., Costantini, F., and Lacy, L. (eds.) (1995) *Manipulating the Mouse Embryo: A Laboratory Manual* 2nd ed. Cold Spring Harbor Laboratory Press, Cold Spring Harbor, NY.
8. Gordon, J. W., Scangos, G. A., Plotkin, D. J., Barbosa, J. A., and Ruddle, F. H. (1980) Genetic transformation of mouse embryos by microinjection of purified DNA. *Proc. Natl. Acad. Sci. USA* **77,** 7380–7384.
9. Sambrook, J., Fritsch, E. F., and Maniatis, T. (eds.) (1989) *Molecular Cloning: A Laboratory Manual,* 2nd ed. Cold Spring Harbor Laboratory Press, Cold Spring Harbor, NY.
10. Schedl, A., Larin, Z., Montoliu, L., Thies, E., Kelsey, G., Lehrach, H., and Schütz, G. (1993) A method for the generation of YAC transgenic mice by pronuclear microinjection. *Nucleic Acids Res.* **21,** 4783–4787.
11. Shizuya, H., Birren, B., Kim, U. J., Mancino, V., Slepak, T., Tachiiri, Y., and Simon, M. (1992) Cloning and stable maintainance of 300-kilobase-pair fragments of human DNA in *Escherichia coli* using an F-factor-based vector. *Proc. Natl. Acad. Sci. USA* **89,** 8794–8797.
12. Ioannou, P. A., Amemiya, C. T., Garnes, J., Kroisel, P. M., Shizuya, H., Chen, C., Batzer, M. A., and de-Jong, P. J. (1994) A new bacteriophage P1-derived vector for the propagation of large human DNA fragments. *Nat. Genet.* **6,** 84–89.
13. Lathe, R., Vilotte, J. L., and Clarke, A. J. (1987) Plasmid and bacteriophage vectors for the excision of intact inserts. *Gene* **57,** 193–201.
14. Murphy, G. (1993) Generation of a nested set of deletions using exonuclease III, in *Methods in Molecular Biology, vol. 23: DNA Sequencing Protocols* (Griffin, H. G. and Griffin, A. M., eds.), Humana, Totowa, NJ, pp. 51–59.
15. Deng, W. P. and Nickoloff, J. A. (1992) Site-directed mutagenesis of virtually any plasmid by eliminating a unique site. *Anal. Biochem.* **200,** 81–88.
16. Brinster, R. L., Allen, J. M., Behringer, R. R., Gelinas, R. E., and Palmiter, R. D. (1988) Introns increase transcriptional efficiency in transgenic mice. *Proc. Natl. Acad. Sci. USA* **85,** 836–840.
17. Choi, T., Huang, M., Gorman, C., and Jaenisch, R. (1991) A generic intron increases gene expression in transgenic mice. *Mol. Cell. Biol.* **11,** 3070–3074.
18. Lewis, B. A. and Orkin, S. H. (1995) A functional initiator element in the human β-globin promoter. *J. Biol. Chem.* **270,** 28,139–28,144.
19. Gossler, A., Joyner, A. L., Rossant, J., and Skarnes, W. C. (1989) Mouse embryonic stem cells and reporter constructs to detect developmentally regulated genes. *Science* **244,** 463–465.
20. Kothary, R., Clapoff, S., Darling, S., Perry, M. D., Moran, L. A., and Rossant, J. (1989) Inducible expression of an *hsp68-lacZ* hybrid gene in transgenic mice. *Development* **105,** 707–714.
21. Gutman, A., Gilthorpe, J. D., and Rigby, P. W. J. (1994) Multiple positive and negative regulatory elements in the promoter of the mouse homeobox gene *Hoxb-4. Mol. Cell. Biol.* **14,** 8143–8154.
22. Allen, N. D., Cran, D. G., Barton, S. C., Hettle, S., Reik, W., and Surani, M. A. (1988) Transgenes as probes for active chromosomal domains in mouse development. *Nature* **333,** 852–855.

23. Goldhammer, D. J., Brunk, B. P., Faerman, A., King, A., Shani, M., and Emerson, C. P., Jr. (1995) Embryonic activation of the *myoD* gene is regulated by a highly conserved distal control element. *Development* **121,** 637–649.

24. Luckow, B. and Schutz, G. (1987) CAT constructions with multiple unique restriction sites for the functional analysis of eukaryotic promoters and regulatory elements. *Nucleic Acids Res.* **15,** 5490.

25. Schöler, H. R., Balling, R., Hatzopoulos, A. K., Suzuki, N., and Gruss, P. (1989) Octamer binding proteins confer transcriptional activity in early mouse embryo-genesis. *EMBO J.* **8,** 2551–2557.

26. Beddington, R. S. P., Morgernstern, J., Land, H., and Hogan, A. (1989) An *in situ* transgenic enzyme marker for the midgestation mouse embryo and the visualization of inner cell mass clones during early organogenesis. *Development* **106,** 37–46.

27. Bronson, S. K., Plaehn, E. G., Kluckman, K. D., Hagaman, J. R., Maeda, N., and Smithies, O. (1996) Single-copy transgenic mice with chosen-site integration. *Proc. Natl. Acad. Sci. USA* **93,** 9067–9072.

28. Gautier, C., Mehtali, M., and Lathe, R. (1989) A ubiquitous mammalian expression vector, pHMG, based on a housekeeping gene promoter. *Nucleic Acids Res.* **17,** 8389.

29. Mehtali, M., LeMeur, M., and Lathe, R. (1990) The methylation-free status of a housekeeping transgene is lost at high copy number. *Gene* **91,** 179–184.

30. Püschel, A., Balling, R., and Gruss, P. (1991) Separate elements cause lineage restriction and specify boundaries of *Hox-1.1* expression. *Development* **112,** 279–287.

31. Fire, A., Harrison, S. W., and Dixon, D. (1990) A modular set of *lacZ* fusion vectors for studying gene expression in *Caenorhabditis elegans*. *Gene* **93,** 189–198.

32. Kam, W., Clauser, E., Kim, Y. S., Kan, Y. W., and Rutter, W. J. (1985) Cloning, sequencing and chromosomal localization of human term placental alkaline phosphatase cDNA. *Proc. Natl. Acad. Sci. USA* **82,** 8715–8719.

33. Fields-Berry, S. C., Halliday, A. L., and Cepko, C. L. (1992) A recombinant retrovirus encoding alkaline phosphatase confirms clonal boundary assignment in lineage analysis of murine retina. *Proc. Natl. Acad. Sci. USA* **89,** 693–697.

34. Chiu, M. I. and Nathans, J. (1994) Blue cones and cone bipolar cells share transcriptional specificity as determined by expression of human blue pigment-derived transgenes. *J. Neurosci.* **14,** 3426–3436.

35. Chalfie, M., Tu, Y., Euskirchen, G., Ward, W. W., and Prasher, D. C. (1994) Green fluorescent protein as a marker for gene expression. *Science* **263,** 802–805.

36. Kain, S. R., Adams, M., Kondepudi, A., Yang, T. T., Ward, W. W., and Kitts, P. (1995) Green fluorescent protein as a reporter of gene expression and protein localization. *Biotechniques* **19,** 650–655.

37. De Giorgi, F., Brini, M., Bastianutto, C., Marsault, R., Montero, M., Pizzo, P., et al. (1996) Targeting aequorin and green fluorescent protein to intracellular organelles. *Gene* **173,** 113–117.

38. Plautz, J. D., Day, R. N., Dailey, G. M., Welsh, S. B., Hall, J. C., Halpain, S., et al. (1996) Green fluorescent protein and its derivatives as versatile markers for gene expression in living *Drosophila melanogaster*, plant and mammalian cells. *Gene* **173,** 83–87.

39. Amsterdam, A., Lin, S., Moss, L. G., and Hopkins, N. (1996) Requirements for green fluorescent protein detection in transgenic zebrafish embryos. *Gene* **173,** 99–103.

40. Ogawa, H., Inouye, S., Tsuji, F. I., Yasuda, K., and Umesono, K. (1995) Localization, trafficking, and temperature-dependence of the *Aequorea* green fluorescent protein in cultured vertebrate cells. *Proc. Natl. Acad. Sci. USA* **92,** 11,899–11,903.

41. Heim, R., Prasher, D. C., and Tsien, R. Y. (1994) Wavelength mutations and post-translational autoxidation of green fluorescent protein. *Proc. Natl. Acad. Sci. USA* **91,** 12,501–12,504.

42. Delagrave, S., Hawtin, R. E., Silva. C. M., Yang, M. M., and Youvan, D. C. (1995) Red-shifted excitation mutants of the green fluorescent protein. *Biotechnology* **13,** 151–154.

43. Ehrig, T., O'Kane, D. J., and Prendergast, F. G. (1995) Green fluorescent protein mutants with altered fluorescence excitation spectra. *FEBS Lett.* **367,** 163–166.

44. Heim, R., Cubitt, A. B., and Tsien, R. Y. (1995) Improved green fluorescence. *Nature* **373,** 663,664.

45. Cormack, B. P., Valdivia, R. H., and Falkow, S. (1996) FACS-optimized mutants of the green fluorescent protein (GFP). *Gene* **173,** 33–38.

46. Heim, R. and Tsien, R. Y. (1996) Engineering green fluorescent protein for improved brightness, longer wavelengths and fluorescence resonance energy transfer. *Curr. Biol.* **6,** 178–182.

47. Anderson, M. T., Tjioe, I. M., Lorincz, M. C., Parks, D. R., Herzenberg, L. A., Nolan, G. P., and Herzenberg, L. A. (1996) Simultaneous fluorescent-activated cell sorter analysis of two distinct transcriptional elements within a single cell using engineered green fluorescent proteins. *Proc. Natl. Acad. Sci. USA* **93,** 8508–8511.

48. Zolotukhin, S., Potter, M., Hauswirth, W. W., Guy, J., and Muzyczka, N. (1996) A "humanized" green fluorescent protein cDNA adapted for high-level expression in mammalian cells. *J. Virol.* **70,** 4646–4654.

49. Aguzzi, A. and Theuring, F. (1994) Improved *in situ* beta-galactosidase staining for histological analysis of transgenic mice. *Histochemistry* **102,** 477–481.

50. Callahan, C. A. and Thomas, J. B. (1994) Tau-β-galactosidase, an axon-targetted fusion protein. *Proc. Natl. Acad. Sci. USA* **91,** 5972–5976.

51. Gorman, C., Padmanabhan, R., and Howard, B. H. (1983) High efficiency DNA-mediated transformation of primate cells. *Science* **221,** 551–553.

52. Brinster, R. L., Chen, H. Y., Trumbauer, M. E., Yagle, M. K., and Palmiter, R. D. (1985) Factors affecting the efficiency of introducing foreign DNA into mice by microinjecting eggs. *Proc. Natl. Acad. Sci. USA* **82,** 4438–4442.

53. Townes, T. M., Lingrel, J. B., Chen, H. Y., Brinster, R. L., and Palmiter, R. D. (1985) Erythroid specific expression of human beta-globin genes in transgenic mice. *EMBO J.* **4,** 1715–1723.

54. Jackson, R. J., Howell, M. T., and Kaminski, A. (1990) The novel mechanism of initiation of picornavirus RNA translation. *Trends Biochem. Sci.* **15,** 477–483.

55. Kozak, M. (1989) The scanning model for translation: an update. *J. Cell Biol.* **108,** 229–241.

56. Kim, D. G., Kang, H. M., Jang, S. K., and Shin, H. S. (1992) Construction of a bifunctional mRNA in the mouse by using the internal ribosome entry site of encephalomyocarditis virus. *Mol. Cell. Biol.* **12,** 3636–3643.

57. Curatola, A. M., Nadal, M. S., and Schneider, R. J. (1995) Rapid degradation of AU-rich element (ARE) mRNAs is activated by ribosome transit and blocked by secondary structure at any position 5' to the ARE. *Mol. Cell. Biol.* **15,** 6331–6340.

58. Sachs, A. B. (1993) Messenger RNA degradation in eukaryotes. *Cell* **74,** 413–421.

59. Hahnel, A. C., Rappolee, D. A., Millan, J. L., Manes, T., Ziomek, C. A.,Theodosiou, N. G., Werb, Z., Pedersen, R. A., and Schultz, G. A. (1990) Two alkaline phosphatase genes are expressed during early development in the mouse embryo. *Development* **110,** 555–564.

60. Wingender, E. (1994) Recognition of regulatory regions in genomic sequences. *J. Biotechnol.* **35,** 273–280.

61. Gundersen, K., Sanes, J. R., and Merlie, J. P. (1993) Neural regulation of muscle acetylcholine receptor epsilon- and alpha-subunit gene promoters in transgenic mice. *J. Cell Biol.* **123,** 1535–1544.

62. Echelard, Y., Vassileva, G., and McMahon, A. P. (1994) *Cis*-acting regulatory sequences governing *Wnt-1* expression in the developing mouse CNS. *Development* **120,** 2213–2224.

63. Gérard, M., Duboule, D., and Zákány, J. (1993) Structure and activity of regulatory elements involved in the activation of the *Hoxd-11* gene during late gastrulation. *EMBO J.* **12,** 3539–3550.

64. Logan, C., Khoo, W. K., Cado, D., and Joyner, A. L. (1993) Two enhancer regions in the mouse *En-2* locus direct expression to the mid/hindbrain region and mandibular myoblasts. *Development* **117,** 905–916.

65. Song, D.-L, Chalepakis, G., Gruss, P., and Joyner, A. L. (1996) Two Pax-binding sites are required for early embryonic brain expression of an *Engrailed-2* transgene. *Development* **122,** 627–635.

# 11

## Application of *lacZ* Transgenic Mice to Cell Lineage Studies

**Paul A. Trainor, Sheila X. Zhou, Maala Parameswaran, Gabriel A. Quinlan, Monica Gordon, Karin Sturm, and Patrick P. L. Tam**

### 1. Introduction

Cell lineage analyses trace the hierarchy of cell types derived from a progenitor population. Critical to these analyses is the ability to track reliably all or defined subsets of the clonal descendants of the progenitor population. This necessitates marking the cells with a heritable and cell autonomous marker. Transgenes encoding molecules that can be visualized directly *in situ* without compromising cell differentiation, such as the reporter, β-galactosidase encoded by *lacZ,* and chloramphenicol acyltransferase, encoded by the CAT gene are the most widely used.

*lacZ* can be readily detected using a sensitive histochemical assay such that cells in which the β-galactosidase gene is transcriptionally active produce a blue stain in tissue sections or in whole mounts *(1)*. Some lineage studies demand the simultaneous detection of the *lacZ* product and other tissue-specific proteins or transcripts. In this chapter, we discuss the experimental strategies in which the *lacZ* transgene can be utilized in the analysis of cell lineages, and we detail assays for detecting β-galactosidase by X-gal histochemistry or immunological localization of the enzyme in combination with mRNA *in situ* hybridization, immunochemistry, and histochemical procedures, such as alkaline phosphatase staining.

### 1.1. Choosing the Appropriate Transgenic Animals for Lineage Analysis

The expression of the *lacZ* transgene varies according to the nature of the regulatory elements (promoter and enhancer) that drive gene expression, the

From: *Methods in Molecular Biology, Vol. 97: Molecular Embryology: Methods and Protocols*
Edited by: P.T. Sharpe and I. Mason © Humana Press Inc., Totowa, NJ

number of active copies and the chromosomal domain where the transgene is localized. The choice of transgene is dictated by the specific questions to be addressed. There are several examples of transgenes that give ubiquitous non-lineage biased expression. For example, in transgenic mice where the *lacZ* gene is regulated by the hydroxymethylglutaryl coenzyme A (HMG-CoA) reductase promoter, the transgene is expressed in all lineages at high levels during development *(2)*. Similarly the ROSA-β-*geo* transgene produced by site-directed mutagenesis *(3)* and the *lacZ* reporter driven by human β-actin promoter *(4)* also provide ubiquitous tagging of multiple cell lineages. Ubiquitously expressed transgenes provide the most ideal marker for unbiased cell lineage analyses provided that the lineage progenitors or founder cells to be studied can be isolated as a pure cell population.

Integration of the transgene into a specific chromosome may be useful in some cases for tracing cell lineages. Transgenes that are expressed in the same pattern as the neighboring genes at the site of integration will provide a ready marker for any tissue-specific pattern of expression. For example, transgenes that are integrated into the X chromosome may behave like the endogenous X-linked gene. Among the known X-linked transgenes, the activity of the *HMG-nls-lacZ* transgene seems to reflect faithfully the activity of the X chromosome *(5)*.

In female mice that carry the *lacZ* transgene only on one X chromosome, X inactivation during embryonic development generates two cell populations, one that expresses the X-linked transgene and another that does not. Since the status of X inactivation is heritable, descendants of either population will be stable for the transgene expression. Several studies on the lineage relationship of cells in tissues, such as the retina, brain, and tongue, have been performed using the mosaicism of transgene expression generated by random X chromosome inactivation *(6–8)*.

Finally, some transgenes display lineage and stage-specific expression through use of a tissue-specific promoter or as a consequence of their unique integration sites. For example, in *Wnt1-lacZ* transgenic mice, the transgene is regulated by the *Wnt1* 3'-cis-acting enhancer element, which directs specific expression to the dorsal part of the neural tube and subsequently in the neural crest cells derived from this region *(9)*. Another example is R197 transgenic mice in which the *lacZ* gene is expressed principally in the muscle lineage from early organogenesis stages onwards as a consequence of its integration site *(10)*. Provided that the expression of the transgene is tissue-specific for a defined period of development, such transgenes offer an excellent tool for tracing the differentiation of cell lineages. A prerequisite for using these mice and any other transgenic lines is to establish the pattern of transmission and to ascertain the tissue and temporal specificity of transgene expression.

## 1.2. Examples of Three Distinct Patterns of Transgenic Reporter Expression

### 1.2.1. Ubiquitous Expression

The H253 transgenic line *(2)* is an example of ubiquitous expression of an Xlinked transgene. Crossing homozygous X*X* females with X*Y males (where X* represents the transgene bearing X chromosome) produces transgenic embryos (X*X*, X*Y) in which β-galactosidase is expressed at high levels in all tissues and progeny during all stages of development. This series of matings produced a ubiquitous non-lineage bias transgenic line that has been used successfully in several lineage studies in postimplantation mouse embryos *(2,11–13)*.

### 1.2.2. X-Inactivation Mosaicism

The other type of expression is one of a mosaic nonlineage biased transgene. Transgenic H253 male and F1 (C57BL/6 × DBA/2) female mice were crossed. Hemizygous H253 female mice (X*X) were then crossed to wild-type F1 (C57BL/6 × DBA/2) males to produce offspring of four possible genotypes and only three distinct patterns of staining: uniform staining (X*Y), patchy staining (X*X), and negative staining (XX and XY). The patchy staining of X*X genotypic mice is owing to random X chromosome inactivation. This results in the expression of the *lacZ* marker in only about half of the population. The ability to recognize clonal populations means that these two cell populations (one that does expresses *lacZ* and the other that does not) have been used in cell lineage analyses to reveal the ancestral history of spatially related clones *(6–8)*.

The mosaicism of transgene expression following X inactivation in chimeric tissues, has been used extensively in the analysis of specific cell lineages *(6–8,14–16)*. The *lacZ*-expressing cells can be identified from nonexpressing cells by X-gal staining, and the composition of the sorted populations may be examined for the expression of cell-specific markers by staining with appropriate antibodies. This system therefore provides an assay for the measurement of the relative proportion of specific cell types at different stages of maturation.

### 1.2.3. Restricted Expression

The R197 transgenic line is an example of a lineage- and stage-specific transgene *(10)*. The *lacZ* transgene is segmentally expressed primarily in the myogenic lineage from about 8.5 d onward. Following the identification of transgenic pups (*see* **Subheading 3.3.**), these animals were crossed with nontransgenic (C57BL/6 × DBA) hybrids or other transgenic mice to produce transgenic embryos for the experiment. In this transgenic line, homozygous embryos die during the immediate postimplantation period. To ascertain lethality, pregnant mice were sacrificed at different ages of gestation so that the litter size and the genotype of the embryos/fetuses could be analyzed. A reduction in litter size of F1 hemizygous matings and the absence of the homozygotes after genotyping indicated that homozygotes did not survive to term. As an essential step for using transgenic mice, the possibility of embryo lethality should be examined particularly when a significant reduction in litter size is observed. Prior genotyping of the entire litter may be necessary to avoid using cells from dying embryos for lineage analysis.

## 2. Materials

All reagents and solutions used in these protocols should be of molecular biology grade and prepared with sterile distilled water. Reagents for some of the protocols are available commercially as kits, and these are indicated.

### 2.1. PCR Preparation of a lacZ Probe

A digoxigenin (DIG)-labeled probe is generated using the PCR reaction and the following reagents:

1. The PCR mix should be prepared as follows:

   | | |
   |---|---|
   | 1 µg Plasmid DNA (does not need to be linearized) | 2 µL |
   | 10X *Taq* buffer (Boehringer Mannheim 1277049, Indianapolis, IN) | 5 µL |
   | 10 m*M* dNTP (*see* **Subheading 2.1., item 2**) | 1 µL |
   | Primer E128 *(lacZ inner 5' primer position 711–730* 5'-CG CTG TAC TGG AGG CTG AAG-3') | 2.5 µL |
   | Primer E129 *(lacZ* inner 3' primer position 1140–1159 5'-CG GCG TTA AAG TTG TTC TGC-3') | 2.5 µL |
   | Water | 35 µL |
   | Paraffin oil | 100 µL |

2. The 10-m*M* dNTP mix is made up of the following nucleotides (Boehringer Mannheim 1277049):
   10 m*M* dCTP;
   10 m*M* dGTP;
   10 m*M* dATP;
   3.5 m*M* dTTP;
   6.5 m*M* DIG-11-dUTP (Boehringer Mannheim 1093088: alkali stock).
3. *Taq* DNA polymerase (Boehringer Mannheim 1435094).
4. Variable-temperature PCR unit (Omingene, Hybaid, Holbrook, NY).

### 2.2. Isolation of Mouse Tail DNA for Screening

1. Tris-EDTA (TE) buffer (pH 7.6): 10 m*M* Tris-HCl, pH 7.6, 1 m*M* EDTA, pH 8.0.
2. TE-SDS (pH 7.6): 10 m*M* Tris-HCl, pH 7.6, 1 m*M* EDTA, pH 8.0, SDS 1%.
3. Proteinase K (20 mg/mL).

4. Potassium acetate (KAc; 8 *M*).
5. Chloroform.
6. Ethanol (100 and 90%).
7. Sealed pipets (prepared by holding glass Pasteur pipets in a flame until the open pore is sealed and a small ball has formed at the end of the pipet).

## 2.3. Screening for lacZ Transgene Expression

1. Positively charged nylon membrane (Boehringer Mannheim 1417240).
2. Buffer 1: 0.5 *M* NaOH, 1.5 *M* NaCl.
3. Buffer 2: 0.5 *M* Tris-HCl, pH 7.5, 1.5 *M* NaCl.
4. Buffer 3 (2X SSC): This should be prepared from a 20X SSC (3 *M* NaCl/0.3 *M* Na-citrate·2H$_2$O) stock solution adjusted to pH 7.0 with 1 *M* HCl.
5. Stratalinker UV crosslinker (Stratagene, La Jolla, CA).
6. 5X SSC.
7. Pre-hybridization buffer: 5X SSC, 0.1% *N*-lauroylsarcosine, 0.02% SDS, 3% blocking reagent (Boehringer Mannheim 1096176).
8. Buffer 4: 2X SSC, 0.1% SDS.
9. Buffer 5: 0.1X SSC, 0.1% SDS.
10. Buffer 6: 0.1 *M* Tris-HCl, pH 7.5, 0.15 *M* NaCl.
11. Anti-digoxigenin (DIG) antibody (Boehringer Mannheim 1093274) diluted 1:10,000 in buffer 6 containing 3% blocking reagent.
12. Tween-20.
13. Buffer 7: 0.1 *M* Tris-HCl, pH 9.5, 0.1 *M* NaCl, 0.05 *M* MgCl$_2$. Prepare fresh on day of use.
14. CSPD (25 m*M*; Di-sodium, 3-(-4 metho-xyspiro[1,2-dioxetene-3,2'-{5' chloro} tricyclo {3.3.1.1} decan]-4-yl)phenyl phosphate; Boehringer Mannheim 1655884) diluted 1: 100 in buffer 7 before use.
15. X-ray film (Kodak, Rochester, NY).

## 2.4. LacZ Expression by β-Galactosidase Histochemistry

1. PBS (Ca$^{2+}$/Mg$^{2+}$-free; 1X).
2. Paraformaldehyde (Sigma P6148). Prepare as a 4% solution in Ca$^{2+}$/Mg$^{2+}$-free PBS by heating to dissolve, but do not boil. Add two to three drops of 0.05 *M* NaOH to clear the solution (pH 7.5), which can then be stored at 4°C for 2 wk. Ideally, however, it should be freshly prepared. (**Caution:** Paraformaldehyde is toxic and readily absorbed through the skin. It is very destructive to the skin, mucous membranes, eyes, and upper respiratory tract. Therefore, glasses, gloves, and a mask should be worn to avoid contact with the dust and work should be performed in a fume hood.)
3. Glutaraldehyde (25% [w/v]; Sigma G5882).
4. X-gal (4-chloro-5-bromo-3-indolyl-β-D-galactopyranoside; Progen, Richlands, QLD, 2000190). X-gal is prepared as a 4% (40 mg/mL) stock solution in dimethylformamide (DMF: Sigma D8654). Store at –20°C and protect from light.

5. Sorensen's phosphate buffer (PB), 0.1 $M$, pH 7.4: 0.02 $M$ NaH$_2$PO$_4$·0.08 $M$ Na$_2$HPO$_4$.

6. X-gal staining solution: 0.1% X-gal, 2 m$M$ MgCl$_2$, 5 m$M$ EGTA, 0.01% (w/v) sodium deoxycholate, 0.02% (w/v) Nonidet P-40 (NP-40: Sigma), 5 m$M$ K$_3$Fe(CN)$_6$ (potassium ferricyanide), and 5 m$M$ K$_4$Fe(CN)$_6$·6H$_2$O (potassium ferrocyanide) in 0.1 $M$ pH 7.4, Sorensen's phosphate buffer. Potassium ferricyanide and potassium ferrocyanide can be made up as 500-m$M$ stock solutions and should be stored at room temperature, protected from light for a maximum of 2 wk.

7. Ethanol (70, 80, 90, 100%).

8. Histolene.

9. Xylene.

10. Paraffin wax.

11. Aluminum sulfate (Sigma, St. Louis, MO, A7523).

12. Nuclear fast red (0.1%; Merck, Rahway, NJ, 15939): Prepare by adding 500 mg of nuclear fast red and 25 g of aluminum sulfate to 500 mL of distilled water, heat/stir to dissolve, and filter when cooled.

13. Avertin: Prepare 100% stock solution by mixing 10 g of 2,2,2tribromethyl alcohol (Aldrich, Castle Hill, NSW, Australia, T4,840-2) with 10 mL of *tert*-amyl alcohol (Aldrich 24,048-6). To use, dilute 100% stock to 2.5% in water. Both the 100 and 2.5% stocks are stored wrapped in foil at 4°C. The correct dose of avertin will vary with each preparation. Therefore, several mice should be test injected within the range of 0.014–0.018 mL/g body wt.

14. Sucrose (0.2 $M$).

15. Glycol methacrylate (2-hydroxyethyl methacrylate; Aldrich 12863-5).

16. Cresyl violet (Fluka, Buchs, Switzerland, 61120).

17. Acetic acid.

## 2.5. LacZ Expression by β-Galactosidase Immunostaining

1. PBS (Ca,$^{2+}$ and Mg$^{2+}$-free).

2. Paraformaldehyde (4%) in Ca$^{2+}$ and Mg$^{2+}$-free PBS (*see* **Subheading 2.4, items 1** and **2** for preparation).

3. Sucrose (0.2 $M$).

4. OTC embedding fluid (Tissue Tek, Elkhart, IN, 4853).

5. Liquid nitrogen.

6. Avertin (2.5%; *see* **Subheading 2.4., item 13** for preparation).

7. Sodium nitrate (Sigma S5506): Make up as 0.5% solution in 0.5X PBS.

8. Vibratome (Campden Instruments, Loughborogh, UK).

9. Sodium azide (Sigma S2002).

10. Normal horse serum (Trace, Sydney, NSW, Australia, 150400100v).

11. Triton X-100.

12. *Escherichia coli* β-galactosidase antibody rabbit IgG (1° antibody; Cappel 55976).

13. Biotinylated goat antirabbit IgG (2° antibody: Vectastain Elite ABC kit; Immunodiagnostic, Bonsheim, Germany, PK6101).

14. Texas red conjugated donkey antirabbit IgG (2° antibody; Jackson Immuno Research, West Grove, PA).
15. Streptavidin–peroxidase complex (Vectastain Elite ABC kit; Vector Laboratories; Immunodiagnostic, Burlingame, CA, PK6101).
16. Diaminobenzedine (DAB: Dako, Carpinteria, CA, S3000). Store at –20°C.
17. Hydrogen peroxide ($H_2O_2$)
18. Streptavidin–FITC complex.
19. Nuclear fast red (0.1%; *see* **Subheading 2.4., item 12**).

## 2.6. Visualizing lacZ by Immunostaining or Histochemistry in Combination with Alkaline Phosphatase Histochemistry

1. PBS ($Ca^{2+}$ and $Mg^{2+}$-free).
2. X-gal staining solution (*see* **Subheading 2.4., item 6**).
3. Glutaraldehyde (25%; Sigma G5882).
4. Ethanol (25% [w/v]), 50, 70, 80, 90, 100%).
5. Polyester wax.
6. Xylene.
7. Hydrogen peroxide ($H_2O_2$).
8. Methanol.
9. Normal goat serum (Sigma G9023).
10. *E. coli* β-galactosidase antibody rabbit IgG (1° antibody; Cappel 55976).
11. Bovine serum albumin (Sigma Fraction V A9647).
12. Biotinylated goat antirabbit IgG (2° antibody; Immunodiagnostic PK6101): Store at 4°C.
13. Streptavidin–peroxidase complex (Vectastain Elite ABC reagent): Add two drops of reagent A to 5 mL PBS (with $Ca^{2+}$ and $Mg^{2+}$), mix well, then add two drops of reagent B, mixing immediately, and allow the ABC reagent to stand for about 30 min before use (Immunodiagnostic PK6101).
14. Diaminobenzedine (Dako S3000): Store at –20°C.
15. Peroxidase substrate solution: Dissolve one DAB tablet in 10 mL PBS to give 1 mg/mL DAB solution. Mix 2 mL of DAB solution with 15 mL of 3% $H_2O_2$ (or 1.5 mL of 30% $H_2O_2$). The substrate solution is stable for 2 h at room temperature. Unused stock DAB solution can be stored in the dark (wrapped in foil) for up to 5 d at 2–8°C, or longer at –20°C.
16. Alkaline phosphatase (ALP) buffer: 0.1 *M* Tris-HCl, 0.1 *M* NaCl, 50 m*M* $MgCl_2$, pH 9.5. Dissolve 12.1 L/g Tris-base and 5.84 g NaCl in d$H_2O$, adjust pH to 9.5 with concentrated HCl, add 50 mL 1 *M* $MgCl_2$, and make up volume to 1 L, (pH 9.5, alkaline phosphatase substrate package, Gibco-BRL, Gaithersburg, MD 8280SA).
17. NBT/BCIP substrate (Gibco-BRL 8280SA): Store at –20°C. Prepare NBT/BCIP substrate immediately prior to staining. Add 44 µL NBT to 10 mL alkaline phosphotase buffer, mix gently, then add 33 µL BCIP, mix gently.
18. Entellan (Merck).
19. 4% (w/v) Paraformaldehyde (*see* **Subheading 2.4., item 2**).
20. X-gal staining solution: 0.1% (w/v) X-gal, 2 m*M* $MgCl_2$, 5 m*M* EGTA, 0.01% (w/v) sodium deoxycholate, 0.02% (w/v) NonidetP-40 (NP-40: Sigma), 5 m*M*

$K_3Fet(CN)_6$ (potassium ferriicyanide) and 5 $M$ $K_4Fe(CN)6\cdot6H_2O$ (potassium ferrocyanide) in 0.1 $M$ pH 7.4 Sorenson's PB.

## 2.7. Visualizing lacZ by Histochemistry in Combination with In Situ Hybridization

1. PBS (without $Ca^{2+}$- and $Mg^{2+}$) (*see* **Subheading 2.4., item 2**).
2. 4% Paraformaldehyde (Sigma: *see* **Subheading 2.4., item 2** for preparation).
3. Ethanol (50, 70, 80, 90, 100% v/v).
4. Xylene.
5. Paraffin wax.
6. TESPA (3-aminopropyltriethoxy-saline; Sigma A3648): TESPA coated slides are prepared as follows. Wash slides in 10% HCl/70% ethanol, followed by 95% ethanol for 1 min each, and then air-dry. Dip the slides in 2% TESPA in acetone for 10 s. Wash twice with acetone and then with distilled water before drying at 37°C.
7. DIG RNA labeling kit (Boehringer Mannheim 1175025).
8. Triton X-100.
9. Proteinase K (20 mg/mL; 6 µg/mL).
10. SSC (20X).
11. SDS (10% w/v).
12. Hybridization buffer: 5X SSC, 50% (v/v) deionized formamide, 0.4% (w/v) SDS, 0.1% (w/v) *N*-lauroylsarcosine, 2% (w/v) blocking reagent (Boehringer Mannheim 10961761).
13. Heparin (50 µg/mL; Sigma H2149).
14. Yeast tRNA (20 µg/mL).
15. NaCl (1 $M$).
16. Maleic acid (1 $M$).
17. Normal sheep serum (Sigma S3771).
18. DIG antibody-sheep conjugated alkaline phosphatase (dilute 1:200 in 150 m$M$ NaCl/100 m$M$ maleic acid/10% sheep serum; Boehringer Mannheim 1093274).
19. Alkaline phosphatase buffer (ALP; *see* **Subheading 2.6., item 16**).
20. NBT/BCIP substrate (*see* **Subheading 2.6., item 17**).
21. *E. coli* β-galactosidase antibody rabbit IgG (5 µg/mL in PBS/20% normal goat serum; Cappel 55976).
22. Normal goat serum (Sigma G9023).
23. Biotinylated goat anti-rabbit IgG (2° antibody; diluted 1:200 in PBS/0.2%BSA; Immunodiagnostic PK6101) for 60 min at room temperature.
24. Streptavidin–β-galactosidase conjugate (diluted 1:200 in PBS/0.2% BSA; Boehringer Mannheim 1112481).
25. X-gal washing buffer (X-gal staining solution without X-gal; *see* **Subheading 2.4., item 6**).
26. X-gal staining solution (*see* **Subheading 2.6., item 20**).
27. CHAPS (3-[{3-cholamidopropyl} dimethylammonio]-1-propansulfonat; Boehringer Mannheim 810118).
28. PBT (0.1% Tween-20 in PBS).
29. Methanol (25, 50, 70, 100% v/v).

30. Hydrogen peroxide ($H_2O_2$).
31. Dimethylsulfoxide (DMSO; Sigma D2650).
32. Glycine.
33. Glutaraldehyde (25% w/v; Sigma G5882).
34. Tween-20.
35. Prehybridization solution consists of the following and should be prepared on the day of use: 50% deionized formamide, 5X SSC, 2% blocking powder, 0.1% Tween-20, 0.5% CHAPS, 50 µg/mL yeast RNA, 5 m*M* EDTA, 50 µg/mL heparin.
36. Bovine serum albumin (Fraction V; Sigma A9467).
37. Mouse embryo powder: This is prepared by homogenizing 12.5–14.5-d mouse embryos in a minimum vol of PBS. Add 4 vol of ice-cold acetone, mix, and incubate on ice for 30 min. Spin at 10,000*g* for 10 min and remove the supernatant. Wash the pellet with ice-cold acetone, and spin again. Spread the pellet out, and grind into a fine powder on a sheet of filter paper. Allow the powder to air-dry, and store in an air-tight tube at 4°C.
38. NTMT: 100 m*M* NaCl, 100 m*M* Tris-HCl, pH 9.5, 50 m*M* $MgCl_2$, 0.1% Tween-20.
39. Levamisole (2 m*M* in NTMT; Sigma L9756).
40. Color reaction mix consists of the following and should be prepared immediately before use: 4.5 mL NBT (Gibco-BRL 8280SA), 3.5 µL BCIP (Gibco-BRL 8280SA) in 1.0 mL NTMT.

## 3. Methods

This protocol for screening gene transmission describes the identification of transgenic animals by the non-radioactive detection of the *lacZ* inserts via a PCR generated *lacZ* probe and cuts down on the time required by more traditional methods, such as slot blotting.

### 3.1. Preparation of a lacZ Probe by PCR

1. Prepare the PCR mix (*see* **Subheading 2.1., item 1**) and incubate at 95°C for 5 min.
2. Incubate the PCR mix at 70°C while adding 2 µL (1.25 U) of *Taq* DNA polymerase.
3. Incubate the sample at 55°C for 1 min, then 72°C for 2 min, followed by 94°C for 1 min, and repeat for 35 cycles on a Hybaid PCR unit.
4. Incubate the sample for the final cycle at 72°C for 10 min and then at 20°C for 10 min before storing at –20°C. The labeled probe is stable for up to 12 mo at –20°C.

### 3.2. Isolation of Mouse Tail DNA for Screening

1. Cut 1–1.5 cm of tail from the mouse and mince with scissors in 400 µL of TE-SDS in a large Eppendorf tube.
2. Add proteinase K (5 mL; 20 µg/mL) to each sample, and incubate at 65°C overnight.
3. Add KAc (75 µL; 8 *M*) and chloroform (500 µL) to each sample, and mix by inversion.

4. Centrifuge for 5 min at 14,000$g$, collect the (upper) aqueous phase into an Eppendorf tube containing 1 mL ethanol (100%), and leave at room temperature for 5–10 min.
5. Spool precipitated DNA onto a sealed pipet, wash by dipping each sample in ethanol (90%) and air-dry (5–10 min).
6. Shake DNA off the pipet in 200 µL of TE in an Epppendorf tube and allow to dissolve overnight before calculating concentration by absorbence at 260 nm.

### 3.3. Screening for lacZ Transgene Expression: Membrane Preparation

1. Spot 10 µg and 5 µg of DNA onto a positively charged nylon membrane, and air-dry the membrane thoroughly (10 min).
2. Denature the DNA by floating the membrane DNA side up in buffer 1, then buffer 2, and then buffer 3 each for 5 min.
3. Expose the wet membrane to UV light using the Stratalinker at 12 J/cm$^2$. The Stratalinker calculates the required time to deliver 1200 J automatically. The membrane can now either be used immediately or be air-dried, sealed in plastic and stored at 4°C until required.
4. Wet the membrane in 5X SSC (1 min), and incubate the membrane in prehybridization buffer for 3 h at 68°C.
5. Hybridize with 10 ng/mL of probe (diluted in prehybridization buffer) overnight at 68°C.
6. Wash the membrane in buffer 4 for 7 min (×2) at room temperature.
7. Wash the membrane in buffer 5 for 15–20 min (×2) at 68°C.

### 3.4. Chemiluminescent Detection of lacZ

1. Equilibrate membranes in buffer 6 for 5 min.
2. Block the membranes at room temperature with 3% (w/v) blocking reagent in buffer 6 for 30–45 min.
3. Incubate membranes at room temperature for 30 min with antibody fragments diluted in 3% blocking reagent in buffer 6.
4. Incubate membranes in buffer 6 + 0.3% (w/v) Tween-20, twice for 15–20 min to wash off excess antibody.
5. Equilibrate membranes in buffer 7 for 5 min.
6. Wash the membrane with CSPD diluted in buffer 7 for 5 min in the dark.
7. Blot the membrane to remove any excess liquid before sealing the membrane in a plastic bag and incubating for 15 min at 37°C in the dark.
8. Expose the membrane to X-ray film for between 15 min and 3 h, and develop (*see* **Notes 1–4**).

### 3.5. Histochemical Staining for lacZ with X-Gal

#### 3.5.1. Small Embryos (7.5–11.5 D Postcoitum [p.c.]) and Fragments of (12.5–18.5 D p.c.) Embryos

1. Fix small embryos or fragments of older embryos in 4% paraformaldehyde/0.2% glutaraldehyde for 5–15 min.

2. Wash the embryos in PBS (2 × 5 min), and stain overnight in X-gal staining solution at 37°C in the dark.
3. Wash the embryos in 70% (10 min) ethanol and dehydrate in 70, 80, and 90% ethanol (10 min each), and then 100% ethanol (3 × 10 min).
4. Clear the embryos in xylene or Histolene (3 × 10 min), and impregnate in paraffin wax (3 × 20 min) before embedding in fresh wax.
5. 5–10 μm sections are cut and counterstained with nuclear fast red (*see* **Notes 5–9**).

### 3.5.2. Adult Embryos and Organs

1. Anesthetize adult animals and perfuse intracardially with 4% paraformaldehyde/ 0.2% glutaralde]hyde (or 2% paraformaldehyde) in 0.1 $M$ phosphate buffer
2. Dissect the organ required and postfix in the same fxative for 1 h, followed by cryoprotection in 30% phosphate buffered 0.2 $M$ sucrose solution overnight.
3. Cut frozen sections at 100–200 μm and briefly postfix with 4% paraformaldehyde for 5 min to preserve the histology.
4. Rinse the sections in phosphate buffer (3 × 20 min), and then incubate in X-gal solution overnight at 37°C. in the dark.
5. Rinse the sections with PB (3 × 20 min) and examine under a dissecting microscope (*see* **Note 10**).

## 3.6. Immunostaining for lacZ

### 3.6.1. Embryos (7.5–18.5 D p.c.)

1. Dissect embryos in 4% paraformaldehyde (in $Ca^{2+}$ and $Mg^{2+}$-free PB) and fix for 4 h.
2. Wash the embryos in 0.2 $M$ sucrose (1 h) and then immerse in OTC briefly and immediately before freezing slowly in liquid nitrogen vapor, followed by immersion in liquid nitrogen.
3. Cut 60-μm thick sections using a cryostat and store at 4°C until required, then go to **Subheading 3.6.3., step 1**.

### 3.6.2. Adult Mice and Body Organs

1. Anesthetize adult mice with 2.5% avertin (0.017 mL/g of body wt) and perfuse via the left ventricle with 0.5% w/v sodium nitrate in half-strength phosphate-buffered saline (PBS).
2. Dissect the required organ, and postfix with the same sodium nitrate fixative (2 h).
3. Section the tissue with a vibratome at 50–100 μm thickness.
4. Store sections for immunohistochemistry in PBS/0.02% sodium azide at 4°C until required. Then go to **Subheading 3.6.3., step 1**.

### 3.6.3. Detection

1. Wash sections from **Subheadings 3.6.1.** and **3.6.2.** selected for detection of β-galactosidase in PBS/10% normal horse serum /0.02% Triton X-100 at room temperature (12 h).
2. Incubate sections with *E. coli* β-galactosidase antibody (diluted 1:1000 in PBS/ 10% normal horse serum/0.02% Triton X-100) overnight at room temperature.

3. Wash sections in PBS (3 × 10 min), and then incubate with 2° antibodies. either biotinylated goat antirabbit IgG (diluted 1:400 in 0.02% Triton X-100 in PBS) or Texas red donkey antirabbit IgG (diluted 1:500 in 0.02% triton X-100 in PBS) for 2 h.
4. Remove unbound antibody by washing in PBS (10 min).
5. If using biotinylated goat anti-rabbit IgG, incubate the sections in streptavidin-peroxidase complex (diluted 1:100 in 0.2% Triton X-100) for 2 h.
6. Visualize the immunochemical reaction by washing sections with diaminobenzidine (0.5 mg/mL) in the presence of 0.01% $H_2O_2$ until the brown color is apparent.
7. If Texas red-conjugated donkey antirabbit IgG is used, then incubate the sections in streptavidin-FITC (diluted to 1:100 in 0.02% Triton X-100 in PBS) for 2 h.
8. Wash sections in PBS (3 × 10 min), and view either by light or fluorescence microscopy. Sections may be counterstained with 0.1% nuclear fast red.

## 3.7. Visualizing lacZ by Immunostaining or Histochemistry in Combination with Alkaline Phosphatase Histochemistry

### 3.7.1. Simultaneous Immunological Detection of lacZ and Alkaline Phosphatase on Sections

This protocol allows for the staining of *E. coli* β-galactosidase antibody and NBT-BCIP histochemical detection of alkaline phosphatase. It is designed to detect alkaline phosphatase in primordial germ cells in *lacZ*-expressing embryos. All staining should be done at room temperature unless otherwise indicated and on 6–8 μm sections.

1. Dissect out embryos in PBS medium and stain either the tail, head, yolk sac, or non-essential body part with X-gal staining solution to screen for transgenic embryos.
2. Fix embryos in 0.5% glutaraldehyde in PBS for 5–15 min, depending on the size of the embryos. As a general rule:

| | |
|---|---|
| <8.5 d p.c. | 5 min |
| 9.5 d p.c. | 5–10 min |
| 10.5–12.5 d p.c. | 10–15 min |
| 13.5–15.5 d p.c. | 15 min |
| >16.5 d p.c. | 20–30 min |

3. Wash embryos 2 × 10 min with PBS. The embryos then can be either stored in 70% ethanol for a short period of time, or dehydrated and embedded as described in **Subheading 3.7.1., step 4**.
4. Dehydrate embryos through 70, 80, 90, and 100% (×3) ethanol according to the times given below:

| | |
|---|---|
| <8.5 d p.c. | 2–5 min |
| 9.5 d p.c. | 5 min |
| 10.5–12.5 d p.c. | 10 min |
| 13.5–15.5 d p.c. | 10–15 min |
| >16.5 d p.c. | 20–30 min |

5. Transfer embryos to fresh polyester wax for 10–15 min. Repeat with fresh wax 2 × 10 min each. Embed with fresh polyester wax at room temperature, and section with either a cryostat or a microtome kept at 15°C (*see* **Note 11**).

6. Dewax sections in xylene (5 min), and rehydrate the sections by incubating them in a 37°C oven for 30 min followed by 2-min washes in 100% ethanol (2× at 37°C), 90, 70, 50, and 25% ethanol. Finally wash the sections for 5 min in dH₂O.

7. Quench the endogenous peroxidase activity with 0.3% $H_2O_2$ in 40% methanol for 20 min, or 3% $H_2O_2$ in $H_2O$ for 5 min.

8. Wash the sections with PBS (3 × 3 min).

9. Block nonspecific binding by overlaying sections with 5–10% normal goat serum for 20 min.

10. Drain excess serum and incubate slides with *E. coli* β-galactosidase antibody (diluted 1–5 µg/mL in PBS) for 60–75 min.

11. Wash sections in PBS + 0.2% BSA (3 × 10 min).

12. Incubate sections with biotinylated goat antirabbit IgG (2° Ab diluted 1:200 in PBS, 0.2% BSA, 3% normal goat serum) for 30 min.

13. Wash sections with PBS + 0.2%, BSA (3 × 10 min).

14. Incubate sections for 30 min with streptavidin peroxidase complex.

15. Wash sections with PBS + 0.2% BSA (3 × 10 min).

16. Incubate sections with peroxidase substrate solution (1 mg/mL DAB + 0.0225% $H_2O_2$) for 2–7 min or until desired staining intensity develops.

17. Wash sections for 5 min with dH₂O.

18. Incubate sections with PBS (10 min).

19. Incubate sections with alkaline phosphatase buffer, pH 9.5 (2 × 10 min).

20. Incubate sections in NBT/BCIP substrate for 10–15 min in the dark at 37°C, or until desired intensity develops.

21. Wash the sections with dH₂O (5 min).

22. Dehydrate and mount in Entellan.

### 3.7.2. Simultaneous Whole-Mount Histochemical Staining for β-Galactosidase and Alkaline Phosphatase

1. Fix 7.5–11.5 d mouse embryos or older embryo (12.5–18.5 d) fragments and organs in 4% paraformaldehyde, 0.5% glutaraldehyde in PBS for 5–10 min.

2. Wash embryos in PBS (2 × 10 min). Incubate embryos for 10 min in alkaline phosphatase buffer.

3. Incubate embryos in NBT/BCIP substrate at 37°C for 10–20 min. Check intensity every 5 min.

4. Wash embryos extensively in X-gal washing buffer (staining solution without X-gal; 2 × 10 min).

5. Incubate embryos overnight in X-gal staining solution at 37°C.

## 3.8. Visualizing lacZ by Histochlemistry in Combination with In Situ Hybridization

### 3.8.1. Simultaneous Detection of Oct 4 mRNA and β-Galactosidase Protein on Paraffin Sections

1. Fix 7.5–11.5 d embryos or fragments and organs of older embryos (12.5–18.5 d) in 4% paraformaldehyde overnight at 4°C, and then wash in PBS (2 × 10 min).

Dehydrate the embryos according to the schedule given in **Subheading 3.7., step 4**, wash in xylene (3 × 10 min), followed by paraffin wax (3 × 15 min), and then embed the embryos in fresh paraffin wax before storing at 4°C. Cut sections at 8 mm, and mount on TESPA-coated slides.

2. A 600 bp *Oct4* RNA digoxigenin probe was generated by in vitro transcription with T3 (antisense strand) and T7 (sense strand) RNA polymerase (DIG RNA labeling kit, Boehringer Mannheim, cat no. 1175025).

3. Dewax the embryos in xylene (5 min), rehydrate (100% x 2, 90, 80, 70, 50, dH$_2$O; for 1 min each), and then wash the sections in PBS (2 min).

4. Refix the sections in 4% paraformaldehyde (10 min), and then wash with PBS (2 × 5 min).

5. Treat the sections with 0.3% Triton X-100 in PBS (15 min), and then wash in PBS (2 × 5 min).

6. Incubate the sections in 37°C prewarmed proteinase K (20 mg/mL; 5 min), wash in PBS (3 × 10 min), and then post-fix the sections in 4% paraformaldehyde (15 min).

7. Prehybridize the sections for 3 h in hybridization buffer (5X SSC, 50% delonized formamide, 0.4% SDS, 0.1% *N*-lauroylsarcosine, 2% blocking reagent.

8. Dilute the probe in hybridization buffer containing 50 μg/mL heparin and 20 mg/mL yeast tRNA to a final concentration of 0.05 ng/mL. Heat the probe at 80°C (10 min) and immediately store on ice until addition to the slides. Incubated with sections with 500 μL of probe at 55°C in a humidified chamber overnight.

9. Wash the sections at room temperature in 2X SSC/0.2% SDS (2 × 10 min); then at 58°C in 0.5X SSC/0.2% SDS (2 × 30 min) followed by 150 m*M* NaCl/100 m*M* maleic acid (pH 7.5; 10 min).

10. Block the section in 150 m*M* NaCl/100 m*M* maleic acid (pH 7.5) containing 10% sheep serum (30 min), and then incubate with diluted alkaline phosphatase-conjugated sheep anti-DIG antibody (1:200 dilution) for 2 h.

11. Wash the sections in 150 m*M* NaCl/100 m*M* maleic acid (pH 7.5; 2 × 15 min), then in alkaline phosphatase buffer (100 m*M* Tris-HCl, 100 m*M* NaCl, 50 m*M* MgCl$_2$, pH 9.5, 10 min), and incubate with NBT/BCIP substrate. Allow the color to develop in the dark for 60–90 min at room temperature.

12. Wash the sections in PBS (3 × 10 min), block with 20% normal goat serum in PBS (30 min), and then incubate with 5 μg/mL *E. coli* anti-β-galactosidase antibody overnight at 4°C.

13. Wash the sections in PBS containing 0.2% BSA (3 × 10 min) and then incubate with biotinylated goat anti-rabbit IgG (2° antibody; diluted 1:200 in PBS/0.2%BSA) for 60 min at room temperature.

14. Wash the sections in PBS containing 0.2% BSA (3 × 10 min), and then incubate with streptavidin–β-galactosidase conjugate for 60 min (1:200 dilution).

15. Wash the sections in PBS (2 × 10 min), then in X-gal washing buffer (10 min), and the incubate in X-gal solution at 37°C (20 min) in the dark.

16. Dehydrate the sections, and mount in Entellan.

### 3.8.2. Whole-Mount In Situ Hybridization Combined with X-Gal Staining

All steps are performed at room temperature unless otherwise stated.

1. Fix embryos in 4% paraformaldehyde for 5 min.
2. Stain embryos in X-gal staining solution for 1 h.
3. Refix embryos in 4% paraformaldehyde overnight.
4. Wash embryos in PBT (2 × 10 min).
5. Wash embryos in 25, 50, and 75% methanol/PBT (5 min), and then in 100% methanol (2 × 10 min). Embryos can be stored in 100% methanol at –20°C.
6. Fix the embryos in methanol:DMSO (4:1) for 1 h and then bleach the embryos in methanol:DMSO:30%$H_2O_2$ (4:1:1) for 1 h.
7. Rehydrate embryos in 75, 50, and 25% methanol/PBT (5 min), and then wash the embryos in PBT (2 × 5 min). It is important to poke a few holes with a small needle into these embryos prior to washing to facilitate the flow of solutions.
8. Treat the embryos with Proteinase K (6 mg/mL; 5 min).
9. Wash the embryos in 0.2% glycine in PBT (2 × 5 min).
10. Refix the embryos in 4% paraformaldehyde/0.2% glutaraldehyde in PBS (30 min).
11. Wash embryos in PBT (2 × 5 min).
12. Wash embryos in 1 mL of prehybridization solution.
13. Prehybridize embryos in 1 mL of prehybridisation solution at 65°C (3 h). Embryos can be stored at this point at –20°C in prehybridization solution.
14. Incubate embryos in 1 mL of hybridization solution at 60°C overnight.
15. Posthybridize wash with each of the following for 5 min at 60°C:
    a. 100% Prehybridization solution;
    b. 75% Prehybridization solution: 25% 2X SSC;
    c. 50% Prehybridization solution: 50% 2X SSC;
    d. 25% Prehybridization solution: 75% 2X SSC.
16. Wash with each of the following for 30 min at 60°C:
    a. 2X SSC: 0.1% CHAPS;
    b. 0.2X SSC: 0.1% CHAPS (×2).
17. Wash embryos in PBT (2 × 10 min),
18. Preblock embryos in 10% normal sheep serum/1% BSA in PBT at 4°C (3 h).
19. Pre-absorb the antibody at 4°C (3 h) with 3 mg mouse embryo powder, 500 µL of 10% normal sheep serum, 1% BSA in PBT, and 1 µL of anti-DIG antibody, rocking the mixture gently. Spin the solution at 2000$g$ for 5 min and remove the supernatant (preabsorbed antibody).
20. Incubate the embryos in preabsorbed antibody overnight at 4°C with constant rocking.
21. Wash embryos in 0.1% BSA in PBT (5 × 1 h).
22. Wash embryos in 2 m$M$ levamisole in NTMT (2 × 10 min).
23. Wash embryos in NTMT (10 min).
24. Incubate embryos in color reaction mixture until sufficient coloration is visible in the specimen.
25. Refix embryos in 4% paraformaldehyde overnight, wash in PBS and examine (*see* **Notes 12** and **13**).

## 4. Notes

1. When crosslinking the DNA to the membrane, UV exposures >5 min will appreciably diminish the signal. As an alternative to UV exposure, baking the membrane at 120°C for 30 min will enable an equally sensitive detection.

2. For detecting high-copy-number transgenics, a positive result can be seen in as little as 1–10 ng of genomic DNA. For low-copy-number transgenics, a positive result can be seen after 15 min for 5–19 µg genomic DNA. A distinct result can be seen after 1.5–2 h.

3. Chemiluminescent detection produces a permanent record. The colorimetric reaction is equally sensitive, but the color reaction will fade with time.

4. The sensitivity of the detection can be improved by changing the wash stringency and probe concentration. The conditions shown here will minimize backgound in relation to signal. However, a clear background may not be possible for low-copy-number transgenic samples.

5. Enzyme activity in embryos from preimplantation up to about 11.5 d p.c. can be directly analyzed in whole mounts. However, for embryos older than 11.5 d penetration of the substrate in whole embryos becomes limiting and this poses a problem for accessibility of staining reagent to internal tissues, especially after a substantial amount of insoluble reaction product has been deposited in more superficial *lacZ*-expressing tissues. Dissection of older embryos (>11.5 d) into smaller fragments facilitates penetration of the substrate and greatly improves staining of deeper tissues.

6. In order to reduce the background associated with endogenous β-galactosidase activity which is normally found in bone, kidney, and brain, the histochemical reaction should be carried out at pH 7.4, which is optimal for bacterial β-galactosidase in contrast to the mammalian enzyme, which is most active at a more acidic pH value of 4.0.

7. Prolonged exposure of embryonic tissues to the fixative diminishes β-galactosidase activity, and enzyme activity becomes undetectable after more than 60 min in paraformaldehyde.

8. If the fixation of the embryo fragment/specimen is insufficient, leakage of the reaction product stains the reaction mixture blue. It is advisable not to leave stained samples in alcohols, and solvents longer than necessary, since this may leach out some of the reaction product. When sections are viewed by dark-fleld microscopy, the *lacZ* stain appears pink and contrasts with surrounding tissues.

9. In order to enhance the discrimination between expressing, nonexpressing cells, and cells expressing endogenous β-galactosidase-like activity, it is useful to target the *lacZ* product to a particular cellular compartment, such as the nucleus. The nuclear localization signal (nls) from the early region of the simian virus (SV40) genome has been used to provide effective nuclear localization of β-galactosidase (nlsLacZ; *17,18*).

10. For more detailed histological studies, small parts of the organ are cut by scalpel blade, dehydrate in alcohol, and embedded either in paraffin wax or glycol methacrylate. Wax sections are counterstained in 0.1% nuclear fast red. Methacrylate

sections are counterstained in 0.1% cresyl violet in 1% acetic acid (pH 3.3) for 50 min at 40°C.

11. When infiltrating with polyester wax, the specimens may float, which is normal, but they will sink as they are infiltrated with wax. If after the first 15 min of incubation in wax the specimens have not sunk, it is still advisable to go to the next wax change.

12. It is important that the embryos are not overfixed with paraformaldehyde for X-gal staining prior to the immunohistochemistry, because it may result in some loss of antigenicity.

13. In each of these procedures, there are a lot of wash steps that require the transfer of embryos. If the dish containing the embryos is placed on a black background, it is easier to see the embryos during transfer.

## References

1. Kothary, R., Clapoff, S., Darling, S., Perry, M. D., Moran, L. A., and Rossant, J. (1988) Inducible expression of an *hsp68-lacZ* hybrid gene in transgenic mice. *Development* **105,** 707–714.

2. Tam, P. P. L. and Tan, S.-S. (1992) The somitogenic potential of cells in the primitive streak and the tail bud of the organogenesis-stage mouse embryo. *Development* **115,** 703–715.

3. Freidrich, G. and Soriano, P. (1991) Promotor traps in embryonic stem cells: a genetic screen to identify and mutate developmental genes in mice. *Genes Dev.* **5,** 1513–1523.

4. Beddington, R. S. P., Morgenstern, J., Land, H., and Hogan, A. (1989) An in situ enzyme marker for the mid-gestation mouse embryo and the visualisation of inner cell mass clones during early organogenesis. *Development* **106,** 37–46.

5. Tan, S.-S., Williams, E. A., and Tam, P. P. L. (1993) X-chromosome inactivation occurs at different times in different tissues of the post implantation mouse embryo. *Nature Genet.* **3,** 170–175.

6. Tan, S.-S. and Breen, S. (1993) Radial mosaicism and tangential cell dispersion both contribute to mouse neocortical development. *Nature* **362,** 638–640.

7. Stone, L. M., Finger, T. E., Tam, P. P. L., and Tan, S.-S. (1995) Taste receptor cells arise from local epithelium, not neurogenic ectoderm. *Proc. Natl. Acad. Sci USA* **92,** 1916–1920.

8. Tan, S.-S. Faulkner-Jones, B., Breen, S. J., Walsh, M., Bertram, J. F., and Reese, B. E. (1995) Cell dispersion patterns in different cortical regions studied with an X-inactivated transgenic marker. *Development* **121,** 1029–1039.

9. Echelard, Y., Vassileva, G., and McMahon, A. P. (1994) Cis-acting regulatory sequences governing Wnt-1 expression in the developing mouse CNS. *Development* **120,** 2213–2224.

10. Tan, S.-S. (1991) Liver specific and position effect expression of a retinol-binding protein-lacZ fusion gene (RBP-*lacZ*) in transgenic mice. *Dev. Biol.* **146,** 24–37.

11. Trainor, P. A., Tain, S.-S., and Tam, P. P. L. (1994).Cranial paraxial mesoderm: regionalisation of cell fate and impact upon craniofacial development in mouse embryos. *Development* **120,** 2397–2408.

12. Quinlin, G. A., Williams, E. A., Tan, S.-S., and Tam, P. P. L. (1995) Neuroecto-dermal fate of epiblast cells in the distal region of the mouse egg cylinder: impli-cation for body plan organisation during early embryogenesis. *Development* **121,** 87–948.

13. Trainor, P. A. and Tam, P. P. L. (1995) Cranial paraxial mesoderm and neural crest cells of the mouse embryo: co-distribution in the craniofacial mesenchyme but distinct segregation in branchial arches. *Development* **121,** 2569–2582.

14. Berger, C. N., Tan, S.-S., and Sturm, K. S. (1994) Simultaneous detection of b-galactosidase activity and surface antigen expression in viable haematopoietic cells. *Cytometry* **17,** 216–233.

15. Berger, C. N., Tam, P. P. L., and Sturm, K. S. (1995) The development of haematopoietic cells is biased in embryonic stem cell chimaeras. *Dev. Biol.* **170,** 651–663.

16. Reese, B. E., Harvey, A. R., and Tan, S.-S. (1995) Radial and tangential disper-sion patterns in the mouse retina are class specific. *Proc. Natl. Acad. Sci. USA* **92,** 2494–2498.

17. Bonnerot, C., Rocancourt, D., Briand, P., Grimber, G., and Nicolas, J. F. (1987) A β-galactosidase hybrid protein targeted to nuclei as a marker for developmental studies. *Proc. Natl. Acad. Sci. USA* **84,** 6795–6799.

18. Kalderon, D., Roberts, B. L., Richardson, W. D., and Smith, A. E. (1984) A short amino acid sequence able to specify nuclear localisation. *Cell* **39,** 499–509.

# 12

# Mouse Primordial Germ Cells

*Isolation and In Vitro Culture*

## Patricia A. Labosky and Brigid L. M. Hogan

## 1. Introduction

Primordial germ cells (PGCs) in the embryo give rise to functional gametes in the adult animal. Considering their importance in the continuation of the species, it is no wonder that there is much interest in understanding the biology of these highly specialized cells. Much has been learned from the analysis of mouse mutants that are defective in germ-cell proliferation and survival. However, given the relative inaccessibility of PGCs in the embryo, the ability to culture these cells in vitro has led to a greater understanding of the mechanism by which growth factors control their proliferation, migration, and differentiation in vivo. This chapter will outline methods to obtain PGCs from various embryonic stages, culture them in vitro, stain them for endogenous alkaline phosphatase activity, and finally generate embryonic germ (EG) cell lines.

Successful in vitro culture systems for PGCs include the use of a mitotically inactivated feeder layer of somatic cells. However, even under these conditions, the number of surviving PGCs in culture decreases dramatically after 5 d. The addition of various growth factors to the culture medium can greatly increase the survival and proliferation of PGCs in vitro and, in some cases, lead to the generation of cell lines resembling blastocyst-derived embryonic stem cells (ES cells). These growth factors include stem cell factor (SCF), leukemia inhibitory factor (LIF), and basic fibroblast growth factor (bFGF). SCF, which is also known as mast cell growth factor, kit ligand, and Steel factor is encoded at the *Steel* locus. Embryos mutant for SCF have PGCs, but they fail to divide and resulting *Steel* mutant mice are sterile. SCF is a transmembrane

From: *Methods in Molecular Biology, Vol. 97: Molecular Embryology: Methods and Protocols*
Edited by: P. T. Sharpe and I. Mason © Humana Press Inc., Totowa, NJ

protein that can be alternatively spliced to produce a soluble form. When PGCs are cultured on feeder cells that express the membrane-bound form of SCF, their survival and proliferation are greatly enhanced *(1–3)*. Another factor made by the feeders is LIF, also known as differentiation inhibiting activity (DIA). It has been shown that by using LIF alone, it is possible to establish and maintain pluripotent blastocyst-derived ES cell lines *(4–6)* and soluble LIF stimulates PGC proliferation in the presence of other factors *(2,3)*. The receptor for LIF is made up of two subunits, one that binds LIF directly and the other, which is a signal transducer called gp130. The gp130 subunit is shared by other cytokines, such as oncostatin M (OSM) and ciliary neurotropic factor (CNTF). OSM and CNTF can substitute for LIF in PGC culture, and neutralizing antibodies against gp130 block PGC survival in culture *(7,8)*, suggesting a role for this receptor in PGC survival and proliferation in vivo. Tumor necrosis factor-$\alpha$ (TNF-$\alpha$) selectively stimulates the proliferation of early PGCs *(9)*, interleukin-4 (IL-4) is a survival factor *(10)*, and retinoic acid is a growth activator of PGCs *(11)*. Also agents, such as dibutyryl cAMP and forskolin, that raise the intercellular concentrations of cAMP also stimulate the proliferation of PGCs in vitro *(12)*. The in vivo functional significance of these growth factors is not yet known.

Long-term proliferation of PGCs was demonstrated by the addition of bFGF to a mixture of SCF and LIF in the in vitro culture system outlined in this chapter *(13–16)*. It has been determined that these growth factors act directly on the PGCs and are not merely an indirect effect on the feeder layers to produce even more growth factors *(7)*. EG cell lines derived from PGCs have many characteristics of ES cells. They can form embryoid bodies in vitro, produce teratomas in nude mice, and even contribute to the germ line of chimeric mice *(13–17)*. However, EG cells are not identical to ES cells with regard to their genomic imprinting. These cell lines may be used in the future to study genomic imprinting, the potency of PGCs at various stages during embryogenesis, and finally as a route to generate pluripotent stem cells from species other than mice.

## 2. Materials

### 2.1. Isolation of Primordial Germ Cells

1. Mice: ICR mice are an outbred strain that can be used to generate large numbers of PGCs. These can be obtained from a number of suppliers, including Taconic Farms (Germantown, NY, 1-800-822-6642) or Harlan Sprague-Dawley (Indianapolis, IN, 1-317-894-7521). Noon on the day of plug is 0.5 d postcoitum (dpc). *See* **Note 1** for discussion of different mouse strains.
2. Dissecting tools: Fine scissors and forceps (watchmaker's #5) for the dissections. Diamond pen for cutting glass pipets. These can be ordered through Fisher (Atlanta, GA, 1-800-766-7000) or VWR (Marietta, GA, 1-800-932-5000). More

specialized dissection equipment is available from Roboz Surgical Instruments (Rockville, MD, 1-800-424-2984) or Fine Science Tools (Foster City, CA, 1-800-521-2109).

3. Tissue-culture equipment, dishes, pipets, and so on, can all be obtained through Fisher or VWR. Manufacturers include Corning, Falcon, and Nunc.

4. Tissue-culture reagents can be purchased from Grand Island Biological Company (Gibco, Grand Island, NY, 1-800-828-6686). Trypsinization solution is 0.25% trypsin, 1 m$M$ EDTA in Hank's balanced salts (cat. no. 25200-056). Dulbecco's phosphate-buffered saline (PBS) is 0.2 g/L KCl, 0.2 g/L $KH_2PO_4$, 8 g/L NaCl, 1.15 g/L $Na_2HPO_4$, and 2.16 g/L $Na_2HPO_4 \cdot 7H_2O$. (cat. no. 14190). Medium is Dulbecco's Modified Eagle's Medium (DMEM) supplemented with 4.5 g/L glucose (cat. no. 11965-084). For the culture of PGCs, this medium is further supplemented with 0.01 m$M$ nonessential amino acids (cat. no. 11140-019), 2 m$M$ glutamine (cat. no. 25030-016), 50 µg/mL gentamycin (cat. no. 15750-011), 0.1 m$M$ 2-mercaptoethanol (cat. no. 21985-023), and 15% fetal bovine serum (ES cell tested, from either Summit Biotechnology, Fort Collins, CO, 1-800-933-0909 or HyClone Laboratories, Logan, UT, 1-800-492-5663). For the routine culture of fibroblast feeder cells, the DMEM with glucose is supplemented with 10% fetal bovine serum, 2 m$M$ glutamine, and 50 µg/mL penicillin-streptomycin (Gibco cat. no. 15070-014). Collagenase/dispase can be purchased from Sigma (1-800-325-3010) (cat. no. C3180), made up to 1% (1 g/100 mL) in either DMEM or PBS, and stored in aliquots at –20°C. This is a 10X concentration. Dimethyl sulfoxide (DMSO) for freezing cell lines is available from Fisher Scientific (cat. no. BP231-1). Bovine serum albumin (BSA) for dissections is available from Sigma (cat. no. A 9647).

5. Feeders for primary cultures are *Sl4-m220* cells (from David Williams, Indiana University School of Medicine), whereas those for the secondary culture are generated from mouse embryos (*see* **Notes 2** and **3** for more details about feeder cells). *Sl4-m220* is a cell line isolated from homozygous *Steel* mutant embryos and stably transfected with the membrane-associated form of SCF. Growth factors necessary for the generation of EG cell lines are LIF (ESGRO™, Gibco cat. no. 13275-011), bFGF (also called FGF2) (Gibco cat. no. 13256-029) and SCF (also known as *Steel* factor, mast cell growth factor, and c-*kit* ligand) (R&D Systems, Minneapolis, MN, 1-800-343-7475, cat. no. 455-MC or Preprotech, Rocky Hill, NJ 1-800-436-9910 and London, 0171 603 8288, cat. no. 250-03). *See* **Note 3**.

6. Feeder layers are mitotically inactivated by γ-irradiation or by mitomycin treatment. Mitomycin can be obtained from Sigma (cat. no. M0503), and a 1 mg/mL stock is stored frozen in PBS in a dark container. Mitomycin C is very toxic, and care should be used when handling (gloves, mask, and so on) (*see* **Note 4**).

7. Dissecting microscope with low-power objectives (10×–80×).

8. A mouth-controlled pipet should be used for most of the manipulations. They can be purchased from Fisher (cat. no. 258616).

## 2.2. Staining for Endogenous Alkaline Phosphatase Activity

1. Fixative (4% paraformaldehyde in PBS, Fisher cat. no. 04042-500) (*see* **Note 5**).

2. Alkaline phosphatase staining solution: 25 m$M$ Tris-malate, pH 9.0, 0.4 mg/mL

Na-α-napthyl phosphate (Sigma cat. no. N7255), and 1 mg/mL fast red TR salt (Sigma cat. no. F2768), 8 m$M$ MgCl$_2$. This staining solution must be made up fresh before use (<30 min). The color reaction is dependent on the high pH of the staining solution.
3. PBS (same as above).

## 2.3. Determination of the Sex of Embryos or EG Cell Lines by PCR for Zfy Gene

1. Lysis buffer, 100 m$M$ EDTA, 50m$M$ Tris-HCl, pH 8.0, 100 m$M$ NaCl, 1% SDS. Proteinase K should be added at 0.5 mg/mL immediately before use. Proteinase K (Sigma cat. no. P2308) is made up in water at 10 mg/mL, self-digested at 37°C for 60 min and stored in aliquots at –20°C.
2. Oligonucleotide primers. 5'-primer AAGATAAGCTTACATAATCACATGGA, 3'-primer CCTATGAAATCCTTTGCTGCACATGT. A stock solution of these primers is 100 μ$M$ and should be kept frozen in aliquots at –20°C. Oligonucle-otides can be purchased from Operon Technologies (510-865-8644).
3. 10X PCR buffer, 100 m$M$ Tris-HCl, pH 8.3, 500 m$M$ KCl, 25 m$M$ MgCl$_2$, 0.1% gelatin. Store frozen at –20°C.
4. dNTPs (Pharmacia 1-800-526-3593, cat. no. 27-2035-01). The purchased dNTP stock is 100 m$M$ and should be kept frozen at –20°C. 10X working solution 10 m$M$ each dNTP) is made by combining 10 μL of each of the four stocks and adding 60 μL of water.
5. *Taq* polymerase (Stratagene 1-800-424-5444, TaqPlus™ DNA Polymerase cat. no. 600203). As with any enzyme, *Taq* should be stored at –20°C in a nonfrost-free freezer.

## 3. Methods
### 3.1. Isolation of Primordial Germ Cells

Primordial germ cells can be obtained from different stages of embryos. **Subheading 3.1.1.** will describe how to dissect early embryos (8.5 d pc) in order to obtain early migratory PGCs, whereas **Subheading 3.1.2.** will describe how to isolate the genital ridge from later embryos (11.5 d pc and 15.5 d pc) in order to obtain later PGCs. **Subheading 3.1.3.** describes how these isolated tissues should be processed for in vitro culture. Further information about the numbers of PGCs and their location in the embryo at various times can be found in **refs. *18*** and ***19***.

Since the ultimate goal of these dissections is to generate PGCs for in vitro culture, care should be taken to maintain sterility at all times. Dissections are done using sterile PBS with 0.1% BSA (0.1 g/100 mL) and sterilized instruments.

### 3.1.1. Dissection of 5–8 Somite Stage Embryos

In the 8.5 d pc embryo, most of the PGCs are located in the posterior third of the embryo at the base of the allantois. After the embryo is removed from the uterus, decidual tissue, and yolk sac, it will appear as in **Fig. 1A**. Details of this

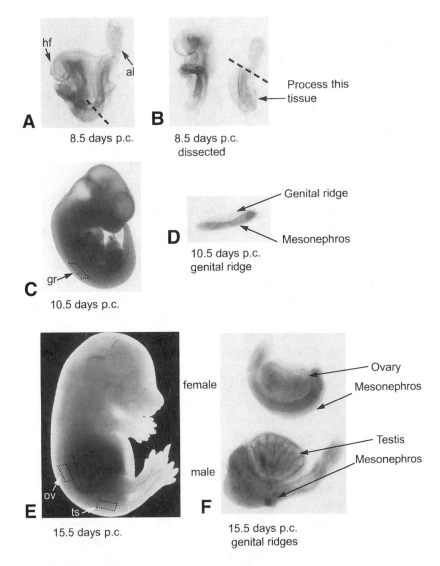

Fig. 1. (**A,B**) A 8.5-d pc embryo dissected free of the uterus, decidual tissue, and extraembryonic membranes. At this stage, the majority of the PGCs are located at the base of the allantois. In order to isolate the PGC-containing tissue, the embryo is cut at the position of the dashed lines and the indicated tissue processed further. (**C**) An embryo at 10.5 d pc with the position of the genital ridges indicated by the box. (**D**) An isolated 10.5-d pc genital ridge showing the position of the mesonephros relative to the developing gonad. (**E**) An embryo at 15.5 d pc with the positions of the developing ovary or testis (depending on the sex of the embryo) illustrated by the boxes. (**F**) Isolated gonads from 15.5-d pc embryos. Note the well-differentiated tubules in the developing testes. Abbreviations: al, allantois; gr, genital ridge; hf, head folds; ov, ovary; ts, testes.

dissection can be found in **ref. 20**. Briefly, the embryos are removed from the uterus, and the broader end of the decidua is cut away. The embryo must then be gently shelled out of the remaining decidual tissue and yolk sac. In order to isolate large numbers of PGCs, the anterior two-thirds of the embryo and the allantois are discarded, as indicated in **Fig. 1A** and **B**. The remaining tissue is then processed as in **Subheading 3.1.3.**

### 3.1.2. Dissection of Genital Ridges and Embryonic Gonads

Alternatively, if later PGCs are desired, they can be obtained from the developing genital ridge. After the 8.5 d pc stage, the PGCs become incorporated into the hindgut of the embryo. They subsequently migrate into the dorsal mesentery and then the genital ridge by 10.5–11.5 d pc. By this stage of embryogenesis, the genital ridges are attached to the dorsal body wall on either side of the spinal column. Their position inside the embryo is illustrated in **Fig. 1C**. At 10.5 d pc, it is often difficult to differentiate between a developing testis and a developing ovary in the light microscope as shown in **Fig. 1D**. In order to remove the genital ridges, the embryo is dissected out of the uterus and the extraembryonic tissues are removed. The embryo should be killed by pinching off the head with forceps. Next, the intestines and liver are removed from the body cavity. The genital ridges lie attached to the dorsal wall of the body cavity, adhered tightly to the developing mesonephros. A genital ridge from a 10.5-d pc embryo is shown in **Fig. 1D**. The genital ridge and the mesonephros are gently teased apart with forceps, and the genital ridge is processed as in **Subheading 3.1.3.** Be careful to avoid disrupting the integrity of the genital ridge, since PGCs can flow out if it is poked or torn.

Using a similar approach, germ cells can be isolated from later developing gonads. Shown in **Fig. 1F** are testes and ovaries from 15.5-d pc embryos. At this stage, it is now easy to distinguish between a testis and an ovary, since the tubules in the testes are well differentiated. Again, the embryo should be sacrificed by removing the head, and then the intestines and liver are removed in order to isolate the gonads. Testes will be located in the caudal region of the pelvis (above the hips), but ovaries will be located further rostrally near the kidneys. Their location is illustrated in **Fig. 1E**.

### 3.1.3. Isolating Germ Cells for Culture

Depending on the experiment, PGC-containing tissues are either pooled at this stage or treated individually. Isolated tissues are kept on ice while the remaining dissections are completed. Using a mouth pipet (*see* **Note 6**), the isolated tissue is rinsed through at least two 30-µL drops of PBS, and then placed into a drop of trypsin, being careful to minimize the amount of PBS transferred so as not to dilute the trypsin. The tissue is mechanically disrupted by tearing with forceps, and the sample is incubated at 37°C for 5 min. Then the tissue is again mechanically disrupted using a finely pulled Pasteur pipet. The internal diameter of the pipet should be approx 75% smaller than the tissue that is being treated. It is often necessary to use a series of pipets, decreasing in diameter, as the tissue is disrupted. The internal diameter of the pipets range between 200 μm for the largest genital ridges down to 30–50 μm for single cells. Depending on how well the cells are disrupted, the plate can be returned to the 37°C incubator for an additional 5 min. This is repeated until the tissue has been reduced to a single-cell suspension. This should be fairly easy for a 8.5-d pc embryo and a 11.5-d pc genital ridge. However, the size and extent of development of a 15.5-d pc genital ridge makes more vigorous enzymatic treatment necessary. In this case, a mixture of collagenase and dispase can be used at 37°C. Again, it is best to disrupt the tissue mechanically as much as possible before enzymatic treatment, and also to check the tissue at 5-min intervals, incubating for up to 30 min total. Despite these efforts, it is often difficult to dissociate the testes tubules, and any remaining large clumps of intact tissue should not be plated out in the next step. However, plenty of germ cells will be released from the tubules and can be collected with a mouth pipet (*see* **Note 7**).

Once the tissue is primarily a single cell suspension of somatic cells and PGCs, the cells are plated onto mitotically inactivated feeder layers. One kind of feeder cells that can be used is *Sl4m220* fibroblasts. Alternatively, a mouse embryonic fibroblast line, STO, can also be used (*see* **Note 3**). Feeder cells are mitotically inactivated with either mitomycin C or irradiation (*see* **Note 4**) and plated onto gelatin coated tissue culture dishes at a density of $1.5 \times 10^5$ cells for each well of a 24-well dish (1 cm in diameter). For convenience, feeder layers are usually prepared the day prior to the dissection. However, most feeder cells have adhered by 2–4 h after plating, so the feeder plates could be prepared the same day as the dissection.

At this stage of the procedure, cells from the equivalent of one 8.5-d pc embryo are plated into each well. Similarly, 1/10 of a 11.5-d pc or 15.5-d pc genital ridge is placed into each well. Growth factors are added as desired. If the goal is to produce EG cell lines, the growth factors should be added in the following concentrations: 20 ng/mL LIF, 60 ng/mL SCF, and 20 ng/mL bFGF. All cultures are maintained in a humidified incubator at 37°C with 5% $CO_2$. Cultures are fed every day with fresh medium and growth factors, and monitored for possible yeast and/or bacterial contamination. It beyond the scope of this chapter to describe the effects of all tested growth factors and cytokines on the growth and differentiation of PGCs. An excellent review of these experiments is contained in **ref. *21*** and references therein.

If the generation of EG cell lines is desired, a secondary culture is generated from the primary culture after 10 d. It is a good idea to stain some portion of the primary cultures for alkaline phosphatase activity to be sure that the PGCs are present and growing before proceeding further. If there are no alkaline-positive cells in the primary culture, the experiment should be stopped at this time. To begin a secondary culture, primary cultures are washed in PBS and trypsinized for 5 min at 37°C. Add an equal volume of serum-containing culture medium to inactivate the trypsin, and pipet up and down 5–10 times to obtain a single-cell suspension. This suspension is then distributed onto fresh feeder plates at a 1:4–1:10 dilution. Secondary cultures need to be monitored daily for the appearance of EG colonies. These colonies can then be isolated and grown as cell lines, although it is not clear that they are clones and they will undergo changes in culture *(16)*. Once EG cell lines are generated, they can be grown in the same manner as ES cell lines on feeder layers from either primary mouse embryo fibroblasts or STO fibroblasts, with only LIF added to the medium, since bFGF and SCF are no longer necessary. Single-cell suspensions of individual cell lines should be frozen at low passage numbers in a freezing medium of 10% DMSO, 20% fetal bovine serum in culture medium, and stored indefinitely under liquid nitrogen. It should be noted that despite many attempts in this laboratory and others *(16)*, EG cell lines have not been generated derived with PGCs from late genital ridges (over 12.5 d pc).

### 3.2. Staining for Endogenous Alkaline Phosphatase Activity

The yield and growth of PGCs can be measured by staining the cultures for endogenous alkaline phosphatase activity. Unfortunately, the cells must be fixed in order to stain for AP, so it is a good idea to have plenty of starting cultures so that a few wells can be stained every day in order to monitor PGC growth.

1. Wash cultures twice in PBS. Washing is done by adding fresh PBS, swirling the plate gently, removing the PBS, and adding the fresh PBS. A long incubation is not necessary.
2. Fix cultures in paraformaldehyde for 20 min at room temperature (*see* **Note 5**).
3. Wash twice in PBS.
4. Wash once in 25 m*M* Tris-malate, pH 9.0.
5. Stain for 15–20 min at room temperature, or until PGCs are a reddish-brown color. This can be monitored in the microscope.
6. Wash twice in PBS, and leave cultures in liquid to photograph.

### 3.3. Determination of the Sex of Embryos or EG Cell Lines by PCR for Zfy Gene

Depending on the nature of the experiments, it may be desirable to know the sex of either the embryos used to culture PGCs or of any EG cell lines generated. This can be done easily by using PCR to amplify sequences found only on the Y chromosome *(16,22)* (*see* **Note 8**).

1. To generate DNA samples from embryos or cell lines, tissue is incubated in lysis buffer with 0.5 mg/mL proteinase K overnight at 56°C. On the next day, samples are extracted once with phenol, once with phenol:chloroform:isoamyl alcohol (25:24:1), and once with chloroform:isoamyl alcohol (24:1), and then ethanol-precipitated. References for these molecular biology techniques can be found elsewhere in this volume or in **ref. 23**.
2. PCR reactions are performed in a temperature cycler with 400 ng of genomic DNA as a template. Reactions are set up as follows:

    $x$ µL DNA (400 ng).

    5 µL dNTPs (10 m$M$ each nucleotide).

    5 µL 10X PCR buffer.

    1 µL each primer.

    1 µL *Taq* (diluted 1:5 in 1X PCR buffer).

    Water to 50 µL.
3. Reactions are cycled at 95°C for 45 s, 62°C for 25 s, and 72°C for 1 min for a total of 30 cycles.
4. Run 1/10 or more of the completed reaction on a 1% agarose gel in order to detect the presence of a 600-bp reaction product indicating the presence of a Y chromosome.

## 4. Notes

1. ICR mice have been used for many PGC studies. However, depending on the objective of the experiment, and especially if EG cell lines are to be generated, it may be desirable to use inbred mouse strains, such as C57BL/6 or 129/Sv, to be assured of a uniform genetic background. Additionally, both of these have been used to generate EG cell lines that will contribute to the germ line of chimeras *(15–17)*. Inbred mouse lines, such as C57BL/6 and 129/Sv, are available from Taconic Farms, Harlan Sprague-Dawley (**Subheading 2.1., item 1**), and The Jackson Laboratories (Bar Harbor, MA, 1-207-288-3371).
2. Primary mouse embryonic fibroblasts (mefs) can be generated from embryos between 12.5 and 14.5 d pc, but with 13.5 d pc embryos giving the highest yield of cells. Embryos are dissected free of their extraembryonic membranes and the heads, limbs, and visceral organs removed. The carcasses are then passed five times through an 18-gage needle (approx 5 embryos/syringe) in approx 3 mL of PBS, and the resulting cell mixture is plated into a 150-mm tissue-culture dish with fibroblast culture medium. After 3 d, the cultures are trypsinized and replated at a dilution of 1:4, and this secondary culture can then be mitotically inactivated or cultured further to generate more cells. These feeders can be frozen in 10% DMSO/20% fetal bovine serum in DMEM and stored indefinitely in liquid nitrogen. *See* **Subheading 2.1., item 4** for a description of tissue-culture reagents and suppliers.
3. It is absolutely essential to use feeder cells that are producing SCF on their cell surface. If *Sl4m220* cells are not available, STO fibroblasts (American Type Culture Collection [ATTC]) can be used *(14)*. STO cells are available from the ATTC (1-800-638-6597, cat. no. ATCC CRL 1503). Primary mefs do not make sufficient amounts of SCF to allow the PGCs to survive and divide *(15)*. It also may

be important to use species-specific growth factors. For example, rat SCF can be used to promote the survival mouse PGCs, but human SCF will not affect the growth of mouse PGCs. This may be a limiting factor for the generation of EG cell lines in species other than mouse, but this information is primarily anecdotal.

4. To inactivate feeder cells mitotically, use >5000 rad of γ-irradiation. Alternatively, mitomycin C can be added to the medium at 10 μg/mL and incubated in the 37°C incubator for 2–3 h. The cells are then washed very well with PBS (five changes of PBS), and trypsinized and plated as usual. Both methods are equivalent, but irradiation requires an expensive and specialized machine, so most investigators use mitomycin C. It is important to take safety precautions with mitomycin C (wear gloves and a mask), since it is very toxic.

5. Paraformaldehyde must be dissolved in PBS and heated to 65°C before it will go into solution. Care should be used when handling paraformaldehyde in order to avoid contact with fumes as well as the liquid or powder form.

6. For many manipulations of the starting cell cultures, a finely pulled Pasteur pipet is used with mouth control. It generally takes some practice both to use the mouth pipet as well as to pull the Pasteur pipets. The long, thin end of the pipet is broken off the wider part by using a diamond pen to score the glass. Then, the center of the long, thin part is softened in the small flame of a microburner or alcohol burner (a traditional Bunsen burner produces too violent a flame for this purpose), removed quickly from the flame, and pulled in one quick motion. Score the thin glass between the two ends with a diamond pen and break the glass, generating blunt ends. If necessary, the ends can be flame polished to make a smoother opening. Glass capillary tubes can also be used for this. The internal diameter of the pipet will vary with the task, so it is often a good idea to pull many pipets of various sizes (from 30 to 200 μm internal diameter). It is also important that these pipets be kept sterile so that the resulting cultures do not become contaminated. This is most easily accomplished by making the pipets shortly before use and keeping the pipets stored in a 150-mm tissue-culture dish during the dissection.

7. Most of the dissections are done with bright-field illumination, but at the end of the tissue dissociation, the use of dark-field illumination makes it easiest to see single cells. This is especially important in order to retrieve single cells from dissociated genital ridges.

8. PCR for the *Zfy* gene detects the presence of a Y chromosome. However, it should be noted that an EG cell line that is negative for this PCR product may actually be an XO cell line, not a normal XX female cell line. It would be necessary to perform karyotype analysis to determine the difference between an XO and an XX cell line accurately.

## References

1. Dolci, S., Williams, D. E., Ernst, M. K., Resnick, J. L., Brannan, C. I., Lock, L. F., et al. (1991) Requirement for mast cell growth factor for primordial germ cell survival in culture. *Nature* **352,** 809–811.
2. Matsui, Y., Toksoz, D., Nishikawa, S., Nishikawa, S.-I., Williams, D., Zsebo, K., et al. (1991) Effect of *Steel* factor and leukemia inhibitory factor on murine primordial germ cells in culture. *Nature* **353,** 750–752.

3. Godin, I., Deed, R., Cooke, J., Zsebo, K., Dexter, M., and Wylie, C. C. (1991) Effects of the *steel* product on mouse primordial germ cells in culture. *Nature* **352,** 807–809.

4. Williams, R. L., Hilton, D. J., Pease, S., Wilson, T. A., Stewsart, C. L., Gearing, D. P., et al. (1988) Myeloid leukaemia inhibitory factor maintains the developmental potential of embryonic stem cells. *Nature* **336,** 684–687.

5. Smith A. G., Heath, J. K., Donaldson, D. D., Wong, G. G., Moreau, J., Stahl, M., et al. (1988) Inhibition of pluripotential embryonic stem cell differentiation by purified polypeptides. *Nature* **336,** 688–690.

6. Pease, S. and Williams, R. L. (1990) Formation of germ-line chimeras from embryonic stem cells maintained with recombinant leukemia inhibitory factor *Exp. Cell Res.* **190,** 209–211.

7. Cheng, L., Gearing, D. P., White, L. S., Compton, D. L., Schooley, K., and Donavan, P. J. (1994) Role of leukemia inhibitory factor and its receptor in mouse primordial germ cell growth. *Development* **120,** 3145–3153.

8. Koshimizu, U., Taga, T., Watanabe, M., Saito, M., Shirayoshi, Y., Kishimoto, T., et al. (1996) Functional requirement of gp130-mediated signalling for growth and survival of mouse primordial germ cells in vitro and derivation of embryonic germ (EG) cells. *Development* **122,** 1235–1242.

9. Kawase, E., Yamamoto, H., Hashimoto, K., and Nakatsuji, N. (1994) Tumor necrosis factor-α (TNF-α) stimulates proliferation of mouse primordial germ cells in culture. *Dev. Biol.* **161,** 91–95.

10. Cooke, J. E., Heasman, J., and Wylie, C. C. (1996) The role of interleukin-4 in the regulation of mopuse primordial germ cell numbers. *Dev. Biol.* **174,** 14–21.

11. Koshimizu, U., Watanabe, M., and Nakatsuji, N. (1994) Retinoic acid is a potent growth activator of mouse primordial germ cells *in vitro. Dev. Biol.* **168,** 683–685.

12. DeFelici, M., Dolci, S., and Pesce, M. (1993) Proliferation of mouse primordial germ cells *in vitro:* a key role for cAMP. *Dev. Biol.* **157,** 277–280.

13. Matsui, Y., Zsebo, K., and Hogan, B. L. M. (1992) Derivation of pluripotent embryonic stem cells from murine primordial germ cells in culture. *Cell* **70,** 841–847.

14. Resnick, J. L., Bixler, L. S., Cheng, L., and Donovan, P. J. (1992) Long-term proliferation of mouse primordial germ cells in culture. *Nature* **359,** 550–551.

15. Labosky, P. A., Barlow, D. P., and Hogan, B. L. M. (1994) Embryonic germ cell lines and their derivation from mouse primordial germ cells, in *Germline Development,* Ciba Foundation Symposium 182, Wiley, Chichester, pp. 68–91.

16. Labosky, P. A., Barlow, D. P., and Hogan, B. L. M. (1994) Mouse embryonic germ (EG) cell lines: transmission through the germline and differences in the methylation imprint of insulin-like growth factor 2 receptor (Igf2r) gene compared with embryonic stem (ES) cell lines. *Development* **120,** 3197–3204.

17. Stewart, C. L., Gadi, I., and Blatt, H. (1994) Stem cells from primordial germ cells can reenter the germ line. *Dev. Biol.* **161,** 626–628.

18. Ginsberg, M., Snow, M. H. L., and A. McLaren. (1990) Primordial germ cells in the mouse embryo during gastrulation. *Development* **110,** 521.

19. Gomperts, M., Garcia-Castro, M., Wylie, C., and Heasman, J. (1994) Interactions between primordial germ cells play a role in their migration in mouse embryos. *Development* **120,** 135–141.
20. Hogan, B. L. M., Beddington, R., Costantini, F., and Lacy, E. (1994) *Manipulating the Mouse Embryo: A Laboratory Manual,* 2nd ed. Cold Sping Harbor Laboratory, Cold Spring Harbor, NY.
21. Donovan, P. J. (1994) Growth factor regulation of mouse primordial germ cell development. *Curr. Topics Dev. Biol.* **29,** 189–225.
22. Nagamine, C. M., Chan, K., Kozak, C. A., and Lau, Y. (1989) Chromosome mapping and expression of a putative testis-determining gene in mouse. *Science* **243,** 80–83.
23. Sambrook, J., Fritsch, E. F., and Maniatis, T. (1989) *Molecular Cloning: A Laboratory Manual.* Cold Spring Harbor Laboratory Press, Cold Spring Harbor, NY.

# II

## CHICKEN EMBRYO

# 13

# The Avian Embryo

*An Overview*

**Ivor Mason**

## 1. Origins of Avian Embryology

The major advantage of the avian embryo for the embryologist is its accessibility for manipulation and observation. Indeed, it is for this reason that, historically, detailed descriptions of normal development were first available for avian embryos, generally chick embryos *(1)*. Artificial incubation and hatching of chicken eggs date to the time of the 18th dynasty in Egypt (ca. 1400 BC), and possibly even earlier in ancient China. The Egyptian practice of egg incubation is well documented in Roman literature, including references by Pliny, Diodorus Siculus, and in the letters of Emperor Hadrian. However, artificial incubation was lost throughout the Middle Ages and was only revived during the 18th century. The first recorded observations of avian embryos is included in the works attributed to Hippocrates (ca. 430 BC), although it was Aristotle who provided the first significant observations (*Historia Animalium* and *De Generatione Animalium*, ca. 350 BC). Detailed and illustrated accounts were published between the late 16th and 18th centuries written by Aldrovanus (*Ornithologia*, 1597) Fabricius (*De Formatione Ovi et Pulli*, 1604), Harvey (*De Generatione Animalium*, 1651), Shrader (*Observations et Historiae*, 1674), Malpighi (*De Ovo Incubato* and *De Formatione Pulli in Ovo*, 1672), Mayow (*De Respiratione Foetus in Utero et Ovo*, 1674), Maître-Jan (*Observations sur la formation du poulet*, 1722), and Haller (*Sur la formation du coeur dans le poulet*, 1767). The first attempts to incubate eggs with part of the shell removed to form a "window" were described by Beguelin (*Memoir sur l'art de couver les oeufs overts*, 1749). From the 19th century, there was an explosion of publications relating to the anatomy, physiology, and biochemistry of the chick embryo to be followed by more invasive procedures during the present century.

From: *Methods in Molecular Biology, Vol. 97: Molecular Embryology: Methods and Protocols*
Edited by: P. T. Sharpe and I. Mason © Humana Press Inc., Totowa, NJ

## 2. Advantages of the Avian Embryo for Experimental Embryology

The avian embryo offers a number of distinct advantages for embryonic investigations. The legacy of its long history of descriptive studies coupled with physiological and biochemical studies of the 19th century and the experimental embryology of the 20th century is an enormous bibliography concerning system and organ development. It is readily accessible between laying (blastoderm stage) and hatching and the comparatively large size of the embryo, even at gastrulation stages, greatly facilitates microsurgical manipulation. Accessibility can be further improved by *in vitro* culture methods pioneered by New (*2; see* Chapter 15) that allow early embryos to be maintained for 3 d on culture rings. Older embryos can be cultured to term by other methods. Birds, like mammals, are amniotes, but are much cheaper to purchase and generally have no associated animal housing costs. Microsurgical manipulations are readily performed until about embryonic d 8, by which time all of the major patterning events are complete and all of the organ systems are established. Tissue grafting has facilitated the determination of instructive and permissive tissue interactions important in sculpting the embryo and facilitated studies of position effects, lineage, commitment, and determination (*see*, e.g., Chapters 16–19, and 22). Many lineage-tracing studies have been undertaken in the chick embryo using engrafted quail tissue, which is distinguished cytologically or immunochemically (*see* Chapter 22), or by application of fluorescent tracers (*see* Chapter 23). The development of avian retroviral vectors has facilitated the ectopic expression of genes (*see* Chapter 35) and the exciting new development of hammerhead ribozyme vectors (*3,4*) offers the prospect of being able to inhibit expression of any gene in this embryo. The field owes a considerable debt to Hamburger and Hamilton (*5*), who provided a detailed staging regime using definitive criteria for the chick embryo; a similarly detailed staged series is not available for the mouse embryo. For stages between fertilization and gastrulation, Eyal-Giladi and Kochav (*6*) provide a further staging regime. The recently published *Atlas of Chick Development (7)* is recommended to those undertaking histological studies and also provides a good first reference source for descriptions of early development and organogenesis.

There are, however, some areas in which other vertebrate embryos offer advantages over avian embryos. First, the chicken (or any other bird) has little genetics, and it is doubtful whether this will change greatly in the foreseeable future. A linkage map is being constructed, but the chicken has about 80 chromosomes; many of them are microchromosomes and are difficult to distinguish one from another. Unfortunately, it seems as if most of the transcribed genes reside on these microchromosomes. It is now clear that transgenic chickens can be generated, but the methods are more laborious than for mice, and the costs of maintaining transgenic flocks will largely preclude this technology except for agricultural purposes. Finally, the very earliest stages of development, from fertilization to formation of the blastoderm, occur prior to laying and are therefore not readily accessible.

## 3. A Brief Overview of Early Chick Development

The following is a brief descriptive account of the early development of the chick embryo. Excellent photographs of embryos at gastrulation and later stages are included in Chapter 15.

### 3.1. Fertilization to Laying

During the first reduction division involved in generation of the oocyte, the latter expands from a diameter of about 1–3.5 cm. Most of the content of this extremely large single cell is yolk surrounded by the plasma membrane. The egg nucleus and associated cytoplasm are located peripherally in a region called the germinal vesicle. The "yolk" beneath this structure is clear and less dense than the remaining yellow yolk, causing the germinal vesicle to lie on the uppermost surface of the oocyte. Follicle cells surround the oocyte and pump yolk, synthesized in the maternal liver, from the blood into the oocyte. The follicle ruptures to release the mature oocyte, but the innermost part of the follicle, which is accellular, remains attached to the oocyte and forms the inner layer of the vitelline membrane. When the follicle-derived vitelline membrane is in place, the plasma membrane of the oocyte breaks down.

Fertilization occurs during the time between release of the oocyte and its entry into the end of the oviduct. Sperm penetrate the follicle-derived vitelline membrane to fertilize the oocyte and the second reduction division occurs. The resultant one-cell stage embryo is called a blastodisc. The requirement for rapid fertilization following follicle rupture is met by the female chicken's ability to store sperm in viable form for a number of weeks.

Peristaltic movements carry the egg down the oviduct, a journey that takes about 22 h and, during which the egg is subject to a number of modifications. First, a thickened outer layer is applied to the vitelline membrane, which has two extensions, chalazae, which act as stabilizers for the yolk. Albumen is then applied to the outer surface providing a source of water, protein, and antibiotic agents. Next, the double-layered shell membrane is applied, the layers of which are closely apposed to one another, except at the blunt end of the egg, where the gap between them eventually becomes the air space. Finally, in the shell gland, calcite crystals are deposited in and over the outer shell membrane layer to form the eggshell.

Cleavage of the blastodisc is rapid. The first two cleavage divisions occur within a few hours of fertilization and are not related to future body axes. The third cleavage coincides with application of shell membrane. The chick is telolecithal (yolk is concentrated at one end of the egg), and cleavage is meroblastic (incomplete). The first three divisions are radial and incomplete with the cells being opened to the yolk ventrally forming a syncytial blastoderm. The fourth cleavage division is horizontal, producing a bilayered blastodisc and occurs as the shell is being applied. Further divisions increase the thickness of the blastoderm, but the diameter of the embryo remains roughly constant at 3 mm during this period and zygotic transcription is activated.

Subsequently, the blastoderm begins to expand over the yolk and marginal cells of the outer region, known as the area opaca, become specialized to engulf the underlying yolk. The more central region, area pellucida, appears dark owing to the underlying translucent "yolk." The area pellucida comprises an upper layer known as the epiblast from which the embryonic tissues derive and a lower layer of large, yolky cells, the hypoblast, which comprises the extraembryonic endoderm. These two layers are separated by a narrow fissure, which is equivalent to the blastocoel. The hypoblast derives in part from ingression of cells from the overlying blastoderm and in part from the posterior marginal zone. The hypoblast forms a triangle posteriorly, the embryonic shield or posterior marginal zone, and is generated particularly from an adjacent region of epiblast known as Koller's sickle. At this stage, the egg is laid (Hamburger and Hamilton [HH] stage 1) and comprises about 60,000 cells.

### 3.2. Gastrulation (HH Stages 2–4)

A major function of hypoblast is to initiate gastrulation through formation of the primitive streak in the overlying epiblast through which epiblast cells migrate to form prospective endoderm and mesoderm. Thus, the initial position of the hypoblast determines the body axis. At the onset of streak formation, the blastoderm is 5–6 mm in diameter. The primitive streak begins to extend forward from the posterior of the area pellucida and ingressing epiblast cells, which mostly have an endodermal fate, migrate anteriorly and centrifugally displacing the hypoblast. Extension of the streak (between HH stages 3 and 4) is by recruitment of anterior cells rather than by cell movement from the existing streak. Hypoblast cells driven to the anterior of the area pellucida will form the primary germ cells. As gastrulation proceeds, the area pellucida changes from being round to becoming pear-shaped with the expanded end anterior.

Lateral epiblast cells converge toward the streak, invaginate, extend, and diverge ventrally. Those cells of the epiblast that do not involute are fated to form the ectoderm and neuroectoderm, and divide to compensate for the loss of ingressing cells. At intermediate streak stages (HH stage 3, 12 h postlaying), prospective mesoderm begins to ingress through caudally, whereas endoderm is still ingressing through rostral streak. At this stage, the anterior end of the streak broadens to form Hensen's node, a structure roughly equivalent to the organizer of *Xenopus* embryos and the Shield of zebrafish. As the streak reaches its longest extent (HH stage 4, 16–20 h), prospective mesoderm cells fated to contribute to lateral plate, somites (posterior streak levels), heart (midstreak), and notochord (head process; derived form the deep node) are migrating through the streak.

## 3.3. Neural Induction and Neurulation

Studies with molecular markers suggest that a neural plate is established by HH stage 5 (20–22 h). After this stage, the streak and node begin to regress posteriorly, a process that continues until about the 20 somite stage, after which they are no longer visible and their remnants become incorporated into the tail bud.

At stage 5, the first formation of the "head process" becomes apparent. It is a short aggregation of mesoderm directly anterior to the node and continuous with prechordal mesoderm, and it derives from the deep part of the node (primitive pit). The head process condenses to form the notochord, which is readily apparent anterior to the node as the latter starts to regress at subsequent stages. The first pair of somites condense either side of the notochord at HH stage 7, and these, and the subsequent four pairs, lie beneath part of the neural plate fated to form the hindbrain. The remainder underlie prospective spinal cord. The cardiac primordium begins to form at the anterior end of the embryo at HH stage 8 (four somites).

The neural plate comprises a pseudostratified epithelium. Between its caudal tip and the prospective infundibulum, its midline overlies the notochord, which induces the overlying neuroepithelial cells to become the floorplate. Anterior to the infundibulum the neural plate overlies prechordal mesoderm. Lateral to the midline, the neural plate overlies paraxial mesoderm: the segmented somitic mesoderm posteriorly (behind the prospective otocyst) and unsegmented cranial paraxial mesoderm anteriorly.

Neural tube closure (neurulation) begins at stage 8 at the level of the midbrain and extends both anteriorly and posteriorly. Closure at the rostral extremity (anterior neuropore) is complete by stage 10, whereas the posterior neuropore remains open until the tail bud develops. Soon after closure, neural crest emerges from the midbrain and hindbrain (HH stages 10–12; 36 h) and, later, from the spinal cord.

Simultaneously with neurulation, the embryo also folds ventrally to enclose the gut and bring the two heart primordia together to fuse.

## 3.4. Later Development

At or before 36 h (HH stage 10), the three germ layers are present, the body axis is established and anteroposterior patterning is well under way, left–right asymmetry is established early during gastrulation, and dorsoventral patterning is ongoing. Somites are being continuously generated and will give rise to the dermis, musculature, and axial skeleton. Neural crest migration has commenced to give rise to derivatives that include the nonplacodal parts of the peripheral nervous system, cranial skeletal elements, some smooth muscle, and melanocytes.

During the third day of development (HH stages 13–19), the head begins to rotate to the left, and the optic vesicle, otic vesicle, nasal pits, branchial arches, and pituitary begin to develop. The limb buds begin to form and project from the trunk during this period, and the amnion extends over the embryo and closes. During the following 48 h or so, the majority of the organ systems have been established such that between the sixth day and hatching, much of development is concerned largely with increase in size of existing organs, although in the case of the nervous system, this is accompanied by considerable increase in complexity.

## Acknowledgments

Work in the author's laboratory is supported by program grants from the MRC and Human Frontier Science Program and project grants form the MRC and the Wellcome Trust.

## References

1. Needham, J. (1934) *A History of Embryology*. Cambridge University Press, UK.
2. New, D. A. T. (1966) *The Culture of Avian Embryos*. Logos, London.
3. Zhao, J. J. and Lemke, G. (1998) Selective disruption of neuregulin-1 function in vertebrate embryos using ribozyme-tRNA transgenes. *Development* **125**, in press.
4. Zhao, J. J. and Lemke, G. (1998) Rules for ribozymes. *Mol. Cell. Neurosci.* **8**, in press.
5. Hamburger, V. and Hamilton, H. (1951) A series of normal stages in the development of the chick embryo. *J. Morphol.* **88**, 49–92.
6. Eyal-Giladi, H. and Kochav, S. (1975) From cleavage to primitive streak formation; a complementary Normal table and a new look at the first stages of development of the chick. I. General morphology. *Dev. Biol.* **49**, 321–337.
7. Bellairs, R. and Osmons, M. (1998) *The Atlas of Chick Development*. Academic, London.

# 14

# Chick Embryos

*Incubation and Isolation*

**Ivor Mason**

## 1. Introduction

The following methods comprise a brief description of chick egg incubation and embryo harvesting. For further details concerning optimizing embryo viability during incubation, the reader is referred to the excellent work of New *(1)*. Embryos are staged according to the staging series devised by Hamburger and Hamilton *(2)*, which has recently been reproduced *(3)*. The reader is also referred to Chapter 15. A further, detailed staging series for very young embryos has also been devised *(4)*.

## 2. Materials

1. Howard's Ringer. (0.12 $M$ NaCl, 0.0015 $M$ CaCl$_2$, 0.005 $M$ KCl (Per liter: 7.2 g NaCl, 0.17 g CaCl$_2$, 0.37 g KCl, pH 7.2, with very dilute HCl) containing Pen/Strep (AAM, Gibco, Paisley, Scotland).
2. Pair of curved, sharp-pointed scissors.
3. Fine watchmakers' forceps (#5 or 55).
4. Pair of spring scissors (Vannas scissors).
5. Stainless-steel spatula, 5- to 8-mm wide tip.
6. Dissecting microscope with magnification to 50× (minimum), preferably greater and transmitted, and incident illumination, preferably from a cold light source.
7. Whatman 3MM filter paper (Millipore, Southampton, UK).
8. Glass (9 cm) and plastic (3 cm) Petri dishes.

From: *Methods in Molecular Biology, Vol. 97: Molecular Embryology: Methods and Protocols*
Edited by: P. T. Sharpe and I. Mason © Humana Press Inc., Totowa, NJ

## 3. Methods

### 3.1. Storage of Unincubated Eggs

At the time an egg is laid, the embryo has already undergone cleavage, and gastrulation has just started; it comprises a bilaminar blastoderm of about 60,000 cells. Conveniently, however, if the egg is not incubated, further development is virtually arrested at this stage, but normal development is resumed when the temperature is raised. Unincubated eggs can remain viable for about 7–10 d if stored at room temperature, although the proportion of nonviable embryos will increase during this time. The number of viable embryos is increased by storage of unincubated eggs at 12°C with the air space (blunt end) upward.

### 3.2. Incubation

1. Take eggs from the cool store, and leave on bench to reach room temperature before setting in the incubator.
2. When embryos are being collected and fixed directly, or used to provide material for culture, the eggs can be incubated vertically. Lay eggs on their sides when incubating for manipulations (*see* Chapters 16–19 and 22), since during incubation, the yolk will rotate such that its least dense region, the embryonic blastoderm, rotates to the highest point: the upper side of the egg, which is where a window will later be cut through the shell. This property of rotation is retained in older embryos.
3. For best viability, incubate eggs at high relative humidity (>50%) and in the range of 37.5–39°C. Times to reach a specific developmental stage depend on the exact temperature (which will vary with position within the incubator) and duration of previous storage (old eggs have an appreciable lag-phase before they resume development). If older embryos (more than 9 d of incubation) are required, rocking of the eggs during incubation improves viability by preventing adhesion between extraembryonic membranes and the shell (egg incubators are commercially available for this purpose). However, rocking does not appreciably increase viability of younger embryos, and these can be incubated in conventional ovens or incubators that function with high internal humidity.

### 3.3. Opening Eggs for Embryo Harvest

#### 3.3.1. Rapid Method

1. Crack the egg on the edge of, and deposit contents into, a 100-mm glass Petri dish keeping the yolk unbroken. The embryo usually lies on the top of the yolk, but if it does not, pour the egg from one dish to another to rotate the yolk until the embryo is visible (a small white patch at 1–1.5 d, later surrounded by blood islands and then by blood vessels).
2. Vital stain as **Subheading 3.3.3.** (*see* **Note 1**), observe under a microscope, and stage.
3. Cut around the perimeter of the area opaca with Vannas scissors, lift out the embryo with a prewetted spatula, and place into fix (usually 4% w/v paraformaldehyde in PBS, but this varies with application).

### 3.3.2. Method for Preserving Shape and Orientation of Young Embryos (up to Stage 14)

This is the preferred method for embryo isolation among members of my group.

1. Crack the egg into a Petri dish as in **Subheading 3.3.1.** above.
2. Prepare a square frame from thick filter paper (Whatman 3MM), with external dimensions of about 1.4 × 1.4 cm and with an internal "window" of about 1 cm$^2$.
3. Lay the frame down on the surface of the yolk such that the embryo is central within the window, and allow the filter paper to become wet; this causes the vitelline membrane to adhere to the paper.
4. Cut around the outside of the frame with spring scissors, and lift the frame up gently using forceps. The embryo will be stretched out within the frame and will remain so throughout processing (*see* **Note 2**).
5. Wash in a dish of Howard's Ringer to remove adherent yolk, cut the embryo from the window, peel away the overlying vitelline membrane, and transfer the embryo to fix.

### 3.3.3. A Careful Method for Observations of Living Embryos

1. Incubate eggs to desired developmental stage *(2)*. Crack eggs against the side of a bowl filled with warm (37°C) Howard's Ringer.
2. Hold the egg with the crack submerged, and gently ease the two halves of the shell apart.
3. Release the contents into the solution. The yolk will float to the surface, and the blastoderm (the least dense region) should be uppermost. If not, carefully rotate the yolk with a spatula.
4. Stain young stage embryos (prestage 15) by applying a drop of neutral red (1% v/v aqueous solution) from the tip of a fine glass rod. The stain will rapidly permeate the vitelline membrane and stain the blastoderm beneath. Stage the embryo *(2–4)*.
5. Carefully cut around the perimeter of the area opaca with spring scissors, without rupturing the yolk and clouding the Ringer.
6. Pull the embryo (plus overlying vitelline membrane) away with #5 watchmaker's forceps to a clear region of the bowl, and lift by immersing a 30-mm Petri dish beneath it and withdrawing it slowly.
7. Grip the area opaca with a pair of forceps, and gently shake the embryo free of the vitelline membrane and adherent yolk.
8. Transfer to another Petri dish containing Howard's Ringer.

## 4. Notes

1. With only a little experience, early embryos can be staged by observation through a dissecting microscope using transmitted light alone. At stages following the onset of somitogenesis, staining is not required for staging, which only requires observation with reference to **refs.** *2* and *3*.

2. This works well for embryos with up to 3 d of incubation, but the filter paper cannot hold the weight of older embryos—these should be simply cut out using the method in **Subheading 3.3.1.**, hand-washed and transferred to fix using a spatula.

## Acknowledgments

Work in the author's laboratory is supported by program grants from the MRC and Human Frontier Science Program, and project grants form the MRC and the Wellcome Trust.

## References

1. New, D. A. T. (1966) *The Culture of Avian Embryos.* Logos, London.
2. Hamburger, V. and Hamilton, H. (1951) A series of normal stages in the development of the chick embryo. *J. Morphol.* **88,** 49–92.
3. Hamburger, V. and Hamilton, H. (1951) A series of normal stages in the development of the chick embryo. *Dev. Dynamics* **195,** 229–272.
4. Eyal-Giladi, H. and Kochav, S. (1975) From cleavage to primitive streak formation; a complementary Normal table and a new look at the first stages of development of the chick. I. General morphology. *Dev. Biol.* **49,** 321–337.

# 15

# New Culture

## Amata Hornbruch

## 1. Introduction

As the title indicates, this culture method was developed by Denis New and first described in 1955 *(1)*. It enables the observer to study the events of gastrulation in the chick embryo in much greater detail than possible until then. It also opens the way to microsurgery without the problem that yolk and the vitelline membrane cause. Studying the effects of treatment with compound affecting morphogenesis and development became meaningful, because precise concentrations and volumes could be administered and successfully washed out again.

Most of the early work on the mechanics of gastrulation was performed on the egg of *Xenopus laevis*, because of its accessibility. New Culture offered a chance to emulate and refine some of those experiments in the chick embryo.

Several attempts have been made to culture blastoderms or fragments thereof in vitro, but none have been very successful. The chorioallantoic membrane (CAM grafts) has been used as a substratum to study self-differentiation, but fragments do not develop consistently true to origin. Spratt *(2,3)* grew blastoderms on a bed of a mixture of yolk and agar. Some of the observations from this culture method resulted in misleading conclusions being drawn because morphogenetic movements were inhibited in the layer that was in contact with the agar.

Gallera and Nicolet *(4)* tried a more fluid culture medium and were slightly more successful. They also tried a modification of New Culture with a double-ring setup in which the vitelline membrane was sandwiched between the two rings, one of which was just small enough to fit into the larger one. This was to facilitate operations to the dorsal aspect of the embryo. The explanted embryo could be turned over without slipping off.

*From:* Methods in Molecular Biology, Vol. 97: Molecular Embryology: Methods and Protocols
Edited by: P. T. Sharpe and I. Mason © Humana Press Inc., Totowa, NJ

The most successful method was pioneered by New *(1)*. Blastoderms can be explanted unincubated or incubated for up to 30 h. This culture system not only allows observation of morphogenetic movements, which are so important for the normal development of the embryo, but also lends itself to a variety of minor and major transplantation experiments, ablations, treatment with compounds in solution or bound on beads, or other carriers and many other experiments. There are two main restrictions to the use of New Culture. First, only the ventral aspect of the blastoderm is accessible to immediate and direct manipulations, although there is no reason not to invade the epiblast as well. However, this can only be done by cutting through hypoblast and mesoderm, and causing perhaps unnecessary damage to these tissues. Second, the incubation time in culture is limited to about 48 h. Consequently, experiments have to be short term.

## 1.1. Embryo Development

At the time of laying, the chick embryo is a flat disk consisting of about 60,000 cells. It is made up of two layers, the hypoblast ventrally, lying in close association with the yolk, and the epiblast dorsally, facing the vitelline membrane, the membrane which surrounds the yolk. The disk has a translucent inner core, the area pellucida, from which the embryo will form, and a denser outer ring of mainly yolk-containing cells, the area opaca. Only the cells at the periphery of the blastoderm are attached to the vitelline membrane, and as they migrate radially, the blastoderm expands.

After about 8–10 h of incubation, the embryonic shield or posterior marginal zone has formed, and the position from which the primitive streak will arise is determined. The disk now has polarity dorso/ventrally and antero/posteriorly as well as left/right.

## 2. Materials

All glassware and instruments should be clean to tissue-culture standards and autoclaved or heat-sterilized the day before use. AnalaR® reagents are recommended for the solutions, which should be made up and autoclaved the day before use. If the technique is used as a classroom exercise, it is quite sufficient to use Howard's Ringer. However, if it is employed as a tool to do serious research, the use of a phosphate-buffered salt solution, such as the one described later, is advised.

Explantation of the cultures may be performed in a quiet draft-free place on the laboratory bench if a tissue culture hood is not available.

## 2.1. Nonsterile Equipment

1. Humidity incubator with fan for egg incubation, preculture.
2. Humidity incubator with fan for new culture incubation.

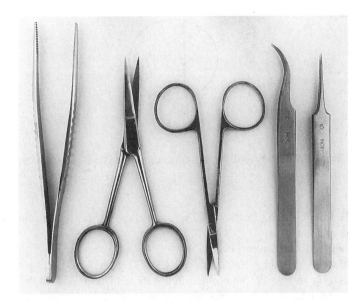

Fig. 1. Set of recommended instruments for New Culture.

3. Stereoscopic binocular dissecting microscope—transmitted light is preferable.
4. Container for egg disposal.

## 2.2. Sterile Equipment

1. Large Pyrex pie dish about 5 cm deep, 20–25 cm in diameter.
2. Watch glasses, 50–55 mm diameter, preferably with flat bottoms, for stability.
3. Glass rings, optimal 25/26 mm inner, 29 mm outer diameter *(1)*.
4. Glass Pasteur pipets, 145-mm, flamed tips and plugged.
5. Disposable Petri dishes, 90 mm—bacteriological grade is sufficient.
6. Kleenex tissue strips for humidity retention in culture dish

## 2.3. Sterile Instruments

1. A pair of coarse forceps with blunt ends, 12 cm (anatomical).
2. A pair of straight scissors, 11–12 cm.
3. A pair of straight fine scissors, 9–10 cm.
4. A pair of straight no. 5 watchmaker's forceps *(2)*.
5. A pair of curved no. 7 watchmaker's forceps *(2)* (**Fig. 1**).

## 2.4. Balanced Salt Solutions (BSS) and Other Solutions

1. Pannett and Compton *(5)* buffered salt solution or Howard's Ringer, autoclaved.
2. Antibiotic/antimycotic solution (100X) (Gibco)
3. Howard's Ringer: 7.2 g NaCl, 0.23 g CaCl·2H$_2$O, and 0.37 g KCl. Dissolve in 1000 mL double-distilled water. Adjust to pH 7.2 with one or two drops of 0.5 *M*

Fig. 2. Cartoon of an egg yolk being cut around the equator to remove the vitelline membrane.

NaOH. Fill in suitable bottles or flasks, autoclave the day before, and keep refrigerated.

4. Pannett and Compton solution:
   a. Solution A: 12.11 g NaCl, 1.55 g KCl, 0.77 g CaCl, and 1.27 g $MgCl_2 \cdot 6H_2O$. Dissolve in 100 mL double-distilled water and autoclave. Make up multiples if required in large amounts, keep in aliquots of 40 mL, and autoclave.
   b. Solution B: 0.189 g $Na_2HPO_4 \cdot 2H_2O$ and 0.019 g $NaH_2PO_4 \cdot 2H_2O$. Dissolve in 100 mL double-distilled water and autoclave. Make up multiples if required in large amounts, keep in aliquots of 60 mL, and autoclave.
   c. Working solution: To make up working solution, autoclave 900 mL double-distilled water with 0.9 g glucose in a 1000-mL Erlenmeyer flask the day before. When cool, add 40 mL of solution A and 60 mL of solution B in that order, and keep at 4°C overnight. Working solution can be kept at 4°C for several days if not used up on the day of transplantation.

## 3. Methods

1. Incubate eggs on their sides for 18–20 h at 38 ±1°C for stages 4 and 5 *(3)*.
2. Wipe or spray eggs with 70% alcohol.
3. Pour cold BSS in pie dish, about 4-cm depth (6–700 mL).
4. Add 1 mL antibiotic/antimycotic solution/100 mL BSS *(4)*.
5. Open eggs with coarse forceps at the blunt end, i. e., air sack.
6. Remove shell to make opening large enough to let yolk out.
7. Pour off thick albumin (into a waste bucket nearby), and save a little of thin albumin in a 90-mm Petri dish or sterile beaker *(5)*.
8. Pour the yolk quickly into BSS *(6)*.
9. Put up to six yolks into the dish, and top up with BSS if they are not completely covered.
10. Remove remaining albumin from all yolks in the dish with coarse forceps or large-mouthed well-flamed pipet.

### 3.1. Cutting the Vitelline Membrane

For right-handed operators (reverse for left-handers) (**Fig. 2**):

1. Place yolk with blastoderm facing east.
2. Hold yolk with coarse forceps at west pole, and cut with larger pair of scissors around the equator or just below it, starting south and moving north, keeping an eye on the position of the blastoderm at all times.
3. When cut right around, take both pairs of watchmaker's forceps and lift vitelline membrane lightly at the cut edge. Position yolk so that blastoderm is on top, facing the operator.
4. Hold the edge of the vitelline membrane with both pairs of forceps, and pull **very slowly** by inverting it over the yolk, keeping the membrane low over the yolk, and pulling towards the bottom of the dish (*see* **Note 7**). Stage 4 and 5 blastoderms adhere fairly well to the vitelline membrane and are easy to handle. Older and younger blastoderms need more care.
5. When the vitelline membrane is freed from the yolk and the blastoderm still attached to it, immerse a watch glass into the BSS and pull the membrane onto it (*see* **Note 8**). As you raise the watch glass above the BSS level, hold onto the membrane with tweezers or it will float off.
6. Place the watch glass with the blastoderm in a 90-mm Petri dish.

### 3.2. Arranging Vitelline Membrane on the Ring

1. Use a stereoscopic, binocular dissecting microscope
2. Stretch vitelline membrane, with the blastoderm in the center, well out on the watch glass. Remember the vitelline membrane ought to be inside-out at this stage.
3. Wet a glass ring in BSS (*see* **Note 9**), and lay it on the vitelline membrane and fold the membrane over the ring, holding ring with straight forceps and pulling membrane with curved tweezers, working all the way around until all the membrane is securely draped over the ring. Hold the membrane **only** at the cut edge to avoid injury to it and the blastoderm (*see* **Note 10**).
4. Pipet off most of the liquid and the adhering yolk. With fine scissors, cut away the surplus membrane close to the inside of the ring.
5. Wash all yolk from under the ring with BSS, and pipet 1 mL of thin albumin onto the watch glass beneath the ring to supply the developing embryo with a source of water and protect it from bacterial infections. Albumin is rich in bactericidal proteins. Little or no liquid should cover the embryo.
6. Lay one or two strips of Kleenex tissue in Petri dish, and wet with BSS to create a moist chamber.
7. Now the embryo is ready for surgery or the designed treatment. If surgery is performed, leave the embryo to heal at room temperature for a couple of hours before further incubation. This helps to close the wound while the blastoderm is not expanding, and there is less tension in the fragile tissues. It may be appropriate to photograph embryos at this stage.
9. Incubate at 38 ($\pm$) 1°C in a tissue-culture incubator. Facilities for an air mix of 95% air and 5% $CO_2$ are not essential.
10. Observe the development of the embryos the next day. They should have reached stage 10–12 (**Fig. 3**) and may be photographed again.

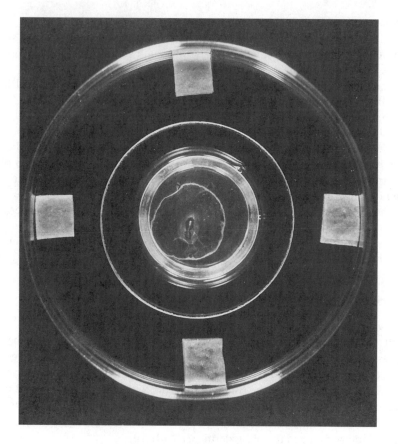

Fig. 3. New Culture after 24 h of incubation.

### 3.3. Analysis of Results

At this stage, the embryos can be fixed to be processed for many different purposes, including whole mounts, histology, immunocytochemistry, *in situ* hybridization, and others. The type of fixative used will depend on the further treatment. It is important **not** to bring the rings and the watch glasses into contact with any fixative to be used. It saves lengthy cleansing processes, which are unpleasant, time-consuming, and not always a great success.

Place the ring with the embryo in a Petri dish of leftover Howard's Ringer or Pannett and Compton solution from the day before, and wash off unwanted albumin and yolk particles. Either peel the blastoderm away from the vitelline membrane along its peripheral edge and stretch out on a piece of thin plastic (Melinex or overhead projection foil is very good) cut into small squares before immersing in fixative, or just cut the embryo out and transport on a spatula to fixative. Embryos not supported tend to curl up, fold, or distort, and are much more difficult to analyze.

## 4. Notes

1. Any glassblower will produce glass rings from glass rods but the best rings are cut from glass tubes with the desired diameter, anything from 22–28 mm inner diameter, 3–4 mm thick. They are oblong when viewed from the side, and are either left rough or better still baked to give a smooth finish on the cut surface.

    **Here is a word of warning:** If any kind of treatment with compounds is envisaged, rings with a rough cut edge should not be used. Despite thorough cleaning they will retain traces of compound and contaminate the next experiment.
    Not all egg yolks are the same size. Particularly eggs from very young hens are small and may not fit onto rings 26 mm in diameter. A few smaller rings with an inner diameter of 22 or 23 mm are recommended. Smaller rings are also easier to handle for the novice operator, and a few larger rings (28 mm inner diameter) come in handy if the eggs happen to be on the generous side.

2. Watchmaker's forceps may be straight or curved to the individual's preference. Curved forceps are handy for the release of the membrane when folded under the ring and in tricky situations.

3. The incubation time depends on the developmental stage required for the intended experiment (*see* **ref. 6**). Incubating the eggs on their sides is preferable to on their points, but it is not necessary. The blastoderm will always rise to the top of the yolk because of its lower specific gravity compared to the yolk. If the eggs are incubated on their tips, the blastoderm will be directly under the air sack and is more likely to be damaged when the egg is opened.

4. Antibiotic/antimycotic solution (100X) from (Gibco) contains
    10,000 U of penicillin (base).
    10,000 µg of streptomycin (base).
    25 µg of amphotericin B/mL.
    Use 1% (1 mL in 100 mL BSS).

5. When separating the albumin from the yolk, **do not pull on the chalazae too hard or the vitelline membrane will break.** The chalazae are two thick gycoprotein threads by which the yolk is tied to the vitelline membrane and which allow it to rotate when the egg is turned so that the blastoderm may rise to the top.

6. Do not try to pour the yolk slowly and carefully from the shell into the BSS. That will certainly tear the yolk.

7. Keep the pie dish on a black surface to show up contours of the yolk, adhering albumin, and so forth.

8. Hold on to the vitelline membrane while lifting the watch glass out of the liquid.

9. A dry ring will instantly stick to the vitelline membrane when put down, and make it difficult to rearrange.

10. As mentioned before, just the cells on the periphery of the blastoderm adhere to the vitelline membrane. It is important not to destroy these delicate contacts.

Carefully pipet small amounts of BSS at the time against the inner edge of the ring to wash away yolk particles, and suck liquid up by slowly moving the pipet backward along the ring. Embryos younger than stage 4 have **very** delicate contacts to the vitelline membrane, and extra care is needed to transplant these stages successfully.

Best results are achieved after incubation times of 20–36 h. Embryos will develop normally and can be compared to embryos of the same stage *in ovo,* though not in the same time. New Culture embryos will be between one and two stages younger having their development interrupted by the explanting procedure and the following manipulation. Longer incubation times than 36 h may be tolerated, but the development will slow down and the incidence of abnormalities increases dramatically. Lack of nutrients and space to expand for the embryo may all play a part.

## References

1. New, D. A. T. (1955) A new technique for the cultivation of the chick embryo *in vitro. J. Embryol. Exp. Morphol.* **3**, 326–331.
2. Spratt, N. T., Jr. (1947) A simple method for explanting and cultivating early chick embryos *in vitro. Science* **106**, 452.
3. Spratt, N. T., Jr. (1963) Role of the substratum, supra cellular continuity and differential growth in morphogenetic cell movements. *Dev. Biol.* **7**, 51–63.
4. Gallera, J. and Nicolet, G. (1961) Quelques cooentaires sur les methodes de culture *in vitro* de jeunes blastodermes de poulet. *Experientia* **17**, 134,135.
5. Pannett, C. A. and Compton, A. (1924) The cultivation of tissue in saline embryonic juice. *Lancet* **206**, 381–384.
6. Hamburger, V. and Hamilton, H. L. (1951) A series of normal stages in the development of the chick embryo. *J. Morphol.* **88**, 49–67.

## Appendix

Here are a few helpful notes and figures for the novice New Culture operator to facilitate easy apprehension of the most commonly applied early stages in the development of the chick embryo for this method. This summary cannot justify the diversity of nuances for all the stages that the operator will recognize with increasing awareness and experience, but serve as a guide only. The stages described here span from an unincubated blastoderm to a range of stages found after 24–30 h in culture.

---

Fig. 4. **(A)** A disk with virtually no landmarks to indicate the future polarity of the embryo. **(B)** Stage 1 after Hamburger and Hamilton (HH), the prestreak, incubation for 4–8 h. After several hours of incubation, the formation of the hypoblast appears as an area of higher density roughly shaped as a triangle rising from the posterior marginal wall. This is called the embryonic shield or Koller's sickle. **(C)** Stage 2 (HH), the

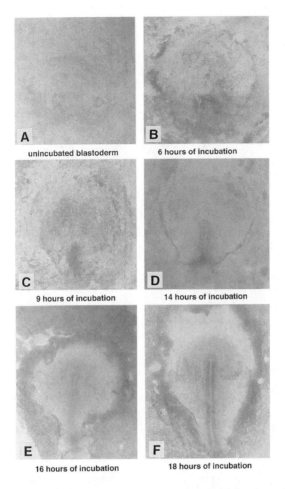

A
unincubated blastoderm

B
6 hours of incubation

C
9 hours of incubation

D
14 hours of incubation

E
16 hours of incubation

F
18 hours of incubation

Fig. 4. *(continued)* initial streak (6–10 h), a short, squat, thickening of the blastoderm at the posterior marginal zone. (**D**) Stage 3 (HH), the intermediate streak (12–16 h), the streak is now extending to one-third of the area pellucida. Invagination of prospective mesoderm proceeds at the posterior part of the primitive streak and ingression of endoderm at more anterior levels of the primitive streak. (**E,F**) Stage 4– and stage 4 (HH), respectively, the definitive streak (15–20 h). The elongation of the primitive streak proceeds over several hours until it has reached its full length extending to two-thirds of the area pellucida from the posterior marginal zone. The average length of the fully extended primitive streak is about 2 mm, but varies enormously from embryo to embryo. It consists of the primitive groove flanked on either side by the primitive folds. The most anterior part of the primitive streak is broadened into a bulb, and is known as Hensen's node, the organizer region of the avian embryo, the homolog to the dorsal lip of the blastopore in amphibian embryos. Somitic mesoderm invaginates through the posterior primitive streak and heart mesoderm through the midstreak regions. The area pellucida is pear-shaped.

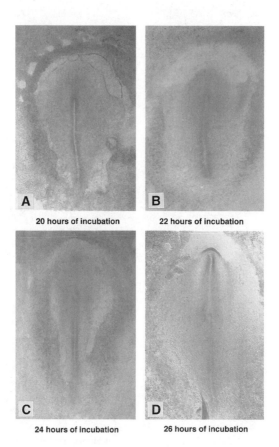

**20 hours of incubation**          **22 hours of incubation**

**24 hours of incubation**          **26 hours of incubation**

Fig. 5. (**A**) Stage 4+ (HH) (18–22 h) shows the first sign of the head process extending cranially from Hensen's node to give rise to the head notochord. (**B,C**) Stage 5–
and stage 5 (HH), respectively, the head process (20–24 h), shows the condensation of
mesodermal cells of the head process more clearly. (**D**) Stage 6 (HH), the head fold
(23–26 h). The anterior rim of the medullary plate is pushed forward by the extending
head process until it folds over under the tension, forming the subcephalic pocket
ventrally. The ensuing fold in the endoderm is to give rise to the foregut. The duration
for this stage is very short. The head fold can flip over within 1 h. Hensen's node is
now beginning to regress leaving in its wake the notochord. Mesenchyme cells form
isolated blood islands in the extraembryonic mesoderm.

All stated times of incubation are very tentative. They will depend on the
time an egg took to pass down from the fimbriated edge of the oviduct, where
it was fertilized, to the vagina, this can take up to 22 h, depending on how long
the eggs were stored before incubation commenced and the temperature at
which they were stored. A further variable is that most laboratory incubators
are set to slightly different temperatures, which will be reflected in the stage of

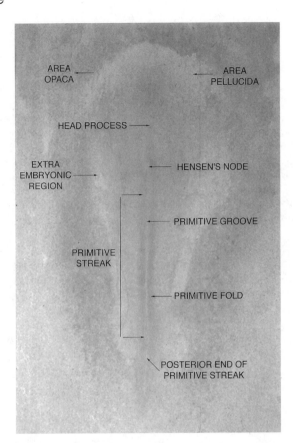

Fig. 6. Stage 5 (HH), the head process has reached its most cranial position and laid down the prospective head territory. The cells of the epiblast overlying the head process form a pseudostratified epithelium, the medullary plate, and the presumptive neural plate. Hensen's node is clearly asymmetrical. The bilateral heart primodia lie lateral and anterior to the node.

development of the embryo. Many investigators make the additional distinction between winter and summer eggs.

All embryos were fixed in 3.5% paraformaldehyde in phosphate-buffered saline (PBS) for at least 4 h and stained in saturated carmine red in 4% Borax and diluted 50/50 in 70% alcohol until they reached the required intensity, differentiated in 70% acid alcohol, dehydrated, cleared and mounted in DPX, and photographed on a Zeiss Axiophot (mag. ×40) or on a Zeiss Stemi SV6 (mag. ×20–30) on Fujichrome 64 T slide film. All embryos were photographed from the dorsal aspect with their antero–posterior polarity running north to south.

When the egg is laid, it has little resemblance to an embryo. The unincubated blastoderm is a flat disk made up of two layers (*see* **Subheading 1.1.**).

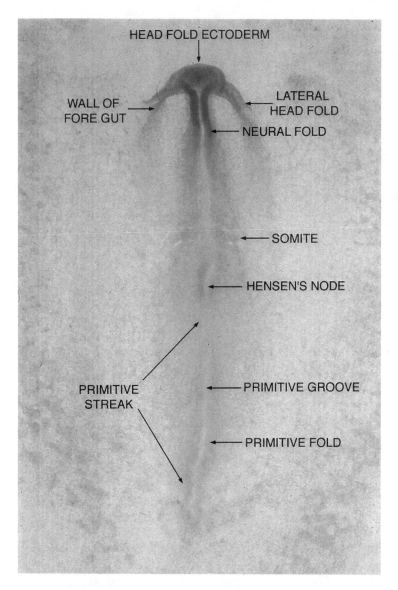

Fig. 7. Stage 7 (HH) (26–30 h) shows the segmentation of the first pair of somites from the *para*-notochrodal mesoderm. This is in fact the second pair of somites. The first somite is a phantom somite. It never forms fully and will disperse again in the next few hours of development. It is a slight mesodermal condensation anterior to the cleavage for the first somite proper. Not all embryos have a visible first somite. Subsequent pairs of somites will segment in two hourly intervals from the *para*-axial mesoderm. Neural folds reach to the mesencephalon. The wall of the foregut emerges close to the lateral head fold.

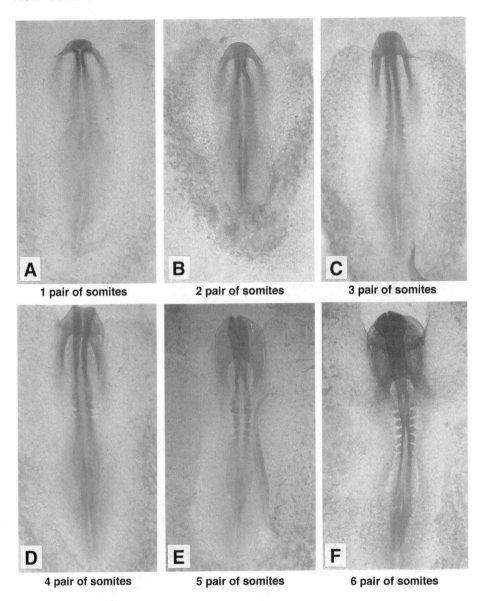

**A** 1 pair of somites  **B** 2 pair of somites  **C** 3 pair of somites

**D** 4 pair of somites  **E** 5 pair of somites  **F** 6 pair of somites

Fig. 8. Comprises stages 7 and 8. (**A**) Stage 7 (HH), 1 pair of somites. (**B**) Stage 7+ (HH), 2 pairs of somites. (**C**) Stage 8– (HH), 3 pairs of somites. (**D**) Stage 8 (HH) 4 pairs of somites. (**E**) Stage 8+ (HH), 5 pairs of somites. (**F**) Stage 9– (HH). 6 pairs of somites, dorsally the first cleft between rhomobomere 5 and 6 can be detected.

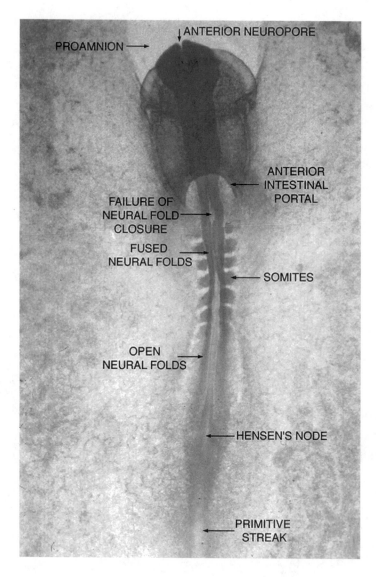

Fig. 9. Stage 8 (HH) (28–34 h), five pairs of somites. The head is raised above the proamnion. The proamnion is a region free of mesoderm between ectoderm and endoderm at the anterior edge of the area pellucida flanked by the lateral horns meeting cranially. Open neuropore, neural folds are beginning to close in the region of the mesencephalon (failure to close will lead to spina bifida in the adult, as seen here rostral to the first somite). Ventrally, the notochord can be seen through the open folds of the neural tube as the node regresses along the primitive streak. Bilaterally heart primodia are developing from the amnio-cardiac vesicle level with the midbrain. More blood islands are developing in the area opaca. (*See* color plate 10 appearing after p. 368.)

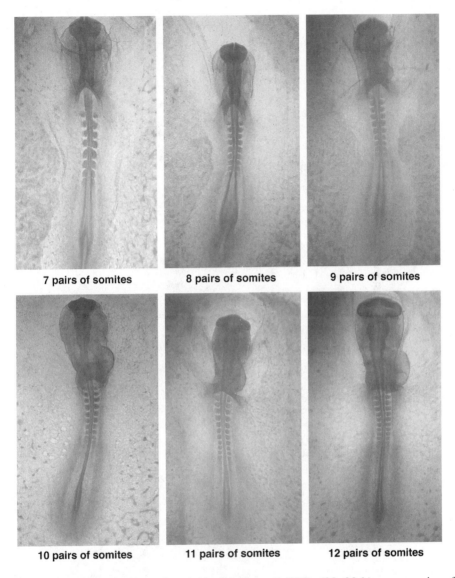

**7 pairs of somites**     **8 pairs of somites**     **9 pairs of somites**

**10 pairs of somites**     **11 pairs of somites**     **12 pairs of somites**

Fig. 10. Comprises stages 9 and 10. (**A**) Stage 9 (HH), (32–38 h), seven pairs of somites, dorsally bilateral primodia of the optic vesicle appearing. Mesencephalon is clearly demarcated from rhombencephalon at the isthmus. Segmentation of rhombomeres continues with boundaries 2/3 and 3/4 appearing. Ventrally the heart primodia fusing medially into a tube. Margin of anterior intestinal portal is level with vitelline vein which is connecting with heart, and splaying out laterally to area opaca. (**B**) Stage 9+ (HH), 8 pairs of somites. (**C**) Stage10– (HH), 9 pairs of somites. (**D**) Stage 10 (HH) 10 pairs of somites. (**E**) Stage 10+ (HH) 11 pairs of somites. (**F**) Stage 11– (HH), 12 pairs of somites.

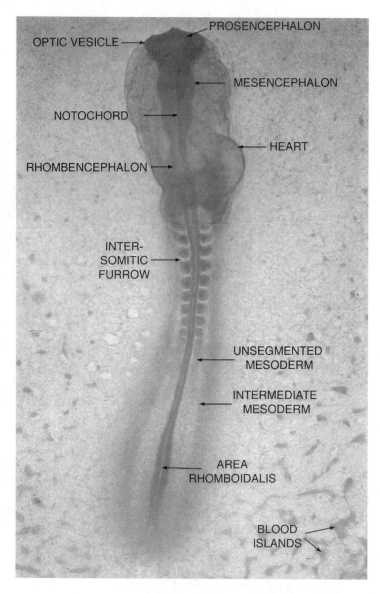

Fig. 11. Stage 10 (HH), 10 pairs of somites (36–42 h) dorsally anterior neuropore closed, prominent optic vesicles, rhombomere boundary 4/5 formed, and neural folds are closed to almost the level of the node. Ventrally, Hensen's node has regressed almost to the end of the primitive streak (the 10th pair of somites has not fully segmented caudally in this illustration), pronephric tubules develop between somites 6 and 10, heart tube turns asymmetrical bulging out to the right and contractions can be seen, and bilateral vitelline veins fan out toward the area opaca, which shows large blood islands to establish circulation.

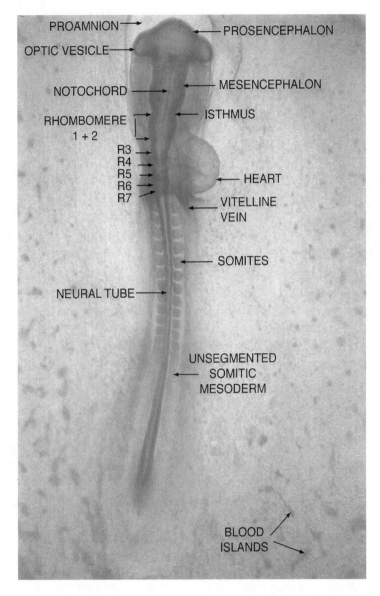

Fig. 12. Stage 11 (HH), 13 pairs of somites (40–46 h), slight flexure of the head, prominent optic vesicles with slight constriction at their base form the lateral parts of the prosencephalon, mesencephalon has clear boundaries, and all rhombomeres can be distinguished. The neural tube is virtually closed along its entire length. On the ventral side, the rostral part of the heart forms the ventricle, whereas the caudal part gives rise to the atrium and the vitelline vein leading from it. The heart beat is rhythmical, but the circulation of blood is not yet fully connected up to all peripheral blood islands. Anterior somites are beginning to differentiate.

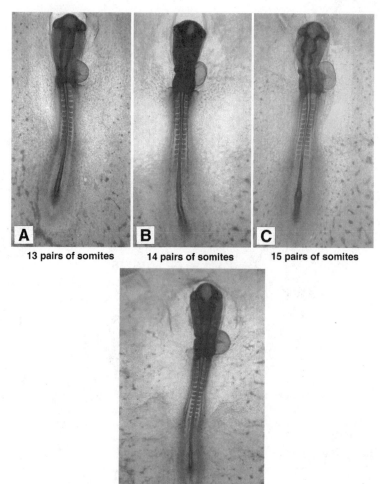

**13 pairs of somites     14 pairs of somites     15 pairs of somites**

**17 pairs of somites**

Fig. 13. **(A)** Stage 11 (HH), (40–46 h), 13 pairs of somites. **(B)** Stage 11+ (HH), 14 pairs of somites. **(C)** Stage 12– (HH), 15 pairs of somites. **(D)** Stage 12 (HH) (48–54 h), 17 pairs of somites. Embryos in New Culture will rarely develop well beyond this stage. It is possible in exceptional circumstances to maintain embryos for up to 48 h in culture. For example, embryos on rings can be left on semipermeable membranes instead of watch glasses and given culture medium containing serum instead of albumin, but development is slowed down and abnormalities increase. The eyes are budding off from the prosencephalon, the bilateral otic vesicles are clearly visible as round indentations adjacent to rhombomere 5. Ventrally the circulation of blood from the heart to the periphery is nearly complete. Somites are segmenting off at about one every 2–3 h. The notochord has regressed to the end of the primitive streak. (*See* color plate 11 appearing after p. 368.)

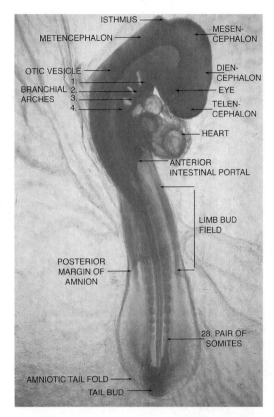

Fig. 14. Stage 16 (HH) (about 72 h), 28 pairs of somites. Very few embryos may get to this stage in development. The vascular system is ill-equipped to function in a two-dimensional space. When the cranial flexure of the embryo reaches 90° or more the main blood vessels fold under the weight of the trunk and can no longer work efficiently. The heartbeat will increase without fullfilling a better function. The embryo fills up with fluid and development slows right down. The lens of the eye has separated from the eye cup, there is a clear demarcation between the telen- and the diencephalon, the isthmus divides the mesencephalon from the metencephalon, and the otic vesicle is a shallow, almost round indentation with a raised epithelial rim. All four branchial arches are clearly structured. The flexure of the body extents to the anterior intestinal portal and the body folds curve ventrally to the tail bud. The wing bud primodia can clearly be seen in a slight thickening of the flank, but there is no such indication for the leg bud yet. The posterior margin of the amnion extends to about the 20 pair of somite. Somite numbers are no longer a reliable aid to stage embryos of this stage or older. The most cranial somites are starting to be incorporated into the base of the scull, and the increase of the flexure of the body right down to the tail bud makes it difficult to give an accurate account. There are many other landmarks for accurate staging. The shape and structure of the branchial arches, the color of the eye, the angle of the head, and the shape and size of the limb buds are just some.

# 16

## Grafting Hensen's Node

### Claudio D. Stern

### 1. Introduction

Soon after Spemann and Mangold's *(1)* famous demonstration in 1924 that the dorsal lip of the blastopore of the gastrulating amphibian embryo has the unique ability to induce a second axis when grafted into an ectopic site in a host embryo, Waddington *(2,3)* showed that Hensen's node is its equivalent in amniotes. After transplanting this region into an ectopic site in interspecific combinations of rabbit, duck, and chick embryos, he found that a second axis developed, where the nervous system was derived from the host ectoderm *(4)*. Hensen's node is situated at the anterior (cranial) tip of the primitive streak during gastrulation, and in chick embryos appears as a bulbous thickening, some 100 µm in diameter, centered around a depression, the primitive pit. At this point, the three germ layers of the embryo are in very close apposition.

In the avian embryo, operations involving Hensen's node at the primitive streak stage (10–20 h incubation, Hamburger and Hamilton *(5)* [HH] stages 3–5) are most easily performed in whole embryo culture, as described by New *(6)* (*see* Chapter 15). For assays of induction, it is essential to be able to distinguish donor from host cells because the change of fate of the host cells is central to the definition of induction. This can be easily achieved by using interspecies chimaeras, for example quail donors and chick hosts, whose cells can be distinguished by either the Feulgen-Rossenbeck technique or using anti-quail cell antibodies (e.g., QCPN) or using species-specific riboprobes in *in situ* hybridization analysis. Another way to trace the fate of the grafted cells is to label the transplanted node with a cell autonomous vital dye, such as the carbocyanine dye DiI (*see* **refs.** *7* and *8*). In a recent study, using these techniques in combination with tissue- and region-specific markers, Storey et al. *(9)* were able to determine that Hensen's node is at the peak of its inducing

From: *Methods in Molecular Biology, Vol. 97: Molecular Embryology: Methods and Protocols*
Edited by: P. T. Sharpe and I. Mason © Humana Press Inc., Totowa, NJ

ability at the primitive streak stage but that this ability quickly declines as soon as the head process begins to emerge (HH stage 5).

One problem with the way in which Waddington originally performed his grafting experiments (2,3) is that he placed the transplanted node into a region now known to be fated to form neural plate, and therefore, although he demonstrated that this was able to initiate the formation of a second axis in the host, it is impossible to conclude from his experiment that the host cells underwent a change in fate. One way to overcome this is to place the grafts into a peripheral ring of the avian blastoderm, the inner third of the *area opaca* (**Fig. 1**). During normal development, this region only contributes to extraembryonic tissues, but is nevertheless able to respond to a graft of Hensen's node by generating a complete embryonic axis, where the host epiblast changes its fate from extraembryonic ectoderm to neural tissue (9,10). The competence of this region to respond to such a graft declines rapidly, such that by HH stage 5 it is no longer able to respond to grafts of nodes derived from donors of any stage (9–11).

Manipulating Hensen's node in its normal position in the embryo is technically very difficult, as are most other microsurgical operations on chick embryos at the primitive streak stage. This is particularly true when the manipulation involves all three germ layers (ectoderm, mesoderm, endoderm), because cutting through the whole thickness of the embryo often leads to holes that expand greatly and eventually destroy the embryo. The main reason for this is that at these stages, the embryo only develops well when it is under some tension. This tension is maintained by the migration of cells at the peripheral edge of the *area opaca* on the vitelline membrane, to which they are attached. There are several ways to overcome this problem, at least in part. One is to remove the embryo from its vitelline membrane and to culture it, epiblast side down, on the surface of agar-albumen or agar-egg extract, as described by Spratt (12). But under these conditions growth of the embryo is stunted and abnormalities of the development of the axis are the rule rather than the exception. Another way is to excise the most peripheral edge of the *area opaca* but leaving the embryo on its vitelline membrane. The excised cells slowly appear to regenerate, while the hole has time to heal, and the embryo gradually develops tension once again in time for normal axial development to occur. In my experience, this is a very successful way to proceed. A third way to prevent large holes from expanding is to keep the newly-operated embryo at room temperature for 2–3 h, followed by a period (3–5 h) at 30°C before placing it at 38°C. The low temperature appears to slow down expansion of the *area opaca* while allowing healing to occur. This is also a successful approach. Whatever the course of action chosen, it is important to consider that the extent of the healing process will probably determine the outcome of the experiment. Healing after excision of a large portion of the embryo will bring new cells into

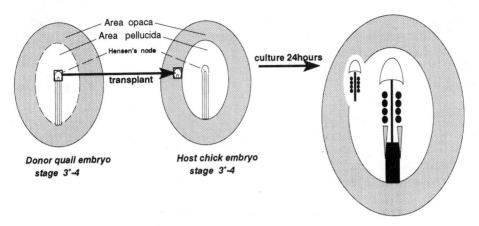

Fig. 1. Diagram illustrating the operation of grafting a quail Hensen's node into a chick host to demonstrate embryonic induction in the inner part of the *area opaca* of the host.

contact with one another, and the result may therefore be different than when these cells are prevented from interacting.

In the following sections, I consider two examples of operations on Hensen's node: excision from a donor quail embryo and transplantation to the inner ring of the *area opaca* of a host chick embryo to demonstrate embryonic induction, as done by Storey et al. *(9)* and rotation of the node about its rostrocaudal axis *in situ*, to demonstrate embryonic regulation as done by Abercrombie *(13)*.

## 2. Demonstration of Embryonic Induction by Transplantation of Hensen's Node

### 2.1. Materials

1. Dissecting microscope, preferrably one with trasmitted light base.
2. Pannett-Compton saline: solution A: 121 g NaCl, 15.5 g KCl, 10.42 g CaCl$_2$·2H$_2$O, 12.7 g MgCl$_2$·6H$_2$O, H$_2$O to 1 L; solution B: 2.365 g Na$_2$HPO$_4$·2H$_2$O, 0.188 g NaH$_2$PO$_4$·2H$_2$O, H$_2$O to 1 L; before use, mix (in order): 120 mL A, 2700 mL H$_2$O, and 180 mL B.
3. Two pairs of watchmakers' forceps, number 4 or 5.
4. One pair of coarse forceps, about 15 cm (6 in.) long.
5. One pair of small, fine scissors, with straight blades about 2 cm (3/4 in.) long.
6. A spoon/spatula or teaspoon.
7. Petri dish (about 10–15 cm diameter) to collect embryos.
8. Container for egg waste.
9. Pasteur pipet with the end cut off at the shoulder, stump flamed to remove sharp edges, and rubber teat.

10. Pasteur pipet (short form), end lightly flamed to remove sharp edges; rubber teat.
11. Pyrex baking dish about 2 in. (5 cm) deep, 2 L capacity.
12. 35-mm Plastic dishes with lids.
13. Watch glasses, about 5–7 cm diameter.
14. Rings cut from glass tubing, approx 27 mm outer diameter, 24 mm inner diameter, 3–4 mm deep.
15. Very fine needles (e.g., entomological size A1 or D1) or sharpened tungsten wire mounted by melting the fine end of a Pasteur pipet (to act as a handle) or into a metal needle holder.
16. 1 Small beaker (50–100 mL).
17. Plastic lunch box with lid for incubating culture dishes.
18. 38°C Incubator.
19. Hens' eggs incubated 12–18 h (depending on stage needed).
20. Quails' eggs incubated 12–18 h (depending on stage needed).

## 2.2. Method

### 2.2.1. Preparation of Donor (Quail) Embryo

1. Remove quail eggs from incubator. With the scissors, gently tap near the blunt end of an egg so as to penetrate the shell. Use the tip of the scissors to cut off a small cap of shell near this end, carefully to avoid damaging the yolk.
2. Allow egg white to pour into waste bucket, assisted by the scissors, taking care to avoid damage to the yolk. You may need occasionally to cut through the rather thick albumen using the scissors.
3. Once most of the albumen has been poured off, make sure the embryo is uppermost; if not, turn the yolk by stroking it very gently with the sides of the scissors.
4. Use the scissors to make four cuts into the vitelline membrane around the embryo. If the embryo does not lie exactly in the center of the egg, make the first cut on the side of the embryo nearest the shell, and proceed in this way until all four cuts have been made. Make sure all the cuts meet each other.
5. Pick up the square of embryo/membrane with the spoon/spatula, trying to collect only a minimal amount of yolk.
6. Transfer the yolk/embryo/membrane with the spoon into the large Petri dish with Pannett-Compton saline under a dissecting microscope. With fine forceps, turn the square of yolk/membrane/embryo so that the embryo is uppermost.
7. After the desired number of donor embryos have been placed into the Petri dish, use two pairs of forceps to separate the embryo from adhering yolk. Working at low magnification, pick up a corner of the square of vitelline membrane with one pair of forceps and slowly but steadily fold it back, steadying the yolk with the other pair of forceps. During the whole procedure the membrane and embryo should remain totally submerged in saline. The embryo should be attached to the membrane. If not, peel the membrane completely and then use foceps gently to remove the embryo from the underlying yolk.
8. Pick up the embryo, with or without adhering membrane, with the wide-mouth Pasteur pipet, and transfer it to a 35-mm dish with clean saline for final cleaning

and dissection. The edges of the extraembryonic membranes will be perfectly circular, provided that the embryo has not been damaged during the explantation procedure. Put this aside while preparing the host embryo.

## 2.2.2. Preparation of Host (Chick) Embryo

The following description is for preparing cultures based on the method of New *(6)* but with some modifications as described in Stern and Ireland *(14)*. The procedure has been adapted from *(15,16)*. You may also follow the method described in Chapter 15 *(7)*. The main difference between these methods and that originally described by New *(6)* is the use of rings cut from glass tubing, rather than bent from a glass rod with circular cross-section. The advantage of these rings, with rectangular profile, is that they grip the vitelline membrane tightly and therefore allow transfer of the assembly to a flat plastic dish. Above, I have recommended rings of 27 mm outer diameter, because it is easier to wrap the membrane around these for a novice. However, if larger (ca. 30 mm diameter) rings are used, the embryos will develop up to 6–9 h longer. The longevity of the cultured embryo appears to depend both on the amount of thin albumen under the ring and on the length of time for which it can be cultured before the edges of the *area opaca* reach the ring.

1. Fill the large Pyrex dish about three-fourths full with saline (about 1.5 L).
2. Open an incubated hen's egg by tapping the blunt egg with coarse forceps, and carefully removing pieces of shell. Tip the egg gently to collect the thin albumen in the small beaker (this is required for culturing), and discard the thicker albumen, assisted with the coarse forceps. Try to remove as much albumen as possible, which will simplify the later steps.
3. When yolk is clean and free from adhering albumen, carefully tip it into the saline container, taking care not to damage the vitelline membrane on the edges of the broken shell. The blastoderm should face upwards. If not, carefully turn the yolk with the side of the coarse forceps.
4. Make a cut into the vitelline membrane enveloping the yolk just below the equator. Continue to cut all the way around the circumference of the yolk.
5. With two pairs of fine forceps, slowly but steadily 'peel' the North Pole of the vitelline membrane, all the way off the yolk. Do not stop during this process. The embryo should come off with the membrane. Let the membrane rest on the bottom of the dish, inner face (containing the embryo) pointing upwards.
6. Lower a watch glass and a glass ring into the container. Slide the vitelline membrane, preserving its orientation, onto the watch glass, and arrange the ring over it so that membrane protrudes around the ring. Pull out the assembly from the saline.
7. With fine forceps, work carefully to fold the cut edges of the vitelline membrane over the edge of the ring, all the way around its circumference. Do not pull too tightly but ensure that the bottom of the membrane is smooth and free from wrinkles as you work around the circumference.

8. Place the watch glass over a black surface. Suck off as much fluid as possible from the outside of the ring with the flamed Pasteur pipet. If there is much yolk remaining over and/or around the embryo, wash it carefully with clean saline. Discard any embryos in which the vitelline membrane has been damaged. Leave the host submerged in saline for the operation.

### 2.2.3. Grafting and Incubation

1. Having prepared both donor quail and host chick embryos, you are ready for the operation. First bring the dish with the donor quail embryo under the microscope and arrange it so that its ventral (endoderm) surface is uppermost. Using two fine needles, carefully cut out the very tip of the primitive streak, cutting through the whole thickness of the embryo but making sure that you do not cut through the vitelline membrane if this is still attached.
2. Lower the magnification of the microscope, keeping track of the excised node, and pick this up with a Gilson P20 fitted with a yellow tip, set to 1–2 μL.
3. Move the dish with donor embryos away and place the watch glass with the host chick embryo under the microscope. While looking down the microscope, insert the tip of the Gilson under the saline covering the host embryo and gently expel the quail node onto its surface, keeping track of it at all times.
4. Use fine needles to manipulate the donor node to close to the desired grafting site (**Fig. 1**). Now carefully lift up a portion of the flap of yolky cells (germ wall margin) that covers the inner margin of the *area opaca*, working outwards from the *area pellucida* and taking care not to penetrate the ectoderm underneath, which is only one cell thick. This will produce a pocket into which the graft can be inserted.
5. Slide the quail node into the pocket, pushing it as deep as possible so that when the flap of germ wall margin is replaced it will cover the graft completely.
6. Working under the microscope, carefully remove any remaining saline, both inside and outside the ring. During this process, keep watching the graft to make sure that it does not become dislodged. It is important that the embryo and the inside of the ring remain completely dry during incubation.
7. Now pour some thin albumen (about 2–3 mm thick layer) on the bottom of a 35-mm plastic dish. Slide the ring with vitelline membrane off the watch glass, and transfer it to the dish, over the pool of egg albumen. Press lightly on the ring with two forceps to allow it to adhere to the dish.
8. If the level of albumen comes close to the edge of the ring, remove the excess. Also aspirate any remaining fluid from inside the ring. It is best if the vitelline membrane bulges upwards, above a good pool of albumen. This will also help to drain off further fluid accumulated during culture to the edges of the ring.
9. Wet the lid of the plastic dish with albumen. Discard the excess, and seal.
10. Place the dish in a plastic box containing a piece of tissue paper or cotton wool wetted in distilled water, seal the box, and place it in an incubator at 38°C.

### 2.3. Analysis of Results

After the desired period of incubation, fix the embryo by flooding with methanol (for most immunological detection procedures), Zenker's fixative (for Feulgen-Rossenbeck staining) or in 4% formaldehyde in PBS (for *in situ* hybridization and most other procedures). In the case of methanol or Zenker's, which are "rapid" fixatives, it is advantageous first to submerge the cultured embryo in saline, then to detach it from the vitelline membrane and to transfer it to a clean dish with saline prior to fixation. Otherwise the embryo will adhere permanently to the membrane and the fixative will denature the albumen, generating threads of protein that will tend to stick to the embryo. In the case of formaldehyde, this can be poured directly onto the embryo provided that the embryo is then detached from the membrane within a few minutes.

Whatever the fixative and subsequent method of processing chosen, it is advantageous to ensure that at the time of fixation the embryo is as flat as possible. If it has been detached from the membrane prior to fixation, place it in a small drop of saline on a plastic surface, then suck off most of the saline with a fine Pasteur pipet so that the embryo flattens on the plastic, and then place the first drop of fixative directly onto the surface of the embryo, taking care not to break it. After this it is safe simply to submerge the embryo in fixative, perhaps transferring it to a glass vial. Embryos fixed in this way usually remain flat through all subsequent manipulations.

Depending on the purpose of the experiment to be performed, embryos operated and cultured as described can be subjected to histochemistry, immunological procedures (as whole mounts), or whole-mount *in situ* hybridization. In many cases, it is possible to combine two or more of these methods. For example, it is possible to fix the embryos in formaldehyde, process them as whole mounts for *in situ* hybridization, postfix in formaldehyde, and then perform whole-mount immunoperoxidase histochemistry with QCPN antibody to detect the quail cells. After this, they can be embedded in wax and sectioned. Methods for this have been published in this volume elsewhere in some detail (*see* Chapter 22 and **ref. *17***).

## 3. Rotation of the Node *In Situ*

The following procedure is "generic", and similar manipulations can be done to operate on smaller or larger portions of embryo. For example, whole large sections of the primitive streak may be rotated as done by Abercrombie *(13)* and others, or very small sub-sectors of Hensen's node transplanted as described by Selleck and Stern *(18)* and Storey et al. *(8)*.

### 3.1. Materials

The materials required are the same as listed in **Subheading 2.1.** In addition you will need fine capillary glass pulled to a very fine tip with an electrode puller, and connected to an aspirator (mouth) tube. The former can be manufactured by pulling the thin end of a Pasteur pipet over a very hot Bunsen flame and pulling rapidly. The latter can be purchased from Sigma (A5177; St. Louis, MO).

## 3.2. Method

1. Follow **steps 1–8** of the method in **Subheading 2.2.2.** to place a chick embryo in modified New (1955) culture as if to receive a graft.
2. Use fine needles to cut out the node, involving the whole thickness of the embryo but being very careful to avoid damaging the vitelline membrane underneath. Even a small hole will cause leakage of albumen and prevent healing, or even displace the graft. It is best to work in steps: first make a very superficial cut of the shape required. Then deepen the cuts, a little at a time, until the node is finally freed all around.
3. It is of advantage to mark one edge of the node with fine carbon (e.g., pencil lead shavings) or carmine (Sigma C1022) particles, using a fine needle. This allows the orientation of the node to be controlled through subsequent manipulations.
4. Manoeuver the excised node, using the fine needles, to the desired orientation, again taking care not to damage the vitelline membrane that is now exposed.
5. Still observing under the microscope, carefully and slowly withdraw as much saline as possible from inside and outside the ring, as described for grafting Hensen's nodes above. In this experiment, where the manipulated node is not secured under a flap of tissue, it is much easier to lose it while sucking off the fluid. If necessary, replace it in position as required, using the fine needles.
6. Once all the fluid has been removed, use the fine capillary and mouth tube (if necessary) to remove all remaining fluid from the site of the graft. It is likely that the excised piece will appear to have shrunk. Sucking the fluid off in this way will close the gap and "knit" the pieces together, if performed with care.
7. Now slide the ring from the watch glass and set up the culture as described in **Subheading 2.2.3., items 7–10**. Finally, if necessary, suck off some more fluid with the mouth capillary assembly to ensure that the graft site is totally dry and appears closed.
8. Before placing in the incubator at 38°C, it is advantageous to keep the operated embryo at room temperature (in the sealed Petri dish) for 2–3 h. After this, place it at 30°C for 3–5 h. Finally, transfer the dishes to an incubator at 38°C. These periods at lower temperature will help the healing process as described above.

## 3.3. Analysis of Results

Operated embryos may be analyzed by histology, whole-mount immunohistochemistry, and/or *in situ* hybridization with appropriate probes, as described for Hensen's node grafts.

## References

1. Spemann, H. and Mangold, H. (1924) Über Induktion von Embryonanlagen durch Implantation artfremder Organisatoren. *Wilh. Roux Arch. EntwMech. Organ.* **100,** 599–638.
2. Waddington, C. H. (1932) Experiments on the development of chick and duck embryos, cultivated in vitro. *Phil. Trans. R. Soc. Lond.* B **221,** 179–230.

3. Waddington, C. H. (1933) Induction by the primitive streak and its derivatives in the chick. *J. Exp. Biol.* **10,** 38–46.
4. Streit, A., Théry, C., and Stern, C. D. (1994) Of mice and frogs. *Trends Genet.* **10,** 181–183.
5. Hamburger, V. and Hamilton, H. L. (1951) A series of normal stages in the development of the chick embryo. *J. Morph.* **88,** 49–-92.
6. New, D. A. T. (1955) A new technique for the cultivation of the chick embryo in vitro. *J. Embryol. Exp. Morph.* **3,** 326–331.
7. Selleck, M. A. J. and Stern, C. D. (1991) Fate mapping and cell lineage analysis of Hensen's node in the chick embryo. *Development* **112,** 615–626.
8. Storey, K. G., Selleck, M. A. J., and Stern, C. D. (1995) Induction by different subpopulations of cells in Hensen's node. *Development* **121,** 417–428
9. Storey, K. G., Crossley, J. M., De Robertis, E. M., Norris, W. E., and Stern, C. D. (1992) Neural induction and regionalisation in the chick embryo. *Development* **114,** 729–741.
10. Stern, C. D. (1994) The avian embryo: a powerful model system for studying neural induction. *FASEB J.* **8,** 687–691
11. Streit, A., Stern, C. D., Théry, C., Ireland, G. W., Aparicio, S., Sharpe, M., and Gherardi, E. (1995) A role for HGF/SF in neural induction and its expression in Hensen's node during gastrulation. *Development* **121,** 813–824
12. Spratt, N. T., Jr. (1947) A simple method for explanting and cultivating early chick embryos in vitro. *Science* **106,** 452.
13. Abercrombie, M. (1950) The effects of antero-posterior reversal of lengths of the primitive streak in the chick. *Phil. Trans. Roy. Soc. Lond.* **234,** 317–338
14. Stern, C. D. and Ireland, G. W. (1981) An integrated experimental study of endoderm formation in avian embryos. *Anat. Embryol.* **163,** 245–263.
15. Stern, C. D. (1993) Avian embryos, in: *Essential Developmental Biology: A Practical Approach* (Stern, C. D. and Holland, P. W. H., eds.), IRL at Oxford University Press, Oxford, UK, pp. 45–54.
16. Stern, C. D. (1993) Transplantation in avian embryos, in: *Essential Developmental Biology: A Practical Approach.* (Stern, C. D. and Holland, P. W. H., eds.), IRL at Oxford University Press, Oxford, UK, pp. 111–117.
17. Stern, C. D. and Holland, P. W. H. (eds.) (1993) *Essential Developmental Biology: A Practical Approach.* IRL at Oxford University Press, Oxford, UK.
18. Selleck, M. A. J. and Stern, C. D. (1992) Commitment of mesoderm cells in Hensen's node of the chick embryo to notochord and somite. *Development* **114,** 403–415.

# 17

## Grafting of Somites

### Claudio D. Stern

## 1. Introduction

The somites are an intriguing invention of vertebrate embryos. They represent the most overtly segmented structures of the body plan, but they give rise to both obviously segmental (e.g., the axial skeleton) as well as not-so-obviously metameric (dermis and skeletal muscle) elements. In addition, they play a key role in controlling several aspects of the organization of the central and peripheral nervous system of the trunk, and appear to participate in several different types of inductive interactions both within themselves and with neighboring tissues like the neural tube, the notochord, the metanephric and lateral plate mesoderm and the ectoderm and endoderm (*see* **refs.** *1* and *2* for reviews).

Questions that can be addressed by manipulating somites range from investigations on the mechanisms by which metameric pattern is established, to their influence on the segmental outgrowth and differentiation of precursors of the peripheral nervous system (neural crest cells, motor axon growth cones), to the control of myogenesis, and patterning and the establishment of regional identities of cells that contribute to the dermis, limbs, and axial skeleton. Previous experiments *(1,2)* have suggested that although many aspects of somite development are controlled by surrounding tissues, many others appear to be remarkably autonomous.

Somites form at the posterior end of the embryo, such that a pair of somites is added every 100 min or so (*see* **refs.** *1* and *2* for reviews). Therefore at any particular stage of development, the embryo contains younger (more recently formed) somites at their caudal end, and older (at more advanced stages following their formation) more rostrally. To indicate this, Ordahl and his colleagues have introduced a "somite stage" numbering system, using Roman numerals to indicate the position of the somite being referred to with respect to

From: *Methods in Molecular Biology, Vol. 97: Molecular Embryology: Methods and Protocols*
Edited by: P. T. Sharpe and I. Mason © Humana Press Inc., Totowa, NJ

its neighbors (*see* **ref. 2**). In this system, somites are counted upwards from the most recently formed one, which is designated as I. The most caudal 4–6 somites are usually epithelial spheres. Stages V/VI and higher designate somites whose dorsolateral surfaces still remain epithelial (the dermomyotome; **Fig. 1**) but whose ventromedial parts have become mesenchymal once more, to form the sclerotome (*1,2*). The neural tube, notochord, ectoderm and endoderm all play a role in determining the dorsoventral polarity of the somites with respect to their ability to form a dermomyotome and a sclerotome.

Despite the simplicity of this numbering system, it is important to remember when investigating somite development that overlapped with this age-structure is also a position-dependent address (reviewed in **refs.** *1* and *2*). This can be demonstrated by transplanting somitic mesoderm from the thoracic level to the neck, where they go on to develop ribs as if they had not been transplanted. However, if a similar experiment is conducted to investigate the nature of the muscles that develop, it is found that any somite will give rise to muscles appropriate for its new position. Thus, some somitic derivatives behave in a cell autonomous way concerning their positional information, while others are subject to cues from their environment. In addition, the most rostral 5 or so somites ("occipital somites") have a different fate from the rest (they do not contribute to the vertebral column, some of their cells appear to contribute to the tongue, and in addition, they do not support the development of dorsal root ganglia from neural crest cells migrating within them).

Experiments in which the somitic mesoderm is manipulated can be done either *in ovo* or in whole embryo (*3–5*) culture. The main advantage of the former method is that it allows embryos to develop for a long time, even up to hatching, but embryos younger than about the 4 somite stage (*see* **ref.** *4* [HH] stage 8) are very difficult to manipulate in this way and their survival is poor. The latter technique allows very young embryos (even before incubation, at preprimitive streak stages) to be operated, but they will only survive for 36–48 h, even in the most expert of hands. In the following sections, I describe two examples. The first is a detailed method for operations on somitic mesoderm *in ovo*, in which the anterior half of the segmental plate (unsegmented paraxial mesoderm; *see* **Fig. 1**) of a quail embryo is grafted into the same position of a host chick embryo. The same procedure is "generic" and can be adapted easily for manipulation of newly formed or older somites at stages 9–15, as well as for manipulations of the notochord, neural tube and other tissues at these stages. The second example, to be used in conjunction with the instructions in the chapter on grafting Hensen's node (*see* Chapter 16), gives advice on manipulating younger embryos to investigate the mechanism of segmentation from the primitive streak stage onward (*see* **refs.** *7* and *8*). The procedures given here have been adapted from those in **ref.** *9*.

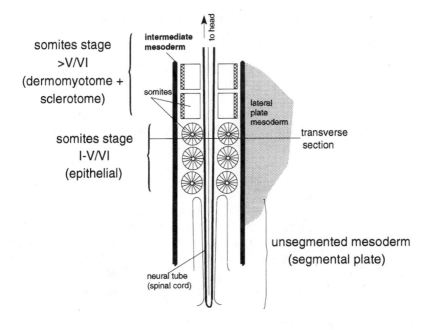

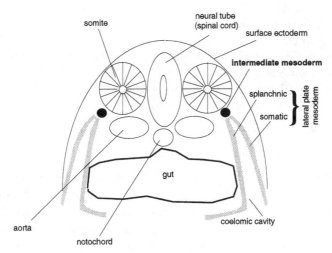

Fig. 1. Schematic diagrams of embryos at about stage 11. The three regions of the somitic mesoderm can be seen along the axis of the embryo in the upper diagram: unsegmented mesoderm toward the bottom of the drawing, followed by epithelial somites (somite stages I through V or VI), followed more anteriorly by somites that have already split into dermomyotome and sclerotome (somite stages V/VI and higher). Although only three epithelial somites are shown, there are usually five or six of this type in embryos at this stage. The lower diagram represents a transverse section at the level of one of the epithelial somites.

## 2. Grafting Paraxial Mesoderm and Somites *In Ovo*

### 2.1. Materials

1. Dissecting kit: 2 pairs small forceps (watchmaker's; number 4 or 5), 1 pair small scissors (about 2 cm straight blades), 1 scalpel (No. 3 handle, No. 11 blade), a Gilson micropipet for 3 µL, yellow tip(s).
2. Eye surgeon's micro-knife, 15° angle. Suitable microknives are: Microfeather micro-surgery scalpels for eye surgery, 15° blade angle, cat. no. 715, manufactured by Feather (Japan) and marketed by pfm GmbH. Sold in boxes of 5, not cheap!
3. 2 Entomological pins, size A1 or D1, or sharpened tungsten needles, each mounted by melting the end of a short Pasteur pipet (to act as handle) or inserted into a metal needle holder.
4. Plasticine (modeling clay) to make ring for resting egg on its side.
5. 100 mL Calcium-/magnesium-free Tyrode's saline (CMF): dissolve: 80 g NaCl, 2 g KCl, 0.5 g $NaH_2PO_4 \cdot 2H_2O$, and 10 g glucose in 1 L $H_2O$. Autoclave for storage. On day of use, dilute 1:10 with distilled water. The working solution may be buffered with bicarbonate, but we usually omit this.
6. Antibiotic/antimycotic solution, 100X concentrate (e.g., Sigma A9909; St. Louis, MO).
7. 70% Ethanol to wipe shell.
8. PVC tape to seal egg.
9. 2 Pasteur pipets and rubber teats.
10. Container for egg waste.
11. 10-mL Plastic syringe with Gilson yellow tip stuck into the end, filled with high vacuum silicon grease or vaseline.
12. 1-mL Syringe, 27-gage (or finer) × 3/4 in. needle (for ink injection).
13. 1-mL Syringe, 21-gage needle (for antibiotics).
14. 5-mL Syringe, 21-gage needle (for withdrawing albumen).
15. Paper tissues.
16. 35-mm Plastic dish coated with Sylgard and steel insect pins, size A1 or D1.

Sylgard 184 (Dow Corning) is clear silicone rubber polymerized by mixing two components (9 parts rubber solution:1 part of accelerator/catalyst). Mix the two well and pour to the desired depth (2–5 mm) into the plastic Petri dish. Allow the dishes to stand for about 1 h at room temperature for air bubbles to leave, then cure at about 55°C until polymerized (3 h to overnight). The dishes can be stored indefinitely. Black Sylgard is also available.

17. India ink (Pelikan Fount India is best; most other makes are toxic), diluted 1:10 with CMF and loaded in the ink injection syringe.
18. 50 mL Trypsin (DIFCO, 1:250), freshly made up to 0.12% (w/v) in CMF.
19. Hens' and quails' eggs incubated 40–44 h so that they are at stages 10–12. The hens' eggs (hosts) should have been resting on their sides at least 30 min.
20. Dissecting microscope, preferrably with transmitted light base.
21. Fiber optic incident illumination.

## 2.2. Method

Here the host embryo is operated *in ovo*, which is only suitable with ease for embryos older than about 36–48 h incubation. As an example, the operation described here consists of a graft of presomitic mesoderm (segmental plate) with a quail embryo as the donor and a chick embryo as the host. An identical method can be followed to transplant paraxial mesoderm that has already formed somites, but it should be remembered that this is much easier when it involves the youngest (most caudal) 5–6 pairs of somites at stages 10–14. Somites situated more anteriorly in the embryo have already separated into dermomyotome and sclerotome (**Fig. 1**) and special care is needed both to keep these components together during grafting and to separate them cleanly. Whatever the operation to be performed, it is generally a good idea to fix for histological analysis some of the pieces like those to be transplanted, as well as some embryos immediately after the operation, to confirm that the graft involves the tissues of interest and no contaminating cells.

### 2.2.1. Preparation of Donor Quail Embryo

1. Remove quail eggs from incubator. With the scissors, gently tap near the blunt end of an egg in order to penetrate the shell. Use the tip of the scissors to cut off a small cap of shell near this end, carefully to avoid damaging the yolk.
2. Allow egg white to pour into waste bucket, assisted by the scissors, taking care to avoid damage to the yolk. You may need occasionally to cut through the rather thick albumen using the scissors.
3. Once most of the albumen has been poured off, make sure the embryo is uppermost; if not, turn the yolk by stroking it very gently with the sides of the scissors.
4. Use the scissors to make four cuts into the vitelline membrane around the embryo. If the embryo does not lie exactly in the center of the egg, make the first cut on the side of the embryo nearest the shell, and proceed in this way until all four cuts have been made. Make sure all the cuts meet each other.
5. Pick up the square of embryo/membrane by grasping one corner with fine forceps and transfer it immediately to the Sylgard-coated 35-mm dish containing about 5 mL CMF-Tyrode's solution.
6. Using two pairs of fine forceps, separate the vitelline membrane from the embryo, but leave the extraembryonic membranes attached to the embryo.
7. Stretch out the embryo (either side up) by placing 4–6 entomological pins (size A1 or D1) through the corners of extraembryonic membranes, into the Sylgard rubber. The embryo should be under some tension. Put the dish aside while preparing the host.

### 2.2.2. Preparation of Chick Host Embryo

The procedure described here differs from that in the chapters on grafting of the notochord and neural tube, AER/ZPA and neural crest (*see* Chapters 18–

20). In the method described here, a small window is cut into the shell and the embryo floated up to the level of this window for the operation. The advantages of this technique are:

a. The embryo lies very close to the operator's hands.
b. It is completely submerged in liquid at all times (which avoids drying out and simplifies some of the manipulations).
c. The liquid above the embryo can be changed several times, for example to wash the embryo.
d. The light illuminating it can be shone tangentially to its surface (the bubble of liquid above acts as a lens that greatly enhances the optical clarity).
e. The tension generated by floating the embryo seems to aid the action of the trypsin, such that tissues almost appear to dissect themselves.
f. It is very easy to notice if the endoderm has been punctured accidentally because ink will fountain out very quickly.
g. After the operation the small hole can be closed very tightly with plastic (PVC, or electrical) tape, which allows the egg to be incubated with the window downwards and turned—both of these greatly enhance survival.

1. Shape plasticine (modeling clay) into a ring about 2 in. (5 cm) in diameter and place it on the stage of the microscope. Place a hens' egg (host) onto the plasticine ring, being careful that it does not rotate with respect to its resting position.
2. Using the 5-mL syringe with 21-gage needle, held nearly vertical, insert needle into blunt end of egg until the shell is felt at the bottom surface. Withdraw 0.5–1 mL egg albumen, which should come up easily.
3. Score a shallow $1 \times 1$ cm square on the top of the shell with the scalpel and lift up the square of shell.
4. With a pair of watchmakers' forceps, pierce and remove the underlying shell membrane, after wetting it with CMF. Avoid damage to the embryo underneath: before air is allowed into the egg, the embryo will lie very close to the membrane.
5. Fill the cavity with CMF so that the embryo floats up to the level of the window.
6. After ensuring that there are no air bubbles in the syringe with Indian ink or in the needle, insert the needle under the vitelline membrane, tangentially, at a position as far away from the embryo proper as possible. Point toward and slightly below the embryo, and inject about 20–50 µL. It is important to minimize movement of the needle after penetrating the vitelline membrane, or the hole will be very large and yolk/ink will leak out. Introduce and withdraw the needle with one clean, decisive movement and do not stir the needle inside the yolk; only one attempt per egg!
7. Draw a shallow, continuous border of silicon grease around the window. This will contain a standing drop in which the operation will be done. Now fill this chamber with CMF saline until there is a standing drop, and adjust the fibre optic light to shine tangentially to the surface of the egg so that the embryo can be seen very clearly with minimal light intensity from the light source.

## 2.2.3. Grafting Procedure

1. Break the vitelline membrane just over the region to be operated with a needle. The hole should be as small as possible. The segmental plates are the rod-like structures lying on either side of the neural tube at the tail end of the embryo, just behind the last somite. The portion to be rotated in this example is the most anterior half of the plate.
2. Replace the bubble of CMF with trypsin/CMF. Increase the magnification of the microscope as much as possible.
3. Operating in the drop of trypsin, use the micro-knife to make initially very shallow cuts in the ectoderm next to the neural tube in the region of the operation.
4. Gradually deepen the cuts using the knife blade more as a spatula, allowing penetration of the trypsin, than as a sharp cutting edge. Once the ectoderm has been penetrated with the tip of the blade, the trypsin does the rest. Find the lateral border of the segmental plate, and do the same there: a shallow cut in the ectoderm first, then separate the tissue gradually. In both cases, make sure you do not penetrate the endoderm (which is 1-cell thick) or ink will pour out. Finally, free the posterior end of the piece and loosen the graft.
5. Remove the piece of segmental plate with a Gilson micropipet set to 1–2 µL. Replace the bubble over the embryo with fresh CMF twice to remove the trypsin solution. Make a new bubble of CMF while obtaining the graft from the donor.
6. Turn to the donor embryo in the Sylgard dish. Replace almost all of the CMF in which it was submerged by trypsin solution. Repeat the trypsin wash and perform the dissection in this solution at room temperature.
7. Cut out an equivalent piece of segmental plate as the one removed from the host, using the same technique.
8. Pick up the graft with the Gilson. With the other hand, place the host under the microscope and, observing under low magnification, carefully place the graft into the CMF bubble over the embryo.
9. Use the knife and work at low magnification to manipulate the graft into the gap made by removal of the host piece of segmental plate.
10. When the graft is in position (approximately), very carefully remove the most of the fluid from above the embryo with a Pasteur pipet while watching under low power. If necessary, reposition the graft with a mounted needle.
11. Insert the 5-mL syringe with 21-gage needle into the original hole in the blunt part of the eggshell, vertically, and carefully withdraw 2.5–3 mL thin egg albumen. This will lower the operated embryo back to its original position. Be careful when you insert the needle, since the pressure could make the graft come out of its site.
12. Add 2–3 drops of antibiotic/antimycotic concentrate (away from the graft site).
13. Wipe the edges of the shell with tissue paper moistened lightly in 70% alcohol to remove the silicon grease.
14. Cut a piece of PVC tape about 6 cm long. Stretch it slightly, and then let it relax. Place it over the window, smoothing out any unevenness carefully to avoid breaking the shell or applying too much pressure on the window.

15. Keeping the egg on its side, place it (window down!) into an egg tray in a humidified incubator at 38°C. If you are worried about the graft falling out, it is a good idea to incubate the embryo with the window upwards until the next day, and then to turn it.

16. Incubate 1–3 d. Embryo survival 2 d after this operation should be 80–100%.

## 3. Operations on Paraxial Mesoderm in Embryos in New Culture

When the aim of the experiments is to investigate the mechanisms that set up the paraxial mesoderm, or when the fates and movements of presumptive somitic cells at or shortly after gastrulation are to be investigated, operations in the egg are very difficult. It is therefore generally necessary to resort to a method of whole embryo culture. For this, set up cultures of embryos at the desired stage (between HH stage 3 [mid-primitive streak], and HH stage 8 [4 somites]), following the protocol in Chapter 15 of this volume. Leave the embryo completely submerged in saline during the operation.

The region of the primitive streak embryo (stages 3–4) that contributes to the somites lies in the most anterior (cranial) one-third or so of the primitive streak and in the ectoderm to either side of this. By stage 4–5, some of the somite progenitors have already left the streak and lie in the middle layer next to the anterior streak and gradually migrate cranially as more cells are added from the streak and node regions.

To operate on these cultured embryos, it is usually better to use fine mounted needles (entomological or sharpened tungsten) rather than knives. Trypsin is generally not necessary, but may be used (at 0.1% w/v) to facilitate separation of tissues if the specific manipulation desired turns out to be difficult because of adherence of the tissues to one another. Operations on the mesoderm are usually easier in the absence of trypsin because the middle layer readily separates into cells in the presence of enzyme. However, for each type of operation, the sequence in which the cuts are made is very important. In some regions of the embryo it is easier to begin with a medial cut and proceed laterally, whereas for some other regions an anterior cut, proceeding caudally, is more effective. You should investigate the relative merits of different ways to dissect the tissue of interest before starting a set of experiments.

As described for operations on Hensen's node, it is important to avoid damaging the vitelline membrane at all costs. Any leakage of the albumen culture fluid to the inside of the ring will diminish survival and may cause the grafted tissue to fall out of its site.

After the operation, follow the steps described for Hensen's node grafts (*see* Chapter 16). Transfer the ring with the embryo to a plastic dish, seal it, place it in a humid chamber, and incubate 38°C for 24–48 h. If cultures are set up using large glass rings (about 30 mm; *see* **ref. 5**) or other methods for extended culture (*see* **ref. 8**), embryos operated at stages 3–6 should survive to stage 15 or even longer.

## 4. Analysis of Results

Embryos, operated as described in **Subheading 3.**, can be studied by a variety of methods, according to the question being addressed. Fix the embryo in methanol (for most immunological detection procedures), Zenker's fixative (for Feulgen-Rossenbeck staining) or in 4% formaldehyde in PBS (for *in situ* hybridization and most other procedures). Details on these procedures may be found elsewhere in this volume and in **ref. 9**.

For embryos operated *in ovo*, it is easiest to crack the egg into a large Petri dish first, cut the membranes around the embryo with scissors, and then lift a corner of these membranes with fine forceps (as described above in **Subheading 2.2.1.**). Then immediately transfer this to a small dish with saline to clean off any adhering yolk. Finally transfer it to a Sylgard dish (*see* **Subheading 2.1.**) containing CMF and pin the embryo so as to straighten out the head and trunk, but avoiding the region close to the operation. Then remove the CMF and replace it with the fixative of choice. In this way, the embryos will be perfectly straight which will simplify subsequent histological sectioning, and they will also be more photogenic if stained as whole mounts.

For embryos operated in New culture *(3–5)*, the glass ring should first be flooded with CMF, and the edges of the *area opaca* then detached from the vitelline membrane. Then pick up the embryo with a wide-mouthed pipet or with fine forceps (from the membranes!) and transfer it to a Sylgard dish for pinning and fixing as described in **Subheading 2.2.1.**

## References

1. Keynes, R. J. and Stern, C. D. (1988) Mechanisms of vertebrate segmentation. *Development* **103,** 413–429.
2. Tam, P. P. and Trainor, P. A. (1994).Specification and segmentation of the paraxial mesoderm. *Anat. Embryol.* **189,** 275–305.
3. New, D. A. T. (1955) A new technique for the cultivation of the chick embryo in vitro. *J. Embryol. Exp. Morph.* **3,** 326–331.
4. Hamburger, V. and Hamilton, H. L. (1951) A series of normal stages in the development of the chick embryo. *J. Morph.* **88,** 49–92.
5. Selleck, M. A. J. and Stern, C. D. (1991) Fate mapping and cell lineage analysis of Hensen's node in the chick embryo. *Development* **112,** 615–626.
6. Selleck, M. A. J. and Stern, C. D. (1992) Commitment of mesoderm cells in Hensen's node of the chick embryo to notochord and somite. *Development* **114,** 403–415.
7. Stern, C. D. (1993) Transplantation in avian embryos, in: *Essential Developmental Biology: A Practical Approach.* (Stern, C. D. and Holland, P. W. H., eds.), IRL at Oxford University Press, Oxford, UK, pp. 111–117.

8. Stern, C. D. (1993) Avian embryos, in: *Essential Developmental Biology: A Practical Approach* (Stern, C. D. and Holland, P. W. H., eds.), IRL at Oxford University Press, Oxford, UK, pp. 45–54.
9. Stern, C. D. and Holland, P. W. H. (eds.) (1993) *Essential Developmental Biology: A Practical Approach*. IRL, Oxford University Press, Oxford, UK.

# 18

## Notochord Grafts

### Andrew Lumsden and Susanne Dietrich

### 1. Introduction

Many patterning structures have been identified by microsurgical manipulation of chick embryos *in ovo*, such as ablation or heterotopic grafting experiments. Among the structures studied, the notochord has received much attention, since it plays a crucial role in the development of the surrounding tissues. In the overlying neural tube (**Fig. 1**), the notochord induces the ventral midline structure, the floor plate, which subsequently specifies neuronal cell types in the ventral half of the neural tube (reviewed in **ref. 2**). In the paraxial mesoderm that flanks the neural tube and notochord, the latter induces the sclerotome and, in synergy with signals from dorsal neural tube or surface ectoderm, the epaxial myotome (*3*, reviewed in **ref. 4**). Studies on zebrafish notochord mutants suggest that the notochord also acts in the formation of the dorsal aorta (*5*). Finally, the proximity of notochord and the subjacent endoderm during early phases of development suggests that the notochord may play a role in the development of the roof of the gut.

Much of the function of the notochord in these tissue interactions has been attributed to the signaling molecule sonic hedgehog (Shh; reviewed in *2,6,7*). However, in mice lacking Shh, notochord-dependent structures develop to some extent (*8*). Thus, the notochord will remain the subject of many embryological studies in the future. Therefore, we will describe notochord heterotopic transplantation in chick embryos as one approach to study the functions of this structure in vivo. It should be noted, however, that many other tissues in the chick embryo can be grafted using similar methods. Similarly, corresponding structures from quail embryos may be grafted into chick hosts, allowing the detection of quail cells with quail-specific markers (*9*). Finally, tissues derived from mouse embryos have successfully been transplanted to chick hosts using this approach (*10*).

*From:* Methods in Molecular Biology, Vol. 97: Molecular Embryology: Methods and Protocols
Edited by: P. T. Sharpe and I. Mason © Humana Press Inc., Totowa, NJ

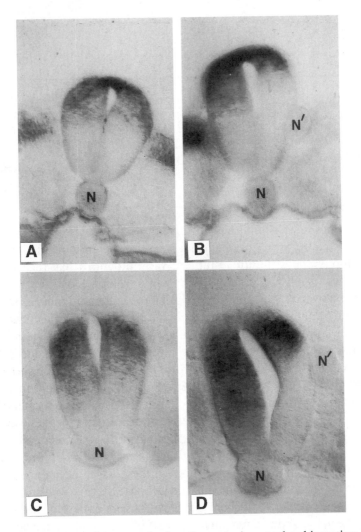

Fig. 1. Repatterning of the neural tube after ectopic notochord insertion *(1)*. **(A,B)** sections of embryos hybridized with the dorsal neural tube marker *Pax3*. **(C,D)** Sections of embryos hybridized with the intermediate marker *Pax6*. A, C, controls, B, D, sections of operated embryos. Where the ectopic notochord (N') has established contact to the neural tube, it induces a wedge-shaped structure normally found ventrally in the neural tube: the floor plate. Simultaneously, the graft suppresses the expression of the dorsal marker *Pax3* and the intermediate marker *Pax6*.

## 2. Materials

1. Roll of clear adhesive tape (Sellotape) and roll of thick plasticized canvas tape, both 25-mm wide and on a dispenser.

2. 27-gage × 21 mm, 19-gage × 35 mm hypodermic needles; 1- and 5-mL syringes.
3. Egg stools, cut from flexible plastic sleeve used for insulating water pipes.
4. 2 Sylgard-bottomed Petri dishes (1 for fixed material and 1 for live). Sylgard (Dow Corning, Wiesbaden, Germany) is a silicone resin that comes in two parts. When mixed, it sets to produce a near-transparent rubbery matrix, ideal for pinning down embryos. To prepare Sylgard-bottom dishes, premix Sylgard, pour into 35- or 60-mm plastic Petri dishes, and allow to set. Store the dishes at room temperature, sealed with Parafilm. Use with stainless-steel minuten pins.
5. Howard's Ringer. 0.12 $M$ NaCl, 0.0015 $M$ CaCl$_2$, 0.005 $M$ KCl/L: 7.2 g NaCl, 0.17 g CaCl$_2$, 0.37 g KCl. Adjust to pH 7.2, with very dilute HCl and autoclave.
6. Penicillin/streptomycin/fungizone (PSF) (Gibco-BRL, Paisley, UK). Aliquot (500 µL) and store at –20°C.
7. Dispase (Boehringer, Lewes, UK). 1 mg/mL in tissue culture medium, such as F12 or L15; 500-µL aliquots. Store at –20°C.
8. DNase I (Boehringer, Lewes, UK) for enzymatic preparation of notochord according to **Subheading 3.3.2.**, **step 2**: 1 mg/mL stocks in Howard's Ringer, aliquot (100 µL). Store at –20°C.
9. Sterile fetal calf serum for coating Gilson tips, 50-µL aliquots. Store at –20°C.
10. L15 tissue culture medium (for enzymatic preparation of notochord according to **Subheading 3.3.2.**, **step 2**).
11. India ink (e.g., Pelikan "Fount"–lacks shellac, nontoxic).
12. #11 Scalpel blade and handle or commercial egg pricker.
13. Pair of curved, sharp-pointed scissors.
14. Pair of Vannas spring scissors.
15. Pair of ultrafine scissors.
16. Pair of coarse forceps.
17. 2 pairs of fine watchmakers' forceps (#5 or 55).
18. Stainless-steel spatula, 5- to 8-mm wide tip. Bend to obtain a 135° elbow.
19. Stainless steel minuten pins (for Sylgard dishes).
20. Pure tungsten wire, 100 µm.
21. Four needle holders (6 × 120 mm glass capillary rod).
22. Sealing wax.
23. 20-µL Gilson tips (sterile).
24. 60-mm Bacterial Petri dishes.
25. Long glass Pasteur pipets.
26. Thin silicone tubing plus adapter.
27. Bunsen burner.

## 3. Methods

### 3.1. Getting Started

1. Incubate eggs. Generally, chick embryos between HH 8–13 can be used for notochord transplantations. To study floor plate induction, use HH 9–10. If you plan to use the mechanical method of notochord preparation (**Subheading 3.3.1.**), incubate some eggs until HH 13–16 as donors. Despite these guidelines, the ideal

stages of hosts and donor has to be determined according to the specific aim of the study. Follow the guidelines for storage and incubation of eggs elsewhere in this volume (Chapter 14). Lay eggs on their sides when incubating for manipulations. Mark this side with a pencil.

2. Microscope: Set up a stereomiscroscope with fiber optics. Check that the light source has a heat filter. Hand-rests to steady the hands may be useful. Have a Bunsen burner within reach.

3. Thaw one aliquot each of PSF, Dispase solution, and FCS. If you are using method II to isolate the grafts, also thaw DNase I.

4. Make up 50 mL of Howard's Ringer with 1:100 PSF.

5. Prepare contrast medium: 1–5 mL of India ink, diluted 1:5 in Howard's Ringer-PSF. Draw into a 1-mL syringe, attach a 27-gage × 21 mm needle, expel any air, and bend the last 1 cm of the needle into a 60° angle.

6. Prepare operation needles. Warm up one end of the glass rod with the Bunsen burner, and then dribble some molten sealing onto the tip of the rod. Do not set fire to the wax. Cut 20-mm lengths of tungsten wire from the reel, and straighten out the curve in the wire. Hold the wire with coarse forceps, and then fasten the wire to the glass rod by pushing it through the soft wax into the hole within the rod. Bend 4 mm at the free end of the wire to form a 135° elbow. Holding this tip into the hottest region of the Bunsen flame (blue color) and keep it there until the last 1–2 mm flies off. Withdraw the needle from the flame immediately—it will have a very fine sharp tip. Be careful not to touch anything hard with your needle, since this will ruin the tip.

7. Prepare a thin glass capillary. Hold capillary of Pasteur pipet over the Bunsen flame, and bend an angle in the middle of it. Warm the portion between this angle and the main part of the pipet, and then pull to obtain a very thin capillary. Break the rest of the pipet off, leaving about 2 cm of the thin capillary behind. The coarse end of this capillary can be attached to the adapter of the silicone tubing to form a mouth pipet.

## 3.2. Opening Eggs for Manipulations (Windowing)

1. Disinfect eggs by spraying with 70% v/v ethanol, allow to dry, and then pierce the blunt end using an egg-pricker or the point of a #11 scalpel blade to open the air space to the atmosphere (*see* **Note 1**). You may prepare eight eggs at a time.

2. Insert a 19-gage needle on a 5-mL syringe through the hole, and direct the tip toward the bottom inside surface of the egg, avoiding the thick albumen that surrounds the yolk. Withdraw 1–2 mL of the thin albumen.

3. With the point of a pair of curved scissors, carefully drill a hole at the middle of the upper side without pushing the scissor tip through the shell membrane. Carefully tear the shell membrane, and leave egg aside for a few moments while the embryo, which is immediately beneath the shell, subsides into the space created by the withdrawal of albumen. A drop of Howard's Ringer over the hole may accelerate the process.

4. Cover the top side of the egg with a piece of 25-mm wide Sellotape, stretched out before sticking down so that it will conform to the double curvature. Press it down firmly, and smooth out creases.
5. Use the curved scissors to make a spiral cut, starting with a small hole and leaving a circular aperture in shell about 12–15 mm in diameter. Direct your cut, as it progresses, so that the aperture overlies the blastoderm (*see* **Note 2**).
6. Examine under the stereoscope and stage the embryo by somite count (use neutral red to visualize if required; *see* Chapter 15).
7. Select those eggs in which the embryo lies more or less centrally in the window as hosts; use the remainder as donors if undertaking isochronic grafting.
8. Hosts: place a drop of Howard's Ringer on the embryo, and reseal egg with loosely applied Sellotape. Keep eggs on the bench—they are happy to be out of the incubator for several hours at this stage.
9. Donors: When doing isochronic grafts, you can economize by using as donors those embryos that are found, on windowing, to be less accessible. Cut around perimeter of the area opaca with Vannas scissors, taking care not to rupture the yolk. Lift embryo with prewetted spatula, and transfer to dish of Howard's Ringer. Float off the vitelline membrane, shake the blastoderm with forceps to free it off adherent yolk, and transfer to fresh Ringer. Subdissect as required (*see* **Subheading 3.3.1.**).

## 3.3. Dissection of Donor Notochords

To explant the notochord from donor embryos, use one of the three methods described below. Method 1 is a rapid procedure, but may leave cells from adjacent tissues sticking to the explant. Methods 2 and 3 use proteolytic enzymes to free the notochord from adhering cells, thus providing clean grafts. The best results are obtained with method 3. Here, the embryos are treated very gently to maintain their structural integrity and to recover additional tissues for further transplantation experiments.

### 3.3.1. Mechanical Method

1. Cut off and discard body from HH 13–16 embryos below the level of the last somites.
2. Grip the embryo with two pairs of forceps, one squeezing around the first branchial arch, and the other gripping the upper trunk. Pull apart. The notochord will stay attached to the head and will slide out intact from the trunk.
3. Trim off the notochord, and set it aside in a dish of fresh Howard's Ringer on ice (*see* **Note 3**).

### 3.3.2. Enzymatic Method 1

1. Transfer embryo to a Sylgard-bottomed dish containing Howard's Ringer. Submerge and pin out through the membranes with stainless-steel minuten pins. Subdissect region using needles or ultrafine scissors under transmitted illumination.
2. Using a serum-coated 20-μL tip, transfer tissues to a solution of 1 mg/mL Dispase, 5 μg/mL DNase I in L15 medium in a 35-mm dish.

3. Treat for 20–25 min at room temperature, and then transfer to fresh Howard's Ringer containing 10% serum. Reflux notochord through tip until free of adherent cells—check under stereoscope. Use fine needles to help separation if necessary.
4. Transfer to fresh Howard's Ringer, and set on ice (*see* **Note 3**).

### 3.3.3. Enzymatic Method 2

1. Pin down the embryo as in **Subheading 3.2.2.** The notochord is reached best if the embryo is ventral side up.
2. Carefully open the embryo along the notochord, using the sharpened tungsten needles.
3. Overlay the embryo with 50–100 µL Dispase solution. Leave at room temperature until structures come loose (15–20 min). Use operating needles to help separation.
4. Cut notochord with needles or ultrafine scissors.
5. Transfer to fresh Howard's Ringer on ice (*see* **Note 3**).

## 3.4. Preparation of Host Embryos

### 3.4.1. Improving Visibility of the Embryo

Have the syringe filled with contrast medium before starting. Push the needle into the yolk outside the area pellucida, and move the tip, as horizontally as possible, into the subblastodermal space. Be careful not to scrape the underside of the embryo. Expel 0.1 mL, taking care not to introduce air bubbles (*see* **Note 4**).

### 3.4.2. Grafting the Donor Notochord into the Host

#### 3.4.2.1. General Remarks

1. Illuminate the embryo with a single fiber optic lens on an oblique path. Use around 50× magnification, hold egg with one hand. Gently press down on the compressible egg stool to keep the object plane in focus. Support wrists on hand-rests and steady the third to fifth fingers of the operating hand on the side of the egg—this will enable you to stabilize the needle holder better. Avoid large movements.
2. Under the dissecting microscope, place one drop of Howard's Ringer over the host embryo. Nick the vitelline membrane with a fine needle, grasp the edge of the membrane with no. 5 forceps and reflect a flap above the embryo; be sure to pull back both layers of the vitelline. The hole in the vitelline should be just large enough to expose the operation site. As the vitelline ruptures, the drop of Ringer will flood in and prevent albumen from covering the embryo (if this should happen, both host and donor tissues become very sticky, and grafting becomes difficult).

#### 3.4.2.2. Grafting Notochord into the Host

1. The manipulation described below will result in ectopic induction of floor plate, since the graft will face the dorsoventrally uncommitted neural plate (**Fig. 1**, *see* **Note 5** for sclerotome induction). Using a very sharp tungsten needle, make a longitudinal slit through the ectoderm at its junction with the neural plate in the region of the open posterior neuropore: push the point through and cut by pulling

sharply upward (50× magnification). The slit should be 500–600 μm long. Periodically clean the needle of tissue or albumen by flaming (*see* **Note 6**).

2. Using the mouth pipet, and suck a small drop of Dispase solution into the fine capillary. Apply to the opened ectoderm of the host embryo. Use the tungsten needle to gently separate the neural plate from the adjacent paraxial mesoderm: roll off or push away the neural plate as it comes loose. Rinse the treated side with Howard's Ringer immediately to wash off the Dispase. The aim of this procedure is to make a pocket beneath the lateral part of the neural plate to house a notochord implant in close apposition with the basal surface of the neuroepithelium. Therefore, do not fully separate neural plate and paraxial mesoderm, as the graft will then settle down too deeply next to the host notochord.

3. Cut donor notochords into 300–400 μm lengths, transfer one piece in minimum fluid using a serum-coated 20-μL tip.

4. Deposit the notochord piece over the posterior end of the embryo, avoiding the albumen.

5. Maneuver the piece over the prepared slit. Open the slit by pulling on the lateral ectoderm with a needle. The notochord should drop into the hole. Push it in more deeply with the point of the needle.

### 3.4.3. Reincubation of Host Embryos

1. After operations, place another drop of Howard's Ringer over the embryo, check that a graft has not been dislodged, return the shell flap, and reseal egg with canvas sticky tape.

2. Stretch the tape before applying, and smooth it down carefully onto the egg. Make sure the tape adheres smoothly to the shell without creases that would admit air. High humidity inside the egg is essential after operations—drying, even slightly, is the major cause of mortality. Mark the shell with name and number/stage of operation using a pencil and place in the recovery incubator (38–39°C, >90% humidity). Do not open the egg until you want to harvest the embryo—checking up on progress causes drying and reduces survival.

## 4. Notes

1. Egg albumen is bactericidal, so instruments need not be sterilized, but they should be thoroughly cleaned after use and may be dipped in 70% v/v ethanol and allowed to air-dry before use. **On no account should instruments (other than tungsten needles) be flamed—it destroys the temper of the steel.**

2. Some workers prefer to cut an incomplete circular aperture, tipping this back and out of the way (like a trapdoor) during manipulations, and then closing it and sealing with tape on completing the procedure. Keep the outside of the egg dry, since the tape will not stick to a wet surface.

3. Instead of storing the explanted notochords on ice, you may directly graft one into a readily prepared host. However, do not keep the grafts at room temperature for long as they will coil up and become intractable.

4. Do not inject the contrast medium until each egg is ready for grafting—it can disperse quite rapidly. You may have to test India ink of several suppliers for toxicity, since the best brand (Pelikan "Fount") is now hard to find.

5. To induce sclerotome ectopically, the procedure in **Subheading 3.4., step 2** may be slightly varied. The paraxial mesoderm gains competence to respond to notochordal signals at the time of somite formation. In addition, newly formed somites remain labile with respect to their dorsoventral and mediolateral pattern. Therefore, it is not critical to use HH 9-10 embryos or operate at the level of the open neuropore. You may find it convenient to insert the graft at the level of the anterior segmental plate.

6. During operations make sure the needle is kept clean of adherent albumen and tissue by passing it quickly through the flame of a Bunsen burner. Keep the needle in a hot flame to remove and resharpen a bent tip.

## Acknowledgment

We thank Ivor Mason for critically reading the manuscript.

## References

1. Goulding, M. D., Lumsden, A., and Gruss, P. (1993) Signals from the notochord and floor plate regulate the region-specific expression of two Pax genes in the developing spinal cord. *Development* **117**, 1001–1016.
2. Lumsden, A. and Graham, A. (1995) A forward role for hedgehog. *Curr. Biol.* **5**, 1347–1350.
3. Dietrich, S., Schubert, F., and Lumsden, A. (1997) Control of Dorsoventral pattern in the chick paraxial mesoderm. *Development* **124**, 3895–3908.
4. Christ, B. and Ordahl, C. P. (1995) Early stages of chick somite development. *Anat. Embryol.* **191**, 381–396.
5. Fouquet, B., Weinstein, B. M., Serluca, F. C., and Fishman, M. C. (1997) Vessel patterning in the embryo of the zebrafish: guidance by notochord. *Dev. Biol.* **183**, 37–48.
6. Ingham, P. W. (1995) Signalling by hedgehog family proteins in Drosophila and vertebrate development. *Curr. Opin. Genet. Dev.* **5**, 492–498.
7. Roelink, H. (1996) Tripartite signaling of pattern: interactions between Hedgehogs, BMPs and Wnts in the control of vertebrate development. *Cur. Opin. Neurobiol.* **6**, 33–40.
8. Chiang, C., Litingtung, Y., Lee, E., Young, K. E., Corden, J. L., Westphal, H., et al. (1996) Cyclopia and defective axial patterning in mice lacking Sonic hedgehog gene function. *Nature* **383**, 407–413.
9. Le Douarin, N. (1973) A feulgen-positive nucleolus. *Exp. Cell Res.* **77**, 459–468.
10. Fontaine-Perus, J., Jarno, V., Fournier le Ray, C., Li, Z., and Paulin, D. (1995) Mouse chick chimera: a new model to study the in ovo developmental potentialities of mammalian somites. *Development* **121**, 1705–1718.

# 19

## Transplantation of Avian Neural Tissue

**Sarah Guthrie**

### 1. Introduction

The transplantation of neural tissue provides a means of addressing many questions in developmental neurobiology and regeneration of the nervous system. Although this technique has been used in fish, amphibia, avian, and mammalian species, this chapter will focus on neural transplantation *in avia*, which has several advantages. The development of the nervous system provides a close parallel to that of mammalian embryos in many of its aspects, and embryos can be accessed at developmental stages impossible in mammals. In addition, avian eggs can be obtained at low cost, and minimal equipment and facilities are required for these types of grafting experiments. Transplantation is possible at both cranial and spinal levels of the neuraxis, but the accessibility of these regions varies with stage. For example, transplantation of brain regions (e.g., **ref.** *1*) is relatively easy at stages in which the neural tube has yet to develop the local expansions of the brain vesicles and has not yet become extensively vascularized, i.e., before E3 in the chick, but becomes more difficult after this time. Stages at which transplantation into the spinal cord (e.g., *2*) can be performed depend on the axial level to be investigated owing to the rostro-caudal order of its generation, but transplants into spinal regions remain feasible at later stages.

Examples of the application of neural grafting in chick embryos have been to explore the role of a signaling region by transplanting it to an ectopic site *(3)* or to investigate the mechanisms of hindbrain segmentation *(4)*. In some experimental paradigms, it is desirable to follow the position and fate of the transplanted tissue in the host environment. For this purpose, chick–quail chimeras have been used extensively to explore a number of developmental questions (*see* Chapter 22 for detailed discussion). Quail tissues transplanted into a chick host can later be recognized using quail cell-specific or axon-specific

From: *Methods in Molecular Biology, Vol. 97: Molecular Embryology: Methods and Protocols*
Edited by: P. T. Sharpe and I. Mason © Humana Press Inc., Totowa, NJ

antibodies. This allows the fate mapping of cells derived from the grafted tissue or of axon projections arising from the graft. In chick-to-chick transplants, it is also possible to track grafted tissues that have been labeled using Hoechst or other fluorescent dyes (e.g., *5*). Transplantation of retrovirally infected avian tissue into a host embryo resistant to viral infection is a possibility in order to analyze the fate of cells carrying a transgene *(6)*. Grafting of mouse tissue into a chick host has also been used to analyze developmental interactions between tissues of transgenic animals and that of the host embryo *(7)*.

This chapter describes the procedure for grafting of neural tissues in general, and provides diagrams depicting grafting of rhombomeres as an example of this technique. Learning neural transplantation is a long and frustrating business, which requires many repetitions and probably weekly practice to achieve any success. However, many of the trivial problems of its execution can be overcome by the strategies outlined here. Grafting experiments can often provide a valuable adjunct to other studies, e.g., molecular biological or tissue-culture experiments, in exploring the role or interactions of a molecule or tissue region. In addition, they may be the method of choice for addressing some developmental problems, such as the state of determination of a tissue, its inductive influence, or to construct a fate map. This technique will therefore continue to provide a valuable tool for experimental embryologists.

## 2. Materials
### 2.1. Microscopes

It is important to use a good-quality stereomicroscope with focussing eyepieces and provision for both epi-illumination and *trans*-illumination. Bear in mind that for much of the day, you will probably need to monopolize two microscopes, one for the grafting itself (epi-illumination) and one with the dish of tissue pieces for transfer (*trans*-illumination). A good light source (preferably a cold light source) is also essential, particularly to locate grafted pieces once you have transferred them into the egg. You will also need a comfortable chair of adjustable height and armrests, so that both forearms are supported while you dissect and do the transplantions. It is virtually impossible to accomplish these types of grafting experiments with quaking hands suspended over the egg. If you are not "sitting comfortably" at the microscope (for what may be many hours), then transplantation becomes a frustrating and ultimately fruitless business.

### 2.2. Dissecting Tools for Preparation
### of Donor Embryo Fragments and Host Embryos

1. Tungsten needles: These can be made by mounting 2- to 3-cm pieces of 100-μm pure tungsten wire (Goodfellow) on needle holders made of aluminum or wood of diameter 30–40 mm. Pieces of wire can be mounted using either sealing wax or adhesive (e.g., Araldite). The end of the needle should be sharpened to a taper-

ing point in the hottest part of the Bunsen flame either before or after mounting. The end of the needle can then be bent over to an angle of 90–135° using forceps. Needles will tend to get bent, and should be kept straight and sharpened in the flame throughout the operation to remove attached tissues and to prevent snagging of the tissues being dissected.

2. Spring scissors and fine scissors for dissection of donor embryos.
3. Small spatula for removal of embryos from donor eggs.
4. Curved scissors or scapel for opening host eggs.
5. Fine forceps (Dumont 5) for dissection of donor embryos.
6. Coarse- and fine-gage needles for withdrawing albumen from eggs and injecting ink.
7. Sellotape and opaque egg tape (Beiersdorf).
8. Carmine dye particles (Sigma) for labeling transplanted tissue fragments.
9. Sterile plastic pipets or fire-polished Pasteur pipets for transferring tissue pieces.

N.B. Sterilize all tools that are to contact the embryos by spraying with 70% ethanol and then allowing them to air-dry. Do not flame fine dissection tools.

## 2.3. Solutions

1. Chick Ringer solution: 7.2 g/L NaCl, 0.23 g/L CaCl, 0.37 g/L KCl, pH 7.2. Immediately before use, add 1 in 50 of penicillin/streptomycin solution (Pen/Strep; Gibco, Grand Island, NY).
2. 1X Hanks' Balanced Salt Solution (HBSS) (Gibco).
3. Dispase (Grade 1; Boehringer Mannheim, Mannheim, Germany). Dilute at 1 mg/mL in HBSS containing 50 µg/mL Deoxyribonuclease (Sigma, St. Louis, MO).
4. Ink solution for subblastodermal injection to visualize embryos. Ink is Pelikan drawing ink A, shade 17 (black). Dilute at 1 in 15 to 1 in 20 in Ringer solution containing Pen/Strep.

## 3. Methods

### 3.1. Preparation of Donor Tissues

1. Remove donor embryos from eggs. This may be done by cracking eggs into a Petri dish or by windowing them (*see* Chapter 14). If you are dealing with embryos of stage 9 or younger, it is more reliable to window eggs to avoid losing embryos by accidentally puncturing the blastoderm.
2. Remove donor embryos by cutting through the vitelline membrane and the blastoderm and around the embryo using spring scissors.
3. Slip a small spatula under the embryo and transfer into a dish of Ringer solution.
4. Dissect away extraneous tissues to isolate the region of interest. In the case of rhombomere transplants, for example, the entire hindbrain from E2 embryos is dissected away from spinal cord, midbrain, and forebrain, and the heart and gut removed (**Fig. 1**).
5. Place the isolated regions in a small volume of Dispase (1 mL) in a 35-mm Petri dish, and leave for 5 min.
6. After this time, use tungsten needles to test whether adjacent tissues are dissociating from the neural tubes. If so, transfer immediately to a fresh dish of HBSS. If

**DONOR**

**HOST**

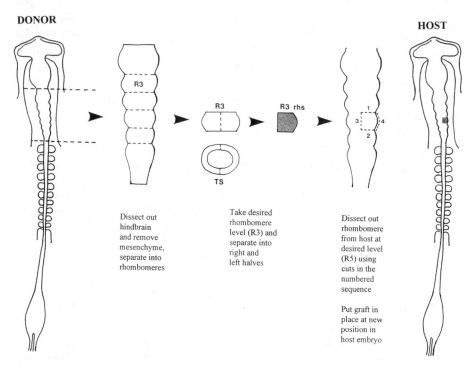

Dissect out
hindbrain
and remove
mesenchyme,
separate into
rhombomeres

Take desired
rhombomere
level (R3) and
separate into
right and
left halves

Dissect out
rhombomere
from host at
desired level
(R5) using
cuts in the
numbered
sequence

Put graft in
place at new
position in
host embryo

Fig. 1. This figure illustrates diagrammatically the procedure for grafting a rhombomere unilaterally from a donor stage 10 chick embryo into an isochronic host embryo. Once the embryo has been removed from the egg, the entire hindbrain region is dissected free of adjacent tissues, and the adhering mesenchyme cells are removed by exposure to Dispase. Then individual rhombomeres are separated from each other, and at the desired axial level (rhombomere 3 in this case), the right and left halves are separated from each other. Meanwhile the region that will receive the graft in the host embryo is prepared by removing the host rhombomere 5 using 4 cuts in the numbered sequence. The graft tissue can then be transferred into the egg and put in place in the ectopic location. The right-hand side piece of the rhombomere has been grafted such that the anteroposterior and dorsoventral polarity have been maintained.

not, leave for a minute or so longer—the exact time will depend on the age of embryo and region to be dissected. In any case, it is crucial that embryos are not exposed to Dispase for too long. Otherwise, neural tissue will become irreversibly damaged.

7. Adjacent mesodermal tissues can be teased away from the neural tubes using tungsten needles and then clean neural tubes transferred into fresh HBSS. Cut away undesired tissues by pressing down on the plastic Petri dish with the tungsten needle. When pulling away mesenchymal or other tissues, avoid contacting the neural tissue directly with the needles.

8. From now on, keep tissues as far as possible on ice, since this prevents deterioration.

9. Transfer individual neural tubes from the dish on ice to the microscope for further subdissection, e.g., dissection of individual rhombomeres (**Fig. 1A**), but keep the rest cold. At this stage or previous stages of the dissection, a few Carmine particles can be brushed against the surface of the neural tissue using a needle in order to label its polarity, if this is required.

10. Different tissue pieces may be stored in individual wells in multiwell plates, and chick tissues can be kept on ice for up to 8 h without detriment. For rhombomere transplantation in which unilateral grafts are to be performed, the bilateral rhombomere fragments can be kept intact until just before grafting and then separated into right and left halves (**Fig. 1**).

## 3.2. Preparation of Host Embryos

1. The outer surface of eggs should be sterilized with alcohol and 1 mL of albumen removed via the blunt end of the egg using a syringe and 21-gage needle.

2. Eggs can then be windowed by cutting a hole in the top using a scalpel or curved scissors (*see* Chapters 14 and 16). Avoid pointing the blades of the scissors down into the egg, since this may damage the embryo.

3. Embryos can then be visualized by subblastodermal injection of ink (about 100 µL) using a 25-gage needle. Inject the ink outside the area opaca at a shallow angle, and minimize the size of the hole.

4. Eggs can be resealed with pieces of Sellotape and left at room temperature (or in the incubator) until they are to be used (*see* **Note 1**).

## 3.3. Grafting of Neural Tissue Pieces

1. Take a host egg of the desired stage, remove the sealing tape, and remoisten the embryo with a few drops of Ringer solution.

2. Using a tungsten needle, lift up the vitelline membrane, and cut a flap above the embryo; keep this to the minimum size required for access to the region of interest.

3. Using one tungsten needle make cuts around the tissue to be removed in order to receive the graft (**Fig. 1**). To do this, insert the needle to a depth of 100 µm or so, and pull gently upward and separate tissues. Then gradually deepen the cut with successive strokes of the needle (*see* **Note 2**). It is preferable to make the hole slightly bigger than the piece you intend to put into it.

4. Then either remove the tissue piece by hooking it out with the needle, or leave it in the egg. If you do the latter and you are using unlabeled donor tissue, you should be confident that you can tell the grafted tissue piece apart from the host tissue you have just removed.

5. Retrieve the neural tissue piece to be grafted from the Petri dish using a P20 pipetman and yellow tip set on 10–20 µL. If there are problems with the tissue getting stuck to the plastic pipet, you can coat the inside of the yellow tip with sterile serum before use.

6. Draw some fluid into the pipet before drawing up the tissue piece. Otherwise the tissue may hit the meniscus and disintegrate. Then transfer the piece into the egg

as quickly as possible. Do this on low magnification, and try to track the piece with your eye to avoid losing it.

7. Maneuver the tissue piece into place adjacent to the grafting site using a tungsten needle (**Fig. 1**).

8. Orient the piece as required, and check for several seconds to see that it does not move out of position. It is possible to place small fragments of sterile drawn glass or other small objects on top of the neural tissue fragment to keep it in place, but this is not normally necessary (*see* **Note 3**).

9. Seal the egg with a piece of opaque egg tape (preferably stretchy). If you intend to incubate the embryo for more than a few days, extra Sellotape around the edges is also a good idea.

10. Eggs should be replaced in the incubator and left until the desired stage for harvesting (*see* **Note 4**). Opening up to check the stage of development and then resealing for further incubation is not recommended, and leads to certain death. When calculating the stage for harvesting, assume a day's delay in the development of embryos.

11. To open the eggs, cut a slightly bigger hole than before in the top of the egg using curved scissors.

12. Cut and lift the embryo out using spring scissors and forceps or a spatula, transfer to Ringer's solution, and wash off excess yolk before further processing.

## 4. Notes

1. For grafting of very young embryos (younger than stage 9), it may be desirable to avoid opening the eggs until immediately before grafting, since dehydration is a major cause of mortality. Replacing the cut piece of shell over the hole and securing with tape may also improve survival rates.

2. Always keep the motion of the needle toward you, rather than down into the egg, since this guards against damage to the embryo. Use more smaller cuts rather than few larger ones to free the region from its surrounding tissues. For rhombomere grafting, this involves cuts at the rostral and caudal ends, the lateral edge where the neural tube abuts the mesenchyme, in the dorsal midline, and the ventral midline (where the ventral neural tube abuts the floor plate).

3. If you have been struggling for many minutes with the graft, you may notice that the surface of the embryo has dried out, and needs the addition of Ringer solution. If so, do this very carefully to avoid dislodging the graft.

4. The major problem of all neural grafting experiments is the low survival rates of the grafted embryos. Even if the operations have been performed flawlessly, mortality can be about 50% after several days, falling to as low as 10% after a week or so. The causes of this are unclear, but the major factors are likely to be dehydration of the embryo and/or contamination. Dehydration can be prevented by addition of Ringer solution, performing the operation in the shortest possible time, and making sure the egg is well sealed at the end. Contamination should not be a problem provided sterile plastics, solutions, and tools are used throughout, and the only tool used inside the egg is a tungsten needle, which is frequently flamed. Excessive damage to the embryo, surrounding membranes and yolk and

the developing vasculature, are also obvious problems, which should diminish with practice. Low survival rates produce the related problem that sample sizes in any one experimental category are very small. It may take a long time to obtain data that are amenable to statistical analysis. Furthermore, experimental results that are obtained within these categories may sometimes be heterogeneous owing to differences in the grafting procedure. For this reason, is may be desirable to label the grafted tissue in some way (*see* **Subheading 1.**).

## References

1. Hallonet, M. E. R., Teillet, M.-A., and Le Douarin, N. M. (1990) A new approach to the development of the cerebellum provided by the quail-chick marker system. *Development* **108,** 19–31.
2. Matise, M. P. and Lance-Jones, C. (1996) A critical period for the specification of motor pools in the lumbosacral spinal cord. *Development* **122,** 659–669.
3. Martinez, S., Marin, F., Nieto, M. A., and Puelles, L. (1995) Induction of ectopic *engrailed* expression and fate change in avian rhombomeres: intersegmental boundaries as barriers. *Mech. Dev.* **51,** 289–303.
4. Guthrie, S. and Lumsden, A. (1991) Formation of regeneration of rhombomere boundaries in the developing chick hindbrain. *Development* **112,** 221–229.
5. Guthrie, S., Prince, V., and Lumsden, A. (1993) Selective dispersal of avian rhombomere cells in orthotopic and heterotopic grafts. *Development* **118,** 527–538.
6. Fekete, D. M. and Cepko, C. L. (1993) Retroviral infection coupled with tissue transplantation limits gene transfer in the chicken embryo. *Proc. Natl. Acad. Sci. USA* **90,** 2350–2354.
7. Itasaki, N., Sharpe, J., Morrison, A., and Krumlauf, R. (1996) Reprogramming *Hox* expression in the vertebrate hindbrain: influence of paraxial mesoderm and rhombomere transposition. *Neuron* **16,** 487–500.

# 20

## Grafting of Apical Ridge and Polarizing Region

### Cheryl Tickle

## 1. Introduction

The apical ectodermal ridge (AER; also known as the apical ridge) and the zone of polarizing activity (ZPA; also known as the polarizing region) are two major signaling regions in developing vertebrate limbs. Limbs arise as small buds of undifferentiated mesenchyme cells encased in ectoderm. The apical ridge and the polarizing region were first identified in limb buds of chick embryos, and functions of these regions were explored by traditional "cut and paste" experiments by Saunders (reviewed in **ref.** *1*). Similar signaling regions are also found in embryonic limb buds of other vertebrate species, including mice, rats, snapping turtles, and humans *(2)*. Signals from apical ridge and polarizing region act together with signal(s) from the ectoderm in a region of undifferentiated cells called the progress zone *(3)*. The progress zone is found at the tip of the bud as it grows out and is maintained by the apical ridge. As cells leave the progress zone, they lay down the structures of the limb in sequence with proximal structures being formed first and distal structures later *(4)*. Cellular responses to polarizing and ectoderm signals in the progress zone establish the limb plan, and subsequently cartilage, bone, muscle, tendon, and other specialized cell types differentiate and become organized into tissues.

The apical ridge is the thickened rim of epithelium at the tip of the limb bud and is required for bud outgrowth (**Fig. 1**). Apical ridge cells are elongated, tightly packed, and linked by extensive gap junctions. When the apical ridge is removed, bud outgrowth ceases and a truncated limb develops; in contrast, grafting a second apical ridge to the surface of a limb bud induces a second outgrowth *(1,5)*. The signal from the apical ridge can be substituted by beads soaked in fibroblast growth factors (FGF-2, FGF-4, FGF-8; *6,7*). Several members of the FGF family are expressed in tissues at the tip of the limb bud, in-

From: *Methods in Molecular Biology, Vol. 97: Molecular Embryology: Methods and Protocols*
Edited by: P. T. Sharpe and I. Mason © Humana Press Inc., Totowa, NJ

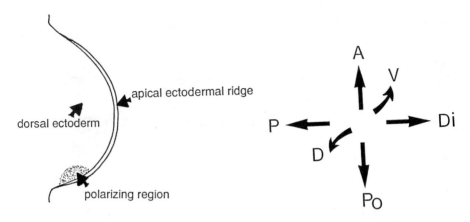

Fig. 1. Diagram to show early chick limb with signaling tissues and the three axes, A-Po (anterior–posterior), P-Di (proximal–distal), D-V (dorsal–ventral).

cluding apical ridge (reviewed in **ref. 8**). The polarizing region is a group of mesenchyme cells at the posterior edge of the bud (**Fig. 1**; the posterior edge is that nearest the tail end of the embryo). Signaling by the polarizing region controls antero-posterior limb pattern (reviewed in **refs. 1** and **5**). Cells of the polarizing region of chick wing buds cannot be distinguished morphologically, but have a remarkable effect on the pattern when grafted to the anterior margin of a second wing bud. The normal chick wing has three digits arranged from anterior to posterior 2 3 4 (**Fig. 2**). The limb develops six digits following a polarizing region graft, and these are arranged in the mirror-image symmetrical pattern 4 3 2 2 3 4 (**Fig. 2**). The additional digits arise from anterior mesenchyme in response to a signal from the graft. Tissues from different regions of the limb bud and from other regions of embryos have been assayed for polarizing activity. This has led to construction of maps showing distribution of cells with polarizing activity or potential polarizing activity (*9–12*; **Table 1**).

The extent of digit duplication depends on both the number of polarizing region cells in the graft and the length of time that the graft is left in place (reviewed in **ref. 5**). With small numbers of polarizing region cells (around 30),only an additional 2 are produced giving the pattern 2234; with more polarizing region cells (60 cells) an additional digit 3 is formed (patterns, such as 32234, 3234, 334) and about 120 cells are required to give an additional digit 4 (**ref. 21**). Signaling by the polarizing region can be attenuated by X-irradiation and various chemicals. All of these treatments probably effectively reduce the number of signaling cells. When the graft is left in place for short periods of time (14 h) and then removed, only an additional digit 2 is formed,

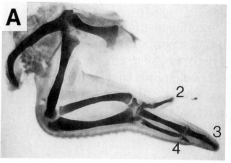 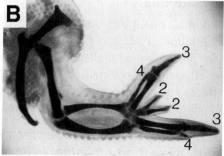

Fig. 2. (**A**) Normal chick wing at 10 d development stained with alcian green to show skeleton laid down in cartilage. Digit pattern <u>234</u>. (**B**) Chick wing with mirror-image duplication following a polarizing region graft to anterior margin. Digit pattern <u>432234</u>.

### Table 1
### Tissues with Polarizing Activity

Posterior margin of limb bud, chicken, quail, mouse,
    snapping turtle, human *(1,2,12)*
Hensen's node, chicken, mouse *(13,14)*
Presumptive flank tissue, chicken *(11,15)*
Floor plate of neural tube, chicken, mouse *(14,16)*
Amnion, chicken *(17)*
Mesonephros, chicken *(1)*
Tail bud mesenchyme, chicken *(18)*
Genital tubercle, mouse *(19)*
Gut epithelium, chick *(20)*

but when the graft is left for longer, then the formation of more posterior digits, such as digit <u>3</u>, results *(22)*. Polarizing signaling can be reproduced by beads soaked in either retinoic acid or sonic hedgehog protein *(23,24)*. Many of the tissues that have polarizing activity have been shown to be able to generate retinoic acid and to express *sonic hedgehog* (reviewed in **ref. 5**).

Limb duplications can be obtained when just the mesenchyme cells from the limb posterior margin are grafted without the normally associated ectoderm and/or apical ridge, provided that grafts are placed in contact with the apical ridge of the host limb. In the normal limb bud, apical ridge and ectoderm maintain polarizing activity of posterior mesenchyme *(25)*. *Sonic hedgehog* transcripts are found in posterior mesenchyme, and the distribution of these transcripts closely follows maps of polarizing activity in developing limbs *(26)*. The posterior part of the apical ridge expresses *Fgf-4*, and FGF-4 maintains *sonic hedgehog* expression in posterior mesenchyme *(27)*.

## 2. Materials

For all manipulations, a good set of dissecting instruments is required. The minimum kit consists of: a pair of small fine scissors, a blunt pair of forceps, two pairs of watchmaker forceps (nos. 4 or 5), two sharpened tunsten needles, small spatula, very small spatula. Other instruments that could be useful include iridectomy scissors and a small spoon (*see* **Note 1**). Also needed are a series of Hamburger/Hamilton stages (reprinted in **ref. 28**).

### 2.1. Separation of Ectoderm/Mesoderm and Dissection of Apical Ridge

1. Sterile phosphate-buffered saline (PBS).
2. Sterile ice-cold tissue-culture medium containing serum (for example, Minimal Essential medium + Hank's salts buffered with HEPES + antibiotics + fetal calf serum).
3. Sterile ice-cold 2% trypsin. (Recently, we have been using crude type II from porcine pancreas made up in calcium-magnesium free saline; pH to 7.5) store in aliquots at –20°C.
4. Ice-cold slab (a freezer block covered with black plastic will do).
5. Sterile pipets for transferring tissue.

### 2.2. Apical Ridge Grafts and Ectoderm-Mesenchyme Recombinations

1. Platinum wire pins. (These can be made by cutting short lengths of platinum wire [0.025 µm in diameter; Goodfellow metals] under 70% alcohol in a Petri dish and bending one end into a hook. The pins are then washed in tissue culture medium to rinse away the alcohol before use.)

### 2.3. Polarizing Region Grafts

1. Sterile trypsin (*see* **Subheading 2.1., item 3**).
2. Sterile ice-cold medium (*see* **Subheading 2.1., item 2**).
3. Ice-cold slab.
4. Sterile pipets or gilson pipet tips.

## 3. Methods

### 3.1. Separation of Ectoderm and Mesoderm Components of Limb Buds (Fig. 3A)

This can be accomplished for limb buds over a range of developmental stages from stage 19 to 29 *(29)*.

1. Remove chicken embryo from egg through window in shell. To do this, first cut a circle into the yolk through the vitelline and chorionic membranes around the embryo, and then lift out embryo on a small spatula.
2. Place embryo in Petri dish containing PBS or tissue-culture medium (that equilibrates with air), and pull away the membranes surrounding embryo with a fine pair of forceps.

**A**

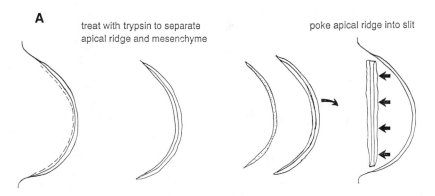

treat with trypsin to separate
apical ridge and mesenchyme

poke apical ridge into slit

Fig. 3. (**A**) (**Subheading 3.2.**) Diagrams to show how an apical ridge is isolated and grafted to the dorsal surface of a limb bud. Remove thin sliver of donor limb bud tip, and soak in 2% trypsin at 4°C for 5–10 min. Separate apical ectodermal ridge, and keep in ice-cold medium. Graft ridge to slit made in dorsal surface of host limb bud, and poke into place with blunt needle.

3. Remove limb buds from embryo using fine forceps to pinch through tissue where buds attach to body wall.
4. Prick ectoderm of limb buds with fine needles ,so that trypsin will penetrate easily to bud tip.
5. Place limb buds in ice-cold 2% trypsin, and leave on ice for between 30 min and 1 h (precise length of time will depend on size of limb bud and batch of trypsin; **ref. *30*).
6. Transfer limb buds to ice-cold tissue-culture medium containing serum and leave on ice for 5 min. (Serum contains trypsin inhibitor and will "stop" trypsin)
7. Place Petri dish containing limbs on a cold slab, and ease loosened ectoderm from mesenchyme by inserting a needle between the tissues and working from cut edge of ectoderm. It should be possible to remove the ectoderm, which looks like a diaphanous mitten, with attached basement membrane *(31)*.

## 3.2. Dissection of Apical Ridge and Grafting to Limb Bud (Fig. 3A)

1–3. As in **Subheading 3.1.** (*see* **Note 3**).
4. Dissect apical ridge of the limb bud away from underlying mesenchyme using fine needles.
5. Place apical ridge with small amount of adherent mesenchyme into ice-cold trypsin (*see* separation section) for 5–10 min. This should be sufficent time to loosen tissues.
6. Transfer tissue to ice-cold tissue-culture medium containing serum and peel apical ridge away from mesenchyme in ice-cold tissue-culture medium.
7. To graft apical ridge to the dorsal surface of a wing bud, cut a shallow slit running from anterior to posterior in the dorsal surface near the tip of right wing bud (good stages are 20–21) with a sharp needle.

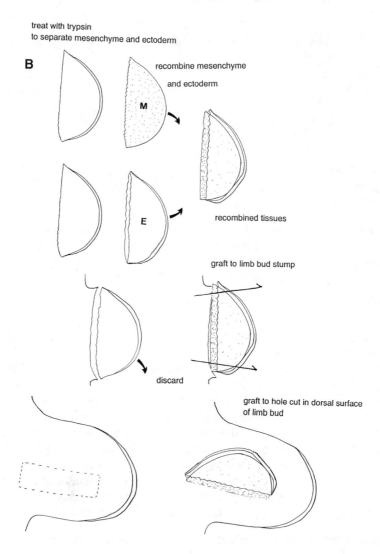

Fig. 3. *(continued)* **(B)** Diagrams to illustrate how mesenchymal and ectodermal components of limb buds are separated, recombined and grafted to allow further development (**Subheadings 3.1. and 3.3.**). Separate ectoderm and mesenchyme of chick limb buds by soaking in 2% trypsin for 1 h at 4°C and then placing in tissue culture containing serum. Recombine mesenchymal core (M) from one limb bud with ectodermal hull (E) from another bud. Allow tissues to reanneal at 37°C for 1 h, and then graft recombined tissues either to wing bud stump and pin in place or to dorsal surface of older wing bud.

8. Transfer isolated ridge into host egg, maneuver over slit, and poke into slit with a blunt needle.

**C**

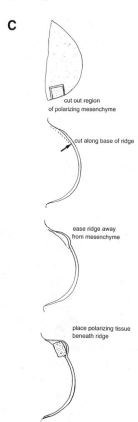

cut out region
of polarizing mesenchyme

cut along base of ridge

ease ridge away
from mesenchyme

place polarizing tissue
beneath ridge

Fig. 3. *(continued)* (**C**) (**Subheading 3.4.**) Diagram to show how a polarizing region is grafted beneath the apical ridge at the anterior margin of a chick limb bud. Cube of posterior mesenchyme cut from donor limb bud. Site for graft prepared in host limb bud by cutting along base of apical ridge, starting at arrow, and then easing ectoderm away from mesenchyme. Cube of posterior mesenchyme placed beneath anterior loop of apical ridge.

### 3.3. Apical Ridge Recombinations with Limb Mesenchyme (Fig. 3B)

1. First remove the ectoderm from a limb bud that will be used to provide the mesenchyme core, and remove this ectoderm from the dish.
2. Now transfer limb bud that will provide the ectodermal jacket for the recombination into the dish containing the isolated mesenchyme core.
3. Separate ectodermal jacket, and immediately push mesenchymal core inside. **Steps 1–3** are to be carried out in ice-cold medium on a cold slab.
4. Take the dish with recombined tissues off cold slab, and leave at room temperature for 10 min. Then place at 38°C for about 1 h to allow the ectodermal jacket to shrink on to the mesenchyme.

5. Graft recombined tissues to prepared site in a new embryo to continue development.
   a. Pin to limb bud stump. The embryo normally lies on its left side, so that dorsal surfaces of right limbs are uppermost. It is therefore easy to manipulate the right limb bud and leave the left as a control. Cut off right wing bud to give a stump (stages 20–22 would be suitable). Pin recombined tissues to stump. It is best to push one pin through one edge of the tissue recombination in a dish so that the hooked end marks the ectoderm surface and the pointed end without a hook protrudes from the naked mesenchyme surface. Then transfer speared tissue into host egg and pin tissue loosely to stump, so that the bare mesenchyme surface abuts the mesenchyme of the stump. Then take a second pin and secure other edge of graft. (Pins can be removed the next day.)
   b. Place recombination in trench on dorsal surface of wing bud. Cut a square away from the tip on dorsal surface of right wing bud (developmental stages 22–24). Lift up ectoderm and a layer of mesenchyme to create a shallow trench, and remove tissue from the egg on a small spatula. (The bleeding will soon stop.) Transfer tissue recombination into host egg and stand tissue recombination in trench with mesenchyme surface downward.

## 3.4. Polarizing Region Grafts (Fig. 3C) (see Note 4)

1. Remove chick embryo from egg, and place in medium (as in **Subheading 3.1.**).
2. Treat chick limb buds (stages 19–21) with cold trypsin and remove ectoderm.
3. Polarizing region in chick wing buds has been mapped from stages 17–29, and with reference to these maps of polarizing activity *(9–11),* cut appropriate region out of mesenchymal core of limb bud with sharp needles. A piece of tissue about $200 \times 200 \ \mu m^2$ from the thickness of the limb bud is easy to handle, although smaller pieces can be used.
4. Prepare host embryo by peeling away vitelline membranes and amnion over right wing bud. (Host embryos can be used between stages 18 and 24 but graft must be placed in contact with the ridge; to obtain complete digit duplications do not use stages later than 20/21; **ref. 32**).
5. Make site for graft by cutting along base of apical ridge over anterior margin of right wing bud with sharpened needle. First make a series a short superficial cuts through the dorsal ectoderm and into the underlying mesenchyme. The cuts are made just below the junction between the transparent ridge and opaque mesenchyme working from the bud apex toward the anterior edge of the bud (**Fig. 3C**). Then push needle through to the ventral side of wing bud, and repeat process working along slit. Pull loosened apical ridge away from mesenchyme to make a loop. The apical ridge can be stretched, but it will gradually shrink back against the mesenchyme.
6. Transfer polarizing tissue graft into host egg using either a glass or gilson pipet or a very small spatula.
7. Maneuver graft under apical ridge loop using the end of a needle. The apical ridge will hold graft in place.

8. Seal egg with sellotape.
9. It is often useful to cut a small hole in the sellotape window the next day, so that graft and limb bud can be inspected. It is usually possible to see whether the graft has remained in place.

## 4. Notes

1. Dissecting instruments need to be kept clean to guard against infection. Wipe instruments before use and before putting away with 70% alcohol. Another useful tip that can cut down on infection, as well as preventing the embryo from drying under the bright lights used to illuminate it, is to add a small drop (up to 10 μL) of medium containing antibiotics to the embryo during or after the operation.
2. For separation of mesenchyme and ectoderm, it is very important that all solutions and the tissue be kept very cold. Otherwise the tissue can become sticky. In early days of tissue disaggregation, the sticky gel associated with cells following trypsin treatment was thought to be glue that held cells together—later this glue was shown to be digested by DNase! Batches of trypsin may vary so it is wise to test out new batches in a trial run.
3. Until one gets used to looking at the embryo in the egg, it can be difficult to make out the apical ridge and to find tissue/cells that have been transferred into the egg. One way of increasing contrast is to use a green filter in the light path, but this reduces the light intensity. It is also possible to use a very weak solution of a vital dye, such as nile blue sulfate. Add a few drops of 0.1–0.2% nile blue sulfate in PBS to the embryo in the egg, or place tissue in a similar solution for a few seconds so that it is stained a very light blue. Be aware that nile blue sulfate at high concentrations can be toxic. We have found that if polarizing regions are stained, their signaling ability is reduced in parallel with the deepness of the shade of blue.
4. For polarizing region grafts to work most effectively, they must be placed in contact with the apical ridge. It is therefore important that the apical ridge is separated as cleanly as possible from the underlying mesenchyme. If the wing bud bleeds while you are lifting the ridge, you are cutting too far away from the apical ridge and into the marginal sinus that runs about 100 μm away from the chick wing ectoderm *(33)*.

## References

1. Saunders, J. (1977) Analysis of chick limb development, in *Vertebrate Limb and Somite Morphogenesis* (Ede, D. A., Hinchliffe, J. R., and Balls, M., eds.), Cambridge University Press, Cambridge, UK, pp. 1–24.
2. Fallon, J. F. and Crosby, G. M. (1977) Polarizing zone activity in limb buds of amniotes, in *Vertebrate Limb and Somite Morphogenesis* (Ede, D. A., Hinchliffe, J. R., and Balls, M., eds.), Cambridge University Press, Cambridge, UK, pp. 55–71.
3. Yang, Y. and Niswander, L. (1995) Interaction between the signalling molecules WNT and SHH during vertebrate limb development: dorsal signals regulate antero-posterior patterning. *Cell* **80**, 939–947.
4. Summerbell, D., Lewis, J., and Wolpert, L. (1973) Positional information in chick limb morphogenesis. *Nature* **224**, 492–496.

5. Tickle, C. and Eichele, G. (1994) Vertebrate limb development. *Ann. Rev. Cell Biol.* **10**, 121–152.

6. Fallon, J. F., Lopez, A., Ros, M. A., Savage, M. P., Olwin, B. B., and Simandl, K. (1994) FGF-2: Apical ectodermal ridge outgrowth signal for chick limb development. *Science* **264**, 104–107.

7. Niswander, L., Tickle, C., Vogel, A., Booth, I., and Martin, G. (1993) FGF-4 replaces the apical ectodermal ridge and directs outgrowth and patterning of the limb. *Cell* **75**, 579–587.

8. Tickle, C. (1995) Vertebrate limb development. *Curr. Opin. Genet. Dev.* **5**, 478–485.

9. MacCabe, A. B., Gasseling, M. T., and Saunders J. W. (1973) Spatiotemporal distribution of mechanisms that control outgrowth and antero-posterior polarization of the limb in the chick embryo. *Mech. Age. Devel.* **2**, 1–12.

10. Honig, L. S. and Summerbell, D. (1985) Maps of strength of positional signalling activity in the developing chick wing. *J. Embryol. Exp. Morph.* **87**, 163–174.

11. Hornbruch, A. and Wolpert, L.(1991) The spatial and temporal distribution of polarizing activity in the flank of the pre-limb-bud stages in the chick embryo. *Development* **111**, 725–731.

12. Wanek, N. and Bryant, S. V. (1991) Temporal pattern of posterior positional identity in mouse limb buds. *Dev. Biol.* **147**, 480–484.

13. Hornbruch, A. and Wolpert, L. (1986) Positional signalling by Hensen's node when grafted to the chick limb bud. *J. Embryol. Exp. Morph.* **94**, 257–265.

14. Izpisúa-Belmonte, J. C., Brown, J. M., Crawley, A., Duboule, D., and Tickle, C. (1992) Hox-4 gene expression in mouse/chicken heterospecific grafts of signalling regions to limb buds reveals similarities in patterning mechanisms. *Development* **115**, 553–560.

15. Yonei, S., Tamura, K., Ohsugi, K., and Ide, H. (1995) MRC-5 cells induce the AER prior to the duplicated pattern formation in chick limb bud. *Devel. Biol.* **170**, 542–552.

16. Wagner, M., Thaller, C., Jessell, T., and Eichele, G. (1990) Polarizing activity and retinoid synthesis in the floor plate of the neural tube. *Nature* **345**, 819–822.

17. McLachlan, J. C. and Phoplonker, M. (1988) Limb reduplicating effects of chorio-allantoic membrane and its components. *J. Anat.* **158**, 147–155.

18. Saunders, J. W. and Gasseling, M. T. (1983) New insights into the problem of pattern regulation in the limb bud of the chick embryo, in *Limb Development and Regeneration Part A* (Fallon, J. F. and Caplan, A. I., eds.), Liss, New York, pp. 67–76.

19. Dollé, P., Izpisúa-Belmonte, J. C., Brown, J. M., Tickle, C., and Duboule, D. (1991) *Hox-4* genes and the morphogenesis of vertebrate genitalia. *Genes Dev.* **5**, 1767–1776.

20. Roberts, D. J., Johnson, R. L., Burke, A. C., Nelson, C. E., Morgan, B. A., and Tabin, C. (1995) Sonic hedgehog is an endodermal signal inducing *Bmp-4* and *Hox* genes during induction and regionalization of the chick hindgut. *Development* **121**, 3163–3174.

21. Tickle, C. (1981) The number of polarizing region cells required to specify additional digits in the developing chick wing. *Nature* **289**, 295–298.

22. Smith, J. C. (1980) The time required for positional signalling in the chick wing bud. *J. Embryol. Exp. Morph.* **60,** 321–328.

23. Tickle, C., Alberts, B., Wolpert, L., and Lee, J. (1982) Local application of retinoic acid to the limb bud mimics the action of the polarizing region. *Nature* **296,** 564–566.

24. Lopez-Martinez, A., Chang, D. T., Chiang, C., Porter, J. A., Ros, B. K., Simandl, B. K., Beachy, P. A., and Fallon, J. F. (1995) Limb patterning by Sonic hedgehog protein fragment. *Curr. Biol.* **5,** 791–797.

25. Vogel, A. and Tickle, C. (1993) FGF-4 maintains polarizing activity of posterior limb bud cells *in vivo* and *in vitro. Development* **119,** 199–206.

26. Riddle, R. D., Johnson, R. L., Laufer, E., and Tabin, C. (1993) *Sonic hedgehog* mediates the polarizing activity of the ZPA. *Cell* **75,** 1401–1416.

27. Niswander, L., Jeffrey, S., Martin, G., and Tickle, C. (1994) Positive feedback loop in signalling in vertebrate limb development. *Nature* **372,** 609–612.

28. Hamburger, V. and Hamilton, H. L. (1951) A series of normal stages in the development of the chick embryo. *J. Morphol.* **88,** 49–92, reprinted (1992) *Developmental Dynamics* **195,** 231–272.

29. Rubin, L. and Saunders, J. W. (1972) Ectodermal-mesodermal interactions in the growth of limb buds in the chick embryo: constancy and temporal limits of the ectodermal induction. *Devel. Biol.* **28,** 94–112.

30. Szabo, G. (1955) A modification of the technique of 'skin splitting' with trypsin. *J. Pathol. Bacteriol.* **70,** 545.

31. Bell, E., Gasseling, M. T., Saunders, J. W., and Zwilling, E. (1962) On the role of ectoderm in limb development. *Devel. Biol.* **4,** 177–196.

32. Summerbell, D. (1974) Interaction between the proximo-distal and antero-posterior co-ordinates of positional values during specification of positional information in the early development of the chick limb bud. *J. Embryol. Exp. Morph.* **32,** 227–237.

33. Feinberg, R. N., Repo, M. A., and Saunders, J. W. (1983) Ectodermal control of the avascular zone of the peripheral mesoderm in the chick embryo. *J. Exp. Zool.* **226,** 391–398.

# 21

## Tissue Recombinations in Collagen Gels

### Marysia Placzek and Kim Dale

## 1. Introduction

The characterization of molecular and antigenic markers that identify specific vertebrate cells has increased dramatically in recent years. As a result, patterns of cell differentiation and development can be observed in vivo, and subsequently, the tissue interactions and differentiation factors that may operate to establish these patterns can be examined in vitro. Three-dimensional collagen gels provide a culture environment in which in vitro assays can be established, and used to assess the biological activity of one tissue or protein in patterning cells within a second potentially responsive tissue. Initially developed as a means to culture embryonic neuronal tissue and examine the effect of trophic factors (1), such gels have been used more recently to identify tissues and molecules responsible for inductive (2–17), chemotropic (18–23), and chemorepulsive (24,25) interactions. The advantages of a three-dimensional culture system are especially marked when the amount of material that is available to assay is limiting, and so, to date, they have been especially useful in the development of functional bioassays for explanted embryonic tissue. When used in conjunction with in vivo assays, such as described in Chapters 17–19, results obtained from such three-dimensional in vitro assays are especially compelling and can be used to extend and evaluate rapidly an observation made initially in an in vivo bioassay. The ability to assay specific tissues in isolation has many advantages. Large numbers of experiments can be set up in a single day, and the effects of tissues and proteins can be examined in the absence of other tissue that would normally be adjacent and potentially interfering in vivo. Furthermore, the isolation and culture of appropriate pieces of tissue enable one to understand the mechanism underlying an observed effect; thus, for

From: *Methods in Molecular Biology, Vol. 97: Molecular Embryology: Methods and Protocols*
Edited by: P. T. Sharpe and I. Mason © Humana Press Inc., Totowa, NJ

instance, one can distinguish trophic from tropic interactions and inductive vs migratory effects in a manner that is impossible in an in vivo assay.

The three-dimensional culture system has a number of advantages compared to a two-dimensional culture system. The collagen gel supports well the integrity of explanted tissue, which develops in culture in a manner appropriate to an equivalent piece developing in vivo: the explanted tissue does not flatten or spread, as occurs when cultured on a two-dimensional substrate. Explanted tissue of very small size often develops appropriately, when similar-sized explanted tissue cultured on a two-dimensional substrate fails to survive. The ability to manipulate the orientation of tissue explants within the three-dimensional gel means that very small quantities of tissue can be used to assay the effect of a protein or one tissue on a second. Consequently, sufficient tissue can be easily gathered to perform experiments and controls that are statistically relevant. Furthermore, the ability to manipulate the orientation of two tissues relative to one another within a three-dimensional gel provides a means to assess whether the action of a tissue is mediated by a factor that is readily diffusible, or by a membrane-associated factor. Finally, subsequent to culture, the development of the explanted tissue can readily be examined by either immunohistochemical or *in situ* hybridization techniques.

Although other types of material, such as fibrin clots or matrigel, can be used in place of collagen, the collagen matrix is less costly and is easy to manipulate. The collagen, which is prepared and maintained under conditions of low pH and low temperature, begins to gel as the temperature and the pH are increased. These two paramenters can be altered to establish optimal conditions in which the collagen sets sufficiently slowly to allow tissues to be manipulated within it. Matrices such as agar or agarose cannot be used as substitutes for collagen—the high density of such gels appears to suffocate the explanted tissue, whereas collagen fibrils are sufficiently spaced to allow the diffusion of nutrients from the culture fluid through the collagen gel to reach the cultured tissue.

The protocol presented here describes how to prepare collagen that is suitable for culturing embryonic tissues, and how to process it to make a three-dimensional gel in which tissue explants can be manipulated and embedded. The techniques used to isolate and embed embryonic tissue are described, as are the ways in which growth and differentiation factors can be presented to tissue explants within these gels. In addition, a summary is provided of the manner in which such cultures can be manipulated to process by immunohistochemical or *in situ* hybridization techniques in order to examine the pattern of cell differentiation within tissue explants.

## 2. Materials

All solutions are made using molecular biology-grade reagents and sterile distilled water.

### 2.1. Tissue Dissection

1. Liebovitz L15-air (Gibco-BRL) *(1)*.
2. Dispase (Grade 1, Boehringer-Mannheim, Mannheim, Germany), 1 mg/mL in L15-air, made fresh.
3. Serum.

### 2.2. Collagen Preparation (2)

1. Hemostats and scissors, sterilized with ethanol *(3)*.
2. Four to six adolescent rat tails.
3. Glacial acetic acid.
4. 0.1X DMEM (Gibco-BRL), pH 4.0 (without bicarbonate, with phenol red, filtered, sterile).
5. Dialysis tubing—sterile.

### 2.3. Tissue Embedding in Three-Dimensional Collagen Gel

1. Collagen.
2. 10X DMEM (without bicarbonate, with phenol red, filtered, sterile).
3. 0.8 $M$ NaHCO$_3$, sterile.
4. Four-well (16-mm) multidishes (Nunc cat. no. 134673).

### 2.4. Assaying Growth and Differentiation Factors

1. LipofectAMINE (Gibco-BRL).
2. DMEM with fetal calf serum (FCS).

### 2.5. Immunohistochemical or In Situ Analysis of Cultures

1. 0.2 $M$ Phosphate buffer, pH 7.4 (PB): 6 g/L NaH$_2$PO$_4$, 21.8 g/L Na$_2$HPO$_4$.
2. 4% Paraformaldehyde in 0.12 $M$ PB, pH 7.4: heat 20 mL H$_2$O to 80°C. Add 2 g paraformaldehyde, invert, and add two to three drops 1 $M$ NaOH. Invert or stir until solution clears. Make up to 50 mL with 0.2 $M$ PB and filter.
3. 30% sucrose in 0.1 $M$ PB.
4. OCT (TissueTek) and gelatin-subbed slides or Superfrost Plus slides (Fisher).
5. Glycergel (DAKO).

## 3. Methods

### 3.1. Tissue Dissection

Tissues from mouse, rat, and chick embryos have been cultured succesfully in three-dimensional collagen gels.

1. Dissect embryos from the decidua, or chick embryos from the eggshell into ice-cold L15-air *(1)*. Dissect embryos away from membranes and adhering yolk, and assess developmental stage.

2. To isolate a specific portion of tissue, first dissect out that tissue surrounded by adjacent tissues, using either electrolytically sharpened tungsten needles or small Vannas scissors. Treatment with an enzyme such as Dispase, frequently is necessary to free the required explant from adjacent tissues. Place the tissue in a small volume (approx 1 mL) of Dispase at room temperature for 5–15 min. Once the reaction has occurred *(4)*, remove tissues into cold L15-air *(5)* with a drop of serum included *(6)*, and allow to rest on ice for approx 15–30 min before beginning to dissect out the appropriate portion of tissue using tungsten needles. Collect tissues into L15-air with a drop of serum, and keep on ice.

## 3.2. Collagen Preparation

1. Collect tails from adolescent rats, and rinse in 95% ethanol. Keep frozen or use fresh.
2. Sterilize frozen tails by rinsing in 95% ethanol, and remove the tendons as follows: Clamp the tail 2 cm from its distal end using a hemostat and clamp the second hemostat immediately adjacent and proximal. Fracture the tail by bending it sharply with the hemostats; the distal portion of the tail will now be held only by the tendons. Slide this piece of tail off the tendons slowly. Cut the pieces of tendon that dangle from the remainder of the tail, so that they drop into a sterile Petri dish. Repeat, working up the tail, until the tendons are completely removed.
3. Wash tendons three times with sterile water. Tease apart using blunt #5 forceps, so they form a fine mesh of fibers. Collect 3 g of wet tendons.
4. Dissolve for 24–36 h, but no longer *(7)*, in 300 mL of 3% (v/v) glacial acetic acid at 4°C in a 200 mL sterile conical flask, stirring as slowly as possible *(7)* (note that the stir bar will stop stirring as the solution becomes viscous; check a few times during first few hours).
5. When most of the tendons have dissolved, spin at 20,000$g$ for 60 min to pellet the remainder.
6. Transfer supernatant to dialysis tubing (boiled in EDTA and washed extensively in sterile water), and dialyze at 4°C against 3 L of sterile 0.1X DMEM, pH 4.0, without bicarbonate. Dialyze for three days, changing medium once a day. Dialysis removes excess acid while keeping the pH low (the collagen gels when alkalinized).
7. Store in 15–50-mL aliquots at 4°C. Do not freeze. Keeps at least 6–12 mo.

## 3.3. Tissue Embedding in Three-Dimensional Collagen Gel

Explants are transferred to a base of gelled collagen, then overlaid with a cover of gelling collagen. Explants are manipulated and positioned appropriately as the top cushion sets.

1. Prepare a desired volume (100–500 µL) of 90% collagen and 10% of 10X DMEM *(8)* in an Eppendorf tube. Vortex. Add enough 0.8 $M$ NaHCO$_3$ to make the solution turn straw yellow after vortexing. For each batch of collagen, the NaHCO$_3$ must be titered: typically 0.7–1.2 µL NaHCO$_3$ is added to 100 µL of collagen and 10X DMEM *(9)*. Collagen can be used within 10–15 min if kept on ice. At room temperature, it will start to set within minutes.

2. In a four-well multidish *(10)*, prepare a collagen cushion by pipeting 20–25 µL of collagen mix onto the bottom of the dish. Stir around to spread and flatten the drop *(11)*, and let it set at room temperature for 20 min.

3. Transfer explant(s) from the medium onto the cushion, using either a Pasteur pipet or a fine-drawn capillary tube with a mouth attachment *(12)*.

4. Remove excess medium with a fine-drawn 25-µL Drummond micro-dispenser capillary tube with a mouth attachment. Leave on a tiny amount of medium so that the explants do not dry out.

5. Overlay with 25–30 µL of gelling collagen *(13)* prepared in the same way as the bottom cushion, position explants using an electrophoretically sharpened tungsten needle *(14)* (with blunted tip), and allow to set for 30–60 min at room temperature *(see* **Fig. 1**). Cover collagen moulds with approx 0.4–0.5 mL medium *(15)*, and transfer to incubator for the desired time of incubation *(16)*.

## 3.4. Assaying Growth and Differentiation Factors

The effect of growth and differentiation factors on explanted tissue can be assayed in different ways. Factors that are available as purified proteins can be assayed simply by adding them to the culture medium *(17)*, or by deriving them to appropriate beads and then implanting the beads within the collagen. Alternatively, if purified protein is not available, but full-length clones encoding a protein are available, attempts can be made to produce protein in COS cell aggregates transiently, which are then implanted in the collagen. The technique is as follows:

1. Seed COS-1 cells at approx $2.5 \times 10^5$ cells/35-mm dish on d 1.

2. On d 2, transfect expression vector containing clone under examination into COS-1 cells, using standard techniques (e.g., lipofection, electroporation, DEAE transfection).

3. On d 3, clump transfected cells. Treat cell layers with either trypsin or enzyme-free dissociation buffer. Wash cells twice with DMEM with 10% HIFCS, and resuspend in DMEM with 1% HIFCS (0.25 mL/35-mm dish *[18]*).

4. Place 20-µL drops of the cell suspension onto the lids of 60-mm culture dishes, and invert over dishes containing 5 mL of DMEM. Incubate the hanging drop cultures for 14–16 h.

5. Harvest cell aggregates into L15-air medium, and trim as required with tungsten needles for use in explant culture.

## 3.5. Analysis of Cultures

Subsequent to culture, the patterns of cell differentiation within explants can be examined either by immunohistochemical or *in situ* hybridization techniques (**Fig. 2**). In either case, explants can be sectioned directly in the gel and then analyzed, or processed for whole-mount labeling. The decision whether to section or process for whole-mount is largely probe-dependent. With a strong

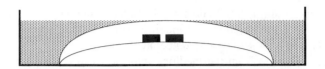

Fig. 1. Schematic diagram of a three-dimensional collagen gel culture. Side view, showing explants positioned at the interface between the collagen bilayer. The explants should lie in the center of the gel and rest on the bottom collagen cushion. The bottom layer of collagen is flat. The top layer of collagen overlays the bottom cushion precisely. Medium is added just to cover the gel.

probe or antibody, whole-mount labeling techniques work well (**Fig. 2C**). With a weaker probe or antibody, sectioning is more appropriate (**Fig. 2D**).

### 3.5.1. Immunohistochemical Analysis of Cultures

3.5.1.1. CRYOSTAT SECTIONING EXPLANTS FOR IMMUNOHISTOCHEMICAL ANALYSIS

1. Remove culture medium, and replace with 0.12 M PB containing 4% paraformaldehyde, so that explants are immersed in the fixative. Leave at 4°C for 2 h, and then replace fixative with 30% sucrose in 0.1 *M* phosphate buffer for 24–48 h at 4°C.
2. Using blunt forceps, tease around the edges of the collagen bilayer containing the explant until it lifts off the base of the four-well dish. Transfer onto a glass plate. Trim the gel, so that the explant is contained in a square of collagen, approx 0.6 cm². Freeze this in OCT *(19)* (TissueTek).
3. Collect 12–20 µm frozen sections *(20)* onto gelatin-subbed slides or Superfrost Plus slides, and process according to standard techniques.

3.5.1.2. WHOLE-MOUNT IMMUNOHISTOCHEMICAL ANALYSIS OF EXPLANT CULTURES

1. Remove culture medium, and replace with 0.12 *M* PB containing 4% paraformaldehyde, so that explants are immersed in the fixative. Leave at 4°C for 2 h.
2. Using blunt forceps, etch around the edges of the collagen containing the explant until it lifts off the base of the four-well dish and transfer to an Eppendorf tube. Process according to standard whole-mount techniques.

### 3.5.2. In Situ Hybridization of Cultures

3.5.2.1. SECTIONING EXPLANTS FOR IN SITU HYBRIDIZATION

1. Remove culture medium, and replace with 0.12 *M* PB containing 4% paraformaldehyde so that explants are immersed in the fixative. Leave at 4°C for 2 h.
2. Transfer explants in collagen to 30% sucrose in 0.1 *M* PB, leave at 4°C overnight, embed in OCT, and process according to standard techniques.

3.5.2.2. WHOLE-MOUNT IN SITU ANALYSIS OF EXPLANT CULTURES

1. Remove culture medium, and replace with 0.12 *M* PB containing 4% paraformaldehyde so that explants are immersed in the fixative. Leave at 4°C for 2 h.

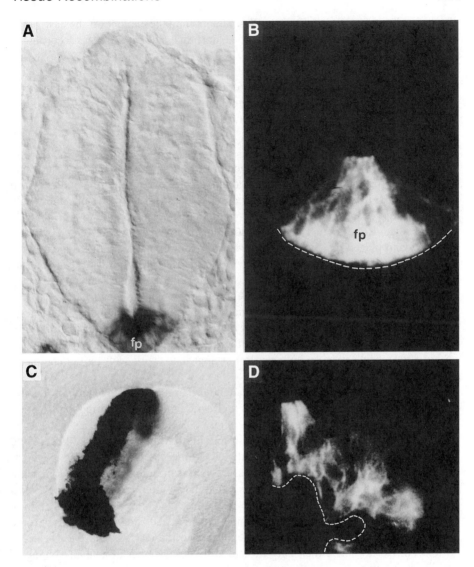

Fig. 2. Analysis of cell differentiation within neural explants cultured in three-dimensional collagen gels. (**A,B**) Transverse sections through an embryonic rat, showing expression of sonic hedgehog mRNA (A) and expression of the FP3 antigen (B) by floor plate (fp) cells of the spinal cord. (**C,D**) Floor plate cells that differentiate within explanted neural tissue in collagen gels can be detected through *in situ* hybridization techniques (C) or immunolabeling techniques (D). (C) Shows a neural explant that has been cultured and processed by *in situ* hybridization with antisense probe to sonic hedgehog. Floor plate cells that express the gene extend in a stripe within the explanted neural tissue. (D) Shows a section through a cultured rat neural explant, labeled to detect FP3. The dotted line in (B) marks the edge of the spinal cord, and in (D), marks the edge of the neural tissue.

2. Using blunt forceps, tease the collagen containing the explant off the base of the four-well dish, transfer to a sterile Eppendorf tube, and then process according to standard techniques *(21)*.

3. Explants can subsequently be viewed directly as whole-mounts or can be sectioned. If sectioning, refix in 4% paraformaldehyde 1–2 h at 4°C, then transfer to 30% sucrose, and process as described above. Optimal signal is maintained on thicker sections (20 µm). Mount sections in Glycergel.

## 4. Notes

1. Although dissections can be performed in PBS, tissues isolated from very young embryos survive far better if dissected into an air-buffered culture medium, such as L15-air. It is of utmost importance to keep tissue ice-cold at all stages (with the exception of the enzymatic treatment step) prior to positioning on the collagen bed. Rat or mouse embryos should be removed from the decidua immediately and placed on ice. Never allow embryos or explanted tissue to remain off ice for more than approx 5–10 min. If working to isolate a specific piece of tissue from an embryo, all the other embryos from the litter should remain on ice. When dissecting small portions of tissue, relax and remember to breathe.

2. Collagen is also available commercially (e.g., Vitrogen). However, the gelling properties of collagen isolated from rat tails are somewhat better than those of commercially obtained samples. Furthermore, the currently tested whole-mount *in situ* hybridization techniques do not appear to work well on commercially obtained collagen.

3. Collagen cannot be sterilized once made. It is therefore critical to prepare it under sterile conditions. Swab instruments well with ethanol, and during the isolation of the collagen, work in an air-filter hood.

4. The optimal time needed for Dispase to work varies according to the age and tissue, and is best assessed empirically. As a rule, the Dispase can be judged to have been effective when sheets of tissue, such as mesoderm and ectoderm, begin to separate. Tissues can be microdissected easily after an acute treatment with Dispase. Thus, material can be more easily dissected after a 5-min incubation in Dispase at 1 mg/mL than after a 10-min incubation at 0.5 mg/mL. Tissues become difficult to dissect after prolonged exposure to the enzyme and die if overexposed. Do not place too many explanted tissues together in a small volume of Dispase—they will become intertwined, sticky, and difficult to separate. Dispase is made up fresh, but the same batch can be used throughout the day.

5. Take care when transferring the tissues from Dispase to cold L15-air: the tissues coming out of Dispase will be relatively warm, i.e., room temperature, and may thus rise to the top of the colder, denser L15-air, and burst at the meniscus. This can be avoided if the tissues are transferred slowly and into a deep amount of L15-air. Small tissues are especially fragile after being incubated in Dispase and must be transferred with care. To prevent the tissue explants sticking to the transfer pipet, use a glass Pasteur pipet or a pulled glass capillary (*see* **Note 12**), and do not draw the tissue explants too far up the pipet. Siliconize the glass pipets if

necessary. Always use the microscope to visualize the tissues as they are transferred between vials.

6. Treatment with Dispase results in the tissue becoming sticky as cells begin to break down. Further dissection is facilitated by the inclusion of a small quantity of serum into medium to which tissue is transfered (1 drop in 5 mL). Serum inhibits further action of the Dispase. In addition, nonspecific binding by the serum to the tissue results in it becoming less sticky.

7. If stirred too vigorously or too long, tendon proteins other than collagen will start dissolving.

8. The purpose of the 10X DMEM is twofold. First, it provides a nutritious medium, enhancing the condition of the explanted tissue. Second it serves as a color indicator, facilitating titration of the bicarbonate—the solution will turn pink if too much bicarbonate is added.

9. If insufficient bicarbonate is added, the collagen will not set; if too much is added, it will set too quickly and unevenly.

10. The four-well dishes are convenient to use, and in these wells, gels can be covered with as little as 0.4 mL of medium (important for testing growth factors and conditioned medium).

11. The beds must be flat, since explants will tend to slide off a convex bed.

12. Fragile tissues can be manipulated far better using a pulled capillary tube than a Pasteur pipet. When picking up and transferring small amounts of tissue, always work through the stereo-dissecting microscope. When transferring tissue in a pulled capillary tube, allow the tissue to enter the tube by gentle sucking, and then remove the mouth attachment from the mouth. The tissue will remain in position. Otherwise, tissues are frequently either expelled or swallowed as the dissector suddenly remembers to breathe. To place the tissue on the collagen bed, it should be sufficient to touch down the capillary tip onto the collagen bed. Otherwise, place mouth attachment back in mouth, and expel very gently. Do not blow hard—the tissue will fragment. Sufficient medium should be transferred with the explant(s), so that they do not dry out. However, the medium should stay on the collagen bed: if too much medium is added and flows off the bed, the second layer of collagen will also flow off the bed, and explants cannot be appropriately positioned.

13. Keep the gelling collagen on ice, so it does not set too quickly. When adding the top layer of collagen, explants often float to the top of the drop. This will be minimized if explants are left in a tiny amount of medium and are not allowed to dry out. Nevertheless, even when explants do float to the top of the drop, they can and must be pushed back down onto the collagen and repositioned. With practice,this can be accomplished with the tungsten needle without tearing even fragile tissue. It is important that the tissue comes to lie on the bottom collagen bed, rather than within the top cushion. Also note that the collagen cushion and top collagen will become alkaline (pink) while setting. This has never been found to affect experiments, although it is possible that some tissues may be sensitive to the alkalinization. The top layer of collagen should overlap the bottom collagen bed precisely (*see* **Fig. 1**).

14. When establishing collagen gel cultures, first practice manipulating only a single explant within a gel. With practice, many (up to about 10) explants (approx 100–200 $\mu m^2$) can be positioned within a single bed. Explants can be positioned in experimental configurations relative to each other as the collagen sets. For instance, explants can be intertwined if contact is required between them or can be positioned apart. Manipulation of the explant(s) is facilitated if it lies in the center of the collagen bed.

15. The type of medium used varies according to the explants being cultured. Neural tissue survives well in Opti-MEM (Gibco-BRL), especially if supplemented with small quantities of sera. Serum screening should be performed, since certain batches adversely affect embryonic tissue in an apparently species-dependent fashion. Medium should be added to the top of the collagen, rather than to one side, to prevent the two layers from separating.

16. Neural explants survive well within collagen cultures for at least 5 d. Often, necrosis is observed early on in culture around the cut edges of the explanted tissue.

17. The collagen gels are composed of fibers that are not so dense that they preclude the diffusion of small growth and differentiation factors. Nevertheless, the isolation of collagen is sufficiently crude that the cultures may be contaminated with extracellular matrix components. Although there are no documented examples, it remains possible that these may affect the properties or presentation of growth factors within the cultures.

18. The resuspension volume can vary considerably. In different experiments, the number of cells per hanging drop (20 $\mu L$) has varied between 500 and 20,000 *(4,9)*.

19. To freeze in OCT, place a small amount of OCT on a chuck and place on dry ice. As soon as the OCT begins to freeze, remove the chuck from the dry ice, and add a small drop of OCT. Quickly place the collagen in the center of this drop, and transfer back to dry ice to freeze. Often it is unecessary to push the collagen into the OCT—sufficient OCT flows over it by capillary action.

20. If the tissue explants are very small (around 50– 200 $\mu m^2$), the entire explant will be collected in only a few sections. Appropriate care must therefore be taken to ensure every section is collected.

21. The variables that must be considered are the same as those considered for standard *in situ* hybridization of embryos—temperature, probe concentration, and washing stringency must all be optimized. For some probes, better results are obtained if the explants are removed from either one-half or all of the collagen after fixation and processed in small vials.

## References

1. Ebendal, T. and Jacobson, C. O. (1977) Tissue explants affecting extension and orientation of axons in cultured chick embryo ganglia. *Exper. Cell. Res.* **105,** 379–387.
2. Basler, K., Edlund, T., Jessell, T. M., and Yamada, T. (1993) Control of cell pattern in the neural tube: Regulation of cell differentiation by dorsalin-1, a novel TGFbeta family member. *Cell* **73,** 687–702.
3. Ericson. J., Muhr. J., Placzek, M. A., Lints, T., Jessell, T. M., and Edlund, T. (1995) Sonic Hedgehog induces the differentiation of ventral forebrain neurons: a

common signal for ventral patterning along the rostrocaudal axis of the neural tube. *Cell* **81,** 747–756

4. Fan, C. M. and Tessier-Lavigne, M. (1994) Patterning of mammalian somites by surface ectoderm and notochord: Evidence for sclerotome induction by a hedgehog homolog. *Cell* **79,** 1175–1186.

5. Fan, C.-M., Porter, J. A., Chiang, C., Chang, D. T., Beachy, P. A., and Tessier-Lavigne, M. (1995) Long-range sclerotome induction by sonic hedgehog: direct role of the amino-terminal cleavage product and modulation by the cyclic AMP signalling pathway. *Cell* **181,** 457–465.

6. Marti, E., Bumcrot, D. A., Takada, R., and McMahon, A. P. (1995) Requirement of 19K form of Sonic hedgehog peptide for induction of distinct ventral cell types in CNS explants. *Nature* **375,** 322–325.

7. Placzek, M., Tessier-Lavigne, M., Yamada, T., Jessell, T., and Dodd, J. (1990) Mesodermal control of neural cell identity: Floor plate induction by the notochord. *Science* **250,** 985–988.

8. Placzek, M., Jessell, T. M., and Dodd, J. (1993) Induction of floor plate differentiation by contact-dependent, homeogenetic signals. *Development* **117,** 205–218.

9. Roelink, H., Augsburger, A., Heemskerk, J., Korzh, V., Norlin, S., Ruiz i Altaba, A., Tanabe, Y., Placzek, M., Edlund, T., and Jessell, T. M. (1994) Floor plate and motor neuron induction by vhh-1, a vertebrate homolog of hedgehog expressed by the notochord. *Cell* **76,** 761–775.

10. Roelink, H., Porter, J. A., Chiang, C., Tanabe, Y., Chang, D. T., Beachy, P. A., and Jessell, T. M. (1995) Floor plate and motor neuron induction by different concentrations of the amino-terminal cleavage product of sonic hedgehog autoproteolysis..*Cell* **81,** 445–455.

11. Ruiz i Altaba, A., Placzek, M., Baldassare, M., Dodd, J., and Jessell, T. M. (1995) Early stages of notochord and floor plate development in the chick embryo defined by normal and induced expression of HNF3-beta. *Devel. Biol.* **170,** 299–313.

12. Tanabe, Y., Roelink, H., and Jessell, T. M. (1995) Induction of motor neurons by sonic hedgehog is independent of floor plate induction. *Curr. Biol.* **5,** 651–658.

13. Yamada, T., Placzek, M., Tanaka, H., Dodd, J., and Jessell, T. M. (1991) Control of cell pattern in the developing nervous system: polarizing activity of the floor plate and notochord. *Cell* **64,** 635–647.

14. Yamada, T., Pfaff, S. L., Edlund, T., and Jessell, T. M. (1993) Control of cell pattern in the neural tube: Motor neuron induction by diffusible factors from notochord and floor plate. *Cell* **73,** 673–686.

15. Hynes, M., Poulsen, K., Tessier-Lavigne, M., and Rosenthal, A. (1995) Control of neuronal diversity by the floor plate: contact-mediated induction of midbrain dopaminergic neurons. *Cell* **80,** 95–101.

16. Hynes, M., Porter, J. A., Chiang, C., Chang, D., Tessier-Lavigne, M., Beachy, P. A., and Rosenthal, A. (1995) Induction of midbrain dopaminergic neurons by sonic hedgehog. *Neuron* **15,** 35–44.

17. Liem, K. F., Tremml, G., Roelink, H., and Jessell, T. M. (1995) Dorsal differentiation of neural plate cells induced by BMP-mediated signals from epidermal ectoderm. *Cell* **82,** 969–979.

18. Tessier-Lavigne, M., Placzek, M., Lumsden, A. G. S., Dodd, J., and Jessell, T. M. (1988) Chemotropic guidance of developing axons in the mammalian central nervous system. *Nature* **336,** 775–778.

19. Kennedy, T. E., Serafini, T., del, T. J., and Tessier, L. M. (1994) Netrins are diffusible chemotropic factors for commissural axons in the embryonic spinal cord. *Cell* **78,** 425–435.

20. Serafini, T., Kennedy, T. E., Galko, M. J., Mirzayan, C., Jessell, T. M., and Tessier, L. M. (1994) The netrins define a family of axon outgrowth-promoting proteins homologous to C. elegans UNC-6. *Cell* **78,** 409–424.

21. Heffner, C. D., Lumsden, A. G. S., and O'Leary, D. D. M. (1990) Target control of collateral extension and directional axon growth in the mammalian brain. *Science* **247,** 217–247.

22. Lumsden, A. G. S. and Davies, A. M. (1983) Earliest sensory nerve fibres are guided to peripheral targets by attractants other than nerve growth factor. *Nature* **306,** 786–788.

23. Lumsden, A. G. S. and Davies, A. M. (1986) Chemotropic effect of specific target epithelium in developing mammalian nervous system. *Nature* **323,** 538,539.

24. Colamarino, S. A. and Tessier-Lavigne, M. (1995) The axonal chemoattractant Netrin-1 is also a chemorepellant for trochlear motor axons. *Cell* **81,** 621–629.

25. Messersmith, E. K., Leonardo, E. D., Shatz, C., Tessier-Lavigne, M., Goodman, C. S., and Kolodkin, A. L. (1995) Semaphorin III can function as a selective chemorepellant to pattern sensory projections in the spinal cord. *Neuron* **14,** 949–959.

# 22

## Quail–Chick Chimeras

**Marie-Aimée Teillet, Catherine Ziller, and Nicole M. Le Douarin**

### 1. Introduction

The understanding of several mechanisms that are essential for embryonic development has greatly benefited from cell-marking techniques that allow tracing of definite cells and their progeny, and thus, the study of their behavior and fate. A cell marker must be precise and stable; it must not interfere with normal development. The quail–chick labeling technique meets these requirements perfectly.

The principle of the method (*1*) is based on the observation that in all embryonic and adult cells of the quail (*Coturnix coturnix japonica*), the heterochromatin is condensed in one (sometimes two or more, depending on the cell types) large mass(es) associated with the nucleolus, thus making this organelle strongly stained after DNA staining, e.g., the Feulgen and Rossenbeck staining (*2*). When quail cells are combined with cells of the chick (*Gallus gallus*) which possess, like most of the animal cells, only small chromocenters dispersed in the nucleoplasm, they are readily recognizable by the structure of their nucleus, which thus provides a permanent natural marker (**Fig. 1A**).

The main purpose of constructing quail–chick chimeras is to follow the fate of definite embryonic territories during development and, in this way, to discover the place and time of origin of the different groups of cells constituting certain organs. The investigations carried out on the neural crest (*3*) and the mapping of the neural primordium at different stages (*4,5*) provide good examples of the possible uses of the quail–chick chimera system for studying developmental problems. This type of study implies that the developmental processes unfold in the chimeras as they do in normal embryos. To achieve this, transplantations of quail tissues into chick embryos (or vice versa) are performed *in ovo* and do not consist of adding the grafts to normal embryos,

From: *Methods in Molecular Biology, Vol. 97: Molecular Embryology: Methods and Protocols*
Edited by: P. T. Sharpe and I. Mason © Humana Press Inc., Totowa, NJ

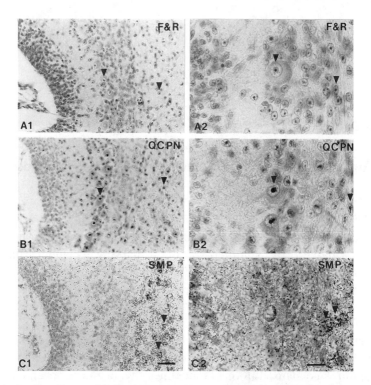

Fig. 1. Quail cells are differentially identified in 5-μm paraffin sections of a chimeric cerebellum (*see* **refs. *16,21,22***). (**A**) With the Feulgen and Rossenbeck staining (F&R), quail neuronal and glial cells show a strongly stained nucleolus (*see* arrowheads), whereas chick cells present almost homogeneously stained nuclei. (**B**) With the QCPN MAb, nucleolus of quail neurons and glial cells are more violently stained, but chick nuclei are not labeled. (**C**) The SMP probe hybridizes specifically with the quail oligodendrocytes. Medial and high magnifications of each staining are presented. A1, B1, C1: bar = 70 μm; A2, B2, C2: bar = 30 μm.

but of removing a given territory in the host and replacing it as precisely as possible by the equivalent region of the donor, which must be at the same developmental stage. Quail and chick are closely related in taxonomy, although they differ by the duration of their incubation time (17 d for the quail, and 21 d for the chick) and their size at birth (about 10 g for the quail and 30 g for the chick). However, during the first days of incubation, when most of the important events in embryogenesis take place, the size of the quail and chick embryos and the chronology of their development differ only slightly. Obviously, it is safe to carry out the grafts not only from quail to chick, but also from chick to quail, and to perform control chronological studies in order to discard any bias owing to differential development processes between chick and quail embryos.

Quail–chick chimeras constructed according to the above principles can hatch and survive in good health for certain time; time of survival is limited by the appearance of an immunological reaction developed by the host. Although no immune reaction against the graft takes place during embryogenesis, when the immune system is immature, the transplant triggers its own rejection, which occurs at various times after birth. For strictly neural grafts, a long delay (1–2 mo) is observed between the onset of immune maturity and the rejection *(6)*. For neural grafts associated with other tissue grafts and for grafts of any nonneural tissue, rejection occurs as soon as maturation of the immune system is achieved *(7,8)*.

Isochronic and isotopic grafts are not the only type of grafts used to construct quail–chick embryonic chimeras. Certain developmental processes can be studied by heterotopic, and heterochronic grafts with or without previous ablation of tissues. This was instrumental in testing the degree of determination of the neural crest cells and their derivatives *(3,9)* and in demonstrating the precise periodicity of the colonization of the primary lymphoid organ rudiments (thymus and bursa of Fabricius) by hemopoietic cells in birds *(10)*.

For many years, the analysis of the quail–chick chimeras was based on the differential staining of the nucleus by either the Feulgen and Rossenbeck nucleal reaction **(Fig. 1A)** or any other method revealing specifically the DNA profiles in light or electron microscopy. Recently, significant progress was accomplished when species-specific antibodies recognizing either quail or chick cells of one or several types were prepared (*see* **Subheading 3.4.3.** and **Fig. 1B**). Nowadays, several specific quail or chick nuclear probes are also available, allowing, at the single-cell level, specific gene activities to be distinguished (*see* **Subheading 3.4.4.** and **Fig 1C**).

This chapter will describe the protocols of several representative examples of neural quail–chick chimeras. These particular protocols can be adapted to the graft of any other type of tissue.

## 2. Materials

### 2.1. Egg Incubation

1. Fertilized chick and quail eggs (*see* **Note 1**).
2. Humidified and ventilated incubators equipped with time programmers. Time programmers are useful in obtaining very precise stages of development (*see* **Note 2**).
3. Egg holders (*see* **Note 3**).
4. Developmental tables of Hamburger and Hamilton for chick embryo development *(11)* and of Zacchei for quail embryo development *(12)* (*see* **Note 4**).

### 2.2. Preparation of Host and Donor Embryos and Grafts

1. 70% Alcohol for sterilization.
2. Disposable syringes (1- or 2-mL) and needles (0.6–0.8 mm in diameter).

3. Transparent Scotch tape (5 cm in width) (*see* **Note 5**).
4. Paraffin 60°C and a thin paint-brush.
5. Physiological liquids: PBS or Tyrode solution.
6. Antibiotics: penicillin and streptomycin (*see* **Note 7**).
7. Proteolytic enzymes (pancreatin) (*see* **Note 7**).
8. Bovine serum (*see* **Note 7**).
9. Pasteur pipets.
10. Glass micropipets hand-drawn from Pasteur pipets and equipped with plastic tubes for mouth use (Tygon) (*see* **Note 8**).
11. Small glass dishes (salières) normal or containing a plastic base for dissections (rhodorsyl, Rhône-Poulenc, France) (*see* **Note 9**) and insect pins.
12. Indian ink (Pelikan for drawing) (*see* **Note 10**).
13. Microscalpels and their holders (*see* **Note 11**).
14. Small curved scissors, iridectomy (Pascheff-Wolff) scissors, and thin forceps (Swiss Dumont no. 5 forceps) (Moria Instruments, Paris).
15. Optical equipment: stereomicroscope equipped with a zoom from ×6 to ×50.
16. Optic fibers for object illumination (*see* **Note 12**).

## 2.3. Histological Analysis of the Chimeric Embryo Tissues and Organs

1. Carnoy fluid or 4% paraformaldehyde for fixation of the tissues (*see* **Note 13**).
2. Materials for paraffin histology.
3. Reagents for Feulgen and Rossenbeck staining *(2,13)* (*see* **Note 14**).
4. Reagents for immunohistochemistry with species-specific and/or cell type-specific antibodies.
5. Reagents for *in situ* hybridization with species-specific nuclear probes.

## 3. Methods
## 3.1. Preparation of the Host and Donor Embryos

1. Incubations: Quail and chick eggs are incubated with their long axis horizontal during the time necessary to obtain the stages adequate for the experiments: 36–48 h for the experiments described here (*see* **Note 4**).
2. Environmental conditions of the experiments: The experiments have to be made under relatively sterile conditions. Eggshells are cleaned with 70% alcohol. Instruments are sterilized in a dry oven (1.5 h at 120°C). Sterile physiological liquids are supplemented with antibiotics (10–20 IU/mL). Experiments are performed in a clean separate room, but never under forced-air apparatus (*see* **Notes 6** and **12**).
3. Opening the eggs: The blastoderm develops on the top of the yolk against the shell membrane. Before opening a window in the shell, a small quantity of albumen (about 0.3 mL) is removed at the small end of the egg, using a syringe in order to separate the blastoderm from the shell membrane. Another method consists of perforating the eggshell at the level of the air chamber and then rolling the egg horizontally several times. These manipulations are sufficient to loosen the

blastoderm from the shell membrane and allow a window through the shell without damaging the embryo. The small holes through the eggshell are obturated with a piece of tape or a drop of paraffin (*see* **Note 5**).

4. Contrasting the embryos *in ovo*: India ink, diluted 1/1 in PBS or Tyrode supplemented with antibiotics, is injected under the blastoderm (donor or recipient) using a glass micropipet mounted with a plastic tube for mouth use (*see* **Note 10**).

5. Gaining access to the embryos *in ovo*: The vitelline membrane, which covers the embryo, is slitted out with a microscalpel just at the place where the microsurgery will be made.

6. Explanting the donor blastoderms: In certain cases, the donor blastoderm is cut out from the egg with Pascheff-Wolff scissors, washed free of vitellus in PBS or Tyrode supplemented with antibiotics, tranferred onto a dish with a black plastic base, and pinned out before dissection.

## *3.2. The Grafts*

### *3.2.1. Neural Tube Transplantations (**Fig. 2**)*

Orthotopic transplantations of fragments of neural tube have allowed the construction of a neural crest fate map *(3)* and the detection of definite crest cell migration pathways *(14)*. The rules of this operation are based on the fact that neural crest cells start migrating first in the cephalic region and then progressively from rostral to caudal when neural tube forms. The interspecific graft is performed at a level where crest cells are still inside the apex of the neural tube, i.e., in the neural folds in the cephalic area, at the level of the last formed somites in the cervical and thoracic regions, and at the level of the segmental plate in the lumbo-sacral region. The operation has to be made on the length of no more than five to six somites to take into account the rostro-caudal differential state of evolution of the neural crest. Donor and host embryos are strictly stage-matched (*see* **Note 4**).

1. Excision of the host neural tube: The selected neural tube fragment is excised from the host embryo by microsurgery *in ovo*. A longitudinal slit through the ectoderm and between the neural tube and the adjacent paraxial mesoderm, at the chosen level, is made bilaterally using a microscalpel. The neural tube is then gently separated from the neighboring mesoderm and cut out transversally, rostrally, and caudally without damaging the underlying notochord and endoderm. The fragment of neural tube is then progressively severed from the notochord and finally sucked out using a calibrated glass micropipet (*see* **Note 8**).

2. Preparation of the graft: The transverse region of the stage-matched donor embryo comprising the equivalent fragment of neural tube plus surrounding tissues (ectoderm, endoderm, and mesoderm) is retrieved with iridectomy scissors and subjected in vitro to enzymatic digestion (pancreatin, Gibco, one-third in PBS or Tyrode) for 5–10 min on ice or at room temperature according to the stage of the embryo (*see* **Note 7**). Then tissues are dissociated using two smooth microscalpels

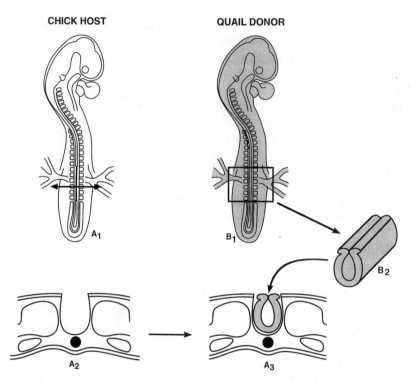

Fig. 2. Scheme of the orthotopic quail/chick neural tube transplantation. A chick embryo *in ovo* (**A1**) is microsurgically deprived of its neural tube (**A2**) at the level of the last formed somites. The corresponding fragment of blastoderm of a quail at the same stage (**B1**) is enzymatically dissociated. The isolated quail neural tube (**B2**) is orthotopically grafted into the chick embryo (**A3**).

    (*see* **Note 11**), and finally, the isolated neural tube fragment is rinsed with PBS or Tyrode supplemented with bovine serum to inhibit the action of the proteolytic enzymes. It is then ready to be grafted.
3. Grafting procedure: The donor neural tube is transferred to the host embryo using a calibrated glass micropipet and placed in the groove produced by the excision, in the normal rostro-caudal and dorso-ventral orientation (*see* **Note 15**).

Heterotopic graftings were instrumental to study whether the fate of the neural crest is specified when the operation is carried out (*9*). The graft is taken at a more rostral or more caudal level than the acceptor level. Depending on the latter, the donor embryo is older or younger than the recipient (*see* **Note 16**).

Partial dorso-ventral orthotopic graftings have also been made in order to localize possible early segregation of precursors in the neural tube (*15*).

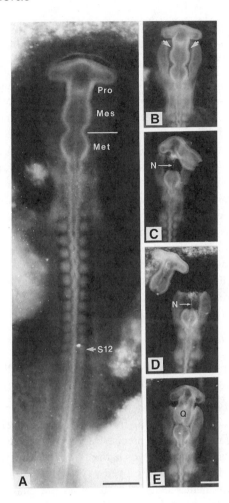

Fig. 3. Scenario of the orthotopic graft of brain vesicles. (**A**) 12-somite chick embryo *in ovo* after injection of a solution of Indian ink under the blastoderm. Brain vesicles are well delineated. (**B**) Longitudinal incisions are made between the cephalic neural tube and the head mesenchyme to delimit the brain excision (arrows). (**C**) After a transversal section at the level of the mesencephalo-metencephalic constriction, the prosencephalon and the mesencephalon are separated from the head mesoderm and endoderm. The notochord (N) is then visible. (**D**) The excised chick brain vesicles are discarded. (**E**) The equivalent quail brain vesicles (Q) are grafted into the chick host. Pro: prosencephalon; Mes: mesencephalon; Met: metencephalon; S12: somite 12. A bar = 0.05 mm; B, C, D, E bar = 0.05 mm.

## *3.2.2. Orthotopic Transplantations of Brain Vesicles (**Fig. 3**)*

This operation has been devised to label defined regions of the neuroepithelium and thus to study cell migrations and morphogenetic movements during brain development *(4,16,17)*. It was also applied to the transfer of a genetic behavioral or functional trait from donor to recipient in either xenogeneic or isogeneic combinations *(8,18,19)*.

1. Excision of brain vesicles from donor and host embryos: Equivalent brain vesicles are excised microsurgically in the same way in stage-matched donors and recipients (*see* **Note 17**). The dorsal ectoderm is slit precisely at the limit between the neural tissue and the cephalic mesenchyme on each side of the selected part of the brain. The neural epithelium is loosened from the cephalic mesenchyme, then cut out transversally at the chosen rostral and caudal levels, and finally severed from the underlying notochord.

2. Exchange of brain vesicles: The transfer of brain vesicles from the quail to the chick and vice versa (or from a mutant to a normal chick embryo) is made using a calibrated glass micropipet. The piece of neural tissue is inserted into the groove made by the excision, with the normal rostro-caudal and dorso-ventral orientation, and then adjusted (*see* **Note 18**).

3. Modifications of the technique consist of orthotopic partial dorsal or dorsolateral grafts of brain vesicles *(17,20–22)*. Heterotopic grafts have also been performed to study specific problems *(20,23–25)*.

### 3.2.3. Neural Fold and Neural Plate Transplantations

Orthotopic and isochronic grafts have been made in order to map the early rostral or caudal neural primordium *(5,26–29)*.

1. Excision of precise pieces of the neural fold or neural plate in the chick host: Very thin microscalpels (made up from insect pins or steel needles sharpened on an oil stone) are used to excise precise fragments of the folds and neural plate in 0- to 5-somite stage chick embryos *in ovo*. An ocular micrometer is used to measure the pieces of tissue to be removed.

2. Excision of equivalent pieces of tissue from the quail donor: The grafts are excised from stage-matched quail in vitro using the same method. They are not subjected to enzymatic treatment.

3. Graft: Pieces of quail tissue are grafted orthotopically into the chick host.

Heterotopic grafts have also been made to establish the degree of autonomy of precise territories of the rhombencephalon *(30,31)*.

Simple excisions combined with adjacent orthotopic grafts have been performed in order to identify neural crest cells differentiating at the level of the excision *(32)*.

### 3.3. Sealing the Eggs and Postincubation

When the grafting operations are achieved, the window in the eggshell is sealed with a piece of tape and the eggs reincubated in a horizontal stable position (*see* **Notes 5** and **19**).

### 3.4. Analysis of the Grafts

#### 3.4.1. Fixation of the Experimental Embryos

Host embryos can be fixed from several hours after the operation to several days after hatching according to the experimental design (*see* **Note 20**). Carnoy fluid is one of the best fixatives utilizable at once for Feulgen and Rossenbeck staining *(13)*, immunohistochemistry, and *in situ* hybridization on paraffin sections. Fixation with 4% paraformadehyde is used for whole-mount immunohistochemistry and *in situ* hybridization, and all treatments on cryostat sections.

#### 3.4.2. Feulgen and Rossenbeck Staining

The Feulgen and Rossenbeck nucleal reaction *(2,13)* (*see* **Note 21**) is applied on 5-μm serial sections.

#### 3.4.3. Immunohistochemistry

1. Two antibodies recognize virtually all cell types in the quail and no one in the chick: the polyclonal antibody raised by Lance-Jones and Lagenaur *(33)* and the monoclonal antibody (MAb) QCPN prepared by B. M. Carlson and J. A. Carlson, which is available at the Developmental Studies Hybridoma Bank (Department of Biology, University of Iowa, 436BB, Iowa City, IA 52242). The use of QCPN is easy and can be combined with other antibodies like HNK1 *(34)*, which recognizes neural crest cells or 13F4 *(35)*, which marks muscle cells and their precursors.
2. Other MAbs are species- and cell type-specific: MB1 and QH1 *(36,37)*, which recognize a glycosylated epitope carried by surface proteins expressed in quail leukocytes and endothelial cells at the exclusion of any cell type of the chick.
3. Neural chimeras can be analyzed with MAbs that recognize either neuronal cell bodies or neurites of quail or chick exclusively *(38,39)*.

#### 3.4.4. In Situ Hybridization

A growing number of species-specific nucleic probes can be used on sections or whole-mount preparations. As examples, the use of a chick probe of the homeobox gene *goosecoid* has demonstrated the induction of this gene in a chick in which goosecoid-producing tissues had been grafted *(40)*. The quail specific SMP (Schwann cell myelin protein) probe *(41)* allows quail oligodendrocytes in chimeric spinal cord to be distinguished *(15)* (**Fig. 1C**). Chick *Wnt1* and quail *Wnt1* probes have been combined to demonstrate the induction of *Wnt1* in quail–chick chimeras *(42)*.

### 4. Notes

1. If possible, select a rapidly growing strain of chickens in which early stages of development will be in phase with the ones of quail embryos. A nonpigmented strain of chickens can be chosen in order to use the heavily pigmented quail melanocytes as a second marker. Freshly laid eggs are stored no more than 1 wk at 15°C.

2. Ideal conditions for chick and quail incubation are 38°C, 45% humidity (first two-thirds of incubation time), 75% (last one-third of incubation time and hatching). Small incubators equipped with time programmers can be placed in a 15°C room in order to obtain very precise stages of develoment.

3. Multiple wire tongs, individual egg holders, and hollowed-out wooden slats can be used for preincubation, microsurgical experiment, and postincubation, respectivley, of the chicken eggs.

4. The operations described here are performed at embryonic d 2 (E2) when somites, which can be easily counted, can serve to stage the embryos.

5. In order to perturb the gas exchanges at the level of the shell as little as possible, avoid using too large pieces of tape, and stick it carefully against the shell without folds, which would cause air entry and consequently progressive drying of the egg content.

6. PBS and Tyrode solutions supplemented with antibiotics (10–20 IU/mL) directly deposited on the blastoderm are commonly used to humidify it at any time during the experiment.

7. Pancreatin (Gibco) is diluted one-third to one-sixth with PBS or Tyrode solution. In this way, the tissue dissociation can be easily controlled. Titer and temperature are adapted to the stage of development of the tissues. The younger the tissues, the lower the titer and temperature; for example, tissues from 10-somite stage embryos will be treated with 20% pancreatin in Tyrode on ice, whereas tissues from 20-somite stage embryos will be treated with 30% pancreatin at room temperature. Tyrode solution supplemented with bovine serum will serve to stop enzyme action.

8. Glass micropipets hand-drawn from Pasteur pipets are curved and calibrated according to use: injection of liquids or transfer of pieces of tissues. Calibration of the micropipet according to the size of the rudiment to be transplanted (for instance, neural tube vs brain vesicle) is an important requirement.

9. The Rhodorsyl base is now preferred over the paraffin base. It can be either black or perfectly transparent depending on whether animal carbon is added to the commercial preparation. Moreover, it can be sterilized as often as necessary in dry oven and supports insect pins without damage.

10. Indian ink has to be tested for toxicity before use and must be used without excess.

11. Microscalpels have to be perfectly adapted to each use. Microscalpels, manufactured by stropping and honing steel needles (sewing needles) on an Arkansas oil stone, are the most convenient for excising fragments of neural tube or brain vesicles, because they can be both extremely thin and resistant. For dissociating tissues after enzymatic treatment, they must have a smooth tip. Tungsten microscalpels (*43*) or microscalpels made up from insect pins are also useful for dissecting very small pieces of tissues. They are quicker to prepare, but more fragile.

12. Formerly used conventional light bulbs with a condensor tend to radiate heat and cause traumatic drying to the embryos during the surgery, so that defects in amnios formation and subsequent death are often observed.

13. Zenker fluids like Carnoy fluid are good fixatives for Feulgen and Rossenbeck staining (*13*) but they do not allow immunohistochemistry or *in situ* hybridization to be made.

14. A modification of the Feulgen and Rossenbeck classical protocol consists of performing the hydrolysis in 5 *N* HCl for 20–30 min at room temperature after Carnoy fixation, instead of 1 *N* HCl for 4–8 min at 60°C as previously recommended in Gabe *(13)*.

15. Dorso-ventral and rostro-caudal orientations of the graft are recognized either by morphological characters or by various labelings, for instance, a minute precisely localized slit.

16. Differences in stage and caudo-rostral level implicate difference of size. If the fragment of neural tube to be grafted is much bigger than the one that has been removed, it should be ressected before grafting.

17. Transplantations of brain vesicles are made at the 12- to 14-somite stages, which are favorable for the following reasons: Brain vesicles, still uncovered by the amnion, are clearly demarcated by constrictions in the absence of brain curvature; the notochord is no longer strongly adherent to the ventral part of the neural epithelium at this level; the neuroepithelium is not yet vascularized. Some neural crest cells and cephalic mesoderm are transferred along with the brain vesicles. Their presence does not interfere with the development of the brain, and presence of melanocytes in the head feathers of the chimera indicates the level of the brain graft.

18. A good adhesion of the graft to the host tissues is favored by sucking out with a micropipet the excess of physiological liquid added during the operation.

19. Daily gentle manual rocking of the operated eggs can enhance embryo survival. Incubator humidity must grow from 45–75% on d 18 of incubation if hatching of the operated embryos is expected.

20. E3–E4 chimeric embryos are fixed as a whole, for 1–3 h at room temperature. Older embryos have to be fixed as fragments and maintained in vacuum, during a growing time according to the stage and the size of the tissue pieces or organs. The same conditions will be applied for dehydratation and paraffin impregnation of the samples.

## Acknowledgments

We thank Marcelle Gendreau and Charmaine Herberts for reference filing, Françoise Viala for photographic work, and Sophie Gournet and Hélène San Clemente for line drawings.

## References

1. Le Douarin, N. (1969) Particularités du noyau interphasique chez la Caille japonaise (*Coturnix coturnix japonica*). Utilisation de ces particularités comme "marquage biologique" dans les recherches sur les interactions tissulaires et les migrations cellulaires au cours de l'ontogenèse. *Bull. Biol. Fr. Belg.* **103,** 435–452.

2. Feulgen, R. and Rossenbeck, H. (1924) Mikroskopisch-chemischer Nachweis einer Nukleinsäure von Typus der Thymonukleinsäure und die darauf beruhende elektive Färbung von Zellkernen in mikroskopischen Präparaten. *Hoppe-Seyler's Z. Physiol. Chem.* **135,** 203–248.

3. Le Douarin, N. (1982) *The Neural Crest.* Cambridge University Press, Cambridge.

4. Le Douarin, N. M. (1993) Embryonic neural chimaeras in the study of brain development. *Trends Neurosci.* **16,** 64–72.
5. Catala, M., Teillet, M.-A., De Robertis, E. M., and Le Douarin, N. M. (1996) A spinal cord fate map in the avian embryo: while regressing, Hensen's node lays down the notochord and floor plate thus joining the spinal cord lateral walls. *Development* **122,** 2599–2610.
6. Kinutani, M., Coltey, M., and Le Douarin, N. M. (1986) Postnatal development of a demyelinating disease in avian spinal cord chimeras. *Cell* **45,** 307–314.
7. Ohki, H., Martin, C., Corbel, C., Coltey, M., and Le Douarin, N. M. (1987) Tolerance induced by thymic epithelial grafts in birds. *Science* **237,** 1032–1035.
8. Balaban, E., Teillet, M.-A., and Le Douarin, N. M. (1988) Application of the quail–chick chimera system to the study of brain development and behavior. *Science* **241,** 1339–1342.
9. Le Douarin, N. M. and Teillet, M.-A. (1974) Experimental analysis of the migration and differentiation of neuroblasts of the autonomic nervous system and neurectodermal mesenchymal derivatives, using a biological cell marking technique. *Dev. Biol.* **41,** 162–184.
10. Le Douarin, N. M., Dieterlen-Lièvre, F., and Oliver, P. D. (1984) Ontogeny of primary lymphoid organs and lymphoid stem cells. *Am. J. Anat.* **170,** 261–299.
11. Hamburger, V. and Hamilton, H. L. (1951) A series of normal stages in the development of the chick embryo. *J. Morphol.* **88,** 49–92.
12. Zacchei, A. M. (1961) Lo sviluppo embrionale della quaglia giapponese (*Coturnix coturnix japonica*, T. e S.). *Arch. Ital. Anat. Embriol.* **66,** 36–62.
13. Gabe, M. (1968) *Techniques Histologiques.* Masson, Paris.
14. Teillet, M.-A., Kalcheim, C., and Le Douarin, N. M. (1987) Formation of the dorsal root ganglia in the avian embryo: segmental origin and migratory behavior of neural crest progenitor cells. *Dev. Biol.* **120,** 329–347.
15. Cameron-Curry, P. and Le Douarin, N. M. (1995) Oligodendrocyte precursors originate from both the dorsal and the ventral parts of the spinal cord. *Neuron* **15,** 1299–1310.
16. Hallonet, M. E. R., Teillet, M.-A., and Le Douarin, N. M. (1990) A new approach to the development of the cerebellum provided by the quail–chick marker system. *Development* **108,** 19–31.
17. Tan, K. and Le Douarin, N. M. (1991) Development of the nuclei and cell migration in the medulla oblongata. Application of the quail–chick chimera system. *Anat. Embryol.* **183,** 321–343.
18. Teillet, M.-A., Naquet, R., Le Gal La Salle, G., Merat, P., Schuler, B., and Le Douarin, N. M. (1991) Transfer of genetic epilepsy by embryonic brain grafts in the chicken. *Proc. Natl. Acad. Sci. USA* **88,** 6966–6970.
19. Batini, C., Teillet, M.-A., Naquet, R., and Le Douarin, N. M. (1996) Brain chimeras in birds: application to the study of a genetic form of reflex epilepsy. *Trends Neurosci.* **19,** 246–252.
20. Alvarado-Mallart, R. M. and Sotelo, C. (1984) Homotopic and heterotopic transplantations of quail tectal primordia in chick embryos: organization of the retinotectal projections in the chimeric embryos. *Dev. Biol.* **103,** 378–398.

21. Martinez, S. and Alvarado-Mallart, R. M. (1989)Rostral cerebellum originates from the caudal portion of the so-called "mesencephalic" vesicle: A study using chick/quail chimeras. *Eur. J. Neurosci.* **1**, 549–560.

22. Hallonet, M. E. R. and Le Douarin, N. M. (1993) Tracing neuroepithelial cells of the mesencephalic and metencephalic alar plates during cerebellar ontogeny in quail–chick chimaeras. *Eur. J. Neurosci.* **5**, 1145–1155.

23. Nakamura, H. (1990) Do CNS anlagen have plasticity in differentiation? Analysis in quail–chick chimera. *Brain Res.* **511**, 122–128.

24. Martinez, S. and Alvarado-Mallart, R. M. (1990) Expression of the homeobox *Chick-en* gene in chick–quail chimeras with inverted mes-metencephalic grafts. *Dev. Biol.* **139**, 432–436.

25. Martinez, S., Wassef, M., and Alvarado-Mallart, R. M. (1991) Induction of a mesencephalic phenotype in the 2-day-old chick prosencephalon is preceded by the early expression of the homeobox gene *en. Neuron* **6**, 971–981.

26. Couly, G. F. and Le Douarin, N. M. (1985) Mapping of the early neural primordium in quail–chick chimeras. I. Developmental relationships between placodes, facial ectoderm, and prosencephalon. *Dev. Biol.* **110**, 422–439.

27. Couly, G. F. and Le Douarin, N. M. (1987) Mapping of the early neural primordium in quail-chick chimeras. II. The prosencephalic neural plate and neural folds: implications for the genesis of cephalic human congenital abnormalities. *Dev. Biol.* **120**, 198–214.

28. Couly, G. and Le Douarin, N. M. (1988) The fate map of the cephalic neural primordium at the presomitic to the 3-somite stage in the avian embryo. *Development* **103(Suppl.)**, 101–113.

29. Couly, G. F., Coltey, P. M. and Le Douarin, N. M. (1993) The triple origin of skull in higher vertebrates: A study in quail-chick chimeras. *Development* **117**, 409–429.

30. Grapin-Botton, A., Bonnin, M.-A., McNaughton, L.A., Krumlauf, R., and Le Douarin, N. M. (1995) Plasticity of transposed rhombomeres: Hox gene induction is correlated with phenotypic modifications. *Development* **121**, 2707–2721.

31. Grapin-Botton, A., Bonnin, M.-A., and Le Douarin, N. M. (1997) Hox gene induction in the neural tube depends on three parameters: competence, signal supply and paralogue group. *Development* **124**, 849–859.

32. Couly, G., Grapin-Botton, A., Coltey, P., and Le Douarin, N. M. (1996) The regeneration of the cephalic neural crest, a problem revisited: the regenerating cells originate from the contralateral or from the anterior and posterior neural folds. *Development* **122**, 3393–3407.

33. Lance-Jones, C. C. and Lagenaur, C. F. (1987) A new marker for identifying quail cells in embryonic avian chimeras: a quail-specific antiserum. *J. Histochem. Cytochem.* **35**, 771–780.

34. Abo, T. and Balch, C. M. (1981) A differentiation antigen of human NK and K cells identified by a monoclonal antibody (HNK-1). *J. Immunol.* **127**, 1024–1029.

35. Rong, P. M., Ziller, C., Pena-Melian, A., and Le Douarin, N. M. (1987) A monoclonal antibody specific for avian early myogenic cells and differentiated muscle. *Dev. Biol.* **122**, 338–353.

36. Péault, B. M., Thiery, J.-P., and Le Douarin, N. M. (1983) Surface marker for hemopoietic and endothelial cell lineages in quail that is defined by a monoclonal antibody. *Proc. Natl. Acad. Sci. USA* **80,** 2976–2980.
37. Pardanaud, L., Altmann, C., Kitos, P., Dieterlen-Lièvre, F., and Buck, C. A. (1987) Vasculogenesis in the early quail blastodisc as studied with a monoclonal antibody recognizing endothelial cells. *Development* **100,** 339–349.
38. Takagi, S., Toshiaki, T., Kinutani, M., and Fugisawa, H. (1989) Monoclonal antibodies against specific antigens in the chick central nervous system: Putative application as a transplantation marker in the quail–chick chimaera. *J. Histochem. Cytochem.* **37,** 177–184.
39. Tanaka, H., Kinutani, M., Agata, A., Takashima, Y., and Obata, K. (1990) Pathfinding during spinal tract formation in quail–chick chimera analysed by species specific monoclonal antibodies. *Development* **110,** 565–571.
40. Izpisúa-Belmonte, J. C., De Robertis, E. M., Storey, K. G., and Stern, C. D. (1993) The homeobox gene *goosecoid* and the origin of organizer cells in the early chick blastoderm. *Cell* **74,** 645–659.
41. Dulac, C., Tropak, M. B., Cameron-Curry, P., Rossier, J., Marshak, D. R., Roder, J., and Le Douarin, N. M. (1992) Molecular characterization of the Schwann cell myelin protein, SMP: structural similarities within the immunoglobulin superfamily. *Neuron* **8,** 323–334.
42. Bally-Cuif, L. and Wassef, M. (1994) Ectopic induction and reorganization of *Wnt-1* expression in quail-chick chimeras. *Development* **120,** 3379–3394.
43. Conrad, G. W., Bee, J. A., Roche, S. M., and Teillet, M.-A. (1993) Fabrication of microscalpels by electrolysis of tungsten wire in a meniscus. *J. Neurosci. Methods* **50,** 123–127.

# 23

## Using Fluorescent Dyes for Fate Mapping, Lineage Analysis, and Axon Tracing in the Chick Embryo

### Jonathan D. W. Clarke

## 1. Introduction

This chapter deals largely with the use of fluorescent dyes in the investigation of the development of the chick embryo. It covers three issues; generating fate maps, lineage (or clonal) analysis from single-cell injections, and axonal tracing techniques to uncover the neuronal organization of the early nervous system. The construction of fate maps in the early embryo is an important step in the process of understanding how an embryo is built. Fate maps tell us about the origin of particular cell groups, the morphogenetic movements that occur as the embryo takes shape, and can reveal the potential for signaling between cells whose proximity may be transient and obscured by subsequent cell rearrangements. Fate maps can be constructed by analyzing the fate of several neighboring cells labeled simultaneously (here the tracking dyes are usually applied to cell surfaces by extracellular injection) or more precisely by following the fate of individual cells (where the tracking dyes are usually injected intracellularly).

## 2. Materials

### 2.1. Fate Mapping with Lipophilic Membrane Dyes

1. DiI (1,1'-dioctadecyl-3,3,3',3'-tetramethylindocarbocyanine) perchlorate, cat. no. D-282 from Molecular Probes (Eugene, OR).
2. Steromicroscope and fiber-optic light source.
3. Micromanipulator attached to base plate or directly to stereomicroscope.
4. Pressure injector and micropipet holder.
5. Micropipet puller.
6. 9-V battery and microelectrode holder.
7. 90% Glycerol in PBS.

From: *Methods in Molecular Biology, Vol. 97: Molecular Embryology: Methods and Protocols*
Edited by: P. T. Sharpe and I. Mason © Humana Press Inc., Totowa, NJ

8. DABCO (and antiquenching agent from Sigma, St. Louis, MO).
9. Paraformaldehyde.
10. Dimethyl formamide (DMF).
11. Dimethyl sulfoxide (DMSO).
12. 1 $M$ lithium chloride.

## 2.2. Lineage Analysis by Single-Cell Injection of Fluorescent Dextran

1. A fixed-stage microscope (i.e., one that focuses by moving the objective lens rather than the specimen platform) with an epifluorescent attachment and an extra-long working distance ×20 objective lens—a condenser and normal microscope stage are not required as a platform to hold the egg, since it can be attached to the fitment designed to hold the condenser. The microscope does not need its own light source. The embryo will be illuminated from the side with a fiber-optic.
2. A high-resolution micromanipulator, for example, a Huxley design. It is helpful to have one that has a fine axial drive on it, i.e., a drive that moves the electrode along the axis of the electrode.
3. An antivibration table or equivalent antivibration device.
4. An oscilloscope.
5. An amplifier suitable for intracellular DC recording and a current injection facility.
6. A fiber-optic light source.
7. A good micropipet-puller (e.g., Flaming/Brown Model P-87 from Sutter Instruments, Novarto, CA).

## 2.3. Neuronal Trancing Using Lipophilic Membrane Dyes, Fluorescent Dextrans, and Horseradish Peroxidase (HRP)

1. Paraformaldehyde.
2. Phosphate-buffered saline (PBS).
3. DiI (*see above*).
4. DMF.
5. Glycerol.
6. DABCO.
7. Physiological saline (137 m$M$ NaCl, 5 m$M$ KCl, 2 m$M$ CaCl$_2$, 1 m$M$ MgCl$_2$, 1 m$M$ NaH$_2$βO$_4$, 5 m$M$ HEPES, 11 m$M$ glucose, pH 7.4).
8. Fluorescent dextran, Molecular Probes product numbers D-3308, D-3306.
9. HRP, Boehringer Grade 1 lyophilized (Boehringer Mannheim, Indianapolis, IN).
10. 2.5% gluteraldehyde in PBS (pH 7.4).
11. Diaminobenzidine (DAB).
12. Hydrogen peroxide.

## 3. Methods

### 3.1. Fate Mapping with Lipophilic Membrane Dyes

The lipophilic membrane dye DiI is the first-choice dye for following the fate of relatively small groups of neighboring cells *in ovo*. It is expensive at first glance ($190 for 100 mg), but 100 mg does go a long way.

DiI, like the other carbocyanine membrane dyes, such as DiO and DiA (Molecular Probes D-275 and D-291), is lipophilic, and thus following application to a tissue, it readily and preferentially diffuses into cell membranes rather than remaining in and diffusing through the aqueous extracellular spaces.

There are a number of ways of applying DiI to tissues, most of which initially involve dissolving the dye in a suitable solvent. Using 3 mg DiI in 1 mL of DMF in the first instance is suggested. DMSO or alcohol can be used as an alternative solvent if you suspect DMF is toxic for your cells.

### 3.1.1. Preparation of Embryos

Incubate eggs on their sides. Punch a small pinhole through the blunt end of the shell, and window eggs. The window needs to be large enough to enable easy access of a micropipet to the embryo. Inject ink beneath the embryo. Make a small hole in the vitelline membrane directly over the part of the embryo to be labeled, and carefully add a small drop of saline onto the embryo to prevent it from drying out. Use alcohol-cleaned instruments and sterile solutions.

### 3.1.2. Pressure Injection

This is the most traditional method of applying DiI to tissues in the chick embryo. It is a good way to label many cells in a small region of the embryo, but is not as suitable for analyzing the fate of small numbers of neighboring cells. Although you can use a mouthpiece to control the application of the dye, it is best to use a pressure injection device, such as the Picospritzer II, made by General Valve Corporation (Fairfield, NJ). You will also need a micromanipulator, which is firmly attached either to a base plate or directly to the microscope being used to visualize the embryo. A good stereomicroscope is essential to target accurately the embryonic tissue. To check the accuracy of application, it is a good idea to use an epifluorescent microscope fitted with a long working distance objective. If the dye deposits are not checked each time at the time of application, then it is essential that some embryos be sacrificed immediately after injection so that the initial spread of the dye can be assessed in fixed and cleared preparations. Dye is delivered via micropipets manufactured on a micropipet puller. The author uses 1-mm diameter, thin-walled borosilicate capillary glass with internal filament and pulls these capillaries to make pipets with a tip opening of approx 1.5 µm. Small aliquots of DiI can be delivered from these pipets using a pressure of 50 psi and a pulse duration of about 5 ms (*see* **Notes 1–4**).

### 3.1.3. Iontophoresis

This is the easiest method for labeling very small numbers of adjacent cells. It can be used to label between 1 and 10 cells. It is not recommended as a way of labeling many cells. DiI is a charged molecule and can thus be driven out of a micropipet by applying a potential difference across the pipet and using it as a microelectrode. In addition to a micropipet puller, good micromanipulator, and stereomicroscope, all you need is a 9-V battery. The negative pole of the battery should be connected to the albumin in the egg. This is easily achieved by a flexible cable attached to a thin silver wire, which is inserted into the hole in the shell created at the time of windowing the egg. The positive pole is attached to the back of the microelectrode via an electrode holder. Contact with the DiI in the electrode tip is acheived via a 1 *M* lithium chloride solution, which is backfilled into the electrode. The author routinely uses an electrode with a 1.5-µm tip diameter to label a few cells in the neural plate although electrodes with submicron tips also work well. The author uses 1.2-mm diameter, thin-walled borosilicate glass with internal filament. First insert the negative silver wire electrode into the albumin, and then manipulate the dye-laden microelectrode tip onto the cells to be labeled before completing the electrical circuit (*see* **Note 4**). Between 1 and 5 s of current are sufficient to label a few cells brightly (*see* **Notes 5** and **6**). The amount of dye expelled is very small and may not be visible without the aid of an epifluorescent microscope, but this technique is very effective and so can be confidently used without epifluorescence. Of course, if you want to check or measure exactly where the dye is deposited, an epifluorescent microscope with long working distance objective becomes essential.

### 3.1.4. Application as a Solid

To label superficial cells exclusively (e.g., ectoderm), rather than deep cells is difficult using micropipets and microelectrodes, because their sharp tips can readily slip through superficial cell layers. A better approach is to manipulate small crystals of DiI directly onto the surface cells. This can be acheived by first recrystalizing the dissolved DiI onto the blunt tip of a tiny glass rod and then touching this glass rod onto the surface to be labeled. Tiny glass rods can be made from micropipets by carefully melting and sealing their tips in a small flame or using a microforge. To recrystalize the dye onto the tip, place a small drop of DiI dissolved in either DMF of alcohol onto a clean plastic surface and wait for most of the solvent to evaporate off. As the solution becomes increasingly more concentrated and sticky, it will simply adhere to the tip of a glass rod when one is dipped into it. Gently manipulate the DiI-loaded glass rod onto the cells, and the dye will redissolve into the cell membranes.

### 3.1.5. Fixation, Mounting, and Viewing

The maximum survival time for the embryo and the dye will depend on the rate of dye dilution by cell division. The fate of rapidly dividing cells can readily be studied for up to 3 d, and less rapidly dividing systems for at least a week. Embryos should then be fixed in 3.5% paraformaldehyde and stored in this solution in the refrigerator until viewing. The material can be viewed as a wholemount if the fluorescent cells are close to the surface, or it can be sectioned either

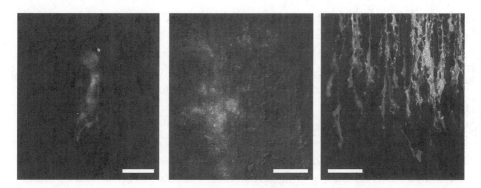

Fig. 1. **(A)** Cells labeled with DiI and DiA in the chick embryo neural plate and viewed immediately after iontophoretic applications. About four cells are labeled with DiI (red) and one cell with DiA (green). Bar is 10 μm. **(B)** After 48 h of development, the descendants of cells labeled as in (A) are still visible and their two colors distinct. Bar is 50 μm. **(C)** Motor neurons retrogradely labeled from their peripheral nerves in fixed tissue. One nerve was labeled with DiI, and the adjacent nerve with DiA. Bar is 40 μm. (*See* color plate 2 appearing after p. 368.)

on a cryostat or vibratome. To increase the transparency of the tissue, the material can be cleared in 90% glycerol in PBS containing 2.5% DABCO. Observe and photograph the material as soon as possible for the best results.

### 3.1.6. Two Color Fate Mapping

If you want to examine the relative movements of two cell populations directly in the same embryo, then each population should be labeled with a differently colored dye. A good combination is to label one set with DiI and the other with DiA (Molecular Probes, cat. no. D-291). DiA fluoresces over a broad range of wavelengths when it is incorporated into cell membranes, but is most intense as a green emission. Using appropriate filter sets, it is thus readily distinguished from the red/orange fluorescence of DiI (**Fig. 1**). DiA performs better than the other commonly used green-fluorescing dye DiO (Molecular Probes, cat. no. D-275), because it is more soluble, it iontophoreses more efficiently, and it fluoresces more intensely.

### 3.2. Lineage Analysis by Single-Cell Injection of Fluorescent Dextran

Labeling single cells *in ovo* and then studying their fate during subsequent development can be a powerful tool in the investigation of how embryos develop. It can be used to construct fate maps at the single-cell level, analyze morphogenetic movements, address issues of tissue specification, and assess the importance of lineage in the determination of regional or cellular identity. One way of labeling single cells is by infection with replication-deficient retroviral vectors. This method allows cells and their descendants to be permanently labeled with a genetic tag, thus allowing for the long-term analysis of clonally related cells. A second method is to inject single cells with tracer dyes using intracellular microelectrodes. The advantages of using the single-cell injection technique are that you can accurately target particular areas of the embryo, there are no doubts about clonality because each injection can be checked visually the time, and by using fluorescent tracers, the sequential analysis of the same clone at different times of development becomes possible. The disadvantage of the single-cell injection technique is that the marker is diluted by cell division and increased cell volume. This effectively limits its usefulness to between about 6 and 9 rounds of cell division, depending on the quality of the initial injection.

Labeling single cells *in ovo* by intracellular injection is not the sort of technique that can easily be learned by simply reading about it. It is much better to see the technique in action. This account will assume a basic understanding of intracellular recording technology and electrophysiological technique. If you do not have this, then make friends with a pharmacologist or neurophysiologist.

### 3.2.1. Method

1. Make microelectrodes—we use 1.2-mm diameter, thin-walled aluminosilicate glass with internal filament (A-M Systems, Everett, WA). Electrode tip should have a nice constant taper and be fine but not wispy. When back-filled with dextran and 1 *M* potassium chloride, they should have a resistance of between 50 and 150 M$\Omega$. In practice, the precise electrode resistance does not matter; if the electrode penetrates a cell, records a stable membrane potential, and passes sufficient dye, then it is a good electrode (*see* **Notes 7** and **8**). Electrodes should first be back-filled with approx 0.5 µL of fluorescent dextran (100 mg/mL in distilled water, Molecular Probes cat. no. D-3308 and D-3306 for tetramethylrhodamine and fluorescein fluorescence, respectively) and then with a little 1 *M* potassium chloride. Don't worry about air bubbles. The capillary action of the internal filament will deal with them.
2. Window egg, inject ink subblastodermally, make a small hole in vitelline membrane over target area, and carefully drop a little saline solution onto the embryo to decrease the risk of drying out and to improve visibility. Do all this under a stereomicroscope.
3. Transfer egg to platform on fixed-stage injection microscope and insert silver/ silver chloride reference electrode into albumin via air hole in shell.
4. Illuminate embryo with fiber-optic directed into the windowed egg. Find and focus embryo under ×20 objective.
5. Manipulate microelectrode tip down into the saline covering the embryo.
6. Switch on amplifier, and check that you are recording a stable baseline potential. You need to work with a "gain" of 10 mV/division and a time base set to 1 s/division. The oscilloscope must be set to monitor DC potentials.

7. Gently manipulate electrode tip down onto the surface of the cells (*see* **Notes 9** and **10**). You probably will not be able to see individual cells. As the electrode tip touches a cell membrane, there will be small change in the appearance of the trace on the oscilloscope. This may be an increase in the thickness of the trace or more usually a small positive deflection. This change is the best indication that the electrode has just touched a cell membrane; it is better than trying to see it happen down the microscope.

8. "Ring," "buzz," or "zap" the capacitance-compensation button or knob on the amplifier to penetrate the cell membrane. If successful, a small but stable negative deflection should be seen on the oscilloscope (anywhere between –5 and –30 mV is common for neural plate cells). This is the most efficient method of getting the electrode tip inside a cell, but it does have one big disadvantage. When you "ring," "buzz," or "zap" an electrode, you simultaneously squirt dye out of the end of the electrode. With a successful penetration, most if not all of this dye will go straight into the cell to be labeled. However, quite often some dye will also be sprayed out onto adjacent cells, and if these have been damaged (by the electrode scraping past them for instance), they will often take up significant amounts of the dye. If this happens, this is the end of your lineage analysis, because you will have labeled more than one cell. It is therefore essential to check briefly each injection visually with epifluorescence in order to ensure only one cell has been labeled. Do not admire your cell for too long, since phototoxicity can kill it. Cell penetration in the absence of simultaneous dye injection can be effected with a piezo-electrode stepper device, but these are not as efficient at getting into small cells as "ringing." No matter how you penetrate your cells, you should then iontophorese more dye into them by using the amplifier's current injection facility (for lysinated rhodamine dextran use positive current pulses of about 4 nA and 250-ms duration for about 30 s).

9. Rapidly withdraw the microelectrode using the axial drive of the manipulator. The oscilloscope trace should spring back up to its original baseline level.

10. Carefully drop a little more saline onto the embryo, and then reseal the egg with tape and reincubate to the appropriate stage.

11. Fix embryo in a solution containing 3.5% paraformaldehyde.

12. Observe results in either sectioned material or whole mounts using conventional epifluorescence or confocal microscopy.

### 3.3. Neuronal Tracing Using Lipophilic Membrane Dyes, Fluorescent Dextrans, and Horseradish Peroxidase (HRP)

The organization of neurons and their axons in the CNS and peripheral nervous system can be examined using anterograde and retrograde tracing techniques with fluorescent or nonfluorescent dyes. The lipophilic membrane dyes, such as DiI, have the advantage that they may be used on fixed as well as living tissue, whereas the intracellular dyes, such as fluorescent dextrans and HRP, rely to some extent on the mechanisms of axonal transport and thus work best on living tissue. The principle is the same for all the dyes; if you are interested in seeing the neuronal cell bodies, then apply the dyes to their distal axons, and if you are interested in the axonal projections, terminations, or growth cones, then apply the dyes to the neuronal cell bodies. The techniques described below work well on chick embryos up to at least Hamburger and Hamilton stage 25.

### 3.3.1. Lipophilic Membrane Dyes

1. Fix embryos in 3.5% paraformaldehyde, and store them in this solution in the fridge until you are ready to label them.
2. Dissect in PBS to reveal the part of the CNS or PNS to be labeled. Do not be too rough with the tissue, since this techniques relies on the dye diffusing through intact cell membranes. If you break the membranes it will not work.
3. Back-fill a micropipet (tip opening approx 2–5 μm) with a small quantity of DiI or DiA (D-282 and D-291 from Molecular Probes, respectively). Use a concentration of 3 mg/mL in DMF. DiI fluoresces intensely red and DiA fluoresces also in the red, but more intensely green, and with appropriate filter sets, can readily be used for double-labeling experiments (**Fig. 1**).
4. Micromanipulate the tip of the micropipet into the tissue, and use a pressure injector (e.g., a Picospritzer II) to deposit the dye. Injections are often made more easily if the pipette is fractionally withdrawn from the full depth of the penetration. If you want to label the cut end of a peripheral nerve, depositing the dye onto the cut surface will efficiently label the axons within.
5. Gently blow away with a Pasteur pipet any excess dye that floats up from the targeted area. If these are allowed to settle onto the embryo, they will label inappropriate areas.
6. Return the specimen to 3.5% paraformaldehyde at room temperature in the dark for 12–48 h.
7. Observe on an epifluorescent microscope either as a whole mount or after sectioning on a cryostat or vibratome. Tissues can be cleared in 90% glycerol in PBS containing 2.5% DABCO (Sigma). The signal in these specimens is not permanent, so they should be analyzed and photographed as soon as possible.

### 3.3.2. Fluorescent Dextrans

Fluorescent dextrans may in some circumstances have advantages over lipophilic membrane dyes. Those that have a lysine residue can be fixed into the cytoplasm and thus form a nice stable signal. Because the fluorescence is in the cytoplasm they may work better in double-staining procedures, which use cell-surface antibodies. The author finds that for labeling axons within the CNS, dextrans often give more intense staining of neuronal projections than the membrane dyes. For efficient labeling of axons in the PNS, the membrane dyes are, however, the best.

1. Carefully remove embryo from egg and transfer to a Sylgard Dow Corning, Midland, MI)-covered dish containing a physiological saline solution.
2. Free embryo of membranes, and pin it down to dissect and reveal the appropriate area of neural tissue.
3. Mix up a small aliquot of fluorescent dextran. Use Molecular Probes product numbers D-3308 and D-3306 for tetramethylrhodamine and fluorescein fluorescence, respectively. The author keeps a small frozen stock of dextran made up at a concentration of100 mg/mL in distilled water from which he transfer a very small drop onto a Sylgard surface. As the water evaporates off, the dextran becomes sticky and is easily picked up on to the tip of a stainless-steel or tungsten micropin or onto the tips of watchmaker's forceps. Simply add more water to the drop of dextran if it dries out too much.
4. To label axons within the CNS, push a micropin laden with dextran into the region of the axon tract. The dextran is taken up into the axons damaged by the pin, so you can regulate the number of axons labeled by controlling the size of the damage. To label axons in peripheral nerves, the author finds it is more efficient to crush the nerve between the tips of dextran-laden watchmaker's forceps.
5. The cut ends of the axons will seal over in about 30 min to 1 h. Thus, if you wait this long and blow away the excess dextran from the first application, a second application with a differently colored dextran can be made nearby without risking contamination of the first axons.
6. Place the embryo in fresh aerated physiological saline at room temperature for up to 12 h (small embryos will need considerably less time than this, about 3–4 h).
7. Fix tissue in 3.5% paraformaldehyde for at least 2 h, and then observe either as a whole mount or after sectioning. Tissue can be sectioned using a cryostat, vibrotome, or by following conventional wax embedding.

### 3.3.3. HRP

HRP is not a fluorescent tracer (it generates a dense brown reaction product as outlined below), but has the advantage of producing permanent preparations and can be used in combination with whole-mount *in situ* hybridization techniques.

1.–4. Follow the procedure outlined for labeling with dextrans, but substitute a thick HRP solution for the dextran mixture. For HRP, the author mixes a small pile of crystals with a small drop of distilled water, and wait for it to evaporate to a sticky consistency.
5. Place the embryo in fresh aerated physiological saline at room temperature for up to 12 h (small embryos will need considerably less time than this, about 3–4 h).
6. Fix tissue in 2.5% gluteraldehyde in PBS (pH 7.4) for 1–2 h.
7. Wash thoroughly in several changes of PBS over a period of at least 2 h.
8. Incubate in diaminobenzidine (DAB) (5 mg in 10 mL of PBS) for at least 1 h. DAB is thought to be carcinogenic and should be handled with care. Wear gloves, and use it only in a fume hood.

9. Add hydrogen peroxide to a concentration 0.003% in the DAB solution for about 5–15 min. Carefully look at the reaction in a covered Petri-dish under a dissecting microscope to check its not going too fast. The reaction can be slowed down by reducing the concentration of hydrogen peroxidase. Stop reaction by washing in excess PBS with azide.

10. Discard DAB solutions into an excess of potassium permanganate before disposal.

11. Observe as a whole mount or after sectioning by vibrotome, cryostat, or after wax embedding. Specimens can be cleared in 90% glycerol in PBS or in nonaqueous mountant.

## 4. Notes

1. Micropipet tips easily become blocked when the DiI solution comes into contact with aqueous solutions and/or is covered in cellular debris. The easiest way to unblock the pipet is to stroke the pipet tip very gently with a small wisp of tissue paper soaked in alcohol. Do this under a dissecting microscope. When the pipet is unblocked, a small deposit of DiI will be drawn out onto the tissue paper.

2. Minimize the time the pipet tip spends in the embryo. The longer it is in contact with embryo or aqueous solution, the more likely it is to block up.

3. If the targeted tissue has sufficient depth, then a small withdrawal of the pipet tip after its initial insertion into the tissue will often facilitate deposition of the dye.

4. If your pipet does not readily push through superficial cell layers to reach deeper cells, then you can first make a small superficial hole with a sharpened tungsten needle and follow this with your DiI pipet.

5. If a small blob of DiI forms on the tip of the electrode as you are trying to inject, first disconnect the battery to stop current flow and wait a short while (30 s or so). The DiI blob may simply dissolve in to the surrounding cells. If it does not dissolve, it will probably stick to the electrode as it is withdrawn from the tissue. Do not worry. You will still have labeled plenty of cells. Use a fresh electrode for the next injection.

6. Your electrode should only ever touch the cells you want to label. Try not to have to reposition it. Its amazing how easily you can inadvertently label cells simply by touching them with the electrode even when the battery is disconnected. This is especially true if the electrode is not a new one.

7. Use a new electrode at least every three embryos. The best electrode is a new one.

8. Poor electrode design is the most likely reason for lack of success.

9. Vibration must be eliminated. Make sure microscope and manipulator are solidly attached to the base plate and that the manipulator only moves when you want it to move, not for instance, when you grab the manipulator, but have yet to turn any knobs.

10. Watch out for embryos that float very slowly across the field of view. This means they are not perfectly balanced within the egg and the egg should be slightly rotated in an appropriate direction to eliminate this drift. You cannot microinject a moving target (well, the author cannot anyway).

## Acknowledgments

The author became familiar with several of these techniques while working with Andrew Lumsden and the single-cell injection procedure is largely inherited from Scott Fraser. They could both have written this chapter better than the author.

# III

# AMPHIBIAN EMBRYO

# 24

# An Overview of *Xenopus* Development

## C. Michael Jones and James C. Smith

## 1. Introduction

Embryos of the amphibian *Xenopus laevis* have been used as a model system for the analysis of developmental mechanisms since the 1950s. As described by Gurdon *(1)*, one of the reasons for the popularity of *Xenopus* is that it is easy to obtain large numbers of embryos. This is illustrated by the fact that until the mid-1950s, *Xenopus* was used as a pregnancy test in humans: injection of urine from a pregnant woman into the dorsal lymph sac of a female *Xenopus* causes the frog to lay eggs. This simple assay can readily be applied only to *Xenopus,* for as Gurdon points out, to persuade *Rana* species to lay eggs, it is necessary to inject homogenized pituitary glands, a procedure that, furthermore, only works at certain times of the year. Thus, *Xenopus* is the easiest amphib'ian species from which to obtain embryos, especially as it is now possible to buy human chorionic gonadotrophin from Sigma (St. Louis, MO), thus circumventing the requirement, in a busy developmental biology laboratory, for a constant supply of pregnant colleagues.

However, there are other reasons for working on *Xenopus*. The animal is completely aquatic, and therefore much easier to keep than other amphibians, which have a tendency to hop away. It is also rather robust, and only rarely succumbs to disease or infection. The early embryo is relatively large in size and, like all amphibian embryos, is readily accessible to the investigator because it develops outside the mother. Dissection and microinjection are therefore easily performed, and because each embryonic cell is provided with yolk reserves to serve as an energy source, tissues can be isolated and cultured for several days in simple salt solutions without the need for poorly characterized serum components. Finally, development is rapid. Together, these virtues have made the *Xenopus* embryo a favorite of vertebrate developmental biologists.

*From:* Methods in Molecular Biology, Vol. 97: Molecular Embryology: Methods and Protocols
Edited by: P. T. Sharpe and I. Mason © Humana Press Inc., Totowa, NJ

This chapter provides a brief overview of *Xenopus* embryogenesis from egg to the formation of the definitive body plan. It concentrates on the stages that are most relevant to the molecular embryologist; other, more detailed, descriptions have recently been published in **refs.** *2* and *3*. In particular, the movements of gastrulation have been extensively described by Keller and colleagues *(3–5)*. The staging series is that of Nieuwkoop and Faber *(6)*.

## 2. The Egg and Fertilization

The *Xenopus* egg is 1.2–1.4 mm in diameter, and consists of a darkly pigmented "animal hemisphere" and a lighter yolky "vegetal hemisphere." When laid, the eggs are oriented randomly with respect to gravity, and held in position by a transparent vitelline membrane inside a jelly coat (**Fig. 1**), but after fertilization, granules located just below the surface of the egg fuse with the plasma membrane and release their contents into the space between the vitelline membrane and the egg. This material provides some lubrication, allowing the egg to rotate such that the less dense animal hemisphere is uppermost. This rotation usually occurs within 20 min of fertilization.

*Xenopus* is monospermic, and the successful sperm enters the egg in the animal hemisphere. The site of sperm entry (the "sperm entry point," or SEP) is often visible as a small aggregation of pigment. The position of sperm entry defines the future dorsal–ventral axis of the embryo: the future "dorsal" side of embryo forms from the side of the egg opposite the SEP. (Dorsal is in quotation marks in the previous sentence, because the dorso–ventral axis of the egg is not directly translated into the dorso–ventral axis of the tadpole; *see below.*) The SEP defines the dorso–ventral axis of the embryo by determining the direction of rotation of a cortical layer of cytoplasm, just beneath the plasma membrane. This rotation, of about 30°, is driven by a transiently aligned microtubule array in the vegetal hemisphere of the egg and begins about 40 min after fertilization. Through mechanisms that are still completely unclear, the rotation establishes a signaling center, often referred to as the "Nieuwkoop Center," which directs the development of the dorso–anterior region of the embryo (*see below*).

## 3. Early Cleavage Stages

It is important to note that early stages of *Xenopus* development rely completely on maternal stores of RNA and protein; transcription does not begin until the so-called midblastula transition, about 7 h after fertilization and when there are 4096 cells (that is, after 12 cleavages) *(7)*.

The first *Xenopus* cell cycle occupies about 90 min at 21°C; subsequent cycles last about 30 min. One of the greatest advantages of *Xenopus* as a tool for developmental biology is that cleavage planes are regular and result in the

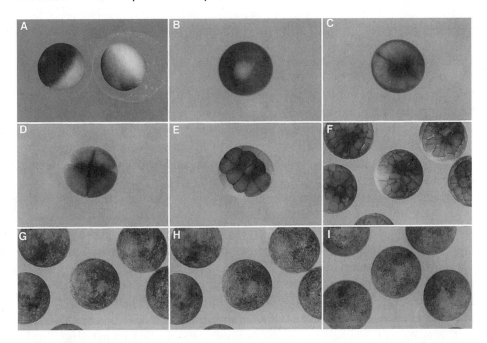

Fig. 1. *Xenopus* eggs and embryos at cleavage and blastula stages. (**A**) Unfertilized eggs. The darkly pigmented animal hemisphere is easily distinguishable from the lighter vegetal hemisphere. The swollen translucent jelly coat is evident around each. (**B**) Animal view of a one-cell embryo after removal of the jelly coat. (**C**) A two-cell embryo. The cleavage plane divides the embryo into the future right and left halves along the animal–vegetal axis. (**D**) Animal view of a four-cell stage embryo. Note the pigmentation differences between the smaller, more lightly pigmented presumptive dorsal blastomeres (the upper two) and the presumptive ventral blastomeres. (**E**) A 16-cell embryo. The smaller animal blastomeres are resting on the larger vegetal cells. (**F**) Animal view of 32- to 64-cell embryos. The lightly pigmented presumptive dorsal side (to the left) is easily recognized in the embryo in the center. (**G**) The same embryos as in (F) at a midblastula stage (stage 8). The embryos consist of approx 4000 cells. (**H**) The same embryos at a later blastula stage (stage 9). Note the smaller size of the cells than in (F) and (G). (**I**) Embryos at a very late blastula stage, just prior to the beginning of gastrulation. The images in (B)–(E) are photographed at the same magnification. The embryos are approx 1.3 mm in diameter.

formation of identifiable blastomeres whose fates may be predicted. It is important to emphasize, however, that the fate maps described here, and by other workers, apply only to embryos in which the cleavages are archetypical. Sometimes (indeed, quite often), cleavages are inegular, and although the embryo develops perifectly normally, it is not possible in these cases to predict the fates of particular cells.

Bearing these provisos in mind, the first cleavage in *Xenopus* separates the future leftand right-hand sides of the embryo. The second, 30 min later, is at right-angles to the first and separates the future dorsal and ventral halves. At this four-cell stage, it is often possible to distinguish the two dorsal blastomeres from their ventral counterparts: if the first two cleavages have occurred in a regular pattern, the dorsal blastomeres are usually slightly smaller and more lightly pigmented than the ventral cells.The third cleavage, which occurs after another 30 min, is orthogonal to the first two, and separates the animal and vegetal poles. Unlike the first two divisions, however, which divide the embryos into (roughly) equal pieces, the plane of the third cleavage is well above the equator of the embryo, reflecting a rule that all cells cleave away from the side with more yolky cytoplasm.

After the fourth (meridional) and fiifth (equatorial) cleavages, the embryo comprises 32 cells, arranged in four tiers of eight cells (stage 6 of Nieuwkoop and Faber). The fates of these cells have been determined by microinjection of cell lineage markers *(8,9)*, and the fate map of Dale and Slack is shown in **Fig. 2**. Notice that individual cells at this stage are far from restricted to a single fate, and that there may be considerable variation from embryo to embryo.

During the early cleavage stages, and beginning with the first, a small space forms becomes larger as cleavages procccd, and eventually becomes the blastocoel.

## 4. Blastula Stages

This rapid series of cleavage divisions continues beyond the 32-cell stage, resulting in the formation of progressively smaller cells. Blastomeres are larger at the vegetal pole than the animal pole, a consequence of the rule that cleavages occur away from the more yolky region of cells. The first tangential cleavage occurs at the early blastula stage (stage 7), and this changes the previously single-cell-layered morula into a double-layered embryo. What was the cleavage cavity is now called the blastocoel, and this enlarges by osmotic uptake of water (*see* **ref. 10**).

This pattern of development—one cleavage every 30 min—lasts until cycle 13, when division becomes asynchronous, and it slows down significantly *(7)*. This point, the midblastula transition (MBT), is also marked by the onset of cell motility and zygotic transcription. It corresponds to stage 8 of *(6)* and presages the next important stage in *Xenopus* development gastrulation.

## 5. Gastrulation

During gastrulation, the blastula, a hollow ball of cells with radial symmetry, is converted into a three-layered structure with a central midline and bilateral symmetry. The highly dramatic movements of gastrulation are preceded, during late blastula stages, by pregastrulation movements of which the most

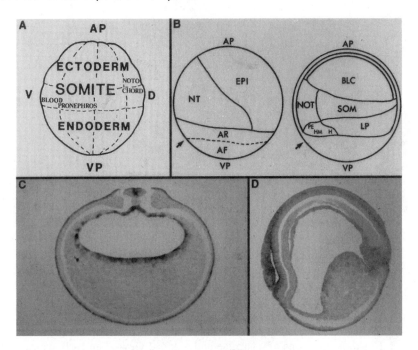

Fig. 2. Fate maps of *Xenopus* embryos, including tracing the outer layer of late blastula embryos by marking the cells with Bolton-Hunter Reagent *(11)*. (**A**) Fate map of the 32-cell embryo *(8)*. (**B**) Fate maps of the superficial (left) and deep layers (right) of an early gastrula embryo. Reproduced from Keller *(12,13)*. (**C**) Transverse section of a neurula stage embryo that was dipped in Bolton- Hunter Reagent at a blastula stage to mark the outer layer of cells. The darkly marked cells have migrated during gastrula stages and can be seen lining the archenteron, in the neural tube and in the epidermis. Dorsal is to the top. (**D**) A sagittal section of a similarly marked embryo showing darkly marked cells in the same tissues. Anterior is to the left and dorsal upward in this photo. AP, animal pole; VP, vegetal pole; V, ventral; D, dorsal; NT, neural tissue; AR, archenteron roof; AF, archenteron floor; EPI, epidermis; BLC, blastocoel; NOT, notochord; SOM, somite; LP, lateral plate; PE, pharyngeal endorlerm; HM, head mesoderm; H, heart.

obvious is epiboly, a vegetally directed movement of animal hemisphere cells. This results in a thinning of the blastocoel roof and the accumulation of prospective mesodermal cells from a position above the equator of the embryo to a subequatorial location.

The first sign of gastrulation proper is the appearance of a pigmented depression in the dorsal—vegetal quadrant of the embryo—the dorsal lip of the blastopore (**Fig. 3**). Formation of this blastopore lip, and the associated line of pigment, reflects the formation of the so-called bottle cells. These are a

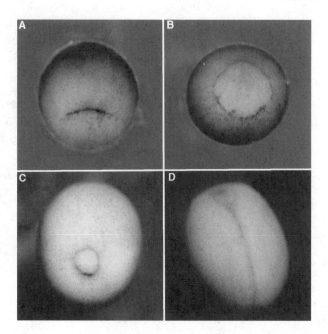

Fig. 3. Blastopore regions of gastrulae and a dorsal view of an early neurula stage embryo. **(A)** Dorso–vegetal view of an early gastrula (stage 10.5). Note the darkly pigmented, crescent-shaped dorsal blastopore lip. The animal hemisphere is upward. **(B)** Vegetal view of a midgastrula (stage 11). The blastopore is now circular. Dorsal is to the top. **(C)** A postero–ventral view of a late gastrula embryo (stage 12.5). The closing circular blastopore is now smaller than at earlier stages. **(D)** Dorsal view of an early neurula (stage 15). The more darkly pigmented closing neural tube is evident in the midline. Anterior is to the upper left.

group of superficial cells that undergo dramatic changes in shape in which their apices contract and cytoplasm is forced away from the surface of the embryo. This causes the cells to adopt their eponymous bottle-like shape, and the accumulation of pigment granules in the apices of the cells results in the formation of the blastoporal pigment line. The first bottle cells to form do so on the dorsal side of the embryo, but they are soon joined by more lateral and ventral cells, so that the lip becomes an arc, then a semicircle, and finally a complete circle. As development proceeds, the bottle cells are propelled to the interior of the embryo by the mesodermal cells. Eventually, by the midgastrula stage, they respread on the roof of the blastocoel and form part of the archenteron, or gut.

At the same time that the blastopore lip becomes visible on the outside of the embryo, prospective mesodermal cells on the inside *(11)* begin to migrate across the roof of the blastocoel toward the animal pole. As with the formation of the blastopore lip, the first mesodermal cells to migrate are those at the dorsal side of the embryo, and the movement then spreads laterally and ventrally, reaching the ventral side by the late gastrula stage. It is only the leading mesodermal cells that migrate; those that follow undergo convergence and extension, processes that occur most vigorously in the prospective notochord and somites. Convergence and extension both result from the directed intercalation of mesodermal cells in such a way that they converge toward the dorsal midline of the embryo and in doing so cause the embryo to extend. An analogy would be to squeeze a tube of toothpaste (not from the bottom, the approach we prefer in a domestic situation, but rather by gripping the whole tube in one's fist). As the toothpaste converges toward the center of the tube, it extends from the hole at the end.

As convergence and extension proceed, the circumference of the blastopore becomes smaller, and eventually is reduced to a slit. The cells of the animal hemisphere now cover the entire embryo, and constitute the ectoderm. The mesoderm lies beneath the ectoderm, having reached this position through active movements, such as migration, convergence, and extension. Finally, the endoderm lies within the mesoderm. The endoderm occupies this position partly by default, because the yolk mass is dragged passively into the middle of the embryo by the mesoderm. However, the superficial cells of the early gastrula that go on to form the endoderm (remember that in *Xenopus*, all the mesoderm is formed from deep tissue *[11]*) also undergo convergent extension, and these foim the roof and walls of the archenteron, or primitive gut (**Fig. 2**).

The closure of the blastoporc marks the end of gastrulation proper, and by this stage the three germ layers have reached their definitive positions. The animal–vegetal and dorso–ventral axes of the blastula-stage embryo are no more, and although at first sight it might seem that they can be translated directly into the antero–posterior and dorso–ventral axes of the late gastrula/early neurula stage embryo, this is not so. The earlier, and more dramatic, gastrulation movements on the "dorsal" side of the early gastrula ensure that these cells form most of the anterior tissues of the embryo as well as the entire notochord and much of the somites. The "ventral" side of the early gastrula forms much of the posterior of the embryo, and, especially toward the tail, gives rise even to substantial amounts of somitic tissue.

Although many region-specific markers are expressed during gastrula stages (which can be used in experiments on inductive interactions), little obvious cytodifferentiation has occurred. The most obvious tissue at this stage is the notochord, which has physically separated from the somites that flank it, and has acquired a "stack-of-coins" appearance.

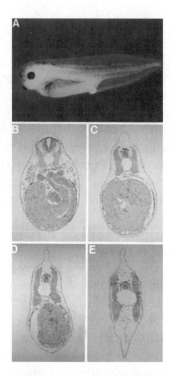

Fig. 4. A late stage tadpole and sections of a similar stage highlighting the tissue organization that results after gastrulation and neurulation are complete. **(A)** A tadpole (stage 40). Anterior is to the left and dorsal to the top. **(B)** A section through the anterior region of the tadpole. The neural tube is stained a magenta color owing to the high concentration of nuclei and is the uppermost structure. Beneath the neural tube is the vacuolated notochord, which is flanked on each side by somitic tissue (stained green). Nephtic tubules are evident lateral and ventral to the somites, and the large yolk-filed gut (endoderm) is stained yellow. **(C)** A slightly more posterior section than (B). The neural tube, notochord, somites, and endoderrn are evident. **(D)** An even more posterior section than (B) or (C). **(E)** A section through the tail region. Note the proportionally smaller neural tube and larger somites than in more anterior regions. The notochord is a similar size along the anterior–posterior axis. The magnification in (E) is slightly greater than in (B)–(D).

## 6. Neurulation and Beyond

Early in neurulation, the posterior neural plate, like the mesoderm, undergoes convergence and extension, thus assisting in blastopore closure. This movement is autonomous and does not require the mesoderm to be underneath the neural plate. At later stages, the process of neurulation in *Xenopus* is much less dramatic than in urodele amphibia, such as *Ambystoma mexicanum*; the neural folds, in particular, are not as obvious, and they do not form the classic, well-defined "keyhole" shape. That said, the first sign of neurulation in *Xenopus* is the thickening of the inner layer of dorsal ectoderm, the so-called censorial layer. This is followed, at the midneurula stage, by the formation of a dark pigment line along the dorsal midline of the embryo. This neural groove arises through the formation of bottle-like cells in the neural midline, which, like the bottle cells of the blastopore lip, contract their apices and concentrate the pigment granules in a smaller area. As neurulation proceeds, the neural groove deepens, and the lateral neural folds converge on the dorsal midline, where they eventually fuse to form the neural tube (**Fig. 3D**).

The most lateral cells of the neural plate do not participate in neural tube closure and are not recruited into the neural tube. These are the future neural crest cells, which in the truck go on to form pigment cells and neural derivatives, such as the dorsal root ganglion, and in the head form a wide variety of structures, such as the cephalic ganglia, the mandibular, hyoid, and branchial arches, and the head mesenchyme.

Finally, during and after neurulation, the mesoderm becomes subdivided into different tissues along the dorso–lventral axis (**Fig. 4**). The most dorsal mesodermal cell type is the notochord, a rod of vacuolated cells running the length of the embryo. Lateral to the notochord are the cells of the somites, which in *Xenopus* form predominantly muscle, and lateral and ventral to the somites are the cells of the pronephros. The lateral mesoderm goes on eventually to form structures such as the limbs, and the most ventral mesoderm forms blood.

## References

1. Gurdon, J. B. (1996) Introductory comments: *Xenopus* as a laboratory animal, in *The Biology of* Xenopus. Zoological Society of London, London, Clarendon, Oxford, UK.
2. Hausen, P. and Riebesell, M. (1991) *The Early Development of* Xenopus laevis*: An Atlas of the Histology.* Springer-Verlag, Berlin, Germany.
3. Keller, R. E. (1991) Early embryonic development of *Xenopus laevis*, in Xenopus laevis: *Practical Uses in Cell and Molecular Biology* (Kay, B. K. and Peng, H. B., eds.), Academic, San Diego, CA, pp. 61–113.
4. Keller, R. E., Danilchik, M., Gimlich, R., and Shih, J. (1985) The function and mechanism of convergent extension during gastrulation of *Xenopus laevis*. *J. Embryol. Exp. Morphol.* **89**, 185–209.
5. Keller, R. and Danilchik, M. (1988) Regional expression, pattern and timing of convergence and extension during gastrulation of *Xenopus laevis*. *Development* **103**, 193–209.
6. Nieuwkoop, P. D. and Faber, J. (1975) *Normal Table of* Xenopus laevis (Daudin). Amsterdam, North Holland.
7. Newport, J. and Kirschner, M. (1982) A major developmental transition in early *Xenopus* embryo: I Characterization and timing of cellular changes at the midblastula stage. *Cell* **30**, 675–686.

8. Dale, L. and Slack, J. M. (1987) Fate map for the 32-cell stage of *Xenopus laevis. Development* **99,** 527–551.

9. Moody, S. A. (1987) Fates of the blastomeres of the 32 cell *Xenopus* embryo. *Dev. Biol.* **122,** 300–319.

10. Gerhart, J. C. (1980) Mechamisms regulating pattern formation in the amphibian egg and early embryo, in *Biological Regulation and Development* (Goldberger, R. F., ed.), Plenum, New York, pp. 133–293.

11. Smith, J. C. and Malacinski, G. M. (1983) The origin of the mesoderm in an anuran, *Xenopus laevis,* and a urodele, *Ambystoma mexicanum. Dev. Biol.* **98,** 250–254.

12. Keller, R. E. (1975) Vital dye mapping of the gastrula and neurula of *Xenopus laevis.* I. Prospective areas and morphogenetic movements of the superficial layer. *Dev. Biol.* **42,** 222–241.

13. Keller, R. E. (1976) Vital dye mapping of the gastrula and neurula of *Xenopus laevis.* II. Prospective areas and morphogenetic movements of the deep layer. *Dev. Biol.* **51,** 118–137.

# 25

## Mesoderm Induction Assays

### C. Michael Jones and James C. Smith

## 1. Introduction

Inductive interactions play a major role in early development, and one of the earliest such interactions in amphibian development, and perhaps the in development of all vertebrates, is mesoderm induction *(1–5)*. Mesoderm induction occurs at blastula stages, when a signal from the vegetal hemisphere of the embryo acts on overlying equatorial cells, causing them to form mesoderm rather than ectoderm. This interaction was first discovered in experiments in which prospective ectodermal tissue of the embryo (the so-called "animal cap") is juxtaposed with future endoderm from the vegetal hemisphere (**Fig. 1**). When cultured alone, the animal caps form epidermis; when cultured adjacent to vegetal pole blastomeres, they form mesoderm.

In the last decade, great advances have been made in coming to understand the signals involved in mesoderm induction as well as in identifying the intracellular signal transduction pathways used by these factors. Progress has also been made in identifying target genes of mesoderm-inducing factors, and even in studying how the transcription of these genes is regulated. This chapter describes three assays for mesoderm induction. The first is the original assay described above, in which animal pole and vegetal pole tissue is juxtaposed. In the second, the inducing factor is supplied to the target tissue not from vegetal pole cells, but as a soluble protein, and in the third, the inducing agent is supplied to the responding cells by microinjecting RNA encoding the protein in question into the developing embryo.

This last method has proven particularly useful for studying proteins that cannot yet be obtained in purified, soluble form *(6,7)* and for studying the activities of intracellular proteins, such as components of the MAP kinase path-

From: *Methods in Molecular Biology, Vol. 97: Molecular Embryology: Methods and Protocols*
Edited by: P. T. Sharpe and I. Mason © Humana Press Inc., Totowa, NJ

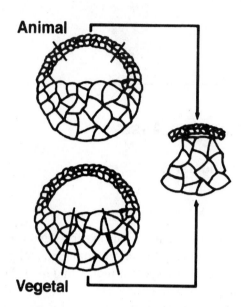

Fig. 1. Mesoderm induction. Cells from the animal hemisphere of the embryo are juxtaposed with cells from the vegetal hemisphere.

way *(8–10)*. The technique can also be adapted to identify novel mesoderm-inducing agents by "expression-cloning" *(11)*. Finally, mesoderm induction can also be used as a convenient read-out for ligand–receptor interactions, which have nothing to do with the in vivo process. For example coexpression of GDNF and the receptor tyrosine kinase Ret in *Xenopus* animal caps causes the formation of mesoderm, whereas expression of either protein alone has no effect, thus suggesting that GDNF signals through Ret *(12)*.

## 2. Materials

All solutions used for culture of *Xenopus* tissue should use distilled water, and should be sterilized by autoclaving or filtration as indicated below.

### 2.1. Obtaining Xenopus *Embryos by Artificial Fertilization*

1. *Xenopus laevis* adult females and males: These may be obtained from a variety of sources; up-to-date details of suppliers throughout the world are available on the *Xenopus* molecular marker resource: URL http://vize222.zo.utexas.edu/. Instructions for keeping *Xenopus* are beyond the scope of this chapter; interested readers may refer to **ref. 13**.
2. Pregnant mare serum gonadotrophin (PMSG; Sigma, St. Louis, MO).
3. Human chorionic gonadotrophin (HCG; Sigma).
4. 2-mL Syringe and 23- to 25-gage needles.

## 2.2. Dissection Dishes and Instruments

1. 35- or 60 mm Petri dishes coated with a thin layer of 1% agarose in water, previously sterilized by autoclaving.
2. Two pairs of no. 5 Dumont watchmaker's forceps, sharpened using an "Arkansas" sharpening stone.
3. Tungsten needles are made by mounting about 5 cm of tungsten wire (0.5-mm diameter. 99.95% purity, Goodfellow metals) in a 7 cm piece of capillary tubing. The needle is sharpened by electrolysis in 1 *M* NaOH using a potential difference of 9–12 V AC. The simplest procedure is to place a small piece of modeling clay on the end of the wire and to dip about 2 mm of naked wire into the NaOH. When the clay falls off, the needle is sharp enough.

## 2.3. Culture Media

1. Embryos and explants are cultured in dilutions of Normal Amphibian Medium (NAM; *14*): 110 m*M* NaCl; 2 m*M* KCl; 1 m*M* $Ca(NO_3)_2$; 1 m*M* $MgSO_4$; 0.1 m*M* $Na_2EDTA$; 2 m*M* sodium phosphate pH 7.5; 1 m*M* $NaHCO_3$; 50 µg/mL gentamycin. It is convenient to prepare a 10X stock solution of all the components of NAM with the exception of the phosphate buffer, the $NaHCO_3$, and the gentamycin. This 10X 'NAM salts' stock may be autoclaved. We then prepare:
   a. 0.1 *M* Sodium phosphate pH 7.5, which may be autoclaved and serves as a 50X stock.
   b. 0.1 *M* $NaHCO_3$, which is filter-sterilized and stored frozen and serves as a 100X stock.
   c. 10 mg/mL gentamycin which is also filter-sterilized and stored frozen and serves as a 200X stock.
   Dilutions of NAM refer only to the "NAM stocks" and $NaHCO_3$; levels of phosphate buffer and gentamycin are held constant.
2. 4% Ficoll in 75% NAM.
3. 0.1% Bovine serum albumin (BSA) in 75% NAM.
4. 60% Leibovitz L-15 medium (for storage of *Xenopus* testes).
5. 2% Cysteine hydrochloride, adjusted to pH 7.9–8.1 with NaOH.
6. Sterile mineral oil.

## 2.4. Microinjection Equipment

As discussed previously *(13)*, there are several different techniques for injecting RNA into *Xenopus* eggs, and it is not possible to describe them all. We therefore describe the methods used in our own laboratory.

1. Microinjection needles are prepared from capillary tubing, such as GC120F-15 made by Clark Electromedical Instruments (Reading, England). We have used two types of needle puller. One consists of nothing more than a heated platinum coil. The capillary is clamped at its top, passed down through the coil, and a piece of modeling clay (about 7 g) is attached to the bottom. The current is then switched on. The shape of the needle and the diameter of the tip depend on the amount of current passed through the coil, the exact weight of the modeling clay, and on whether the investigator is able to catch the tubing before it hits the bench.

The second option is to use more expensive pullers, such as those made by Campden Instruments or Kopf. Whatever kind of puller is used, the diameter of the needle tip should not exceed 15 μm. Needles should be heated to 180°C to destroy RNAase activity, and they should be stored in a dust-free atmosphere.

2. Singer Instrument Company Micromanipulator, Mark I.
3. Inject + Matic air-driven injector (Supplied by Micro Instruments Ltd., Long Hanborough, Oxford, UK).

### 2.5. Lineage Labeling

This is performed with fluorescein-lysine-dextran (Fluoro-emerald: Molecular Probes, Eugene, OR).

### 2.6. RNA Transcription

Stable transcripts and high yields of protein are produced from cDNAs cloned into the vector pSP64T *(15)* or pCS2+ *(16,17)*. The original version of pSP64T is slightly inconvenient, because although it has several sites for linearization, the only cloning site is *Bgl*II. Several laboratories have, however, produced more user-friendly versions of pSP64T, which include a multiple cloning site, and two such vectors are available from our colleague Masazumi Tada. Plasmids should be linearized before transcription with a restriction enzyme that generates a blunt end or a 5'-overhang.

1. Linearized plasmid at 1 mg/mL.
2. 10X Transcription buffer: 400 m$M$ Tris (pH 7.5), 60 m$M$ MgCl$_2$, 20 m$M$ spermidine HCl, 50 m$M$ NaCl.
3. 0.25 $M$ DTT.
4. 10 m$M$ ATP.
5. 10 m$M$ CTP.
6. 10 m$M$ UTP.
7. 1 m$M$ GTP, 10 m$M$ GTP.
8. 2.5 m$M$ GpppG.
9. RNasin.
10. SP6 RNA polymerase.
11. RNase-free water.
12. RQ1 DNase I (Promega).
13. Phenol/chloroform.
14. Ethanol.
15. 1% Agarose TBE gel (RNase-free).

## 3. Methods

### 3.1. Artificial Fertilization of Xenopus *Eggs*

*Xenopus* females are induced to lay eggs by the injection of 500–1000 IU human chorionic gonadotrophin (HCG) into their dorsal lymph sacs. This technique is defined as a "procedure" under the UK Animals (Scientific Procedures) Act of 1986, and requires a Home Office license. Yields of eggs may be improved if the frogs are "primed" a few days before HCG injection with 30 IU of pregnant mare serum gonadotrophin (PMSG). It is prudent to inject two or three frogs in order to be sure of obtaining enough embryos for an experiment. In what follows, it is assumed that the investigator is right-handed.

1. Dissolve 30 IU PMSG in 0.5 mL water and load a 2-mL syringe.
2. Select a female frog (fat ones are usually good). Hold her still on a flat surface. We use the left hand to hold the frog, with the heel of the hand pressed firmly against the bench in front of the frog's head, and with the middle and index fingers between the frog's back legs. If the muscles of the hand are tensed, as if making a "bridge" for a snooker or pool cue, the hand forms a "cage," which restrains the frog without exerting significant pressure on her.
3. Injections are made into the dorsal lymph sac of the animal. Using the right hand, the injection needle should be inserted just beneath the skin on the dorsal surface of the frog's right hindleg. The needle tip should then be moved beneath the row of stitches of the lateral line organs toward the dorsal midline. After injection, the needle tip should be withdrawn slowly along the path of insertion. There should be little or no bleeding, and the frog should not experience any discomfort. It should be returned gently to a marked tank.
4. Two or 3 d after injection of PMSG, frogs should receive injections of 500–1000 IU HCG, using the same technique. They should be kept overnight in the dark, at 18–22°C.
5. To obtain testes, male *Xenopus* should be killed by heavy anesthesia followed by decapitation and rostral and caudal pithing. The testes are pale, curved structures about 1 cm long, which are positioned on either side of the spine. They may be removed by making an incision in the ventral surface of the animal, and they can be stored in 60% Leibovitz L-15 medium at 4°C for up to 1 wk.
6. Female *Xenopus* should start to lay eggs the morning after injection. Fresh eggs can be "squeezed" from the frogs by gentle peristalsis of their ventro-lateral surfaces. The eggs should be transferred to a Petri dish and rinsed with distilled water. Then, using a Pasteur pipet, as much liquid as possible should be removed from the eggs.
7. Fertilize the eggs by rubbing them lightly with a piece of dissected testis.
8. Wait 5 min, and flood the eggs with 10% NAM. After about 15 min, the eggs should rotate so that their heavily pigmented animal hemisphere is uppermost. This is a reliable sign of successful fertilization.
9. "Dejelly" the fertilized eggs by incubating them for about 5 min in 2% cysteine hydrochloride. Rinse thoroughly in 10% NAM. Transfer to an agarose-coated Petri dish. Embryos should begin to cleave about 90 min after fertilization. They can be cultured at temperatures between 14 and 23°C. At the warmer temperature, it takes about 5 h for embryos to reach stage 8 (the midblastula stage), which is when mesoderm induction assays are usually carried out.

## 3.2. RNA Synthesis

1. Set up transcription reaction at room temperature:

| | | |
|---|---|---|
| Linearized DNA (1 mg/mL) | 5 | µL |
| 10X transcription buffer | 5 | µL |
| 0.25 $M$ DTT | 2.5 | µL |
| 10 m$M$ ATP | 5 | µL |
| 10 m$M$ CTP | 5 | µL |
| 10 m$M$ UTP | 5 | µL |
| 1 m$M$ GTP | 5 | µL |
| 2.5 m$M$ GpppG | 10 | µL |
| RNasin | 2.5 | µL |
| SP6 RNA polymerase | 2.5 | µL |
| RNase-free water | 2.5 | µL |

2. Mix gently, spin briefly to get the components to the bottom of the tube, and incubate at 37°C for 30 min.
3. Add 2.5 µL 10 m$M$ GTP, and incubate for an additional 1 h.
4. Add 5 µL RQ1 DNase I, and incubate at 37°C for 30 min.
5. Extract twice with phenol/chloroform, and ethanol-precipitate twice. Wash the pellet twice with 75% ethanol.
6. Redissolve RNA in 30 µL DEPC-treated water. Measure $A_{280}$, and adjust RNA concentration to 1 mg/mL.

RNA can be tested for integrity by taking 1 µL of the reaction after **step 4** and running it on a 1% RNaase-free TBE gel. There should be a strong RNA band of the appropriate size.

## 3.3. Microinjection

Microinjection of RNA and of the lineage tracer fluorescein-lysine-dextran uses the same technique, except that it is essential to bake injection needles used for RNA to destroy RNase activity. As stated above, there are several different kinds of injection apparatus, and our description is therefore rather general.

1. Transfer embryos to be injected to 4% Ficoll in 75% NAM.
2. Attach the injection needle to the micromanipulator and injection system. Dispense about 2 µL of water onto a small piece of parafilm, and suck it up into the pipet. This rinses the pipet and allows one to calibrate it.
3. Calibrate the injector/needle combination by injecting into a dish of sterile mineral oil. Calculate the volume injected by the formula $V = 4/3\pi r^3$. A volume of about 10 nL can be safely injected into a fertilized egg.
4. Expel water, and load the injection sample.
5. Inject embryos. This is most easily done by supporting the embryo with a pair of forceps held in the left hand (this is for a right-hander again) and using the right hand to control the micromanipulator. Injections can be made into any part of the egg, but for uniform distribution of the injected material, it is best to aim for the

equatorial region, where the pigmented animal hemisphere meets the paler vegetal hemisphere. A few hundred embryos can be injected in an hour.

6. Culture embryos in 4% Ficoll in 75% NAM until stage 8.

### 3.4. Animal/Vegetal Conjugates

1. When they reach stage 8, remove the vitelline membranes from a group of
   embryos using sharpened forceps. Half the embryos should have been lineage-
   labeled by injection of fluorescein-lysine-dextran injection, and half should be
   uninjected. Transfer embryos to 75% NAM.
2. Dissect animal pole regions from the center of the pigmented regions of the
   labeled embryos. This can either be done using a pair of forceps as "scissors" or
   by using a "picking" motion with a tungsten needle. During dissection, the
   embryo can be kept still using a pair of forceps held in the other hand.
3. Dissect vegetal pole regions from uninjected embryos as described above for
   animal pole regions.
4. Place an animal pole region with its originally outer surface down, and position a
   vegetal pole region on top of it such that its originally outer surface is up. The
   two pieces of tissue will adhere to each other quickly, but care should be taken
   not to disturb the conjugate for at least 10 min.
5. Culture in 75% NAM at 18–22°C.

### 3.5. Animal Cap Assay with Soluble Inducer

1. Dissect animal pole regions from Xenopus embryos as described above.
2. Transfer animal caps to 75% NAM containing 0.1% BSA and the putative
   inducer. It is sensible to use a range of different inducing factor concentrations.
   Molecules like activin are active at 0.1–10 ng/mL.
3. Culture in 75% NAM at 18–22°C.

### 3.6. Animal Cap Assay Following RNA Injection

1. Dissect animal pole regions from *Xenopus* embryos, which have been injected
   with RNA encoding the putative inducer.
2. Culture in 75% NAM at 18–22°C.

### 3.7. Scoring the Results

The result of a mesoderm induction assay may be score in several ways. In
increasing order of complexity, these are:

1. Observation of gastrulation-like movements. Normally, isolated animal pole tissue "rounds up" after being dissected from the embryo, and it forms a sphere.
   Treatment with a mesoderm-inducing factor causes the animal pole cells to
   undergo gastrulation-like movements. However, and the animal caps elongate in
   a characteristic fashion (*18*; **Fig. 2**). Activin and other members of the TGF-β
   family cause more dramatic gastrulation movements than do members of the FGF
   family (*6,18,19*). Elongation movements are visible after approx 4 h of culture.

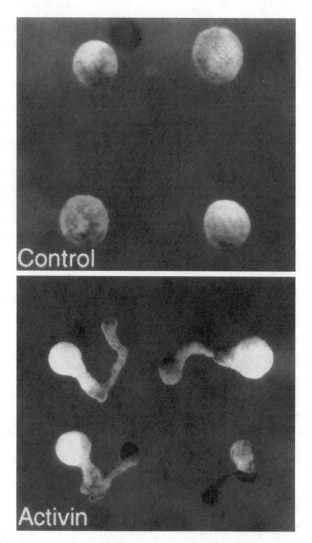

Fig. 2. Gastrulation movements induced by activin treatment of animal pole cells. Control explants on the top form spheres; those treated with activin, on the bottom, elongate.

When animal pole–vegetal pole conjugates have been made, observation with a fluorescent microscope should reveal that the cells undergoing gastrulation movements are those derived from the animal cap.

2. Observation of later morphology: After 3 d of culture, when sibling embryos are at tadpole stages, animal caps treated with mesoderm-inducing factors form characteristic structures. In the case of activin, these have been termed "embryoids," because they resemble miniature embryos *(20)*. Treatment with FGF causes the formation of

translucent balloon-like structures with a layer of smooth muscle beneath epidermis. Uninduced animal caps remain as opaque darkly pigmented spheres.

3. Numerous molecular markers are available that allow the identification not only of mesoderm, but also of which region of the mesoderm has been induced. Suitable markers are listed on the *Xenopus* molecular marker resource: URL http://vize222.zo.utexas.edu/. Expression of these marker genes can be studied by RNase protection or by reverse-transcription PCR (*see* Chapter 39).

## 4. Notes

Mesoderm induction assays are usually very reliable. However:

1. If fertilization of eggs is unsuccessful, it is usually the fault of the eggs rather than the sperm. The sperm can be checked by observation under an inverted phase-contrast microscope. If they are healthy, they should wiggle.

2. Sometimes, animal caps just seem to fall apart into their constituent cells for no reason. There is nothing you can do about this—just repeat the experiment.

## References

1. Nieuwkoop, P. D. (1969) The formation of mesoderm in Urodelean amphibians. I. Induction by the endoderm. *Wilhelm Roux's Arch. EntwMech. Org.* **162,** 341–373.
2. Gurdon, J. B., Fairman, S., Mohun, T. J., and Brennan, S. (1985) Activation of muscle-specific actin genes in Xenopus development by an induction between animal and vegetal cells of a blastula. *Cell* **41,** 913–922.
3. Sive, H. L. (1993) The frog prince-ss: A molecular formula for dorsoventral patterning in Xenopus. *Genes Dev.* **7,** 1–12.
4. Slack, J. M. W. (1994) Inducing factors in *Xenopus* early embryos. *Curr. Biol.* **4,** 116–126.
5. Smith, J. C. (1995) Mesoderm-inducing factors and mesodermal patterning. *Curr. Opin. Cell Biol.* **7,** 856–861.
6. Jones, C. M., Kuehn, M. R., Hogan, B. L. M., Smith, J. C., and Wright, C. V. E. (1995) Nodal-related signals induce axial mesoderm and dorsalize mesoderm during gastrulation. *Development* **121,** 3651–3662.
7. Smith, W. C., McKendry, R., Ribisi, S. J., and Harland, R. M. (1995) A *nodal*-related gene defines a physical and functional domain within the Spemann organizer. *Cell* **82,** 37–46.
8. Gotoh, Y., Masuyama, N., Suzuki, A., Ueno, N., and Nishida, E. (1995) Involvement of the MAP kinase cascade in *Xenopus* mesoderm induction. *EMBO J.* **14,** 2491–2498.
9. LaBonne, C., Burke, B., and Whitman, M. (1995) Role of MAP kinase in mesoderm induction and axial patterning during *Xenopus* development. *Development* **121,** 1475–1486.
10. Umbhauer, M., Marshall, C. J., Mason, C. S., Old, R. W., and Smith, J. C. (1995) Mesoderm induction in *Xenopus* caused by activation of MAP kinase. *Nature* **376,** 58–62.

11. Baker, J. C. and Harland, R. M. (1996) A novel mesoderm inducer, Madr2, functions in the activin signal transduction pathway. *Genes Dev.* **10,** 1880–1889.
12. Durbec, P., Marcos-Gutierrez, C. V., Kilkenny, C., Grigoriou, M., Wartiowaara, K., Suvanto, P., Smith, D., Poner, B., Costantini, F., Saarma, M., Sariola, H., and Pachnis, V. (1996) GDNF signalling through the Ret receptor tyrosine kinase. *Nature* **381,** 789–793.
13. Smith, J. C. (1993) Purifying and assaying mesoderm-inducing factors from vertebrate embryos, in: *Cellular Interactions in Development–A Practical Approach* (Hartley, D., ed.), Oxford University Press, Oxford, UK, pp. 181–204.
14. Slack, J. M. W. (1984) Regional biosynthetic markers in the early amphibian embryo. *J. Embryol. Exp. Morph.* **80,** 289–319.
15. Krieg, P. A. and Melton, D. A. (1984) Functional messenger RNAs are produced by SP6 in vitro transcription of cloned cDNA. *Nucleic Acids Res.* **12,** 7057–7070.
16. Rupp, R. A. W., Snider, L., and Weintraub, H. (1994) *Xenopus* embryos regulate the nuclear localization of XMyoD. *Genes Dev.* **8,** 1311–1323.
17. Turner, D. L. and Weintraub, H. (1994) Expression of achaete-scute homolog 3 in *Xenopus* embryos converts ectodermal cells to a neural fate. *Genes Dev.* **8,** 1434–1447.
18. Symes, K. and Smith, J. C. (1987) Gastrulation movements provide an early marker of mesoderm induction in *Xenopus*. *Development* **101,** 339–349.
19. Green, J. B. A., Howes, G., Symes, K., Cooke, J., and Smith, J. C. (1990) The biological effects of XTC-MIF: quantitative comparison with *Xenopus* bFGF. *Development* **108,** 229–238.
20. Sokol, S. and Melton, D. A. (1991) Pre-existent pattern in *Xenopus* animal pole cells revealed by induction with activin. *Nature* **351,** 409–411.

# 26

# Experimental Embryological Methods for Analysis of Neural Induction in the Amphibian

**Ray Keller, Ann Poznanski, and Tamira Elul**

## 1. Introduction

Our objective is to describe and critique some of the experimental embryological preparations used to analyze tissue interactions involved in neural induction in amphibians. The molecular basis of neural induction and the tissue interactions that carry the inductive signals are areas of active research, stimulated by the recent identification of several potential neural inducers *(1–6)*, availability of regional molecular markers easily visualized with a good whole-mount RNA *in situ* hybridization method *(7)*, and the work on Hox genes that may have a role in specifying regional differentiation of the vertebrate nervous system *(8)*. These advances demand more of and make more useful the classical embryological manipulations used to characterize the tissue interactions involved in neural induction.

We will first describe the location and movements of the inducing and induced tissues, since misunderstanding of these aspects remains the major source of confusion in experimental design and interpretation in this area of research. Then we will describe several classical embryological methods that have been useful to us and to others in studying neural induction and development, pointing out the problems, difficulties, and liabilities of each of these methods.

### 1.1. The Origins and Movements of the Inducer and the Induced

A detailed description of the early embryological development of *Xenopus laevis*, with illustrations of anatomical features and key landmarks as seen under the stereomicroscope, has been presented elsewhere *(9)*. Here we describe those features particularly relevant to analysis of neural induction.

From: *Methods in Molecular Biology, Vol. 97: Molecular Embryology: Methods and Protocols*
Edited by: P. T. Sharpe and I. Mason © Humana Press Inc., Totowa, NJ

### 1.2. Fate Maps of Organizer and Neural Tissue in X. laevis

The fate map of the prospective neural tissue and of the Spemann Organizer, which is thought to be the major neural-inducing tissue, are shown as they appear in the very early gastrula stage of *X. laevis* (**Fig. 1**, stage 10–) *(10)*. The prospective nervous system forms a crescent on the dorsal side, lying mostly above the equator at this stage. It consists of prospective fore- and midbrain (F, M), hindbrain or rhombencephalon (RH), and spinal cord (SC). The prospective RH and SC are very wide in the mediolateral direction and very short in the anterior–posterior direction, extending about five to seven cells along the dorsal midline *(10)* (**Fig. 1**). These regions converge (narrow) mediolaterally, and extend (lengthen) anterior–posteriorly greatly during gastrulation and neurulation, described in detail below. This equatorial–subequatorial annulus of neural tissue undergoing convergent extension closes around the blastopore (**Fig. 1**, stage 12) and is often called the "noninvoluting marginal zone" (NIMZ) *(11)*.

The "involuting marginal zone" (IMZ) lies vegetal to the prospective neural tissue, and its dorsal sector (about 30°) contains the "Spemann Organizer" *(12)*. The Organizer contains much of the neural-inducing activity, based on the experiments of Spemann and Mangold in urodeles *(13)*, as well as others since on *Xenopus (14)*, in which this tissue was grafted to the ventral side of another embryo where it induced the formation of a second axis, including a nervous system. For simplicity, only the tissues of the Organizer lying in the dorsal midline are shown in **Fig. 1**. These tissues are prospective prechordal mesoderm (PM) or "head mesoderm," and prospective notochord (N), both in the deep, nonepithelial region (**Fig. 1**, stage 10–). The superficial epithelial layer of the Organizer consists of prospective endoderm (E) (**Fig. 1**, stage 10–), a fact that is often overlooked. This "suprablastoporal" endoderm lies above the site of blastopore formation and involutes to form the archenteron roof (**Fig. 1**, stages 10– to 17). It is distinct from the "subblastoporal" or vegetal prospective endoderm (VE), which lies below the site of blastopore formation. During gastrulation, the VE is covered over by the IMZ, including the archenteron roof, and becomes the archenteron floor (**Fig. 1**, stage 17). The vegetal end of the suprablastoporal epithelium consists of specialized endodermal cells, prospective bottle cells (BC), which play a role in initiating gastrulation (*see* **Subheading 1.3.1.**). At its lateral regions, the organizer includes some prospective anterior somitic mesoderm (not shown in **Fig. 1**).

### 1.3. Movements of the Inducing and Induced Tissues.

#### 1.3.1. Early Events

Gastrulation nominally begins as the cuboidal prospective BC constrict their apices and elongate in the apical-basal direction, acquiring their definitive

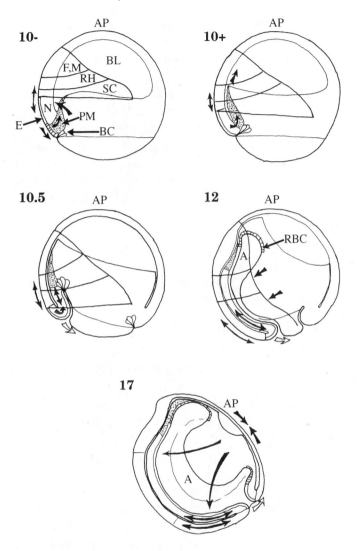

Fig. 1. These diagrams show important aspects of the fates and movements of the prospective neural tissue and the neural-inducing or "Organizer" tissue during gastrulation and neurulation. The embryos are viewed from the right sides such that dorsal is to the left, the AP is at the top, and vegetal pole is at the bottom. The Organizer occupies the dorsal sector of the involuting marginal zone, spreading laterally about 30° on either side of the midline, but for simplicity, only the prospective tissues at the dorsal midline are shown, including prospective prechordal (PM) and prospective notochord (N), covered on the outside by the prospective endoderm (E) of the archenteron roof. Both the midline and the lateral aspects of the prospective neural plate, including the prospective fore and midbrain (F, M), rhombencephalon (RH), and spinal cord (SC) are shown. Movements are indicated by arrows. Also indicated are the BC, the blastocoel (BL), the archenteron (A), and the respread bottle cells (RBC).

shape *(15,16)* (**Fig. 1**, stage 10–, 10+). This process begins about 30 min before stage 10, as described by Nieuwkoop and Faber *(17)*, and is progressive. The number of cells with constricted apices and the degree of constriction of individual cells increas, resulting in an increasing concentration of black pigment in the apices of these cells, transforming them from yellow to gray, and finally to black (*see* **Fig. 2** of **ref. *16***). As a result, IMZ rotates, the outside being pulled vegetally by the constricting BC, and the deep prospective PM, formerly associated with the deep ends of the prospective BC, is displaced upward *(16)* (**Fig. 1**, stages 10–, 10+). At this time, radial intercalation of cells occurs in the dorsal region of the gastrula, including both the prospective nervous system and the Organizer, and these regions become thinner and extend vegetally (*see* *18,19*) (*see* arrows in **Fig. 1**, stages 10–, 10+). This vegetal extension and the formation of BC act together to rotate the vegetal edge of the IMZ inward and upward, such that it comes to lie beneath and in contact with the overlying prospective neural ectoderm (**Fig. 1**, stage 10+). This process of early, precocious movement of the deep prospective mesoderm was described in detail by Nieuwkoop and Florshutz *(20)*.

### 1.3.2. A Stage of Rapid Transition

Because this early rapid movement of PM beneath the prospective neural ectoderm is a critical event in neural induction (*see* **Subheading 4.1.**), we distinguish among and characterize several forms of this stage. Stage 10– is characterized by a small area of prospective BC initiating constriction, their apices darkening slightly, to gray (**Fig. 2**, stage 10–, top). In most 10– embryos, the PM has neither rotated completely nor reached the overlying prospective neural ectoderm. The dorsal blastocoel wall is connected to the blastocoel floor by a smooth concave surface (**Fig. 2**, stage 10–, bottom). In the transition from stage 10– to stage 10, the BC apices constrict and darken further to form the blastoporal pigment line. Initially, this line is straight and extends about one-fifth to one-quarter the diameter of the embryo (stage 10 of Nieuwkoop and Faber, **ref. *17***) (**Fig. 2**, stage 10, top). The interior of the embryo at this stage is variable (**Fig. 2**, stage 10, bottom). It may be identical to that of the previous or the following stages, or intermediate between the two. As more BC are recruited, and those that have already formed constrict more, the field of BC widens and a shallow groove forms (**Fig. 2**, stage 10+, top; cf **Fig. 1**, stage 10+). Inside, the involuting cells have rotated upward, contacting the inner surface of the prospective neural tissue (arrows, **Fig. 2**, stage 10+, bottom). What was previously the smooth concave surface at the margin of the blastocoel has become a cleft along which the PM and the overlying neural tissue are in contact, a fissure called the "cleft of Brachet" (pointer, **Fig. 2**, stage 10+, bottom). Thus the involuted tissue may come into initial apposition with the

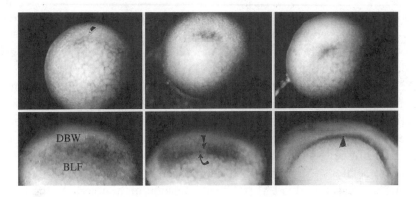

| Stage 10- | Stage 10 | Stage 10+ |

Fig. 2. Video frames taken through the stereomicroscope during dissections to make explants show the external view of the progressive formation of BC (top panels) and the involution of the leading edge of the Organizer mesoderm beneath the prospective neural tissue, on the inside (bottom panels). At stage 10–, only a few of the prospective BC have begun to undergo apical constriction, and these appear in a broad area of gray cells (pointer, stage 10–, top). Internally, the dorsal blastocoel wall (DBW) is continuous with the blastocoel floor (BLF) across a smooth curvature at the lateral edge of the blastocoel (stage 10-, bottom). By stage 10, more BC have formed, and the ones that have formed are darker, indicating greater contraction of their apices (stage 10, top panel). Internally, the curvature of the lateral margin of the blastocoel decreases in radius as the blastocoel floor rises and the blastocoel wall is pulled vegetally (arrows, stage 10, bottom). By stage 10+, more BC have formed, the constriction of those already formed is greater, and a shallow invagination has developed (stage 10+, top). Internally, the leading edge of the Organizer has rotated upward and outward, contacting the overlying neural tissue across the cleft of Brachet (pointer, stage 10+, bottom).

overlying neural tissue in stage 10 embryos, and this has almost always occurred by stage 10+. Thereafter, the cells spread on, and begin to migrate anteriorly on, the inner surface of the prospective neural tissue.

The prospective fate of this early involuting tissue is heterogeneous. A tongue of smaller, grayish or brownish PM cells extends animally from the bottle cells along the inner surface of the neural ectoderm (stippled, **Fig. 1**, stage 10+). By stage 10+ to 10.25, this tongue of PM cells reaches the leading edge of the involuted material in some embryos. In others, the tongue falls short of the leading edge. Central to these PM cells, and anterior to them as well (in embryos in which they do not reach the leading edge of the involuted tissue), larger, light-colored cells with coarse yolk platelets are found. These larger cells are indistinguishable from the rest of the central endodermal cells extending upward from the vegetal pole. Because cells in this region map to the pharyngeal region of the embryo, or anterior to the pharynx, they are thus referred to as pharyngeal endoderm *(20–22)*.

According to Nieuwkoop and Faber *(17)*, these definitive mesodermal mantle (PM) cells are "delimited" from the central endoderm at approx stage 10.25. This appears to be the case in our experience, although the timing of their appearance and the ease with which they can be distinguished from the remaining central endodermal cells varies between spawnings and sometimes between embryos in a single spawning. Nakatsuji *(23)* distinguished these populations of cells on the basis of the size distribution of their yolk platelets, as seen in sections. Markers, such as goosecoid *(24)*, noggin *(4)*, and Otx2 *(25)*, are expressed in this PM or "head" mesoderm, although on the basis of location, the "pharyngeal endoderm" may also express these markers (*see also* **ref. 25a**). Vodicka and Gerhart *(26)* correlated the early regions and movements of the organizer tissues with the expression of molecular markers, specifically, Xbra, noggin, goosecoid, and XNR3, in fate maps (*also see* the fate map in **ref. 27**). Whether the PM, the pharyngeal endoderm, or both, contribute to the inductive activity of the anterior organizer, and whether their contributions are the same or different from one another is a matter of speculation, since no assays of induction have clearly distinguished between them.

Since stage 10 involves rapid internal transitions, the external staging criteria by which it is defined is not a reliable predictor of internal events. Nieuwkoop and Faber *(17)* describe the dorsal gastrula wall of this stage as having one epithelial layer from two or three to five or six layers of deep cell. This range reflects the rapid radial intercalation (*see 19*) and vegetal extension (*see* **ref. 18**) in the dorsal gastrula wall from stage 10– through stage 10+, a process that appears not to be strictly correlated with the BC formation on which external staging is based. Thus, in cases where the presence or absence of this early contact with the neural tissue appears to be important, we directly determine the amount of involution that has occurred at a given external stage (*see* **Subheading 4.2.3.**), particularly during the transitional stage 10, which is less consistent internally, than stage 10– or 10+.

### 1.3.3. Convergent Extension: a Good Way to Make an Axis (and a Good Way to Fool the Investigator!)

The posterior regions of both the neural ectoderm and the dorsal mesoderm are originally very wide and very short, and they acquire their final form by extreme convergence (narrowing) and extension (lengthening) during gastrulation and neurulation *(10)*. One can be seriously misled in designing and interpreting neural induction experiments if these convergent extension movements and their contribution to embryonic development are misunderstood. The common impression of events following BC formation (**Fig. 1**, stage 10.5) is that the length of the dorsal axial structures (the nervous system, the notochord, and the archenteron) is generated as the involuted mesoderm crawls anteriorly a long distance, and finally comes to rest beneath the appropriate part of the prospective neural ectoderm. It is assumed that at that time, or perhaps before, neural induction occurs as signals are passed from the mesoderm to the neural ectoderm.

In fact, this is not what occurs. After BC formation, the mesoderm contacts the neural ectoderm at or slightly above the equator (**Fig. 1**, stage 10+), and it only migrates approximately to the animal pole (**Fig. 1**, stages 12–17). Thus the leading edge of the involuted mesoderm actually migrates only about a quarter of the circumference of the embryo. Although this migration is very important, it contributes little to the elongation of the dorsal aspect of the embryo. The length of the dorsal embryonic tissues is generated as the nervous system extends posteriorly, across the VE, and converges transversely (mediolaterally), squeezing the blastopore shut over the ventral aspect of the yolk plug (see arrows, **Fig. 1**; **Fig. 3A**). As this occurs, the prospective PM/endodermal tissues of the IMZ involute and converge, and extend on the inside, more or less in concert with the overlying posterior neural tissue (*see* **refs.** *28* or *29*). As the dorsal sector of the embryo elongates through neurulation, the ventral sector shortens. The prospective ventral epidermis diverges around both sides of the ventral midline and moves dorsally with the rise of the neural folds (*see* arrows, **Fig. 1**, stages 12–17) *(21)*. These shortening movements on the ventral side contribute to the overall dominance of the sagittal profile at the end of neurulation by the dorsal, extending tissues (**Fig. 1**, stage 17). The powerful and continuing role of convergent extension in elongating the posterior axis means that in the early fate maps, the prospective RH and SC, the two neural regions undergoing most of the convergent extension, appear very broad and very short (**Fig. 3B**). They elongate and narrow (converge and extend) greatly in the course of gastrulation and neurulation *(10)*.

In summary, we present a list of facts important for designing and interpreting experiments on neural induction:

> The potential inducing tissues of the Organizer involute and make contact with the inner surface of the potential responding tissues earlier than previously thought, usually during stage 10, and nearly always by 10+.
>
> Because the prospective neural tissues are very short in the animal–vegetal axis at the early gastrula stage, the first, early contact of the inducing tissues on the undersurface of the ectoderm is in the anterior neural region.
>
> Inducing and responding tissues shear relatively little, since the first contact of the two is relatively anterior, and because their posterior regions extend more or less together. Thus, corresponding anterior–posterior regions spend more time together than previously thought.

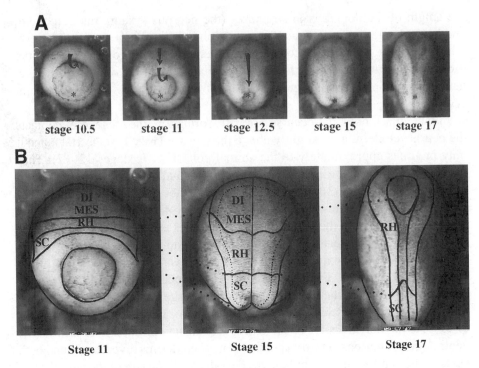

Fig. 3. Video frames from a time-lapse recording of the vegetal view of an embryo during gastrulation and neurulation show the extreme extension of the neural tissue across the yolk plug of the embryo. This embryo was held in clay with its AC fixed. Dorsal is at the top of the frames. In (**A**), the Spemann Organizer involutes (curved arrows, stage 10.5–11), whereas the prospective neural plate converges and extends across the yolk plug (straight arrows, stage 10.5–12.5), to close over the ventral part of the VE (asterisk). After blastopore closure, the neural tissue continues to extend posteriorly, pushing the blastopore posteriorly (stages 12–17). The involuted PM converges and extends coordinately with the overlying neural plate. In (**B**), the fate map of the stage 15 neural plate, redrawn from Eagleson and Harris *(91)* and projected onto the video image of the stage 15 neural plate, was mapped backward to stage 11 and forward to stage 17 by tracing individual cells at the junctions of map coordinates. Shown are the diencephalon (DI), mesencephalon (MES), RH, SC. This is the same embryo shown in panel (A).

Since the prospective nervous system is initially very short, most of it, particularly its posterior region, is very close to the inducing, Organizer tissue. Inducing signals passing through the plane of the tissue do not have to travel very far.

## 1.4. Regionalization of the Inducing Tissue

There is evidence that different regions of the mesoderm produce different inducing signals, which induce the regional differences that pattern the neural tissue (*see* **refs.** *30–37*). Accordingly, it is useful to know how to identify different regions of the mesoderm. The detailed correspondence of mesodermal and neural fates throughout their relative movements, in both anterior–posterior and mediolateral axes, is complex and best visualized with color diagrams (consult Fig. 4 in **ref.** *9*). However, the archenteron cavity and the BC at its edge are useful landmarks for identifying prospective mesodermal regions until they take on a recognizable character of their own. The leading edge of the mesodermal mantle, the first tissue to involute, consists of migratory mesoderm that spreads across the blastocoel roof in dorsal, lateral, and ventral sectors of the gastrula. It consists of prospective PM dorsally, heart mesoderm dorsolaterally, lateral plate mesoderm laterally, and ventral (blood island) mesoderm ventrally. In general, these early involuting tissues lie ahead of the BC, located at the periphery of the archenteric cavity. The late involuting tissues, the prospective notochord in the dorsal sector and the prospective somites in the lateral and ventral sectors lie behind the BC, immediately above the archenteron roof. For example, in the dorsal midline of the early gastrula, the PM lies ahead of the dorsal BC, found at the tip of the archenteron, and the notochordal and somitic mesoderm lie behind the bottle cells (**Fig. 1**, stages 10–, 10+). However, the dorsal BC respread in late gastrulation, when they expand the archenteron anteriorly, beneath the PM *(15,16)*, changing this relationship (**Fig. 1**, stages 11.5–17). At the late gastrula and neurula stages, the PM, notochordal, and somitic mesoderm lie dorsal to and approximately coincident with the roof of the archenteron, and the neural plate lies dorsal to and approximately coincident with this mesoderm. If one cuts through the body wall at the peripheral aspect of the archenteron at the late gastrula through neurula stages, the explant contains the entire neural plate, underlaid with PM, somitic, and notochordal mesoderm, and archenteron roof endoderm.

As one cuts and manipulates different regions of the mesoderm, one should be aware of the dramatic differences in their properties. The posterior, dorsal (notochordal and somitic) mesoderm differs in behavior from the anterior leading edge (PM, heart, lateral plate, and ventral mesoderm) in that the former undergo convergent extension, whereas the latter spread and migrate *(38)*. The cell motility in the posterior mesoderm consists of an intrinsic bipolar, mediolaterally directed protrusive activity *(39)*, whereas in the anterior mesoderm, it consists of monopolar, animally directed, substrate-dependent protrusive activity *(40,41)*. These two types of mesoderm also behave differently with respect to extracellular matrix *(40,41)*, and they express different molecular markers. Goosecoid *(24)*, Otx2 *(25)*, and noggin *(4)* are expressed in the leading edge, PM/pharyngeal endoderm, whereas brachyury is expressed in the postinvolution notochord *(42)*. Other useful markers for regions of mesoderm include the antibody tor-70 *(43)*, which marks notochord from stage 17 onward (*see* **ref. 44**), and 12–101, a monoclonal antibody (MAb) *(45)* that marks somitic mesoderm from stage 18 onward.

The regional character of the mesoderm is not fixed at the onset of gastrulation, but is patterned during gastrulation. The precise anterior to posterior progression of the cell behaviors driving mediolateral cell intercalation *(46)* is actually organized during gastrulation *(47)*. Likewise, regional neural-inducing properties appear in the early neurula *(36)*.

### 1.5. Regionalization of the Animal Cap (AC)

By the onset of gastrulation, if not before, the AC shows several dorsal-ventral differences relevant to neural induction. Early in the blastula stage, only the ventral half of the AC expresses the epidermal-specific antigen, Epi-1 *(48,49)*, possibly owing to early signals acting through the plane of the tissue *(50)*. After treatment with activin, the dorsal half of the AC is more likely than the ventral to form mesodermal tissue *(51)*. The dorsal half can be induced to converge and extend *(52)* and to express neural molecular markers *(53)* when placed in planar apposition to the Organizer, whereas the ventral side shows little or no response. Dorsal and ventral halves differ in expression of protein kinase C isozymes, which is thought to have a role in biasing the dorsal half to form neural tissue *(54)*. Bone morphogenetic protein (BMP) signaling between ventral cells (prospective epidermal) may prevent neuralization and ensure epidermalization, while blocking such signaling results in neuralization (*55*; *see also* **ref. 56**). Thus, the dorsal side of the AC is biased in the direction of neural development by the early gastrula stage. Later ventrally derived epidermal tissue takes an active role in patterning the dorsal aspects of the neural tube in amphibians *(57–59)* and in other vertebrates *(60,61)*. Using ACs from UV "ventralized" embryos, which lack an Organizer and a nervous system *(62)*, would presumably avoid this dorso-ventral bias of the AC.

## 2. Materials

### 2.1. Solutions

A variety of saline solutions are commonly used to culture amphibian embryonic tissues *(63)*. Our favorites are modified Barth's solution (MBS) *(64)*, a general-purpose saline, and modified Danilchik's solution (DFA), a saline specialized for supporting the normal motility and behavior of deep, nonepithelial cells *(65)*.

MBS (modified Barth's solution, from Gurdon, **ref. 64**):

| Stock | Per 1 L | Final concentration, m$M$ |
| --- | --- | --- |
| 4 $M$ NaCl | 22.0 mL | 88.0 |
| 50 m$M$ MgSO$_4$ | 16.4 mL | 0.82 |

| | | |
|---|---|---|
| 0.8 *M* NaHCO$_3$ | 3.0 mL | 2.40 |
| 0.1 *M* KCl | 10.0 mL | 0.01 |
| 33 m*M* Ca(NO$_3$)$_2$ | 10.0 mL | 0.03 |
| 1 *M* CaCl$_2$ | 0.41 mL | 0.41 |
| Hepes | 2.38 g/L | 5.00 |

Adjust pH to 7.4 with NaOH.

DFA ("Danilchik's for Amy" from Sater et al., **ref. 65**):

| Stock | In 500 mL | In 1 L | Final concentration |
|---|---|---|---|
| 4 *M* NaCl | 6.625 mL | 13.25 mL | 53.0 m*M* |
| 1 *M* Na$_2$CO$_3$ | 2.5 mL | 5.0 mL | 5.0 m*M* |
| K gluconate | 0.525 g | 1.05 g | 4.5 m*M* |
| Na gluconate | 3.49 g | 6.93 g | 32.0 m*M* |
| 1 *M* CaCl$_2$ | 0.5 mL | 1.0 mL | 1.0 m*M* |
| 1 *M* MgSO$_4$ | 0.5 mL | 1.0 mL | 1.0 m*M* |

Adjust pH to 8.3 with 1 *M* bicine.

The original Danilchik's (*66,67*) as well as subsequent versions (*see* **refs. 39,65**) were developed to mimic the ionic composition of the blastocoel fluid of *X. laevis* (*68*).

Embryos are kept in third-strength saline, and then transferred to full-strength for microsurgery. DFA is the appropriate solution for "open-faced" explants, which contain exposed deep cells. Although DFA allows more normal behavior of deep cells, it interferes with some functions of the epithelial cells, such as neural fold fusion (*see* **refs. 11,66,67**). MBS or an equivalent should be used for all other microsurgery. After healing, the embryos should be transferred to 3rd- or 10th-strength saline. We make two types of full-strength salines, one plain and one with 0.1% bovine serum albumin (BSA). BSA reduces adhesion of the cells to the surfaces of the dish and coverslips. Embryos should be transferred to plain saline before fixation, or the BSA may be fixed or precipitated on the surface of the specimens, making them appear dirty in a number of staining procedures. Solutions are filtered through a 0.22-μm Millipore filter, and aliquoted into 50-mL plastic centrifuge tubes, and frozen at –20°C until needed. We also freeze 1.0-mL aliquots of 100X antibiotic/antimycotic solution (10,000 U penicillin, 10 mg streptomycin, 25 μg amphotericin B, per mL, in 0.9% NaCl, Sigma Chemical, St. Louis, catalog #A9909) at –20°C. These are thawed and added at 0.5 mL/50 mL tube of solution at the time of use. Culture solutions are replenished every 12 h. Embryos are kept at

16–24°C. We try to avoid temperatures below or above this range, though two more degrees in either direction may be fine with good spawnings.

## 2.2. Tools and Dishes

The vitelline envelops are removed with Dumont #5 watchmakers forceps. For cutting embryos, we use eyebrow hairs embedded in wax in the tips of pulled, disposable "Pasteur" pipets. Hairloops, which are used to position and hold embryonic tissues, are also embedded in pipets. Pipets are pulled to a diameter several times that of a hair and scored with a diamond pencil; the vibration of scoring will usually fracture the pipet transversely. An eyebrow hair with a fine, evenly tapered point is placed in the pipet, butt first, such that its natural curvature brings the distal third to an angle of about 120° with respect to the pipet. In order to make a hairloop, both ends of a long hair are placed into the pipet until the smallest possible loop is formed. (It is important to use a long piece of hair, since the friction of the hair on the side of the pipet, a key factor in making a small loop, is increased with length.) Pipets are then dipped in wax, which has been melted in a spoon. The excess wax is removed by heating a steel spatula in an alcohol burner until warm, covering it with small piece of laboratory tissue, and transiently touching the eyebrow hair or hairloop to the tissue. Once melted, the excess wax will soak into the tissue without releasing the eyebrow hair or hairloop from the pipet. These instruments are sterilized by dipping them in 70% ethanol for a few minutes. We have found that these "low-tech" instruments outperform more pretentious instruments; they are sharp, flexible, and relatively nonadhesive to cells.

For microsurgery, a plastic disposable 60-mm Petri dish is used. Grafts between embryos are most easily made in a dish containing black modeling clay (Pastalina, Certified Nontoxic, Van Aken International, Rancho Cucamonga, CA 91730). The clay, available from many toy and art stores, is pressed into the bottom of the dish to a depth of 2–3 mm. To sterilize the clay, the dish is flooded for 20–30 s with 70% alcohol and rinsed in five changes of sterile saline.

To hold grafts or explants in place, small fragments of coverslips are cut ahead of time in rectangular shapes, about 2–4 mm by 10–15 mm, with a diamond pencil, and kept in plastic Petri dishes. These coverslip fragments are handled with watchmaker's forceps (*see* **Subheading 3.6.**).

## 2.3. Microscopes

Proper setup of a stereo-dissecting-type microscope is important for good microsurgery. In addition to to good posture, eyepiece height, and working height of the bench, it is very important to adjust the eyepieces according to the manufactuer's instructions, such that both eyes focus at the same level and have stereoptic vision. We work on a large, flat, cooled surface, rather than on a raised stage. We buy the microscope with a boom stand or without a stand, and mount the microscope to a large cooling plate, of dimensions 40 × 80 cm. The cooling plate is made by milling a continuous, 2-cm-wide channel in a sheet of Plexiglas of this size, about a centimeter thick, beginning at the back right hand corner, coming forward, turning towards the back again, and so on, in a continuous, sinuous pattern, staying about 3 cm from the edge, and ending at the back left. This channeled plate is glued to a second Plexiglas plate. Finally the pair of plates are glued and screwed, with machine screws, to the aluminum plate. The screw heads are recessed into the bottom Plexiglas plate, forming a smooth surface, and are screwed into tapped holes on the underside of an aluminum plate. The aluminum plate is then drilled and tapped for a water inlet and outlet. These are located over the ends of the channels in the Plexiglas at the back right and left of the plate in order to maximize water flow and to be out of the way. The inlet is made smaller (3/8 in.) than the outlet (1/2 in.) to reduce buildup of internal pressure. Water is circulated through the plate from a large water bath/chiller, and adjusted to 16–18°C. A boom stereoscope is swung out over the plate, or a pillar matching the diameter of that required for the stereoscope is mounted on the aluminum plate, using a 3/8 in. machine screw-tapped into the base of the pillar and into the top of the plate. Illumination is by fiber optic illuminator, which is cool and does not heat the embryos.

## 3. Methods

All the operations described below, and others as well, can be done with a combination of a few basic manipulations.

### 3.1. Cutting Tissues

Large pieces that are to be cut along straight lines from an intact embryo are cut out by pressing an eyebrow hair directly through the embryo, along the line of the cut until the substratum is reached. Explants are trimmed to size with the same type of movement (**Fig. 4A**). It is best to use a long, stiff eyebrow hair for this type of operation. In order to remove only a superficial part of the embryo without damaging deeper tissues, a type of "stitching motion" is used. The eyebrow hair is inserted into the wall of the embryo at an angle of 45° with the surface until the tip is at the desired depth; the hairloop is placed next to the eyebrow hair, along the line of cut on the surface of the embryo, and the tissue is cut by quickly raising the tip of the eyebrow hair (**Fig. 4B**). The operation is repeated very quickly, advancing the eyebrow hair and the retaining hairloop only a small distance at each cycle, never cutting through more than a few cells at a time, to avoid undue strain in the tissue (**Fig. 4B**). By varying the depth of insertion of the eyebrow hair, a cut of any depth can be made, ranging from the epithelial layer alone (about 10–15 µ thick) to the epi-

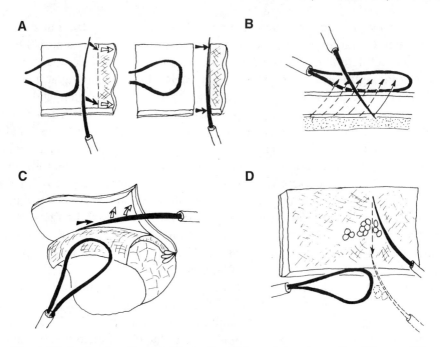

Fig. 4. Methods of cutting (**A,B**), peeling (**C**), and shearing cells off layers (**D**) are shown with diagrams. In order to trim an explant usin a long, straight type of cut (A), the tissue is held lightly with a hairloop, while a long, stiff eyebrow hair knife is brought down on the explant along the desired line of the cut (dashed line, A, left), and pressed through the tissue until it hits the substratum. The eyebrow hair is then quickly moved to the right, cleanly separating the two pieces of tissue (A, right). A method of cutting through a limited and controlled depth of tissue is shown in a face-on view of the cutting surface, as if the tissue on the viewer's side of the cut had been removed (B). The hairloop is used to hold the tissue along the line of the cut, and the eyebrow hair tip is inserted to the desired depth and lifted repeatedly and quickly (arrows, B), cutting only a few cells at a time. A method of peeling tissue from underlying tissue is shown (C). In this diagram, the postinvolution PM (below restraining hairloop) and the neural ectoderm (above the eyebrow hair) of an early gastrula are being separated. The tip of the eyebrow hair is moved along the interface between the tissues (solid arrow), whereas the heel of the curved hair is rocked against the prospective neural tissue (open arrows). Adherent mesodermal cells are sheared off from the inner surface of an explant of the neural tissue (D). The hairloop is placed at the bottom of the tissue to prevent the tissue from moving toward the operator. The eyebrow hair is then swept at low angles straight toward the operator, just above the surface of the explant, shearing off the adherent cells. For clarity, the size of the eyebrow hair and hairloop is reduced relative to the size of the tissues.

thelial and deep region together (about 25–100 μm, depending on the stage and region). A small, sharp-tipped eyebrow hair is best for this type of stitching cut. An alternative method for cutting tissue at a limited depth is to insert a long, stiff eyebrow hair along a line of tissue that is to be cut, and then rub the hairloop back and forth against it, on the outside, cutting the intervening tissue. This method is good for making straight cuts, but is not conducive to making curved cuts, and it does not allow precise control of depth below the curved surface of the embryo.

## 3.2. Peeling Layers Apart

One will often want to peel layers of tissue apart. Examples include removing the epithelium from the underlying deep cells (*69*) or removing the outer wall of the gastrula from the underlying involuted cells (*see* **Subheadings 4.2.** and **4.2.2.**). The "peel" is begun by probing with the tip of an eyebrow hair for the interface between the tissues to be separated; the interface reveals itself as a line of easier separation. Once the interface is discovered, the background tissue is held with the side of the hairloop. The tissue to be peeled off is pulled away by running the tip of the eyebrow hair along the interface, and angling the butt or heel of the eyebrow hair against the surface of the tissue (**Fig. 4C**). Initiation of a peel at the proper interface is difficult if the cut at the edge of the peel has gone too deep. In this case, the area of the underlying tissue beneath the peel will be more likely to come along with the peeled tissue.

## 3.3. Shearing Cells Off Layers

In order to remove remaining unwanted cells on layers that were peeled apart, one can perform a shearing operation. For example, when making sandwich explants from older gastrula-stage embryos, mesodermal cells will often adhere to the inner surface of the blastocoel wall. The following method can be used to remove them. The explant is positioned with its deep surface uppermost. The eyebrow hair is turned such that a considerable length of its tip lies parallel to the surface and just above the explant, to the right of the area of contamination (**Fig. 4C**). The tip of the eyebrow hair should be inclined at an angle of about 20–30° off a line running vertically from the bottom to the top of the field of view. The hairloop is then placed against the near edge of the explant, just to the left of the eyebrow hair; this will prevent the explant from moving toward the experimenter. The eyebrow hair is repeatedly and rapidly brought down until it just touches the surface of the explant, and back towards the experimenter, always remaining in its original plane. The restraining hairloop is moved slowly to the left and followed with the eyebrow hair, shearing off any recalcitrant, adherent cells. The eyebrow hair should be oriented at a low angle of attack, since a low attack angle does not exert large forces on the explant (**Fig. 4C**). The eyebrow hair and hairloop can be switched right to left, and the process begun at the other edge of the explant, if one prefers.

## 3.4. Open-Faced Explants

"Open-faced" explants are ones having the inner, deep cells exposed to the medium, often for the purpose of studying deep cell motility *(39,46,66,67,70)*, and thus Danilchik's (DFA) medium is used. A culture chamber for these explants is made by drilling a 20-mm hole in a 60-mm plastic Petri dish and gluing a #1.5, 24-mm coverslip over the hole with silicone high vacuum grease (Dow Corning, Midland, MI). The tissue is placed with the inner or deep surface down on the coverslip, and restrained with another coverslip, supported at each end with silicone high vacuum grease. This allows high-resolution optics to be used with an inverted microscope. An upright microscope can be used by overfilling the dish with media, covering the top with a large glass slide, and then inverting it, such that the coverslip and the deep surface of the explant are at the top, facing the objective.

## 3.5. Sandwich Explants

Sandwiches consist of abutting the deep, inner surfaces of two identical tissues together, such that the epithelial sheets covering each half heal across the exposed edges, forming a physiological barrier surrounding the deep cells. The two components are excised and their inner surfaces apposed, the quicker the better, since delay will result in curling of the two halves, which makes apposition difficult. Dabs of silicone high vacuum grease are placed on both ends of a precut rectangular coverslip, considerably larger than the explant, and the coverslip is rested on the grease, straddling the explant and some distance above it. Then, the coverslip is tapped with a forceps, bringing it down on top of the sandwich, pressing the two components together lightly. Healing should occur in 15–20 min, and then the explant should be removed. The high vacuum grease does not appear to be overtly toxic, but the less time the explant spends near it, the better.

## 3.6. Grafting Tissues

The easiest method of grafting from embryo to embryo is to use a clay-bottomed dish. A hole the size of the embryo is made in black modeling clay with a ball tip, formed at the end of a disposable pipet by holding it in a flame until the correct size molten ball is formed. The host embryo is placed in the hole, graft site uppermost, if possible. If the embryo rotates, small nibs of clay can be pushed against it with forceps, holding it in position. The donor embryo is placed alongside the host, also in a depression (necessary only if one wants to keep the same levels of both embryos in focus at higher magnifications). A graft is cut out of the donor and a graft site cut in the host; which operation is done first depends on which is the most troublesome to do, and which preparation gives the most trouble when left alone for a time. Speed is crucial. If left alone too long, the graft will curl, and the hole made to accept it will first gape and then heal. The boundaries of graft and graft site should be closely matched, since healing will occur fastest under these conditions. If the graft is slightly larger than the site, it is important to tuck the edges of the graft into the graft site. If the edges of the graft overlap the host epithelium, healing will be delayed. The graft epithelium will turn back on itself and ultimately find the edge of the host epithelium, but this will take a while. In order to hold the graft in place, forceps are used to push up two long ridges of clay on each side of the embryo, some distance (2–4 mm) away. A small coverslip fragment, preferably a precut rectangle, is bridged across the two ridges, above the embryo. The coverslip is pushed down on the graft with forceps tips, taking care to align the surface of the glass exactly perpendicular to the graft surface, pushing it directly into the graft site. If the embryo or graft moves one way or the other as pressure is applied, the glass bridge can be tilted to counter these movements. Healing should occur in 15–20 min at which time the coverslip is removed and embryo is removed from the clay. Most batches of clay appear to be somewhat deleterious to the embryo over long periods, although embryos from many spawnings will develop normally to advanced stages on clay. We prefer black clay for contrast, but other colors can be used according to the experimenter's taste. Some colors appear to affect the embryos more than others.

The method described above is the easiest, but not necessarily the best way to graft tissues. With practice, one can make grafts between free, unconstrained embryos lying in a dish. This is the least disruptive and yields the best results, if done correctly. It is important that the edges of grafts are mated precisely. The grafts should be relatively small, and they must stick in the graft site without external pressure, all of which require much practice.

## 3.7. Tricks

Many tricks can be used to facilitate grafting and explanting embryonic tissues. Although these tricks provide advantages in some situations, they usually come with liabilities. High pH and or low calcium makes tissues easier to separate or cut, but also retards healing. Hypertonicity and hypotonicity may facilitate separation of layers of cells, by shrinking or swelling the cells, but may retard healing, and kill or damage cells. In cutting through the epithelial surface of the embryo, either the epithelial or the deep cell populations will be at a disadvantage, depending on what solution is used. The deep cells and basolateral surfaces of the epithelial cells require high salt, whereas the outer surface of the epithelium is normally exposed to low salt. External solutions of high pH and/or high salt cause lesions in the epithelial layer and increased cell motility at these lesions, particularly in the case of high pH. Conversely, low pH and low salt cause sluggish or abnormal motility or swelling of deep cells, respectively. These facts, discovered by Holtfreter (*see* **refs. 71–73**), make any grafting or explantation operation suboptimal for the health and function of one cell population or the other.

### 3.8. Handling Albinos

It is advantagous to use albino embryos for experiments involving whole mount, RNA *in situ* hybridization, as well as for some antibody staining of tissues. However, because these embryos have low contrast, they are difficult to stage and orient. Recognition of BC and other pigment-dependent landmarks is very difficult. Albino embryos can be staged and oriented more easily if are stained a pale, "baby" blue, achieved by soaking the embryos for a few minutes in third-strength saline containing a few drops of 1.0% Nile blue. With this light background stain, the constricting BC apices will appear darker, much as they would in normal embryos. Staging and orientation during gastrulation can be done much like that of normal embryos, and at later stages, the staining will enhance the neural plate and the neural folds.

### 3.9. Tipping and Marking

"Tipping and marking" embryos *(74)* facilitate identification of the dorsal, lateral, or ventral sides of the embryo at early cleavage and late blastula stages, which is essential for regional injection of dyes, RNAs, or plasmids. This procedure makes use of the fact that movement of the cortex of the egg relative to the deep cytoplasm during the first cell cycle determines where the dorsal side will form *(75)*. If the equator is rotated uppermost at any meridian early in the first cell cycle and left there, the dorsal side will form at that meridian, overriding the influence of the sperm entry site on specifying the dorsal side. The embryos are first placed on a Nitex grid (about 1 mm mesh size) in 6% ficol in saline. After fetilization, but before first cleavage, the equator is tilted uppermost until after first cleavage. The site that is uppermost is marked with a wand bearing vital dye. The wand is made by pulling a disposable pipet to a small diameter and then melting a small glass ball at the end. Dye is precipitated by placing a bit of Nile blue sulfate (1% in water) and 100 m$M$ sodium carbonate on a microscope slide, and mixing the two with the wand. A precipitate of dye will form, which is picked up on the tip of the wand and placed against the embryo for a couple of seconds. Care should be taken, since the dye will appear darker later, and it is easy to overstain. The dorsal side will develop at the stained site.

## 4. Issues and Specific Experimental Preparations

We will describe in detail several methods of analyzing some current and difficult problems in neural induction, including the relative roles of planar and vertical signaling in neural induction and patterning. However, many of

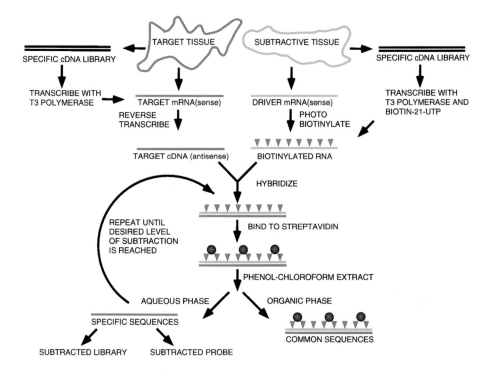

**Plate 1, Fig. 1** (*see* full caption on pg. 560, and discussion in Chapter 37). Schematic view of the steps in constructing a subtracted cDNA library or subtracted probe.

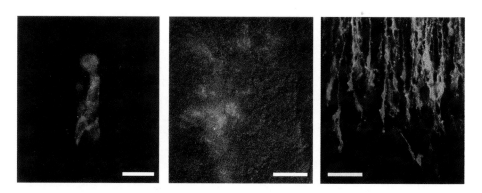

**Plate 2, Fig. 1** (*see* discussion in Chapter 23). **(A)** Cells labeled with DiI and DiA in the chick embryo neural plate and viewed immediately after iontophoretic applications. About four cells are labeled with DiI (red) and one cell with DiA (green). Bar is 10 μm. **(B)** After 48 h of development, the decendants of cells labeled as in (A) are still visible and their two colors distinct. Bar is 50 μm. **(C)** Motor neurons retrogradely labeled from their peripheral nerves in fixed tissue. One nerve was labeled with DiI, and the adjacent nerve with DiA. Bar is 40 μm.

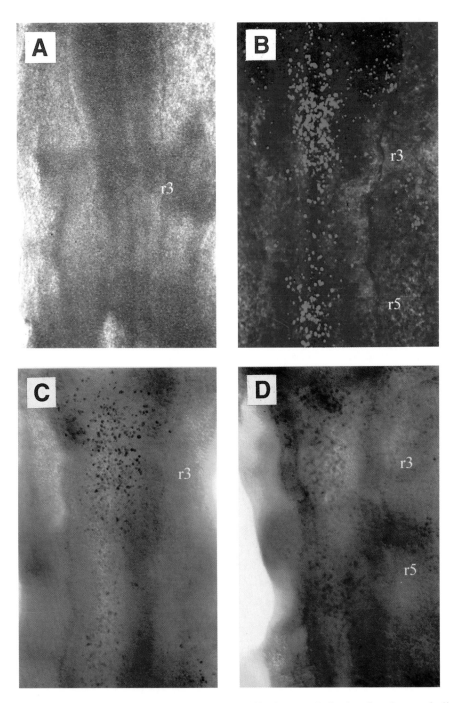

**Plate 3, Fig. 1** (*see* discussion in Chapter 48). Apoptosis in the rhombencephalic neural crest of the chick embryo as revealed by acridine orange staining **(A,B)** and TUNEL **(C,D)**. The embryos shown in (A) and (C) are at stage 10, but those in (B) and (D) are older, stage 11.

**1) Colour temperature**

Blue filter

No filter

**2) Pseudocolour**

colour look-up tables

Output Intensity

Input Intensity

Raw image

Sharpened

Pseudocolour

combining different fluorescence wavelengths

488 nm

568 nm

Combined

combining different optical sections

Ventricular

Pial

Combined

**Plate 4, Fig. 4** (*see* full caption on pp. 726 and 727, and discussion in Chapter 51). **(1)** Color temperature: effects of a blue filter on a specimen under tungsten illumination. **(2)** Pseudocolor: three examples of the use of pseudocolor with black-and-white digital images. (Middle) Black-and-white images of the same specimen illuminated for different fluorescent dyes are color-coded and merged to produce an image equivalent to a photographic double exposure. This image is then merged with a DIC picture of the same specimen. (Bottom) Confocal microscope optical sections are color-coded and merged to show differences in cell dispersal at different depths in the tissue.

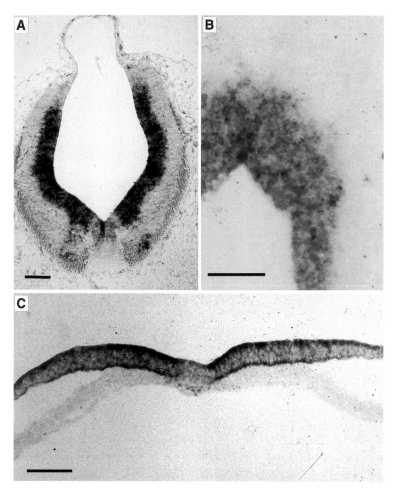

**Plate 5, Fig. 1** (*see* full caption on p. 648 and discussion in Chapter 45).

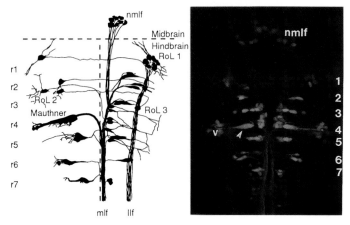

**Plate 6, Fig. 1** (*see* full caption on p. 434 and discussion in Chapter 29).

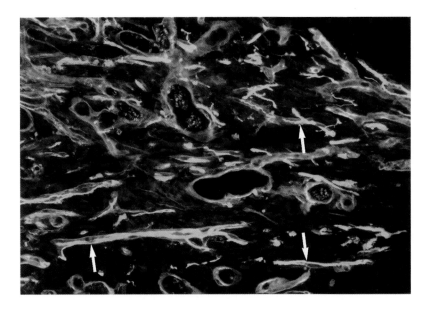

**Plate 7, Fig. 1** (*see* full caption on p. 656 and discussion in Chapter 46). Regrowing axons from a transected peripheral nerve. Migrating Schwann cells are labeled with antilaminin and detected with antirabbit IgG-FITC. Axons are labeled with anti-(-tubulin isoform III and detected using biotinylated antimouse IgG and Extravidin-TRITC.

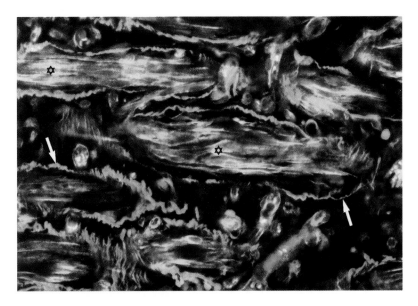

**Plate 8, Fig. 2** (*see* full caption on p. 657 and discussion in Chapter 46). Schwann cells of a transected peripheral nerve. The sarcolemmal basement membrane is labeled with antilaminin and detected with antirabbit IgG TRITC.

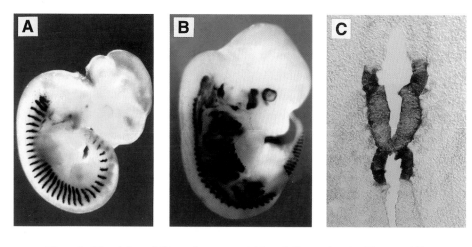

**Plate 9, Fig. 1** (*see* full caption on p. 165 and discussion in Chapter 10).

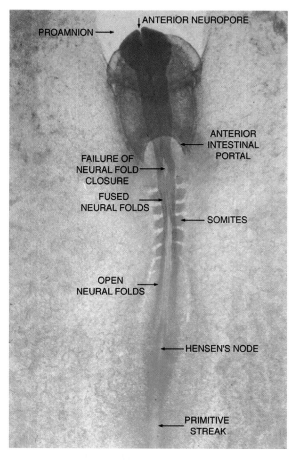

PROAMNION

ANTERIOR NEUROPORE

ANTERIOR INTESTINAL PORTAL

FAILURE OF NEURAL FOLD CLOSURE

FUSED NEURAL FOLDS

SOMITES

OPEN NEURAL FOLDS

HENSEN'S NODE

PRIMITIVE STREAK

**Plate 10, Fig. 9** (*see* full caption on p. 235 and discussion in Chapter 15).

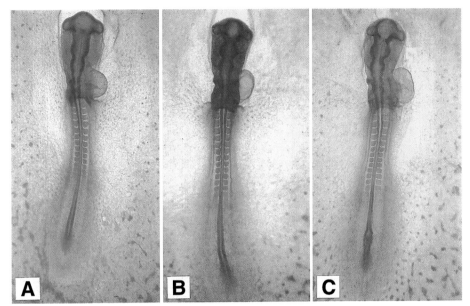

**13 pairs of somites**        **14 pairs of somites**        **15 pairs of somites**

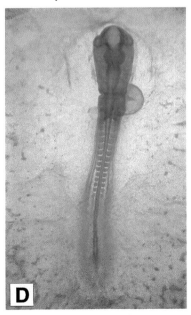

**17 pairs of somites**

**Plate 11, Fig. 13** (*see* full caption on p. 236 and discussion in Chapter 15). **(A)** Stage 11 (HH) (40–46 h), 13 pairs of somites. **(B)** Stage 11 + (HH), 14 pairs of somites. **(C)** Stage 12– (HH), 15 pairs of somites. **(D)** Stage 12 (HH) (48–54 h), 17 pairs of somites.

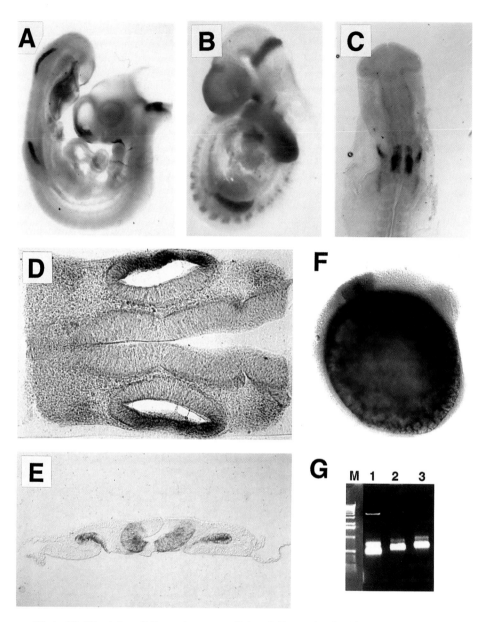

**Plate 12, Fig. 1** (*see* full caption on p. 624 and discussion in Chapter 42). **(A)** *In situ* hybridization of Fgf-8 to a stage 20 chicken embryo using a mouse cDNA sequence as probe. **(B)** The same probe as in (A) hybridized to a 9.5 d-mouse embryo. **(C)** *In situ* hybridization of a chicken Fgf-3 probe. **(D)** Vibroslice section of a chicken embryo. **(E)** Thin section of the specimen in (C). **(F)** "Two-color" *in situ* hybridizations showing expression in adjacent sections of zebrafish hindbrain. **(G)** A typical gel showing DIG-labeled riboprobes before (lane 1) and after (lanes 2 and 3) DNase treatment.

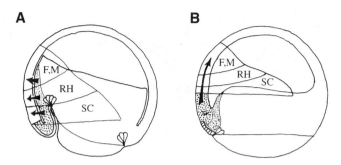

Fig. 5. Diagrams illustrate the concepts of vertical induction (**A**) and planar induction (**B**). In vertical induction, the organizer mesoderm (stippled) first involutes beneath the prospective neural tissue, and then sends signals radially (outward or vertically) to the overlying neural tissue (arrows, A). In planar induction, the Organizer mesoderm sends signals animally, from its posterior edge, into the prospective neural tissue (arrow, B). One difference is that vertical signaling requires involution of the Organizer and thus must occur at a relatively older stage (stage 10.5 shown in A), whereas planar signaling can occur early (stage 10– shown in B). The prospective fore- and midbrain (F, M), rhombencephalon (RH), and spinal cord (SC) are shown. The components of the organizer in the sagittal plane, the prospective prechordal mesoderm (PM) and notochordal mesoderm, are indicated with fine and course stippling, respectively.

the issues, problems, and solutions described below are generic ones, applicable to other experiments as well.

### 4.1. Planar and Vertical Signaling

The traditional view of neural induction is that the mesoderm comes to lie beneath the prospective dorsal neural ectoderm and sends signals vertically, inducing the ectoderm to realize its neural fate (**Fig. 5A**). In the past 10 years, there has been a revival of interest in edgewise or "planar" induction. Planar induction is thought to involve a signal passing from the posterior edge of the Organizer, through the plane of the tissue, into the prospective neural region (*see* **refs. 25,50,52,53,65,76–78**) (**Fig. 4B**). This idea is not new, originally favored by Spemann, but discarded because of Holtfreter's work *(79)*, demonstrating that exogastrulae, which supposedly have planar, but not vertical apposition of the inducing and responding tissues, underwent no obvious neural development. With the revival of interest in planar induction, the traditional route of neural induction is now called "vertical" induction.

Several experimental preparations have been used to study the relative roles of planar and vertical signaling. We will describe these preparations and examine their usefulness and limitations, including difficulties in making them, problems of interpretation, and unknown factors that might make them misleading.

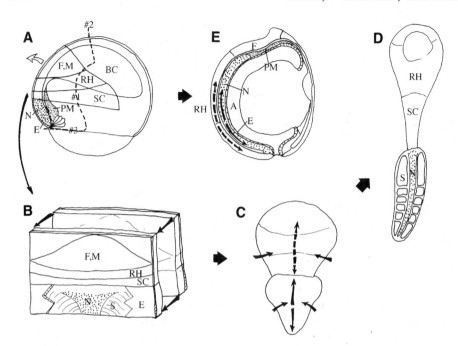

Fig. 6. The method of making a sandwich ("Keller") explant of the dorsal sector of the gastrula is shown (**A,B**). The dorsal sector of the early gastrula (A) is excised by cutting 45–60° on both sides of the dorsal midline, and across the AC (heavy dashed lines, #1 and #2, respectively). This flap of tissue is then peeled outward, away from any involuted material (open arrow), and cut off where it joins the VE, above or below the BC (#3). Two such explants are trimmed at the edges to match one another, and sandwiched, with their inner, deep surfaces together (B) by placing them between a coverslip fragment and the bottom of the dish. The prospective areas of such an explant are shown, including prospective fore- and midbrain (F, M), hindbrain (RH), SC, endoderm (E), notochord (N), and somite (S). Prospective PM may or may not be included, depending on whether the explant was made early or late in stage 10. After 15 min or so, of healing, the explant is taken from under the coverslip and allowed to develop. Such an explant undergoes convergent extension (**C,D**) of both the involuting marginal zone (solid arrows) and the prospective posterior neural tissue (broken arrows). The involuting marginal zone differentiates into notochord and somitic tissue, whereas the neural region differentiates into regions that express early markers of the SC, RH, and F and M *(12,25,78)* (D). If the explant is large enough in the animal direction, cement gland is also formed. In the intact embryo (E), the neural and mesodermal tissues converge and extend in parallel (arrows, E), the later beneath the former, rather than serially, as in the explant (C).

## 4.2. The Sandwich Explant of the Dorsal Marginal Zone and Prospective Neural Tissue

This explant is made by sandwiching the inner deep surfaces of two dorsal sectors, taken from two individual gastrulae, together (**Fig. 6**). Both the involuting (PM/endodermal) and noninvoluting (prospective neural) regions of this explant converge and extend, whereas the animal-most ectodermal region, corresponding to the prospective mid- and forebrain, remains bulbous *(52,66,67)* (**Fig. 6**). Since Organizer tissues of this explant are in edgewise or planar apposition to the ectoderm, it is assumed the planar signals must induce the ectoderm to acquire a neural fate. However, some thought on the requirements for demonstrating planar induction suggests that the interpretation of the behavior of this sandwich explant is more complex than might first appear.

### 4.2.1. Is the Assayed Behavior Already Autonomous When the Test for Induction Is Done?

The neural convergent extension or marker expression in sandwich explants of early gastrulae is not necessarily induced by planar signals from the Organizer. They may have been induced previously or patterned by cytoplasmic localizations, and may be autonomous behaviors by gastrulation. Testing for autonomous expression at the early gastrula is not a trivial problem, particularly in the case of the SC and hindbrain, since these regions are short and difficult to manipulate *(52)*.

### 4.2.2. Do Contaminating Cells Provide Vertical Signaling?

The second problem is that mesodermal cells may contaminate the interior of the noninvoluting, prospective neural tissue, providing vertical signals in a preparation supposedly devoid of vertical signals. In sandwiches made at stage 10 to 10+, the leading edge of the mesodermal mantle likely has begun involution, as described in **Subheading 1.3.2.** and **Fig. 2**. If these mesoderm cells are not removed from the inner surface of the explant, they have access to the fibronectin-rich inner surface of the blastocoel roof. Under these conditions, these cells are highly invasive *(41)* and will invade the core of sandwich and provide vertical signals (**Fig. 7A**). If sandwich explants are made early, at stage 10–, PM will be found at the vegetal end of the explant. These PM cells can also migrate animally beneath the neural tissue, both as individuals and in streams, and even lead the more posterior cells in an invasion of the core of the explant (**Fig. 7B**) (*see also* Fig. 10 in **ref. 9**). As few as four or five notochordal cells can change the expression of neural genes in the overlying ectoderm (*see* **ref. 80**).

There are several methods one can use to solve the problem of invasion. The best method, of course, is to remove all the potential invading cells from the inner surface of the vegetal end of the explant, as well as adherent cells higher in the marginal zone. They are easy to identify (**Fig. 7A**; *see also* Fig. 10 in **ref. 9**). One could also monitor such invasions with markers expressed by the potentially invading tissue. However, one must know what markers this tissue should express, and the test for marker expression must be sufficiently sensitive at the appropriate stage.

**A**

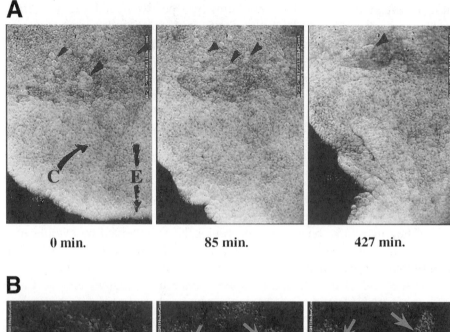

0 min.                    85 min.                    427 min.

**B**

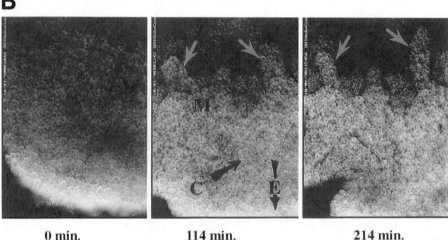

0 min.                    114 min.                   214 min.

Fig. 7. Time-lapse video microscopy of the inner surface of an open-faced explant shows the tendency of mesodermal cells to migrate animally on the inner surface of the prospective neural tissue. As the bulk of the mesoderm converges and extends vegetally (arrows, **A**), a number of individual mesodermal cells break off from the posterior edge of the mesoderm and migrate on the inner surface of the prospective neural ectoderm (pointers, A). In some cases, large tongues of mesodermal cells may migrate animally (white arrows, **B**) while the rest of the mesoderm converges (C) and extends (E) vegetally (black arrows, B). Photos courtesy of John Shih.

A better way to document the absence of such invasion and to demonstrate convincingly that only planar induction has occurred is to appose an Organizer from a fluorescently labeled embryo next to an unlabeled sandwich of ectoderm that, in isolation, does not show neural development. Any labeled cells from the inducing tissue that invade, regardless of phenotype, will be noticed (*see* Fig. 7C in **ref. 52**).

Finally, a positive control for the effect of contamination can be done. Contamination is allowed, both a little of it and a lot of it, to determine if it makes a difference in neural convergent extension or marker expression, relative to cases thought to have no contamination. For example, neither a little nor a lot of contamination increased the amount of convergence and extension of the neural region, and thus we concluded that vertical signals had no detectable effect on convergent extension induced by planar signals alone *(52)*.

### 4.2.3. Has Vertical Signaling Already Occurred? Use of Skewered Sandwiches

One must also be certain that vertical signals have not occurred prior to testing the effect of planar signals alone. As pointed out above, the prospective nervous system is very wide and very short, and *Xenopus* shows early, cryptic movement of the leading edge mesoderm inside. This means that the distance the leading edge of the organizer must move to come beneath a good part of the prospective nervous system, and the time it takes to do so are very short. If staging is not absolutely precise, or too late a stage is used, some opportunity for vertical signaling will already have occurred prior to experimental manipulations used to study solely planar signals (*see* **refs. 52,80**).

We use the following method to determine how far beneath the prospective neural ectoderm the Organizer has reached at the time of making explants. Explants are made at stages 10- through 10+. Also, as the dorsal sector is dissected, the highest point of contact of any involuted mesoderm/endoderm is marked by piercing the prospective neural ectoderm with an eyebrow hair (**Fig. 8A**, A'). Two similarly marked explants are selected and sandwiched together. A small glass bead (glass beads, acid-washed, 106 μ and finer, US Sive 140, cat. no. G4649, Sigma Chemical Co.) is inserted into the hole made by the eyebrow hair (**Fig. 8B**, B'), marking the upper limit of mesodermal/endodermal contact. The two halves are then held together beneath a glass coverslip for healing. After healing, the coverslip and glass bead are removed and the explant is skewered to a bed of 2% agarose, with a hair pushed through the hole left by the bead. The hair permanently marks the site of mesodermal/endodermal contact (**Fig. 8C**, C'). Video microscopy shows convergent extension is not affected by the presence of the bead and hair. Convergent extension occurs, moving the explant with respect to the anchor point provided by the skewer. Using this method, we have found that expression of the homeobox gene *Hoxb-1* is affected by the early vertical contact of PM with neural tissue, whereas

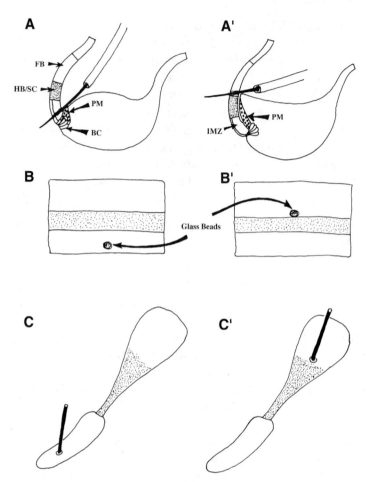

Fig. 8. The skewer method of marking the anterior limit of contact of the involuting mesoderm with the overlying neural ectoderm is shown in diagrams in cases of no contact (**A–C**) and substantial, anterior contact (A'-C'). When the AC is cut open, in preparation for making the explant, an eyebrow hair knife is used to puncture the outer gastrula wall at the point of inflection of the VE to gastrula wall at early stages (A), or at the top edge of contact of the involuted material with the gastrula wall (A'). Two explants with holes in matched or nearly matched positions are then sandwiched and the holes marked with small glass beads (**B**, B') before pressing them together under a coverslip. After the coverslip is removed, the beads are removed and the explants skewered to the bottom of a dish covered with 2% agarose. As the explants converge and extend, the position of the original contact is marked by the eyebrow hairs (C, C').

convergent extension is not *(80)*. Expression of other neural properties and genes may or may not be affected by these early, transient signals.

### 4.3. The Edgewise (Planar) Apposition of Sandwiched Tissues

The best way to demonstrate planar induction is to place the Organizer in edgewise apposition to a responding tissue that normally does not undergo neural development. We have done such an experiment to demonstrate the planar induction of convergent extension *(52)* by grafting a labeled piece of Organizer to the edge of a piece of dorsal ectoderm, taken from far above the blastoporal lip, so that it did not include the original converging and extending neural tissue (**Fig. 9A**). In another experiment, a second Organizer was abutted to the animal end of the dorsal sector of the gastrula, inducing a second extension of polarity opposite that of the original (**Fig. 9B**). These explants can be made as sandwiches of two full thickness of tissue, grafted with their inner surfaces together, as shown above (**Fig. 6**). In this case, the explant consists of the outer epithelium, the apposed deep layers of each, and the outer epithelium of the second.

A somewhat easier way to make these explants is to excise one-half of such an explant from one embryo and graft to its inner, deep surface the corresponding epithelium from a second embryo, as shown in **Fig. 9A,B**. This provides a complete epithelial covering, maintaining the advantage of a controlled internal environment. In addition, these thinner explants are somewhat easier to abut edgewise than their thicker counterparts.

Multistep preparations such as these are easier done if several rules are followed. They should be done fast. Otherwise, problems will arise, such as rolling of the epithelial sheets, migration of the epithelium over the cut edges to be apposed, and "bowing" of the tissues, which results in tissues "saddling" crosswise to one another when the sandwich is made. Also, the preparations should be done in stages, at least until one develops expertise. For example, this experiment is easier if two sandwiches, one of the Organizer and one of the responding ectoderm, are made first. Each of these sandwiches consist of an explant and the corresponding epithelium from another embryo (steps #1 and #2, **Fig. 9**). When these sandwiches are slightly healed, they are removed from beneath their coverslips and abutted to one another edgewise (step #3, **Fig. 9**). A fresh wound surface should be made at the surfaces to be mated so that they will join and heal.

In any case, these "blunt-end ligations" of tissue are difficult. Epithelial sheets abhor a free edge, and immediately after being cut, their margins will migrate across the exposed deep cells at the ends of the explants, until they meet another epithelium. If the edges of the explants do not match precisely and a gap appears where the deep cells are abutted, the epithelial sheets will not span the breach. Rather, each epithelium will migrate down along its own deep cells to the bottom of the gap, where it will meet and heal with its counterpart. At this point, the continuous epithelium will pop up out of the groove and

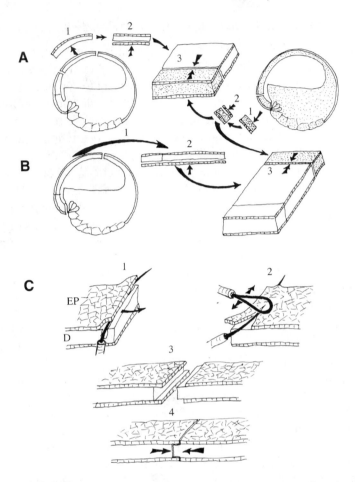

Fig. 9. A method for edgewise apposition of inducing and responding tissue is shown. In the first case, Organizer tissue is placed in edgewise (planar) contact with AC ectoderm at the early gastrula (**A**). In the second case, a second organizer is grafted in planar apposition to the opposite end of the AC ectoderm, whereas the original Organizer is left in place (**B**). In both cases, the inducing and responding tissues are excised and sandwiched with an epithelial layer from the corresponding region of another embryo (steps #1 and #2, A and B), to provide a protective covering. The inducing and responding tissues are then abutted edgewise (step #3, A and B). A method of making a slight lap joint, which will aid in abutment and healing of tissues joined edgewise, is shown (**C**). One explant is cut such that flaps of epithelial tissue are exposed (#1, C), and the other is cut such that the deep nonepithelial cells are exposed (#2, C). The two are laid facing one another (#3, C) and then pushed together and held there with a coverslip until healed (# 4, C).

form a smooth surface. This takes some time, and such slow or incomplete healing could block any induction that might have occurred.

To avoid these problems, we often make a slight lap joint, or "sticky-end ligation", in which the cut edges of the deep cells and epithelial cells do not coincide. One end of an explant is cut off squarely, and several rows of epithelial cells are lifted off the deep region at the edge, with the eyebrow hair (#1, **Fig. 9C**). This operation is repeated on the opposite side. A cut is then made downward, through most of the deep layer; the operation is repeated from the opposite side, removing a short section of deep layer (#3, **Fig. 9C**). On the other explant, the opposite operation is performed; an eyebrow hair is pushed back under the epithelium a couple of cells, and a hairloop is rubbed across the top, cutting off a short piece of epithelium (#2, **Fig. 9C**); the operation is repeated on the other side, making a notched explant (#3, **Fig. 9C**). If one works slowly, it is best to reverse the order of these operations, since the free epithelial sheets in the second type of explant will curl faster than their counterparts will advance across the deep cells in the first type of explant. Next, the edges of the free epithelial sheets are teased outward and the exposed deep cells of the other explant are stuffed in between them (#4, **Fig. 9C**). Finally, the entire preparation is pressed lightly between a coverslip and the bottom of the dish for the duration of healing, about 10–15 min.

## *4.4. Exogastrulae*

The last, and the least useful, of the experimental preparations for analyzing planar induction in *Xenopus* is the exogastrula. Holtfreter *(79)* found no obvious sign of neural development in the ectoderm in exogastrulae, in which the endomesoderm was extruded outward and thus made only planar or edgewise contact with the ectoderm. In *Xenopus,* exogastrulae can be made by culturing the blastulae in hypertonic salt solution, usually 1.3–2.0× normal culture medium. Under these conditions, the endoderm and mesoderm "evaginate" rather than involute, resulting in a specimen in which the notochord and somites, covered by endoderm, are connected by a narrow stalk to an ectodermal sack. *Xenopus* exogastrulae have been used to argue that planar signals induce neural tissue properties, on the assumption that they, like the exogastrulae used by Holtfreter, have no vertical signaling *(81,82)*.

Holtfreter, however, worked with the axolotl and several species of anurans, other than *Xenopus* (*Rana fusca* and *Hyla arborea*), for which he described exogastrulation in some detail. In contrast, the movements and mechanism of exogastrulation in *Xenopus* have never been described in adequate detail. What we do know about normal gastrulation and exogastrulation in *Xenopus* suggests that its exogastrula is unsuitable for studies of neural induction. Video microscopy of *Xenopus* "total exogastrulae" shows that they form BC and progress as far as the equivalent of normal stage 10.25–10.5 before exogastrulating (R. Keller, unpublished observations). This fact, together with the fact that head mesoderm undergoes early cryptic movement in *Xenopus*, and the fact that much of the prospective neural tissue is very close to the mesoderm in the early gastrula, make it very likely that early and transient, and perhaps permanent vertical neural-inducing signals occur in *Xenopus* exogastrulae. In addition, the blastocoel shrinks and the blastocoel roof often collapses on the VE in high-salt solutions, making it likely that both the morphogenetic movements and inductive interactions in the blastula stages are abnormal (*see* **refs.** *83,84*). Unless more work is done on the movements and anatomy of the *Xenopus* exogastrulae, it is useless for the study of neural (and mesodermal) induction. It may be very useful in situations where the possible early vertical contact of mesoderm with ectoderm is not an issue.

Interestingly, Holtfreter's work on the axolotl may not suffer from any of these weaknesses. Urodele embryos of the type he studied have a very long prospective neural plate and Organizer, and do not have cryptic, early involution of mesoderm (*see* **ref.** *85*), thus negating the problem of early vertical induction at anterior levels. Moreover, these urodele embryos have a large blastocoel, which tends to wrinkle on itself rather than against the VE (R. Keller, unpublished observations), making fortuitous mesoderm inductions less likely.

## 5. Planar Signals in the Context of Prospective Neural and Epidermal Interactions

The sandwich explants described above consist of the dorsal 90–120° of the embryo, which includes little or none of the prospective epidermal ectoderm. Thus, in these explants, planar signals from the Organizer act on the prospective neural tissue in the absence of any influences from the prospective epidermis. Since there is evidence that epidermal–neural interactions pattern the neural tissue and regulate its morphogenesis *(57–61)*, we have developed several types of explants that allow planar signals from the Organizer to act in the context of also having planar neural–epidermal interactions.

### 5.1. The "Giant" Sandwich

In the giant explant, the entire AC, NIMZ, and IMZ are excised and sandwiched (**Fig. 10**). The vegetal region of the embryo is turned uppermost, and single-stroke cut is made (**Fig. 4A**), from the center of vegetal region toward the midventral line, through the entire embryo all the way to the AC (cut #1, **Fig. 10A**). The embryo is then turned over, and the AC, NIMZ and IMZ are pulled away from any mesoderm/endoderm that has involuted (**Fig. 10A**; *see* **Subheading 3.2.** and **Fig. 4C**). These tissues are separated from the involuted tissue, as well as the VE by a second cut at the boundary of the VE (cut #2, **Fig. 10A**). The explant is then laid out, inner surface uppermost. Any adherent, postinvolution cells are removed from its inner surface (*see* **Subheading 3.3.**

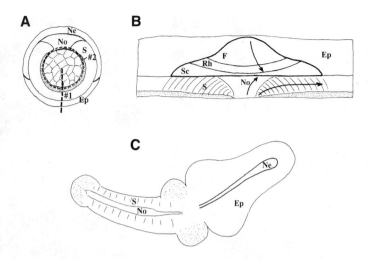

Fig. 10. To test the results of planar signaling in the context of the entire AC, a giant sandwich is used. A giant sandwich is made by cutting through the entire gastrula, from the ventral pole animally, along a line that extends from the center of the vegetal pole along the ventral midline (step #1, **A**). The embryo is then turned over, the entire gastrula wall peeled off from any involuting material, and the entire involuting marginal zone and AC cut away from the VE (step #2, **A**). This large piece, spanning the entire dorsoventral extent of the embryo, is then trimmed at the animal end to form a rectangle and sandwiched with a similar piece from another embryo to form a giant explant (**B**). As is the case with the standard sandwich explant, this explant shows convergent extension of both the neural and mesodermal regions. It differentiates neural and epidermal regions in the former, and notochord and somitic mesoderm in the latter (**C**). Prospective areas or the corresponding differentiated tissues are shown, including notochord (N), somitic mesoderm (S), epidermis (Ep), forebrain/midbrain (F), RH, SC, and neural tissue (Ne). The mesodermal prospective fates are shown as they would appear if the overlying epithelial endoderm were removed. The arrows in A and B show the anterior to posterior polarity of the neural and mesodermal tissues.

and **Fig. 4D**), and it is trimmed at the animal end to make a clean rectangle. The explant consists of the entire array of prospective tissues of the gastrula, its free edges at the ventral midline (**Fig. 10B**). Two of these giant explants are sandwiched together and allowed to develop. Both the neural and mesodermal/endodermal regions of the explant converge and extend (**Figs. 10C, 11A,B**). Moreover, the cells of the neural ectoderm columnarize, wedge, and attempt to roll the neural plate into a neural tube. Generally, a neural groove, rather than a tube, forms on both sides of the explant (pointers, **Fig. 11B**). Patterning in such explants can be visualized by RNA whole-mount *in situ* hybridization according to the method of Harland *(7)*. Otx2, a marker for forebrain and head meso-

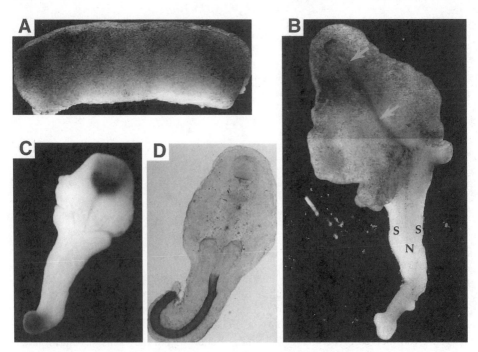

Fig. 11. Giant sandwiches are shown shortly after construction (**A**) and at stage 27
(**B**). The vegetal end at the bottom and the AC is at the top of all figures. The neural
region has converged, extended, and rolled into a neural trough (pointers, B), whereas
the lateral regions form epidermis. The mesodermal/endodermal region converges and
extends in the opposite direction, and consists of notochord (N) centrally and somitic
mesoderm (S) on both sides (B). A whole-mount RNA *in situ* hybridization (**C**) shows
an explant stained for Otx-2, expressed in the forebrain region of the giant explant (top
of explant) and in the prechordal mesoderm (bottom of explant). An explant stained
with the antibody tor 70 shows the notochord in the deep region of the mesodermal/
endodermal component of the explant (**D**).

derm (**Fig. 11C**), and staining with tor 70 antibody, a marker for notochord
(**Fig. 11D**), shows the locations of these tissues in the giant sandwich explant.

### 5.2. The "Pita" Sandwich

The giant explant preserves the normal dorsoventral planar tissue relationships in the gastrula, except at the ventral midline, where it has a free edge.
Moreover, the corresponding inner surfaces of the explant are apposed, and the
rolling of the neural tubes occurs opposite one another, both abnormal configurations that might affect morphogenesis. Another type of explant, the "pita"
sandwich, both retains the normal planar tissue relationships of the gastrula by
keeping the midventral line intact and avoids the problem of basal surface

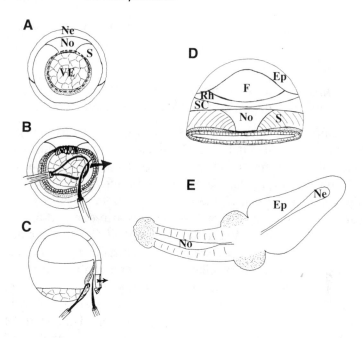

Fig. 12. The "pita" sandwich is similar to the giant sandwich, but consists of tissues taken from only one embryo, sandwiched together in the dorsoventral direction. This explant is made by cutting around the VE (dashed lines, **A**), and then sticking the eyebrow hair animally, along the interface between postinvolution and preinvolution tissues. The hairloop is used to hold the VE and postinvolution material, and the eyebrow hair is moved outward, pulling the gastrula wall away from the postinvolution tissues (**B,C**). The hairloop is then used to extract all the postinvolution material and VE out through this artificial "blastopore," leaving a pita-like affair, which is compressed in the dorsal-ventral direction (**D**). This explant develops much like the giant sandwich, but has a neural groove only on one side (**E**). Prospective areas or the corresponding differentiated tissues are shown, including notochord (N), somitic mesoderm (S), epidermis (Ep), forebrain/midbrain (F), rhombencephalon (RH), spinal cord (SC), and neural tissue (Ne). The mesodermal prospective fates are shown as they would appear if the overlying epithelial endoderm were removed.

apposition. The pita sandwich is made by cutting through the IMZ, along the blastoporal lip, as described above (**Fig. 4B**). The IMZ is teased outward and an eyebrow hair inserted into the blastocoel, toward the animal pole (**Fig. 12A**). The IMZ-NIMZ-AC is then pulled away from the involuted tissue (**Fig. 12B,C**). By gently stretching the tissue, the entire IMZ-NIMZ-AC can be removed from the involuted mesodermal tissues at any stage from 10 to 11.5, and flattened or cultured as a vesicle (**Fig. 12D**). The pita sandwich converges, extends, and differentiates dorsal neural and mesodermal tissues as well as epidermis and other tissues (**Fig. 12E**). This explant, though difficult to make, preserves some relationships of the normal embryo not maintained in the giant or standard explant.

## 5.3. Variations of Age, of Amount of Involution, and of Prospective Areas of the Organizer Included

Varying amounts of Organizer tissue (prospective PM and pharyngeal endoderm) can be included in the giant, pita, and standard sandwich explants. The amount of Organizer tissue included is determined by the age of the embryos from which the explants are made and where, with respect to the blastoporal lip, the anterior end of the explant is trimmed. For example, when explants are made from stage 10 or 10+ embryos, trimming off the vegetal end of the explant just above the VE, at the level of the BC, will leave behind with the embryo most of the prospective PM, which has already involuted (**Fig. 13A**). In order to include prechordal mesoderm in explants at this stage, the following method should be used. The wall of the gastrula (the preinvolution IMZ-NIMZ-AC) should be peeled outward, as described above, and the postinvolution prospective PM cut away from the central endoderm of the gastrula (**Fig. 13B**). The explant should be bent at the point of involution such that the AC-NIMZ-preinvolution IMZ and postinvolution IMZ material are spread open and apart, like the leaves of a strap hinge (**Fig. 13C**). Finally, two such explants are flattened and sandwiched together. The PM or head mesoderm will form a ball of tissue that does not converge and extend at the anterior end of the IMZ portion of the explant. It should stain with PM markers, such as Otx2 (**Fig. 11C**).

## 6. Vertical Signaling

Analysis of vertical signaling might appear simple, because it is relatively simple to combine parts of the dorsal organizer mesoderm with ectoderm. Unfortunately, it is not very straightforward to evaluate or control the amount of planar signaling that has happened or is continuing along with the vertical signaling. One can explant the prospective neural plate at the early through late gastrula stages, using the fate maps as guides. At progressively later stages, a number of properties of the neural plate become established and can be studied in such an explant. If one makes the explant from late gastrula stages, the dorsal mesoderm can be included with the prospective neural plate. The development of the neural plate with its corresponding mesodermal tissues can be compared with development of the neural plate alone. Although these types of explants have not been used much to date in *Xenopus*, they were used in the pioneering experiments of Jacobson and his associates on the axolotl and the newt, work that defined many of the morphogenetic issues involved in con-

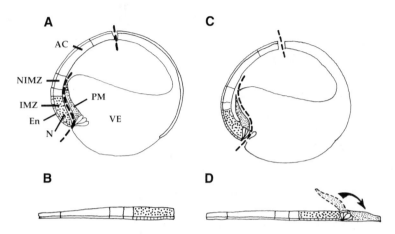

Fig. 13. Two methods of dissecting a stage 10+ gastrula for the purpose of making Organizer-prospective neural explants are shown. On the left (**A**), the involuting marginal zone, noninvoluting marginal zone, and animal cap (IMZ-N IMZ-AC) is separated from the postinvolution PM and cut from the VE at the level of the BC (heavy dashed line), such that the PM is left with the embryo. In this case, the vegetal end of the explant (**B**) consists of prospective notochord (N, coarse stippling) and the overlying superficial endodermal layer (En, cuboidal layer). On the right (**C**), the IMZ-NIMZ-AC is peeled away from the underlying involuted prechordal mesoderm along the interface (Cleft of Brachet) between the two, but instead of cutting this flap of tissue off the VE at the BC, a cut is made between the PM and the VE (heavy dashed line, C). The involuted, PM is folded vegetally to make the explant (arrow, **D**).

structing a neural plate and neural tube *(86,87)*. To study regionalization of the neural plate, various parts of the neural plate and mesoderm can be recombined (*see* **ref. 36**).

### 6.1. Neural Plate Explants

The neural plate can be excised cleanly from the underlying dorsal mesoderm as diagrammed in **Fig. 14**. The tip of the eyebrow hair should be pushed only through the neural plate, which consists of an epithelium and one layer of deep cells at these stages (stages 11–12.5), using the method described above (**Fig. 4B**). The depth of the neural plate is determined by experience, judging at what depth the eyebrow hair is two cells deep, staying on the shallow side. The eyebrow hair tip is used to tease away the edges of the two layers of the neural plate, searching for the easily separable interface that lies between the neural ectoderm and the underlying mesoderm. If only one layer comes up, the cut is too shallow. If the edge of the patch is lifted and more than two layers come up, the vertical cut likely was too deep, traversing into the mesoderm. In this case, the underlying mesoderm on the neural plate side of the cut will come

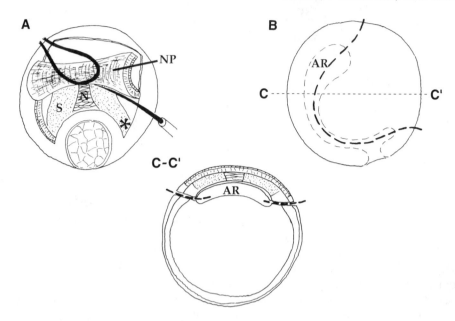

Fig. 14. The neural plate can be removed from later-stage gastrulae by first cutting through the two layers of the neural plate (NP) completely surrounding the area to be removed, using method diagrammed in **Fig. 4B**. A separation is then begun at the interface of the neural plate and underlying mesoderm, using the tip of the eyebrow hair. We prefer to begin at the posterior, right corner (asterisk). Once a sufficiently large area is lifted, the hairloop is used to deflect the neural plate animally, and the tip of the eyebrow hair is run along the interface, separating it from the underlying meso- derm. From stage 11.5 onward, the notochordal mesoderm (N) can be distinguished from somitic mesoderm (S) lying on both sides. The neural plate and underlying dor- sal mesoderm can be removed together at late gastrula stages **(B)**, by cutting (heavy dashed lines) through the body wall, into the archenteron (AR, outlined with light dashed lines), from its lateral aspect. A cross-section **(C-C')** shows the line of cut (dashed lines) through the body wall, into the archenteron (AR).

along with the neural plate. If this occurs, one should use the tip of the eyebrow hair to start a cleft at the interface of neural and mesodermal tissues. The neural cells are usually smaller and grayer than those of the mesoderm. However, once the integrity of the mesoderm has been violated, the forces of peeling off the neural plate may tear it further; if so, one must start over with a new embryo.

This procedure is easy to do at stages 10.5–11.5, but at stage 11.5 onward, the region lying above the notochord, the notoplate *(87)*, progressively adheres more tightly to the underlying notochord. As one peels off the neural plate, the notoplate cells will tend to stay with the notochord, or vice versa. When the undersurface of the notoplate is inspected, gaps will appear, reflecting notoplate cells that adhered to the notochord. Conversely, cells sitting above the general surface of the notoplate are notochord cells that have adhered to the notoplate and were pulled out of the notochord. These cells must be stripped off with an eyebrow hair, one by one (*see* **Fig. 4D**) to assure a pure neural plate cell population. It is best to check the explant after fixation with tor 70 staining for notochord cells, since just a very few notochord cells can affect development of the neural plate (*see* **ref. 80**). Laterally, contamination of the neural plate with the underlying somitic cells is less of a problem, since these cells adhere weakly to the neural plate.

Another difficulty is controlling whether or not mesoderm is left in planar apposition with the posterior neural plate. This requires knowing precisely where the posterior boundary of the neural plate lies. If one desires mesoderm to be present, the explant should be cut at the blastoporal lip. If not, one must err on the neural side of the boundary to be assured that no mesoderm will be included. This presents a dilemma. Although considerable convergent extension has occurred at the mid- to late-gastrula stages, the posterior neural tissue is still relatively short and wide. Cutting a little too far animally will remove all the prospective posterior neural tissue that would have converged, extended, and made the spinal cord. Likewise, lateral epidermis can be included in the explant, or not, depending on where the lateral cuts isolating the explant are made with respect to the boundaries shown in the fate maps. This type of explant has been used to study the effect of the lateral epidermis on morphogenesis of the neural tube in the newt and the axolotl *(59)*.

### 6.2. Neural Plate/Dorsal Mesoderm Explants

The neural plate can be explanted together with the underlying dorsal mesoderm/endoderm at the same stages by cutting completely through the anterior, lateral, and ventral walls of the archenteron *(88)*. This type of explant was first used to analyze *Xenopus* mesodermal cell behavior by stripping off the endoderm of the archenteron roof, exposing the deep somitic and notochordal cells (*see* **ref. 88**). However, it can also be used to compare autonomous development of the neural plate isolated and cultured alone, with development under the continued influence of the underlying mesoderm. In this explant, the ventral mesoderm that is destined to contribute to posterior somitic tissue moves around both sides of the blastopore, which is pushed posteriorly by the extending notochord (*see* **refs. 88,89**). This is the only explant preparation that allows the posterior somitic mesoderm, which lies at the ventrolateral and even ventral IMZs of the gastrula, to move beneath the neural plate in a relatively normal fashion (*see* Fig. 5 in **ref. 29**).

### 6.3. Recombination of Neural Plate and Organizer Mesoderm

A major issue in neural development is determining to what degree detailed regional patterning of the nervous system is owing to self-organizing processes within the neural plate, elicited by general signals from the mesoderm, and what is owing to direct induction by pre-existing regionalized pattern within the mesoderm. This issue can be addressed by recombining parts of the prospective neural plate and Organizer mesoderm in vertical apposition (*see* **ref.** *36*). In this type of preparation, as in the neural plate isolates described above, care should be taken to separate mesodermal and neural areas cleanly before recombining, so that one does not have crosscontamination of regional tissues. One must also carefully define the regions involved, which is not a trivial task, considering the dramatic movements occurring in both neural and mesodermal tissues. Although the published fate maps can serve as a guide, it is best to check the prospective fates with control marking of tissues. Similar concerns apply to recombination of mediolateral components of the neural and mesodermal tissues.

### 6.4. Visualizing Cell Behavior in the Neural Plate

High-resolution video microscopy of neural plate explants similar to the type described above can be used to characterize the cell behaviors driving neural convergent extension. Protrusive activity of cells in a tissue is best visualized if a scattered population of the cells is labeled with a vital dye. To achieve such a labeling in the neural plate, we inject 20–30 nL of rhodamine dextran amine (RDA, Molecular Probes Inc., Eugene, OR) into the "A" or "B" dorsal tiers of 32 cell-stage embryos. Dorsal is identified by the tipping and marking method described above. By the midgastrula stage (stage 11.5), cell rearrangement associated with convergence and extension has occurred among the derivatives of the injected blastomere, so that labeled cells are scattered along the length of the neural plate. At stage 12.5, the outer epithelial layer of the neural plate is removed and discarded, exposing the labeled deep cells of the neural plate. The remaining layer of neural plate deep cells, with or without the underlying mesoderm, is removed and cultured under a restraining coverslip, oriented with the neural deep cells facing the bottom of the culture chamber. In this configuration, the neural deep cells can be visualized with time-lapse video recording, using fluorescence illumination, a low-light camera, and image processing, as described previously *(39)*. We have used this method to study the cell behavior underlying convergence extension of the posterior neural plate *(89)*.

### 7. Combination of Vertical and Planar Signaling: The "Kintner" Sandwich Explant

If the sandwich explant is properly made, it should allow only planar signaling. Dixon and Kintner *(77)* tested the effect of vertical signals in addition to planar signals by making standard "Keller" explants and adding head/PM to their inner, anterior surfaces (*see* Fig. 5 in **ref. 77**). In this case, they found that neural tissue, assayed by expression of several markers, was induced poorly when only vertical apposition of tissues was allowed. Planar apposition was more effective, but the strongest response came when vertical apposition was coupled with planar apposition, implying that a synergistic effect of the two signaling routes exists.

## 8. Other Preparations

A large number of other preparations have been used to study induction and neural development, including classical Organizer grafts, pioneered by Mangold and Spemann in urodeles *(13,30)*, and used more recently in studies of *Xenopus* neural development *(14)*, and placing the Organizer into the blastocoel (the "Einsteck") *(90)*. The general rules that can be taken from the above discussion for any embryological manipulation of these types are:

> Know the fates and locations of the tissues being used.
> Understand the intrinsic morphogenetic movements of the tissues involved.
> Mark the tissues involved and follow their movements, instead of assuming what they will do.
> Do the operations quickly so that healing is rapid.

## Acknowledgments

This work was supported by NIH HD25594 and NSF92-20525 to Ray Keller. Tamira Elul is supported by a Howard Hughes Medical Institute Predoctoral Fellowship. We thank the other members of the laboratory, both past and present, for their direct and indirect contributions to the development of the methods described in this chapter.

## References

1. Hemmati-Brivanlou, A. and Melton, D. A. (1992) A truncated activin receptor inhibits mesoderm induction and formation of axial structures in *Xenopus* embryos. *Nature* **359,** 609–614.
2. Hemmati-Brivanlou, A. and Melton, D. A. (1994) Inhibition of activin receptor signaling promotes neuralization in *Xenopus*. *Cell* **77,** 273–281.
3. Hemmati-Brivanlou, A., Kelly, O. G., and Melton, D. A. (1994) Follistatin, an antagonist of activin, is expressed in the Spemann organizer and displays direct neuralizing activity. *Cell* **77,** 283–295.
4. Smith, W. and Harland, R. (1992) Expression cloning of noggin, a new dorsalizing factor localized to the Spemann Organizer in *Xenopus* embryos. *Cell* **70,** 829–840.
5. Lamb, T. M., Knecht, A. K., Smith, W. C., Stachel, S., Economides, A. N., Stahl, N., Yancoplous, G. D., and Harland, R. M. (1993) Neural induction by the secreted polypeptide noggin. *Science* **266,** 650–653.

6. Harland, R. (1994) Neural induction in *Xenopus*. *Curr. Opin. Gen. Devel.* **4,** 543–549.
7. Harland, R. M. (1991) In situ hybridization: An improved whole-mount method for *Xenopus* embryos, in *Methods in Cell Biology,* vol. 36 (Kay, B. and Peng, B., eds.), Academic, San Diego, pp. 685–695.
8. Krumlauf, R. (1995) Hox genes in vertebrate development. *Cell* **78,** 191–201.
9. Keller, R. E. (1991). Early embryonic development of *Xenopus laevis*, in *Xenopus laevis: Practical uses in Cell and Molecular Biology,* vol. 36 (Kay, B. and Peng, H. B. eds.), Academic, San Diego, pp. 59–111.
10. Keller, R. E., Shih, J., and Sater, A. K. (1992) The cellular basis of the convergence and extension of the *Xenopus* neural plate. *Dev. Dyn.* **193,** 199–217.
11. Keller, R. and Danilchik, M. (1988) Regional expression, pattern and timing of convergence and extension during gastrulation of *Xenopus laevis*. *Development* **103,** 193–209.
12. Gerhart, J., Doniach, T., and Stewart, R. (1991) Organizing the Xenopus organizer, in *Gastrulation* (Keller, R., Clark, W., and Griffin, F., eds.), Plenum, New York, pp. 57–77.
13. Spemann, H. and H. Mangold (1924) Über Induktion von Embryonalanlagen durch Implantation artfremder Organisatoren. *Arch. Mikr. Anat. Entw. Mech.* **100,** 599–638.
14. Gimlich, R. and Cooke, J. (1993) Cell lineage and induction of second nervous systems in amphibian development. *Nature* **306,** 471–473.
15. Keller, R. E. (1981) An experimental analysis of the role of bottle cells and the deep marginal zone in gastrulation of *Xenopus laevis*. *J. Exp. Zool.* **216,** 81–101.
16. Hardin, J. and Keller, R. (1988) The behaviour and function of bottle cells during gastrulation of *Xenopus laevis*. *Development* **103,** 211–230.
17. Nieuwkoop, P. D. and Faber, J. (1967) *Normal Table of* Xenopus laevis *(Daudin).* North-Holland Publishing, Amsterdam.
18. Keller, R. E. (1978) Time-lapse cinemicrographic analysis of superficial cell behavior during and prior to gastrulation in *Xenopus laevis*. *J. Morph.* **157,** 223–248.
19. Keller, R. E. (1980) The cellular basis of epiboly: An SEM study of deep-cell rearrangement during gastrulation in *Xenopus laevis*. *J. Embryol. Exp. Morph.* **60,** 201–234.
20. Nieuwkoop P. and Florshutz, P. (1950) Quelques caractères spéciaux de la gastrulation et de la neurulation de l'oeuf de *Xenopus laevis*, Daud. et de quelques autres Anoures. 1ère partie. - Étude descriptive. *Arch. Biol. (Liège)* **61,** 113–150.
21. Keller, R. E. (1975) Vital dye mapping of the gastrula and neurula of *Xenopus laevis*. I. Prospective areas and morphogenetic movements of the superficial layer. *Develop. Biol.* **42,** 222–241.
22. Keller, R. E (1976) Vital dye mapping of the gastrula and neurula of *Xenopus laevis*. II. Prospective areas and morphogenetic movements of the deep layer. *Develop. Biol.* **51,** 118–137.
23. Nakatsuji, N. (1975) Studies on the gastrulation of amphibian embyros: cell movement during gastrulation in *Xenopus laevis* embryos. *Wilhelm Roux' Arch.* **178,** 1–14.
24. Cho, K. W. Y., Blumberg, B., Steinbeisser, H., and De Robertis, E. M. (1991) Molecular nature of Spemann's Organizer: the role of the *Xenopus* homeobox gene goosecoid. *Cell* **67,** 1111–1120.

25. Blitz, I. and Cho, K. (1995) Anterior neuroectoderm is progressively induced during gastrulation: the role of the *Xenopus* homeobox gene orthodenticle. *Development* **121,** 993–1004.

25a. Bouwmeester, T., Sung-Hyun, K., Sasai, Y., Lu, B., and DeRobertis, E. (1996) Cerberus is a head-inducing secretal factor expressed in the anterior endoderm of Spiemann's Organizer. *Nature* **382,** 595–601.

26. Vodicka, M. and Gerhart, J. (1995) Blastomere contributions and domains of gene expression in the Spemann Organizer of *Xenopus laevis*. *Development* **121,** 3505–3518.

27. Bauer, D. V., Huang, S., and Moody, S. (1994) The cleavage stage origins of Spemann's Organizer: analysis of the movements of blastomere clones before and during gastrulation in *Xenopus*. *Development* **120,** 1179–1189.

28. Keller, R., Shih, J., and Wilson, P. (1991) Cell motility, control and function of convergence and extension during gastrulation of *Xenopus,* in *Gastrulation: Movements, Patterns, and Molecules* (Keller,R, Clark, W., and Griffin, F. eds.), Plenum Press, New York, pp. 101–119.

29. Keller, R., Shih, J., Wilson, P., and Sater, A. (1991) Pattern and function of cell motility and cell interactions during convergence and extension in *Xenopus*, in *Cell-Cell Interactions in Early Development, 49th Symp. Soc. Develop. Biol.* (Gerhart, J. C., ed.), Wiley-Liss, New York, pp. 31–62.

30. Spemann, H. (1938) *Embryonic Development and Induction.* Yale University Press, New Haven

31. Nieuwkoop P. D., Boterenbrod, E. C., Kremer, A., Bloemsma, F., Hosessels, E., and Verheyen, F. (1952) Activation and organization of the central nervous system in Amphibians. *J. Exp. Zool.* **120,** 1–108.

32. van Stratten, H. M. V., and Hekking, J. W. M., Wiertz-hoessels, E. J. L. M., Thors, F., and Drukker, J. (1988) Effect of the notochord on the differentiation of the floorplate area in the neural tube of the chick embryo. *Anat. Embryol.* **177,** 317–324.

33. Smith, J. L. and Schoenwolf, G. C. (1989) Notochordal induction of cell wedging in the chick neural plate and its role in neural tube formation. *J. Exp. Zool.* **250,** 49–62.

34. Sive, H., Hattori, K., and Weintraub, H. (1989) Progressive determination during formation of the anteroposterior axis in *Xenopus laevis*. *Cell* **58,** 171–180.

35. Placzek, M., Tessier-Lavigne, M., Yamada, T., Jessell, T., and Dodd, J. (1990) Mesodermal control of neural cell identity: floor plate induction by the notochord. *Science* **250,** 985–988.

36. Yamada, T., Placzek, M., Tanaka, H., Dodd, J., and Jessell, T. M. (1991) Control of cell pattern in the developing nervous system: Polarizing activity of the floor plate and notochord. *Cell* **64,** 635–647.

37. Saha, M. and Grainger, R. (1992) A liabile period in the determination of the anterior-posterior axis during early neural development in *Xenopus*. *Neuron* **8,** 1003–1014.

38. Keller, R. and R. Winklbauer (1992) The cellular basis of amphibian gastrulation, in *Current Topics in Developmental Biology*, vol. 27 (Pedersen, R., ed.), Academic, New York, pp. 39–89.

39. Shih, J. and Keller, R. E. (1992) Cell motility driving mediolateral intercalation in explants of *Xenopus laevis*. *Development* **116,** 901–914.

40. Winklbauer, R. Mesodermal cell migration during *Xenopus* gastrulation. *Dev. Biol.* **142**, 155–168.

41. Winklbauer, R., Selchow, A., Nagel, M., Stoltz, C., and Angres, B. (1991) Mesoderm cell migration in the *Xenopus* gastrula, in *Gastrulation: Movements, Patterns, and Molecules* (Keller, R., Clark, W., and Griffin, F., eds.), Plenum, New York, pp. 147–168.

42. Smith, J. C., Price, B. M. J., Green, J. B. A., Weigel, D., and Herrmannn, B. (1991) Expression of the *Xenopus* homolog of Brachyury (T) is an immediate-early response to mesoderm induction. *Cell* **67**, 79–87.

43. Kushner, P. D. (1984) A library of monoclonal antibodies to *Torpedo* cholinergic synaptosomes. *J. Neurochem.* **43**, 775–786.

44. Bolce, M. E., Hemmati-Brivanlou, A., Kushner, P. D., and Harland, R. M. (1992) Ventral ectoderm of *Xenopus* forms neural tissue, including hindbrain, in response to activin. *Development* **115**, 673–680.

45. Kintner, C. R. and Brockes, J. (1984) Monoclonal antibodies identify blastemal cells derived from differentiating muscle in newt limb regeneration. *Nature* (London) **308**, 67–69.

46. Shih, J. and Keller, R. E. (1992) Patterns of cell motility in the organizer and dorsal mesoderm of *Xenopus laevis*. *Development* **116**, 915–930.

47. Domingo, C. and Keller, R. (1995) Induction of notochord cell intercalation behavior and differentiation by progressive signals in the gastrula of *Xenopus laevis*. *Development* **121**, 3311–3321.

48. Akers, R., Phillips, C., and Wessels, N. (1986) Expression of an epidermal antigen used to study tissue induction in the early *Xenopus* embryo. *Science* **231**, 613–616.

49. London, C., Akers, R., and Phillips, C. (1988) Expression of Ep-1, an epidermis-specific marker in Xenopus laevis embryos, is specified prior to gastrulation. *Devel. Biol.* **129**, 380–389.

50. Savage, R. and Phillips, C. (1989) Signals from the dorsal blastopore region during gastrulation bias the ectoderm toward a nonepidermal pathway of differentiation in *Xenopus laevis*. *Dev. Biol.* **132**, 157–168.

51. Sokol, S. and Melton, D. (1991) Pre-existent pattern in *Xenopus* animal pole cells revealed by induction with activin. *Nature* **351**, 409–411.

52. Keller, R. E., Shih, J., Sater, A. K. and Moreno, C. (1992) Planar induction of convergence and extension of the neural plate by the organizer of *Xenopus*. *Dev. Dynam.* **193**, 218–234.

53. Doniach, T., Phillips, C. R., and Gerhart, J. C. (1992) Planar induction of antero-posterior pattern in the developing central nervous system of *Xenopus laevis*. *Science* **257**, 542–545.

54. Otte, P. and Moon, R. (1992) Protein kinase C isozymes have distinct roles in neural induction and competence in *Xenopus*. *Cell* **68**, 1021–1029.

55. Wilson, P. A. and Hemmati-Brivalou, A. (1995) Induction of epidermis and inhibition of neural fate by BMP-4. *Nature* **376**, 331–334.

56. Harland, R. (1995) The transforming growth factor b family and induction of the vertebrate mesoderm: bone morphogenetic proteins are ventral inducers. *Proc. Natl. Acad. Sci. USA* **91**, 10,243–10,246.

57. Moury, D. and Jacobson, A. (1989) Neural fold formation at newly created boundaries between neural plate and epidermis in the axolotl. *Dev. Biol.* **133**, 44–57.

58. Moury, D. and Jacobson, A. (1990) The origins of the neural crest cells in the axolotl. *Devel. Biol.* **141**, 243–253.

59. Jacobson, A. and Moury, J. D. (1995) Tissue boundaries and cell behavior during neurulation. *Dev. Biol.* **171**, 98–110.

60. Liem, K., Jr., Tremmi, G., Roelink, H., and Jessell, T. (1995) Dorsal differentiation of neural plate cells induced by BMP-mediated signals from epidermal ectoderm. *Cell* **82**, 969–979.

61. Selleck, M. and Bronner-Fraser, M. (1995) Origins of the avian neural crest: the role of neural plate-epidermal interactions. *Development* **121**, 525–538.

62. Scharf, S. and Gerhart, J. (1983) Axis determination in eggs of *Xenopus laevis*: a critical period before first cleavage, identified by the common effects of cold, pressure, and ultraviolet irradiation. *Devel. Biol.* **99**, 75–87.

63. Kay, B. K. and Peng, H. B. (1991) Xenopus laevis: *Practical Uses in Cell and Molecular Biology,* vol. 36, Academic, San Diego.

64. Gurdon, J. (1977) Methods for nuclear transplantation in amphibia. *Meth. Cell Biol.* **16**, 125–139.

65. Sater, A. K., Steinhardt R. A., and Keller R. (1993) Induction of neuronal differentiation by planar signals in *Xenopus* embryos. *Devel. Dynam.* **197**, 268–280.

66. Keller, R. E., Danilchik, M., Gimlich, R., and Shih, J. (1985) Convergent extension by cell intercalation during gastrulation of *Xenopus laevis*, in *Molecular Determinants of Animal Form*, UCLA Symposia on Molecular and Cellular Biology, New Series 31 (Edelman, G. M., ed.), Liss, New York, pp. 111–141.

67. Keller, R. E., Danilchik, M., Gimlich, R., and Shih, J. (1985) The function and mechanism of convergent extension during gastrulation of *Xenopus laevis*. *J. Embryol. Exp. Morphol.* **89(Suppl.),** 185–209.

68. Gillespie, R. (1983) The distribution of small ions during the early development of *Xenopus laevis* and *Ambystoma mexicanum* embryos. *J. Physiol.* **344**, 359–377.

69. Shih, J. and Keller, R. (1992) The epithelium of the dorsal marginal zone of *Xenopus* has organizer properties. *Development* **116**, 887–899.

70. Wilson, P. A. and Keller, R. E. (1991) Cell rearrangement during gastrulation of *Xenopus*: direct observation of cultured explants. *Development* **112**, 289–305.

71. Holtfreter, J. (1943) Properties and function of the surface coat in amphibian embryos. *J. Exp. Zool.* **93**, 251–323.

72. Holtfreter, J. (1943) A study of the mechanics of gastrulation. Part I. *J. Exp. Zool.* **94**, 261–318.

73. Holtfreter, J. (1944) A study of the mechanics of gastrulation. Part II. *J. Exp. Zool.* **95**, 171–212.

74. Kirschner, M. and Hara, K. (1980) A new method of local vital staining of amphibian embryos using ficoll and "crystals" of Nile Red. *Mikroskopie* **36**, 12–15.

75. Gerhart, J., Ubbels, G., Black, S., Hara, K., and Kirschner, M. (1981) A reinvestigation of the role of the grey crescent in axis formation in *Xenopus laevis*. *Nature* **292**, 511–516.

76. Kintner, C. R. and Melton, D. A. (1987) Expression of Xenopus N-CAM RNA in ectoderm is an early response to neural induction. *Development* **99**, 311–325.
77. Dixon, J. and Kintner, C. R. (1989) Cellular contacts required for neural induction in *Xenopus* embryos: evidence for two signals. *Development* **106**, 749–757.
78. Papalopulu, N. and Kintner, C. (1993) *Xenopus* Distal-less related homeobox genes are expressed in the developing forebrain and are induced by planar signals. *Development* **117**, 961–975.
79. Holtfreter, J. (1933) Die totale Exogastrulation, eine Selbstablosung des Ektoderms vom Entomesoderm. *Roux' Arch. Entw. Mech.* **129**, 669–793.
80. Poznanski, A. and Keller, R. (1997) The role of planar and early vertical signaling in patterning and expression of Hoxb-1 in *Xenopus*. *Dev. Biol.* **189**, 256–269.
81. Ruiz i Altaba, A. (1990) Neural expression of the *Xenopus* homeobox gene Xhox3: evidence for a patterning neural signal that spreads through the ectoderm. *Development* **108**, 67–80.
82. Ruiz i Altaba, A. (1992) Planar and vertical signals in the induction and patterning of the *Xenopus* nervous system. *Development* **115**, 67–80.
83. Keller, R. E. (1986) The cellular basis of amphibian gastrulation, in *Developmental Biology: A Comprehensive Synthesis*, vol. 2 (Browder, L., ed.), Plenum, New York, pp. 241–327.
84. Lamb, T. M. (1995) Neural induction and patterning in *Xenopus*: The role of the dorsal mesoderm and secreted molecules derived from it. Ph.D. Thesis, University of California, Berkeley, CA.
85. Vogt, W. (1929) Gestaltanalyse am Amphibienkein mit ortlicher Vitalfarbung. II. Teil. Gastrulation und Mesodermbildung bei Urodelen und Anuren. *Wilhelm Roux Arch. EntwMech. Org.* **120**, 384–706.
86. Jacobson, A. and Gordon, R. (1976) Changes in the shape of the developing vertebrate nervous system analyzed experimentally, mathematically and by computer simulation. *J. Exp. Zool.* **197**, 191–246.
87. Jacobson, A. (1981) Morphogenesis of the neural plate and tube, in *Morphogenesis and Pattern Formation* (Connelley, T. G., Brinkley, L., and Carlson, B, eds.), Wiley, New York, pp. 223–263.
88. Wilson, P. A., Oster, G. M., and Keller, R. (1989) Cell rearrangement and segmentation in *Xenopus*: direct observation of cultured explants. *Development* **105**, 155–166.
89. Elul, T., Koehl, M., and Keller, R. (1995) Cellular mechanism of neural convergence and extension. *J. Cell Biol.* H-49 (abstract).
90. Lehman, F. E. (1932) Die Beteiligung von Implantats- und Wirtsgewebe bei der Gastrulation und Neurulation inducierter Embryonalanlagen. *Wilhelm Roux Arch. Entw.-Mech. Org.* **125**, 566–639.
91. Eagleson, G. and Harris, W. (1989) Mapping the presumptive brain regions in the neural plate of *Xenopus laevis*. *J. Neurology* **21**, 427–440.

# 27

# A Method for Generating Transgenic Frog Embryos

## Enrique Amaya and Kristen L. Kroll

## 1. Introduction

### 1.1. Summary

The early amphibian embryo has been widely used as a model organism for studying early vertebrate development. This chapter describes in detail a new and very efficient method for generating transgenic *Xenopus* embryos. At the end of the chapter, a new method for fertilizing in vitro matured oocytes is also introduced.

### 1.2. Background

The elucidation of the molecular basis of pattern formation and differentiation in frog embryos has been hindered by the lack of a system for temporal and tissue-specific expression of wild type and mutant forms of developmentally important genes. RNA injection, the most common transient expression method in *Xenopus*, has been effectively used to study maternally expressed genes. However, since RNAs are translated immediately after injection, this method is unfavorable for the study of zygotic gene products that are expressed only after the midblastula transition. Direct injection of DNA can be used to express genes behind temporal and tissue specific promoters after the midblastula transition. However, in frog embryos this approach has only been marginally successful for two reasons: (1) injected DNA does not integrate into the frog chromosomes during early cell cycles, and therefore, the embryo expresses the genes in a highly mosaic pattern, and (2) many promoters lack adequate temporal fidelity and tissue specificity of expression when the DNA is not integrated into the genome of the frog. To overcome these technical problems, we have developed the nuclear transplantation-based approach to transgenesis

From: *Methods in Molecular Biology, Vol. 97: Molecular Embryology: Methods and Protocols*
Edited by: P. T. Sharpe and I. Mason © Humana Press Inc., Totowa, NJ

described in this chapter. The approach enables stable expression of cloned gene products in *Xenopus* embryos, allowing a broader range of feasible experimentation than that previously possible by transient expression methods.

Unlike plasmid injection, transgenesis allows stable, temporally and spatially controlled expression of gene products in desired cells of the *Xenopus* embryo. We have used transgenesis to express genes of interest ectopically, to direct expression of modified gene products which dominantly interfere with the function of their endogenous, wild-type counterparts, and to analyze the spatial regulation of promoters in the embryo (*1*). We have been able to obtain large numbers of transgenic embryos readily for these purposes and to interpret reliably the effect of transgene expression without the cell-to-cell variability of expression within an embryo, which plagues many studies using plasmid-injected embryos.

## 1.3. Overview of Transgenesis Procedure

In the transgenesis approach described here, DNA is integrated into isolated sperm nuclei in vitro, followed by transplantation of the nuclei into unfertilized eggs, thus generating transgenic embryos. Nuclear transplantation of transfected cultured cells was previously used by one of us to produce transgenic *Xenopus* embryos, which expressed promoter-reporter plasmids nonmosaically (*2*). However, the cultured cells used as nuclear donors for these transplantations were aneuploid and rarely promoted development of the pseudo-triploid embryos to tadpole stages. To overcome these problems, we now use sperm nuclei to generate transgenic embryos. These nuclei offer many advantages over cultured cell lines. First, since sperm nuclei are haploid, there is no need to destroy the egg nucleus before transplantation to generate a normal diploid embryo. Second, sperm nuclei have been used for many years to investigate the processes of chromosome decondensation, nuclear assembly, and cell-cycle progression (*3–5*). These studies have provided us with valuable information regarding the manipulation of sperm nuclei in vitro. Indeed we have discovered that we can introduce DNA into sperm nuclei swelled and decondensed in cell-free egg extracts using restriction enzyme-mediated integration (REMI) (*6,7*). When these nuclei are transplanted into unfertilized eggs, we obtain large numbers of normal diploid tadpoles, which develop to advanced stages and express inserted genes at high frequency.

The protocol for *Xenopus* transgenesis described here involves the following steps:

1. Sperm nuclei are incubated with linearized plasmid DNA.
2. After a short incubation, a high-speed interphase egg extract and a small amount of the restriction enzyme used for plasmid linearization are added to the sperm nuclei/plasmid mixture. The extract partially decondenses sperm chromatin, but does not promote replication.

3. After the plasmid-treated nuclei are incubated for a brief period in the interphase extract, the mixture is diluted 50- to 100-fold (or 500- to 1000-fold total dilution of sperm nuclei).

4. Approximately one nucleus is transplanted into an unfertilized egg in a 5–15 nL volume. This procedure is schematically represented in **Fig. 1**.

### 1.4. Efficiency of Transgenesis Procedure

After transplantation with swelled sperm nuclei, 20–40% of the eggs cleave and develop normally. One person can transplant about 500 sperm nuclei/h to produce several hundred to one thousand embryos in a typical experiment. As with embryos produced by in vitro fertilization, the frequency of normal, advanced development varies somewhat, depending on the overall quality of the eggs; typically, 5–40% of the cleaving eggs develop normally beyond feeding tadpole stages. We commonly obtain 1–2 mo old tadpoles and are currently raising transplantation-derived metamorphosed froglets to sexual maturity.

Embryos from gastrula through tadpole stages derived from sperm nuclear transplantation express plasmids nonmosaically at high frequency. We have used transgenesis to introduce plasmids containing the simian cytomegalovirus (CMV) *(8)* or the *X. borealis* cytoskeletal actin (CSKA) *(9,10)* promoter into embryos *(1)*. Transgenic embryos express genes from these promoters in every cell starting at the late blastula and early gastrula stages, respectively, as expected for these ubiquitously expressed promoters. When whole-mount *in situ* hybridization is used to detect plasmid expression, 20–50% of transplantation-derived embryos express CMV and CSK promoter-containing plasmids in every cell. In contrast, embryos injected with these plasmids never express reporter genes in all cells and typically express in only a small fraction (about 5–20%) of cells in the embryo.

We have also made transgenic embryos that express promoters that are spatially restricted *(1)*. For example, we have used this method to introduce into embryos plasmids containing a muscle-specific actin promoter *(11)* linked to chloramphenicol acetyltransferase (pRLCAR) or green fluorescent protein (pCARGFP). We find that 40–60% of tadpoles derived from sperm nuclear transplantations with these plasmids show stable, nonmosaic expression. Expression is restricted to the somites and heart tissue, as expected for this regionally restricted promoter. We have also generated transgenic embryos with plasmid DNA that contains a neural specific β-tubulin promoter driving chloramphenicol acetyltransferase (CAT) (provided by Paul Krieg). These embryos express CAT in the primary neurons of the embryo as expected for this promoter. This correct expression is significant, since embryos injected with plasmids not only express CAT mosaically, but also ectopically, suggesting that integrating DNA into the genome is likely to give better regulation of cloned promoters than expression from nonintegrated DNA.

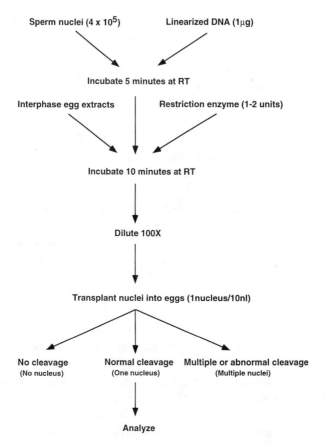

Fig. 1. Overview of transgenesis procedure. Sperm nuclei are incubated with linear DNA for a brief period of time. Interphase egg extracts and a restriction enzyme are then added. The egg extracts partially decondense the chromosomes, and the restriction enzyme very lightly cleaves them. These events facilitate the eventual integration of the linear DNA into the chromosomes. After incubation of nuclei in a mixture of extract, restriction enzyme, and plasmid DNA, the nuclei are diluted, and approximately one nucleus is transplanted per egg. Each activated egg requires a nucleus (or at least the centriole introduced with a nucleus) to divide; therefore, only eggs receiving a nucleus develop into embryos. Eggs that receive more than one nucleus (polyspermic eggs) divide abnormally into multiple cells at the first cleavage division. Embryos developing from monospermic eggs cleave normally during early divisions; only these embryos are isolated and analyzed. Generally, between 20 and 50% of these embryos will be transgenic.

## 1.5. Analysis of DNA Integration in Transgenic Embryos

We have analyzed genomic DNA from transplantation-derived tadpoles to determine whether early integration of introduced plasmids into sperm or egg chromosomes is responsible for the nonmosaic expression observed *(1)*. The pCARGFP plasmid was introduced into embryos by transplantation of sperm nuclei, and tadpoles 2.5 wk to 1 mo old were scored for nonmosaic expression of GFP in the somites; the presence and arrangement of pCARGFP were then determined by probing Southern blots of genomic DNA from these tadpoles with a 1-kb probe consisting of GFP sequences from one end of the linearized pCARGFP plasmid. We found that transplantation-derived tadpoles that did not express pCARGFP did not contain the plasmid, whereas each tadpole that expressed GFP contained between 5 and 35 copies of the plasmid/cell. The probe recognized four to eight bands in each GFP-expressing tadpole which were of unique sizes and were not found in other GFP-expressing tadpoles. These fragments represent putative junction points at which pCARGFP was integrated into the genome of each tadpole. Additionally, the probe recognized two common bands in all of the GFP-expressing tadpoles, corresponding to products formed by tandem and back-to-back concatemerization of the pCARGFP plasmid. By comparing the intensity of these bands relative to the putative junction fragments, we estimate that pCARGFP was integrated into the genome as single copies in some instances and as short (two to six copy) concatemers in other instances. Since the plasmid is expressed in all expected cells in the embryo, it is likely that most of these integrations occurred prior to the first cleavage division, assuring that all cells of the embryo would inherit several copies of the plasmid.

### 1.6. Overview of a New Technique for Fertilizing In Vitro Matured Oocytes

In addition to describing a procedure for making transgenic frog embryos, we are also including a new protocol for using sperm nuclear transplantation to fertilize oocytes that have been matured in vitro. A number of investigators have successfully depleted maternal RNAs from *Xenopus* embryos by injecting oocytes with antisense deoxyoligonucleotides for a message of interest, and then maturing and fertilizing these injected oocytes to produce embryos *(12–15)*. However, the approaches currently available for obtaining embryonic development from oocytes matured in vitro are quite labor-intensive and technically demanding *(16)*. The difficulty arises in large part because the surface of matured oocytes must be altered by proteolytic enzymes, and covered with a jelly coat in the oviduct of the female frog in order to become competent for fertilization. When oocytes are matured in vitro, these processes must either be simulated, or the matured oocytes must be reimplanted into the body cavity of a female frog so that these processes occur as it travels back outside through the oviduct of the host female. However, we have found that direct transplantation of sperm nuclei into matured oocytes bypasses these requirements and has allowed us to very simply produce normal embryos from oocytes handled in vitro. A similar approach has also been developed to overcome male infertility in humans *(17–20)*.

## 2. Materials

### 2.1. Sperm Nuclei Preparation

1. 1X Nuclear Preparation Butter (NPB): 250 m$M$ sucrose (1.5 $M$ stock; filter-sterilize and store aliquots at –20°C), 15 m$M$ HEPES (1 $M$ stock; titrate with KOH so that pH 7.7 is at 15 m$M$, filter-sterilize, and store aliquots at –20°C) (*see* **Note 4**), 0.5 m$M$ spermidine trihydrochloride (Sigma S-2501; 10 m$M$ stock; filter-sterilize and store aliquots at –20°C), 0.2 m$M$ spermine tetrahydrochloride (Sigma S-1141; 10 m$M$ stock; filter-sterilize and store aliquots at –20°C), 1 m$M$ dithiothreitol (Sigma D-0632; 100 m$M$ stock; filter-sterilize and store aliquots at –20°C) (*see* **Notes 4** and **5**).

    For steps requiring protease inhibitors, add: 10 µg/mL leupeptin (Boeringer Mannheim 1 017 101; 10 mg/mL stock in DMSO, store aliquots at –20°C), 0.3 m$M$ phenylmethylsulfonyl fluoride (PMSF) (Boeringer Mannheim 837 091; 0.3 $M$ stock in EtOH, stored at –20°C).

2. 1X Marc's Modified Ringer (MMR): 100 m$M$ NaCl, 2 m$M$ KCl, 1 m$M$ MgCl$_2$, 2 m$M$ CaCl$_2$, 5 m$M$ HEPES, pH 7.5. Prepare a 10X stock, and adjust pH with NaOH to 7.5. Sterilize 10X and 1X solutions by autoclaving.
3. Lysolecithin: 100 µL of 10 mg/mL L-α-lysophosphatidylcholine (Sigma Type I, L-4129); dissolve at room temperature just before use. Store solid stock at –20°C.
4. Bovine serum albumin (BSA) (store at –20°C in 1 mL aliquots): 10% (w/v) BSA (fraction V, Sigma A-7906; prepare stock in water, titrate to pH 7.6 with KOH).
5. Sperm dilution buffer (store at -20°C in 0.5-mL aliquots): 250 m$M$ sucrose, 75 m$M$ KCl, 1 m$M$ EDTA (0.5 $M$ stock, pH8), 0.5 m$M$ spermidine trihydrochloride (Sigma S-2501; 10 m$M$ stock; filter-sterilize and store aliquots at –20°C), 0.2 m$M$ spermine tetrahydrochloride (Sigma S-1141; 10 m$M$ stock; filter-sterilize and store aliquots at –20°C), 1 m$M$ dithiothreitol (Sigma D-0632; 100 m$M$ stock; filter-sterilize and store aliquots at –20°C).
6. Hoechst No. 33342 (Sigma B-2261): 10 mg/mL stock in dH$_2$O; store in a light-tight vessel at –20°C.

### 2.2. High-Speed Egg Extract Preparation

1. 20X Extract buffer (XB) salt stock: 2 $M$ KCl, 20 m$M$ MgCl$_2$, 2 m$M$ CaCl$_2$, filter-sterilize and store at 4°C.
2. Extract buffer (XB): 1X XB salts (100 m$M$ KCl, 0.1 m$M$ CaCl$_2$, 1 m$M$ MgCl$_2$; from 20X XB salts stock solution), 50 m$M$ sucrose (1.5 $M$ stock; filter-sterilize and store in aliquots at –20°C), 10m$M$ HEPES (1 $M$ stock, titrated with KOH such that pH is 7.7 when diluted to 10 m$M$; should require about 5.5 mL of 10 $N$ KOH for 100 mL; filter-sterilize, and store in aliquots at –20°C) (*see* **Note 4**). Prepare about 100 mL.

3. 2% (w/v) Cysteine (Sigma C-7755): Made up in 1X XB salts within 1 h of use and titrated to pH 7.8 with NaOH. Prepare about 300 mL.

4. CSF-XB: 1X XB salts (100 m$M$ KCl, 0.1 m$M$ CaCl$_2$, 1 m$M$ MgCl$_2$), 1 m$M$ MgCl$_2$ (in addition to MgCl$_2$ present in XB salts; final concentration 2 m$M$), 10 mM potassium HEPES, pH 7.7, 50 m$M$ sucrose, 5 m$M$ EGTA, pH 7.7. Prepare 50 mL.

5. Protease inhibitors: Mixture of leupeptin (Boeringer Mannheim 1017 101), chymostatin (Boeringer Mannheim 1 004 638), and pepstatin (Boeringer Mannheim 600 160), each dissolved to a final concentration of 10 mg/mL in dimethyl sulfoxide (DMSO). Store in small aliquots at –20°C.

6. 1X MMR Prepare as described in **Subheading 2.1., item 2**.

7. 1 $M$ CaCl$_2$ (filter-sterilize and store at 4°C).

8. Versilube F-50: Made by General Electric. Can be purchased from Andpak-EMA (1560 Dobbin Drive, San Jose, CA 95133; Tel. [408]-272-8007).

9. Energy mix (store in aliquots at –20°C): 150 m$M$ creatine phosphate (Boeringer Mannheim Biochemicals 127 574), 20 m$M$ ATP (Boeringer Mannheim Biochemicals 519 979), 20 m$M$ MgCl$_2$, Store in 0.1-mL aliquots at –20°C.

10. Pregnant mare serum gonadotropin (PMSG).: 100 U/mL PMSG (367222; Calbiochem, San Diego, CA). Made up in water and stored at –20°C.

11. Human chorionic gonadotropin (HCG): 1000 U/mL HCG (Sigma CG-10 from human pregnancy urine). Made up in water and stored at 4°C.

## 2.3. Nuclear Transplantation Reagents and Equipment

1. 1X MMR Prepared as described in **Subheading 2.1., item 2**.

2. 2.5% Cysteine in 1X MMR (titrate to pH 8.0 with NaOH). Make up fresh each day.

3. Sigmacote (Sigma SL-2).

4. 100 m$M$ MgCl$_2$.

5. 0.4X MMR + 6% (w/v) Ficoll (Sigma Type 400; F-4375) Sterilize by filtration.

6. 0.1X MMR + 50 µg/mL gentamycin (a 10 mg/mL stock solution may be purchased from Gibco-BRL; cat #15710-015). Add 6% (w/v) Ficoll for culturing embryos prior to gastrulation. Culture embryos in 0.1X MMR without Ficoll after gastrulation. Sterilize by filtration.

7. Progesterone (Sigma P-0130; SmM stock in EtOH).

8. Linearized plasmid (200–250 ng/µL): Although we have primarily used plasmid linearized with *Xba*I or *Not*I in transplantation reactions, we have also used *Xho*I, *Bam*HI, and *Eag*I successfully. We think that most enzymes that function in the moderately high salt conditions of the egg extract are likely to work. We commonly purify linearized plasmid for transgenesis using the Geneclean kit by Bio 101, Inc. (cat. no. 1001-200; 1070 Joshua Way, Vista CA 92083; 1-800-424-6101). Plasmid DNA can be eluted in either dH$_2$0 or in TE (10 m$M$ Tris-HCl, pH 8.0; 1 m$M$ EDTA). If linearized plasmid DNA needs to be concentrated, standard precipitation with 0.1 vol sodium acetate (3 $M$ stock; pH 5.2) and 2.5 vol absolute ethanol, followed by a 70% ethanol wash, can be used. We have found that plasmid DNA purified in this manner works well for making transgenic embryos and does not adversely effect embryonic development.

We have used enzymes purchased from Boeringer Mannheim or New England Biolabs for transplantation reactions. Some calibration may be required to determine the optimal amount of enzyme to add to each reaction, since additions of 0.5 µL of undiluted enzyme to reactions can adversely affect the development of nuclear transplant embryos. We generally test several dilutions of enzyme (1:2.5, 1:5; 1:10) to identify a dose that has no apparent deleterious effects on transplant embryo development when compared with embryos produced with no enzyme addition.

9. Agarose-coated injection dishes: 2.5% agarose in dH$_2$O is poured into 35-mm or 60-mm Petri dishes. Before the agarose solidifies, a well template (a rectangular square of Dow-Corning Sylgard 184 elastomer) is laid onto it. After the agarose has solidified and the Sylgard templates have been removed, 1X MMR is poured into each dish to prevent dehydration. The dishes are then wrapped in parafilm and stored at 4°C (weeks to months) until use.

10. Transplantation needles: 30-µL Drummond micropipets (Fisher, cat. #: 21-170J) are pulled to produce large needles with long, gently sloping tips (**Fig. 2**). A micropipet (1 mm wide; 8 cm long) is first heated in a Bunsen burner flame and drawn by hand to make the bore of the needle (200–400 µm wide). This drawn pipet should be 10–15 cm in length and should remain fairly straight when held by one end. To produce a gently sloping needle tip, this pipet is drawn again. We use a gravity-driven needle puller for this: the upper end of the needle is fixed in a brace, the center of the needle bore of the drawn pipet is placed within a small heating coil, and a weight is attached to the lower end of the needle. The gravity driven pullers we have used are home-built and about 10–20 yr old, but similar vertical pullers are commercially available from Narishige (i.e., Model PB-7). The second pull can also be performed with a horizontal needle puller available from Sutter Instrument Co. (Model P-87; Flaming/Brown micropipet puller) using settings like those used to make other injection needles. In limited trials of the Sutter puller using a standard setting, we have found that the needles produced had a steeper slope near the tip and were slightly more difficult to use than those drawn with our vertical puller; however, settings can probably be adjusted on this and other commercially available pullers to produce long, gently sloping tips that will work well for transplantation. Needles are clipped with a forceps to produce a beveled tip of 60–75 µm diameter (*see* inset in **Fig. 2**), using the ocular micrometer of a dissecting microscope for measurement.

11. Transplantation apparatus: We have found most commercial injection apparatuses commonly used for RNA and DNA injections unsuitable for nuclear transplantation. This is largely due to the difference in needle tip size. Flow through the 5–10 µm needle tips used for fluid injections can be controlled at fairly high pressures. However, with standard air-injection systems, we have been unable to obtain the extremely low positive pressure, and gentle, controlled flow required to deliver an intact nucleus in a small volume (10–15 nL) through the 50–70 µm tips of nuclear transplantation needles. Oil-filled injection systems (Drummond) are likely to work, since they are based on a positive displacement mechanism that should not be affected by the tip size of the needle. At this writing, though,

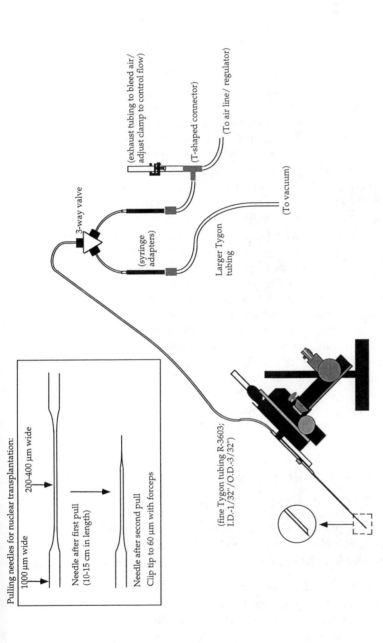

Fig. 2. Diagram of injection apparatus. A pressure regulator is set up on the house air, and a line connecting the regulator is split with a T-shaped connector into an exhaust tube and line to a three-way valve. Another line connects the house vacuum to the three-way valve. Finally, another line connects the three-way valve to the needle. By adjusting the valves in the three-way valve, the air pressure, and the clamp on the exhaust tube, one can very finely control the level of positive or negative pressure going into the needle. The rectangular inset shows how the needle should appear after the first and second pulls. The circular inset shows how the point of the needle should appear after it is clipped.

we have not tried one of these injection apparatuses for nuclear transplantations. Instead, we will describe how to make a home-made air injection apparatus that works extremely well for nuclear transplantions on a large scale and that costs very little (approx $200).

The transplantation apparatus that has given us the most success is shown in **Fig. 2**. A line connects the house vacuum outlet to a three way valve. Another line connects the house air outlet to a T-connector that splits the air flow into an exhaust line and another line connecting to the three way valve. Finally, another line connects the three way valve to the needle. For fine control of the positive pressure into the needle a screw clamp is placed on the exhaust line. Screwing down on this clamp increases the positive pressure into the system, while opening the clamp decreases the positive pressure. Negative pressure is established by opening slightly the valve (on the three way valve) connected to the house vacuum line. A more rough adjustment of positive pressure also can be obtained by opening or closing the valve (on the three way valve) connected to the house air line. By using a combination of these adjustments, we are able to obtain a very slow, controllable flow through a 50–70-$\mu$m needle. As flow is continuous, transplantations can usually be done more rapidly than injections of RNA or DNA, since it is only necessary to move from egg to egg to deliver nuclei. Parts needed to build the transplantation apparatus shown in **Fig. 2** are listed in **Table 1**.

Alternatively, a transplantation apparatus like the one shown in **Fig. 3** can be constructed. For this apparatus, a large, air-filled Hamilton Syringe (30 cc Multifit Interchangeable syringe with Luer-Lok tip; Fisher) is connected to a length of Tygon tubing. A metal plunger removed from a Syringe Microburet (Model # SB2; Micro-metric Instrument Co., Cleveland, OH) is used to control injection of the nuclei. We have found this apparatus usable although it is not controlled as easily as the one shown in **Fig. 2**.

## 3. Methods

### *3.1. Transgenesis Method (see Note 3)*

#### *3.1.1. Sperm Nuclei Preparation*

We have generally followed the standard protocol of Murray *(4)*, but have omitted the protease inhibitors leupeptin and phenylmethylsulfonyl fluoride from many steps to avoid transfer into the final mixture, which is diluted for egg injections.

1. Dissect and isolate the testes from a male:
   a. Anesthetize a male in a bucket containing a liter of 0.1% Tricaine (MS222, aminobenzoic acid ethyl ester, Sigma A-5040) and 0.1% sodium bicarbonate for at least 20 min (immersion of the animal in ice water for 20 min may also be used), and pith it.
   b. Cut through the ventral body wall and musculature, and lift the yellow fat bodies to isolate the two testes, which are attached to the base of the fat bodies, one on each side of the midline.

**Table 1**
**Part for Transplantation Apparatus**[a]

| Company | Product | Catalog number | Price | Phone/address |
|---|---|---|---|---|
| Newport Corp. (This may not be necessary; if your house air pressure is fairly low, you can attach the tubing directly to house air.) | Air regulator/filter | ARF | $79.00 | 1-800-222-6440 1791 Deere Ave. Irvine, CA 92714 |
| Western Analytical Products, Inc. (They are the US distributor for Omnifit Ltd. Phone: 01223-69841 Fax: 01023-61106 51 Norfolk St. Cambridge CB12LE) | Three-way valve Trivalve caps (mixed colors) | 001102 001310 | $105.51 $5.25 | 1-800-541-8421 25407 Blackthorn Murrieta, CA 92563 |
| Fisher | Hoffman open-side tubing clamps | 05-875A | $20.00 for 10 | |
| | T-shaped connectors | 15-319C | $10.44 for 12 | |
| | Tygon tubing 1/32 in. | 12-169-1A | $4.10 | |
| | Tygon tubing 3/16 in. | 14-169-3B | $7.50 | |
| Fine Science Tools, Inc. | Precision micro- manipulator MM33 | 25033-10 | $690.00 | 1-800-521-2109 373-G Vintage Park Dr. Foster City, CA 94404 |
| | Magnetic base | 25810-00 | $84.95 | |

[a]A base somewhat heavier and more stable than the magnetic model available from Fine Science Tools was previoulsy available from Brinkman. To our knowledge, however, a comparable product is no longer available either from Brinkman or from other manufacturers; we have had a copy built by the local machine shop.

    c. Remove the testes with dissecting scissors, and place them in a 35-mm tissue-culture dish containing cold 1X MMR.
       Rinse the testes in three changes of cold 1X MMR and two times in cold 1X NPB, removing any attached pieces of fat body or debris with a fine forceps. Take care not to puncture the tissue pouches, since this releases the sperm.
  2. Move the cleaned testes to a dry 35-mm tissue-culture dish, and macerate the tissue well (until clumps are no longer visible to the naked eye) with a pair of clean forceps.

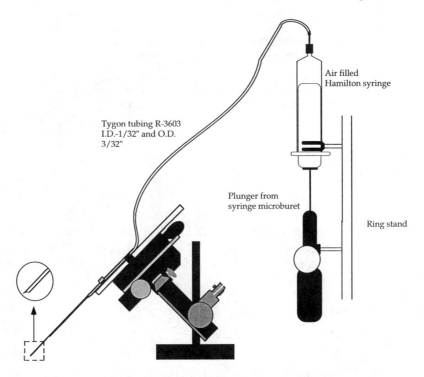

Fig. 3. Diagram of alternative injection apparatus. A line connects the needle to a 30-mL Hamilton syringe that is held by a clamp to a ring stand. The glass syringe plunger is pushed in using a microburet that is also clamped to the ring stand. A slow, controlled flow of liquid through the needle can be obtained by leaving a large cushion of air inside the syringe between the glass plunger and the end attached to the tubing. The circular inset shows how the point of the needle should appear after it is clipped.

3. Resuspend the macerated testes in 2 mL of 1X NPB by gently pipeting the solution up and down through a fire-polished, truncated Pasteur pipet with an opening of about 3 mm in diameter.
4. Squirt the sperm suspension through two to four thicknesses of cheesecloth placed into a funnel, and collect the solution into a 15-mL tube (we use round bottom polypropylene tubes; Fisher, cat. #: 14-956-1J). Rinse the forceps and dish with 8 mL of 1X NPB, and force this through the cheesecloth into the 15-mL tube. With a gloved hand, fold the cheesecloth and squeeze any remaining liquid through the funnel into the 15-mL tube.
5. Pellet the sperm by centrifugation at 1500*g* for 10 min at 4°C (we use a Sorvall HB-6 or similar swinging bucket rotor fitted with the appropriate adapters). Resuspend sperm in 8 mL NPB and repellet by centrifugation at 1500*g* for 10 min at 4°C.
6. Resuspend pellet in 1 mL NPB with a cut plastic pipet tip, warm the suspension to room temperature, and add 50 μL of 10 mg/mL lysolecithin. Mix gently and incubate for 5 min at room temperature.

7. Add 10 mL cold 1X NPB + 3% BSA (with protease inhibitors; 1:1000 dilution of leupeptin and PMSF stock solutions) to the suspension, mix gently by inversion, and centrifuge at 1500$g$ for 10 min at 4°C. Decant the supernatant.

8. Resuspend the pellet in 5 ml cold 1X NPB + 0.3%BSA (no protease inhibitors), mix gently by inversion, and repellet as before.

9. Resuspend the pellet in 500 μL of sperm dilution buffer, and transfer suspension into a 1.5-mL Eppendorf tube. Count the sperm density using a hemacytometer (Fisher, cat. #: 02-671-5): dilute a small amount of the concentrated sperm 1:100 in sperm dilution buffer, and add 1μL of 1:100 Hoechst stock to visualize the sperm heads under a fluorescence microscope. For a 1:100 dilution of our sperm stock, we typically obtain counts of 75–125 ($\times 10^4$ cells/mL). At this concentration, the undiluted stock contains 75–125 sperm/nL. If your sperm stock is substantially less concentrated (i.e., a count of <50 for a 1:100 dilution), repellet the sperm and resuspend in a smaller volume of sperm storage buffer. Sperm can be stored at 4°C and used for transplantations for up to 48 h.

### 3.1.2. High-Speed Extract Preparation

This protocol is an adaptation of Murray *(4)*. Briefly, a crude cytostatic factor (CSF) arrested egg extract (cytoplasm arrested in meiotic metaphase) is prepared. Calcium is then added to allow the extract to progress into interphase, and a high-speed spin is performed to obtain a purer cytoplasmic fraction. Cytochalasin is omitted from the protocol, since carryover of cytochalasin into the final extract used for sperm incubations interferes with normal development of transplant embryos. Use of high-speed rather than crude cytoplasmic extracts is advantageous, because high-speed extracts promote swelling of added sperm nuclei (and some chromatin decondensation), but do not promote DNA replication. Replication of sperm DNA incubated in these extracts occurs after transplantation of the nucleus into the egg rather than in the extract. High-speed extract can be stored frozen in small aliquots (at –80°C) and thawed before use.

1. Prime 8–12 female adult *X. laevis* about 24 h prior to HCG injection by injecting 25–100 U of PMSG into the dorsal lymph sac. Maintain at room temperature. The evening before the extract preparation begins, inject each frog with 500–800 U HCG, and place 2 frogs/container into 2 L 1X MMR. Since one frog with lysing or activating eggs can compromise the whole extract preparation, we prefer to separate the frogs into pairs for the ovulation. The frogs are then placed at 15–18°C overnight (12–14 h). On the next morning, the egg quality from each container is screened before mixing all the eggs and starting the extract preparation. All the eggs released from a frog that lays mottled, lysing, or dying eggs are left out of the extract preparation.

2. All solutions should be prepared before beginning the extract preparation, since the procedure should be carried through all steps promptly once it is initiated; opti-

mally, the high-speed spin should begin within 45–60 min of dejellying the eggs. Gently expel eggs manually from each frog into a large dish of 1X MMR, and collect unbroken eggs with even pigmentation. Good eggs can also be collected from the 1X MMR in the frog buckets. Total volume of eggs should be 100 mL or greater before dejellying.

3. Remove as much MMR as possible from the eggs. Dejelly eggs in 2% cysteine in XB salts (no HEPES/sucrose). Add a small amount at a time, swirl eggs, and partially replace with fresh cysteine several times during dejellying. Remove broken eggs with a pipet during dejellying. Dejellying can be performed separately for different batches of eggs, and batches that show breakage or egg activation are discarded.

4. Wash eggs in XB (with HEPES/sucrose). We use about 35 mL for each wash, and do four washes.

5. Wash eggs in CSF-XB with protease inhibitors. We do two 25-mL washes.

6. Using a wide-bore Pasteur pipet, transfer eggs into Beckman ultraclear tubes. For these volumes, we typically use 14 × 95 mm tubes (cat. no.: 344060; Beckman, Fullerton, CA 344057). If multiple tubes will be used, try to transfer an equal volume of eggs per tube. Remove as much CSF-XB as possible, and replace with about 1 mL of Versilube F-50.

7. Spin in a clinical centrifuge at room temperature for about 60 s at 1000 rpm (150$g$) and then 30 s at 2000 rpm (600$g$). Eggs should be packed after this spin, but unbroken. Versilube should replace the CSF-XB between the eggs, and an inverted meniscus between the Versilube and displaced CSF-XB should be clearly visible. Remove the excess CSF-XB and Versilube, and then balance the tubes.

8. Spin the tubes in rubber adapters for 10 min at 16,000$g$ at 2°C in Sorvall HB-4 or similar swinging bucket rotor to crush the eggs. The eggs should be separated into three layers: lipid (top), cytoplasm (center), and yolk (bottom). Collect the cytoplasmic layer from each tube with an 18-gage needle by inserting the needle at the base of the cytoplasmic layer and withdrawing slowly. Transfer cytoplasm to a fresh Beckman tube on ice. If large volumes of darkly pigmented eggs are used, the cytoplasmic layer may be grayish rather than golden at this step. After a second spin to clarify this extract, it should be golden.

9. Add protease inhibitors to the isolated cytoplasm (do not add cytochalasin); recentrifuge the cytoplasm in Beckman tubes for an additional 10 min at 16,000$g$ to clarify, again using a swinging bucket rotor. Collect the clarified cytoplasm as before. Expect to get obtain 0.75–1 mL cytoplasm/batch of eggs collected from one frog.

10. Add 1/20 vol of the ATP-regenerating system (energy mix). Transfer the clarified cytoplasm into TL100.3 thick-wall polycarbonate tubes (Beckman 349622). Tubes hold about 3 mL each and should be at least half full.

11. Add CaCl$_2$ to each tube to a final concentration of 0.4 m$M$; this inactivates CSF and pushes the extract into interphase. Incubate at room temperature for 15 min and then balance for the high-speed spin.

12. Spin tubes in a Beckman tabletop TL-100 ultracentrifuge in a TL100.3 rotor (gold top; fixed angle) at 70,000 rpm for 1.5 h at 4°C.

13. The cytoplasm will fractionate into four layers, top to bottom: lipid, cytosol, membranes/mitochondria, and glycogen/ribosomes. Remove the cytosolic layer from each tube (about 1 mL if 2–3 mL were loaded into the tube) by inserting a syringe into the top of the tube through the lipid layer. Transfer this fraction to fresh TL-100 tubes, and spin again at 70,000 for 20 min at 4°C.

14. Aliquot the high-speed cytosol supernatant into 25-µL aliquots in 0.5-mL Eppendorf tubes. Quick-freeze aliquots in liquid nitrogen, and store at –80°C until use. We typically obtain 1–2 mL of high-speed cytosol from preparations of this scale. Sperm nuclei should be incubated in an aliquot of extract and stained with Hoechst as previously described to determine whether extract is effective. If active interphase extract has been prepared, nuclei should swell visibly (thicken and lengthen) within 10 min of addition to extract at room temperature.

### 3.1.3. Transgenesis by Sperm Nuclear Transplantation into Unfertilized Eggs (see **Notes 1–3**)

1. Inject two to four adult female frogs in the dorsal lymph sac with 500–800 U HCG, and incubate at 15°C for 12–16 h before transplantations.

2. Set up injection area: Coat inside of transplantation needles with Sigmacote to prevent shearing of sperm nuclei flowing through the needle (needles can be coated 10 min to several months before use). Attach approx 1 cm Tygon tubing (R-3603 1/32 in.; Fisher, cat. #: 14-169-1A) to the end of a plastic pipetman (200 µL) tip, and use the pipetman to draw up Sigmacote; then attach the other end of the tubing to the injection needle. Depress the pipetman plunger to force Sigmacote through the needle until a few drops emerge from the tip, and then release the pipetman plunger to withdraw most of the solution. Rinse needle with water before using for transplantations.

3. Adjust the transplantation apparatus. Fill a Petri dish with water, attach a needle to the tubing and micromanipulator, and establish a very low positive needle pressure through the needle. To do this, set the air flow valve to partially closed (so that it can be opened further), and then open the vacuum valve until liquid is drawn into the needle. When the needle is filled to the wide bore, partially close the vacuum until the flow is either stopped or just slightly outward owing to the partial air flow pressure. The pressure should be so low that it should not be possible to see the meniscus moving at all.

   When you are finished adjusting the system and are ready to load for transplants, put a finger over the exhaust tube to discharge the liquid from the needle, back-load the needle, and add positive pressure to the needle just slightly to begin injecting. This is done either by increasing the air flow or screwing done on the clamp fitted on the exhaust tube.

4. Fill agarose-coated injection dishes with 0.4X MMR + 6% Ficoll.

5. Set up a reaction. Sample reaction (~1:10 dilution of sperm stock): 4 µL sperm stock (~4 × 10⁵ nuclei) and 5 µL linearized plasmid (150–250 ng/µL). Incubate for 5 min.

6. Add: 0.5 µL of an ~1:5 dilution of *Xba*I or *Not*I, 2 µL 100 m*M* MgCl₂ (add to 5 m*M* final at all steps to aid enzyme action), and 25 µL high-speed extract.

7. Mix the reaction by gentle pipeting (using a clipped yellow tip). Incubate for 10 min at room temperature; sperm will now be visibly swelled if diluted into Hoechst as before and observed with a 10X–20X objective.

8. While sperm are swelling in reaction mixture, collect eggs from individual frogs and dejelly them in 2.5% cysteine hydrochloride in 1X MMR (pH 8.0 with NaOH).

9. Under the dissecting microscope, inspect the eggs released from each frog for general health (eggs with even pigmentation and that remain round after dejellying). Draw the healthiest eggs into a wide-bore Pasteur pipet and transfer them to the square space in the injection dish. We generally fill the square space with eggs such that no space is left between the eggs. After about 5 min in 0.4X MMR + 6% Ficoll, the eggs will pierce easily.

10. Dilute the sperm into sperm dilution buffer (SDB) at 1:25–1:100 (such that the final dilution is 1:250–1:1000 or a concentration of 1–2 sperm/10–15 nL injection volume). For some enzymes, such as *Not*I or *Xba*I, add $MgCl_2$ to 5 m$M$ to aid enzyme action.

   Before removing sperm from the stock tube or from the dilution used for injection, **always mix thoroughly** with a cut yellow tip, since sperm will rapidly settle out of the suspension.

11. Use a piece of Tygon tubing attached to a yellow tip (as previously described for Sigmacoting needles) to draw up the dilute sperm suspension and back-load the needle. Reattach the needle to the micromanipulator, and turn the air pressure up just slightly so that solution begins to flow from the needle tip (seen under the microscope as a schlearing solution of a different density). Owing to the low air pressure, solution will flow out of the needle only when the tip enters the liquid.

12. Transplant sperm nuclei into unfertilized eggs. The rate of flow should be robust enough that the needle does not reverse flow or clog with cytoplasm during injections and slow enough to be manageable. At the flow and injection rates we generally use, about 10-nL vol is delivered in each injection, so a 1:500–1:1000 dilution of the original sperm stock allows approximately one sperm to be injected in that volume. Move the needle fairly rapidly from egg to egg, piercing the plasma membrane of each egg with a single, sharp motion and then drawing the needle out more slowly. The angle of the needle should be perpendicular to the membrane surface (rather than glancing) to avoid tearing the plasma membrane.

   A hole about the diameter of the needle tip should be visible on the egg and should remain open for about 5 s after injection; when the flow is too low, the hole created in the egg by the needle instantly closes over after injection and little or no volume is delivered. When the flow is too rapid, the surface of the egg near the injection site may ripple or the site of injection may expand in size significantly. If the needle becomes clogged with cytoplasm, bring the tip to the air-liquid interface of the dish. Sometimes the surface tension of the interface removes the cytoplasm plug in the end of the needle. If a needle tip is too narrow, or if it becomes partially clogged with debris during transplantations, the injected nuclei will be damaged during transplantation, and haploid embryos will result.

Haploid tadpoles have shortened trunks and tails, are thicker than normal through-out the trunk region with a "pigeon-chested" appearance, and often have heads and tails that curl toward the dorsal side; these tadpoles will live for a while, but usually become edemic and die around the time of feeding *(21)*.

You can determine whether your sperm dilution and the flow rate used for injections were appropriate by watching the first cleavage of the transplanted eggs. If few of the eggs received a nucleus, the frequency of cleavage will be low; one-fifth to one-third of our transplantations typically result in normally cleaving embryos. If too much volume was injected into the eggs, they may also fail to cleave; in this case, the animal hemisphere pigmentation may appear mottled or "marbleized," or have other signs or unhealthiness owing to overinjection. Eggs that were injected with more than one nucleus will divide at the time of first cleavage abnormally into three or four (or more) cells. Many of these embryos will develop to blastula stages, but most fail during gastrulation; in some, a region of the embryo will fail to cellularize and die. Eggs injected with multiple nuclei that do gastrulate usually do so abnormally; typically, blastopore closure is incomplete, resulting in embryos that form two wings of somites and neural tissue on each side of the exposed yolky tissue lying in the center of the trunk. This type of gastrulation failure is common to stressed or unhealthy embryos (particularly embryos derived from "soft" eggs).

13. When the cleaving transplant embryos have reached the 4- to 16-cell stage, gently separate them from uncleaved eggs and move with a wide-bore Pasteur or Spemann pipet to a separate dish of 0.1X MMR + 6%Ficoll + 50 µg/mL gentamycin. We commonly culture transplanted embryos in 6- or 12-well tissue-culture dishes with about 10-embryos/well, since culturing embryos at high density can compromise their health. It is also important to remove dying embryos promptly, since they also can compromise the health of their siblings.

14. When embryos are around stage 12, media is replaced with 0.1X MMR + 50 µg/mL gentamycin without Ficoll. Because of the large needle tip used for transplantations, embryos often develop large blebs at the site of injection. These blebs occur when cells are forced out of the hole left in the vitelline membrane at the injection site, but they generally do not affect development. The blebs usually fall off on their own at the neurula or tailbud stages, but they can be removed manually once the embryos have reached the late blastula stage.

## 3.2. Fertilization of In Vitro Matured Oocytes by Nuclear Transplantation

1. Prime female adult *X. laevis* with 50 U PMSG, and leave at 18°C for about 48 h before isolating oocytes. Techniques for obtaining ovary tissue, isolating oocytes by manual defolliculation and culturing and injecting oocytes have been described elsewhere *(16,22)*. Maintain oocytes in 1X modified barth's saline (MBS) + 1 mg/mL (w/v) BSA, and inject and manipulate as desired. Alternatively, oocytes can be maintained in oocyte culture medium (OCM). Prepare fresh for each experiment. This medium is preferable if oocytes will be cultured for an extended

period (>24 h) before maturation. Transfer defolliculated oocytes to fresh media, and change these media several times to remove traces of yolk and debris.

2. Prepare sperm nuclei as previously described (**Subheading 3.1.1.**)

3. Add 1–5 μm progesterone to oocytes maintained in MBS + 1 mg/mL BSA to begin maturation.

4. Determine when sperm nuclei should be transplanted. A general rule to follow is that oocytes should be ready for fertilization in about 2X the amount of time taken to get from progesterone addition to germinal vesicle breakdown (GVBD; appearance of white spot in the animal hemisphere). We typically add 5 μm progesterone in the evening after defolliculating oocytes (5–7 PM), incubate oocytes at 18°C overnight and during the next day, and inject sperm 20–25 h after progesterone addition (*see* **Note 2**).

    The most common mistake made is not allowing oocytes sufficient time after maturation before injecting the sperm nuclei. Oocytes must be able to respond to pricking by a needle with a vigorous cortical contraction before sperm are transplanted, or no development will occur. Even after oocytes first become responsive to pricking, they are probably not fully competent to support embryonic development immediately and should be incubated an additional 3–4 h at 18°C. Since there is probably quite a bit of variability between batches of oocytes from different frogs and between frogs from different colonies, the optimal timing should be determined by prick-activating a small number of test oocytes at several times during the incubation period to determine when they become responsive.

5. Dilute and transplant sperm nuclei as described in the transgenesis protocol. There is no need to swell the nuclei in interphase extract. We have used slightly lower dilutions of sperm than are used for transgenesis for this protocol (such that two to three sperm may be deposited into some eggs) and have done these injections in 0.4X MMR without Ficoll. Use a 40–60 μm wide needle tip to transplant the sperm as described for transgenesis. When successful, oocytes should pierce very easily for injection, and membrane texture should not seem at all rubbery. There should be a normal cortical contraction in the animal hemisphere after activation, and the injected, matured oocytes (eggs now) should look and later cleave like fertilized eggs. When testing this method, approx 25% of the in vitro matured oocytes developed into blastula-gastrula stages. Of these, the majority developed into tadpoles, and were apparently morphologically normal and raised for months.

# 4. Notes

## 4.1. Factors Affecting the Success of Nuclear Transplantation-Based Transgenesis

1. Egg quality is a major factor that contributes to the level of postgastrula development, which is obtained from sperm nuclear transplantation. To obtain good postgastrula development, eggs must be generally healthy. In particular, they should have even pigmentation and should be firm enough to hold their shape well after dejellying. In addition, it is important that they do not become acti-

vated before they are injected with nuclei. When egg quality is poor, a fraction of the embryos will show morphogenetic defects, resulting in incomplete blastopore closure during gastrulation. This problem is often compounded by expressing genes at high levels during the gastrula stages. Therefore, embryos expressing genes from the CMV promoter are more likely to show nonspecific gastrulation defects than embryos expressing genes from strong promoters that are turned on after gastrulation.

2. Transplantations should be performed and eggs incubated after transplantation at temperatures no higher than 22°C. We have found that transplantation and early incubation of activated eggs at elevated temperatures (24-25°C) lowers the frequency with which plasmids are expressed in batches of nuclear transplantation-derived embryos. Embryos in these batches also frequently express plasmids in only one-quarter to one-half of the expected cells. We believe that acceleration of the first cell cycle, which occurs at 24–25°C, may give these embryos inadequate opportunity to integrate introduced plasmids prior to first cleavage, thus resulting in more chimeric and nonexpressing embryos.

3. It may be important to note that although this technique is very efficient and workable, it involves several steps; all of which are critical for its success. Therefore, we suggest that anyone trying to learn the technique does so in steps, rather than all at once. For example, one should first learn to isolate sperm nuclei and transplant them into eggs. Once this can be done successfully, resulting in normal development, then one can determine whether sperm nuclei, swollen in extracts, gives normal development. If swelling of sperm in extract has no adverse effects on the level of development obtained after transplantations, one can add plasmid and enzyme to the reaction, thus reconstituting the whole transgenesis procedure.

4. Dilution drastically changes the pH of HEPES, making it impossible to pH the stock directly.

5. For each sperm nuclei prep, it is most convenient to make about 30 mL of 2X NPB form stock solutions. Then use this 2X stock to make all subsequent solutions (i.e., 1X NPB + protease inhibitors, 1X NPB + 3% BSA, and so forth).

## 4.2. The Frog as a Mouse

6. The transgenesis procedure described here compares very favorably with those developed for mouse or zebrafish. Embryos expressing plasmids nonmosaically can be obtained in high numbers directly, and since the embryos are not chimeric, breeding of animals is not required. In addition, the cost of studies involving transgenic frog embryos will be considerably lower than that required for similar studies in mammals. In fact, the frog may prove useful for the study of regulation of mouse promoters. Limited studies have shown that mouse promoters are regulated appropriately in other systems and vice versa *(23–28)*. Therefore, initial promoter mapping and analysis may be done more easily and effectively in the frog than in the mouse.

## 4.3. Future Prospects in the Frog: Knockouts and Genetics

7. Except for studies where gene function has been inhibited by the expression of dominant negative mutations *(29–31)* or maternal mRNAs where degraded following injection with oligonucleotides *(12–15)*, it has been difficult to inhibit the function of genes in the early embryo specifically. In the future, we also hope to combine transgenesis with antisense *(32–34)* and ribozyme *(35–37)* technologies in order to deplete specific gene products from *Xenopus* embryos.

The advantages of the frog system are numerous, but one major disadvantage is that it has not been exploited at the genetic level. The method for transgenesis we have developed can be adopted for an insertional mutagenesis scheme. Since *Xenopus laevis* is pseudotetraploid and has a long generation time, we suggest using *Xenopus tropicalis*, which is diploid and has a generation time of around 4–6 mo *(38)*. For similar reasons, *Xenopus tropicalis* will also be the species of choice for doing targeted mutations.

## Acknowledgments

This transgenesis procedure was developed while we were both at the Department of Molecular and Cell Biology at the University of California at Berkeley. Above all we would like to thank Ray Keller and John Gerhart for their endless encouragement and patience. Also we thank John Bradley and Nancy Papalopulu for helpful comments on the manuscript. E. A. was a fellow of The Jane Coffin Childs Memorial Fund for Medical Research.

## References

1. Kroll, K. L. and Amaya, E. (1996) Transgenic Xenopus embryos from sperm nuclear transplantation reveal FGf signaling requirements during gastrulation. *Development* **122,** 3173–3183.
2. Kroll, K. L. and Gerhart, J. C. (1994) Transgenic *X. laevis* embryos from eggs transplanted with nuclei of transfected cultured cells. *Science* **266,** 650–653.
3. Leno, G. H. and Laskey, R. A. (1991) DNA replication in cell-free extracts from *Xenopus laevis*, in *Methods in Cell Biology,* vol. 36 (Kay, B. K. and Peng, H. B., eds.), Academic, San Diego, CA, pp. 561–579.
4. Murray, A. W. (1991) Cell cycle extracts, in *Methods in Cell Biology,* vol. 36 (Kay, B. K. and Peng, H. B., eds.), Academic, San Diego, CA, pp. 581–605.
5. Newmeyer, D. D. and Wilson, K. L. (1991) Egg extracts for nuclear import and nuclear assembly reactions, in *Methods in Cell Biology,* vol. 36 (Kay, B. K. and Peng, H. B., eds.), Academic, San Diego, CA, pp. 607–634.
6. Schiestl, R. H. and Petes, T. D. (1991) Integration of DNA fragments by illegitimate recombination in *Saccharomyces cerevisiae. Proc. Natl. Acad. Sci. USA* **88,** 7585–7589.
7. Kuspa, A. and Loomis, W. F. (1992) Tagging developmental genes in *Dictyostelium* by restriction enzyme-mediated integration of plasmid DNA. *Proc. Natl. Acad. Sci. USA* **89,** 8803–8807.

8. Turner, D. L. and Weintraub, H. (1994) Expression of *achaete-scute homolog 3* in *Xenopus* embryos converts ectodermal cells to a neural fate. *Genes Dev.* **8,** 1434–1447.

9. Cross, G. S., Wilson, C., Erba, H. P., and Woodland, H. R. (1988) Cytoskeletal actin gene families of *Xenopus borealis* and *Xenopus laevis. J. Mol. Evol.* **27,** 17–28.

10. Harland, R. M. and Misher, L. (1988) Stability of RNA in developing *Xenopus* embryos and identification of a destabilizing sequence in TFIIIA RNA. *Development* **102,** 837–852.

11. Mohun, T. J., Garrett, N., and Gurdon, J. B. (1986) Upstream sequences required for tissue-specific activation of the cardiac actin gene in *Xenopus laevis* embryos. *EMBO J.* **5,** 3185–3193.

12. Kloc, M., Miller, M., Carrasco, A. E., Eastman, E., and Etkin, L. (1989) The maternal store of the xlgv7 mRNA in full-grown oocytes is not required for normal development in *Xenopus. Development* **107,** 899–907.

13. Torpey, N., Wylie, C. C., and Heasman, J. (1992) Function of maternal cytokeratin in *Xenopus* development. *Nature* **357,** 413–415.

14. Heasman, J., Crawford, A., Goldstone, K., Garner-Hamrick, P., Gumbiner, B., McCrea, P., Kintner, C., Noro, C. Y., and Wylie, C. (1994) Overexpression of cadherins and underexpression of beta-catenin inhibit dorsal mesoderm induction in early *Xenopus* embryos. *Cell* **79,** 791–803.

15. Vernos, I., Raats, J., Hirano, T., Heasman, J., Karsenti, E., and Wylie, C. (1995) Xklp1, a chromosomal *Xenopus* kinesin-like protein essential for spindle organization and chromosome positioning. *Cell* **81,** 117–127.

16. Heasman, J., Holwill, S., and Wylie, C. (1991) Fertilization of cultured *Xenopus* oocytes and use in studies of maternally inherited molecules, in *Methods in Cell Biology,* vol. 36 (Kay, B. K. and Peng, H. B., eds.), Academic, San Diego, CA, pp. 213–230.

17. Palermo, G., Joris, H., Devroey, P., and van Steirteghem, A. C. (1992) Pregnancies after intracytoplasmic injection of single spermatozoon into an oocyte. *Lancet* **340,** 17,18.

18. Van Steirteghem, A. C., Liu, J., Joris, H., Nagy, Z., Jassenswillen, C., Tournaye, H., Derde, M., Van Assche, E., and Devroey, P. (1993) Higher success rate by intracytoplasmic sperm injection than subzonal insemination. Report of a series of 300 consecutive treatment cycles. *Human Reprod.* **8,** 1055–1060.

19. Van Steirteghem, A. C., Nagy, Z., Joris, H., Liu, J., Staessen, C., Smitz, J., Wisanto, A., and Devroey, P. (1993) High fertilization and implantation rates after intracytoplasmic sperm injection. *Human Reprod.* **8,** 1061–1066.

20. Payne, D., Flaherty, S. P., Jeffrey, R., Warnes, G. M., and Matthews, C. D. (1994) Successful treatment of severe male factor infertility in 100 consecutive cycles using intracytoplasmic sperm injection. *Human Reprod.* **9,** 2051–2057.

21. Gurdon, J. B. (1960) The effects of ultraviolet irradiation of the uncleaved eggs of *Xenopus laevis. Q. J. Microsc. Sci.* **101,** 299–312.

22. Smith, L. D., Xu, W., and Varnold, R. L. (1991) Oogenesis and oocyte isolation, in *Methods in Cell Biology,* vol. 36 (Kay, B. K. and Peng, H. B., eds.), Academic, San Diego, CA, pp. 45–60.

23. McGinnis, N., Kuziora, M. A., and McGinnis, W. (1990) Human *Hox-4.2* and *Drosophila Deformed* encode similar regulatory specificities in *Drosophila* embryos and larvae. *Cell* **63,** 969–976.

24. Brakenhoff, R. H., Ruuls, R. C., Jacobs, E. H., Schoenmakers, J. G., and Lubsen, N. H. (1991) Transgenic *Xenopus laevis* tadpoles: a transient *in vivo* model system for the manipulation of lens function and lens development. *Nucleic Acids Res.* **19,** 1279–1284.

25. Dillon, N., Kollias, G., Grosveld, F., and Williams, J. G. (1991) Expression of adult and tadpole specific globin genes from *Xenopus laevis* in transgenic mice. *Nucleic Acids Res.* **19,** 6227–6230.

26. Awgulewitsch, A. and Jacobs, D. (1992) Deformed autoregulatory element from *Drosophila* functions in a conserved manner in transgenic mice. *Nature* **358,** 341–344.

27. Westerfield, M., Wegner, J., Jegalian, B. G., DeRobertis, E.M., and Puschel, A.W. (1992) Specific activation of mammalian Hox promoters in mosaic transgenic zebrafish. *Genes Dev.* **6,** 591–598.

28. Morasso, M. I., Mahon, K. A., and Sargent, T. D. (1995) A *Xenopus* distal-less gene in transgenic mice: conserved regulation in distal limb epidermis and other sites of epithelial-mesenchymal interaction. *Proc. Natl. Acad. Sci. USA* **92,** 3968–3972.

29. Herskowitz, I. (1987) Functional inactivation of genes by dominant negative mutations. *Nature* **329,** 219–222.

30. Christian, J. L., Edelstein, N. G., and Moon, R. T. (1990) Overexpression of wild-type and dominant negative mutant vimentin subunits in developing *Xenopus* embryos. *New Biol.* **2,** 700–711.

31. Amaya, E., Musci, T. J., and Kirschner, M. W. (1991) Expression of a dominant negative mutant of the FGF receptor disrupts mesoderm formation in *Xenopus* embryos. *Cell* **66,** 257–270.

32. Harland, R. and Weintraub, H. (1985) Translation of mRNA injected into *Xenopus* oocytes is specifically inhibited by antisense RNA. *J. Cell Biol.* **101,** 1094–1099.

33. Melton, D. A. (1985) Injected antisense RNAs specifically block messenger RNA translation *in vivo. Proc. Natl. Acad. Sci. USA* **82,** 144–148.

34. Nichols, A., Rungger-Brändle, E., Muster, L., and Rungger, D. (1995) Inhibition of *Xhox1A* gene expression in *Xenopus* embryos by antisense RNA produced from an expression vector read by RNA polymerase III. *Mech. Dev.* **52,** 37–49.

35. Cotten, M. and Birnstiel, M. L. (1989) Ribozyme mediated destruction of RNA *in vivo. EMBO J.* **8,** 3861–3866.

36. Zhao, J. J. and Pick, L. (1993) Generating loss-of-function phenotypes of the *fushi tarazu* gene with a targeted ribozyme in Drosophila. *Nature* **365,** 448–451.

37. Bouvet, P., Dimitrov, S., and Wolffe, A. P. (1994) Specific regulation of *Xenopus* chromosomal 5S rRNA gene transcription *in vivo* by histone H1. *Genes Dev.* **8,** 1147–1159.

38. Tymowska, J. and Fischberg, M. (1973) Chromosome complements of the genus *Xenopus. Chromosoma* **44,** 335–342.

# 28

# Axolotl/newt

## Malcolm Maden

## 1. Overview

Limb regeneration is one of the oldest topics within developmental biology, since the first experiments were reported in 1768 by Spallanzani, a Roman Catholic priest and Professor of Natural History at Padua, Italy. In his work *Prodromo di un opera da imprimersi sopra la riproduzioni animali,* he showed that regenerative ability was widespread throughout the animal kingdom, and that the legs and tails of newts could regenerate perfectly time after time. These studies provided much fuel to the contemporary debate between the preformationists and epigeneticists, and won for Spallanzani his election as a foreign correspondent of the Royal Society.

The Italian connection was crucial to the early development of regeneration research, because in 1823 Todd (*1*) showed that the regeneration of newt limbs was completely inhibited when the nerve supply to the limb was cut. Todd was a physician who served in the Royal Navy and apparently performed his experiments while at the British naval base in Naples. This was the first demonstration of the neurotrophic requirement for limb regeneration. As the 19th century progressed, the pace of regeneration research heated up, and by the turn of the century, there were hundreds of papers being published each year, mostly in German. Limb regeneration, along with studies on frog and newt eggs, were the driving forces in vertebrate developmental biology at that time. This pace continued throughout the first half of this century and from the 1930s researchers in the US made dominant contributions. In the last 20 yr, the pace has considerably slackened, since very few researchers have entered what has become an unfashionable field. Modern developmental biology is now dominated by *Xenopus* eggs and mouse genetics.

From: *Methods in Molecular Biology, Vol. 97: Molecular Embryology: Methods and Protocols*
Edited by: P. T. Sharpe and I. Mason © Humana Press Inc., Totowa, NJ

Nevertheless, virtually all of the fundamental and fascinating questions that have intrigued students of limb regeneration for more than a century remain unanswered and are still there to inspire today's enquiring minds. Why can newts regenerate limbs and mammals cannot? What factors do the nerves supply? What factors does the epidermis supply? How do cells dedifferentiate and undergo transformation into another cell type? Are the mechanisms by which redifferentiation takes place the same as those that were used to develop the limb in the first place? Is the newly regenerated limb a different age from its contralateral unregenerated partner, and if so, is this a way to reverse the deleterious processes of aging?

Imagine what benefit the answers to these questions would bring to humankind. Indeed, there is already one good example of how limb regeneration studies have been of direct practical benefit to medicine. After limb amputation, the epidermis migrates over the stump to close the wound resulting in epidermal/mesenchymal interactions, which are crucial to the induction of regeneration (*see* **Subheading 2.2.**). It has been known since 1906 *(2)* that if full thickness skin is sewn over the amputation plane regeneration is inhibited. This knowledge led to a change in surgical procedure in dealing with the amputated fingertips of children. Prior to Illingworth's report *(3)*, such cases were dealt with by suturing the wound to make a cosmetically acceptable product, but one which was missing the terminal phalanx. Illingworth showed that if such amputations are simply covered and left to heal normally, then the terminal phalanx and nail will regenerate perfectly. Thus young children up to about the age of 11 yr have considerable capacity for regeneration of fingertips, provided the principles of regeneration are followed.

For more details on the subject of limb regeneration than can be included here, the reader is referred to a book by Wallace *(4)*, which also contains much information on the older literature.

## 1.1. Which Species Regenerate?

Limb regeneration occurs after amputation of most larval and adult urodeles (newts and salamanders) and in the tadpole stages of anurans (frogs and toads). After metamorphosis in frogs and toads, regeneration is inhibited with the maximal response being the growth of a long spike. After metamorphosis in urodeles, regeneration is noticeably retarded, but still occurs. It is often said that all urodeles can regenerate their limbs, but this is not so *(5)*, and in general, postmetamorphic regenerative ability declines with age and size.

A host of different species have been used over the past century for limb regeneration studies, including the European crested newt, *Triturus cristatus*, which is now a protected species. Today, however, four species predominate. One is the eastern spotted newt of North America, *Notophthalmus viridescens*, which is used as an adult. The adults of this species are common, adaptable to laboratory conditions, and can be purchased from commercial suppliers or collected in the wild. The spotted salamander, *Ambystoma maculatum*, is used at larval stages having been collected from ponds in the spring. The other two species, the ribbed newt, *Pleurodeles waltl*, and the axolotl, *Ambystoma mexicanum*, breed readily in captivity, and there are several established breeding colonies. These two offer several advantages—they breed more than once a year, the eggs can also be used for studies of early development, the larvae regenerate limbs remarkably rapidly, they are available in large numbers, and comparative studies can be performed between limb development and limb regeneration not only in the same species, but in the same animal. In the latter case, because the hindlimbs develop many days after the forelimbs, grafts can be exchanged between forelimb regeneration blastemas and hindlimb developing limb buds *(6)*. The axolotl offers the peculiar advantage in having suppressed metamorphosis to retain its larval form and aquatic habit throughout life. Consequently, they have become worldwide curiosities not only for the very large size of the elderly adults, but also because by adding thyroxine to the water, they can be induced to metamorphose and transform into a different species! In addition, there is a very large axolotl colony established at Indiana University, Bloomington, IN (run by Susan T. Duhon, IU Axolotl Colony, Jordan Hall 407, Bloomington, IN 47405), which will supply eggs and larvae to anywhere in the world. They also keep all the available mutant lines, some of which are useful for regeneration research, e.g., Short toes *(7)*.

## 2. Limb Regeneration

Young axolotl larvae will regenerate their limbs in 2–3 wk, whereas adult newts take 2–3 mo, but the processes they go through are identical.

### 2.1. Amputation of Limbs

1. Anesthetize the animals by placing them in a solution of 3-aminobenzoic acid ethyl ester (MS222) (Sigma). A concentration of 1 in 10,000 is required for young larvae, but older animals need a stronger solution, 1 in 1000. The acidic solution should be returned to neutrality with NaOH. Five to 10 min in this solution are usually enough.
2. Remove the animals from the anesthetic, place them on a wet paper towel under a dissecting microscope, and amputate the limbs at the desired level with a scalpel, razor blade, or scissors for larger animals.
3. Limb bones will protrude almost immediately owing to retraction of the skin and muscles from the wound surface. Trim these protruding bones, since they interfere with wound healing.
4. Return the animals to their tanks. It is better to keep experimental animals in individual bowls, since this prevents cannibalism, and they can be individually identified.

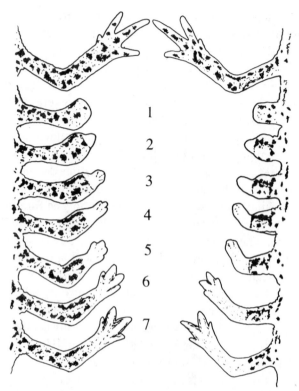

Fig. 1. Stages in the regeneration of adult newt limbs following amputation through the midlower arm (left side) or the midupper arm (right side). At the top are the original limbs. Stage 1—7 d; stage 2—21 d, stage 3—25 d, stage 4—28 d, stage 5—32 d, stage 6—42 d; stage 7—70 d after amputation. Redrawn from **ref. 8**.

## 2.2. Stages and Major Features of Limb Regeneration

**Figure 1** shows a series of drawings following amputation either through the midradius and ulna (the zeugopodium level) or the midhumerus (stylopodium level) of an adult newt limb. This emphasizes several principles—whatever the level of amputation, the regenerate is a perfect copy of what was removed; the stages are the same whatever the level of amputation; regeneration from the lower arm is completed more quickly than from the upper arm. The latter occurs because there is less tissue to be replaced from lower arm amputations and the rate of cell division of blastemal cells is the same at lower arm levels as at upper arm levels (*9*).

Staging systems for various species have been published: adult *N. viridescens* (*10*), large axolotls (*11*), and larval *A. maculatum*, also valid for small larval axolotls (*12*). In the following description, the major events of regeneration are described rather than a precise staging system, which can be obtained from the above references. Also, times are not included, since this depends on the ambient temperature, age of the animal, and species. As mentioned above, in small larval axolotls, this is a very rapid process, being completed in 2–3 wk.

1. Wound healing (**Fig. 1**, stage 1; **Fig. 2A**): This is achieved in a matter of hours after amputation by migration of the epidermis from the cut edges of the stump. It is slower if the protruding bones are not trimmed away. There are large numbers of damaged mesodermal cells and blood clots that the wound epidermis has covered, and during the first few days after amputation, phagocytes accumulate to remove this cell debris. This debris can also be seen between the cells of the wound epithelium, so some must be ejected via this route. Fluid often accumulates at the tip, but a circulating blood supply is rapidly re-established. Mitosis in the wound epithelium is inhibited for several days after amputation.

2. Dedifferentiation (**Fig. 2B,C**): Once the local damage has been repaired, the net effect of wound healing can be appreciated—the juxtaposition of a naked epidermis and mesodermal tissues. Normally the dermis and basal lamina are present to prevent interactions. Indeed, if full thickness skin is sewn over an amputated stump, then limb regeneration is inhibited, as described above *(2,13)*. The wound epithelium seems to be active in inducing histolysis of the mesodermal tissues it is now in contact with and the process of dedifferentiation begins. The cytoplasm of the myotubes fragments and surrounds individual nuclei, and osteocytes appear along the cut ends of the bones breaking down the matrix and releasing individual chondrocytes (**Fig. 2B,C**). This process is clearly an organized one unrelated to simple phagocytic breakdown, because the result of dedifferentiaion is the appearance of embryonic cells with large pale nuclei and active cytoplasms rich in rough endoplasmic reticulum. These dedifferentiated cells now begin cell division and accumulate under the apical cap.

3. Apical cap (**Fig. 2D**): After closure of the wound, the epidermis continues to migrate and piles up at the tip forming a very thick apical epidermis or apical cap. This structure has been likened to the apical ectodermal ridge (AER) of the chick limb bud, since it is responsible for the accumulation of dedifferentiated cells beneath it, for the direction of outgrowth of the blastema, and it behaves like the AER in inducing an accessory outgrowth when transplanted *(14)*. Perhaps it also generates fibroblast growth factor *(15)*, as the AER does.

4. Blastema: The accumulation of dedifferentiated cells beneath the apical cap results in the appearance of a small, conical structure known as the early bud blastema (**Fig. 2D**). The blastemal cells now begin to divide rapidly and increase in number by proliferation rather than by continued dedifferentiation, the latter now ceasing. As more and more cells are generated, the blastema increases in size, forming a larger and larger cone at the limb apex (stage 2 in **Fig. 1**) and is known as the medium bud blastema (**Fig. 2E**). Nerve fibers are present throughout the blastema, but the vasculature appears sparse.

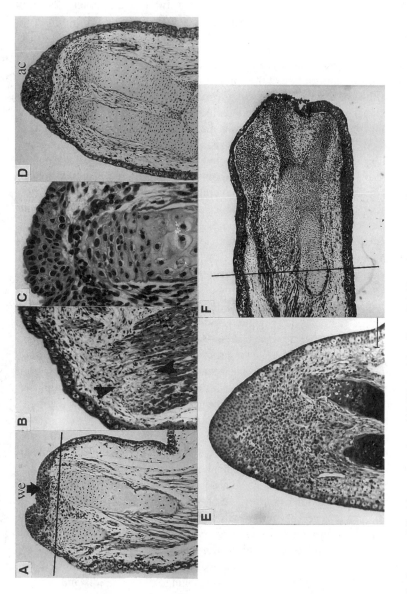

Fig. 2. Stages of limb regeneration. (**A**) Wound healing. Within a few hours after amputation a wound epithelium (we) migrates over the stump to heal the wound completely. Solid line marks the level of amputation through the radius and ulna. (**B,C**) Dedifferentiation. In B, muscle dedifferentiation is shown. The loss of myofiber structure can clearly be seen stretching back a considerable distance from the amputation plane, between the two arrowheads. In C, cartilage dedifferentiation is shown.

5. Redifferentiation (**Fig. 1**, stage 3; **Fig. 2F**): As the blastema enlarges and becomes an elongated cone, the cells in the proximal region adjacent to the stump begin to segregate into precartilage and premuscle masses. This stage is known as the late bud or palette stage. As redifferentiation commences, the first new cartilage element (e.g., the distal half of the humerus if amputation was through the midhumerus level) appears as a continuation from the stump, as if the remaining piece of humerus serves as a model. The same applies to the newly differentiating muscles. Redifferentiation is occurring proximally, while distally, proliferation of the blastemal cells continues. Clearly then, redifferentiation progresses in a generally proximal-to-distal fashion (except for a blip in the wrist), since the appearance of the humerus is soon followed by the appearance of a completely new radius and ulna, followed by the digits, followed by the wrist elements. This sequence is a repeat of the sequence of appearance of elements during development, even down to the wrist elements and digits, which develop and regenerate in an anterior to posterior sequence. When the digits first appear (**Fig. 1**, stage 4 on the left and stage 5 on the right), the regenerate is at the early digit stage, and when all the digits have appeared (**Fig. 1**, stage 5 on the left and stage 6 on the right), the regenerate is at the late digit stage.

## 2.3. Control of Limb Regeneration

1. Nerves: Since the pioneering experiments of Todd *(1)*, it has been known that a denervated limb will not regenerate. The incredibly detailed series of experiments by Singer and colleagues in the 1940s *(16)* in which he partially denervated limbs, counted the remaining nerve fibre numbers at various limb levels and then recorded the resulting frequency of regeneration, led to the hypothesis that a threshold number (between 30 and 50% of normal) was required for regen-

---

Fig. 2. *(continued)* At the bottom of the micrograph are typical small, darkly staining chondrocytes. In the middle of the micrograph, the nuclei of the chondrocytes can be seen to be enlarging and staining less intensely. At the cut tip of the cartilage, cells are released into the blastema as the cartilage matrix has been degraded. (**D**) Apical cap. As dedifferentiation progresses and an early bud blastema begins to accumulate, the wound epithelium piles up thickly, and the apical cap (ac) forms at the tip of the stump. Cartilage dedifferentiation can be clearly seen from the cut ends of the radius and ulna. (**E**) Blastema. Dedifferentiation followed by proliferation of the released cells generates a blastema consisting of a mass of rapidly dividing, embryonic cells covered by an epithelium. This is at the medium bud blastema stage. (**F**) Redifferentiation. After the blastema has reached a certain size, redifferentiation begins proximally and spreads distally until all the elements that were removed by amputation are replaced. Here the solid line marks the amputation plane, and the new distal radius can be seen to be fused perfectly with the remaining proximal radius in the stump. The first two digits can clearly be seen as well as a large cartilage mass between the radius and the digits, which will form the wrist elements. This is at the early digit stage.

eration to occur. Either nerve type, motor or sensory, will suffice provided they are present in sufficient quantity. It was subsequently assumed that the nerves provide a neurotrophic factor and that it was concentration of this factor that Singer had been quantitating. The neurotrophic factor is responsible for stimulating blastema cell division. A strong candidate for the neurotrophic factor is glial growth factor (17), although no one has managed to replace completely the function of the nerves by a defined compound. This is an extremely difficult experiment to perform because of the problem of administering minute quantities of a test substance over a prolonged period of many weeks and the problem of keeping the limb denervated as amphibian nerves readily regrow after crushing or severing.

This well-founded neurotrophic theory also encompasses a fascinating paradox. Limbs that have never been innervated, so-called aneurogenic limbs, regenerate perfectly in the absence of nerves. When nerves are allowed to enter the aneurogenic limb, they gradually become dependent on innervation—the blastemas cells are thought to have become "addicted" to the neurotrophic factor. More recent experiments have cast some light on these rather vague concepts by showing that the nerve controls the molecular phenotype of blastemal cells (18).

2. Hormones: It is generally taken for granted that the appropriate "hormonal mileau" is required for regeneration. This mileau includes the adrenal corticosterioids, somatotrophin, thyroxine, insulin, and prolactin. However, it has been particularly difficult to demonstrate specific requirements for several reasons. First, removal of the gland under consideration is usually so severe an operation either physically or physiologically that the animals do not survive. Second, the successful removal of an organ, such as the pituitary, has such a profound effect on many other hormonal systems that it is impossible to dissect out individual requirements for regeneration. With the advent of cloned products, however, specific requirements are being demonstrated, e.g., for growth hormone (19). Third, removal of a gland followed by replacement therapy has either involved mammalian preparations whose similarity of action in amphibia is unknown, or impure preparations have been used. Nevertheless, we would expect that hormones, such as growth hormone or prolactin, which have such an important role in the control of basic cell metabolism, should be involved in regeneration, but not nesessarily in a controlling capacity.

3. Origin of blastemal cells: The cells that form the blastema arise from within 1–2 mm of the amputation plane by the process of dedifferentiation of the mesodermal tissues as described above. The epidermis cannot contribute to any internal tissues, but only forms the apical cap. Meticulous studies recording cell and mitotic counts have concluded that all mesodermal tissues contribute to the blastema (dermis, muscle, connective tissue, periosteum, bone and even Schwann cells) roughly in proportion to the number of cells in each tissue in a cross-section of the stump (20, although see 21). There is a tendency to assume that the majority of cells revert back to their former differentiated state, but this is not necesarily so, as two recent studies have conclusively demonstrated. In the first, cells were marked by grafting between diploid and triploid animals followed by

meticulous cell counting. It transpired that there was an overrepresentative contribution of dermis from the stump and an underrepresentative contribution from the cartilage *(21)*. In the second study, cultured myotubes were labeled with lysinated dextran, implanted into blastemas, and when redifferentiation began, labeled cells were occasionally seen in the cartilage of the regenerate *(22)*. Thus, a proportion of cells seem to undergo a metaplastic transformation during normal regeneration.

Another type of experiment in which cells are forced to do more than they naturally would has also demonstrated the metaplastic potential of blastemal cells. In these experiments, limbs are irradiated with X-rays, which permanently prevents cell division, and then a graft of unirradiated tissue is provided, which supplies new cells with regenerative potential. If the graft is from a white axolotl and the host is a black axolotl, then the regenerate will be white confirming the origin of the tissues of the regenerate. In these experiments, it was shown that grafts of muscle could provide all the tissues of the regenerate, including cartilage (as revealed in the labeling experiment described above). However, grafts of dermis and grafts of cartilage could provide all the tissues of the regenerate, except muscle. Thus, it seems that most metaplastic transformations are possible, except that only myoblasts can generate new myoblasts.

This type of tissue transformation is readily demonstrable in a simple experiment where the cartilage of the stump is removed prior to amputation. Even though there is no cartilage at the amputation plane and no chondrocytes dedifferentiate to supply the blastema, perfect cartilage elements are produced distal to the amputation plane (**Fig. 3A**).

4. Regional and axial determination: That individual blastemal cells remember the region of the body from which they come is the general conclusion from a long history of grafting studies. Tail tissue grafted to limbs or vice versa or forelimb tissues grafted to hindlimbs or vice versa results in the regeneration of organs specific to the graft type, not the host type. Until recently, it had never been possible to change the organ specificity of cells, but this has now been done. The regenerating tail blastema of frogs can be homeotically transformed into tails by treatment with retinoids prior to metamorphosis *(23,24)*. The same general conclusion is also true of axial determination. From the very beginning, blastemal cells carry a knowledge of their axial position. This has been demonstrated many times in experiments in which blastemas are cut off the stump and then either rotated 180° and put back on or grafted from left to right (or vice versa). The former manipulation reverses both anteroposterior and dorsoventral axes and results in the appearance of supernumerary limbs (**Fig. 3B**). The same is true if either the anteroposterior or dorsoventral axis is reversed (**Fig. 3C**). Thus, axes cannot be respecified. The interaction between cells whose axes conflict results in the generation of extra tissue to resolve the conflict.

In the proximodistal axis level-specific information is similarly present within the blastemal cells. Clearly, it must be or the limb would not know how much of itself to regenerate. If a proximal blastema is grafted onto a distal amputation stump, then the result will be a limb that has serially duplicated elements (**Fig.**

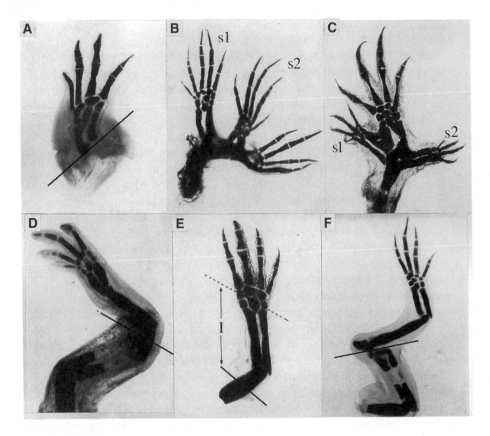

Fig. 3. Victoria blue-stained limb regenerates to show the cartilage patterns after various treatments. (**A**) The structure of the regenerate after removing all the cartilage and bone from a forelimb and then amputating through the lower arm level. A solid line marks the amputation plane. Proximal to the amputation plane, there is only muscle, which does not stain with Victoria blue. Distal to the amputation plane, the ends of the radius and ulna, wrist elements, and digits have regenerated perfectly despite the absence of cartilage in the stump from which the dedifferentiated tissues of the blastema were derived. This is a simple demonstration of tissue metaplasia. (**B**) The result of cutting off a blastema from a regenerating hindlimb, rotating it 180° and putting it back on the stump. Two supernumerary limbs (S1 and S2) have been produced in addition to the original limb. (**C**) The result of putting a left forelimb blastema on a right stump to reverse the anteroposterior axis. As in B, two supernumerary limbs are generated (S1 and S2) in addition to the original limb. (**D**) The result of grafting a proximal blastema (from the shoulder level) onto a distal amputation stump (through the ends of the radius and ulna). The result is two limbs in tandem. A solid line marks the amputation plane. (**E**) The result of grafting a distal blastema (from the wrist level) onto a proximal stump (the midupper arm level). The dotted line marks the level of the grafted distal blastema and the solid line marks the amputation plane through the mid upper arm. The result is a

**3D**). In this case, there seems to be no interaction between stump and blastema in reaction to the disparity in level-specific information. When the converse experiment is performed, a distal blastema grafted onto a proximal stump, then interaction does take place because the disparity is recognized and the gap filled in to generate a normal limb (**Fig. 3E**). The gap is filled in by proximal cells, not by distal cells, and these observations led to the formulation of the "law of distal transformation" which only allows cells within the limb field to become more distal, never more proximal. Significantly, just like the respecification of regional determination by retinoids, the only instance where the law of distal transformation has been broken is when distal blastemas are treated with retinoids. In this case, a complete limb, including proximal elements, can be regenerated from a distal level amputation after retinoid treatment (**Fig. 3F**) *(25)*.

## 2.4. Recent Techniques

Throughout this century, many techniques have been used in an attempt to discover the mechanisms of limb regeneration—denervation, transplantation of blastemas to ectopic sites, organ culture, radiolabeling, gel electrophoresis, administration of compounds, such as retinoic acid, and so forth. However, recently three technical advances have been made that are certain to have an important influence on future discoveries.

One is the advent of whole-mount *in situ* hybridization for use with probes to genes expressed at very low levels, such as the homeobox genes *(26)*. Gardiner et al. have cloned 17 homeobox genes from the axolotl, and work such as this is already beginning to answer some question, such as: Are these genes expressed in the same domain and in the same sequence during regeneration as they were during development?

The second advance is the establishment of long-term cultures of blastemas cells, allowing for their genetic manipulation by transfection and grafting back into the blastema. In contrast to normal, untransformed cells from mammalian tissues, both blastemal cells and a muscle cell line established from a dissociated limb have generated permanent cell lines, showing no senescence *(27)*.

---

Fig. 3. *(continued)* complete limb as the missing portion, the intercalary regenerate represented by the gap between the solid line and the dotted line (I), is filled in by cells from the proximal level stump according to the law of distal transformation. This phenomenon is also shown here because the distal blastema was from a black animal (melanophores can be seen in the digits and wrist) and it was grafted to a white animal. The intercalary regenerate (I) is white, since it has no melanophores showing that it came from stump tissue. (**F**) The result of treating a distal level regenerate (solid line marks the amputation plane through the wrist) with retinoids. Instead of just regenerating the missing elements, a complete limb has been produced from distal level blastemal cells. Retinoids respecify blastemal cells in a manner that breaks the law of distal transformation.

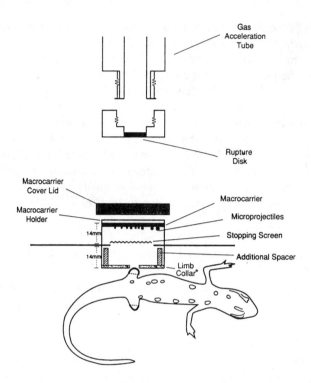

Fig. 4. Drawing of a Biolistics Particle Delivery System for transfecting a regenerating limb. Helium at high pressure is passed into the gas acceleration tube, and the ruptured disk breaks. This releases a shock wave, which propels the macrocarrier disk and DNA-coated gold microprojectiles downward. The screen is halted by a stopping screen, but the gold particles continue on toward the target. The anesthetized animal is supported against the bottom of the assembly and the regenerate inserted into the hole, so that it can serve as a target (from **ref. 29**).

This again emphasizes a unique feature of blastemal cells, and the ability of limb tissues to dedifferentiate and turn into blastemal cells.

The third advance has to some degree obviated the need for cultures, because it is a remarkable technique by which cells can be transfected both in vitro and in vivo. It involves the use of a biolistic gun (*28*), which fires 1.6-μm gold particles coated with the DNA of one's choice into cells in a Petri dish or into the cells of the regenerate (**Fig. 4**). If the gun is fired at the external surface of the blastema then cells in the epidermis are transfected at a frequency of about 10%, and their altered behavior can be assessed in various ways. If the blastema is cut off the limb stump and the cut edge exposed to the gold particles, then the mesenchymal cells of the blastema are transfected. Transfection with constructs, such as chimeric retinoic acid receptors (RARs), has provided a wealth of valuable data on which receptors perform which functions

*(29)*. At least six different receptors are expressed in the limb blastema and in addition to the positional respecification referred to above, RA induces many other phenotypic changes in the cells. In order to examine which receptor mediates which effect, Schilthuis et al. *(30)* constructed chimeric receptors by exchanging each of the newt RAR ligand binding domains with that from the *Xenopus* thyroid hormone receptor α. The genes activated by each RAR then became responsive, not to the RA, but to thyroid hormone. In this way, they showed, for example, that the RARα was responsible for growth inhibition, one of the established effects of RA on blastemal cells. It is hard to imagine how such a result would have been obtained without these important technical advances.

Hopefully, these and further advances will ultimately provide answers to those fundamental questions posed at the beginning of this chapter, the most profound of which is: Why can urodeles regenerate and mammals cannot and will we ever be able to stimulate regeneration in humans?

## References

1. Todd, T. J. (1823) On the process of reproduction of the members of the aquatic salamander. *Quart. J. Sci.* **16,** 84–96.
2. Tornier, G. (1906) Kampf der Gewebe im Regenerat bei Begunstigung der Hautregeneration. *Roux's Arch. Dev. Biol.* **22,** 348–369.
3. Illingworth, C. M. (1974) Trapped fingers and amputated finger tips in children. *J. Pediatr. Surg.* **9,** 853–858.
4. Wallace, H. (1981) *Vertebrate Limb Regeneration.* Wiley, Chichester, UK.
5. Scadding S. R. (1977) Phylogenetic distribution of limb regeneration capacity in adult amphibia. *J. Exp. Zool.* **202,** 57–67.
6. Muneoka, K. and Bryant, V. (1982) Evidence that patterning mechanisms in developing and regenerating limbs are the same. *Nature* **298,** 369–371.
7. Rio-Tsonis, K. D., Washabaugh, C. H., and Tsonis, P. (1992) The mutant axolotl Short toes exhibits impaired limb regeneration and abnormal basement membrane formation. *Proc. Natl. Acad. Sci. USA* **89,** 5502–5506.
8. Goss, R. J. (1969) *Principles of Regeneration.* Academic, New York.
9. Maden, M. (1976) Blastemal kinetics and pattern formation during amphibian limb regeneration. *J. Embryol. Exp. Morph.* **36,** 561–574.
10. Iten, L. E. and Bryant, S. V. (1973) Forelimb regeneration from different levels of amputation in the newt, *Notophthalmus viridescens*:length, rate and stages. *Roux's Arch. Dev. Biol.* **173,** 263–282.
11. Tank, P. W., Carlson, B. M., and Connelly, T. G. (1976) A staging system for forelimb regeneration in the axolotl, *Ambystoma mexicanum. J. Morph.* **150,** 117–128.
12. Stocum, D. L. (1979) Stages of forelimb regeneration in *Ambystoma maculatum. J. Exp. Zool.* **209,** 395–416.
13. Mescher, A. L. (1976) Effects on adult newt limb regeneration of partial and complete skin flaps over the amputation surface. *J. Exp. Zool.* **195,** 117–128.

14. Thornton, C. S. and Thornton, M. T. (1965) The regeneration of accessory limb parts following epidermal cap transplantation in urodeles. *Experientia* **21**, 146–148.
15. Poulin, M. L., Patrie, K. M., Botelho, M. J., Tassava, R. A., and Chiu, I.-M. (1993) Heterogeneity in the expression of fibroblast growth factor receptors during limb regeneration in newts *(Notophthalmus viridescens). Development* **119**, 353–361.
16. Singer, M. (1952) The influence of the nerve in regeneration of the amphibian extremity. *Q. Rev. Biol.* **27**, 169–200.
17. Brockes, J. P. and Kintner, C. R. (1986) Glial growth factor and nerve-dependent proliferation in the regeneration blastema of urodele amphibians. *Cell* **45**, 301–306.
18. Fekete, D. M. and Brockes, J. P. (1987) The aneurogenic limb: a puzzle in cell interactions. *Trends Neurosci.* **10**, 364–368.
19. Landesman, R. H. and Copeland, K. C. (1988) Bioengineered human growth hormone supports limb regeneration in the hypophysectomized newt Notophthalmus viridescens. *J. Exp. Zool.* **248**, 247–250.
20. Chalkley, D. T. (1954) A quantitative histological analysis of forelimb regeneration in *Triturus viridescens. J. Morph.* **94**, 21–70.
21. Muneoka, K., Fox, W. F., and Bryant, S. V. (1986) Cellular contribution from dermis and cartilage to the regenerating limb blastema in axolotls. *Dev. Biol.* **116**, 256–260.
22. Lo, D. C., Allen, F., and Brockes, J. P. (1993) Reversal of muscle differentiation during urodele limb regeneration. *Proc. Natl. Acad. Sci. USA* **90**, 7230–7234.
23. Mohanty-Hejmadi, P., Dutta, S. K., and Mahapatra, P. (1992) Limbs generated at site of tail amputation in marbled balloon frog after vitamin A treatment. *Nature* **355**, 352,353.
24. Maden, M. (1993) The homeotic transformation of tails into limbs in *Rana temporaria* by retinoids. *Dev. Biol.* **159**, 379–391.
25. Maden, M. (1982) Vitamin A and pattern formation in the regenerating limb. *Nature* **295**, 672–675.
26. Gardiner, D. M., Blumberg, B., Komine, Y., and Bryant, S. V. (1995) Regulation of *HoxA* expression in developing and regenerating axolotl limbs. *Development* **121**, 1731–1741.
27. Ferretti, P. and Brockes, J. P. (1988) Culture of newt cells from different tissues and their expression of a regeneration-associated antigen. *J. Exp. Zool.* **247**, 77–91.
28. Pecorino, L. T. and Lo, D. C. (1992) Having a blast with gene transfer. *Curr. Biol.* **2**, 31,32.
29. Pecorino, L. T., Lo, D. C., and Brockes, J. P. (1994) Isoform-specific induction of a retinoid-responsive antigen after biolistic transfection of chimaeric retinoic acid/thyroid hormone receptors into a regenerating limb. *Development* **120**, 325–333.
30. Schilthuis, J. G., Gann, A. A. F., and Brockes, J. P. (1993) Chimeric retinoic acid/thyroid hormone receptors implicate RARα1 as mediating growth inhibition by retinoic acid. *EMBO J.* **12**, 3459–3466.

# IV

## ZEBRAFISH

# 29

## The Zebrafish

*An Overview of Its Early Development*

**Nigel Holder and Qiling Xu**

### 1. The Emergence of the Zebrafish as a Model System for the Study of Vertebrate Development

During the past 10 yr, the zebrafish has emerged as an important model system for the study of vertebrate development. This is primarily because of the promise of the system for developmental genetic studies. but, in addition to the necessary features of an animal that can be used for genetics, there are a range of experimental approaches that have proven successful in studies of tissue interactions, gene function, and early neural development. Such methods include embryonic cell transplantation, the analysis of gene function by injection of RNA or antibodies into the fertilized egg, and the analysis of identified neurons in the developing central nervous system. The object of this chapter is to outline the main features of early zebrafish development and to provide details of the methods for injecting the fertilized egg with nucleic acid or protein.

It is important to point out that analysis of gene function in the zebrafish, whether by mutational screens or DNA/RNA injection, is carried out against an increasingly extensive knowledge of basic embryology. Thus, the transparency and rapid development of the embryo have been exploited to great effect in establishing fate maps *(1–3)* produced using cell marking experiments in which fluorescent dyes are injected into single or small groups of cells. This kind of analysis has also led to an understanding of the lineage relationships of cells in the blastula and an assessment of the timing of commitment. It is now clear, despite some data to the contrary *(4,5)*, that there is no clear restriction of cell fate in the zebrafish until gastrulation begins *(6–8)*. Consistent with this is the dem-

From: *Methods in Molecular Biology, Vol. 97: Molecular Embryology: Methods and Protocols*
Edited by: P. T. Sharpe and I. Mason © Humana Press Inc., Totowa, NJ

onstration that the dorsoventral axis, despite being established during blastula stages, is positioned randomly with respect to the initial blastomere divisions *(9)*.

## 1.1. Developmental Genetics in the Zebrafish

The zebrafish owes its elevation from common pet shop aquarium fish to one of the few model systems for the study of vertebrate development largely to the "Oregon School." George Streisinger, at the University of Oregon at Eugene, first settled on the zebrafish for his pioneering genetic studies. The medaka may have some advantages over the zebrafish for certain procedures and is still used for developmental studies; for example, it has been used recently for the analysis of mesoderm formation *(10)*.

Streisinger established a number of methods for studying zebrafish, including the generation of isogenic homozygous diploid lines and the screening of haploid embryos for developmentally interesting mutations *(11)*. The haploid zebrafish embryo develops for several days and eventually dies; a screening program based on haploid embryos has the advantage that recessive mutations present in the female can be revealed in a single generation. This method has been used effectively by Kimmel's laboratory at the University of Oregon to reveal several mutants either induced or present in the genetic background of the zebrafish stock held in Eugene. Such mutants include *spadetail (12)*, *cyclops (13)*, and *no-tail (14)*; lines that have subsequently been extensively studied, the last two being important for studies of patterning of the axial midline of the embryonic axis. Mapping of candidate genes has allowed a number of mutants to be identified. These include *spadetail (15)*, cyclops *(16)*, and other important regulatory genes, such as fleating head *(17)* and FgF8 *(18)*.

Since the first attempts to establish zebrafish for developmental genetics the single most important step has been the selection of this system for a number of major screens using chemical mutagenesis and analysis of sib-crosses in the F2 generation to reveal mutations originally induced into male germ cells by chemical mutagenesis using ethyl nitrosourea (ENU) *(19)*. The strategy, methodology, and results from these screens have recently been published by these two laboratories, and are the subject of Chapter 30. Together with the continuing screening in Oregon and elsewhere, these two large-scale screens are now at the stage where mutations are being characterized genetically and morphologically. The screens have revealed many mutants of developmental interest affecting the major organ systems and structures of the embryo (*Development*, vol. **123,** 1996).

Considerable progress has been made in generating a genetic recombination map for genes relative to the chromosomes. Methods based on polymerase chain reaction (PCR) (so-called RAPD method) have been used to identifiy polymorphic short sequences which, in conjunction with generation of haploids, have been applied by John Postlethwait's laboratory *(20–22)* to generate

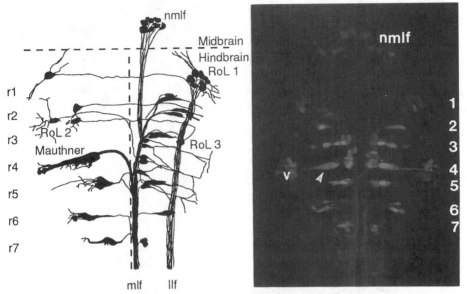

Fig. 1. The reticulospinal complex of the zebrafish. (**A**) The cells of the complex indicating rhombomeres (rl-7) and the principal identified cells (based on **ref. 25**). (**B**) Back-filled reticulospinal neurons in a normal embryo. The Mauthner cell is arrowed, v indicates the vestibulospinal neurons, and nmlf indicates the midbrain nucleus of the longitudinal vesiculus. (*See* color plate 6 appearing after p. 368.)

a recombination map *(21,22)*. Genes newly isolated by mutation and genes cloned by homology to those in other species are now being added to the map. It is clear the map will have many markers within a short time and will be a valuable resource for molecular identification of novel mutations either by mapping of cloned genes to the same site as an identified mutation *(15–17)* or by positional cloning *(23)*.

## 1.2. The Presence of Identified Neurons Allows Analysis of Neural Differentiation at the Single-Cell Level

In addition to genetic studies, there are other advantages to using the zebrafish for studies of early development. Principal among these is the presence in the developing nervous system of identified neurons, cells that differentiate in the same place and at the same time in every embryo *(24)*. The identified neurons that have been most heavily studied, in terms of their time of origin, the mechanisms leading to their determination, and subsequent differentiation are the primary motor neurons of the spinal cord *(25,26)*, and the reticulospinal neurons of the hindbrain *(27–29)*. The presence of identified neurons within the central nervous system allows analysis at the level of the single cell and provides a level of

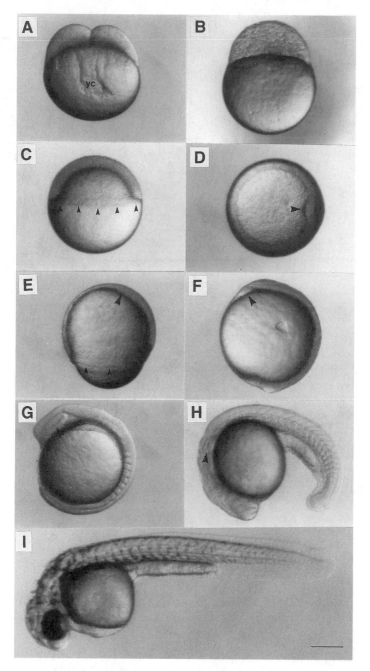

Fig. 2. *(opposite page)* Stages in the development of the zebrafish. This figure is adapted, with permission of the authors, from the definitive zebrafish staging series published by Kimmel et al. *(32)*. (**A**) Two-cell staged embryo showing the large yolk cell as an incomplete cleavage. (**B**) The blastula stage illustrated with an embryo at 3.3 h of development. The ever-smaller blastomeres are piled up on the yolk cell. (**C**) Gas-

analysis not possible with higher vertebrates. An example of such a study is the identification of rhombomere respecification in the hindbrain following retinoic acid treatment of the gastrula *(30)*. In this study, the reticulospinal complex, which is illustrated in **Fig. 1**, *(shown on p. 433)* was used to identify respecifi-cation of the Mauthner cell, normally present in rhombomere 4, in rhombomere 2 *(30)*.

## 1.3. What the Future Holds

The near future holds great promise for work on the zebrafish because of the flood of interesting mutations that emerged from the large-scale screens (*Development*, vol. **123**, 1996). The continued generation of a genetic map is a crucial development and has allowed the first positional cloning projects to be started *(23)*. Other techniques are being tackled, such as the establishment of a mouse ES cell equivalent, which is yet to be obtained in any other vertebrate, and the use of viruses for transgenesis and insertional mutagenesis. The future is becoming increasingly more promising for the studies of developmentally regulated genes in this organism.

## 2. The Zebrafish Is a Rapidly Developing Embryo: Stages and Key Events

A detailed description of the stages of zebrafish development has recently been published *(32)* and the reader who is interested in a detailed look at the key features of this developmental series should read this paper and refer to the *Zebrafish Book (33)*. What follows is a brief description of the key stages of zebrafish development over the first 2-s period—all of the terms and stages are consistent with those used by Kimmel et al. *(32)* and relate to embryos growing at 28.5°C.

## 2.1. The First Five Hours

This includes the zygote, cleavage, and blastula periods *(32)*. Within 10 min of the egg being fertilized, streaming cytoplasmic movements cause a cap of

---

trulation begins at 50% epiboly (5.25 h) by which time the blastomeres have moved down over the yolk cell to cover half its extent (arrows). **(D)** Animal pole view of the shield stage (6 h) at which the position of axial involution is evident (arrow). **(E)** At 70% epiboly, the edge of blastomeres (small arrows) has spread further round the yolk cell. The anterior extent of the axially involuting hypoblast can be seen approaching the animal pole (large arrow). **(F)** The bud stage (10 h) is reached at the end of gastrulation. The yolk cell is completely covered by cells and the anterior extent of the axial hypoblast can be seen as the polster (arrow). **(G)** The eighth somite stage (13 h). The arrow marks the eye. **(H)** 22 somite stage (19 h). The arrow marks the otic vesicle. **(I)** The prim 20 stage (33 h). Pigmentation is now clear as melanocytes spread over the body and the retinal pigment epithelium of the eye differentiates. Scale bar = 250 μ.

clear cytoplasm to emerge at the animal pole of the otherwise yolky cell. After 45 min, the first cleavage occurs incompletely separating the clear cytoplasmic region (**Fig. 2A**). Subsequent cleavages are more rapid, occurring at 15-min intervals, generating the 64-cell stage by 2 h. The cells adjacent to the yolk cell communicate by intracellular bridges. The blastula stage (**Fig. 2B**) begins with the generation of 128 cells. Divisions initially remain largely synchronous, but this synchrony begins to be lost at the time of the midblastula transition *(34)*, which begins at the 10th cell cycle (512-cell stage). Other key events that occur during the blastula stage are the formation of the yolk syncytial layer (YSL) and the beginning of the process of epiboly. The YSL is formed as the marginal blastomeres sink into the yolk cell distributing their cytoplasm and nuclei as a layer across its superior edge. Epiboly is the spreading of the blastodisc and the YSL over the surface of the yolk cell. By the end of gastrulation, the yolk cell is completely engulfed. In addition to the YSL and the deep cells (DEL) of the blastodisc, during the blastula stage, the single outer layer of cells flattens to form the enveloping layer (EVL).

## 2.2. The Second Five Hours

To be accurate, gastrulation begins at $5^{1}/_{4}$ h; this process comprises the cell movements of involution, convergence and extension *(35)*, which are initiated when epiboly has reached 50% (50% coverage of the yolk cell; **Fig. 2C**). At the onset of gastrulation, the marginal zone becomes thickened to form the germ ring, one region of which, at the site of dorsal axial involution, becomes distinctively thickened. This localized thickening, caused by convergent movement of cells to the axial (dorsal) location, is called the shield (**Fig. 2D**). Gastrulation movements lead initially to the formation of two layers, the inner hypoblast (mesendoderm),and the epiblast (ectoderm) (**Fig. 2E**). Understanding the cell movements of gastrulation is aided by appreciation of the fate maps that are available for this period *(1,3)*.

During gastrulation, cells in the axial midline extend anteriorly to create the notochord and the most anteriorly located prechordal plate (**Fig. 2F**). This anterior region produces the hatching gland and the pharyngeal endoderm. More laterally, the rostral hypoblast forms the muscles of the head and, more caudally, the somites. By the end of gastrulation epiboly is complete and the tail bud has formed (**Fig. 2F**); in addition,the ectodermally derived neural plate is beginning to be formed on the dorsal side of the gastrula *(36)*.

## 2.3. The Segmentation Period Runs from Gastrulation Until 24 Hours

One of the striking features of this time period is the growth of the tail, which extends to give the embryo its characteristic shape. In addition to tail elongation, the major organ systems, including the somites, the pronephros, the heart, and the central nervous system, form. Segmentation is seen in the forming hindbrain and in the mesodermally derived somites (**Fig. 2G**). The number of formed somites is a good check on the exact stage that the embryo has reached. Eventually there will be over 30, but, for example, 20 have formed by 19 h of development (**Fig. 2H**). Spontaneous contractions of the somites begin at the 17 somite stage as the axons from the primary motor neurons first reach the differentiating muscle fibers. Within the central nervous system, 10 distinctive brain regions or neuromeres are evident by 18 h (18 somites). The most anterior of these is the telencephalon, which lies adjacent to the diencephalon. These two regions form the forebrain. The midbrain or mesencephalon separates these from the more clearly segmented hindbrain or rhombencephalon. The hindbrain segments are termed rhombomeres of which seven clearly form in the zebrafish.

### 2.4. The Second Day: The Pharyngula Period

During this period, the body continues to lengthen and the pharyngeal arches (after which the stage is named) become evident. The arches are a series of structures located in the head ventral to the midhindbrain region, beneath the otic vesicle (the structure that makes the inner ear). The first arch forms the mandible, the second the hyoid cartilage, and the subsequent arches the gills. An accurate way to stage the embryo during this period is to use Nomarski optics to locate the tip of the caudally migrating lateral line primordium *(37)*. This has reached a precise somite location by a specific time. In addition, the fins begin to grow, and the brain undergoes a compression such that the otic vesicle lies closer to the eye. The heart begins to beat at the beginning of the period, and the vascular system develops. Finally, pigment cells begin to differentiate and are evident over the body and in particular sites, such as the pigmented retinal epithelium (**Fig. 2I**).

### References

1. Kimmel, C. B., Warga, R., and Schilling, T. (1990) Origin and organisation of the zebrafish fate map. *Development* **108,** 581–594.
2. Woo, K. and Fraser, S. E. (1995) Order and coherence in the fate map of the zebrafish nervous system. *Development* **121,** 2595–2609.
3. Shih, J. and Fraser, S. E. (1995) The distribution of tissue progenitors within the shield region of the zebrafish gastrula. *Development* **121,** 2755–2765.
4. Strehlow, D. and Gilbert, W. (1993) A fate map for the first cleavage stages of the zebrafish. *Nature* **361,** 451–453.
5. Wilson, E., Helde, K., and Grunwald, D. (1993) Something's fishy here rethinking cell movements and cell fate in the zebrafish embryo. *Trends Genet.* **10,** 348–352.
6. Kimmel, C. B. and Warga, R. (1986) Tissue specific cell lineages originate in the gastrula of the zebrafish. *Science* **231,** 365–368.

7.  Kimmel, C. B. and Warga, R. (1988) Cell lineage and developmental potential of cells in the zebrafish embryo. *Trends Genet.* **4,** 68–74.

8.  Helde, K., Wilson, E., Cretekos, C., and Grunwald, D. (1994) Contribution of early cells to the fate map of the zebrafish gastrula. *Science* **265,** 517–520.

9.  Abdelilah, S., Solnica-Krezel, L., Stainier, D., and Driever, W. (1994) Implications for dorsoventral axis determination from the zebrafish mutant janus. *Nature* **370,** 468–471.

10. Wittbrodt, J. and Rosa, F. (1994) Disruption of mesoderm and axis formation in fish by ectopic expression of activin variants: the role of maternal activin. *Genes Dev.* **8,** 1448–1462.

11. Kimmel, C. B. (1989) Genetics and early development of the zebrafish. *Trends Genet.* **5,** 283–288.

12. Kimmel, C. B., Kane, D., Walker, C., Warga, R., and Rothman, M. (1989) A mutation that changes cell movement and cell fate in the zebrafish embryo. *Nature* **337,** 358–362.

13. Hatta, K., Kimmel, C. B., Ho, R., and Walker, C. (1991) The cyclops mutation blocks specification of the floor plate of the zebrafish CNS. *Nature* **350,** 339–341.

14. Schulte-Merker, S., van Eeden, F., Halpern, M., Kimmel, C., and Nusslein Volhard, C. (1994) *No tail (ntl) is* the zebrafish homologue of the mouse T *(brachyury) gene. Development* **120,** 1009–1015.

15. Griffin, K., Amacher, S., Kimmel, C., and Kimelman, D. (1998) Molecular identification of spadetail: regulation of zebrafish trunk and tail mesoderm formation by T-box genes. *Development* **125,** 3379–3388.

16. Sampath, K., Rubinstein, A., Cheng, A., Liang, J., Fekany, K., Solnica-Krezel, L., Korzh, V., Halpern, M., and Wright, C. (1998) Induction of the zebrafish ventral brain and floorplate requires cyclops/nodal signalling. *Nature* **395,** 185–189.

17. Talbot, W., Trevarrow, B., Halpern, M., Melby, A. E., Farr, G., Postlethwait, J. H., et al. (1995) A homeobox gene essential for zebrafish notochord development. *Nature* **378,** 150–157.

18. Reifers, F., Bohli, H., Walsh, E., Crossley, P., Stainer, D., and Brand, M. (1998) Fgf 8 is mutated in zebrafish acerebellar mutants and is required for maintenance of midbrain–hindbrain devlopment and somitogenesis. *Development* **125,** 2381–2395.

19. Haffter, P., Granato, M., Brand, M., Mullins, M. C., Hammerschmidt, M., Kane, D. A., Odenthal, J., et al. (1996) The identification of genes with unique and essential functions in the development of the zebrafish, Danio rerio. *Development* **123,** 1–36.

20. Postlethwait, J. H., Johnson, S., Midson, C., Talbot, W., Gates, M., Ballinger, E., et al. (1994) A genetic linkage map for the zebrafish. *Science* **264,** 699–702.

21. Postlethwait, J. H. and Talbot, W. (1997) Zebrafish genomics: from mutants to genes. *Trends Genet.* **13,** 183–190.

22. Postlethwait, J., et al. (1998) Vertebrate genome evolution and the zebrafish gene map. *Nature Genet.* **18,** 345–349.

23. Zhang, J., Talbot, W., and Shier, A. (1998) Positional cloning identifies zebrafish one-eyed pinhead as a permissive EGF-related ligand required during gastrulation. *Cell* **92,** 241–251.

24. Eisen, J. (1991) Developmental neurobiology of the zebrafish. *J. Neurosci.* **11,** 311–317.

25. Eisen, J. (1991) Determination of primary motoneuron identity in developing zebrafish emblyos. *Science* **252,** 569–572.

26. Myers, P., Eisen, J., and Westerfield, M. (1986) Development and axon outgrowth of identified motoneurons in the zebrafish. *J. Neurosci.* **6,** 227,228.

27. Mendelson, B. (1986) Development of reticulospinal neurons of the zebrafish. I Time of origin. *J. Comp. Neurol.* **251,** 160–171.

28. Mendelson, B. (1986) Development of reticulospinal neurons of the zebrafish. II. Early axon outgrowth and cell body position. *J. Comp. Neurol.* **251,** 172–184.

29. Kimmel, C. B., Metcalfe, W., and Schabtach, E. (1985) T reticular interneurons: a class of serially repeating cells in the zebrafish hindbrain. *J. Comp. Neurol.* **233,** 365–376.

30. Hill, J., Clarke, J. D. W., Vargesson, N., Jowett, T., and Holder, N. (1995) Exogenous retinoic acid causes alterations in the development of the hindbrain and midbrain of the zebrafish embryo including positional respecification of the Mauthner neuron. *Mech. Devel* **50,** 3–16.

31. Gaiano, N., Amsterdam, A., Kawakami, K., Allende, M., Becker, T., and Hopkins, N. (1996) Insertional mutagenesis and rapid cloning of essential genes in zebrafish. *Nature* **383,** 829–832.

32. Kimmel, C., Ballard, W., Kimmel, S., Ullmann, B., and Schilling, T. (1995) Stages of embryonic development of the zebrafish. *Develop. Dyn.* **203,** 253–310.

33. Westerfield, M. (1995) *The Zebrafish Book*, 3rd ed. University of Oregon Press, Eugene, OR.

34. Kane, D. and Kimmel, C. (1993) The zebrafish midblastula transition. *Development* **119,** 447–456.

35. Warga, R. and Kimmel, C. (1990) Cell movements during epiboly and gastrulation in the zebrafish. *Development* **108,** 569–580.

36. Papan, C. and Campos-Ortega, J. (1994) On the formation of the neural keel and neural tube in the zebrafish Danio rerio. *Roux Archiv Dev Biol.* **203,** 178–186.

37. Metcalfe, W. (1985) Sensory neuronal growth cones comigrate with posterior lateral line primordial cells in the zebrafish. *J. Comp Neurol.* **238,** 218–224.

# 30

# Small-Scale Marker-Based Screening for Mutations in Zebrafish Development

## Peter D. Currie, Thomas F. Schilling, and Philip W. Ingham

## 1. Introduction

We describe a standardized mutagenic protocol and a methodology for small-scale directed screening of the zebrafish genome for mutations in specific developmental processes. The methods are based primarily on those developed for large-scale screens in Tubingen, Germany; Boston, MA; and Eugene, OR as well as our experiences with a smaller facility. By combining a marker-based screening protocol with both haploid and diploid screening methods, one can efficiently recover mutants in specific processes.

Random mutagenesis provides the ability to survey the genome of an organism, without bias, for genes that function in particular processes. For many years, geneticists have been reaping the rich harvest of mutations produced by such a mutagenic approach, directed against particular developmental processes of the fruit fly *Drosophila melanogaster (1)*. Analysis of the genes uncovered by this approach has revolutionized our understanding of the genetic control of animal development.

Researchers eager to see a similar mutagenic approach applied to the vertebrate genome have been stymied by the genetic intractability of classical vertebrate developmental models. The mouse is the only vertebrate organism in which large-scale screens for mutations have been performed. These screens have, in the main, been limited to identification of defects in visible morphological traits after birth, since screening for embryonic mutant phenotypes is difficult because development occurs *in utero (2)*. Biologists interested in using a mutagenic approach to study early aspects of vertebrate development have been forced to search for an alternative.

From: *Methods in Molecular Biology, Vol. 97: Molecular Embryology: Methods and Protocols*
Edited by: P. T. Sharpe and I. Mason © Humana Press Inc., Totowa, NJ

The fishes represent the largest group of vertebrate taxa and perhaps the largest uncharted waters in terms of vertebrate developmental studies. Until recently, debates on the molecular mechanisms that govern vertebrate development have largely ignored fish. However, with the rediscovery that certain teleosts allow the application of both sophisticated embryological manipulations and classical genetic analysis, the stock of fish as a developmental system has risen. Specifically, studies on the zebrafish, *Brachydanio rerio*, have indicated its promise as both a genetic and an embryological model *(3,4)*.

Zebrafish are small and have a short life cycle, reaching sexual maturity in three months. Females produce a large brood size, typically hundreds of eggs, which are fertilized externally. They are inexpensive to maintain and can be bred to great numbers easily. Embryonic stages are completely transparent, allowing most structures of the developing fish to be viewed without the aid of sophisticated microscopy. This last attribute has allowed researchers to gain an intricate knowledge of cell movements and behavior during early development. This information can now be coupled with detailed fate maps and a precise staging series, making zebrafish a sophisticated embryological model *(5–7)*. As with some other teleosts, zebrafish can be induced to undergo either gynogenetic or haploid development, abilities that greatly enhance its stock as a genetically manipulable model and make it unique among animal developmental systems. These attributes have attracted a number of laboratories to undertake large-scale mutagenic screens in an attempt to reach saturation for lethal mutations within the zebrafish genome *(8,9,* and **Note 1**).

The data published by these laboratories describe the general classes of mutations found using this approach *(8,9)*. These include general necrosis, edema, brain necrosis, and general retardation. Mutations of these common classes account for up to two-thirds of the mutations uncovered and have, for the most part, been discarded by the large screens, since they represent an unwieldy task in the determination of individual complementation groups. Small-scale directed screening may reveal that a significant number of such mutations disrupt in specific developmental processes that can only be revealed by the use of gene or protein-specific markers, that is, marker-based screening.

The efficiency of different mutagenic agents in inducing mutations within the germline of zebrafish has also been assessed. These studies have relied largely on the vast literature concerning the efficient induction of mutations in mice by ethyl nitrosourea (ENU). Similarly, Mullins et al. *(8)* and Solnica-Krezel et al. *(9)* demonstrated that ENU induces mutations within male spermatogonial cells at rates between 1/450 and 1/1000 for specific pigmentation genes in zebrafish. This range is typical for mutagenic rates at given loci for a number of mutations, and the rate is significantly higher than that produced in similar treatments with the other most frequently used alkylating agent, EMS

*(8,9)*. Large-scale mutagenic screens have tended to focus on the use of chemical mutagens as their action is not site-specific and they usually induce lesions that are limited to single genes. However X-and γ-rays are also efficient in inducing mutations within the zebrafish germline and are more applicable to other screening rationales (*see* **ref. *10***).

The two large scale mutagenic screens have both used a similar general methodology (*see* **Fig. 1**). Mutations are induced by ENU in $G_0$ males and detected in the diploid offspring of the intercrossed F2, where the mutation has been driven to homozygosity. A "family" of F2 fish, half of which is heterozygous for a given mutation, is a significant advantage of a F2 diploid screen. It provides an immediate working stock from which to rescreen and recover mutants. Its major disadvantage is the large numbers of fish that have to be maintained and screened. The large screens have chosen not to use F1 screening via the induction of haploid or gynogenetic diploids because: (1) some strains do not consistently produce eggs for fertilization in vitro, and (2) the manipulations produce a high background of developmental defects.

Although these disadvantages may preclude the detection of every mutable locus within the zebrafish genome via F1 screening, those laboratories not concerned with such a goal can still effectively screen for mutations in a given developmental process.

Large-scale diploid screens require facilities and support beyond the scope of most laboratories. The flexibility of zebrafish as a genetic model makes it suitable for use in small-scale mutagenic screens directed to specific developmental processes. In particular, these attributes include:

1. The ability to screen for defects in haploid or gynogenetic diploid embryos. As outlined below, this greatly reduces the number of fish needed, since the offspring of F1 founders can be screened directly. Thus, even small facilities can survey a significant amount of the genome for particular types of mutations *(3)*.
2. The optical transparency of the embryo allows the detection of gene-specific protein or mRNA markers in whole mounts and, therefore, allows a greater efficiency of screening by the direct detection of mutations that are involved in a given developmental process. The three major advantages of such an approach are:
   a. The detection of subtle defects not visible by morphological screening.
   b. The ability to direct a screen toward mutations that affect a process of interest. Thus, the use of tank space can be maximized by immediately discarding mutations in which phenotype is unrevealing, and there are no alterations to a given marker's distribution.
   c. The ability to interpret better mutants identified by morphological inspection in which phenotype is unrevealing regarding gene function, but developmental markers are affected.

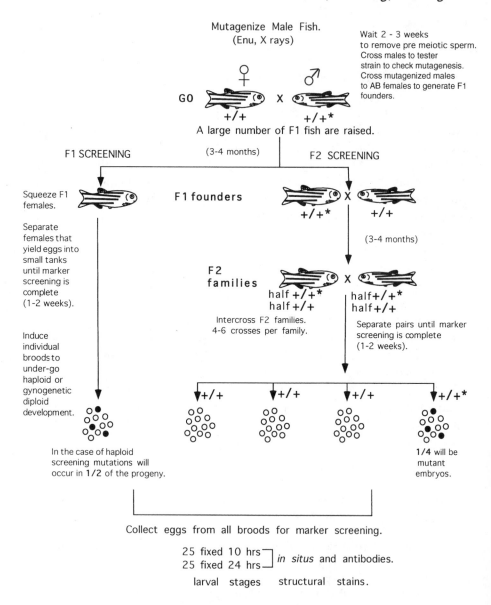

Fig. 1. Flowchart of procedures for F1- and F2-based screening. Screening the progeny of F1 females saves the researcher one generation time and the space required for large families of F2 fish in diploid screening. Single mutations (*) are generated in the spermatogonia of $G_0$ males, and these are mated (**X**) to wild-type females. Similar pairwise matings of F1 or F2 offspring are used to identify fish, male or female, carrying the mutation, and these fish must be kept separate during the screening. Time intervals are indicated alongside arrows, which are not drawn to scale.

## 2. Materials

### 2.1. Mutagenesis

1. Fish strains (*see* **Notes 3** and **4**): Strains are available from laboratories on request. The choice of strain to mutagenize is extremely important, since particular strains have been used for mapping and some are more amenable to genetic manipulations. The inbred strain recommended for mutagenesis is one derived originally by the late George Streisinger (University of Oregon) and is designated AB, although other laboratories have developed their own inbred strains *(8,9)*. The advantages of AB over other strains are to be found primarily in haploid screening. AB fish have been selected over many generations for their ability to provide eggs for fertilization in vitro and for the lack of embryonic lethal mutations within the genetic background. Further, the most extensive genetic map for zebrafish, using polymorphic variation in the PCR products generated by random primers, has used AB as its inbred strain *(11)*. In the absence of information about the amount of polymorphic variation between other strains and AB, it would seem useful to induce mutations in this strain to aid in mapping.

2. Mutagens: The choice of mutagen will determine the type and number of mutants you find. ENU (Sigma, St. Louis, MO) has proven to be the most efficient mutagen for the induction of mutations within the zebrafish germline *(8,9)*. It is highly mutagenic and carcinogenic, and must be handled with extreme care. Whenever possible, dedicated sets of equipment and tanks must be used, and all mutagenic procedures carried out in a suitable fume hood. All solutions and equipment contaminated with ENU are inactivated by incubation in a 10% solution of sodium thiosulfate, adjusted to pH 10.0 with sodium hydroxide, for at least 24 h at room temperature. To create the mutagenic solution, ENU is dissolved directly in the ampule in 10 m$M$ acetic acid to a concentration of 100 m$M$ and stored at –20°C as a stock solution (ENU activity is highly pH-dependent; it is higher with increasing pH and a correspondingly higher lethality). Immediately prior to use, the stock solution is thawed and diluted to 3–10 m$M$ sodium phosphate buffer (pH 6.6) to make the final mutagenic ENU solution (*see* **Notes 5** and **6**).

### 2.2. Fish Raising and Embryo Manipulation

1. Tanks and water (*see* **Note 2**): Systems for rearing and maintaining zebrafish range from commercially available modular systems to small-scale, self-built facilities *(8,12)*. What is of most importance, however, is water quality. To manage a diploid screen efficiently, fish must lay in a high percentage of pair matings. This seems to relate, at least in part, to the water quality. Tap water quality can be greatly improved by the use of filtration systems. Charcoal filters seem to be neccesary with sand filters proving a useful second-step filtration, although neither of these can buffer against regional differences in water quality, such as heavy metal content. Water quality should be assayed before facility designs are too advanced to accommodate any specialized needs. One valuable addition to consider is a UV sterilizer, which can greatly aid in keeping stocks disease-free.

It is also important that pH is kept neutral or slightly high, since zebrafish are sensitive to an acidic environment, but seem to tolerate mild alkaline water.

There are many varieties of tank designs. In a diploid screen where many pair matings are performed, it is vital to have an efficient system to set up breeding pairs to collect embryos. We use small plastic boxes with wire mesh replacing the bottom of one box and stacked inside another. Eggs fall through the mesh, and parents are restrained from eating them. Pairs or individual females are kept separate in small plastic boxes. Larger 4-L mouse cages are used for rearing of juvenile fish, and adult fish are kept in larger permanent glass tanks of the facility. A serial system of glass tanks with a commom water-flowthrough is the most effient design for facilities with space constraints (*see* **Note 7**).

2. Food: There are also many diets that can be used. Adults do well on most commercial flake diets, and we use Tetramin flakes finely ground with a mortar and pestle to a powder. Juveniles from about 3 wk of age are fed hatched live brine shrimp, and these can be fed to the adults also. Care must be taken to remove all unhatched brine shrimp, since these are not easily digested. Fry from a week old are fed alternatively filtered paramecium cultures and a commercial protein diet of liquifry (Tetramin) and Tetramin fry food mixed. Fish of all stages are fed twice a day. Sexual maturity can be achieved much more quickly if the number of feedings and food density is kept high to juveniles. However, adult fish fed in this manner may have a shorter life cycle and stop breeding more quickly. This can be a major problem at the end of a screen lasting a year or so, since tanks of females identified as carrying mutations may have stopped breeding before they can be out-crossed or sperm collected for freezing, in the case of the males.

3. Hank's solution: Full-strength Hank's solution is $0.137\ M$ NaCl, $5.4\ mM$ KCl, $0.25\ mM$ $Na_2HPO_4$, $0.44\ mM$ $KH_2PO_4$, $1.3\ mM$ $CaCl_2$, $1.0\ mM$ $MgSO_4$, $4.2\ mM$ $NaHCO_3$. Bicarbonate is made fresh.

4. Embryo medium: This is 10% (v/v) Hank's with magnesium and calcium at full strength.

5. Methylene blue: A weak solution of methylene blue acts as an antifungal agent. A stock solution is made by diluting methylthione chloride powder (Sigma) to 2% (w/v) in $dH_2O$. One or 2 drops are added/400–500 mL just until water turns pale blue. Higher concentrations result in uptake of the blue dye by developing embryos.

6. 1-Phenyl-2-thiourea (PTU): An active solution of PTU is made by dissolving the powder to 0.003% (w/v) in 10% Hank's. Fish must be raised in this solution from fertilization to inhibit pigment synthesis. PTU is also neurotoxic, and solutions and powder should be handled with care. Fish raised in this solution should be separated from the rest of the fish to avoid the possibility of PTU being placed into the system.

7. Tricaine: This is the most readily used fish anesthetic and, if applied correctly, seems to have no adverse side effects. A stock solution of tricaine is made by dissolving 3-amino benzoic acidethylester (Sigma) powder to 0.4% (w/v) in 20 $mM$ Tris (pH 9.0) and adjusting the pH to 7.0. This solution is stored at $-20°C$

and diluted to approx 0.015–0.020% (v/v) in water to provide a working solution. The stock solution loses its efficacy over time and should be made up monthly. Tricaine is also a mild carcinogen. Although it is impossible to avoid contact with the solution in manipulations involving its use, care should be taken to avoid overexposure.

8. Methyl cellulose: A 3% (w/v) solution of methyl cellulose can be used to mount and photograph live anesthetized embryos. Powder is dissolved in water via gentle heating to a viscous solution. The solution is refrigerated to remove air bubbles and then brought to room temperature before use. Fish can simply be removed from the methyl cellulose by sequential rinses in water.

9. UV source: Any source of shortwave UV light can induce haploidization in zebrafish embryos as long as this effect is calibrated. The source used in our laboratory is Ultraviolet Products Incorporated UVL-56.

10. Fish Ringers. Fish Ringers is 0.11 $M$ NaCl, 0.0034 $M$ KCl, $CaCl_2(H_2O)_2$, and 0.0024 $M$ $NaHCO_3$. Bicarbonate is added last and immediately before use.

## 2.3. Detection of Mutant Phenotypes

1. Microscopy: All morphological and marker screening utilizes simple light microscopy via a dissecting scope. We use Zeiss Stemi SR microscopes. However, any microscope with the capability of translucent illumination and up to 5X objectives would be adequate. Translucent lighting is essential for clear optical inspection of structures of the embryo.

2. *In situ* hybridizations: *In situ* probes and protocols use standard methods (*see* Chapter 31 and **refs. *12,13***).

3. Antibodies: Antibodies are either bought commercially or are available from different laboratories on request. We utilize standard methods for protein detection *(12)*.

4. Alcian blue: A stock solution of alcian blue stain is made by dissolving powder to 0.1% (w/v) in 70% (v/v) EtOH/30% (v/v) glacial acetic acid.

## 3. Methods

Many of the methods listed below are derived from **refs. *8,9,12***.

### 3.1. Mutagenic Methods: ENU Mutagenesis

Tests of different mutagenic regimes in zebrafish favor mutagenizing premeiotic germ cells in adult males (**Fig. 1**). This is because, in a classical diploid screen, use of sperm derived from postmitotic stages of spermatogenesis results in a mosaic germline in F1 founder fish and increases the number of crosses required between F2 siblings to drive mutations to homozygosity. Mutagenizing $G_0$ males involves the following procedures:

1. Pretest males (4–8 mo optimal) for high fertility by single pair matings. Place males directly in ENU solution.

2. Mutagenize in ENU solution for 1 h at room temperature in a darkened fume hood. The fish are stressed and tend to leap out of containers. A darkened quiet environment seems to minimize this behavior.

3. Transfer to a similar volume tank containing water from the fish facility at room temperature. Allow to recover for 6–8 h at this temperature.
4. Finally, transfer to a large 5-L tank in the fish facility.
5. Following a 1-wk recovery period, the treatment is repeated two further times, each followed by 1 wk of recovery. This recovery period may be shortened, but may lead to an increase in lethality. In a typical mutagenesis, only 30–50% of mutagenized fish survive all three treatments with increasing lethality per treatment.
6. Fish are then mated several times in a 4–6 wk period to remove mutant, mosaic postmeiotic sperm cells.
7. To test the efficacy of the mutagenesis, these $G_0$ males (**Fig. 1**) are then crossed to a tester strain female, usually a strain carrying a homozygous viable pigment mutant, such as *golden* (*gol*). An effective mutagenesis should reveal a pigmentless embryo at a specific locus frequency of approx 1/500 (*8,9*).
8. These embryos may be raised as F1 founder fish or their $G_0$ fathers crossed to wild-type females of a required strain to generate F1 founders.

## 3.2. Screening Methods

The ability to generate haploid or gynogenetic diploid development provides the researcher with a number of different options for screening mutagenized F1 founder fish. More than one of these can be performed in parallel, and each of these will be discussed in turn.

### 3.2.1. Haploid-Based Screen

This type of screen saves one generation by screening F1 fish directly and reduces the number of fish required to survey a given number of haploid genomes. Haploid development is initiated by inactivating sperm by UV irradiation and using this sperm to fertilize eggs in vitro. Haploid embryos develop essentially as normal for the first few days, but have a few consistent defects, including short tails, deformed ear capsules, and edema, by early larval stages, after which they die. Therefore, a haploid-based screen for mutations in organogenesis and late differentiating structures would be inadvisable. However, mutations that alter the basic structure of the zebrafish body plan are easily identifiable, and most of the mutations previously identified in zebrafish have been identified by haploid screening of X-ray-induced mutations (*14–16*).

The ease with which haploids may be produced depends entirely on the ability of mutagenized F1 females to yield their stored eggs, a trait that seems highly strain-dependent. Typically, in a strain considered accessible to this manipulation, sexually mature gravid females will yield eggs in at least 20–30% of squeeze attempts. Each wild-type strain should be assessed for its ability to undergo this procedure. Ideally, fish of the AB strain should be used, since they have been selected over several generations for the ability to allow in vitro fertilization in this manner. Tricks that seem to lead to a more consistent yield from females are:

1. The separation of males from females the night before squeezing to prevent the males from inducing spawning in the females.
2. Squeezing females early in the morning, since they are induced to spawn when the lights come on. Since sperm has to be collected and inactivated prior to females being squeezed, placing female fish in an enclosed light cycle that comes on later than the main facility may help this procedure.

Production of haploid embryos requires little or no specialized equipment *(4)*. A shortwave UV source and anesthetic are all that is required. Sperm may be collected from males and used to fertilize eggs in vitro in the following way:

1. Anesthetize sexually mature males by placing them in a 0.002% (w/v) tricaine solution until gill movement has stopped. Rinse anesthetized fish in water.
2. Place ventral side up in a moist sponge bed designed to keep the fish wedged upright. The genital opening is located between the pelvic fins and should be wiped dry before sperm collection.
3. With Millipore forceps, gently squeeze this area while collecting sperm in a glass capillary pipet attached by thin-gage tubing to a mouth pipet or other suction device.
4. Place sperm in a solution of full-strength Hank's on ice, and store until enough has been collected in the vial to turn the solution opaque and slightly milky. Approximately 0.5 mL are needed for 20 fertilizations.
5. Transfer the Hank's/sperm solution on a watch glass or small plastic Petri dish, and UV-irradiate. Inactivated sperm should be stored on ice. Since individual sources vary in their output of UV irradiation, a time-course of irradiation should be performed. However, as a reference point, we find that irradiation for 5 min at a distance of 20 cm by our UV shortwave source produces inactivated sperm still able to fertilize. Shorter periods fail to inactivate and lead to diploid embryos, easily discernable by their longer tails. Longer periods highly reduced fertility.
6. Anesthetize females, and then dry initially by flipping the fish on a clean paper towel.
7. Place in a clean, small plastic Petri dish. Apply even pressure to the belly, pushing toward the genital opening. Hold the back of the fish gently with the other hand. Undue pressure will lead to internal hemorrhaging and death, so a desire to yield eggs must be tempered against the need to keep the fish alive, no matter how gravid a female may appear. Increased pressure will never lead to an increased percentage of fish that yield eggs. Also, the success of a haploid-based screen relies on the female surviving, since she carries the only recoverable chromosomes in the screen. This risk can be circumvented by separating expressed eggs into dishes, fertilizing one with wild-type, unirradiated sperm, and keeping these out-crossed embryos until the results of a screen are known.
8. Females often yield eggs that cannot be fertilized. These eggs appear milky and opaque when they are squeezed into the Petri dish. Fertile eggs appear shiny yellow and complete in appearance. Separate eggs from the female using a clean spatula, and add 50–100 µL of the Hank's/sperm solution, followed immediately with 0.5 mL of water. Eggs must not be left to dry and should be covered. Sperm solution should be added as quickly as possible after the eggs have been expressed.

9. After 1 min add 3–4 mL of water, and swirl the fertilized eggs gently to avoid clumping.
10. Monitor eggs to see if the chorions raise. Separate from infertiles, and allow to develop.

## 3.2.2. Gynogenetic-Based Screen

This type of screening provides the space and time-saving aspects of haploid screening, but has the added advantages of allowing the observation of the diploid phenotype combined with viable resultant embryos. Gynogenetic diploid embryos can be produced by subjecting eggs fertilized with UV-inactivated sperm to either heat-shock or high hydrostatic pressure. Heat-shock administered 13 min after fertilization suppresses the first mitotic division and, therefore, induces gynogenesis. However, a large percentage of heat-shocked embryos exhibit atypical development possibly owing to structural damage to the fertilized eggs, limiting its effectiveness as a methodology for random screening. High hydrostatic pressure delivered just after fertilization restricts the second meiotic anaphase. A significant number of early pressure- (EP) induced gynogenetic embryos also develop abnormally owing to damage to the eggs, but this is seen at a lower rate than in heat-shock treated embryos. Since crossingover occurs at the first meiotic division, only mutations that occur close to the centromere will exhibit a high rate of homozygosity in the gynogenetic embryos. This somewhat hampers the use of this technique, since the percentage of mutant embryos is directly proportional to the distance that a mutation maps from the centromere and, therefore, is unpredictable. It has consequently been used as a measure of gene distance from the centromere of known mutations. This technique also has the drawback of requiring the use of controlled pressure equipment.

### 3.2.2.1. Production of Gynogenetic Embryos via Heat-Shock

1. Collect and UV-inactivate sperm as described in **Subheading 3.2.1.**
2. Collect eggs and fertilize in vitro as described in **Subheading 3.2.1.** Keep fertilized eggs at 28°C.
3. Approximately 10 min after fertilization, transfer the developing eggs to a 42°C water bath for 2 min.
4. Return embryos to 28°C water, and allow to develop normally.

### 3.2.2.2. Production of Gynogenetic Embryos via Early Pressure

This requires a hydraulic press or French press. These are not easily located but perhaps one place to look is amongst the discarded equipment of a microbiology department.

1. Collect and UV-inactivate sperm as described in **Subheading 3.2.1.**
2. Collect eggs and fertilize in vitro as in **Subheading 3.2.1.**

3. Place fertilized eggs in embryo water in pressure vials within pressure cylinder.
4. Apply pressure to fertilised embryo 1.5 min after fertilization to 8000 lbs./sq. in. for 4.5 min.
5. Remove embryos from vials, distribute into dishes, and allow to develop normally. With both gynogenetic diploid and haploid screening, we find it useful to separate embryos with mechanically induced defects from normal developing embryos as soon as possible. All embryos are kept and screened, but are also ranked by quality. Any consistent defect that occurs in a percentage of the embryos predicted by the screening rationale is considered mutant and rescreened. Researchers learn defects that are induced by manipulating the embryos and those that reflect a true genetic lesion.

### 3.2.3. Classical Diploid Screening

Alternatively, those researchers that possess a large enough facility to generate and maintain large numbers of mutagenized families may opt for a classical diploid screen approach. In such a screen, mutations are homozygosed in the F3 generation. Advantages include a very low incidence of abnormal development per a given embryo in a brood, and mutations typically show Mendelian segregation. Recessive mutations will be a quarter of the progeny of the F2 intercross of the raised family (**Fig. 1**). Another drawback of haploid screens is that two generations are required before you recover the mutation as a diploid, because the squeezed female must be out-crossed and then the resulting progeny intercrossed before homozygous mutant embryos are produced. The F2 families raised in the diploid screen also provide a large number of fish immediately for analysis, since 50% of these fish should be heterozygous for the induced mutation (**Fig. 1**).

Diploid screening simply involves pair-mating fish of the F2 families, so that the statistical likelihood that two heterozygous fish for an induced mutation have been crossed is high. It is important to raise families of sufficient size and numbers of males and females to achieve this. Ideally 30–50 fish should be raised. Small families should be discarded as soon as it is realized that insufficient fish are present in the brood, since they will take up tank space and not provide sufficient fish to ensure detection of any mutations. Because 50% of fish should carry the mutation as heterozygotes in a given family, four to six crosses are usually sufficient to uncover mutations within a family. If the number of families is limited, more can be crossed to increase the certainty of detecting mutations. Pairs need to be kept separate until all embryo screening methods have been performed and, if no mutation is detected, can be returned to original tanks. Once it has been determined which mutations are to be kept, fish carrying those mutations must be identified from individual families by crossing to already identified fish. Noncarriers can be discarded, and tanks used for new F2 families.

### 3.3. Marker-Based Detection of Mutant Phenotypes

Irrespective of how mutant embryos are produced, a similar protocol can be employed for the use of markers to detect mutant phenotypes. Zebrafish offer some unique alternatives in the detection of specific phenotypes. The ease with which embryos can be manipulated, the large brood size, optically clear embryos and reliability of procedures to detect mRNA or protein in fixed embryos allow the use of "markers" of specific developmental processes. Several markers may be used in combination. A typical protocol for marker-based screening is outlined below. Obviously, the stage of embryos fixed will vary between those markers used (*see* **Note 8**).

### 3.3.1. Strategy for Marker-Based Screening

1. Collect and sort 100 fertile embryos from each pair mating. Separate into two dishes of 50 embryos. Individual pairs or F1 females must be labeled and kept separately. Space constraints will determine the number of pairs or females that can be screened at any given time, since the fish must be kept separate until marker screening is done and it may take a number of weeks until enough individual crosses or squeezes are generated to perform an *en masse* marker-based screen.
2. Allow the embryos to develop to the stages to be screened morphologically (*see* **Subheading 4.** for specific details on morphological screening). One dish of 25– 50 embryos is fixed for use in a marker-based screen. At least half of the embryos are kept for morphological screening and structural stains at later stages. We fix 25 embryos at bud and early pharyngula stages for use in an *in situ*-based screen. Embryos of different stages from the same parent(s) are combined in small baskets and incubated through the *in situ* protocol in 24-well tissue-culture plates. A similar approach can be used for screening with antibodies to specific antigens. We routinely screen embryos from 30–50 individual broods at a time. It is possible to scale up the *in situ* protocol to use larger well number microtiter plates, such as 96-well plates, moved through common solution baths. However, it is difficult to keep large numbers of pairs or individual females separate for any length of time in a small facility. Thus, the number of broods screened at any given time will be dictated by this space constraint.
3. Allow to develop to 3 d, and screen for defects in organogenesis if a diploid screen is being employed. Collect embryos for structural stains at appropriate stages. If no phenotype has been detected at any stages, pairs or females should be returned to original tanks in diploid screens, or to a separate tank in haploid- or gynogenetic-based screens.

    Choosing markers: A large number of cloned genes and cell-type-specific antibodies have been generated that allow visualization of specific regions or molecular processes within the zebrafish embryo. These markers can often reveal subtle changes in the molecular phenotype of a given mutation that may not be detectable morphologically. They may also implicate a mutation that has an otherwise uninteresting phenotype in a specific molecular process. Some examples are given in **Table 1**.

**Table 1**
**In Situ and Antibodies Availiable for Marker Based Screening of Early Segmentation and Pharyngula Stages in Zebrafish[a]**

| Germ layer | Marker | Region expressed | Ref. |
| --- | --- | --- | --- |
| Neuroectoderm | | | |
| | *zash 1a* and *1b* | tel. and di., r1-6 | *18* |
| | *emx 1* and *2* | tel. and di. | *19* |
| | *otx 1, 2,* and *3* | tel., di., and mes | *20* |
| | *pax 2* | optic stalk, mes/rhomb, and dorsal spinal neurons | *21* |
| | *pax 6* | retina, lens, and dorsal spinal neurons | *22* |
| | *nk 2.2* | di. | *23* |
| | *shh* | fp, ventral di. | *24* |
| | *dlx2* and *4* | tel. and di. | *25* |
| | *wnt 1* | mes/rhomb | *26* |
| | *eng 1, 2, 3* | mes/rhomb | *27* |
| | *krx 20* | r3 and 5 | *28* |
| | *rtk 1* | r2 and 4 | *29* |
| | *pou 2* | r2 and 4 | *30* |
| | *axial* | fp, ventral di | *31* |
| | *collagen type 2* | fp | *32* |
| | ZN1* | 1° motor neurons, most differentiating neurons | *33* |
| | ZN12* | reticulospinal and Rohon-Beard neurons | *33* |
| Mesoderm | | | |
| | *axial* | noto., pcp | *31* |
| | *znot* | noto | *34* |
| | *gsc* | pcp | *35* |
| | *shh* | noto., pcp | *24* |
| | *Brachyury* | noto | *36* |
| | *snail 1* | paraxial | *37* |
| | *twist* | noto., somite | *38* |

*a*This is by no means an exhaustive list and is meant to provide the reader with a starting point for designing marker screens. Abbreviations: di—diencephalon, fp—floorplate, mes—mesencephalon, noto—notochord, pcp—prechordal plate mesoderm, r—rhombomere, rhomb—rhombencephalon, tel—telencephalon.

### 3.3.2. Selection of Markers for Use in Screens

The techniques for whole-mount *in situ* hybridization and antibody staining have been comprehensively reviewed elsewhere and, therefore, will not be discussed (Chapter 31, *12,13*). However, some general pointers and pitfalls for the selection of markers can be outlined.

1. Care must be taken to use *in situ* probes or antibodies that are robust in their application. If this is the case, most protocols can be modified and considerably shortened without much loss of sensitivity.
2. Reagents can often be reused many times without loss of sensitivity of detection, and this sensitivity can actually increase with use. Since cost is a major factor in designing such a screen, with dioxygenin and antibodies being the major expense, this is an important consideration.
3. Since it is envisaged that a screen based on revealing gene patterns or specific proteins would utilize several markers at a time, care must be used to select individual markers that are within each other's range for sensitivity of detection. Also, each individual marker should mark distinct subpopulations of cells that do not obscure visualization via other markers.
4. For antibody stainings, fixation times for individual antigens may be incompatible. These parameters can often only be determined empirically.
5. Importantly, detection must be via simple inspection utilizing light microscopy, without the need for sophisticated mounting and visualization techniques, since this is preclusive to the ability to screen large numbers of individual broods.
6. Choose markers that reveal some aspect of developmental regulation that is not obvious on inspection. For example, somite formation is clearly visible within the transparent embryo, and markers that stain the whole of the developing somite may reveal less than simple inspection of the developing somite. However, a marker, such as *krox 20*, that stains specific rhombomeres in the hindbrain of the developing embryo that are difficult to visualize under the dissecting microscope may be more useful in a marker-based screen.

A large number of genes and proteins have been identified in zebrafish that fit these basic criteria. Examples of markers that can be utilized in such a screen are given in **Table 1**. This is by no means an exhaustive list, with the number of possible markers growing fast, and it is meant only as a guide to demonstrate the possibilities of this approach. Although the number of mutations that directly affect the expression pattern or distribution of gene transcript and proteins of interest will only be a small subset of the mutations revealed, this type of approach immediately allows the assessment of a particular mutation and its possible involvement in a process of interest.

### 3.3.3. Use of Stains in Screening

Structural stains: A number of stains have been described in vertebrate systems that bind particular cellular or extracellular components. These types of stains can be used to reveal particular aspects of fish anatomy that may not be visible on inspection, such as the vertebral column and head skeleton (*17*). A number of fluorescent stains are available that stain nuclear or cell membrane components that are detectable by fluorescent microscopy. These are included for reference, but clearly may not be accessible techniques for every lab and may not be suitable for use in a larger screen.

### 3.3.3.1. ALCIAN BLUE

This stain binds to a matrix component of cartilage that begins to develop during the hatching period. Staining in embryos and early larvae reveals the pattern of the developing fin cartilage and head skeleton. Subtle changes in the patterning of these structures can be revealed by using simple light microscopy without the need for dissection.

1. Fix early larvae (3–5 d) in 10% buffered formalin (pH 7.0) for 3 h to overnight.
2. Rinse once in 50% EtOH/50% (v/v lX) PBS.
3. Place in 0.1% (w/v) alcian blue overnight at room temperature.
4. Rinse in EtOH and rehydrate with PBS.
5. To clear tissue, place in dilute trypsin solution (~0.05% w/v) and leave at room temperature until embryos become soft and transparent. The eyes can then be removed, and the embryos mounted in glycerol. (This step is optional and required only when looking in detail at cartilage.)
6. Pigmentation may be bleached by placing stained specimens in 0.35% $H_2O_2$ dissolved in 0.1 $M$ KOH.

### 3.3.3.2. ALIZARIN RED

This stain binds to forming bone *(17)*. Although most bone develops relatively late, researchers may wish to analyze the affects of mutation on early ossifications.

1. Fix larvae in 3% (w/v) KOH for 24 h.
2. Stain in 0.001% (w/v) alizarin red/1% (w/v) KOH for 3 h.
3. Rinse and store in glycerol.

### 3.3.3.3. ACRIDINE ORANGE

This stain can be used to access the amount of cell death in a given mutation.

1. Soak live embryos in ~0.1 mg/mL solution for 2–3 min.
2. Examine immediately with fluorescein filter set.

## 4. Notes

1. An in-depth guide to zebrafish raising and techniques has been provided in *The Zebrafish Book: A Guide for the Laboratory Use of Zebrafish* (Brachydanio rerio), University of Oregon Press *(12)*. This book provides an invaluable reference for most experimental manipulations in zebrafish.
2. Raising of zebrafish embryos has traditionally been done in a defined salt solution or embryo medium. We find that this is not necessary as long as an antifungal agent, such as methylene blue, is added to system's water. This may of course vary with the quality of individual water sources.
3. It may be advantageous for some researchers to screen for mutations in pigmentless embryos. In wild-type embryos, pigment develops during the early pharyngula period of development. By early hatching, pigment can significantly

reduce the information gained from optical inspection of the embryo. Although a number of fully viable pigment mutations exist in the appropriate genetic background, the researcher has the option of using PTU to inhibit pigment synthesis (*see* **Subheading 2.**). This may be of use when rescreening mutations to determine effects later in development.

4. It is important to realize that many different regulations govern the importation of fish into different countries, and obtaining import licences can often be costly and time-consuming. It is important to investigate these procedures to avoid fish being impounded and delays that could lead to death of transported fish.

5. Many regulations also govern the mutagenesis of vertebrate animals. In many instances mutagenesis of zebrafish requires a licence from the appropriate authority. Mutagenesis without this licence may be illegal.

6. X- or γ-ray mutagenesis and PCR-based screening: A number of studies have indicated the efficacy of X- and γ-rays in inducing mutations within the germline of zebrafish, although at a significantly lower efficiency than the above-described treatment with ENU *(8)*. X- and γ-rays induce mutation by producing double-strand breaks within the DNA, and often result in deletion or rearrangement of large regions of the chromosome. This complexity may reduce overall viability of F1 embryos or affect the segregation of mutations in the adult germline. This may explain the lower frequency of recovered mutations. Inducing translocations and deletions, however, is desirable in screens that utilize PCR to detect gene-specific lesions or as a reference point for cloning the gene represented by a given mutation. Mutations can be easily induced in adult fish or in collected sperm, and a brief method for mutagenesis is described below. It is important to use a calibrated X- or γ-ray source with a wide beam to ensure whole-sample irradiation.

   a. Sperm can be collected as outlined above, irradiated while being kept on ice and resuspended in 10% Hank's solution in small glass vials. Alternatively, adult male fish, pretested for high fertility, can be anesthetized in tricane and placed on a moist sponge bed to constrain the fish within the path of the beam.

   b. The ability of a given source to induce mutations should ideally be calibrated, but 200–300 rads seem to be sufficient to induce mutations.

   c. Sperm after irradiation can be added directly to eggs obtained from squeezed females, or irradiated fish can be treated similarly to those treated with ENU to produce F1 founder fish.

It is theoretically possible to search for specific DNA lesions generated by X- or γ-rays via the use of PCR. Primers that span a region of interest can be used to amplify target sequences from haploid embryos derived from F1 founder fish generated from X or γ-rays mutagenized fish or sperm. Although the chances of locating a rearrangement that breaks specifically in a known gene are small, it is possible to identify deletions that span a gene region seemingly with fairly high frequency. This is dependent on the dose of radiation that is used to induce such rearrangements, and it is quite likely that complex rearrangements are induced by high doses of X- or γ-rays. Such complex rearrangements invariably led to complex phenotypes, and it is difficult to say with certainty that a given phenotype results from rearrangement in a specific gene region. It is thus yet to be determined if PCR can be used to identify gene-specific lesions efficiently by this method. It can be used to check quickly if an identified phenotype results from the deletion of a gene of interest, and it may be in this application that PCR-based screening is most useful, rather than its systematic application to every brood of a random screen.

The ability to generate a number of gene-specific PCR markers in any one PCR reaction allows multiple genes to be screened at any one time for a given phenotype. The screen procedure first involves the preparation of DNA from fish demonstrating a given phenotype. PCR is performed on this DNA and scored for the presence or absence of a DNA marker. PCR can also be used to map quickly the limits of a deletion and assess its suitability for further genetic analysis, such as the saturation mutagenesis of a specific region.

7. Stock maintenance: The protocols required for general fish raising and maintenance have been listed in detail elsewhere *(12)*, but it is worthwhile to mention a few tips that specifically impinge on the efficiency of a small-scale mutagenic screen. Mutations will be identified in an ENU screen on average once per one to two haploid genomes screened *(8,9)*. There is a great temptation to keep every mutation that is identified from such a screen. However, the maintenance of these mutations as stocks can quickly cut into available tank space in a small facility. It is far more efficient to make a decision about mutations when they are identified. Alternatively, mutations can be stored as frozen sperm and can be recovered by in vitro fertilization of squeezed wild-type eggs with the thawed sperm. Care should be taken to assay the protocol for the ability of thawed sperm to fertilize by initially utilizing wild-type sperm in test fertilizations.
   Freezing sperm:
   a. Sperm is collected using the methods outlined above. Sperm from at least four males identified as heterozygotes for the mutation of interest should be used.
   b. Sperm is added to 4–5 vol of fish Ringers containing 10% methanol and 15% powdered nonfat milk.
   c. The solution is frozen in capillary tubes placed in 10-mL plastic centrifuge tubes (Sorvall) by incubation in dry ice for 20 min.
   d. Capillary tubes are stored under liquid nitrogen after freezing.

8. Morphological screening: Marker-based screening is most efficient when coupled with a morphological screen. Zebrafish embryos are amenable to screening at most stages of development. If a researcher is interested in processes during organogenesis at stages later than 2 d it may be advisable to perform mutagenesis in a strain mutant for pigment synthesis, such as *golden* or *albino*, since this greatly increases the ability to detect subtle defects in organogenesis. The rapid development of the zebrafish embryo can often make screening at some stages inconvenient. The ability to screen some stages may rely on placing the fish "off cycle" by utilizing time-regulated light sources for tanks so that fish are induced

to lay at specific times. Researchers should make a checklist of structures and processes that are of interest and screen accordingly. After early pharyngula stages embryos should be anesthetized in tricaine to stop twitching and movement to best reveal structure. A subset of embryos should be dechorinated, but the dorsal aspects of the embryo are best revealed by rolling the embryo within the chorion to present the dorsal surface. Later stages require anesthetizing also, since by 3 d fry are able to swim rapidly and can be frustrating to screen.

## References

1. Nusslein-Volhard, C. and Wieschaus, E. (1980) Mutations affecting segment number and polarity in *Drosophila. Nature* **287,** 795–801.
2. Rinchink, E. M. (1991) Chemical mutagenesis and fine-structure functional analysis of the mouse genome. *Trends Genet.* **7,** 15–21.
3. Kimmel, C. B. (1989) Genetics and early development of zebrafish. *Trends Genet.* **5,** 283–288.
4. Streisinger, G., Walker, C., Dower, N., Knauber, D., and Singer, F. (1981) Production of clones of homozygous diploid zebrafish (*Brachydanio rerio*). *Nature* **291,** 293–296.
5. Kimmel, C. B., Ballard, W. W., Kimmel, S. R., Ullmann, B., and Schilling, T. F. (1995) Stages of Embryonic Development of the Zebrafish. *Dev. Dyn.* **203,** 253–310.
6. Warga, R. M. and Kimmel, C. B. (1990) Cell movements during epiboly and gastrulation in zebrafish. *Development* **108,** 569–580.
7. Kimmel, C. B., Warga, R. M., and Schilling, T. F. (1990) Origin and organization of the zebrafish fate map. *Development* **108,** 581–594.
8. Mullins, M. C., Hammerschmidt, M., Haffter, P., and Nusslein-Volhard, C. (1994) Large scale mutagenesis in the zebrafish: in search of genes controlling development in a vertebrate. *Curr. Biol.* **4,** 189–202.
9. Solnica-Krezel, L. A., Schier, F., and Driever, W. (1994) Efficent recovery of ENU induced mutations from the zebrafish germline. *Genetics* **136,** 1401–1420.
10. Walker, C. and Streisinger, G. (1993) Induction of mutations by γ-rays in the pregonial germ cells of zebrafish embryos. *Genetics* **103,** 125–136.
11. Postlethwaite, J. H., Johonson, S. L., Midson, C. N., Talbolt, W. S., Gates, M., Ballinger, E. W., Africa, D., Andrews, R., Carl, T., Eisen, J. S., et al. (1994) A genetic map for the zebrafish. *Science* **264,** 699–703.
12. Westerfield, M. (1993) *The Zebrafish Book: A Guide for the Laboratory Use of Zebrafish* (Brachydanio rerio). The University of Oregon Press, Eugene, OR.
13. Jowett, T. and Lettice, L. (1994) Whole-mount in-situ hybridizations on zebrafish embryos using a mixture of digoxigenin-labeled and fluorescein-labeled probes. *Trends Genet.* **10,** 73,74.
14. Kimmel, C. B., Kane, D. A., Walker, C., Warga, C. M., and Rothman, M. B. (1989) A mutation that changes cell fate in the zebrafish embryo. *Nature* **337,** 358–362.
15. Halpern, M. E., Ho, R. K., Walker, C., and Kimmel, C. B. (1993) Induction of muscle pioneers and floor plate is distinguished by the zebrafish *no tail* mutation. *Cell* **75,** 99–111.

16. Hatta, K., Kimmel, C. B., Ho, R. K., and Walker, C. (1991) The cyclops mutation blocks specification of the floor plate of the zebrafish central nervous system. *Nature* **350,** 339–341.
17. Kelly, W.L. and Bryden, M. M. (1983). A modified differential stain for cartilage and bone in whole mount preparations of mammalian fetuses and small vertebrates. *Stain Technol.* **58,** 131–134.
18. Allende, M. L. and Weinberg, E. S. (1994) The expression pattern of two zebrafish *achaete scute homolog (ash)* genes is altered in the embryonic brain of the cyclops mutant. *Devel. Bio.* **166,** 509–530.
19. Morita, T., Nitta, H., Kiyama, Y., Mori, H., and Mishina, M. (1995) Differential expression of 2 zebrafish emx homeoprotein messenger-RNAs in the developing brain. *Neurosci. Lett.* **198,** 131–134.
20. Mori, H., Miyazaki, Y., Morita, T., Nitta, H., and Mishina, M. (1994) Different spatio-temporal expression of three *otx* homeoprotein transcripts during zebrafish embryogenesis. *Mol. Brain. Res.* **27,** 221–231.
21. Krauss, S., Johansen, T., Korzh, V., and Fjose, A. (1991) Expression of the zebrafish paired box gene *pax [zf-b]* during early neurogenesis. *Development* **113,** 1193–1206.
22. Krauss, S., Johansen, T., Korzh, V., Moens, U., Ericson, J. U., and Fjose, A. (1991) Expression pattern of zebrafish pax genes suggests a role in early brain regionalization. *Nature* **353,** 267–270.
23. Barth, K. A. and Wilson, S. W. (1995) Expression of zebrafish *nk2.2* is influenced by the *sonic hedgehog/vertebrate hedgehog-1* and demarcates a zone of neuronal differentiation in the embryonic forebrain. *Development* **121,** 1755–1768.
24. Krauss, S., Concordet, J.-P., and Ingham, P. W. (1993) A functionally conserved homolog of the *Drosophila* segment polarity gene *hh* is expressed in tissues with polarizing activity in zebrafish embryos. *Cell* **75,** 1431–1444.
25. Aikimenko, M.-A., Ekker, M., Wegner, J., Lin, W., and Westerfield, M. (1994) Combinatorial expression of three zebrafish genes related to *distal-less*: Part of a homeobox gene code for the head. *J. Neurosci.* **14,** 3474–3486.
26. Krauss, S., Korzh, V., Fjose, A., and Johansen, T. (1992) Expression of 4 zebrafish *wnt*-related genes during embryogenesis. *Development* **116,** 249–259.
27. Ekker, M., Wegner, J., Akimenko, M. A., and Westerfield, M. (1992) Coordinate embryonic expression of 3 zebrafish *engrailed* genes. *Development* **116,** 1001–1010.
28. Oxtoby, E. and Jowett, T. (1993) Cloning of the zebrafish *krox-20* gene (*krx-20*) and its expression during hindbrain development. *Nucleic Acids Res.* **21,** 1087–1095.
29. Xu, Q. L., Holder, N., Patient, R., and Wilson, S. W. (1994). Spatially regulated expression of 3 receptor tyrosine kinase genes during gastrulation in the zebrafish. *Development* **120,** 287–299.
30. Hauptmann, G. and Gerster, T. (1995) *Pou-2*—a zebrafish gene active during cleavage stages and in the early hindbrain. *Mech. Devel.* **51,** 127–138.
31. Strahle, U., Blader, P., Henrique, D., and Ingham, P. W. (1993) *Axial*, a zebrafish gene expressed along the developing body axis, shows altered expression in cyclops mutant embryos. *Genes Devel.* **7,** 1436–1446.

32. Yan, Y. L., Hatta, K., Riggleman, B., and Postlethwait, J. H. (1995) Expression of a type-II collagen gene in the zebrafish embryonic axis. *Dev. Dyn.* **203,** 363–376.

33. Trevarrow, B., Marks, D. L., and Kimmel, C. B. (1990) Organization of hindbrain segments in the zebrafish embryo. *Neuron* **4,** 669–679.

34. Talbot, W. S., Trevarrow, B., Halpern, M. E., Melby, A. E., Farr, G., Postlethwait, J. H., Jowett, T., Kimmel, C. B., and Kimelman, D. (1995) A homeobox gene essential for zebrafish notochord development. *Nature* **378,** 150–157.

35. Stachel, S. E., Grunwald, D. J., and Myers, P. Z. (1993) Lithium perturbation and *goosecoid* expression identify a dorsal specification pathway in the pregastrula zebrafish. *Development* **117,** 1261–1274.

36. Schulte-Merker, S., Hammerschmidt, M., Beuchle, D., Cho, K. W., DeRobertis, E. M., and Nusslein-Volhard, C. (1994) Expression of zebrafish *goosecoid* and *no tail* gene-products in wild-type and mutant *no tail* embryos. *Development* **120,** 843–852.

37. Thisse, C., Thisse, B., Schilling, T. F., and Postlethwait, T. F. (1993) Structure of the zebrafish *snail 1* gene and its expression in wild-type, *spadetail* and *no tail* mutant embryos. *Development* **119,** 1203–1215.

38. Riggleman, B. and Grunwald, D. J. (1991) Molecular cloning and analysis of three developmentally regulated genes in zebrafish N-Cam, myogenin and twist. *J. Cell Biol.* **115,** 146.

# 31

## Transgenic Zebrafish

### Trevor Jowett

### 1. Introduction

The zebrafish has become an important model for studying early vertebrate development. This is largely because of the pioneering work of
George Streisinger and the later work of his colleagues in establishing much
of the methodology in rearing and studying the animal *(1)*. The large numbers
of progeny and the ease of performing genetic crosses have also allowed large-
scale screens for mutations in genes affecting early development *(2–6)*. The
development of the genetic linkage map *(7,8)* means that the genes identified
by these mutation screens can now be cloned.

The number of zebrafish genes cloned by different methods is growing rapidly, and there is a need to exploit transgenic methods to study gene regulation
and interactions in the zebrafish. There are three ways in which transgenic
zebrafish can be used to answer important biological questions.

1. Transgenes are expressed transiently in living embryos to study gene regulation
   and interactions.
2. Stable transgenic lines are generated to study cis-acting regulatory elements, perform rescue of mutant phenotypes, and study genetic regulatory pathways.
3. Integration of foreign DNA provides a means of generating new mutations.
   Insertional mutations allow cloning of the DNA sequences at the point of insertion by transgene tagging or plasmid rescue.

### 1.1. Major Considerations in Performing Transgenic Experiments

When plasmid DNA is injected into zebrafish embryos, it may meet with
several different fates. It may replicate and persist in the cell and its descendants for several cell divisions. It may integrate into the chromosomal DNA of

From: *Methods in Molecular Biology, Vol. 97: Molecular Embryology: Methods and Protocols*
Edited by: P. T. Sharpe and I. Mason © Humana Press Inc., Totowa, NJ

the cell and generate a clone of transgenic somatic cells. It may integrate into the chromosomal DNA of a progenitor germ cell and generate a group of germ cells that will pass the transgene on to the F1 progeny. Alternatively, it may be lost from the embryo.

The first two fates lead to embryos and fish that are mosaic with respect to the presence of the plasmid DNA. If the plasmid carries a reporter gene fused to a promoter, which is active in zebrafish, then injected embryos can be assayed for the presence of transgene product. The promoter may be one that is active in all cells or one that will only drive expression of the reporter gene in a subset of cells in a spatially and temporally restricted manner. In either case, the pattern of transgene expression is transient and will be different in each injected embryo, since it depends on which cells in the embryo retain the plasmid DNA.

If the injected DNA is incorporated into the chromosomal DNA of the germ cells, the fish has the potential to pass the transgene on to its progeny. Such founder fish are expected to be mosaic in their germline, since not all germ cell precursors may have integrated the injected DNA. The fish may also be mosaic for somatic integration of the DNA. If all the germ cells carry a single insertion of the injected DNA, then 50% of the progeny are expected to inherit the transgene. However, if the germline is mosaic, the proportion of transgenic F1 progeny depends on the degree of mosaicism. Germline transmission may be detected by sampling DNA from either somatic or germ cells of F1 fish, since the fish should be hemizygous for the transgene in all cells. This is done by performing a polymerase chain reaction (PCR) amplification with DNA extracted from a fin biopsy using primers specific to the transgene construct. It is important to note that this alone does not confirm the DNA has integrated into a chromosome. Southern analysis will show the number of transgene copies present and also the presence of junction fragments derived from the point of insertion. Although the presence of junction fragments is indicative of integration into the genome and subsequent germline transmission, this must be confirmed by showing the Southern pattern of the integrated transgene is inherited by 50% of the offspring in the F2.

A transgene construct that expresses a reporter gene in a transient manner in the injected embryo may not show expression in the transgenic fish of the next generation. This may be because the transgene has undergone a rearrangement at some stage during integration into the genome, or the integrated DNA may be transcriptionally inactive.

## 1.2. Transient Transgene Expression

The analysis of transient expression in the injected embryos provides a way of checking whether a particular promoter will drive transcription in zebrafish. This allows potential transgenic vectors to be tested before embarking on a major F1 screen. Several viral promoters have been tested in this way. Stuart et al. *(9)* injected a plasmid containing the SV-40 early region promoter upstream of the Rous sarcoma virus long-terminal repeat (RSV-LTR) fused to the chloramphenicol acetyltransferase (CAT) gene and showed that CAT activity could be detected as early as 8 h and up to 12 d after injection. This was a complex promoter, but expression appeared to be primarily driven by the RSV-LTR.

Bayer and Campos-Ortega *(10)* injected two plasmids, one containing the RSV-LTR promoter fused to *lac*Z, and the second containing a truncated mouse heat-shock promoter linked to *lac*Z. Injected embryos were fixed and stained for β-glactosidase (β-gal) activity about 24 h later. Embryos injected with the RSV construct exhibited staining in cells of mesodermal origin, whereas the heat-shock construct gave β-gal activity in most tissues.

The most powerful promoter allowing expression in all cell types is probably the immediate early promoter and enhancer of human cytomegalovirus, CMV *(11)*. This promoter fused to *lac*Z drives expression in most tissues of the zebrafish embryo *(12)*. Gibbs et al. *(13)* compared the CMV promoter with the SV-40 early promoter and the RSV-LTR promoter, each fused to the luciferase gene. The CMV promoter gave the strongest signal with the lowest concentration of luciferase reaction components, allowing positive fish to be raised to adulthood.

This type of transient assay can be exploited to examine the effect of ectopic expression of one gene on the regulation of a potential downstream gene. The CMV promoter is fused to a full-length cDNA of the first gene, and the plasmid injected into the embryo. The embryos are subsequently subjected to double *in situ* hybridization first to locate those cells that are expressing the CMV-driven gene and then to determine whether the presumptive target gene is also active in the same cells. This transient assay is a powerful technique, since it is a quick way of determining potential regulatory cascades and causes far less disruption to the embryo than injecting RNA.

Westerfield et al. *(12)* used the transient assay to test the ability of mammalian Hox gene promoters to drive the expression of *lac*Z in specific regions and tissues of the zebrafish. Linear DNA fragments injected at the one cell stage gave expression in one to several hundred cells when examined about a day later. The expression patterns were always mosaic in that expressing cells were mixed with nonexpressing cells in a variegated manner. Mosaicism in these transient assays probably occurs through the uneven distribution and replication of the DNA during subsequent cell divisions and through the failure of some cells to activate the promoter. A similar study has been performed with promoter lacZ fusions for the mammalian GAP-43 gene *(14)*. This approach provides a way of examining the conservation of signaling pathways and *cis*-acting regulatory elements in evolutionary-related vertebrate genes.

## 1.3. Stable Transgenic Lines to Study Gene Regulation and Genetic Interactions

Since the pioneering work of Stuart et al. *(9,15)*, several groups have reported the generation of stable transgenic lines of zebrafish. Although, the frequency with which transgenic fish were produced was low, some important features have emerged about generating stable transgenic lines.

The preferred method of introduction of the DNA sequences is injection *(9,10,13,15–17)*. An alternative is by electroporation *(18–21)*, but these studies have not demonstrated that the transgenic lines are genetically stable or that the transgenes are expressed. Therefore, microinjection of the DNA into the one-cell stage blastula is the most reliable method.

The amount of DNA injected is difficult to estimate and can vary with each needle used. In general, solutions ranging from 25–50 µg/mL give the highest number of survivors, which may also be transgenic in their germline *(9,15,16)*. Phenol red may be added to the solution at 0.1–2.0% as a visual guide to the amount injected. Alternatively, the volume may be estimated by using a radioactive nucleotide in the injection mix and counting the embryo in a scintillation counter. The volume injected should be about 1–3 nL *(14)*.

It appears to make little difference to transient transgene expression or the frequency of generating transgenic germlines whether injected DNA is linear or circular DNA. In all cases reported, all the founder fish (F0), which transmit the transgene to their progeny, were found to be mosaic. In most cases where the transgenic lines have been subjected to Southern analysis, the transgene is inserted at a single chromosomal location, but in multiple copies probably in a tandem array. However, the single enhancer trap insert reported by Bayer and Campos-Ortega *(10)* appears to be a single copy, and 12 out of the 15 integrations reported by Gibbs et al. *(13)* were of one to two copies. The reason for this lower copy number in the latter study is not clear. The DNA was injected as supercoils at 10 µg/mL in 0.01X TE containing 0.05% phenol red. They estimated injecting a total of 100 pg of DNA/embryo and obtained a survival of 10–20% to sexual maturity.

A problem with transgenic technology in zebrafish is the high level of mosaicism in the germlines of the founders. This means that only a small proportion of the F1 will inherit the transgene (frequencies reported range from 2–40%). This makes the identification and live isolation of the relatively rare transgenic F1s very time-consuming if they are screened by PCR. Founder fish with mosaic germ-lines can be identified by extracting DNA from a proportion of embryos from each individual spawning. If the frequency of transgenic F1 fish is low, there is chance that transgenic fish may be missed or that the siblings that are raised to sexual maturity do not carry the transgene. The alternative is to raise all F1 fish until they are near maturity and then take a fin biopsy. This means that many fish must be reared before they can be tested. These problems can be overcome if transgenic fish carry a dominant marker to identify individual transgenic embryos. One possibility is to include a *lac*Z reporter gene in the injected DNA and screen the F1 progeny with a substrate that can be used on living cells. The use of two such substrates has been reported. Westerfield et al. *(12)* used the Imagene Green™ substrate to identify transient expression of injected hox-*lac*Z constructs. Lin et al. *(22)* used a similar fluoresceinated substrate (FDG) on transgenic embryos expressing a *lac*Z fusion construct with the *Xenopus* elongation factor 1a (EF1a) transcriptional regulatory element. In the latter study, four out of five different transgenic lines expressed the *lac*Z gene in early embryos. However, the pattern of expression was distinct for each line, with two lines first beginning expression at the midblastula transition. One of these expressed solely in motor neurons, whereas the other showed patchy expression. Since the different lines contain the same transgene construct, the different expression patterns must be a consequence of the site of insertion in the genome. Positive embryos of the most highly expressing transgenic lines were identified by permeabilizing with dimethylsulfoxide and soaking the embryos for 2–3 min in the FDG substrate.

The *lac*Z gene as the reporter gene has the disadvantage that weakly expressing embryos may not be identified. Prolonged incubation in the substrate gives rise to false positives, probably through the activity of endogenous β-gal. This problem may be overcome with transgene constructs with luciferase, which is not normally found in zebrafish. Gibbs et al. *(17)* injected CMV-luciferase plasmids into embryos, and showed transient expression by incubating the embryos in luciferin solution and detecting the luminescence on X-ray film. However, the transgenic lines in this study failed to express the luciferase enzyme.

The most promising alternative to the β-gal substrates and to luciferase described above is the green fluorescent protein (GFP) from *Aequorea victoria* *(23)*. GFP produces fluorescence without the need to add an exogenous substrate. Amsterdam et al. *(24,25)* have fused a truncated version of the 4.6 kb *Xenopus* elongation 1a enhancer/promoter to the GFP cDNA. Expression of such a construct starts at about 4 h after injection at the one-cell stage when incubated at 28°C. This corresponds to onset of transcription at the midblastula transition *(26)*. Fluorescence continues to be detected for at least 3 wk after injection.

A further three different constructs were made, which all had the EF1a enhancer/promoter in different fusion configurations with GFP cDNA *(25)*. One, with an intron from rabbit β-globin, yielded five transgenic lines, each of which expressed the fluorescence in the F1 fish and, where tested, in the F2. Most expressing lines showed a similar pattern of expression. In transgenic embryos, fluorescence was not seen until 20 h after fertilization, when it was seen uniformly expressed throughout the embryo. Fluorescence was greatest between 24 and 36 h, but persisted for up to 5 d being particularly strong in the eye. A single copy of the transgene was sufficient to detect fluorescence.

The absence of expression in the F1 and subsequent generations is a serious problem with zebrafish transgenic technology. The study of Gibbs et al. *(17)* demonstrated that the CMV-luciferase transgenes were highly methylated in the F1 and subsequent generations, which could account for the lack of activity. Partial activity could be recovered by incubating in 0.3 m*M* 5-azacytidine, which can inhibit methylation in vivo. However, survival was very much reduced, and many of the embryos were deformed. It is possible that incorporation of extra sequences into the vectors to "buffer" the promoter–gene fusions may allow subsequent expression of the transgenes after passage through the germline. The incorporation of the intron from rabbit β-globin in the GFP construct described by Amsterdam et al. *(25)* appears to be beneficial with respect to the ability of the transgenes to be transcribed. However, they report that when included with additional insulating sequences flanking the same construct, only one out of three transgenic F1 lines expressed the GFP. The number of transgenic lines are low, so these results should be treated with caution. However, they suggest that transcription of the transgenes may depend on some inherent property of the particular DNA sequences.

Caldovic and Hackett *(27)*, have tested the ability of border elements incorporated into transgenic constructs to affect integration and expression of a carp β-actin gene enhancer/promoter–CAT reporter gene fusion. Border elements are sequences of DNA that are part of the scaffold attachment regions involved in forming the nuclear scaffold and can impart position-independent expression of a gene in transgenic mice *(28)*. Caldovic and Hackett *(27)* made transgenic constructs with border elements from the *Drosophila* heat-shock 87A7 locus and the attachment-element sequence from the chicken lysozyme locus. These constructs were problematical to make, since plasmids containing two copies flanking the transgene construct failed to amplify in *Escherichia coli*. To overcome this problem they generated linear concatamers by ligation of separate purified fragments in vitro. Transgenic fish carrying the border elements showed uniform expression at about the same level in all tissues. Without the border elements, the expression was restricted and variable between different transgenic lines.

The recent work of Higashijima et al. *(29)* has shown that transgenic zebrafish, which express GFP reliably, can be generated at high frequencies. Unlike the previous studies, which used promoter constructs of heterologous origins, they have made constructs in which the expression of GFP is driven by a zebrafish muscle-specific actin promoter (α-actin) promoter. Transgenic zebrafish carrying the α-actin-GFP promoter constructs reliably expressed the protein muscle. Further constructs with a cytoskeletal β-actin promoter produced transgenic zebrafish that expressed GFP throughout the body.

The use of GFP as the reporter gene allows embryos, injected at the one-cell stage, to be examined for transient expression of the protein a day later. Only those embryos expressing the GFP need to be raised to adulthood. These founder fish are crossed with wildtype fish and the F1 progeny examined for expression of GFP. In the case of α-actin-GFP constructs, 41/194 (21%) of the founder fish transmitted the GFP expression to their progeny. Of the 41 lines generated, 40 expressed the GFP in the same spatially restricted pattern as the normal α-actin gene. It also appears that the α-actin-GFP construct can be incorporated into a plasmid that also carries other transgenes driven by other promoters. Thus the α-actin-GFP transgene can be used as a transformation marker and will be invaluable in generating transgenic zebrafish carrying different gene constructs. Important points arising from this work are that the DNA to be injected should be linearized and the constructs made such that any plasmid sequence is 3' to the α-actin-GFP transgene and that those embryos expressing GFP in only a few cells, a day after injection, should be rejected (usually about 25%). In a similar study by Lin's group *(30)* tissue-specific expression of GFP has been achieved using a modified promoter of the zebrafish GATA-1 gene.

### 1.4. Integration and Germline Transmission of Retroviral Vectors

Control over the number of copies inserted at any one locus can be achieved by using retroviral vectors. Lin et al. *(31)* reported the generation of transgenic zebrafish using a pseudotyped retroviral vector. This vector, developed by Burns et al. *(26)*, can be concentrated to very high titers and can infect cultured cells derived from zebrafish embryos. The virus contains an Moloney murine leukemia virus (MoMLV)-based genome surrounded by an envelope containing the glycoprotein (G-protein) of the vesicular stomatis virus (VSV), completely replacing the retroviral *env* glycoprotein. The presence of the VSV G-protein confers the same broad host range, which is characteristic of VSV, on the pseudotyped vector. On entry into a permissive cell, the vector integrates retroviral sequences into the host genome.

The pseudotyped virus, LZRNL(G), was injected into the blastoderm cells at the 2000–4000-cell stage. About 50–100 infectious units were injected/embryo. Embryos were raised to sexual maturity, and DNA from pools of their 24-h-old F1 progeny was tested for the presence of LZRNL sequences by PCR. Eight of 50 founder fish showed germline transmission. Two transgenic F1 fish transmitted the viral sequences at 44% (11 out of 25) and 47% (8 out of 17) to their F2 progeny. Southern analysis showed that all founder fish had single copies at the point of insertion and that some founder fish had more than one retroviral insertion, which segregated independently in the F1. The viral construct had two reporter genes: *lacZ* fused to Moloney LTR, and *neo* driven by RSV-LTR. The former is not active in zebrafish cells. The *neo* gene has similarly not been shown to be expressed in these transgenic lines.

Gaiano et al. *(33)* constructed modified retroviral vectors that give high titers. When injected into 2000- to 4000-cell blastula, they give very high transgenic frequencies. In three experiments, 110 of 133 (83%) of the embryos injected were found to transmit proviral insertions through their germ line into the F1. In one of these three experiments, all 50 founder fish transmitted the proviral insertions to their F1 progeny. The difference being that the embryos were incubated at 28°C rather than 26°C. In two experiments, on average each positive founder transmitted transgene insertions to 29% of its F1 progeny. Each positive founder on average transmitted 11 proviral insertions to its F1 progeny. A second experiment with a different retroviral stock with a quarter of the titer of the first gave lower rates of transmission (13% compared with 29%). These significantly higher rates of generation of transgenic zebrafish reported by Gaiano et al. *(33)* are probably as a result of the approx 100-fold higher titers of retroviral vectors used.

## 1.5. Insertional Mutagenesis Retroviral Vectors

Compared to previous results involving either DNA microinjection or retroviral infection, the latest results of Gaiano and Hopkins represent a 20- to 30-fold increase in the efficiency of generating transgenic zebrafish. If the retroviral constructs insert at random into the genome, then they may be potent insertional mutagens for screening the zebrafish genome. If the same high frequency of integration can be achieved by a *lacZ*- or GFP-expressing retroviral vector, then living F1 embryos can screened directly, and positive ones raised to sexual maturity and subsequently bred to homozygosity. This makes it feasible to breed thousands of retroviral insertions to homozygosity in zebrafish, allowing a large-scale insertional mutagenesis screen. The great advantage of such a screen is that any interesting mutation that is found will be readily cloned by transgene tagging or plasmid rescue.

Hopkins and her colleagues suggest that 4–6 people can generate 10,000 to 20,0000 founder fish that will carry 100,000–200,000 transgenes in about 3 mo *(33)*. Given the size of the zebrafish genome ($1.6 \times 10^9$ bp), then 200,000 random insertions would on average represent an insertion every 8 kb. This potenially could be an insertion in every gene. This assumes that the insertions are indeed random, which is as yet unknown. Analysis of some of the insertions already generated does suggest that the integrations of the retroviral sequences are in, or near, transcribed regions *(34,35)*. Determining the mutant phenotypes of different insertions either involves generating homozygotes, which will be labor-intensive, or screening haploids generated from the F1 transgenic fish. These fish may carry multiple insertions and so could have complex phenotypes. An alternative is to modify the retroviral vectors so that they can be used as gene traps. These would contain a reporter gene which can only be expressed after integration into a transcribed endogenous gene.

## 2. Materials

### 2.1. Chemicals and Solutions

1. 100% Hank's (*1*): 0.137 $M$ NaCl, 5.4 m$M$ KCl, 0.25 m$M$ Na$_2$HPO$_4$, 0.44 m$M$ KH$_2$PO$_4$, 1.3 m$M$ CaCl$_2$, 1.0 m$M$ MgSO$_4$, 4.2 m$M$ NaHCO$_3$. It is made from stock solutions that can be kept for several months. Sodium bicarbonate solution is made fresh. Solution 1: 8.0 g NaCl, 0.4 g KCl in 100 mL water. Solution 2: 0.358 g Na$_2$HPO$_4$, 0.6 g KH$_2$PO$_4$ in 100 mL water. Solution 3: 0.72 g CaCl$_2$ in 50 mL water, Solution 4: 1.23 g MgSO$_4$·7H$_2$O in 50 mL water. Hank's premix: combine the following in order, 10 mL solution 1, 1 mL solution 2, 1 mL solution 3, 86 mL water, 1 mL solution 4. Store in refrigerator. Dissolve 0.35 g NaHCO$_3$ in 10 mL water. Final 100% Hank's is 9.9 mL Hank's premix, 0.1 mL fresh sodium bicarbonate solution.
2. 10X Digoxigenin nucleotide labeling mix: 10 m$M$ each ATP, GTP, CTP, 6.5 m$M$ UTP, 3.5 m$M$ digoxigenin-11-UTP. Store in aliquots at –20°C (available from Boehringer Mannheim #1277 073).
3. 10X Fluorescein nucleotide labeling mix: 10 m$M$ each ATP, GTP, CTP, 6.5 m$M$ UTP, 3.5 m$M$ fluorescein-12-UTP. Store in aliquots at –20°C (Boehringer Mannheim #1685 619).
4. 10X TBE: dissolve 109 g Tris base, 55 g boric acid, 9.3 g diNaEDTA in 1 L of water. pH should be 8.3. 1X TBE is 90 m$M$ Tris, 2.5 m$M$ EDTA, 90 m$M$ boric acid.
5. 10X Transcription buffer: 400 m$M$ Tris-HCl pH 8.0, 60 m$M$ MgCl$_2$, 100 m$M$ dithiothreitol, 20 m$M$ spermidine, 100 m$M$ NaCl, RNase inhibitor 1 U/µL. Store in aliquots at –20°C. (Supplied with RNA polymerase from Boehringer Mannheim.)
6. 20X SSC for hybridization and washing: (20X SSC is 3 $M$ NaCl, 300 m$M$ tri-sodium citrate). Dissolve 175.3 g NaCl and 88.2 g sodium citrate in 800 mL water. Adjust the pH with 1 $M$ citric acid to 6.0. Adjust the volume to 1 L and sterilize by autoclaving.
7. Alkaline phosphatase (AP) inactivation buffer 0.1 $M$ glycine-HCl, pH 2.2, 0.1% Tween-20.
8. Antidigoxigenin–AP Fab fragments (Boehringer Mannheim #1093274): 150 U/200 µL.
9. Antidigoxigenin–horseradish peroxidase (POD) Fab fragments (Boehringer Mannheim #1207733): 150 U lyophilized.
10. Antifluorescein–AP Fab fragments (Boehringer Mannheim #1426338): 150 U/200 µL.
11. Blocking solution for antibodies: 1X PBS, 0.1% Tween-20, 2 mg/mL BSA (BDH/Merck #44155), 5% sheep serum (Gibco-BRL Life Technologies #035-6070H), 1% dimethylsufoxide, DMSO (Merck-BDH #28216).

12. DetectaGene™ Blue CMCG *lacZ* gene expression kit Molecular Probes D-2921.
13. DetectaGene™ Green CMFDG *lacZ* gene expression kit Molecular Probes D-2920.
14. DNase I, RNase-free (Boehringer Mannheim #776785).
15. Durcupan ACM (Sigma D0166).
16. ELF™ stop reaction buffer: 25 m$M$ EDTA, 0.05% Triton X-100 in PBS, pH 7.2. Dissolve EDTA in PBS, and check pH.
17. ELF™-AP substrate kit (Molecular Probes Inc. Eugene #E-6601): dilute the substrate 1:100 in ELF™ reaction buffer supplied with the kit. Use within 30 min. Tissues should be equilibrated with pre-reaction wash buffer (30 m$M$ Tris, 150 m$M$ NaCl, pH 7.5) prior to adding the diluted substrate solution. The reaction is stopped with 25 m$M$ EDTA, 0.05% Triton X-100 in PBS, and the final pH should be 7.2 (addition of 1.0 m$M$ levamisole is optional, but is not necessary for zebrafish embryos).
18. Fast Red tablets, AP substrate (Boehringer Mannheim #1496549): each tablet contains 0.5 mg naphthol substrate, 2 mg Fast Red chromogen, and 0.4 mg levamisole. Store tablets for a short term at 4°C or at –20°C for long term. Wear gloves, and use plastic forceps to handle the tablets. Dissolve one tablet in 2 mL of 100 m$M$ Tris-HCl, pH 8.2. Use the solution within 30 min.
19. Fluorescein di-β-D-galactopyranoside (FDG): Molecular Probes F-1179.
20. Glutaraldehyde: 50% glutaraldehyde for EM (BDH/Merck #36218) stock solution of glutaraldehyde is stored in aliquots at –20°C.
21. Gentamicin (Sigma G 1272) 10 mg/mL in water stock solution.
22. Hybridization mix: 50% formamide, 5X SSC, 0.1% Tween-20, 500 μg/mL yeast RNA (Sigma #R6750), 50 μg/mL heparin (Sigma #H9399), pH to 6.0, with 1 $M$ citric acid.
23. Imagene Green™ $C_{12}$FDG *lacZ* gene expression kit Molecular Probes I-2904.
24. Imagene Red™ $C_{12}$RG *lacZ* gene expression kit Molecular Probes I-2906.
25. Paraformaldehyde/phosphate-buffered saline (PBS) fixative: paraformaldehyde is dissolved in PBS at 65°C. If it does not readily dissolve, add a drop or two of 1 $M$ NaOH solution to pH 7.5. It should be cooled to 4°C and used within 2 d.
26. PBS: 130 m$M$ NaCl, 7 m$M$ $Na_2HPO_4$, 3 m$M$ $NaH_2PO_4$. For a 10X PBS, mix 75.97 g NaCl, 12.46 g $Na_2HPO_4 \cdot 2H_2O$, 4.80 g $NaH_2PO_4 \cdot 2H_2O$. Dissolve in less than 1 L of distilled water, adjust to pH 7.0 and to final volume of 1 L, sterilize by autoclaving.
27. PBT: PBS, 0.1% Tween-20.
28. Penicillin/streptomycin solution (Sigma P0781): 10,000 U penicillin and 10 mg/mL streptomycin in 0.9% sodium chloride.
29. Proteinase K (Boehringer Mannheim #1000144): make up a stock solution of proteinase K at 20 mg/mL in 50% glycerol 10 m$M$ Tris, pH 7.8, and store at –20°C.
30. RNA polymerase SP6 (Boehringer Mannheim #810274).
31. RNA polymerase T3 (Boehringer Mannheim #1031163).
32. RNA polymerase T7 (Boehringer Mannheim #881767).
33. RNase inhibitor (Boehringer Mannheim #799017).
34. Sigma *FAST*™ Fast Red TR/Naphthol AS-MX (Sigma F 4648). Store at –20°C. Each tablet set contains the following when dissolved in 1 ml of water: 1.0 mg/mL

Fast Red TR, 0.4 mg/mL Naphthol AS-MX, 0.1 M levamisole, 0.15 mg/mL Tris buffer, pH 8.2. Remove required tablets from freezer, and allow to reach room temperature. Do not touch tablets with your fingers. Open Tris tablet (gold foil), and drop into a 1.5-mL microfuge tube. Add 1 mL of distilled water. Vortex until dissolved. Add Fast Red TR/Naphthol AS-MX tablet (silver foil) to Tris buffer. Vortex to dissolve. The solution should be used within 1 h. Reactions can be stopped by rinsing in PBS. Dispose of contaminated materials by incineration.

35. TE: 10 m$M$ Tris HCl, 1 m$M$ EDTA, pH 8.0. Make up with RNase-free water and autoclave.
36. Tricaine, 3-amino benzoic acid ethyl ester (Sigma A-5040) make a stock solution by dissolving 400 mg tricaine in 97.9 mL water with 2.1 mL 1 $M$ Tris-HCl pH 7.0. Store in aliquots in freezer. Use 4.2 mL tricaine stock solution in 100 mL tank water.
37. True Blue™ peroxidase substrate (Kirkegaard & Perry Laboratories, #71-00-64).
38. Vector™ Red alkaline phosphatase substrate (Vector Labs SK-5100): Mix stock solutions as described with kit.
39. X-GAL, 5-bromo-4-chloro-3-indolyl-β-D-galactopyranoside, (Sigma B 4252): Dissolve in dimethylformamide at 20 mg/mL. Store at –20°C.

## 2.2. Equipment

Joystick manipulator from Narishige (Tokyo, Japan), electrode puller from Campden Instruments (Sileby, UK), glass capillaries with filament from Clark Electromedical Instruments (Reading, UK), and Pneumatic Picopump PV830, World Precision Instruments (Sarasota, FL) are the necessary equipment.

## 3. Methods

The following sections describe methods for injection of DNA into zebrafish and subsequent analysis of the embryos for transient transgene expression. Methods for identifying founder fish with mosaic germ lines and F1 fish that are hemizygous for transgene constructs are also described. A recent improvement in screening for transgenic F1 fish has been developed by Kawakami and Hopkins *(36)*. Their strategy is to identify founder fish with mosaic germlines by raising injected embryos to adulthood and mating them with a wildtype fish. DNA is them extracted from 100 24-h-old F1 embryos and a PCR amplification performed. Only 2–20% of the F1 progeny will carry the transgene so DNA from a large number must be tested. If the PCR is positive for a transgene insertion, then new embryos must be raised and rescreened individually to identify transgenic F1 fish. Previously this was done by taking a fin biopsy, but this requires raising the F1 fish to adulthood. If only 2% carry the transgene then this is uneconomical and labor intensive. The improved strategy is to treat 72–75 h postfertilization F1 embryos with proteinase K (20 µg/mL) for a short time (12 min at room temperature). This is sufficient to release enough DNA into solution for PCR without killing the embryo. Use of 96-well culture plates and a multichannel pipet can facilitate performing 96 PCR amplifications. Analysis of the amplification products by gel electrophoresis means that the positive transgenic F1 fish are identified when still young and the nontransgenic can be disposed.

## 3.1. Fish Maintenance and Egg Collection

1. Fish are raised in a temperature-controlled room maintained at 26–28.5°C and on a 14-h light:10-h dark cycle. The main room lights switch on at 9:00 AM, but some tanks are blacked out with respect to the room lighting and have their own lights, which are kept on a similar light/dark cycle but with their lights set to come on at 10:00 AM, 11:00 AM, and 12:00 AM.
2. Eggs are collected using a polyethylene sandwich box with a rectangular hole cut out of the lid. The hole is covered with nylon mesh with a pore size of about 3 mm. The mesh is fixed to the lid with aquarium sealer. The mesh also has a few 10-cm strands of green plastic garden twine attached to simulate foliage. One of these boxes is placed in each 50-L tank, and the fish spawn among the plastic twine. The box is weighted down with a few glass marbles on the lid.
3. Ten to 25 fish are placed in the tanks at least a day before egg collections. They should be 6- to 12-mo old, and approximately twice as many females as males. The collection boxes are placed in the tank the evening before the eggs are required. Spawning occurs when the lights go on and can continue for 2–3 h.
4. The collecting boxes are removed and replaced at 15–30 min intervals.

## 3.2. Injection of Embryos

1. DNA for injection should be the purest available, preferably prepared by Qiagen midi- or maxipreps. DNA is made up in 0.1 *M* KCl or water at about 25–50 µg/mL final concentration (*see* **Note 1**). The DNA solution contains phenol red (0.25%) to give a visual guide regarding how much solution is injected and which embryos have received DNA.
2. Embryos are injected through their chorions at the one-cell stage (*see* **Note 2**). Eggs persist at this stage for 10–30 min after fertilization. Cooling to 16°C will delay the first cleavage, providing more time for injection of large collections of embryos.
3. Freshly harvested eggs are rinsed several times in 10% Hank's before transferring to a Petri dish containing 1% agarose made with 10% Hank's and covered with 10% Hank's (*see* **Fig. 1** and **Note 3**).
4. Align the embryos at the bottom of the ramp in the agarose as shown in **Fig. 1C**.
5. Injection can be done using a dissecting microscope equipped with transmitted and/or incident light. The injection needle is mounted in a joystick manipulator and connected to a gas pressure injection apparatus (*see* **Note 4**).
6. The needles are pulled on a microelectrode puller (*see* **Note 5**). The shape of the needle is critical. It should have a sharp tip with a very small aperture, and the shank length should be short so that it is strong enough to penetrate the chorion.

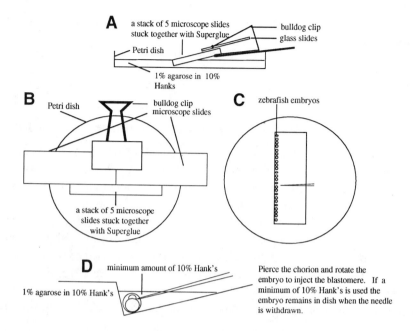

Fig. 1. (**A,B**) Five microscope slides are stuck together with Superglue and held in a bulldog clip. Two other slides are held together on the top surface of the five slides such that their ends protrude at either side. These are rested on the rim of the bottom part of a 90-mm Petri dish so that the five stuck slides extend a few millimeters into the dish, but do not touch the bottom surface. A 1% agarose solution, made by heating in 10% Hank's, is poured into the dish and allowed to set. The slides are removed, leaving an indentation in the surface of the agarose in the form of a ramp. (**C**) Zebrafish embryos are added to the dish in 10% Hank's solution with a wide-mouthed, heat-polished glass Pasteur pipet. (**D**) The embryos are injected by piercing the chorion and rotating each embryo so the needle can be inserted into the single cell above the yolk. If a minimum of Hank's solution is used, the embryos remain in the well when the needle is withdrawn.

7. The needles are made from glass capillaries, which contain a glass filament (*see* **Note 6**). The filament aids back-filling the needle. The needle is held vertically in a micropipet storage jar with a layer of water in the bottom to keep the chamber humid. A 1–2 μL drop of solution is placed on the blunt end of the capillary, and the solution is slowly drawn down into the capillary along the filament (*see* **Note 7**). The tip of the needle is sealed when the capillary is drawn out by the needle puller. The tip must be broken by touching against a fine pair of watchmakers forceps. The tip is best kept immersed in liquid to avoid it becoming blocked.

8. After injection, remove the uninjected or dead embryos, and transfer the remaining embryos to a 35-mm Petri dish with a cushion of 1% agarose covered with 10% Hank's solution containing gentamicin at 20 μg/mL (*see* **Note 8**).

9. Incubate the embryos for several hours at 28.5°C (*see* **Note 9**). Remove those that have failed to develop or are dead.
10. Some embryos can be examined for transient gene expression 4 h after injection onward.
11. Raise the embryos at low density until they hatch as fry, and then rear to sexual maturity.

### 3.3. Double Fluorescence and Two Color In Situ Hybridization

The following method was developed specifically for identifying transcripts with overlapping or colocalized expression domains *(37)*. It is therefore useful for identifying ectopically expressed transgene products in transient transgene experiments along with upregulated transcripts of downstream genes. Double *in situ* hybridizations are performed by hybridization with a mixture of fluorescein- and digoxigenin-labeled antisense RNA probes followed by sequential incubation in AP-conjugated antibodies to fluorescein and digoxigenin, respectively. The first antibody is stained with Fast Red (*see* **Note 10**) and the second with enzyme labeled fluorescence (ELF™) substrate. The signals are viewed on a fluorescence microscope with rhodamine and DAPI filter sets, respectively.

### 3.3.1. RNA Probe Synthesis

1. Digest linearized plasmid equivalent to 1 µg of insert DNA (*see* **Note 11**) with proteinase K (*see* **Note 12**) (0.05 µg/µL) for 30 min at 37°C
2. Phenol/chloroform-extract and ethanol-precipitate.
3. Redissolve the DNA in TE so that 1 µg of insert DNA is in 4 µL.
4. Mix together 4 µL of linearized plasmid (1 µg of insert DNA), 2 µL of 10X transcription buffer, 2 µL of 10X nucleotide mix (with digoxigenin-UTP or fluorescein-UTP), and 20 U of RNase inhibitor. Add water to give a final reaction volume of 20 µL and 40 U of the appropriate T7, T3, or SP6 RNA polymerase.
5. Incubate the mixture for 2 h at 37°C.
6. Add 40 U of DNase I, and incubate at 37°C for 15 min to remove the plasmid DNA.
7. Stop the reaction by adding 2 µL of 200 m$M$ EDTA, pH 8.0.
8. Precipitate the RNA with 2.5 µL 4 $M$ LiCl, and 75 µL prechilled ethanol (*see* **Note 13**).
9. Spin down the pellet, and redissolve in 100 µL of RNase-free water containing 40 U of RNase inhibitor.
10. Check the probe by running 2.5 µL on an 0.8% (w/v) agarose, 1X TBE minigel. Wash the apparatus thoroughly before preparing the gel, and run the samples quickly to avoid problems with RNase (*see* **Note 14**).
11. The probe can be used without further treatment (*see* **Note 15**). It should be split into aliquots and stored at –20°C.

### 3.3.2. Fixation

1. Fix embryos in 4% paraformaldehyde in phosphate-buffered saline (PBS) at 4°C for 4 h to overnight.

2. Wash twice in PBT (PBS with 0.1% Tween-20) at 4°C.
3. Dehydrate with a series of methanol/PBT solutions (25:75%, 50:50%, 75:25% methanol:PBT) and then twice with 100% methanol. Incubate for 10 min in each solution.
4. The tissues can conveniently be stored at this stage at –20°C.

### 3.3.3. Pretreatments Prior to Hybridization

1. Rehydrate the embryos through a methanol/PBT series (75:25%, 50:50%, 25:75% methanol:PBT) finishing with three washes of PBT (*see* **Note 16**).
2. Fix the embryos with 4% paraformaldehyde in PBT for 20 min at room temperature. This second fixation helps prevent the embryos from falling apart.
3. Wash twice for 5 min in PBT.
4. Add 0.2–0.4 mL of hybridization buffer, 50–65% formamide, 5X SSC, 500 µg/mL yeast RNA, 0.1% Tween-20, 50 µg/mL heparin, and 1 *M* citric acid to pH 6.0. The volume used depends on the number and size of embryos.
5. Incubate for 5 min, and then add fresh hybridization solution and incubate at 65–70°C for a minimum of 2 h.
6. The embryos are now safe from RNases.

### 3.3.4. Hybridization

1. Replace the prehybridization solution with preheated hybridization mix containing probes. Total probe concentration should not exceed 1 µg/mL (*see* **Note 17**).
2. Incubate overnight at 60–70°C. If background proves to be a problem, then the temperature and formamide concentration can be increased or the salt concentration in the hybridization buffer reduced.

### 3.3.5. Posthybridization Washes

**Steps 1–7** are performed at the hybridization temperature. **Steps 8–11** are performed at room temperature.

1. Wash the embryos for 10 min in 50% formamide, 5X SSC.
2. Wash for 10 min with 0.5 mL of a 3:1 mixture of 50% formamide, 5X SS :2X SSC.
3. Wash for 10 min with 0.5 mL of a 1:1 mixture of 50% formamide, 5X SSC:2X SSC.
4. Wash for 10 min with 0.5 mL of a 1:3 mixture of 50% formamide, 5X SSC:2X SSC.
5. Wash for 10 min in 1 mL of 2X SSC, 0.01% Tween-20 alone at the hybridization temperature.
6. Wash for 30 min in 1 mL of 0.2X SSC, 0.01% Tween-20 at the hybridization temperature. This is best done on the shaker in the hybridization oven. Alternatively, use a heating block or water bath, but gently invert the tube every 5–10 min.
7. Repeat the stringent wash in **step 6**.
8. At room temperature, wash for 10 min in 1 mL of a 3:1 mixture of 0.2X SSC:PBT.
9. At room temperature, wash for 10 min in 1 mL of a 1:1 mixture of 0.2X SSC:PBT.
10. At room temperature, wash for 10 min in 1 mL of a 1:3 mixture of 0.2X SSC:PBT.
11. Replace the solution with PBT.

### 3.3.6. Detection of Signals with Fast Red and ELF AP Substrates

1. Replace the PBT after the posthybridization washes with blocking solution (5% sheep serum, 2 mg/mL BSA, 1% DMSO in PBT).
2. Incubate for at least 60 min.
3. Incubate in 1:5000 dilution (0.15 U/mL) of sheep antifluorescein-AP Fab fragments (Boehringer Mannheim) in blocking solution overnight at 4°C.
4. Wash embryos with PBT for 2 h (8 × 15 min).
5. Equilibrate with 100 m$M$ Tris-HCl, pH 8.2, 0.01% Tween-20 at room temperature by washing 3 × 5 min.
6. Stain embryos with Fast Red™ (Boehringer Mannheim).
7. Stop reaction by washing several times with PBT.
8. Inactivate the alkaline phosphatase activity by incubating in 100 m$M$ glycine, pH 2.2, 0.1% Tween-20 for 2 × 15 min at room temperature with gentle shaking. Do not incubate in acidic glycine for longer than is necessary, since prolonged incubation reduces the second signal presumably by masking the second antisense probe.
9. Wash four times for 5 min with PBT.
10. Fix in 4% paraformaldehyde in PBS for 20 min.
11. Wash five times for 5 min with PBT.
12. Incubate embryos in blocking solution for 60 min.
13. Incubate in a 1:5000 dilution (0.15 U/mL) of sheep antidigoxigenin–AP Fab fragments (Boehringer Mannheim) in blocking solution overnight at 4°C.
14. Wash in 0.5% Triton X-100 in PBS for 2 h. Note that either DMSO or Tween-20 in the wash solutions causes the final ELF crystals to be large, so they are substituted with Triton X-100.
15. Wash 3 × 5 min at room temperature with the ELF™ prereaction buffer (150 m$M$ NaCl, 30 m$M$ Tris-HCl, pH 7.5).
16. For staining, incubate in 1:100 dilution of the ELF™ substrate in ELF substrate buffer (provided with the ELF™ kit) at room temperature for 5 h to overnight (*see* **Note 18**).
17. Monitor the staining reaction using a UV fluorescent microscope with a DAPI filter set at intervals after starting the reaction.
18. Stop the reaction by washing with 25 m$M$ EDTA, 0.05% Triton X-100 in PBS, pH 7.2.
19. Fix the embryos in 4% paraformaldehyde in PBS for 20 min.
20. Mount the tissue with the special aqueous mounting medium supplied with the kit from Molecular Probes (*see* **Note 19**).
21. Specimens should be flat-mounted underneath a coverslip for best resolution. View the Fast Red precipitate with a rhodamine filter set and the ELF precipitate with a DAPI filter set.
22. Caution should be taken in interpreting colocalized fluorescent signals with sequential AP antibodies. If the first enzyme is not completely inactivated by the acid-glycine treatment, then any residual activity may give rise to false-positive signals with the ELF substrate. Therefore, controls should be performed where some embryos are not incubated with the second antibody, but otherwise treated identically with the ELF substrate.

### 3.3.7. Detection of Signals with AP-Conjugated and Horseradish Peroxidase-Conjugated Antibodies

As an alternative to sequential AP-conjugated antibodies, one can use antifluorescein–Fab fragments conjugated to AP and Vector Red to identify the first signal followed by antidigoxigenin Fab–conjugated to horseradish peroxidase. The second staining reaction is performed with TrueBlue™ (Kirkegaard Perry). TrueBlue™ is an alternative peroxidase substrate to diaminobenzidine (DAB). It is 50 times more sensitive than DAB, so the antidigoxigenin Fab–conjugated to horseradish peroxidase must be used at 10- to 50-fold greater dilution. This method overcomes the problem of possible residual AP activity giving rise to a false-positive colocalized signal. However, the TrueBlue™ precipitate is not stable in aqueous solution, so specimens must be photographed immediately or rapidly dehydrated and transferred to organic mountant.

Proceed as in **Subheading 3.3.6.** up to **step 5**, and then as follows. Note that the glycine inactivation step is omitted.

1. Incubate in Vector Red™ staining solution for a few hours to overnight. When color has developed to the desired extent, wash 3X with PBT.
2. Replace PBT with blocking solution, and incubate for a further 1 h.
3. Incubate for at least 2 h with sheep antidigoxigenin–horseradish peroxidase Fab fragments at a concentration of 0.015–0.0375 U/mL.
4. Wash embryos with PBT for 2 h ($8 \times 15$ min).
5. Transfer to a embryo dish, and add TrueBlue™ staining solution.
6. Monitor staining and photograph.
7. If overstained, replace staining solution with PBT to destain.

## 3.4. Detection of β-Gal Activity in Fixed Embryos (see Note 20) (12)

1. Fix embryos in 4% paraformaldehyde, 0.2% glutaraldehyde, 4% sucrose, 0.15 m$M$ $CaCl_2$, 0.1 $M$ sodium phosphate, pH 7.3, for 30 min at 4°C.
2. Rinse the embryos in 0.1 $M$ sodium phosphate.
3. Incubate in 4% 5-bromo-4-chloro-3-indolyl-β-D-galactopyranoside, 150 m$M$ NaCl, 1 m$M$ $MgCl_2$, 1.5 m$M$ $K_4(Fe_3[CN]_6)$, 1.5 m$M$ $K_3$ $(Fe_2[CN]_6)$ in 5 m$M$ sodium phosphate buffer at 37°C for 3 h.
4. Rinse the embryos in 0.1 $M$ sodium phosphate.
5. Mount in glycerol or dehydrate through alcohol series, xylene, and mount in Durcupan.

## 3.5. Detection of β-Gal Activity in Live Embryos

### 3.5.1. FDG (see *Note 21*) Visualization of β-Gal Activity in Live Embryos (22)

1. Embryos are dechorionated and stained in a solution containing 2 m$M$ FDG and 20% DMSO in ddH$_2$O for exactly 2 min and 30 s at 33°C. Note that older embryos

that have muscular movements may be anesthetized in tricaine to aid viewing and photography.
2. Wash extensively with fish water.
3. Examine with a low-power objective (5X) on a fluorescence microscope with an FITC filter set.
4. False positives can develop after 30–40 min. Some *lac*Z-expressing embryos may be overlooked if expression is not very strong.

### 3.5.2. ImaGene Green™ Visualization of β-Gal Activity in Live Embryos (see **Note 22**) (12)

Embryos injected at the one-cell stage may be examined 20–24 h later.

1. Dechorionate the embryos and transfer them to 20 µ*M* Imagene β-gal substrate for 30 min at 28.5°C.
2. Rinse and view with fluorescence optics (*see* **Note 23**).

### 3.5.3. DetectaGene™ (see **Note 24**) Visualization of β-Gal Activity in Live Embryos

Embryos injected at the one-cell stage may be examined from 4–24 h later.

1. Dechorionate the embryos and transfer them to 200 µ*M* Detectagene Green™ β-gal substrate solution containing 20% DMSO. for 2–3 min at 28.5°C.
2. Rinse and view with fluorescence optics.

### 3.6. Visualization of Luciferase Expression in Live Embryos (17) (see Note 25)

1. Embryos are placed in the individual wells of a modified 96-well round-bottom microtiter plate (*see* **Note 26**) over orthochromatic X-ray film.
2. Incubate embryos in 100 µL of embryo-rearing solution (a 1:1 mixture of tap water, pH 7.8, hardness 350 ppm, with deionized water, 25 µg/mL neomycin sulfate, 0.5 µg/mL methylene blue, 17.5 µg/mL sodium thiosulfate) containing 0.1–1.0 mg/mL luciferin.
3. Cover the plate with a lid and tape it over a piece of film (TMAT-G orthochromatic X-ray film) (*see* **Note 27**) supported on a rigid piece of plastic.
4. The embryos are incubated in a light-tight box for 1–5 d, and the film is then developed.

### 3.7. Green Fluorescent Protein Expression in Live Embryos (24,25)

The cDNA encoding GFP from *Aequorea victoria*, provides an alternative, more convenient marker for living embryos than the β-gal substrates described above (*20*). GFP produces fluorescence without the need to add an exogenous substrate. GFP absorbs blue light (maximally at 395 nm with a minor peak at 470 nm) and emits green light (peak emission at 509 nm with a shoulder at 540 nm).

The fluorescence is very stable with virtually no photobleaching. Embryos can be viewed directly with a fluorescein filter set.

## 3.8. Histology (10)

Sections can be made of stained embryos by embedding in Durcupan (Sigma).

1. Fish embryos are incubated in a 1:1 mixture of xylene and Durcupan (Sigma) overnight at 4°C.
2. After evaporation of the xylene at room temperature in a fume cupboard, the embryos are embedded in fresh Durcupan.
3. Incubate for 24 h at 45°C.
4. Incubate for 30 h at 60°C.
5. Cut 2–3 µm sections with a microtome equipped with a glass knife (*see* **Note 28**).

## 3.9. DNA Extraction for Southern Blot Analysis and PCR (see Note 29)

### 3.9.1. Taking Fin Biopsies (see **Note 30**)

1. Anesthetize fish by incubating in 100 mL tank water containing 4.2 mL of tricaine solution.
2. Rinse the fish in fresh tank water.
3. Place fish on a moist sponge, and with a fine pair of dissecting scissors, cut off about 25% of the caudal fin.
4. Place fish in fresh tank water, and allow to recover.
5. The fin biopsy can be stored at –80°C prior to DNA extraction.

### 3.9.2. DNA Extraction from Fin Biopsies for PCR or Southern Analysis

1. Place fin biopsy in 500 µL of 10 m$M$ Tris-HCl, pH 8.0, 10 m$M$ EDTA, 100 m$M$ NaCl, 0.5% SDS, and 200 µg/mL proteinase K at 55°C for 12 h.
2. Add an equal volume of phenol:chloroform (1:1). Mix gently.
3. Spin in microfuge for 5 min.
4. Remove the upper, aqueous layer to a fresh microfuge tube.
5. Repeat **steps 2–4**.
6. Extract with an equal volume of chloroform.
7. Add 1/10 vol of 2.5 $M$ sodium acetate, pH 5.5, and 2.5 vol of ethanol.
8. Spin down in microfuge for 10 min.
9. Rinse pellet in 80% ethanol and respin.
10. Dry the pellet.
11. Resuspend in TE. Use 0.5–1 µg per PCR.

### 3.9.3. DNA Extraction from Embryos for PCR (31)

This method is quicker than the previous one and is suitable for embryos. The DNA is suitable for PCR, but is sheared so it is not suitable for Southern analysis.

1. Total nucleic acid is extracted from pools of 40 to several hundred dechorionated embryos at 16–24 h.
2. Add 10 vol of 4 $M$ guanidinium isothiocyanate, 0.25 m$M$ sodium citrate, pH 7.0, 0.5 % Sarkosyl, 0.1 $M$ β-mercaptoethanol.
3. Vortex for 1 min.
4. Extract once with phenol:chloroform:isoamyl alcohol (25:24:1).
5. Precipitate nucleic acid by addition of 3 vol ethanol and 1/10 vol sodium acetate (3 $M$, pH 5.5).
6. Wash the pellet with 70% ethanol.
7. Dry and redissolve in TE, pH 8.0.

## 4. Notes

1. Do not use DNA in 1X TE or other Tris-based buffers.
2. Several workers prefer to dechorionate the embryos before injection. This is done *en masse* with pronase (0.5 mg/mL) followed by extensive washes to remove the residual pronase before proceeding. However, it is not necessary, since the embryos can be injected through the chorion *(1)*.
3. The embryos are best handled with a glass Pasteur pipet that has been cut and polished in a Bunsen flame to give an aperture of about 2.0 mm.
4. In *The Zebrafish Book*, Westerfield *(1)*, describes a simple gas pressure injection apparatus that may be made by a competent technical workshop. Alternatively, a pneumatic picopump air pressure supply system is supplied World Precision Instruments.
5. The needle puller is Model 753 from Campden Instruments, UK.
6. The capillaries are from Clark Electromedical Instruments (GC-100F 500 pcs).
7. If filament capillaries are not used, then the needles can be back-filled with a finely drawn out glass Pasteur. Use a rubber bulb with a small hole in the end. Cover the hole with your index figure to draw up or expel liquid into the pipet. (This is the same technique as used for filling Drummond Scientific microcaps.)
8. Alternatively, penicillin and streptomycin can be added at 20–50 µg/mL.
9. If left at 16°C the embryos will fail to gastrulate properly.
10. Three different Fast Red substrates have been compared: Vector™ Red alkaline phosphatase substrate *(38,39)*, Vector Laboratories; Fast Red tablets, Boehringer Mannheim *(40)*, Sigma Fast™ Fast Red TR/Naphthol AS-MX). The different Fast Red substrates have different characteristics. All are made up in Tris-HCl, pH 8.2. Vector Red is made up from three solutions provided in the kit from Vector Laboratories. Staining is quite rapid, appearing within a few minutes to 1–2 h. Prolonged incubation tends not to intensify the signal. The yolk stains yellow, but background in the embryo remains yellow. The Fast Red from Boehringer Mannheim is in the form of tablets. The solution throws down a precipitate after prolonged incubation. The signal develops less quickly than Vector™ Red, but produces a more intense red precipitate. Backgrounds can be quite orange in both the yolk and the embryo. Sigma *Fast*™ Fast Red is supplied as tablets for Tris buffer and the red substrate. It gives a similar result to Boehringer Fast™ Red in that the signal is intense and develops slowly. Backgrounds are a little less than

with the Boehringer Fast™ Red, but greater than with Vector™ Red. It also produces less precipitate in the staining solution. The AP can be inactivated by heat treating at 65°C for 30 min. with the Vector™ Red precipitate but both the other precipitates are heat-labile and are lost if heat-treated. Thus, the AP activity must be inactivated by incubating in 100 m$M$ glycine, pH 2.2, for 30 min.

11. It is important that the plasmid DNA is free of contaminants. Plasmid prepared by the various commercial resins is suitable (Magic™ or Wizard™ Minipreps Promega, Qiagen Plasmid Mini or Midi kit).

12. Treat with proteinase K to ensure that the DNA is free of RNase prior to riboprobe synthesis.

13. LiCl/ethanol precipitation does not completely remove unincorporated rNTPs, but this is not normally a problem.

14. In the case of fluorescein probes, the pellet is yellow, and unincorporated fluorescein-UTP runs at the front during electrophoresis and is easily seen on the UV-transilluminator. There should be a single, sharp band migrating behind the front. If it is smeared, then there is a problem with RNase contamination or with the RNA polymerase.

15. Previous protocols required that the probe should be partially hydrolyzed by heating in a sodium bicarbonate buffer (*see* **ref. 42**). This generates shorter lengths of probe, which more easily penetrate the tissue. However, this hydrolysis is difficult to control and tends to give lower signals. Now many laboratories omit this step. Perhaps with larger probes (>6 kb), partial hydrolysis may be worth considering.

16. Treat with proteinase K (10 µg/mL in PBT) to increase the permeability of the membrane. This is performed for 10–20 min at room temperature, depending on the stage of development. Embryos up to 24 h postfertilization do not need to be treated with proteinase K.

17. Use 1/100 of a standard digoxigenin or fluorescein riboprobe reaction in 200 µL hybridization solution as a starting point (equivalent to a final concentration of 0.5 µg/mL). Probes which give strong signals may be effective at 0.1 µg/mL.

18. This is a quarter of the dilution and longer incubation than recommended by Molecular Probes. However, this gives smaller crystals and less background. With zebrafish embryos, it is also not necessary to add 1.0 m$M$ levamisole. Stain in 100–200 µL of staining solution in a 1.5-mL microfuge tube. Lay the tube almost horizontal, so that the embryos spread out and are equally exposed to the staining solution.

19. Note that other antifade mounting media may cause the ELF precipitate to fade.

20. Slight variations on this method are found in the following: Bayer and Campos-Ortega *(10)*. Fix in 12.5% glutaraldehyde 10 min in 1X PBS, pH 7.2.Wash several times in PBT (0.3% Triton X-100, in 1X PBS, pH 7.2). Stain in 0.2% 5-bromo-4-chloro-3-indolyl-β-D-galactopyranoside in dimethylformamide, 5 m$M$ $K_3$FeIII (CN)$_6$, 5m$M$ $K_4$Fe II (CN)$_6$ in 0.1 $M$ citrate buffer, pH 8.0, overnight at 37°C. Dehydrate embryos in ethanol, incubate for 5 min in xylene and mount in Durcupan. Prolonged incubation in xylene will remove the blue stain.

21. FDG is a substrate of β-gal. The compound is nonfluorescent, but on enzymatic cleavage, releases fluorescein. The embryos must be permeabilized before they

can take up FDG. This is achieved by incorporating DMSO into the staining solution. The released fluorescein is freely diffusible, so the embryos must be examined immediately after staining.

22. The Imagene product was developed by Molecular Probes to overcome the problems associated with the necessity to permeabilize cells before they could take up FDG. The Imagene products are fatty acyl-modified FDG substrates. They are taken up without cellular permeabilization, and they yield fluorescent enzymatic hydrolysis products, which are retained in the cells. Once inside the cell, the substrate is cleaved by β-gal, producing a green fluorescent lipophilic product that is retained by the cells, probably by incorporation of the fatty acyl tail within the cellular membranes. Molecular Probes produce a variety of ImaGene kits with acyl-modified FDG of various chain lengths. The Imagene Green™ C12FDG *lacZ* Gene Expression kit includes a stock solution of $C1_2FDG$, chloroquine (for reducing acid hydrolysis and endogenous β-gal activity) and phenylethylthio-β-D-galactopyranoside, which is a broad-spectrum galactosidase inhibitor for stopping the reaction. The Imagene Red™ $C1_2RG$ *lacZ* Gene Expression kit (I-2906) contains a 12-carbon fatty acyl analog of the nonfluorescent *lacZ* substrate resorufin-β-D-galactopyranoside, which yields a red fluorescing enzymatic hydrolysis product.

23. A low-light video camera (SIT-57, General Electric) with an image processor *(12)* can be used or normal fluorescent optics.

24. DetectaGene™ Green *lacZ* substrate is an FDG analog that has been modified to react with intracellular glutathione. Once inside the cell, chloromethylfluorescein di-β-D-galactopyranosidase (CMFDG) is cleaved by intracellular β-gal, and then its chloromethyl moiety reacts with glutathione to form a tripeptide-fluorescein analogue. Since peptides do not readily cross the plasma membrane, the resulting fluorescent tripeptide cleavage product is much better retained than fluorescein. Although cells must be permeabilized to take up the DetectaGene™ Green substrate, its enzymatic cleavage products are brighter and more photostable than those of the ImaGene Green™ substrate; 200 μ*M* CMFDG are as effective as 2 m*M* FDG. The fluorescent product stays in the cells for 3 d, whereas fluorescein diffuses out after 30 min. If cells are permeabilized by DMSO, in order for them to take up the substrate, they must undergo repair on returning to fish water, and presumably the enzymatic product does not diffuse out. DetectaGene™ Blue *lacZ* Expression kit contains 4-chloromethylcoumarin-β-D-galactopyranoside (CMCG), which forms a bright blue fluorescent membrane-impermeant product after reacting with intracellular glutathione. Visualization of CMCG fluorescent product requires a fluorescent microscope equipped with a DAPI filter set.

25. This method has been used to detect luciferase activity in embryos injected with a CMV-luciferase transgene construct. In preliminary experiments, Gibbs et al. *(17)* coinjected luciferin at 1 mg/mL with the DNA solution. However, injected embryos may also be incubated in a luciferin solution of 0.1–1 mg/mL. A lower concentration of 0.01 mg/mL detected only 10% as positives. Since F1 transgenic fish failed to express luciferase, it has not been used successfully to iden-

tify stable transgenic lines. Homogenized embryos can be assayed for luciferase activity using a scintillation counter assay *(41)*.

26. The modified plates are made by removing the outer rim with a hot knife and sanding the bottom of the plate with a belt sander until a hole appears in the bottom of each well. The hole is covered with a strip of Mylar sealing tape (Fisher Scientific No. 14-245-20). Embryos incubated in the plate are therefore close to the underlying film.

27. Orthochromatic film (TMAT-G) is used, since it is sensitive to 560 nm light, whereas normal X-ray film is not (XAR-OMAT). TMAT-H film gives similar results.

28. This requires a knifemaker suitable for making long-edge, extremely sharp "Ralph"-type disposable histological knives. Such knives can be used on a conventional microtome fitted with a suitable holder and will produce quality semithin sections on the order of 0.5 μm. A suitable knifemaker is available from Taab: K058 Taab Histoknifemaker complete with tools and glass.

29. PCR and Southern blotting are standard procedures, so they are not included here.

30. Taking fin biopsies followed by recovery of the fish requires a Home Office License in the UK. You should be familiar with your country's regulations for animal research.

## Acknowledgments

T. J. was supported by grants from EMBO, the Wellcome Trust and the North of England Cancer Research Campaign. Thanks are given to Yi-Lin Yan, Peter Currie, Philip Ingham, Daniel Alexander, Jeremy Wegner, and Uwe Strähle for sharing their expertise and to Nancy Hopkins for providing her preliminary results.

## References

1. Westerfield, M. (ed.) (1995) *The Zebrafish Book*, University of Oregon Press, Eugene, OR.

2. Mullins, M., Hammerschmidt, P., Haffter, P., and Nüsslein-Volhard, C. (1994).Large-scale mutagenesis in the zebrafish: in search of genes controlling development in a vertebrate. *Curr. Biol.* **4,** 189–202.

3. Driever, W., Stemple, D., Schier, A., and Solnica-Krezel, L. (1994) Zebrafish: genetic tools for studying vertebrate development. *Trends Genet.* **10,** 152–159.

4. Solnica-Krezel, L., Schier, A., and Driever, W. (1994) Efficient recovery of ENU-induced mutations from the zebrafish germline.*Genetics* **136,** 1401–1420.

5. Haffter, P., Granato, M., Brand, M., Mullins, M. C., Hammerschmidt, M., Kane, D. A., et al. (1996) The identification of genes with unique and essential functions in the development of the zebrafish. *Development* **123,** 1–36.

6. Driever, W., Slonica-Krezel, L., Schier, A. F., Neuhauss, S. C. F., Malicki, J., Stemple, D. L., et al. (1996) A genetic screen for mutations affecting embryogenesis in zebrafish. *Development* **123,** 37–46.

7. Postlethwait, J. H., Johnson, S. L., Midson, C. N., Talbot, W. S., Gates, M., Ballinger, E. W., et al.. (1994) A genetic map for the zebrafish. *Science* **264,** 699–701.

8. Postlethwait, J. H. and Talbot, W. S. (1997) Zebrafish genomics: from mutants to genes. *Trends Genet.* **13,** 183–190.

9. Stuart, G. W., Vielkind, J. R., McMurray, J. V., and Westerfield, M. (1990) Stable lines of transgenic zebrafish exhibit reproducible patterns of transgene expression. *Development* **109,** 557–584.

10. Bayer, T. A. and Campos-Ortega, J. A. (1992) A transgene containing *lacZ* is expressed in primary sensory neurons in zebrafish. *Development* **115,** 421–426.

11. MacGregor, G. R. and Caskey, C. T. (1989) Construction of plasmids that express *E. coli* b-glactosidase in mammalian cells. *Nucleic Acids Res.* **17,** 2365.

12. Westerfield, M., Wegner, J., Jegalian, B. G., DeRobertis, E. M., and Püschel, A. W. (1992) Specific activation of mammalian *Hox* promoters in mosaic transgenic zebrafish. *Genes Develop.* **6,** 591–598.

13. Gibbs, P. D. L., Gray, A., and Thorgaard, G. (1994) Inheritance of P element and reporter gene sequences in zebrafish. *Mol. Marine Biol. Biotech.* **3(6),** 317–326.

14. Reinhard, E., Nedivi, E., Wegner, J., Skene, J. H. P., and Westerfield, M. (1994) Neural selective activation and temporal regulation of a mammalian GAP-43 promoter in zebrafish. *Development* **120,** 1767–1775.

15. Stuart, G. W., McMurray, J. V., and Westerfield, M. (1988) Replication, integration, and stable germ-line transmission of foreign sequences injected into early zebrafish embryos. *Development* **103,** 403–412.

16. Culp, P., Nüsslein-Volhard, C., and Hopkins, N. (1991) High frequency germ-line transmission of plasmid DNA sequences injected into fertilised zebrafish eggs. *Proc. Natl. Acad. Sci. USA* **88,** 7953–7957.

17. Gibbs, P. D. L., Peek, A. and Thorgaard, G. (1994) An *in vivo* screen for the luciferase transgene in zebrafish. *Mol. Marine Biol. Biotech.* **3(6),** 307–316.

18. Buono, R. J. and Linser, P. J. (1992) Transient expression of RSVCAT in transgenic zebrafish made by electroporation. *Mol. Marine Biol. Biotech.* **1,** 271–275.

19. Zhao, X., Zhang, P. J., and Wong, T. K. (1993) Application of Baekonization: a new approach to produce transgenic fish. *Mol. Marine Biol. Biotech.* **2(1),** 63–69.

20. Müller, F. Ivics, Z. Erdélyi, F., Papp, T., Váradi, L., Horváth, L., Maclean, N., and Orbán, L. (1992) Introducing foreign genes into fish eggs with electroplated sperm as a carrier. *Mol. Marine Biol. Biotech.* **1(4/5),** 276–281.

21. Powers, D. A. Hereford, L., Cole, T., Chen, T. T. Lin, C. M., Kight, K., Creech, K., and Dunham, R. (1992) Electroporation: a method for transferring genes into the gametes of zebrafish (*Brachydanio rerio*), channel catfish (*Ictalurus punctatus*), and common carp (*Cyprinus carpio*). *Mol. Marine Biol. Biotechnol.* **1(4/5),** 301–308.

22. Lin, S., Yang, S., and Hopkins, N. (1994) *lacZ* expression in germ line transgenic zebrafish can be detected in living embryos. *Dev. Biol.* **161,** 77–83.

23. Chalfie, M., Tu, Y., Euskirchen, G., Ward, W. W., and Prasher, D. C. (1994) Green fluorescent protein as a marker for gene expression. *Science* **263,** 802–805.

24. Amsterdam, A., Lin, S., and Hopkins, N. (1995) Transient and transgenic expression of green fluorescent protein (GFP) in living zebrafish embryos. *CLONTECHniques* **X(3),** 30.

25. Amsterdam, A., Lin, S., and Hopkins, N. (1995) The *Aequorea victoria* green protein can be used as a reporter in live zebrafish embryos. *Dev. Biol.* **171,** 123–129.

26. Kimmel, C. B., Ballard, W. W., Kimmel, S. R., Ullmann, B., and Schilling, T. F. (1995) Stages of embryonic-development of the zebrafish *Develop. Dynam.* **203,** 253–310.

27. Caldovic, L. and Hackett, P. B. (1995) Development of position-independent expression vectors and their transfer into transgenic fish. *Mol. Mar. Biol. Biotech.* **4,** 51–61.

28. McKnight, R. A., Shamay, A., Sankaran, L., Wall., R. J., and Henninghausen, L. (1992) Matrix-attachment regions can impart position-independent regulation of a tissue-specific gene in transgenic mice. *Proc. Natl. Acad. Sci. USA* **89,** 6943–6947.

29. Hagashijima, S.-I., Okamoto, H., Ueno, N., Hotta, Y., and Eguchi, G. (1997) High-frequency generation of transgenic zebrafish which reliably express GFP in whole muscles or the whole body by using promoters of zebrafish origin. *Dev. Biol.* **192,** 289–299.

30. Long, Q., Meng, A., Wang, H., Jessen, J. J., Farrell, M. J., and Lin, S. (1997) GATA-1 expression pattern can be recapitulated in living transgenic zebrafish using a GFP reporter gene. *Development* **124,** 4105–4111.

31. Lin, S., Gaiano, N., Culp, P., Burns, J. C., Friedmann, T., Yee, J.-K., and Hopkins, N. (1994) Integration and germ line transmission of a pseudotyped retroviral vector in zebrafish. *Science* **265,** 666–669.

32. Burns, J. C., Friedmann, T., Driever, W., Burrascano, M., and Yee, J.-K. (1993) Vesicular stomatitis virus G glycoprotein pseudotyped retroviral vectors: Concentration to very high titer and efficient gene transfer into mammalian and nonmammalian cells. *Proc. Natl. Acad. Sci. USA* **90,** 8033–8037.

33. Gaiano, N., Allende, M., Amsterdam, A., Kawakami, K., and Hopkins, N. (1996) Highly efficient germ-line transmission of proviral inertions in zebrafish. *Proc. Natl. Acad. Sci. USA* **93,** 7777–7782.

34. Allende, M. L., Amsterdam, A., Becker, T., Kawakami, K., Gaiano, N., and Hopkins, N. (1996) Insertional mutagenesis in zebrafish identifies two novel genes, pescadillo and dead eye, essential for embryonic development. *Genes Devel.* **10,** 3141–3155.

35. Gaiano, N., Amsterdam, A., Kawakami, K., Allende, M., Becker, T., and Hopkins, N. (1996) Insertional mutagenesis and rapid cloning of essential genes in zebrafish. *Nature* **383,** 829–832.

36. Kawakami, K. and Hopkins, N. (1996) Rapid indentification of transgenic zebrafish. *Trends Genet.* **12,** 9,10.

37. Jowett, T. and Yan, Y.-L. (1996) Double fluorescent *in situ* hybridisation to zebrafish embryos. *Trends Genet.* **12,** 387,388.

38. Jowett, T. and Lettice, L. (1994) Whole-mount *in situ* hybridization on zebrafish embryos using a mixture of digoxigenin and fluorescein-labelled probes. *Trends Genet.* **10,** 73,74.

39. Strähle, U., Blader, P., Adam, J., and Ingham, P. W. (1994) A simple and efficient procedure for non-isotopic *in situ* hybridization to sectioned material. *Trends Genet.* **10,** 75,76.

40. Hauptmann, G. and Gerster, T. (1994) Two-color whole-mount *in situ* hybridisation to vertebrate and *Drosophila* embryos. *Trends Genet.* **10,** 266.

41. Nguyen, V. T., Morange, M., and Bensaude, O. (1988) Firefly luciferase luminescence assays using scintillation counters for quantification in transfected mammalian cells. *Anal. Biochem.* **171,** 404–408.

42. Wilkinson, D. G. (1992) The theory and practice of *in situ* hybridisation, in *In Situ
Hybridisation: A Practical Approach* (Wilkinson, D. G., ed.), IRL, Oxford, UK,
pp. 1–13.

# 32

# Microinjection of DNA, RNA, and Protein into the Fertilized Zebrafish Egg for Analysis of Gene Function

## Nigel Holder and Qiling Xu

## 1. Introduction

The use of microinjection to study gene function in the zebrafish has become widespread in recent years. This includes ectopic expression of genes by introducing DNA *(1)* or RNA into embryos or injection of blocking molecules, such as RNA encoding truncated proteins *(2)* or antibodies, to perturb the function of endogenouse gene products *(3,4)*. The method involves microinjection of DNA and RNA molecules into the cytoplasm of one-cell-stage embryos using pressure microinjector and micromanipulator as described below.

## 2. Materials

1. Breeding pairs of adult zebrafish are maintained at 28.5°C with a 14-h light and 10-h dark cycle. Embryos are obtained by natural spawning or by in vitro fertilization (refer to the *Zebrafish Book [5]* for the method).
2. The microinjection system consists of a stereo microscope, a three-dimensional micromanipulator, and an automatic pressure microinjector.
3. An injection chamber is made by placing a glass microscope slide into a 85-mm diameter disposable plastic Petri dish. Flood the injection chamber with embryo medium *(5)*, and allow a thin film of liquid to form between the slide and the dish. Remove the excess liquid using a glass pipet.
4. Microinjection pipets are prepared as follows: Pull fine glass capillaries with inner filament (Clark Electromedical Instruments, Reading, UK, 1.0-mm od, and 0.58-mm id) with a micropipet puller (Model P-87, Sutter Instrument Co. Novato, CA, or any equivalent). The tip of the micropipet should be about 0.05-mm. If the tip is too thin, it will lack the tensile strength to penetrate the chorion. Thicker

From: *Methods in Molecular Biology, Vol. 97: Molecular Embryology: Methods and Protocols*
Edited by: P. T. Sharpe and I. Mason © Humana Press Inc., Totowa, NJ

pipets do not easily withdraw from the chorion without dragging the embryo out of the chamber after injection. Just before microinjection, break the micropipet tip under a microscope with a pair of blunt forceps to produce a sharp end.

5. Molecular biology grade reagents should be used to prepare the solutions, DNA, and RNA samples.

## 3. Methods

### *3.1. Preparation of DNA*

For the chromosomal integration and germline transmission of transgenes, the DNA should be linearized or isolated from the plasmid sequences. Supercoiled DNA, however, can be used for transient gene expression work. After phenol/chloroform extraction to remove the restriction enzyme, the DNA is further purified on a microconcentrator (Microcon 100, Amicon Inc., Bedford, MA) as follows:

1. Equilibrate the microcon with 250 µL sterile $dH_2O$ and spin at 3000 rpm for 5 min.
2. Load the DNA solution into the microcon, and spin at 3000 rpm for 5–10 min until most of the liquid has gone through the membrane.
3. Wash the microcon three times each with 500 µL sterile $dH_2O$, and spin as above.
4. Invert the microcon, and spin briefly to recover the DNA solution into a fresh Eppendorf tube. Estimate the DNA concentration by running an aliquot of the purified DNA on an agarose gel, and store the DNA at –20°C.
5. Before microinjection, dilute the DNA to 0.1 µL in 0.2 $M$ KCl, 0.25% phenol red. The dye facilitates the estimation of the injection volume.

### *3.2. Preparation of RNA*

The RNA for microinjection is synthesised by in vitro transcription. The structure of the 5'- and 3'-termini of synthetic mRNA can affect its stability and the efficiency of translation. Capping of the 5' terminus with GpppG and addition of the poly(A) tail improve the stability of synthetic mRNAs. The constructs used widely to generate the DNA templates for in vitro transcription are pSP64T and CS2 vectors (*6–9*). The synthetic RNA should be tested in an in vitro translation system to ascertain that a protein product is made.

1. Prepare 10 µg of template DNA by digestion with a suitable restriction enzyme. Make sure the reaction is completed by analyzing an aliquot of the DNA digest in an agarose gel.
2. Purify the linear DNA by extraction with phenol/chloroform and precipitation with ethanol. Redissolve the DNA at 0.5 mg/mL in RNase-free water (Sigma, W-4502) or DEPC-treated water.
3. Mix the following components at room temperature in the order shown:
   4.0 µL DNA template.
   19.5 µL RNase-free water.
   2.0 µL 0.1 $M$ dithiothreitol.

4.0 µL 10X transcription buffer.

4.0 µL each of 50 m*M* rATP, rUTP, rCTP.

2.0 µL 50 m*M* rGTP.

2.5 µL 40 m*M* cap analog (GpppG).

2.0 µL placental RNase inhibitor (10 U/mL).

2.0 µL DNA-dependent RNA polymerase.

4. Incubate for 2 h at 37°C. Analyze an aliquot of the reaction (2 µL) on an agarose gel to check the transcription efficiency.
5. Add 1 µL of RNase-free DNase I (1 mg/mL), mix, and continue the reaction for 15 min at 37°C.
6. Add 160 µL of RNase-free water and purify the RNA by phenol/chloroform extraction twice.
7. Further purify the RNA by the microconcentrator as described above.
8. Check the size and quality of RNA by gel electrophoresis. Aliquot the RNA, and store at –80°C.
9. The amount of RNA for microinjection varies according to the nature of its encoded protein. It is, therefore, necessary to titrate the RNA.

### 3.3. Microinjection

1. Fill the micropipet with the DNA or RNA using a microloader (Eppendorf, Germany). The capillary reaction of the inner microfilament should direct the solution to the very tip of the micropipet.
2. Attach the micropipet to the needle holder connected with the micromanipulator (Narishige, Tokyo, Japan). The system is operated using a compressed air cylinder, which delivers pulses controlled by a microinjector (Picospritzer II, General Valve Corp., USA, or any equivalent).
3. Precalibrate the amount injected by counting the number of pulses to expel 1 µL of solution under fixed pressure and duration (e.g., 50 psi and 30 ms). For 200 pulses to deliver 1 µL of solution indicates 5 nL injection volume. Inject about 2–5 nL DNA or RNA.
4. Use a wide-mouth glass pipet to transfer fertilized embryos (*see* **Notes**) into the injection chamber and align them (30–40 embryos) along the trough created – between the slide and the Petri dish. Tilt the injection chamber slightly to collect and remove the excess liquid as much as possible. With a pair of blunt forceps, orient embryosm such that the germinal disk is facing the trough.
5. Injection can be done through the chorion and directly into the cytoplasm. To prevent the germinal disk rotating away from the micropipet tip, it is important to have a relatively steep angle (about 45°) between the pipet and embryo; use the micromanipulator to move the micropipet vertically to penetrate the chorion and the cell membrane. Injection through the yolk and then into the cytoplasm can result in blockage of the micropipet and difficulty in withdrawing the needle from injected embryo.
6. Following injection, tilt the injection chamber, and add embryo medium to the embryos. Transfer the embryos into a clean dish, and incubate them in a humidified incubator at 28.5°C. Later, remove all dead or uncleaved embryos.

## 4. Notes

It is not necessary to dechorionate the embryos for microinjection. Dechorionation is time-consuming and dechorionated embryos are much more fragile and susceptible to infection. If one has to work with dechorionated embryos for microinjection, 1.5% (w/v) agarose-coated dishes should be used to incubate embryos and prepare the injection chamber as follows:

1. Pour 15 mL of hot 1.5% agarose in embryo medium into a 85-mm Petri dish, and leave to set.
2. Add another 15 mL of the agarose solution to the solidified dish. Place the glass slide into the liquid agarose overlay at an angle such that a trough is created. Remove the slide after it is set, and flood the chamber with the embryo medium. Alternatively, store the injection chamber at 4°C for later use.

## Acknowledgments

We are indebted to our colleagues in the fish group at the Randall Institute and University College, London for helping with developing any techniques outlined in this chapter.

## References

1. Reinhard, E., Nedivi, E., Wegner, J., Skene, J., and Westerfield, M. (1994) Neural selective activation and temporal regulation of a mammalian GAP-43 promotor in zebrafish. *Development* **120,** 1767–1775.
2. Krauss, S., Maden, M., Holder, N., and Wilson, S. (1992) Zebrafish pax (b) is involved in the formation of the midbrain/hindbrain boundary. *Nature* **360,** 87–89.
3. Durbin, L., Brennan, C., Shiomi, K., Cooke, J., Barrios, A., Shanmugalingam, S., Guthrie, B., Lindberg, R., and Holder, N. (19998) Eph signalling is required for segmentation and differentiation of somites. *Genes Dev.* **12,** 3096–3109.
4. Xu, Q., Alldus, G., Holder, N., and Wilkinson, D. (1995) Expression of truncated Sek- 1 receptor tyrosine kinase disrupts the segmental restriction of gene expression in the *Xenopus* and zebrafish hindbrain. *Development* **121,** 4005–4016.
5. Westerfield, M. (1995) *The Zebrafish Book*, 3rd ed. University of Oregon Press, Eugene, OR.
6. Melton, D. A., Krieg, P. A., Rebagliati, M. R., Maniatis, T., Zinn, K., and Green, M. R. (1984) Efficient *in vitro* synthesis of biologically active RNA and RNA hybridisation probes from plasmids containing a bacteriophage SP6 promoter. *Nucleic Acids Res.* **12,** 7035–7056.
7. Krieg, P. A. and Melton, D. A. (1984) Functional messanger RNAs are produced by SP6 *in vitro* transcription of cloned cDNAs. *Nucleic Acids Res.* **12,** 7057–7070.
8. Rupp, R. A. W., Snider, L., and Weintraub, H. (1994) *Xenopus* embryos regulate the nuclear localisation of XMyoD. *Genes Dev.* **8,** 1311–1323.
9. Turner, D. L. and Weintraub, H. (1994) Expression of achaete-scute homolog 3 in *Xenopus* embryos converts ectodermal cells to a neural fate. *Genes Dev.* **8,** 1434–1447.

# 33

## Retinoids in Nonmammalian Embryos

**Malcolm Maden**

## 1. Introduction

The family of retinoids comprise an enormous number of compounds related to vitamin A. Many of these compounds are naturally occurring substances generated during the biological functioning of retinoids: the conversion of dietary sources of retinoids (β-carotenes, retinyl esters) to those that can be taken up by the absorptive epithelium of the gut (retinol); the conversion of absorbed forms to stored forms in the liver (retinyl esters); the conversion of stored forms to active forms as mediators of vision (retinals), skin differentiation (retinoic acids), and general cell differentiation and proliferation (retinoic acids). However, more of these compounds are synthetic, for example, the retinobenzoic acids, and have been generated in the desire to find more potent and less teratogenic retinoids for pharmaceutical use. This chapter will be concerned only with a very few retinoids, ones that have been used in an embryological context.

The original definition of a retinoid was based on its chemical structure, but this soon became obsolete when these new compounds were synthesized. The current definition of a retinoid is a substance that can elicit specific biological responses by binding to and activating specific receptors or a set of receptors. This perceptive definition was formulated by Sporn and Roberts *(1)* before the discovery of the retinoid receptors and remains pertinent today with our knowledge of the retinoic acid receptors (RARs), whose ligand is all-trans-retinoic acid (tRA), and the retinoid X receptors (RXRs), whose ligand is 9-cis-retinoic acid (9-cis-RA) *(2)*. In **Table 1**, the structure of tRA and 9-cis-RA is shown, as well as several other retinoids of embryological interest, and in the following section are a few comments on each of the compounds.

From: *Methods in Molecular Biology, Vol 97: Molecular Embryology: Methods and Protocols*
Edited by: P. T. Sharpe and I. Mason © Humana Press Inc., Totowa, NJ

**Table 1**
**Chemical Formulae of the Retinoids That Are of Embryological Interest**[a]

| | |
|---|---|
| Retinol | |
| Retinal | |
| All-trans-Retinoic Acid | |
| 9-cis-Retinoic Acid | |
| 3,4-Didehydro-Retinoic Acid | |
| 4-oxo-Retinoic Acid | |
| Retinyl Palmitate | |
| TTNPB | |
| Ch 55 | |
| CD 367 | |

[a]The numbering of the carbon atoms is shown only on the tRA.

### 1.1. Embryologically Interesting Retinoids

#### 1.1.1. Retinol

This is the "parent" vitamin A molecule from which tRA is derived. All-trans-retinol is first converted to retinal and then tRA by two dehydrogenase enzymes, presumably specific. It is found at high levels endogenously in mammalian and chick embryos *(3–5)*. Usually it is about 10-fold less potent than tRA in most biological assays *(6)*, including mammalian teratogenicity *(7)*, but it is inactive in inducing pattern duplication when applied locally to the regenerating amphibian limb *(8)*, the chick limb bud *(9)*, or developing chick skin *(10)*.

#### 1.1.2. Retinal

This is the intermediate metabolite between retinol and tRA and is used in the visual cycle where 11-cis-retinal binds with the protein opsin to form rhodopsin, the visual pigment. It is undetectable in mammalian and chick embryos, but present at very high levels in amphibian *(11,12)* and zebrafish eggs *(13)*. It is not active in inducing pattern duplication in the regenerating amphibian limb *(8)*.

#### 1.1.3. tRA

This was formerly the most active naturally occurring metabolite of *t*-retinol in biological assays, including pattern respectfication in embryos, although perhaps it is now superseded by 9-cis-RA, at least in some assays *(14)*. It is the ligand for the RARs and is present endogenously in varying concentrations throughout the mammalian and chick embryo *(3–5,15)*, in the regenerating amphibian limb *(16)*, and the *Xenopus* and zebrafish embryo *(12,13)*.

#### 1.1.4. 9-cis-RA

This is the ligand for the RXRs. Presumably, it is generated from tRA via a specific isomerase, although this is not known. It is present in adult mouse liver and kidney *(17)*, but undetectable endogenously in chick and mouse limb buds *(4)*. It is present in *Xenopus* embryos *(12)* and the regenerating amphibian wound epidermis generates 9-cis *(18)*. It is 25 times more potent than tRA in inducing duplications of the chick wing bud *(14)*.

#### 1.1.5. 3,4-Didehydroretinoic Acid

This is found in the chick limb bud at 5–6 times higher amounts than tRA *(4,19)*. It is generated by a parallel pathway to tRA from 3,4-didehydroretinol via 3,4-didehydroretinal. It is equipotent with tRA at inducing duplications in the chick wing bud *(19)*.

### 1.1.6. 4-oxo-RA

This is a more polar metabolite of tRA, originally thought to be a breakdown product, but present in *Xenopus* embryos *(20)*. It is more potent than tRA in inducing neural defects in *Xenopus* and a more potent mammalian teratogen *(21)*. It binds and activates RARβ, but only weakly activates RXRα, resembling tRA in this respect *(20)*.

### 1.1.7. Retinyl Palmitate

This is a typical esterified form of retinol used as storage in the liver. It is convenient for administration to amphibians, since it is available in a soluble form attached to corn starch and can be simply added to the tank water. It is eightfold less potent than tRA at inducing limb duplications during regeneration as assayed by this method of administration *(22)*.

### 1.1.8. TTNPB, Ch-55, CD-367

These are examples of synthetic derivatives of tRA, which usually have far higher potencies in biological assays than tRA, including embryological assays, such as duplication of the chick limb bud *(23–25)*, the regenerating amphibian limb *(8,25)*, and feather transformation in chick skin *(10)*. This may be because they are much more stable in cells than tRA and are not broken down by the tRA metabolism pathway.

## 2. Materials

### 2.1. Preparation of Retinoid Solutions and Stability

Retinoids are insoluble in water, but soluble in organic solvents, such as ethanol and dimethylsulfoxide (DMSO), which are the most commonly used solvents for administration of tRA. Although one always performs control experiments, it is important to be aware of the fact that DMSO acts as a differentiating agent to embryonal carcinoma cell cultures in the same way that tRA does, so perhaps ethanol is a "better" solvent. The maximum solubility of tRA in DMSO is about 50 mg/mL.

Retinoids are stable in powder form for many months at –20°C. Light and oxygen cause isomerization and breakdown, so they are often stored in brown vials in an atmosphere of nitrogen. Solutions of retinoids are unstable, especially at room temperature, and tRA in DMSO begins to isomerize to 13-cis-RA, 11-cis-RA and 9-cis-RA in a few minutes when left on the laboratory bench. This can be very useful for generating standards for HPLC, but a problem when administering retinoids to animals over periods of several hours! The latter tends to negate the emphasis on absolute purity that many retinoid researchers insist on—as soon as one administers these compounds to the embryo, either the light or the embryonic cells will change the nature of thesubstance you have given it. Nevertheless, for experimental purposes it is better to make up one solution and aliquot it as follows;

1. Prepare 1 mL of a 10 mg/mL solution of tRA (Sigma, Dorset, UK) in DMSO (Spectrosol or Analar-Grade, BDH, Poole, UK) or ethanol.
2. Split the 1 mL into 5-µL aliquots in capped tubes.
3. Freeze at –20°C.
4. Use one aliquot for each experiment, never refreeze and use again.

## 2.2. Preparation as a Solid

An alternative to administering retinoids as a solution is to administer them as neat powder in a depot. In this case, a rubber compound called Silastic 382 Medical Elastomer (Dow Corning Corporation, Midland, MI) acts as the depot. tRA is released from this compound at a constant rate over the first 24 h and then the release plateaus by which time about 65–70% of the total content has been released *(26)*.

1. Measure 1 mL of Silastic, either as a volume from a syringe or by weight, into a small Petri dish.
2. Add the required amount of tRA, e.g., 10 mg to the Silastic.
3. Mix thoroughly for 10 min until an even consistency is obtained—easy to judge with a yellow compound like tRA.
4. Add one drop of catalyst, stannous octoate, mix, and leave to set for an hour at room temperature.
5. The silastic can now be cut up into any shape and size for administering to the embryo. This is performed down a dissecting microscope with a graticule on the stage plate. The required size is normally determined empirically by the dose or by the size of the embryo or embryonic field to be treated. The amount of retinoid administered can be calculated from the size of the block, e.g., a 200-µm cube has a volume of $8 \times 10^6$ $\mu m^3$. If the concentration of the mixture is 10 mg/mL = 10 mg/$10^{12}$ $\mu m^3$ then this block will have $(8 \times 10^6)/10^{12} \times 10$ mg = 80 ng of RA in it.

## 2.3. Preparation on a Bead

This method was designed for adminstering RA to the anterior side of the chick limb bud *(27,28)*.

1. Use AG 1-X2 chromatography beads, formate form, 100–200 mesh (Bio-Rad, Hemel Hempstead, UK).
2. Select a number of uniformly sized beads (approx 200 µm) in a small Petri dish and add an aliquot of tRA in DMSO at a concentration of, for example, 1 mg/mL. Leave for 20 min. The beads are very difficult to see once in this solution, and the angle of the light source has to be altered to be able to visualize them.
3. Rinse the beads by removing the RA solution and putting on a few drops of medium, e.g., MEM. The beads will take up phenol red from the medium and become perfectly visible.

4. Repeat the rinse.
5. Incubate the beads in medium for 20 min at 37°C. They are then ready for implantation.

## 3. Methods

### 3.1. Administration to Chick Embryos

Chicks have been treated with RA administered in solution to cultured embryos, with RA on beads, with RA in Silastic, or directly injected into the embryo.

#### 3.1.1. Administration in Silastic

1. Cut a block of the appropriate size.
2. Window eggs at the appropriate stage.
3. Tear the vitelline membrane adjacent to the part of the embryo to be treated with a tungsten needle.
4. Place the block adjacent to the embryo.
5. Seal the egg and replace in the incubator.

This method is useful for early embryos, e.g., treating a stage 9/10 hindbrain where the embryo is relatively immobile, but not good for late embryos, which will turn and begin to move. In this case, the block is displaced and the experiment fails.

#### 3.1.2. Administration on Beads

These can be placed over the embryo or adjacent to the embryo exactly as described in **Subheading 3.1.1.**, but beads are also ideal for inserting into the embryo and were originally used for inserting into the anterior margin of the chick limb bud *(27)*. In this case:

1. Window the eggs at the appropriate stage, e.g., stage 20.
2. Tear all the membranes above the limb bud. This is usually the right limb bud owing to the turning of the embryo, but not always.
3. With a tungsten needle, make a slit between the AER and the mesenchyme from the apex of the bud to the anterior margin.
4. Stretch the AER away from the mesenchyme to expand it, but do not break it.
5. With the AER pulled away with a needle, take a bead with a pair of fine forceps, place it into the gap between the AER and the mesenchyme, and release the AER. The bead will stay in place as the AER contracts.
6. Seal the eggs and replace in the incubator.

#### 3.1.3. Administration of RA on Paper

An alternative to beads is to use the RA soaked onto paper *(29)*.

1. A piece of absorbent paper (blotting paper, newsprint, and so forth) is placed into a solution of RA in DMSO (*see* **Subheading 2.1.**) for 10 min and then washed in medium to remove excess RA.

2. Cut the paper into small pieces (500-μm squares).
3. Insert a square into a slit made in the anterior margin of the limb bud with a tungsten needle. The slit is made perpendicular to the base of the limb opposite intersomite 16/17.

### 3.1.4. Exposure of Embryos in New Culture to RA

Chick embryos in culture have been treated with RA in several instances *(30,31)*. RA is dissolved in DMSO or ethanol and added to the culture medium at the required concentration for a particular period of time. At the end of that time, embryos are washed and fresh medium without RA is added.

### 3.1.5. Injection of RA into the Yolk

A further variation on the theme of administering retinoids to the chick embryo is to inject them directly into the yolk through the air sac. In this case the required amount of tRA dissolved in DMSO is taken up into a syringe and injected after making a hole in the egg at the end with the air sac with a 20-gage needle.

## 3.2. Administration to Amphibians

As in the case of chicks, various methods have been used to treat amphibians with retinoids and for two experimental purposes: first, to study early development, and second, to study limb development and regeneration. Obviously, these two types of experiments involve treatment at widely differing stages, and a large number of different species of amphibians have been used for limb regeneration studies.

### 3.2.1. Early Development

Treatment with retinoids at gastrulation stages results in the progressive loss of head structures *(11)*. In this case, stage 10 embryos are treated by taking an aliquot of tRA in DMSO and simply adding it to the medium in which the embryos are cultured to produce a final concentration of $10^{-5} M$ or $10^{-6} M$. A typical treatment time would be for 30 min, at the end of which the medium is changed.

Virtually all studies on early development of amphibians have used this technique because of its speed and simplicity. One experiment, however, involved the injection of RA suspended in corn oil *(32)*. Stage 10–11 embryos had their fertilization envelope manually removed, and were injected with a 1-nL droplet between the surface and deep ectodermal layers by means of an air pressure injection system. The droplet also contained a fluorescent dye, DiI, so its presence could be traced throughout further development of the embryo. The additional unique feature of this system was that the effects of RA were asymmetric because only one side of the embryo was injected.

### 3.2.2. Limb Regeneration Studies

Adult newts, or axolotls, or *Pleurodeles* at various stages of larval growth are usually used for these studies. The limbs are amputated, and then the animals are treated with retinoids in various ways. One method of administration is simply to add 10–60 mg retinyl palmitate (Sigma, complexed with corn starch) to 1 L of water in which the animals are kept. The water turns cloudy because of the corn starch and should be changed every day. Treatment times vary from 1–14 d *(22)*. It is also possible to treat animals in this way with tRA and other retinoids. In this case the appropriate amount of retinoid is used dissolved in DMSO (e.g., 300 µL of a 1 mg/mL solution of tRA), and this is put into the water. The retinoid immediately precipitates out, being insoluble in water, forming a yellow scum on the surface. Nevertheless, this method works perfectly well.

It is also common to administer retinoids by ip injection (e.g., *33*). Animals are anesthetized and placed on their backs on a wet towel. A 27-gage needle is used to make a puncture through the skin and abdominal muscles into the ip cavity of each animal, a few millimeters anterior to one hindlimb and lateral to the midline. The volume of solution containing the desired dose to tRA is then injected into the abdomen with a Hamilton microsyringe using the previously made puncture hole. tRA percipitates at the injection site, being visible through the skin as a yellow mass, and gradually dissipates over the next 24–48 h. Animals are returned to their water as soon as possible. Control animals are injected with an equal amount of DMSO. This method tends to induce some mortality among the animals, since DMSO injected into the peritoneum is very unpleasant for the liver.

A very reliable and nontoxic method of administration is to use retinoids in Silastic as described above. Pieces of a uniform size of Silastic, e.g., 1 mm$^3$ containing tRA at the appropriate dose, e.g., 100 mg/mL, are prepared. Animals are anesthetized and a tunnel made in the limb under the skin by inserting the points of a pair of watchmaker forceps through the skin and pushing them along the limb between the skin and the underlying musculature. In this way, the piece of Silastic can be placed directly adjacent to the regeneration blastema on either the dorsal, ventral, anterior, or posterior side. Animals are returned to the water, and the Silastic remains in place for several weeks. With a colored compound, such as tRA, one can see the color gradually disappear from the Silastic. The diffusion characteristics of tRA from Silastic have been described *above* *(26)*. Only very rarely are the blocks ejected from the limb. One example of this we have found is when the Siastic contains retinal *(22)*. Uniquely, this retinoid induced the epidermis to eject the block.

Finally, gastric intubation has been used to administer retinoids to adult newts *(34)*. In this case, the retinoid solution in DMSO is squirted into the stomach of the anesthetized animal through a thin plastic tube attached to a syringe.

### 3.3. Adminstration to Fish

Zebrafish have been used in studies on the effects of CNS development *(35)*, heart development *(36)*, eye development *(37)*, and fin regeneration *(38)*. In all these cases, tRA is administered to the animals in the tank water. A tRA in DMSO aliquot is used at an appropriate concentration to give a final concentration of $10^{-6}$ *M* or $10^{-7}$ *M* when diluted in, say, 5 mL of zebrafish water. Many embryos can be treated at once in a 5-mL Petri dish. For studies of CNS and heart development, embryos are treated for 1 h at 50% epiboly; for eye development for 4 h at 10.5 h of development, and for fin regeneration, adults are treated for 4 or 7 d.

### 3.4. Reporter Cells

Several stably transfected F-9 or L-cell lines have been generated that can be used as reporter cells for the presence of RA when embryonic tissue is placed on top of them. The line produced by Wagner, et al. *(39)* contains a retinoic acid-response element (RARE) from the human β-retinoic acid receptor gene, which functions as an inducible enhancer that responds to the α, β, and γ RAR subtypes. A single copy of this RARE was placed upstream of the *Escherichia coli lacZ* or firefly *luciferase* genes, conferring retinoid responsiveness to these genes. The *lacZ* gene is used for histochemical detection of retinoid responsive cells and quantitation by cell counting, and the *luciferase* gene is used to provide a direct quantitative assay. The inclusion of the NEO$^r$ gene permits the establishment of transfected cell lines that stably maintain the reporter gene constructs.

### 3.4.1. Preparation of Media and Solutions

1. Medium: 10 mL 10X DMEM with NEAA without L-glutamine, sodium pyruvate, or NaHCO$_3$ (Gibco-BRL, Paisley, Scotland, UK, no. 12501-011): 7 mL 5.3% NaHCO$_3$, 1 mL penicillin/streptomycin, 58.4 mg L-gluamine (Sigma), 82 mL autoclaved milli-Q H$_2$O. Filter-sterilize and adjust pH to 7.4 with 800 µL 1 *N* NaOH.
2. Medium with serum: 15 mL medium (*see* **item 1**), 3 mL heat-inactivated fetal bovine serum, 2 mL heat-inactivated horse serum.
3. Growth medium: Medium with serum, add 0.8 mg/mL G418 (Geneticin G-418 sulphate; Gibco-BRL no. 11811-01E), filter-sterilize.
4. Coated dishes: Make fresh a 0.1% gelatin solution (Sigma) in PBS by gentle heating and shaking. Filter-sterilize while still warm. Completely cover the base of tissue-culture dishes. Leave for 2 h at room temperature, and then aspirate off and wash 1X with sterile PBS. Plate cells or cover with fresh PBS and store at 4°C (maximum 2 wk).

5. Serum-free medium: 10 mL media, 20 µL N3 (serum supplement—*see* **item 6**).
6. N3 serum supplement: 246 µL Hank's balanced salt solution (HBSS) without calcium and magnesium, 50µL 10 mg/mL bovine serum albumin in HBSS (store at 4°C), 100 µL 100 mg/mL human transferrin in HBSS (store at –20°C), 20 µL 80 mg/mL putrescine hydrochloride in HBSS (store at –20°C), 50 µL 10 m$M$ sodium selenate in HBSS (store at –20°C), 5 µL 20 mg/100 mL triodothyronine, sodium salt in 0.01 $M$ NaOH (store at –20°C), 20 µL 25 mg/mL bovine pancreatic insulin in 20 m$M$ HCl (store in plastic at –20°C), 5 µL 12.5 mg progesterone in absolute alcohol (store in glass at –20°C), 1 µL 2 mg/mL corticosterone in absolute alcohol (store in glass at –20°C).
7. X-gal stain: 3.08 mL 0.5 $M$ Na$_2$PO$_4$ 0.46 mL 1 $M$ NaH$_2$PO$_4$, 0.026 mL 1 $M$ MgCl$_2$ 1.2 mL 50 m$M$ K$_3$Fe(CN)$_6$ 1.2 mL 50 m$M$ K$_4$Fe(CN)$_6$, 1.0 mL X-gal stock = a 2% solution of 5 bromo-4 chloro-3-indoxyl-β-D galactopyranoside (Novabiochem) made up in *N*-*N*-dimethylformamide (Sigma), 13 mL milli-Q H$_2$O.

### 3.4.2. Preparation of Cells from a Frozen Aliquot

1. Thaw the frozen cells in 20 mL of medium with serum to 37°C.
2. Plate in a gelatin-coated 80-cm$^2$ tissue-culture flask (Nuclon) with 20 mL of growth medium. The latter is added to enrich the plated culture in transfected F-9 cells; the transfected cells; are geneticin-resistant, but untransfected F-9 cells are not.
3. Grow the culture at 37°C 5%CO$_2$ for 3–5 d until just confluent.

### 3.4.3. Preparation of Cells for Explant Coculturing

1. For culturing to assay explants, cells are grown on gelatin-coated 35-mm diameter tissue-culture dishes.
2. Starting with a just confluent flask, wash the cells with PBS, and then loosen the cells by covering with with 5 mL trypsin/EDTA (Gibco-BRL no. 610-5300AG) (store at –20°C in 5-mL aliquots) for 3 min.
3. Triturate the cell suspension thoroughly with a plastic pipet to give a complete single cell suspension. F-9 cells are tough and will tolerate much up and down pipeting.
4. Transfer the cell suspension to a 15-mL tube and add 5 mL of medium with serum. Add 40–50 drops of this suspension to 20 mL of growth medium. Put 1 mL of this suspension into each of 20 35-mm dishes. This gives a cell concentration which 5% of which will give near confluence in 2 d when incubated at 37°C.

### 3.4.4. Explant Coculturing

The cell plating and the incubations of embryonic tissue must be synchronized to give confluent F-9 cells on the day embryonic tissue is ready for explanting. Superconfluent F-9 cultures should be avoided, since the cells pile up and this makes it more difficult to keep the explant attached to the monolayer of cells.

1. Dissect the tissue explants in a balanced salt solution. If the ectoderm is trypsinized off, for example, or any other enzymes used to separate tissues, it is very important

to wash the tissue and neutralize the enzymes with medium with serum prior to explanting, since any remaining trypsin will destroy the monolayer.

2. Immediately before placing the explant on the cell monolayer, remove the growth medium from the F-9 cells and replace with 1 mL serum-free medium. Explants are always cultured in serum-free medium, since serum has retinoids in it.

3. Transfer the explants onto the cell monolayer with a pipet.

4. The dishes should be moved as little as possible to allow the explants to settle and attach to the cell monolayer. If attachment proves to be a problem, minimizing the medium over the cells when first introducing the explant can increase cell contact. After a hour or so, gently add more medium to these cultures to ensure that they do not become dry overnight.

5. Culture at 37°C, in a 5% $CO_2$ atmosphere overnight.

6. Wash the cultures twice very gently (in order not to disturb the F-9 cells) with room temperature PBS. Then fix for 5 min at 4°C with 2% paraformaldehyde, 0.2% glutaraldehyde in PBS.

7. Rinse gently twice more with PBS, then cover with 1 mL of X-gal stain/dish, and incubate at 37°C overnight.

8. Count the number of blue cells around the explant as a percentage of all cells.

Each time an experiment is performed with tissue explants, a series of control RA dilutions need to be run in parallel, which is why one needs about 20 dishes. A typical series would be $10^{-6}$–$10^{-11}$ $M$ in steps of 10-fold dilutions. In the case of F-9 cells with the *lacZ* gene, the results are determined by taking a field of view down an inverted microscope and counting the number of blue cells as a percentage of the total cells. The plate is then moved to another field and the counts repeated. The result of such an experiment is shown in **Fig. 1A**. When the explants are counted, then an approximation of the amount of retinoid present can be made by comparing % blue cells with the control curve.

Several explants of the same tissue type can be placed in one Petri dish. Usually, the explants remain in place all through the β-gal staining process, but if they do not, that is rarely a problem because a ring of blue cells marks the place where they were and counts can still be performed. Counts are made of the percent blue cells within a fixed distance from the explant.

Obviously, 50 cell diameters away from the explant, none of the cells are blue, whereas the majority of those directly touching the explant will probably be blue, so a compromise in distance much be reached and adhered to for all explants. A typical result for chick limb buds cut into two halves is shown in **Fig. 1B**. By comparison with the standards in **Fig. 1A** it is possible to suggest that the posterior halves of chick limb buds contain in the region of $10^{-9}$ M RA and that the anterior halves contain considerably less. This result fits well with the original HPLC data of Thaller and Eichele *(3)*.

Finally, it is important to remember that this assay does not measure tRA alone because several retinoids are equally efficient at activating the RARs,

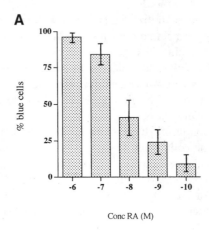

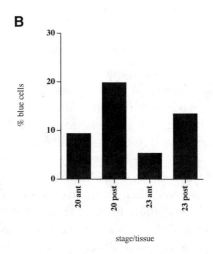

Fig. 1. (**A**) Calibration curve generated by counting the number of F-9 reporter cells that have turned blue after treatment with increasing concentrations of tRA. A straight line is usually obtained. From these data, an approximation of the concentration of RA present in explanted tissues can be made from the % blue cells recorded in the explant dishes. Bars mark standard deviations. (**B**) Example of data obtained after explanting anterior (ant) and posterior (post) half chick limb buds at two different stages of development, stage 20 and stage 23. It is clear that posterior halves contain more RA than anterior halves. A value of 20% blue cells from stage 20 posterior halves gives a concentration of close to $10^{-9}$ $M$ RA from the standard curve in A.

including didehydroretinoic acid and 9-cis. Therefore, what one measures here is the combined concentration of the retinoic acids.

## 3.5. High Pressure Liquid Chromatography (HPLC)

The use of HPLC to identify precisely individual retinoids is a methodology that is subject to infinite variability (*see*, for example, **ref. 40**). The variables include normal-phase or reverse-phase, type of column, composition of buffers, mix of buffers, single-step or multistep elution, step or gradient elution, and rates of flow. The idea is to use a system such that each of the retinoids one is interested in will come off at a different and highly repeatable elution time. Reverse-phase is generally preferred over normal-phase chromatography, the latter suffering from difficulty in reproducing precise elution times. Therefore, which method should one use?

The method that was used to identify tRA in the chick limb bud by Thaller and Eichele *(3)* is described below with only minor modifications, since it has been successfully used by several groups, including ourselves, to identify retinoids in the mouse embryo *(5)*. An example of this type of chromatography is given in **Fig. 2**. **Figure 2A** shows the separation of a mixture of standards and **Fig. 2B**, the retinoids present in a mouse embryo. However, this method was developed before the discovery of 9-cis-RA, and it fails to separate this isomer because 9-cis coelutes with retinol, so this is by no means a perfect method. To identify 9-cis, we have developed a method *(18)* that involves first running the sample on a normal-phase system with an $NH_2$ column that retains only the acids and then running the acid fractions on a reverse-phase system as described below.

Because there is no contaminating retinol left, all the tRA isomers can be detected, but one does not obtain any quantitation for retinol by this method. However, Kraft et al. *(12)* describe a method involving a gradient elution of a mixture of ammonium acetate and methanol whereby 9-cis *is* identifiable as a separate peak.

Most embryological interest has been centered on the retinoic acids and retinol, but there are other retinoids of biological interest, such as the retinyl esters, that these methods do not identify, because they come off the column after very long elution times. Another method that was designed to separate 14 different retinoids, including those of greatest interest, such as 4-oxo-RA, tRA, and retinol, as well as retinyl acetate, palmitate, and stearate, was described by Cullum and Zile *(41)*. This involves a multistep, three solvent (water, methanol, chloroform) gradient system eluting for 50 min. Thus, the choice of method is dependent on which retinoids one wants to identify.

### 3.5.1. Extraction Procedure

Ideally this should be done in red or gold light, but at the very least it should be done in subdued lighting (turn the lab lights off) and in lightproof (brown) microcentrifuge vials. All chemicals should be of the highest quality.

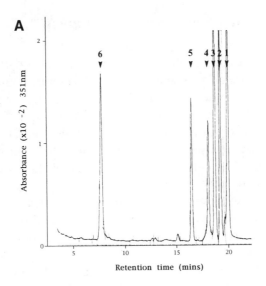

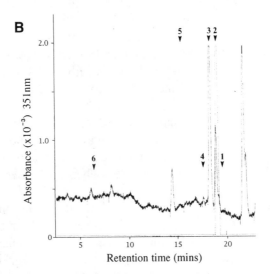

Fig. 2. Examples of HPLC chromatograms using the methods described in the text. Solid lines show the UV absorbance at 351 nm, and dotted lines the cpm measured with an on-line radioactivity detector. (**A**) A mixture of six retinoid standards separated out into individual peaks with different elusion times. Peak 1 = retinal, peak 2 = tRA, peak 3 = retinol, peak 4 = 13-cis-RA, peak 5 = didehydroretinol, peak 6 = 4-oxo-RA. The dotted line marks the cpm of [$^3$H] tRA, which was also added to the mixture and this coeluted with the cold tRA. (**B**) The retinoid extracted from a whole 10.5-d mouse embryo. The same six peaks of known standards are marked. Only retinol

1. Collect batches of embryonic tissue on ice. Not <500 mg can usually be used. They can be stored for short periods at −70°C to allow the collection of enough material.
2. Add an equal volume of ice-cold stabilizing buffer and sonicate. Stabilizing buffer = 5 mg/mL ascorbic acid + 5 mg/mL $Na_3$ EDTA dissolved in PBS, with pH adjusted to 7.3 with NaOH.
3. Take 10 μL of homogenate for protein estimation, so that at the end the amount of retinoid can be given in ng/mg protein. Alternatively, use this sample for DNA determination, and express the data in ng/μg of DNA.
4. Add a known amount, 1–2 nC, of [$^3$H] tRA, so that the recovery ratio can be determined and the tRA peak identified on the chromatograph.
5. Extract in 2 vol of extraction solvent. Extraction solvent = ethyl acetate:methyl acetate 8:1 + 50 μg/mL butylated hydroxytoluene as an antioxidant. Extract for 20 min while continuously mixing on a vibromax.
6. Separate the solvent phase by microcentrifugation at low speed. Keep the solvent phase.
7. Repeat the extraction on the homogenate, and separate again by centrifugation. Remove the solvent phase, and pool it with that from the first extraction.
8. Dry down the combined solvent phases under a stream of nitrogen.
9. Resuspend the extract in 100 μL methanol, and microcentrifuge at high speed to remove the particulate matter. This is now ready to be injected onto the HPLC column.

### 3.5.2. Chromatography

The ideal HPLC hardware is one that measures both the UV absorbance of the eluate and the radioactivity with an on-line isotope detector. The two sets of data are then superimposed by the computer to allow for the identification of individual peaks (**Fig. 2**). Chemicals should be HPLC grade and buffers degassed before use.

1. Use a 5-μm encapped $C_{18}$ column (Lichrospher, Merck, Darmstadt, Germany) with an equivalent precolumn. The precolumn prevents any particulate matter, such as phospholipids, from entering the column and is changed at regular intervals.
2. Inject the sample onto the column.
3. Elute at 1 mL/min using the following mobile phases. Solvent A, 1% acetic acid in MilliQ water. Solvent B, acetonitrile/methanol 3:1. The initial conditions are 60% solvent B rising linearly to 100% solvent B over 25 min.
4. Monitor the eluent at 351 nm with a UV detector and the radioactivity with an isotope detector.

---

Fig. 2. *(continued)* (peak 3) and tRA (peak 4) are clearly identifiable. Peak 2 coelutes with authentic [$^3$H] tRA, suggesting that the mouse embryo contains significant quantities of tRA and a good deal of retinol. The peak on the extreme right of the chromatogram is an unknown, and the peak to the left of arrow 5 is butylated hydroxytoluene, the antioxidant added to the extraction solvent.

### 3.5.3. Quantitation and Identification of Peaks

The radioisotope detector will give a readout of the cpm of the tRA, which was added to the initial tissue prior to extraction. By knowing how much was added, one can thus work out a recovery ratio that will allow a correction factor to be introduced when working out the endogenous levels of retinoids. Typical values are in the 60–80% range. The UV detector will give a readout of the peak areas and the amount in nanograms for each peak it has recorded. It can do this, because when setting up the machine, one establishes a calibration curve for each retinoid, which the machine uses to give a value in nanograms for each of the UV peaks.

The identification of individual peaks can be done in several ways. First, an indication is given simply by comparing the elusion time of each peak with the known elution time of each standard. This one has done when setting up the machine to establish calibration curves as described above. Thus, in **Fig. 2A**, the elution profile of a set of retinoid standards is shown. The elution time of peak 2 (IRA) is 19.25 min. In **Fig. 2B**, the retinoids extracted from a 10.5-d mouse embryo is shown, and in this chromatograph, a peak is present at exactly the same elution time (peak 2) suggesting that tRA is present in mouse embryos. By comparing the standards with the mouse embryo extract, it can be seen seen that some peaks are present (2 and 3) and the others are absent. Second, a further identification is provided by coelution of the radioactivity in the spiked sample with an endogenous UV peak (**Fig. 2B**). This is shown here for tRA, but can clearly be done for all the other retinoids of interest, or a cocktail of radioactive retinoids can be added to the sample. Third, the sample can be spilt into two, and a spike of one cold retinoid, e.g., tRA is added to one of the samples. They are then run on the HPLC, one after the other, and in the spiked case, only one peak should increase in height, i.e., the tRA one. Fourth, the individual peaks can be collected in a fraction collector and then peaks of interest can be rechromatographed on a different system. For example, if this was done with the retinol peak obtained from the reverse-phase chromatograph in **Fig. 2B** (peak 3) and the sample was then run on a normal-phase column, this would reveal whether there really were two peaks here (retinol and 9-cis), since the elution times of each retinoid changes on a different chromatography system. Fifth, the peaks can be collected and then derivatized. This involves methylating the material in the peak, e.g., the suspected tRA and then running it on a normal-phase system to confirm that it now coeluted with authentic tRA methyl ester or performing gas chromatography/mass spectorscopy to the same end (*3*).

Amounts as low as 1 ng of tRA can readily be detected by the method described above. However, it is to be emphasized that the figures one generates after quantitation of retinoids in tissue are only estimates. In addition to the corrections described above (recovery ratio, comparison with calibration curves), there is the further problem of UV detection. The suggested wavelength of 351 nm is an average value for several retinoids. In fact, each retinoid has a different absorption maximum, e.g., all-trans-retinol = 325 nm; all-trans-retinal = 383 nm; tRA = 350 nm; 9-cis- RA = 345 nm; ddRA = 370 nm; 4-oxo-RA = 360 nm. So were the wavelength to be set differently then a different value for the quantitation of each retinoid would be obtained. The use of a scanning wavelength detector would solve this problem, but such detectors are an order of magnitue less sensitive than fixed-wavelength detectors. Nevertheless, within these limitations, good results have been obtained, particularly where differences in retinoid content between different parts of the embryo have been identified.

## References

1. Sporn, M. B. and Roberts, A. B. (1985) What is a retinoid? *Ciba Found. Symp.* **113,** 1–5.
2. Mangelsdorf, D. J., Umesono, K., and Evans, R. M. (1994) The retinoid reception, in *The Retinoids. Biology, Chemistry and Medicine,* 2nd ed. (Spore, M. B., Roberts, A. B., and Goodman D. S., eds.), Raven, New York, pp. 319–349.
3. Thaller, C. and Eichele, G. (1987) Identification and spatial distribution of retinoids in the developing chick limb bud. *Nature* **327,** 625–628.
4. Scott, W. J., Walter, R., Tzimas, G., Sass, J. O., Nau, H., and Collins, M. D. (1994) Endogenous status of retinoids and their cytosolic binding proteins in limb buds of chick vs mouse embryos. *Dev. Biol.* **165,** 397–409.
5. Horton, C. and Maden, M. (1995) Endogenous distribution of retinoids during normal development and teratogenesis in the mouse embryo. *Dev. Dynam.* **202,** 312–323.
6. Sporn, M. B., Dunlop, V. M., Newton, D. L., and Henderson, N. R. (1976) Relationships between structure and activity of retinoids. *Nature* **263,** 110–113.
7. Lee, Q. P., Juchau, M. R., and Kraft, J. C. (1991) Microinjection of cultured rat embryos: Studies with retinol, 13-*cis* and all-*trans*-retinoic acid. *Teratology* **44,** 313–323.
8. Keeble, S. and Maden, M. (1989) The relationship among retinoid structure, affinity for retinoic acid-binding protein and ability to respecify pattern in the regenerating axolotl limb. *Dev. Biol.* **132,** 26–34.
9. Summerbell, D. and Harvey, F. (1983) Vitamin A and the control of pattern in developing limbs, in *Limb Development and Regeneration,* part A. (Fallon, J. and Caplan, A., eds.), Liss, New York, pp. 109–118.
10. Cadi, R., Pautou, M.-P., and Dhouailly, D. (1984) Structure-activity relationships in the morphogenesis of cutaneous appendages in the chick embryo. *J. Investig. Dermatol.* **83,** 105–109.
11. Durston, A. J., Timmermans, J. P. M., Hage, W. J., Hendriks, H. F. J., de Vries, N. J., Heideveld, M., and Nieuwkoop, P. D. (1989) Retinoic acid causes an antero-

posterior transformation in the developing central nervous system. *Nature* **340,** 140–144.

12. Kraft, J. C., Schuh, T., Juchau, M., and Kimelman, D. (1994) The retinoid X receptor ligand, 9-*cis*-retinoic acid is a potential regulator of early Xenopus development. *Proc. Natl. Acad. Sci USA* **91,** 3067–3071.

13. Costaridis, P., Horton, C., Zeitlinger, Holder, N., and Maden, M. (1995) Endogenous retinoids in the zebrafish embryo and adult. *Dev. Dynam.* **205,** 41–54.

14. Thaller, C., Hofmann, C., and Eichele, G. (1993) 9-cis-retinoic acid, a potent inducer of digit pattern duplications in the chick wing bud. *Development* **118,** 957–965.

15. Chen, Y., Huang, L., Russo, A. F., and Solursh, M. (1992) Retinoic acid is enriched in Hensen's node and is developmentally regulated in the early chicken embryo. *Proc. Natl. Acad. Sci. USA* **89,** 10,056–10,059.

16. Scadding, S. R. and Maden, M. (1994) Retinoic acid gradients in limb regeneration. *Dev. Biol.* **162,** 608–617.

17. Heyman, R. A., Mangelsdorf, D. J., Dyck, J. A., Stein, R., Eichele, G., Evans, R. M., and Thaller, C. (1992) 9-*cis*-retinoic acid is a high affinity ligand for the retinoid X receptor. *Cell* **68,** 397–406.

18. Viviano, C. M., Horton, C., Maden, M., and Brockes, J. P. (1995). Synthesis and release of 9-*cis*-retinoic acid by the urodele wound epidermis. *Development* **121,** 3753–3762.

19. Thaller, C. and Eichele, G. (1990) Isolation of 3,4-didehydroretinoic acid, a novel morphogenetic signal in the chick wing bud. *Nature* **345,** 815–819.

20. Pijnappel, W. W. M., Hendriks, H. F. J., Folkers, G. E., van den Brink, C. E., Dekker, E. J., Edelenbosch, C., van der Saag, P. T., and Durston, A. J. (1993) The retinoid ligand 4-oxo-retinoic acid is a highly active modulator of positional information. *Nature* **366,** 340–344.

21. Kraft, J. C., Bechter, R., Lee, Q. P., and Juchau, M. R. (1992) Microinjections of cultured rat conceptuses: Studies with 4-oxo-all-trans-retinoic acid, 4-oxo-13-cis-retinoic acid and all-*trans*-retinoyl-β-glucuronide. *Teratology* **45,** 259–270.

22. Maden, M. (1983) The effect of vitamin A on the regenerating axolotl limb. *J. Embryol. Exp. Morph.* **77,** 273–295.

23. Eichele, G., Tickle, C., and Alberts, B. M. (1985) Studies on the mechanism of retinoid-induced pattern duplications in the early chick limb bud: temporal and spatial aspects. *J. Cell Biol.* **101,** 1913–1920.

24. Tamura, K., Kagechika, H., Hashimoto, Y., Shudo, K., Oshugi, K., and Ide, H. (1990) Synthetic retinoids, retinobenzoic acids, Am80, AmS80 and Ch55 regulate morphogenesis in chick limb bud. *Cell Differ. Dev.* **32,** 17–26.

25. Maden, M., Summerbell, D., Maignan, J., Darmon, M., and Shroot, B. (1991) The respecification of limb pattern by new synthetic retinoids and their interaction with cellular retinoic acid-binding protein. *Differentiation* **47,** 49–55.

26. Maden, M., Keeble, S., and Cox, R. A. (1985) The characteristics of local application of retinoic acid to the regenerating axolotl limb. *Roux's Arch. Dev. Biol.* **194,** 228–235.

27. Tickle, C., Alberts, B., Wolpert, L., and Lee, J. (1982) Local application of retinoic acid to the limb bud mimics the action of the polarizing region. *Nature* **296,** 564–565.

28. Eichele, G., Tickle, C., and Alberts, B. M. (1984) Microcontrolled release of biologically active compounds in chick embryos: beads of 200 µm diameter for the local release of retinoids. *Anal. Biochem.* **142,** 542–555.

29. Summerbell, D. (1983) The effect of local application of retinoic acid to the anterior margin of the developing chick limb. *J. Embryol. Exp. Morph.* **78,** 269–298.

30. Osmond, M. K., Butler, A. J., Voon, F. C. T., and Bellairs, R. (1991) The effects of retinoic acid on heart formation in the early chick embryo. *Development* **113,** 1405–1417.

31. Sundin, O. and Eichele, G. (1992) An early marker of axial pattern in the chick embryo and its respecification by retinoic acid. *Development* **114,** 841–852.

32. Drysdale, T. A. and Crawford, M. J. (1994) Effects of localized application of retinoic acid on *Xenopus laevis* development. *Dev. Biol.* **162,** 394–401.

33. Thoms, S. D. and Stocum, D. L. (1984) Retinoic acid-induced pattern duplication in regenerating urodele limbs. *Dev. Biol.* **103,** 319–328.

34. Koussoulakos, S., Sharma, K. K., and Anton, H. J. (1988) Vitamin A induced bilateral asymmetries in *Triturus* forelimb regenerates. *Biol. Structures Morphogenesis* **1,** 43–48.

35. Hill, J., Clarke, J. D. W., Vargesson, N., Jowett, T., and Holder, N. (1995) Exogenous retinoic acid causes specific alterations in the midbrain and hindbrain of the zebrafish embryo including positional respecification of the Mauthner neuron. *Mech. Dev.* **50,** 3–16.

36. Stainier, D. Y. R. and Fishman, M. C. (1992) Patterning the zebrafish heart tube: acquisition of anteroposterior polarity. *Dev. Biol.* **153,** 91–101.

37. Hyatt, G. A., Schmitt, E. A., Marsh-Armstrong, N. R., and Dowling, J. E. (1992) Retinoic acid induced duplication of the zebrafish retina. *Proc. Natl. Acad. Sci. USA* **89,** 8293–8297.

38. Geraudie, J., Brulfert, A., Monnot, M. J., and Ferretti, P. (1994) Teratogenic and morphogenetic effects of retinoic acid on the regenerating pectoral fin in zebrafish. *J. Exp. Zool.* **269,** 12–22.

39. Wagner, M., Han, B., and Jessell, T. M (1992) Regional differences in retinoid release from embryonic neural tissue detected by an in vitro reporter assay. *Development* **116,** 55–66.

40. Packer, L. (1990) Retinoids. Part A Molecular and metabolic aspects. *Meth. Enzymol.* **189.**

41. Cullum, M. E. and Zile, M. H. (1986) Quantitation of biological retinoid by high-pressure liquid chromatography: primary internal standardization using tritiated retinoids. *Anal. Biochem.* **153,** 23–32.

# V

## Nonvertebrate Chordates

# 34

## Protochordates

### Peter W. H. Holland and Hiroshi Wada

## 1. Introduction

The protochordates (amphioxus and tunicates) occupy a pivotal position in chordate phylogeny, being the closest living invertebrates to the vertebrates. In spite of their evolutionary significance, these animals do not feature commonly in modern developmental biology research. This has not always been the case; indeed, amphioxus ranked as one of the principal animals for embryological description in the early part of this century. The ascidia (one group of tunicates) have received intensive study as a model for determinative development, and considerable experimental and molecular data have been accumulated over the past few decades *(1)*.

The realization that many genes playing key roles in early development have been widely conserved in animal evolution has helped bring protochordates back toward the mainstream of developmental biology research. The existence of homologous control genes in divergent species is a starting point for investigating evolutionary changes in developmental control; ascidia and amphioxus are a natural choice for inclusion in such studies, since they occupy such important phylogenetic positions.

Here we give protocols for obtaining embryos and larvae of one ascidian species, *Ciona intestinalis*, and one amphioxus species, *Branchiostoma floridae*.

## 2. Materials

### 2.1. Spawning Ascidia

1. Fine scissors and watchmakers forceps.
2. Dechorionating solution: 1% sodium thioglycolate, 0.05% protease type I (Sigma) in filtered seawater; adjust to pH 10.0–11.0 using NaOH.
3. Plastic dishes coated with 1% agar (dissolved in filtered seawater).

From: *Methods in Molecular Biology, Vol. 97: Molecular Embryology: Methods and Protocols*
Edited by: P. T. Sharpe and I. Mason © Humana Press Inc., Totowa, NJ

## 2.2. Spawning Amphioxus

1. Strong shovels with 1–2 m handles.
2. 40-cm Sieves with 1 mm mesh size.
3. Electrical stimulator (e.g., Grass) fitted with platinum electrodes.

## 3. Methods

### 3.1. In Vitro Fertilization of Ciona intestinalis (Ascidia)

Many species of ascidian have been studied by developmental biologists; here we cover only *C. intestinalis*. This species is cosmopolitan and common; in the United Kingdom it is abundant in the vicinity of several major marine biology laboratories, including Plymouth, Southampton, and Millport. As with many other ascidian species, if adults are kept in constant illumination, they can be induced to spawn by moving them to the dark *(2)*; animals can also be kept in the dark, and spawned by moving to the light. An alternative procedure to obtain embryos involves in vitro fertilization using gametes obtained by dissecting gonoducts. This method is convenient, since embryos can be obtained immediately after collection of animals. They can be spawned throughout the year or throughout summer in northern Europe.

1. Adults should be kept in seawater aquaria (12–18°C or temperature at collection site), in constant dark or constant illumination to prevent spontaneous spawning.
2. Select several mature adults; these can be recognized by the white sperm duct visible through the body wall. The yellow or brown oviduct is also sometimes visible.
3. Using sharp scissors, cut the test and body wall open longitudinally. The white sperm duct is usually obvious in mature animals; the oviduct, which lies parallel, is sometimes less clear. A dissecting microscope is sometimes required to see eggs within the oviduct.
4. Carefully cut through the oviduct, and squeeze eggs out using fine forceps. Transfer eggs to a Petri dish using a Pasteur pipet. Only use eggs from the oviduct; do not dissect from the ovary.
5. Now cut through the sperm duct and squeeze out sperm in a similar manner. Transfer sperm to a separate Petri dish, before it becomes dilute.
6. Repeat **steps 3–5** for at least one further animal.
7. Add a drop of sperm from one animal to several hundred eggs collected from another. After 10 min, wash away excess sperm with several changes of filtered seawater.
8. At 16°C, first cleavage occurs after 1 h, gastrulation at 5 h, and hatching after about 24 h. *Ciona* has a nonfeeding tadpole larva that swims for 12–24 h before settling and metamorphosis.
9. If prehatching stages are required, e.g., for *in situ* hybridization to embryos, the chorion must be removed before fixation. This membrane surrounding the egg is associated with large follicle cells projecting from its outer surface and test cells on its inner surface. The chorion and associated cells can be removed manually

using sharp tungsten needles; however, this is difficult and time-consuming owing to the small perivitelline space.

10. An alternative method is to place fertilized eggs, prior to first cleavage, into dechorionating solution. Observe under a dissecting microscope until most eggs are dechorionated (10–20 min), and then gently wash with several changes of filtered seawater. This procedure can also be performed prior to fertilization.

11. After dechorionation, embryos must be cultured in agar-coated Petri dishes.

## 3.2. In Vitro Fertilization of B. floridae (Amphioxus)

Several species of amphioxus can be found in temperate and tropical seas; the protocol given is designed for *B. floridae*. These can be collected in large numbers from Old Tampa Bay, Florida, or other sandy bays around the Gulf of Mexico. They can only be spawned between July and September; even then, successful spawning is obtained only once every 10–14 d.

1. Locate a suitable population of adult amphioxus, by sieving sand from subtidal regions. Sites around Old Tampa Bay include south of the Courtney Campbell Causeway and St. Petersburg beach—in each case, in around 1 m of water.

2. After 4 PM, collect several hundred amphioxus by digging sand and sieving. Transfer animals to clean seawater for transport and keep in the shade. Collect seawater from the same site.

3. Transfer ripe animals into beakers containing 50 mL seawater (up to 5 animals/ beaker). Keep males and females separate; the serially repeated gonads are white in ripe males, and yellow in females. Keep in the light at 25°C.

4. Attempts to collect eggs and sperm should be made between 9 PM and 1 AM. Either place beakers of animals in the dark, or (more reliably) use a stimulator to give a brief (2-s) nonlethal shock of direct current (50 V in 10-ms pulses) to a beaker of animals.

5. If gametes are released from the gonopore, they should be transferred immediately to Petri dishes (eggs) or microfuge tubes (sperm) using clean Pasteur pipets. Use separate pipets for eggs and sperm, and keep sperm as concentrated as possible. Not every animal will spawn, even on a successful spawning night.

6. If no gametes are obtained within 30 min after electrical stimulation of 50–100 animals, it is likely that spawning cannot be induced on that night. In this case, return animals to the collecting site on the next day, and repeat **steps 1–5**.

7. In vitro fertilization should be set up with freshly obtained eggs and sperm. Add one drop of sperm suspension to several hundred eggs in a 9-cm Petri dish of seawater filtered through Whatman No. 1 paper. After 5 min, observation under a dissecting microscope should reveal the presence of an elevated membrane around each fertilized egg.

   Change the water twice to flush away excess sperm.

8. At 25°C, first cleavage occurs after 45 min, second cleavage after 75 min, and gastrulation after 5 h.

9. Hatching from the fertilization membrane occurs at 10–12 h; at this stage, the embryo is a bean-shaped neurula actively swimming using epidermal cilia. Pour

hatched cultures into 50-mL centrifuge tubes and direct an angle-poise lamp at the water surface for 20 min to attract swimming neurula away from debris. Transfer active neurulae from the top 10 mL water to fresh Petri dishes, using a P1000 Gilson pipetman (Gilson Co., Worthington, OH) fitted with a cut-off disposable tip.

10. Embryos are readily raised from 12 h (5 somite stage neurula) to 60 h in Petri dishes of filtered seawater, with little attention.

11. At 36 h, the larval mouth opens; feeding commences around 60 h. Beyond this stage, culture is considerably more difficult; **ref.** *2* should be consulted.

12. If embryos are to be fixed, e.g., for *in situ* hybridization, the embryo cultures may need to be concentrated prior to fixation. Pour cultures into 15-mL centrifuge tubes, spin at low speed (2000$g$) for 5 min, and dispose of all except the bottom 1 mL of water. The embryos in 1 mL seawater are then transferred to a microfuge tube and concentrated further by centrifugation at 6000$g$.

## Acknowledgments

We thank Nicholas Holland and Linda Holland (Scripp's Institution of Oceanography, San Diego, CA) for extensive and invaluable advice on collection and spawning of amphioxus, and Noriyuki Satoh, members of his laboratory (Kyoto University, Japan), and Yusuke Marikawa (University of Toronto, Canada) for advice on ascidia.

## References

1. Satoh, N. (1994) *Developmental Biology of Ascidians.* Cambridge University Press, Cambridge, UK.

2. Holland, N. D. and Holland, L. Z. (1993) Embryos and larvae of invertebrate deuterostomes, in: *Essential Developmental Biology: A Practical Approach* (Stern, C. D. and Holland, P. W. H., eds.), IRL Press at Oxford University Press, Oxford, UK, pp. 21–32.

# VI

## Retroviruses

# 35

## Gene Transfer to the Rodent Embryo by Retroviral Vectors

**Grace K. Pavlath and Marla B. Luskin**

### 1. Introduction

Over the last 15 years investigators studying vertebrate development have capitalized on the use of retroviral-mediated gene transfer to determine the lineage relationships of diverse cell types, particularly in many regions of the mammalian central nervous system (for review, *see* **ref. 1**). Whereas an intraperitoneal or intrauterine injection of cell proliferation markers such as tritiated thymidine or bromodeoxyuridine suffices to examine the birthdates of cells, many fundamental questions dealing with the formation of a structure in the mammalian embryo often necessitate performing intrauterine surgery to introduce genes by retroviral vectors. As retroviral vectors can be used not only to study lineage *(2)*, but also to introduce genes to perturb development *(3–5)*, the methods for delivering retroviruses into the developing mammalian embryo will be in increasing demand. This chapter describes a set of procedures to generate and introduce retroviral vectors into rodent embryos.

Retroviruses only integrate into the DNA of replicating cells *(6)*. As a result of this chromosomal integration, the progeny of the infected cell inherit the retroviral DNA. Recombinant retroviruses have been engineered to eliminate the viral structural genes. Thus, unlike wild-type retroviruses, these recombinant retroviruses deliver exogenous genes to target cells, but cannot replicate on their own and infect unrelated cells. These attributes make replication-defective recombinant retroviruses ideal for use as tracers of cell lineage during development. In addition, a replication-defective recombinant retrovirus encoding an exogenous marker gene is a useful tool because it can be introduced into dividing cells at virtually any stage of development and, unlike many

From: *Methods in Molecular Biology, Vol. 97: Molecular Embryology: Methods and Protocols*
Edited by: P. T. Sharpe and I. Mason © Humana Press Inc., Totowa, NJ

other cell proliferation markers, no dilution of the exogenous gene occurs as a function of increasing numbers of cell division.

To generate infectious retroviral particles for use in experiments, retroviral producing cell lines are utilized. These cell lines are made by introducing the retroviral plasmid DNA into a packaging cell line which has been engineered to synthesize all the proteins required for viral assembly (7). Packaging cell lines are either ecotropic or amphotropic depending on the type of viral envelope proteins they produce. The envelope proteins are critical for determining the host range of cells which a retrovirus can infect; ecotropic retroviruses infect only rat and mouse cells, whereas amphotropic retroviruses infect cells from a broader variety of species. To generate retroviral particles, one can either use a stable producer cell line or utilize a transient transfection technique of highly transfectable packaging cells (8). A stable producer cell line offers the simplicity of thawing identical cell aliquots, but generates infectious retroviruses with relatively low titers ($10^4$–$10^6$ infectious particles/mL). The transient technique generates short-lasting producer cells, but production of higher titers (>$10^7$ infectious particles/mL). In either case, ecotropic packaging cell lines are used for the techniques described in this chapter.

The first retroviral lineage tracers to be used contained the reporter gene *Escherichia coli* β-galactosidase (*lacZ*), whose expression was detected histochemically at the light microscopic level (9,10). The subsequent detection of the lacZ histochemical reaction product at the ultrastructural level offered several advantages (11). Among them, it could be used to determine more definitively cell phenotype. The detection of *lacZ* at the ultrastructural level in combination with the use of other antibodies further expanded the types of questions that could be addressed (12). Furthermore, as better antibodies to *lacZ* became available, the expression of *lacZ* could be routinely detected immunohistochemically at the light microscopic level (13), opening up the possibility of using it in conjunction with other antibodies in double- and triple-labeling procedures (14). To facilitate the use of retroviral vectors encoding the gene for *lacZ* to answer more questions, highlights of procedures used to detect *lacZ* histochemically and immunohistochemically, at the light and ultrastructural level, are provided. These techniques can be applied to studies in vivo as well as in vitro.

## 2. Materials

All solutions should be prepared to either tissue culture or molecular biology standards. Prepare solutions using double-distilled water unless otherwise noted. For cell culture, all solutions and materials need to be sterile. Wherever possible, reagents that have been tissue culture or molecular biology tested by the manufacturer are recommended. The use of sterile solutions when performing surgical procedures is also advantageous.

## 2.1. Production of Supernatant Containing Infectious Retroviral Particles

### 2.1.1. If Using Stable Producer Cell Line

1. Producer cell line for generation of ecotropic retrovirus encoding *lacZ*: CRE BAG 2 (American Type Culture Collection, Accession Number CRL 1858).
2. Growth media (GM, store at 4°C): The following are added to DMEM (formulation containing 4500 mg/L glucose and 2 m$M$ L-glutamine) to give the final indicated concentrations: 10% fetal bovine serum (FBS); 100 U/mL penicillin; 1000 µg/mL streptomycin.
3. Passaging cells: Phosphate-buffered saline-calcium and magnesium free (PBS-CMF), pH 7.4, Trypsin (0.05% trypsin/0.53 m$M$ EDTA).
4. Freezing media: 90% calf serum (*see* **Note 1**) and 10% dimethylsulfoxide (DMSO).

### 2.1.2. If Using Transient Producer Cells

1. Purified retroviral vector DNA: A number of retroviral vectors are available. Appropriate vectors should be chosen according to the needs of the individual investigator. These vectors differ in the promoter driving the gene of interest, and the presence or absence of drug resistance genes. Similarly, different reporter genes are available, such as *lacZ* and alkaline phosphatase. The use of a *lacZ* reporter gene is detailed here.
2. Ecotropic envelope-expressing packaging cell line: Bosc 23 (American Type Culture Collection, Accession Number CCRL 11270). For culture of these cells, the same reagents as detailed in **Subheading 2.1.1., items 2–4** are needed.
3. Transfection reagents:
   a. 25 m$M$ Chloroquine stock solution in PBS-CMF (*see* **Note 2**). Filter through a 0.2-µm disposable sterile filter unit and store at –20°C.
   b. 2X $N,N$-bis[2-Hydroxyethyl]-2-aminoethanesulfonic acid (BES): 50 m$M$ BES, 280 m$M$ NaCl, 1.5 m$M$ $Na_2HPO_4$. The final pH should be 7.0 ± 0.05. Filter through a 0.2-µm disposable sterile filter unit and store at 4°C.
   c. 2 $M$ $CaCl_2$: Filter through a 0.2-µm disposable sterile filter unit and store at –20°C.
   d. 8 mg/mL polybrene stock solution in PBS. Filter through a 0.2-µm disposable sterile filter unit and store at –20°C.

### 2.1.3. For Either Stable or Transient Methods

1. Harvest of supernatant containing retrovirus:
   a. 0.45 µm Disposable sterile syringe filters with cellulose acetate membrane.
   b. Cryogenic vials with gasket seal closure.
2. Enzymatic stain for *lacZ* activity:
   a. 0.1 $M$ sodium phosphate buffer, pH 7.2.
   b. Fixative: 4% paraformaldehyde, 0.5% glutaraldehyde in 0.1 $M$ sodium phosphate buffer, pH 7.2. For 400 mL: Heat 200 mL of 0.1 $M$ sodium phosphate buffer and 16 g of paraformaldehyde in a fume hood with stirring until dis-

solved (*see* **Note 3**). Add 8 mL of 25% glutaraldehyde stock and the remaining 192 mL of 0.1 *M* sodium phosphate buffer. Store at 4°C.

   c. 40 mg/mL 5-bromo-4-chloro-3-indolyl-β-D-galactopyranoside (X-gal) in DMSO. Store at –20°C.

   d. Staining buffer (*see* **Note 4**): 5 m$M$ K$_3$Fe(CN)$_6$; 5 m$M$ K$_4$Fe(CN)$_6$·3H$_2$O; 2 m$M$ MgCl$_2$ in PBS, pH 7.2–7.4 (*see* **Note 5**). Store at 4°C.

## *2.1.4. Test For Wild-Type Retrovirus*

Occasionally DNA recombination occurs between the replication-defective retroviral DNA construct and the genome of the packaging cell line that encodes viral proteins resulting in the generation of wild-type retrovirus. Wild-type retrovirus is capable of replicating on its own and infecting unrelated cells making lineage experiments impossible to interpret.

1. 3T3 Fibroblast cells (American Type Culture Collection, Accession Number CCL 92).
2. 3T3 Growth Media (store at 4°C): The following are added to DMEM (formulation containing 1000 mg/L glucose and 2 m$M$ L-glutamine) to give the final indicated concentrations: 10% FBS; 100 U/mL penicillin; 1000 µg/mL streptomycin.
3. 8 mg/mL Polybrene stock solution in PBS.
4. 0.45-µm Disposable sterile syringe filters with cellulose acetate membrane.

## *2.2. Injection of Retrovirus into Fetuses*

Accurately establish the gestational age of the embryonic animals to be injected with retrovirus. The question to be addressed will dictate at which gestational age the retrovirus should be injected, and will be related to the developmental time course of the organ or structure under examination. The time point under analysis should coincide with the period of cell genesis for the structure under examination since the retroviral vectors only infect dividing cells.

Pregnant dams of a known gestational age can be obtained from a supplier. Alternatively, the investigator can establish a breeding colony and determine when insemination has occurred according to routinely used procedures. By convention, the day a vaginal plug is detected is designated embryonic d 0 (E0). Gestation usually lasts 20 d for mice and 22 d for rats.

1. A long-acting anesthetic, such as chloral hydrate should be used (*see* **Note 6**). The effective intraperitoneal dose of chloral hydrate is approx 400 mg/kg for rodents. Dissolve chloral hydrate in 0.1 *M* PBS or 0.9% NaCl. A 3-cc syringe and 25 gage needle are needed to administer the chloral hydrate.
2. Electric shaver.
3. A bifurcated fiber optic light source to transilluminate the uterine swellings (*see* **Note 7**).
4. Pneumatic picopump for controlled delivery of retrovirus. There are several pneumatic picopumps available from various vendors, including one manufactured by World Precision Instruments (Sarasota, FL) (Model # PV 820) that is suitable for

injections of small volumes. The operation of a pneumatic picopump also requires a dedicated nitrogen gas tank and vacuum line.

5. Supplies for the laparoscopic surgery and intraventricular retroviral injections:
   a. 3- × 3-in. Sterile gauze pads.
   b. Sterile PBS.
   c. 5 cc Syringe.
   d. Capillary tubing (Omega dot borosilicate glass tubing, # 30-30-1, FHC, ME).
   e. Tape.
6. Surgical platform and collecting tray for supporting the experimental animal above the reservoir of irrigation fluids during surgery. Any acrylic or metal tray large enough to accommodate the mouse or rat with extended limbs and that is raised in some fashion 1–3 cm above the base can serve as a platform. The designated platform should fit inside a tray, of slightly larger dimensions than the platform, so that the irrigation fluid (PBS) will not run all over the operating surface. A shallow metal tray will work well for this purpose.
7. Surgical instruments for laparoscopic surgery (*see* **Note 8**):
   a. Small scissors.
   b. 2 Hemostats (rat surgery only).
   c. 2 Dog ear clips (mouse surgery only).
   d. Small toothed forceps.
   e. Forceps (# 5 or 3).
   f. # 10 Scalpel holder.
   g. #10 Blades.
   h. Needle holder.
   i. #5 (rat) or # 6 (mouse) suture thread and needles (Ethicon Inc., Somerville, NJ).
8. Retroviral supernatant
   a. Ice bucket filled with ice.
   b. 1% Fast green dye (F7252, Sigma, St. Louis, MO).
   c. 1 mg/mL polybrene, prepare as in **Subheading 2.1.2., item 3d**.
   d. Aliquot of frozen viral supernatant.

The viral supernatant, fast green dye, and polybrene should all be kept on ice when not in use.

9. An infrared heat lamp should be used to warm the pregnant dam while recovering from surgery (*see* **Note 9**). The infrared light source, with shield, can be bought at a hardware store, and attached on an adjustable stand, if desired.

## 2.3. Detection of lacZ Positive Cells at the Light Microscopic Level

1. Perfusion instruments:
   a. Cannula for directing the flow of fixative into the left ventricle.
   b. 2 Hemostats.
   c. Scissors.
   d. Small toothed forceps.

    e. # 10 Scalpel blade holder.

    f. # 10 Blades.

    g. Two # 5 forceps.

2. Peristaltic pump for delivery of the fixative (Masterflex Pump #7520-24, Cole-Parmer Instrument Co., Vernon Hills, IL).

3. Ether chamber: Any container with a lid large enough to accommodate the experimental animal during the anesthetization procedure and resistant to ether fumes is suitable.

4. Fixative: 4% paraformaldehyde, 0.2% glutaraldehyde in 0.1 *M* sodium phosphate buffer, pH 7.2 prepared as in **Subheading 2.1.3., item 2b**.

5. Histochemical detection of *lacZ*-positive cells: In addition to the chemicals listed in **Subheading 2.1.3., item 2**:

    a. Sodium deoxycholate.

    b. Nonidet P-40 (NP-40).

    c. 24-Well tissue culture plates.

    d. Vibratome (*see* **Note 10**).

    e. Superfrost slides (*see* **Note 11**, #12-550-15, Fisher Scientific, Pittsburgh, PA).

6. Immunohistochemical detection of *lacZ*-positive cells:

    a. Blocking serum: 10% normal goat serum, 0.2% Triton X-100 in PBS, pH 7.4.

    b. Primary antibody: mouse anti-β-galactosidase (Five Prime-Three Prime, Inc., CO), diluted 1:500 in blocking serum.

    c. Secondary antibody: Anti-mouse IgG conjugated to fluorophore, diluted in blocking serum according to manufacturer's recommendations.

7. Detection of the phenotype of *lacZ*-positive cells: The investigator can double- or triple- label tissue sections to detect the expression of *lacZ* immunohistochemically, in conjunction with other cell-type specific antibodies of choice *(14)* (*see* **Note 12**).

## *2.4. Detection of* lacZ *Positive Cells at the Ultrastructural Level*

1. Fixative: 4% paraformaldehyde, 0.5% electron microscopy grade glutaraldehyde in 0.1 *M* phosphate buffer, pH 7.2–7.4.

2. Post fixation solution: 4% paraformaldehyde, 2% glutaraldehyde in 0.1 *M* phosphate buffer, pH 7.4.

3. Staining and dehydration of tissue samples:

    a. 1% Osmium tetroxide in 0.1 *M* phosphate buffer, pH 7.4.

    b. 6- and 24-Well plastic tissue culture plates.

    c. 0.1 *M* Sodium acetate.

    d. 1% Uranyl acetate.

    e. 0.1 *N* Sodium acetate.

    f. 25, 50, 70, 95, and 100% ethyl alcohol.

4. Embedding compounds:

    a. Araldite.

    b. 3- × 2-in. glass slides.

    c. High-quality acetate paper (*see* **Note 13**).

    d. Hole-punch.

    e. Small weights.

## 3. Methods

### 3.1. Safety Procedures for Working with Retroviruses

These procedures are carried out at Biosafety Level 2. The proper institutional guidelines should be followed before using retroviruses. When handling retroviruses observe the following safety precautions:

1. Wear a lab coat and gloves. Minimize touching surfaces with gloved hands.
2. All work is performed in a biological safety cabinet.
3. If using pipets to transfer small volumes of retroviral supernatant, use disposable tips which contain aerosol resistant barriers.
4. Liquid waste is aspirated into a flask which contains approx 100 mL of liquid bleach at full strength.
5. All the solid waste which is generated (pipets, tips, gloves, dishes, and so forth) must be disinfected by bleaching or autoclaving before disposal. When beginning, put a small autoclavable bag in the hood and place solid waste into it as it is generated. When finished, tape the bag shut in the hood and autoclave.
6. Clean any spills of supernatant with 10% bleach.
7. When finished working with retroviral producer cells, wipe the biological safety cabinet and pipetor with 70% ethanol.
8. Expose the biological safety cabinet to UV light for a minimum of 30 min after working with retroviruses in the biosafety hood.

### 3.2. Production of Retroviral Supernatant Using Stable Producer Cell Line

#### 3.2.1. Growth of CRE BAG2 Producer Cells

CRE BAG 2 producer cells are grown in Growth Media (GM, *see* **Subheading 2.1.1., item 2**) in a humidified 37°C incubator containing 5–7% $CO_2$.

1. Remove a cryogenic vial from liquid nitrogen storage and rapidly thaw it in a 37°C water bath.
2. Transfer the contents of the vial to a 100-mm dish containing 10 mL of GM.
3. On the following day, aspirate the spent media and refeed the cells with 10 mL of fresh GM.
4. Feed the cultures with fresh GM every 2–3 d, subculture the cells 1:4 using trypsin/EDTA when cells reach approx 80% confluence (*see* **Note 14**).

#### 3.2.2. Subculturing CRE BAG 2 Cells

1. Aspirate the spent media and gently rinse the cells twice with PBS-CMF.
2. Add 1 mL of trypsin (0.05% trypsin/0.53 m$M$ EDTA)/100 mm dish. Incubate for a few min at room temperature until the cells round up or lift off the dish.
3. Add 9 mL of GM to the dish and gently pipet the cells to disperse.
4. Seed 1/5 of the cells into each of five 100-mm dishes in 10 mL of GM.

### 3.2.3. Collection of Supernatant Containing Retroviral Particles

1. When the cells in the desired number of 100-mm dishes are 70–80% confluent, refeed the cultures with 5 mL of GM (*see* **Note 15**).
2. Leave the GM on the CRE BAG 2 cells for 6–12 h (*see* **Note 16**).
3. Collect the supernatant, pool, filter through 0.45-μm syringe filters, aliquot into cryogenic vials at 50 μL and 1 mL each, and freeze in a liquid nitrogen cryogenic tank.

## 3.3. Production of Retroviral Supernatant Using Transient Transfection of Producer Cells

The method given is modified from that of **ref. 8**.

1. Plate $7 \times 10^6$ Bosc23 cells/100-mm tissue culture dish the day prior to transfection. Cells are grown in a humidified 37°C incubator with 5% $CO_2$.
2. On the day of transfection: Refeed Bosc23 cells with 8 mL GM + 10 μL of 25 m$M$ chloroquine (*see* **Note 17**).
3. In a 5 mL tube, mix 35–40 μg of retroviral plasmid DNA with water to give a final volume of 875 μL.
4. Add 125 μL 2 $M$ CaCl$_2$ and mix well.
5. Add 1 mL of 2X BES solution to the DNA/ CaCl$_2$ mixture dropwise with bubbling or vortexing.
6. Gently add the DNA mixture to the Bosc23 cells.
7. Return the cells to the 37°C incubator at 5% $CO_2$ for 6–12 h.
8. Aspirate the media and refeed with 27 mL of fresh GM.
9. Twenty-four hours after the start of transfection, refeed the Bosc cells with 9 mL of GM and transfer to a humidified 32°C incubator at 5% $CO_2$ (*see* **Note 18**).
10. Forty-eight hours after the start of transfection, collect the supernatant and filter through a 0.45-μm filter. Aliquot the supernatant to cryogenic vials (50 μL and 1 mL) and rapidly freeze using dry ice or liquid N$_2$. Store in a liquid nitrogen cryogenic tank.
11. Refeed the dish with 9 mL of GM and keep at 32°C.
12. Sixty hours after the start of transfection, collect the supernatant again. Filter, aliquot, and freeze as above.
13. Refeed the dish with 9 mL of GM and keep at 32°C.
14. At 72 h after the start of transfection, collect the supernatant for the last time (*see* **Note 19**). Filter, aliquot, and freeze as before.

### 3.3.1. Determining the Efficiency of Transfection if Using a lacZ Expressing Retrovirus

1. Fix the dish for 4 min in cold fixative (*see* **Note 21**).
2. Wash the dish, two times for 5 min, each in a large volume of PBS.
3. Dilute the X-gal to 1 mg/mL in staining buffer and incubate the cells for 3 h overnight at either room temperature or 37°C (*see* **Note 22**).
4. Determine the percentage of *lacZ*-positive cells.

## 3.4. Determination of Retroviral Titer

1. On the day before the retroviral infection is to be done: Seed a 6-well tissue culture plate with 3T3 fibroblasts at a density of $2 \times 10^4$ cells/well.
2. On the day of infection add 100, 200, or 400 µL of the retroviral supernatant in a final volume of 2 mL GM to different duplicate wells. Add 8 µg/mL polybrene. The control well receives 2 mL of GM + 8 µg/mL polybrene.
3. Incubate in a humidified, 5–7% $CO_2$ incubator for 2 d.
4. Fix and stain with X-gal as in **Subheading 3.3.1.**
5. Count the number of *lacZ*-positive cells in each well.
6. Plot volume vs number of *lacZ*-positive cells. The *y*-intercept yields the approximate number of infectious retroviral particles per microliter.

## 3.5. Test for Presence of Wild-Type Retrovirus

Retroviral supernatants should always be checked for the presence of wild-type retrovirus. Lineage studies can only be properly interpreted if replication-defective retroviruses are used. If replication competent wild-type virus is present, throw away all of the stored supernatant and the batch of retroviral producer cells used and obtain a new lot of cells.

1. On the day before the infection is to be done: Seed a 6-well tissue culture plate with 3T3 fibroblasts at a density of $2 \times 10^4$ cells/well.
2. On the day of infection: Infect the 3T3 cells with 2 mL of the test supernatant using 8 µg/mL polybrene (*see* **Note 23**). The negative control dish receives fresh 3T3 GM + polybrene.
3. After allowing 3–4 h for absorption, aspirate the media (*see* **Note 24**).
4. Refeed the cultures with fresh 3T3 GM + 2 µg/mL polybrene (*see* **Note 25**).
5. The day that the 3T3 cells from **step 2** become 90% confluent, seed a new 6-well plate with 3T3 fibroblasts at a density of $2 \times 10^4$ cells/well for infection the following day. Also, replace the media on the initial set of 3T3 cells infected with test supernatant with half the volume of fresh 3T3 GM.
6. Harvest the supernatant from the 3T3 cells 6–12 h later. Filter the supernatant through a 0.45 µm filter before doing the infection of the second set of 3T3 cells (*see* **Note 26**).
7. Infect the second set of 3T3 cells with 2 mL of the filtered supernatant from the original 3T3 cells using 8 µg/mL polybrene. The negative control dish receives fresh 3T3 GM+ polybrene.
8. After allowing 3–4 h for absorption, aspirate media as before.
9. Refeed cultures with fresh 3T3 media + 2 µg/mL polybrene.
10. For testing the presence of a wild-type retrovirus expressing the *lacZ* gene (*see* **Note 27**): Stain with X-gal when the 2nd batch of 3T3 cells become confluent. If the test supernatant contains only replication defective virus, there should be no *lacZ*-positive cells in the 2nd batch of 3T3 cells.

## 3.6. Intraventricular Injection of Retrovirus

The procedure for making injections of retrovirus into the telencephalic ventricles of embryonic mice and rats will be described. The basic procedures can be adapted to inject retrovirus into other structures, such as the eye *(15)* or olfactory epithelium *(16,17)*. Intraventricular injections of retrovirus can be made most readily between embryonic d 11 and 15 of mice and between embryonic d 13 and 17 of rats. Injections earlier and later than the stated times are difficult because the uterine membranes are somewhat opaque, making it difficult to resolve the enclosed embryo. Furthermore, at earlier times the embryo, as well as its brain is very small and difficult to aim for with the tip of a micropipet (*see* **Note 28**). Nevertheless, the particular question under investigation will determine the precise developmental age at which the rodent embryo is injected with recombinant retroviruses.

1. Sterilize the instruments and work area: Place the instruments in a tray or beaker with 70% alcohol for approx 30 min; subsequently allow them to air-dry on sterile gauze or paper. Clean the surgical work space with 70% alcohol and allow the surface to air-dry.
2. The use of a pneumatic picopump simplifies the procedure for delivering retrovirus into the chosen location of the developing embryo, and lateral ventricles of the forebrain in particular (*see* **Note 29**). Turn on the picopump and set the flow rates for the vacuum and the nitrogen gas tank. Attach a pipet to the electrode holder connected to the picopump. The pipet tip should be broken with forceps; the optimum diameter will need to be determined by the investigator. However, the pipet tip needs to be rigid enough to penetrate the uterine membranes and fetal skull without breaking, but sufficiently thin to prevent untoward tissue damage. Pipets that gradually taper to a sharp point usually work the best. To check the rate of intake, outflow, and hold pressure of the picopump before filling the pipet with retrovirus, first draw up colored water through pipet. Adjust rates if necessary, according to manufacturers suggestions (*see* **Note 30**).
3. Anesthetize the pregnant rat or mouse by an intraperitoneal injection of chloral hydrate (or another suitable anesthetic) (*see* **Note 31**). After the pregnant dam is completely anesthetized and no longer responds to forelimb pinch or displacement of the eyelid, immobilize the dam with limbs extended on the surgery platform. The limbs should be loosely extended and taped in a fixed position on the platform or secured in some other comparable way (*see* **Note 32**).
4. Shave the abdomen of the immobilized dam from the pubic bone to the xiphoid process (*see* **Note 33**). Clean the region of the shaved skin with 70% ethyl alcohol followed by sterile PBS. In circular movements, wipe from the center of the surgical field to the margins with both the alcohol and PBS.
5. After illuminating the abdomen with the fiber optic light, gently lift the skin at the midline of the anesthetized pregnant dam using dull or small toothed forceps. Incise the superficial skin layer of the abdomen (*see* **Note 34**), using a #10 scalpel blade, along the linea alba, which will be apparent as a longitudinally running

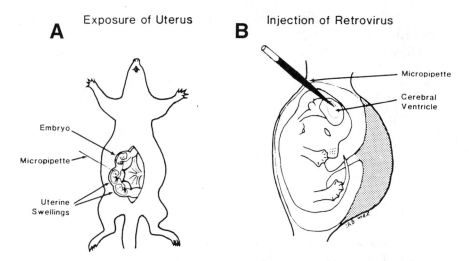

Fig. 1. Method for introducing recombinant retrovirus into the mid-gestation rat embryo. (**A**) To expose the uterine horns a midline incision is made through the skin and muscle layers of the abdomen of an anesthetized rat. In this figure three uterine swellings of the right uterine horn are diagrammed. *In utero* the orientation of the embryos is variable. (**B**) An enlargement of one uterine swelling illustrating the placement of a micropipet in the right lateral ventricle of an embryonic d 16 rat. Shaded area represents the placenta. (Reproduced with permission from Luskin, 1992.)

white line where the external and internal oblique muscles of each side of the body meet. Extend the opening anteriorly and posteriorly; do not extend the opening rostral to the rib cage or more caudal than the position of the bladder. The smaller the abdominal opening, the better.

6. Use dull or small toothed forceps to lift the exposed abdominal muscles away from body, and then with care use scissors to make a small midline, longitudinal opening in the muscle wall. Use an index finger or blunt instrument to ensure that the abdominal organs are not adhered to the muscle wall before extending the incision rostrally and caudally about the same extent as the overlying skin layer was opened.

7. Surround the entire surgical field with a bed of gauze. Open 3 × 3 in. gauze pads so that they become 3 × 6 in., and lay the strips longitudinally on both sides of the incision. Place multiple layers of gauze along the opening.

8. Constantly use sterile PBS or physiological saline to keep the area of the incision and the exposed abdomen moist. The abdominal tissue and uterine membranes must not be allowed to dry out (*see* **Note 35**).

9. Use hemostats (for rats) or dog ear clips (for mice) to retract the skin and abdominal muscle wall. The two layers can be retracted together, or separately.

10. Mice and rats have both a right and left uterine horn, each comprised of several uterine swellings, that meet at the cervix (*see* **Fig. 1**). Gently remove the uterine

horns from the abdominal cavity, and accurately count all the uterine swellings on each side (*see* **Note 36**). Dull forceps can be used to manipulate the uterine swellings. Make sure that all uterine swellings are accurately counted by identifying the rostral to the uppermost swelling, the ovary rostral to the uppermost swelling on each side. Diagram the arrangement of the uterine swellings and give each a number. By visual inspection, determine if all swellings contain a viable embryo (*see* **Note 37**).

11. Tuck the uterine swellings of one side back inside the abdominal cavity, and cover all the remaining uterine swellings with moist gauze to prevent them from becoming dry (*see* **Note 38**).

12. Thaw an aliquot of the retroviral supernatant. To every 50 µL of supernatant, add 2.5 µL of 1 mg/mL polybrene and 2.5 µL of 1% fast green dye; the final concentration of the polybrene and fast green dye is 0.5 mg/mL and 0.5%, respectively (*see* **Note 39**).

13. With careful positioning of the fiber optic light guides the uterine membranes are somewhat transparent. The embryo's body and head should be detectable. The embryonic sagittal and transverse dural sinuses demarcating the lateral ventricles should be recognizable. Using dull forceps, or some other comparable instrument, the position of the embryo and its head within the uterine swelling can be maneuvered (*see* **Note 40**). Use one motion, if possible, to penetrate the uterine membranes, skull and telencephalon (*see* **Note 41**). Once the pipet tip is in the lateral ventricle, the virus can be slowly released. The presence of the blue dye can be used to verify the placement of the pipet tip. The volume of the retrovirus to be injected will depend on the design of the experiment. Volumes of up to 5 µL can be safely injected.

14. After each embryo has been injected, replace all the uterine swellings back into the abdominal cavity. Suture closed the muscle and skin layers separately. First, close the inner muscle wall using silk suture thread; be sure that the fascial coverings of muscles do not slip into the abdominal cavity. Use interrupted sutures spaced about 1/4–3/8 in. apart. Similarly, suture closed the outer skin layer.

15. Place the pregnant mouse or rat in its cage under the infrared heat lamp. Cover half of the cage with aluminum foil to give the pregnant dam a choice of environmental temperature upon arousal and recovery; the tin foil will reflect the heat.

16. Discard all items that touched retrovirus in a dilute solution of bleach.

## 3.7. Localization of β-Galactosidase Positive Cells in the Central Nervous System

### 3.7.1. Histochemical and Immunohistochemical Detection of lacZ-Positive Cells at the Light Microscopic Level

The injected embryos can be examined for the presence and distribution of *lacZ*-positive cells at a later embryonic age or at some perinatal or postnatal time point. Since it is beyond the scope of this chapter to describe individual perfusion protocols for animals of every developmental age, a protocol that can be followed for adult animals will be given in the greatest detail.

The total number of pups should be counted after the pregnant dam has given birth to the injected embryos. Often, fewer pups are born than the number of embryos that were injected. Moreover, one can not correlate individual pups with the number they were assigned prenatally at the time of surgery unless the injection site was substantially different (i.e., right hemisphere vs left hemisphere, or big differences in the volume of retrovirus injected). If the injected animals are retrieved at a later embryonic time point, individual injected animals can be distinguished by their position along each uterine horn.

### 3.7.1.1. PERFUSION

1. Anesthetize the experimental animal by placing it in a chamber lined with gauze that has been treated with ether.
2. Extend and immobilize the limbs of the fully anesthetized animal and follow standard intracardiac perfusion procedures. Make a longitudinal incision over the sternum from neck to xiphoid process (*see* **Note 42**).
3. Lift the skin and fur over the xiphoid process and cut laterally with scissors several centimeters in both directions.
4. Grasp the xiphoid process with toothed forceps or a hemostat and then free the lower margins of the ribs from any attachments to the abdominal organs. Carefully snip the diaphragm at its midline attachment to the sternum. Free the diaphragm from the lower margin of the rib cage.
5. Free the rib cage by cutting along the axillary line on each side; begin at the lateral margins of the abdominal incision. Make sure the mediastinal structures underlying the rib cage are detached from the ribs.
6. Retract the severed rib cage; be careful not to put any tension or pressure on the structures in the neck which will interfere with the flow of fixative to the brain (*see* **Note 43**).
7. In rapid succession, snip the right atrium so the animal begins to exsanguinate and insert a cannula or syringe needle, connected to tubing of the peristaltic pump, into the left ventricle (*see* **Note 44**).
8. Turn on the peristaltic pump and perfuse the anesthetized rodent with 4% paraformaldehyde, 0.2% glutaraldehyde in 0.1 $M$ phosphate buffer (*see* **Note 45**).
9. After completion of the perfusion, remove the brain from the skull and transfer it to a container with fixative for not more than 1 h (*see* **Note 46**).
10. Section the brain or part of brain of interest on a Vibratome at a chosen thickness (e.g., usually 50–100 µm). Collect the sections in 0.1 $M$ phosphate buffer in a 24-well culture dish or some other partitioned tray.

### 3.7.1.2. FOR HISTOCHEMICAL DETECTION OF *lacZ*

1. Incubate the sections overnight at room temperature in staining buffer containing 1 mg/mL X-gal, 0.02% sodium deoxycholate and 0.01% Nonidet P-40 (NP-40)
2. Rinse the sections in 0.1 $M$ phosphate buffer and mount on Superfrost slides for examination. Alternatively, the sections can be processed for the ultrastructural

visualization of the *lacZ* histochemical reaction product (*see* **Subheading 3.7.2.**).
**Fig. 2A** demonstrates the appearance of two histochemically stained *lacZ*-positive cells.

### 3.7.1.3. For Immunohistochemical Detection of *lacZ* (After Collection of the Sections in 0.1 *M* Phosphate Buffer)

1. Incubate the sections in blocking serum for 1–2 h at room temperature.
2. Incubate the sections in the anti-β-galactosidase antibody diluted in blocking serum overnight on a rotator at 4°C.
3. Rinse the sections three times in PBS.
4. Incubate the sections in the secondary antibody conjugated to the desired fluorochrome (*see* **Note 47**) and proceed with standard immunohistochemical procedures. **Fig. 2B** demonstrates the appearance of an immunohistochemically stained *lacZ*-positive cell.

## 3.7.2. Histochemical Detection of lacZ-Positive Cells at the Ultrastructural Level

1. Day 1: As described in **Subheading 3.7.1.1.**, perfuse the experimental animals, and remove and fix the brains. Collect 100 μm tissue sections and carry out the X-gal reaction.
2. Day 2: Aspirate the X-gal reaction mixture from each well, rinse the sections with 0.1 *M* phosphate buffer and post-fix the tissue sections for a few hours in 4% paraformaldehyde/2% glutaraldehyde in 0.1 *M* phosphate buffer.
3. Rinse the tissue sections three times in 0.1 *M* phosphate buffer, incubating for 5 min each time
4. Transfer the sections from each well of the 24-well plate to a fresh well of a 6-well tissue culture plate (*see* **Note 48**).
5. Aspirate the phosphate buffer and add 0.5 mL of 1% $OsO_4$ in 0.1 *M* phosphate buffer (*see* **Note 49**) to each well (*see* **Note 50**). Place the 6-well plate containing the tissue sections on a rotator in the fume hood for 30 min.
6. Carefully remove the $OsO_4$ to a toxic waste container and add the following solutions to counterstain (uranyl acetate) and dehydrate (ethyl alcohol) the tissue sections in each well for the stated times:

    | | |
    |---|---|
    | 0.1 *N* sodium acetate | 5 min, twice |
    | 1% aqueous uranyl acetate | 30 min |
    | 0. 1 *N* sodium acetate | 5 min |
    | 25% ethyl alcohol | 2 min |
    | 50% ethyl alcohol | 2 min |
    | 70% ethyl alcohol | 10 min |
    | 95% ethyl alcohol | 15 min, twice |
    | 100% ethyl alcohol | 15 min, twice |

7. Aspirate the alcohol from the wells and replace with a 1:1 mixture of 100% ethyl alcohol:Araldite resin. Leave overnight on a rotator at room temperature.

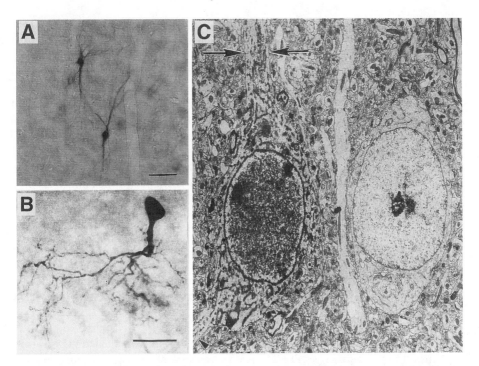

Fig. 2. Representative examples of the appearance of *lacZ*-positive cells in the central nervous system at the light and electron microscopic levels. (**A**) Representative examples of histochemically stained *lacZ*-positive neurons in the rat cerebral cortex resulting from an injection of retrovirus into the cerebral ventricles of the embryonic d 16 rat telencephalon. Note that the cell bodies and proximal dendrites are intensely stained (modified with permission from Luskin et al., 1993). (**B**) Representative example of an immunohistochemically stained neuron in the olfactory bulb resulting from a perinatal injection of retrovirus. The *lacZ*-positive cell was revealed using a primary antibody to β-galactosidase; diaminobenzidine was used as a chromogen to visualize the secondary antibody conjugated to horseradish peroxidase. In most instances immunohistochemistry can be used to label the fine processes of *lacZ*-positive cells (modified with permission from Luskin, 1993). (**C**) An electron micrograph showing a *lacZ*-positive (left) and unlabeled (right) neuron in the visual cortex from a rat that received an intraventricular injection of retrovirus at E16. The nucleus and nuclear membrane are conspicuously stained and in the cytoplasm the reaction product is associated preferentially with the endoplasmic reticulum, which extends into the apical dendrite (large arrows) and basal dendrite (arrowhead) of the labeled cell (modified with permission from Luskin et al., 1993).

8. Day 3: Remove the alcohol/Araldite mixture and replace with 100% Araldite. Change resin every 2 h so that the sections incubate in the resin at room temperature for at least 6 h.

9. Embed the tissue sections in Araldite resin as follows: Trim two acetate sheets for each slide. One piece should be the same size as the slide or slightly smaller and the other should be slightly larger.

10. Using a paper punch, make small holes in the corners of each piece of the smaller acetate sheets. These will become the bottom sheets of the microscope slide sandwich.

11. Place the lower acetate sheet on a microscope slide and place one small drop of plastic in each corner hole. These plastic drops will secure the acetate film to the slide. Place 6 small drops of embedding plastic on the acetate film (two rows of 3 drops evenly spaced) or one for each section that will be embedded. Carefully scoop the section out of the dish where it has been processed and place it on top of one of the drops of plastic. Cover each tissue section with one more small drop of embedding plastic.

12. Cover the tissue sections with a piece of the larger sheets of cut acetate film. Place weights or additional slides on top of each acetate sandwich to flatten the plastic during baking.

13. When ready to bake the tissue embedded between the acetate sheets, place them on cardboard and then place the cardboard on the oven rack for backing. This way the microscope slides will not stick to the metal oven rack if some of the plastic oozes out from between the slides.

14. Bake in the oven at 60°C 48–60 h.

15. The tissue is ready for trimming, cutting on an ultramicrotome and viewing with the electron microscope. **Fig. 2C** demonstrates the appearance of a *lacZ*-positive cell at the ultrastructural level.

## 4. Notes

1. All sera (fetal bovine, calf, horse) are suitable for use in freezing media.

2. Chloroquine increases the retroviral titer *(8)*.

3. Paraformaldehyde will dissolve in approx 20 min with heating and stirring.

4. $K_3Fe(CN)_6$ and $K_4Fe(CN)_6$ are commonly used at concentrations between 5 and 20 m$M$. The higher concentrations may result in increased staining intensity.

5. This pH range is optimal for detection of the bacterial *lacZ* activity. The use of higher pHs should be avoided so as to minimize the endogenous mammalian β-galactosidase activity.

6. The chloral hydrate should be made fresh on the day of surgery; stored chloral hydrate can be used for anesthetizing rats or mice prior to perfusion. Supplementary doses of chloral hydrate should be avoided because death of the experimental animal may occur. Before giving any additional chloral hydrate allow additional time for the chloral hydrate to achieve its effect. As an alternative to using chloral hydrate, equithesin can be used.

7. A heat-absorbing lens should be attached to the end of each fiber optic light guide to reduce the desiccation of exposed tissue, and in particular the uterine swellings, during surgery.

8. Different instruments should be used for surgical and perfusion procedures; traces of fixative may be deleterious to living tissue.

9. The experimental animals must be kept warm post surgery so that they do not succumb to the temperature lowering effects of anesthesia. Cover half of the cage with aluminum foil to reduce the temperature generated by the infrared heat source. Therefore, upon awakening from anesthesia the experimental animal can choose to remain under the heat source or to go under the aluminum foil heat shield. Typically the anesthetized pregnant dams awaken within 30–45 min.

10. As an alternative to sectioning with a Vibratome, the tissue can be sectioned on a cryostat and mounted on glass slides.

11. The use of Superfrost slides eliminates the need to sub the slides.

12. A retrovirus with a nuclear localization signal can be used instead so that *lacZ* expression is restricted to the nucleus *(18)*. This allows the cytoplasmic expression of other antigens to be detected more easily.

13. Choose the acetate sheets carefully. Baking will ruin many kinds of acetate and make then useless for embedding. Avoid acetate sheets that become opaque when baked or that stick permanently to the polymerized Araldite.

14. Growing cells to confluence results in a number of unwanted changes in the producer cells, among them aneuploidy and increased frequency of DNA recombination events.

15. The volume is decreased in order to increase the relative retroviral titer.

16. The half-life of retroviral particles at 37°C is approx 4 h.

17. Care should be taken during media changes as the Bosc23 cells are loosely adherent.

18. Decreasing the temperature to 32°C during collection of the supernatant increases the retroviral titer 5- to 15-fold *(19)*.

19. After 72 h from the start of transfection the titer of infectious retrovirus greatly drops.

20. The efficiency of transfection should approach 50%.

21. The length of fixation is critical. *LacZ* activity drops dramatically if the cells are fixed for long periods.

22. The diluted X-gal in staining buffer can be reused several times.

23. The cells should be subconfluent before infecting with virus. Retroviruses only infect replicating cells.

24. Use a new pipet to aspirate each well so as not to transfer any retrovirus into neighboring wells.

25. If the titer of wild-type retrovirus in the test supernatant is low, very few 3T3 cells will become infected. Continued growth of the cultures allows time for the wild-type retrovirus to infect other 3T3 cells in the dish increasing the likelihood of detecting its presence. Low levels of polybrene in the media help the viral infection of neighboring cells.

26. If the test supernatant contained wild-type retrovirus, then the 3T3 cultures now produce the wild-type retrovirus themselves after several days. Thus, the supernatant from the 3T3 cells is capable of infecting a naive set of 3T3 cells.

27. For testing the presence of a wild-type retrovirus expressing a neomycin resistance gene, after 2–3 d, split the second set of 3T3 cells at a 1:20 dilution into GM with 1 mg/mL G418. Change the media after 3 d and count colonies after 10 d. If the retroviral producer cells express only replication defective virus, there should be no colonies.

28. Retrovirus can also be injected into the amniotic cavity to label structures that can not be seen, but are in direct contact with the amniotic fluid, such as skin.

29. If a pneumatic picopump is not available, then alternatively, a Hamilton syringe can be loaded in the usual way, and used to deliver retrovirus.

30. The pipet tip can be backfilled with the colored water or viral supernatant. Alternatively, 5–10 µL of the viral supernatant can be deposited on a small plastic boat or piece of parafilm, and then with extreme care, the pipet tip gently placed in the center of the droplet at an angle in order to load the pipet by vacuum.

31. Work as quickly as possible once the animal has been anesthetized because the anesthesia can deleteriously lower body temperature. The investigator should strive to finish the surgery within 1.5 h.

32. The surgery platform serves two principal purposes. First, it aids in restricting the pregnant dam in a position in which her abdomen is accessible to surgery. Second, it raises the anesthetized dam above the base or tray, where the irrigation fluids will drain. This helps to keep the dam from laying in a pool of fluid, which would adversely lower the body temperature and curtail recovery.

33. While shaving the fur off the ventral surface of the abdomen, carefully shave around the nipples so that they are not injured. If the nipples are malfunctioning because of surgical trauma, the neonatal animals cannot suckle.

34. Be sure the alcohol has dried before incising the abdomen because alcohol can be irritating to an open wound.

35. If the uterine membranes dry out, they become opaque and resistant to penetration by the pipet tip. If the area of the abdominal incision dries out, a clean closure of the tissue is difficult to achieve, and subsequently the pregnant dam is more at risk of infection.

36. The number of uterine swellings present varies from animal to animal. Mice usually have between 7 and 14, whereas rats usually have between 11 and 17. Be careful not to mistake the bladder for a uterine swelling or to maneuver it any more than necessary.

37. A uterine swelling containing a dead fetus or one that is being resorbed is usually gray rather than pink, and considerably smaller than the healthy looking uterine swellings. The vascularization of the healthy uterine swellings are more prominent. Commonly one or more of the uterine swellings may contain a dead or resorbing fetus.

38. The end of the fiber optic light source can emit heat and surprisingly quickly dry out the tissues in its proximity if the tissues are not repeatedly moistened.

39. When not in use, immerse the vial of retrovirus in the ice bucket. The virus will only remain viable for a couple of hours if maintained on ice.

40. While tilting or raising the embryo within the uterine swelling, be careful not to puncture the uterine membranes or to apply excessive pressure to the embryo.

41. Do not apply excess force to enable the pipet tip to penetrate the uterine wall. If the tip will not go through easily, greater pressure will just cause the pipet tip to snap off. Instead, try to puncture the uterine wall in a slightly different place, at a slightly different angle, or with a faster motion.

42. The following procedure can be modified to perfuse injected embryos by first anesthetizing the pregnant dam and making a midline abdominal incision, similar to that done in advance of the laparoscopic surgery for retroviral injections. Expose the uterine horns. Incise the uterine membranes and remove the injected embryo by disconnecting it from placental circulation. The injected embryos can be perfused as described below or fixed by immersion in the fixative. Remove an embryo and fix it before recovering and fixing additional embryos.

43. If desired, the descending aorta can be clamped.

44. If perfusing embryos, then use a small syringe needle to deliver the fixative to the left ventricle. Butterfly needles of various gages connected to the tubing of the peristaltic pump work well to perfuse embryos as young as E12 mice and E13 rats.

45. The peristaltic pump can be turned on in advance of snipping the right atrium so that the fixative is flowing at the time the cannula is inserted into the left ventricle.

46. If the brain is left in the fixative for more than 1–1.5 h, the *lacZ* may start to decay thereby reducing the intensity of the histochemical reaction product.

47. Alternatively, a secondary antibody conjugated to an enzyme such as horseradish peroxidase or alkaline phosphatase followed by incubation in the appropriate substrate can be substituted.

48. Multiple sections can be placed in the same well.

49. Check to make sure that the osmium tetroxide is fully dissolved. Osmium crystals must be left overnight in distilled $H_2O$ to dissolve, and then mixed the following morning with 0.2 $M$ phosphate buffer to make a working solution of 1% $OsO_4$. Remember to wear gloves and work in the hood when handling osmium tetroxide.

50. Make sure the sections are flat and not overlapping. The $OsO_4$ solution will turn the tissue sections a dark color and make them somewhat brittle.

51. All experimental procedures must comply with legal restriction of the appropriate country (*see also* Notes in Chapter 4).

## Acknowledgments

We thank Dr. John Parnavelas for advice pertaining to electron microscopy. The authors' research has been supported in part by grants from the National Institutes of Health to G. K. P (#AR43410) and to M. B. L (#NS28380) and by a Biomedical Scholars Award from the Pew Charitable Trusts and a Basic Research Grant from the March of Dimes Birth Defects Foundation to M. B. L.

## References

1. McDermott, K. W. and Luskin, M. B. (1995) The use of retroviral vectors in the study of cell lineage and migration during the development of the mammalian central nervous system, in *Viral Vectors: Tools for Study and Genetic Manipulation of the Nervous System* (Kaplitt, M. G. and Loewy, A. P., eds.), Academic, New York, pp. 411–433.

2. Sanes, J. R. (1989) Analysing cell lineage with a recombinant retrovirus. *Trends Neurosci.* **12,** 21–28.

3. Compere, S. J., Baldacci, P. A., Sharpe, A. H., and Jaenisch, R. (1989) Retroviral transduction of the human c-Ha-ras-1 oncogene into midgestation mouse embryos promotes rapid epithelial hyperplasia. *Mol. Cell. Biol.* **9,** 6–14.

4. Galileo, D. S., Majors, J., Horwitz, A. F., and Sanes, J. R. (1992) Retrovirally introduced antisense integrin RNA inhibits neuroblast migration in vivo. *Neuron* **9,** 1117–1131.

5. Lillien, L. (1995) Changes in retinal cell fate induced by overexpression of EGF receptor. *Nature* **377,** 158–162.

6. Stoker, A. W. (1993) Retroviral vectors, in *Retroviral Vectors* (Davidson, A. J., and Elliot, R. M., eds.), Oxford Press, Oxford, UK, pp. 171–197.

7. Danos, O. and Mulligan, R. C. (1988) Safe and efficient generation of recombinant retroviruses with amphotropic and ecotropic host ranges. *Proc. Natl. Acad. Sci. USA* **85,** 6460–6464.

8. Pear, W. S., Nolan, G. P., Scott, M. L., and Baltimore, D. (1993) Production of high-titer helper-free retroviruses by transient transfection. *Proc. Natl. Acad. Sci. USA* **90,** 8392–8396.

9. Sanes, J. R., Rubenstein, J. L., and Nicolas, J. F. (1986) Use of a recombinant retrovirus to study post implantation cell lineage in mouse embryos. *EMBO J.* **5,** 3133–3142.

10. Price, J., Turner, D., and Cepko, C. (1987) Lineage analysis in the vertebrate nervous system by retrovirus-mediated gene transfer. *Proc. Natl. Acad. Sci. USA* **84,** 156–160.

11. Luskin, M. B., Parnavelas, J. G., and Barfield, J. A. (1993) Neurons, astrocytes, and oligodendrocytes of the rat cerebral cortex originate from separate progenitor cells: an ultrastructural analysis of clonally related cells. *J. Neurosci.* **13,** 1730–1750.

12. Mione, M. C., Danevic, C., Boardman, P., Harris, B., and Parnavelas, J. G. (1994) Lineage analysis reveals neurotransmitter (GABA or glutamate) but not calcium-binding protein homogeneity in clonally related cortical neurons. *J. Neurosci.* **14,** 107–123.

13. Luskin, M. B. (1993) Restricted proliferation and migration of postnatally generated neurons derived from the forebrain subventricular zone. *Neuron* **11,** 173–189.

14. Menezes, J. R. L., Smith, C. M., Nelson, K., and Luskin, M. B. (1995) The division of neuronal progenitor cells during migration in the neonatal mammalian forebrain. *Mol. Cell. Neurosci.* **6,** 496–508.

15. Turner, D. L. and Cepko, C. L. (1987) A common progenitor for neurons and glia persists in rat retina late in development. *Nature* **328,** 131–136.

16. Caggiano, M., Kauer, J. S., and Hunter, D. D. (1994) Globose basal cells are neuronal progenitors in the olfactory epithelium: a lineage analysis using a replication-incompetent retrovirus. *Neuron* **13,** 339–352.

17. Schwob, J. E., Huard, J. M., Luskin, M. B., and Youngentob, S. L. (1994) Retroviral lineage studies of the rat olfactory epithelium. *Chem. Senses* **19,** 671–682.

18. Bonnerot, C., Rocancourt, D., Briand, P., Grimber, G., and Nicolas, J. F. (1987) A beta-galactosidase hybrid protein targeted to nuclei as a marker for developmental studies. *Proc. Natl. Acad. Sci. USA* **84,** 6795–6799.

19. Kotani, H., Newton III, P. B., Zhang, S., Chiang, Y. L., Otto, E., Weaver, L., Blaese, R. M., Anderson, W. F., and McGarrity, G. J. (1994) Improved methods of retroviral vector transduction and production for gene therapy. *Hum. Gene Ther.* **5,** 19–28.

# 36

## Gene Transfer in Avian Embryos Using Replication-Competent Retroviruses

### Cairine Logan and Philippa Francis-West

## 1. Introduction

A series of replication-competent, avian-specific retroviral vectors known as RCAS or RCAN have been developed by Hughes et al. *(1)* and used successfully by a rapidly expanding number of groups to assess gene function directly (e.g., **refs.** *2–14*). These proviral vectors are derived from the Rous sarcoma virus and contain a unique *Cla*I restriction site in place of the region normally encoding the *src* oncogene, into which foreign DNA fragments of up to approx 2.4 kb can be inserted. An *Escherichia coli* plasmid backbone allows the gene of choice to be introduced by standard subcloning techniques, whereas retention of the viral long terminal repeat (LTR) sequences together with sequences encoding the viral *gag*, *pol*, and *env* genes facilitates viral replication and transmission. RCAN is a variant of RCAS from which the splice acceptor immediately upstream of the *Cla*I site has been removed preventing translation of the inserted gene and acts as a control for nonspecific effects owing to viral infection.

These proviral vectors can be efficiently transfected into cultured chick embryo fibroblasts (CEFs) using standard techniques, and supernatant containing infectious virions easily collected and concentrated to yield high-titer viral stocks, which can then be used to infect chick embryos of susceptible strains. Alternatively, transfected CEF cells can be grafted directly into host embryos. To increase their host range and to make it possible to introduce two vectors carrying different genes into the same cell, vectors containing different subgroup *env* genes have been constructed. The viral surface glycoprotein encoded by the *env* gene primarily determines the host range specificity of the virus, and vectors are currently available that contain *env* genes derived from subgroups A, B, D (**ref.** *1*) as well as subgroup E (**ref.** *15*).

From: *Methods in Molecular Biology, Vol. 97: Molecular Embryology: Methods and Protocols*
Edited by: P. T. Sharpe and I. Mason © Humana Press Inc., Totowa, NJ

The accessibility of the avian embryo coupled with its amenability to micro-surgical manipulation means that expression of the retrovirally introduced gene in ovo can be easily manipulated simply by varying the site and time of infection and/or limited to a specific region by transplantation of infected tissue or CEF cells into a host not susceptible to infection. Specific examples of such manipulations are to be found in the literature cited herein. Detailed protocols for the numerous steps involved in the production of high-titer viral stocks and/or transfected CEF cells that can subsequently be used to assess gene function directly within the developing chick embryo are outlined below. A protocol for the grafting of transfected cell pellets is also included. Detailed information and protocols for injection of high-titer viral stocks can be found in an excellent review by Morgan and Fekete *(16)*.

## 2. Materials

All solutions should be made to the standard required for molecular biology and/or tissue culture using appropriate molecular biology and/or tissue-culture-grade reagents and sterile distilled water. For tissue culture, all solutions should be sterilized by autoclaving (where possible), or by filtering and sterile technique used throughout.

### 2.1. Construction of Retroviral Vectors

1. Adaptor plasmids:
     pCla12Nco (contact S. Hughes; National Cancer Institute, Frederick, MD).
     pSlax12 (contact B. Morgan; Harvard Medical School and Massachusetts General Hospital, Chestnut Hill, MA).
2. Viral vectors:
     RCASBP (Subgroups A, B, and D) (contact S. Hughes; as above).
     RCANBP (Subgroups A, B) (contact S. Hughes; as above).
     RCASBP (Subgroup E) (contact D. Fekete; Department of Biology, Boston College, Chestnut Hill, MA).

### 2.2. Preparation of Primary Cultures of CEFs

1. One dozen eggs (appropriate strain), incubated for 10 d. Remember to turn the eggs 1/4 turn every 2 d to increase viability.
2. Sterilized surgical instruments (scissors, watchmaker's forceps, razor blade).
3. Sterilized 250-mL Erlenmeyer flask.
4. Sterile 10-mL wide-mouth pipets.
5. Sterile 50-mL plastic centrifuge tubes.
6. Sterile 1.5-mL cryovials.
7. 10-cm Petri and tissue-culture dishes.
8. 70% v/v Ethanol.
9. 1X Trypsin/EDTA (0.5 mg/mL trypsin, 0.2 mg/mL EDTA), available as 1X stock in Modified Puck's Saline A from Gibco-BRL (Gaithersburg, MD).

10. Fetal calf serum (FCS), available from Gibco-BRL.
11. Chicken serum (CS), available from Gibco-BRL.
12. Penicillin/streptomycin (pen/strep), available from Gibco-BRL.
13. Dulbeco's Modified Eagle's Medium (DMEM; high-glucose, L-glutamine, sodium pyruvate), available from Gibco-BRL.
14. Dimethyl sulfoxide (DMSO), available from Sigma. DMSO is an irritant requiring adequate safety precautions.

## 2.3. Transfection of Proviral Constructs

1. 6- and/or 10-cm tissue-culture dishes.
2. Sterile 4-mL polypropylene round bottom tubes.
3. TE pH 8.0 (10 m$M$ Tris-HCl, pH 8.0; 1 m$M$ EDTA).
4. 70% v/v Ethanol.
5. 2X HEPES buffered saline (HBS): 280 m$M$ NaCl, 10 m$M$ KCl, 1.5 m$M$ Na$_2$HPO$_4$, 12 m$M$ dextrose, 50 m$M$ HEPES. Adjust pH to 7.05 with 0.5 $N$ NaOH and filter-sterilize. Aliquot and store at –20°C.
6. 2 $M$ CaCl$_2$: filter-sterilize, and store in aliquots at –20°C.
7. CEF media (*see* **Subheading 3.2.**).
8. 15% (v/v) glycerol in HBS: filter-sterilized.
9. Stock solution of 2 mg/mL polybrene (Sigma, St. Louis, MO) in H$_2$O: filter-sterilize, and store in aliquots at –20°C. Polybrene is harmful by inhalation. Therefore, adequate safety precautions should be taken.

## 2.4. Collection and Concentration of Viral Stocks

1. 10-cm Tissue-culture dishes.
2. Sterile 50-mL plastic centrifuge tubes.
3. Sterile 1.5-mL microfuge tubes.
4. 40 mL polyallomer ultracentrifuge tubes (Beckman, Fullerton, CA): sterilize by rinsing well in 70% ethanol and air-drying in the tissue-culture hood.
5. Phosphate-buffered saline (PBS). PBS tablets are available from Sigma.
6. CEF media (*see* **Subheading 3.2.**).
7. Reduced serum CEF media (DMEM + 5% FCS + 1% CS).

## 2.5. Preparation of Infected Cells for Grafting

1. 250-mL tissue-culture flasks.
2. Sterile 15- or 50-mL plastic centrifuge tube.
3. Sterile 1.5-mL Eppendorf tubes.
4. Sterile scissors.
5. Sterile 3.5-cm Petri dish.
6. Sterilized tungsten needle.
7. 1X Trypsin/EDTA (0.5 mg/mL trypsin, 0.2 mg/mL EDTA). Available as 1X stock in Modified Puck's Saline A from Gibco-BRL.
8. CEF media (*see* **Subheading 3.2.**).

## 2.6. 3C2 Immunostaining of Virally Infected CEF or QT6 Cells

1. PBS: PBS tablets are available from Sigma.
2. Triton X-100: available from Sigma.
3. Goat Serum: available from Gibco-BRL.
4. 3C2 hybridoma supernatant: available from the Developmental Studies Hybridoma Bank, Department of Biological Sciences, University of Iowa, Iowa City, IA.
5. Peroxidase-conjugated antimouse IgG, IgM secondary antibody: available from Jackson Laboratories, West Grove, PA.
6. 30% v/v Hydrogen peroxide: available from Sigma. Hydrogen peroxide can cause burns and is harmful by inhalation. Therefore adequate safety precautions should be taken.
7. 3.5% w/v Paraformaldehyde: dissolve 3.5 g paraformaldehyde in 100 mL of PBS (pH 7.4) by gently heating in a fume hood. Aliquot and store at –20°C. Paraformaldehyde is extremely toxic. Therefore adequate safety precautions should be taken.
8. 0.5 mg/mL 3,3'-diaminobenzidine tetraaminobiphenyl (DAB) in PBS. DAB tablets (10 mg) are available from Sigma. Activate by adding 1 μL/mL of 30% (v/v) $H_2O_2$ to DAB just prior to use. DAB is extremely toxic. Therefore adequate safety precautions should be taken.
9. Nickel chloride: available from Sigma. Nickel chloride is a known mutagen and is extremely toxic. Therefore adequate safety precautions should be taken.

## 3. Methods

### 3.1. Construction of Retroviral Vectors

For optimum expression of the inserted gene, there are several important points to consider when cloning:

First, the level of expression of the inserted gene is not only influenced by proviral sequences within the LTR, but also by sequences within the *pol* region. The more recently developed RCAS vectors containing the *pol* region from the Bryan high-titer strain of RSV, known as RCASBP (BP stands for Bryan *pol*), express introduced genes at higher levels *(17)* than do the original RCAS vectors and may be preferable for cloning (*see also* **Note 1**).

Second, the gene of interest should be cloned into the unique *Cla*I restriction site of the RCASBP retroviral vector via an intermediate cloning step using the adapter plasmids pCla12Nco *(1)* or pSLAX12 *(16)*. pSLAX12 was made by transferring the *Cla*I to *Cla*I fragment from pCla12Nco into a pBluescript (Stratagene) vector and has the increased advantage of having a higher copy number than does pCla12Nco. In addition, commercially available T3 and T7 primers can readily be used for sequencing pSLAX12 subclones. These adapter plasmids contain several restriction sites flanked by two *Cla*I sites, and not only facilitate cloning, but also contain part of the 5'-untranslated region of the *src* oncogene, the inclusion of which has been found to enhance considerably the level of expression of the inserted gene.

Third, the coding sequences of the gene of interest should be inserted in frame with the ATG of the *Nco*I site within the adapter polylinker to avoid altering the N-terminus of the encoded protein. Inclusion of untranslated 5'- and 3'-regions of the gene of interest should be avoided, since they can adversely affect subsequent transcriptional and/or translational efficiency.

1. Appropriate insert fragments to be subcloned can be generated by either restriction digests or by PCR, and subcloned into the adapter plasmid by standard cloning procedures. When using pCla12Nco, recombinants must be identified by radioactive screening and/or PCR as *LacZ* blue-white selection is not possible. If the insert fragment is generated by PCR, restriction sites can be added to the oligos to facilitate cloning. In the latter strategy, an *Nco*I site can be conveniently introduced (if not already present) at the initiator ATG site, thus allowing the fragment to be directly cloned in frame into the *Nco*I site of the adapter polylinker. Providing that the second codon begins with a guanine, such mutagenesis will not alter the encoded sequence. If the second codon does not begin with a guanine, a more complicated cloning strategy, such as that outlined by Morgan and Fekete *(16)* is required. For efficient cutting of the PCR product, at least two bases (preferably four for *Nco*I) should be included in the oligo 5' to the restriction sites. To avoid errors in the PCR product, the oligos should be very pure and a thermostable polymerase, which has proof-reading ability (e.g., PFU polymerase, Promega), should be used. Finally, once subcloned, the sequence of the PCR product and/or inserted fragment should be thoroughly checked to ensure that no mutations have been introduced (*see* **Note 2**).
2. The *Cla*I to *Cla*I fragment from the adaptor plasmid containing the inserted coding sequences is then isolated by partial or complete restriction enzyme digestion and subcloned into the appropriate RCASBP and/or RCANBP vectors by standard cloning procedures. Again recombinants must be identified by radioactive screening and/or PCR analysis. Orientation of the insert may be determined by restriction digest analysis. Alternatively, the inserts may be sequenced using appropriate oligonucleotide primers *(16)* (*see* **Note 3**).

## 3.2. Preparation of Primary Cultures of CEFs

Primary cultures of CEFs, compatible with the retroviral subtype being used, are required for the production of high-titer viral stocks and/or infected cells for grafting. All retroviral subtypes, except E, can be grown on CEFs made from line 0 embryos, which contain no endogenous retroviruses. Line $15b_1$ embryos are susceptible to infection by E subgroup viruses (*see* **Note 4**).

All instruments should be sterilized by autoclaving (where possible), or by rinsing in alcohol and sterile technique used throughout. Wear gloves (ethanol-sterilized).

### 3.2.1. Day 1

1. Wash eggs well with 70% (v/v) ethanol.
2. Open the eggs by cutting a hole in the shell. Pluck embryos out with forceps, and place in a sterile 10-cm Petri dish.

3. Cut off and discard the head and limbs, and remove and discard the viscera.
4. Mince the torso into small fragments with a sterile razor blade.
5. Collect minced embryos using a wide-mouth 10-mL pipet. Place in a sterile 250-mL Erlenmeyer flask with 10 mL of 1X Trypsin/EDTA for every four embryos.
6. Place flask on rotator, and gently swirl for 12–15 min at room temperature. You should see small clumps disappear, but you should not wait until they all disappear or the cells will start to die.
7. Take the flask off the rotator, and allow large clumps to settle. Transfer supernatant (i.e., leave clumps behind) to a sterile 50-mL plastic centrifuge tube.
8. Add an equal volume of fetal calf serum (FCS) to inhibit trypsin. Mix gently, and let stand for approx 5 min again to allow any big clumps to settle. Decant supernatant to a new tube.
9. Spin at low speed (approx 1000–2000$g$) for 5 min at room temperature to pellet cells.
10. Discard supernatant, and resuspend pellet in 25 mL of FCS. Count cell number to obtain an approximate idea of the number of cells/mL.
11. Pellet cells again (5 min at low speed as above).
12. Discard supernatant and resuspend cell pellet in 20 mL of CEF media (DMEM + 10% v/v FCS + 2% v/v CS + 1X pen/strep).
13. Plate on 10-cm tissue-culture dishes (*see* **Note 5**) in CEF media using a range of concentrations: e.g., $10^7$ cells/dish, $3 \times 10^6$ cells/dish, $10^6$ cells/dish, and $3 \times 10^5$ cells/dish. Incubate at 37°C in an 5% $CO_2$ incubator.

### 3.2.2. Day 2

Change media to remove cell debris. It is somewhat usual to see many floating and/or dead cells at this point.

### 3.2.3. Day 3

Cells plated at the appropriate density should be reaching confluence. You now have two choices: you can freeze them immediately (P0), or pass the cells 1:3 or 1:4 and then freeze them 1–2 d later when they look almost confluent (P1). Following trypsinization, cells are frozen in 1-mL aliquots in CEF media (without the pen/strep) + 12% v/v DMSO at a ratio of one 10-cm tissue-culture dish in three 1.5-mL cryovials. The cells are frozen overnight at –70°C before transferring to liquid nitrogen for long-term storage.

## 3.3. Transfection of Proviral Constructs

Being replication-competent, virus will rapidly spread horizontally throughout the culture of CEF cells, and therefore, almost any transfection protocol will suffice. The following protocol works well and minimizes cell death.

### 3.3.1. Day 1

1. Plate cells at 10–20% confluence in a 6-cm (or 10-cm) dish. This can be achieved by splitting a newly confluent dish 1:4 or 1:5. Cells plated at this density should be approx 50–60% confluent by the following day.

2. Precipitate DNA, rinse in 70% (v/v) ethanol, air-dry in the tissue-culture hood and resuspend in sterile TE, pH 8.0, at an appropriate concentration (e.g., 0.3 µg/µL).

### 3.3.2. Day 2

1. For each 6-cm (or 10-cm) dish to be transfected, place 0.3 mL (or 0.6 mL) of 2X HBS into a sterile 4-mL polypropylene tube.
2. Add 3–6 µg (10–20 µg) plasmid DNA and mix well.
3. Add 18 µL (32 µL) of filter-sterilized 2 $M$ CaCl$_2$ dropwise while gently mixing (i.e., tapping the tube) or slowly vortexing.
4. Flick or slowly vortex tube for 20 s.
5. Let stand at room temperature for 45 min to allow a precipitate to form.
6. Remove media from each dish to be transfected by thorough aspiration. Alternatively, this media can be saved and replaced later.
7. Add DNA precipitate to the center of the dish, and rock gently by hand. Incubate at room temperature in the tissue-culture hood for 20 min (rock gently after 10 min).
8. Add 5 mL CEF media (or replace original medium), and incubate at 37°C for 3–4 h.
9. Remove media, and aspirate well. Carefully add 1 mL (2.5 mL) of sterile 15% (v/v) glycerol in HBS, and rock gently. Incubate at room temperature for exactly 90 s.
10. Wash with 5 mL of CEF media, rocking gently to mix.
11. Remove media, and repeat the same wash once or twice.
12. Remove final wash, add fresh CEF media, and incubate at 37°C.

### 3.3.3. Day 3

The media should be changed, and if appropriate, the cells should be split. The cells should be passaged as usual, making sure they are split at least every other day. For all viral subtypes, except A, 2 µg/mL of polybrene should be included in the media at this stage to enhance infection. Viral supernatant should be collected (*see* **Subheading 3.4.**) 7–10 d following transfection. Similarly, transfected cells can be used for grafting experiments (*see* **Subheading 3.5.**) after 7–10 d in culture (*see* **Note 6**).

## 3.4. Collection and Concentration of High-Titer Viral Stock

Prior to viral collection, transfected cells should be checked by immunostaining with the 3C2 monoclonal anti-GAG antibody (*18*) to ensure that the retrovirus has spread throughout the culture (*see* **Subheading 3.6.**). To optimize retroviral titers, it is important that the supernatant is collected from newly confluent or near-confluent dishes of dividing cells. Viral titers can be enhanced by minimizing both the volume and serum concentration of the media used for collection. Collection of supernatant containing infectious viral particles using reduced serum CEF media facilitates resuspension of the viral particles after concentration by centrifugation and does not adversely affect viral titer.

### 3.4.1. Collection of Viral Supernatant

1. Split transfected CEF's 1:5 from newly confluent plates into an appropriate number of 10-cm tissue-culture dishes. Incubate cells until they are 80–90% confluent (1 or 2 d).
2. Discard medium and rinse once with sterile PBS.
3. Replace with 5 mL of reduced serum CEF media containing 5% FCS and 1% CS. Do not include pen/strep or polybrene in the culture media. Incubate cells overnight.
4. Collect supernatant and put into sterile 50 mL plastic centrifuge tubes. Supernatant can be stored at –70°C at this point.
5. If the monolayer of cells is still intact, collect a second aliquot of supernatant by placing an additional 5 mL of reduced-serum CEF media on plates and incubating for a further 4–24 h (*see* **Note 7**).

### 3.4.2. Concentration of Viral Stock

1. Thaw viral supernatant on ice (or in cold water). Be careful to keep thawed supernatant on ice at all times.
2. Spin supernatant at approx 2 K for 10 min at 4°C to pellet nonviral, cellular debris.
3. Carefully decant supernatant into 40 mL polyallomer ultracentrifuge tubes (Beckmann), which have been rinsed well with 70% ethanol and left to air-dry in the tissue-culture hood.
4. Spin at 20 K for 2 h at 4°C in a SW28 rotor to pellet viral particles.
5. Pour off supernatant, and aspirate well.
6. Allow the tubes to stand upright on ice for a few minutes.
7. Carefully resuspend the pellet in the supernatant left in the tube (usually about 50–100 µL) by gently pipeting up and down. Avoid making bubbles, since this can denature the viral proteins. Resuspension may take up to 2 h if the pellet is fairly sticky. It is convenient to leave the tubes in an ice bucket in the tissue-culture hood, and pipet the suspension approx once every 15 min.
8. Pool identical samples. To avoid repeated freeze–thawing, concentrated viral supernatant should be aliquoted into an appropriate volume convenient for injections (e.g., 20 µL) and stored at –70°C (*see* **Note 8**).
9. To determine the number of infectious virions/mL, concentrated viral supernatant stocks can then be titered by infecting either CEFs or the chemically transformed quail embryonic fibroblast cell line, QT6 (*19*). Viral titers are determined by infecting cells with serial dilutions of concentrated viral supernatant. Following an 18–48 h incubation, the cells are fixed using 3.5% paraformaldehyde, and the number of infected cells is determined by immunostaining using the monoclonal 3C2 anti-GAG antibody (*18*) (*see* **Subheading 3.6.**). Titers of concentrated viral stocks should be between $5 \times 10^7$ and $10^9$ infectious virions/mL (*see* **Note 9**). Where possible, QT6 cells are used in preference to CEFs, since although they are initially infected as well as CEFs, the virus does not spread as rapidly from cell to cell, thus making it easier to identify clonal isolates. QT6 cells can be infected with viruses of all subtypes, except B and E.

## 3.5. Preparation of Infected Cells for Grafting

Transfected CEF cells expressing the gene of interest can be easily grafted into developing chick embryos (e.g., **ref. 3**). By using a host embryo from an infection-resistant strain, this technique also provides an effective method whereby infection can easily be limited to the implanted cells. Although it saves having to prepare high-titer viral stocks, infected cells must be cultured for each experiment.

1. To obtain enough cells for grafting, cells should be grown to confluence in a 250-mL tissue-culture flask. This can be achieved within 2 d after transferring cells from a confluent 10-cm tissue-culture dish.
2. Trypsinize cells and transfer to a sterile 15- or 50-mL plastic centrifuge tube. Spin at approx 3 K for 5 min. Remove supernatant, and resuspend cell pellet in 0.5 mL of CEF media.
3. Transfer the cells to a 1.5-mL Eppendorf tube, and pulse briefly (20 s) in a microfuge to pellet the cells.
4. Dislodge the cells carefully using a sterile tungsten needle, keeping the cell pellet as intact as possible (*see* **Note 10**).
5. Using a micropipet (*see* **Note 11**), transfer the cell pellet to a sterile 3.5-cm Petri dish containing 2 mL of CEF media. To avoid fragmenting the cell pellet, cut the end off the yellow or blue tip using sterile scissors. If the cell pellet does fragment while attempting to transfer it, respin and try again.
6. Place the Petri dish containing the cells in a 5% $CO_2$ tissue-culture incubator at 37°C, and allow cells to consolidate for at least 1 h.
7. For grafting, dissect the cell pellet into suitable-size pieces under a dissecting microscope. Transfer the pellet to be grafted onto the embryo (*see* **Note 12**) using a micropipet, and position pellet appropriately using sterile tungsten needles (*see* **Notes 13** and **14**).

## 3.6. 3C2 Immunostaining of Virally Infected CEF or QT6 Cells

The monoclonal antibody (MAb) 3C2 *(18)* recognizes the viral GAG protein and can be used to assess the extent of infection following transfection of line 0 CEF cells (**Subheading 3.3.**) and/or to determine the viral titer of concentrated supernatant (**Subheading 3.4.**).

1. Rinse cells twice with 1X PBS.
2. Fix for 15–30 min at room temperature with 3.5% (v/v) paraformaldehyde in PBS.
3. Rinse cells twice with 1X PBS.
4. Block 3X 30' at room temperature in CEF media containing 0.05% Triton X-100.
5. Incubate in 3C2 1° antibody (diluted 1:4 in CEF plus 0.05% Triton X-100) overnight at 4°C. The 1° antibody can be reused several times, and between uses should be kept at 4°C in the presence of 0.02% (w/v) sodium azide.
6. Rinse 3X 30' with 1X PBS containing 10% serum (e.g., goat serum).
7. Incubate in 2° antibody (peroxidase-conjugated antimouse IgG, IgM diluted 1:200 in 1X PBS plus 10% serum) for at least 1 h at room temperature.

8. Rinse 3X 15' in 1X PBS.
9. Develop with activated DAB. To enhance the signal, 0.5 mg/mL nickel chloride can be added to the DAB solution giving a black precipitate, which is easier to visualize (*see* **Fig. 1**).
10. The reaction is stopped by rinsing several times with 1X PBS.

## 4. Notes

1. The choice of viral subgroup should also be considered. Although subgroup A retroviral vectors have been routinely used for limb studies *(2,3,5–7,12–14)*, subgroup B vectors have been found to infect developing neural tissue more efficiently *(20)*.
2. Before starting the cloning procedure, it is also important to consider how expression of the inserted gene will be determined. If antibodies are not available with which to detect the encoded protein, it may be appropriate to label the protein with an expression tag, such as Myc *(21)* or FLAG *(22)*, provided such a tag does not interfere with protein function. However, if such expression tags are used, it is important first to ensure that the antibody against the tag does not crossreact with endogenous chick proteins.
3. Where possible, expression of the encoded protein should be assayed immunohistochemically via Western blots of total protein extracts and/or supernatant (if secreted) from transfected CEF cells. Protein function should also be assayed if possible. Alternatively, Northern blot analysis of mRNA from transfected CEF cells could be used to determine if the inserted sequences are being properly spliced to produce the transcript required for translation of the exogenous gene. Three alternatively spliced mRNA viral transcripts should be produced in transfected cells. It is the smallest splice variant, whose abundance should be >10% of the total viral RNA, which is used for translation of the exogenous gene. In addition, the pSLAX12 adapter plasmid containing the inserted sequences could be used in an in vitro transcription/translation assay to determine if protein of the correct size can be translated. Kits are readily available from several manufacturers. We routinely use the TNT-coupled wheat germ extract system (L4120) and tRNAnscend nonradioactive translation detection system (L5080) from Promega.
4. The lack of endogenous retroviruses makes line 0 CEFs desirable for growing viral stocks, since the possibility of recombination between the exogenous virus being introduced and endogenous viruses is minimized and viral spread can be readily determined by immunostaining using the monoclonal 3C2 anti-GAG antibody *(18)*.
5. CEF cells are very sensitive to the type of plasticware used. In our experience, they preferentially adhere to and grow best on tissue-culture dishes supplied by Griener, Corning, and/or Nunc.
6. To introduce two vectors carrying different genes into the same CEF cells, separate transfections (using vectors containing different subgroup *env* genes) should be done sequentially. CEF cells should be passaged two to three times between transfections.

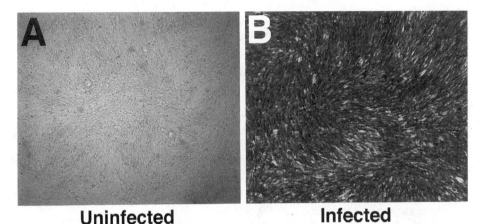

**Uninfected**                    **Infected**

Fig. 1. Immunohistochemical analysis using the 3C2 anti-GAG monoclonal antibody
(**18**) of (**A**) uninfected and (**B**) RCASBP(A)-infected CEF cells (courtesy of I. Campbell).

7. The titers obtained after 4–6 h of incubation are almost as high as those obtained
after 16 h (i.e., O/N). Therefore, multiple collections of supernatant could also be
done at 4–6 h intervals.

8. Concentrated viral stocks can maintain their titer for months and/or years when
stored at –70°C.

9. The titer of concentrated viral stocks varies considerably, depending on the virus,
confluence, and fitness of the cells used for collection and the volume of supernatant
used to resuspend the viral pellet. In general, titers below $10^8$ infectious virions/mL
do not produce high-efficiency infection and would thus not be considered useful.

10. The cell pellet should be compact. If it is too dispersed, the cells cannot be used
for grafting, since the cell pellet will disintegrate during transfer and/or subsequent manipulations.

11. Plastic disposable pipets should not be used to transfer the cell pellet, since we
have found that the pellet often sticks to the side of the pipet.

12. When adding the pellet to the embryo, it is very important to watch where the
pellet goes, since it is often invisible on the embryo.

13. Manipulations, such as the grafting of cell pellets, can sometimes delay development.

14. Although compact cell pellets can be implanted into a specific region of the
embryo, cells often disperse within 24–48 h (e.g., *see* **ref. 3**).

## References

1. Hughes, S. H., Greenhouse, J. J., Petropoulos, C. J., and Sutrave, P. (1987) Adaptor plasmids simplify the insertion of foreign DNA into helper-independent
retroviral vectors. *J. Virol.* **61**, 3004–3012.

2. Morgan, B. A., Izpisúa-Belmonte, J.-C., Duboule, D., and Tabin, C. J. (1992)
Targeted misexpression of *Hox-4.6* in the avian limb bud causes apparent
homeotic transformations. *Nature* **358**, 236–239.

3. Riddle, R. D., Johnson, R. L., Laufer, E., and Tabin, C. (1993) *Sonic hedgehog* mediates the polarizing activity of the ZPA. *Cell* **75**, 1401–1416.

4. Johnson, R. L., Laufer, E., Riddle, R. D., and Tabin, C. (1994) Ectopic expression of *Sonic hedgehog* alters dorsal-ventral patterning of somites. *Cell* **79**, 1165–1173.

5. Riddle, R. D., Ensini, M., Nelson, C., Tsuchida, T., Jessell, T. M., and Tabin, C. (1995) Induction of the LIM homeobox gene *Lmx1* by WNT7a establishes dorsoventral pattern in the vertebrate limb. *Cell* **83**, 631–640.

6. Vogel, A., Rodriguez, C., Warnken, W., and Izpisúa-Belmonte, J. C. (1995) Dorsal cell fate specified by chick *Lmx1* during vertebrate limb development. *Nature* **378**, 716–720.

7. Yang, Y. and Niswander, L. (1995) Interaction between the signalling molecules WNT7a and SHH during vertebrate limb development: dorsal signals regulate anteroposterior patterning. *Cell* **80**, 939–947.

8. Itasaki, N. and Nakamura, H. (1996) A role for gradient *en* expression in positional specification of the optic tectum. *Neuron* **16**, 55–62.

9. Logan, C., Wizenmann, A., Drescher, U., Monschau, B., Bonhoeffer, F., and Lumsden, A. (1996) Rostral optic tectum acquires caudal characteristics following ectopic *Engrailed* expression. *Curr. Biol.* **6**, 1006–1014.

10. Friedman, G. and O'Leary, D. (1996) Retroviral misexpression of *engrailed* genes in the chick optic tectum perturbs the topographic targeting of retinal axons. *J. Neurosci.* **16**, 5498–5509.

11. Yuasa, J., Hirano, S., Yamagata, M., and Noda, M. (1996) Visual projection map specified by topographic expression of transcription factors in the retina. *Nature* **382**, 632–635.

12. Rodriguez-Esteban, C., Schwabe, J. W. R., De La Peña, J., Foys, B., Eshelman, B., and Izpisúa-Belmonte, J. C. (1997) *Radical fringe* positions the apical ectodermal ridge at the dorsoventral boundary of the vertebrate limb. *Nature* **386**, 360–366.

13. Laufer, E., Dahn, R., Orozco, O. E., Yeo, C.-Y., Pisenti, J., Henrique, D., Abbott, U. K., Fallon, J. F., and Tabin, C. (1997) Expression of *Radical fringe* in limb-bud ectoderm regulates apical ectodermal ridge formation. *Nature* **386**, 366–373.

14. Logan, C., Hornbruch, A., Campbell, I., and Lumsden, A. (1997) The role of Engrailed in establishing the dorsoventral axis of the chick limb. *Development* **124**, 2317–2324.

15. Fekete, D. M. and Cepko, C. L. (1993) Replication-competent retroviral vectors encoding alkaline phosphatase reveal spatial restriction of viral gene expression/transduction in the chick embryo. *Mol. Cell. Biol.* **13**, 2604–2613.

16. Morgan, B. A. and Fekete, D. M. (1995) Manipulating gene expression with replication competent retroviruses. *Methods Cell Biol.* **51**, 185–218.

17. Petropoulos, C. J. and Hughes, S. H. (1991) Replication-competent retrovirus vectors for the transfer and expression of gene cassettes in avian cells. *J. Virol.* **65**, 3728–3737.

18. Potts, W. M., Olsen, M., Boettiger, D., and Vogt, V. M. (1987) Epitope mapping of monoclonal antibodies to *gag* protein p19 of avian sarcoma and leukaemia viruses. *J. Gen. Virol.* **68**, 3177–3182.

19. Moscovici, C., Moscovici, M. G., Jimenez, H., Lai, M. M., Hayman, M. J., and Vogt, P. K. (1977) Continuous tissue culture cell lines derived from chemically induced tumors of Japanese quail. *Cell* **11**, 95–103.
20. Homburger, S. A. and Fekete, D. M. (1996) High efficiency gene transfer into the embryonic chicken CNS using B-subgroup retroviruses. *Dev. Dyn.* **206**, 112–120.
21. Evan, G. I., Lewis, G. K., Ramsay, G., and Bishop, J. M. (1985) Isolation of monoclonal antibodies specific for the human c-myc oncogene product. *Mol. Cell Biol.* **75**, 3610–3616.
22. Hopp, T. P., Prickett, K. S., Price, V., Libby, R. T., March, C. J., Cerretti, P., Urdal, D. L., and Conlon, P. J. (1988) A short polypeptide marker sequence useful for recombinant protein identification and purification. *Biotechnology* **6**, 1205–1210.

# VII

# MOLECULAR TECHNIQUES

# 37

## Subtractive Hybridization and Construction of cDNA Libraries

**Bruce Blumberg and Juan Carlos Izpisúa Belmonte**

### 1. Introduction

Genes that are differentially expressed both in time and space are the basis for how single cells, through the process of embryonic development, give rise to animals with an extraordinary diversity of cell types. As a first step in understanding differential gene expression, many researchers seek to identify those genes whose transcripts are temporally or spatially restricted to particular cells, tissues, or embryonic stages. Although there are a variety of methods suitable for identifying moderately to highly expressed genes, the isolation of the most interesting class of mRNAs, those that are not abundant, but that may be cell- or tissue-specific, remains the most difficult task.

Several basic types of methods have been employed to identify low-abundance, tissue-specific transcripts. The more classical differential hybridization techniques (e.g., *1*) are mostly limited to the detection of moderately abundant transcripts representing >0.05% of the mRNA population *(2)*. Subtractive hybridization techniques can increase the detection sensitivity by 10- to 100-fold and make the identification of quite rare genes possible *(2)*. A specialized form of subtractive hybridization, the "Gene Expression Screen" *(3)*, can detect both upregulated and downregulated transcripts. A number of protocols have been devised in recent years to simplify and expedite the process of transcript identification by subtractive hybridization *(4–8)*. Here we present a comprehensive set of methods that have proven quite successful in our laboratories and that may serve as an entry point for future refinement.

In the following protocol, we utilize the commonly available and widely used phage vector λZAPII, since one can produce oriented libraries in phage or

From: *Methods in Molecular Biology, Vol. 97: Molecular Embryology: Methods and Protocols*
Edited by: P. T. Sharpe and I. Mason © Humana Press Inc., Totowa, NJ

phagemids, and subsequently utilize the libraries to produce essentially unlimited quantities of sense or antisense RNAs for subtraction and screening. Furthermore, Stratagene (San Diego, CA) and other suppliers provide a number of high-quality, premade cDNA libraries that can save considerable time if one happens to exist for the tissue of interest. Recent advances in vector technology and hybridization allow one to produce subtracted probes and libraries even when starting material is quite limited.

## 2. Materials

### 2.1. cDNA Synthesis

1. 10X First-strand buffer: 0.5 $M$ Tris-HCl, pH 8.9 (this will be 8.3 at 42°C), 0.1 $M$ MgCl$_2$, 0.2 $M$ KCl, 50 m$M$ dithiothreitol (DTT).
2. 10X Second-strand buffer: 0.2 $M$ Tris-HCl, pH 7.5, 0.05 $M$ MgCl$_2$, 0.1 $M$ (NH$_4$)$_2$SO$_4$, 1.0 $M$ KCl.
3. 10X *Eco*RI methylase buffer: 1.0 $M$ Tris-HCl, pH 8.0, 1.0 $M$ NaCl, 0.01 $M$ EDTA. Make as a stock solution and then add S-adenosyl methionine to 0.8 m$M$ to a small aliquot of this stock before use.
4. 10X T4 polynucleotide kinase buffer: 0.5 $M$ Tris-HCl, pH 8.0, 0.1 $M$ MgCl$_2$, 50 m$M$ DTT.
5. 10X T4 ligase buffer: 0.3 $M$ Tris-HCl, pH 7.4, 0.04 $M$ MgCl$_2$, 0.1 $M$ DTT, 2 m$M$ ATP. Make ~10 mL of this buffer and store in 100-µL aliquots at –20°C. Repeated freeze-thaw cycles rapidly deplete the buffer of both DTT and ATP.
6. dNTP mix with 5-methyl-dCTP: 20 m$M$ dATP, 20 m$M$ 5-methyl-dCTP, 20 m$M$ dGTP, 20 m$M$ dTTP.
7. dNTP mix: 20 m$M$ dATP, 20 m$M$ dCTP, 20 m$M$ dGTP, 20 m$M$ dTTP.
8. 10X *Eco*RI buffer—supplied by the manufacturer: 0.5 $M$ Tris-HCl, pH 7.5, 0.1 $M$ MgCl$_2$, 1.0 $M$ NaCl, 10 m$M$ DTT.
9. Column buffer: 10 m$M$ Tris-HCl, pH 8.0, 1 m$M$ EDTA, 0.15 $M$ NaCl.
10. TBE for electrophoresis (10X): 0.89 $M$ Tris base (Boehringer Mannheim [BMB], Indianapolis, IN), 0.89 $M$ boric acid, 0.02 $M$ EDTA. Filter and autoclave (prevents precipitate from forming with time).

### 2.2. Enzymes for cDNA Library Construction

1. AMV reverse transcriptase (Seikagaku America #120248-1 [St. Petersburg, FL]).
2. DNA polymerase I, endonuclease-free (BMB # 642-711).
3. RNase H (Pharmacia #27-0894-02 [Milwaukee, WI]).
4. T4 DNA polymerase (Pharmacia #27-0718-02).
5. T4 polynucleotide kinase (Pharmacia #27-0734-01).
6. T4 DNA ligase (Pharmacia #27-0870-01).
7. T4 RNA ligase (Pharmacia #27-0883-01).
8. *Escherichia coli* DNA ligase (BMB #862 509).
9. *Eco*RI methylase (New England Biolabs, NEB #211S [Tozier, MA], S-adenosyl methionine is supplied with the enzyme.

10. Human placental ribonuclease inhibitor (Pharmacia RNA guard #27-0815-01).
11. *Eco*RI high concentration (BMB #200-310).
12. *Xho*I—high concentration (BMB #703-788).

## 2.3. Vectors and Packaging Extracts

1. UniZAP-XR—ZAPII, digested with *Xho*I and *Eco*RI and dephosphorylated (Stratagene #237211).
2. *E. coli* (ER-1647 NEB #401-N).
3. Gigapack II Gold packaging extract (Stratagene #200216).

## 2.4. Miscellaneous Reagents and Supplies for cDNA Synthesis

1. A primer consisting of ~15 Ts followed by at least an *Xho*I site and preferably several other rare sites. The one we use is 5'(ACTAGTGCGGCCGCCTAG GCCTCGAGTTTTTTTTTTTTTTTT)3'.
   This has the following restriction sites (in 5' -> 3' order): *Spe*I, *Not*I, *Eag*I, *Sfi*I, *Avr*II, *Stu*I, *Xho*I.
2. *Eco*RI linkers—octamer (GGAATTCC) (Pharmacia 5'-OH, #27-7726-01, 5'-PO$_4$, #27-7428-01).
3. 5-methyl-dCTP (Pharmacia #27-4225-01). Store as a 100-m$M$ stock.
4. dNTPs (Pharmacia #27-2035-01). 100 m$M$ solutions of each.
5. 10 m$M$ ATP (Pharmacia #27-2056-01) (dilute from 100 mM stock).
6. 0.1 $M$ CH$_3$HgOH (ALFA products #89691).
7. 5.8 $M$ 2-mercaptoethanol (BME; Sigma M6250 [St. Louis, MO]).
8. 100% Ethanol (Rossvile Gold Shield or equivalent).
9. $\alpha[^{32}$P]-dATP 800 Ci/m$M$ (New England Nuclear, Wilmington, DE, #NEG-012A).
10. $\gamma[^{32}$P]ATP 6000 Ci/m$M$ (New England Nuclear #NEG-002Z).
11. 1-kb Ladder (BRL #5615SA).
12. Agarose (Pharmacia #17-0554-02 or Bio-Rad, Hercules, CA, #162-0126).
13. 40% acrylamide (19:1 acrylamide:bisacrylamide) acrylamide (Bio-Rad #161-0101), bis-acrylamide (Bio-Rad #161-0201).
14. β-NAD (BMB #775-7). The stock solution is 0.045 $M$ in H$_2$O.
15. Sepharose Cl-4B (Pharmacia #17-0150-01) equilibrated in column buffer.
16. Tris-base (BMB #604-205).
17. X-gal (BMB #745-710).
18. IPTG (BMB #724-815).
19. Ultrapure, recrystallized phenol (BMB #100-300).
20. DTT (BMB #100-032).
21. Sodium dodecyl sulfate (SDS) (BMB #100-155).
22. Guanidine HCl (optional) (BMB #100-173 or BRL #5502UA).
23. Guanidine thiocyanate (BMB #100-175 or BRL #5535UA).
24. Ultrapure urea (BMB #100-164).
25. LiCl (Sigma L-0505).
26. Sephadex G-50 spun columns, equilibrated in TE Sephadex G-50 medium (Pharmacia #17-0043-01), or Sephadex G-50 spun columns (Pharmacia #17-0855-01).

27. LB media and plates.
28. Minimal media and plates (use maltose as carbon source).
29. 20% Maltose (Difco).
30. All other chemicals and reagents should be at least ACS-reagent grade. You cannot go wrong by buying small quantities of ultrapure chemicals (e.g., from Aldrich) and reserving them for making cDNA buffers and solutions.

## 2.5. Reagents for Biotinylation and Subtraction

1. Photobiotin acetate (Clontech K1012-1 [Palo Alto, CA]).
2. Reflector sunlamp (Clontech 1131-3).
3. Biotin-21-UTP (Clontech 5024-1).
4. Streptavidin (BRL Life Technologies #15532-013 [Gaithersburg, MD]).

## 2.6. Reagents for Isolation of Total RNA

1. Guanidine thiocyanate solution (GuSCN): 4.0 $M$ guanidine thiocyanate (BMB #100-175), 0.01 $M$ Na-acetate, pH 7.0, 0.1 $M$ 2-mercaptoethanol (Sigma), 0.1% (w/v) $n$-lauryl sarcosine (Sigma), 0.5% (v/v) antifoam C (Sigma).
2. CsCl-EDTA cushion: 6.0 $M$ CsCl (BMB), 0.1 $M$ EDTA, pH 7.0.
3. Phenol:chloroform:isoamyl alcohol: 25 parts ultrapure phenol (BMB #100-300), 24 parts chloroform (Fisher ACS grade, Fisher Scientific, Pittsburgh, PA), 1 part isoamyl alcohol (Fisher ACS grade). Prepared as described in **ref. 9**.
4. Chloroform:isoamyl alcohol: 24 parts chloroform, 1 part isoamyl alcohol, store at room temperature in a dark bottle.
5. Ethanol-sodium acetate: 0.04 $M$ Na-acetate, 60% v/v ethanol, prepare by mixing 3 parts of 80% ethanol with 1 part of 0.15 $M$ Na-acetate, pH 7.0.
6. Ammonium acetate for precipitations: 7.5 $M$ NH4-acetate pH 7.0—treat with 0.2% DEPC and autoclave. Store at –20°C.

## 2.7. Reagents for Poly A⁺ Selection

1. 2X Loading buffer: 40 m$M$ Tris-HCl, pH 7.6, 1 $M$ NaCl, 2 m$M$ EDTA, 0.2% SDS.
2. 0.1 $N$ NaOH.
3. Loading buffer (low-salt): 40 m$M$ Tris-HCl, pH 7.6, 0.1 $M$ NaCl, 2 m$M$ EDTA, 0.2% SDS.
4. Elution buffer: 10 m$M$ Tris-HCl, pH 7.5, 1 m$M$ EDTA, 0.05% SDS.
5. 3 $M$ sodium acetate, pH 5.2.
6. 100% Ethanol (reserved for RNA use).
7. 70% Ethanol (reserved for RNA use).
8. Oligo (dT) cellulose, Type 2 (Collaborative Research, Los Altos, CA, cat. # 20002).
9. Quik-Sep columns (Isolab, Akron, OH, cat. # QS-Q).

## 2.8. Reagents for In Vitro Transcription (Optional)

1. Megascript T7 kit (Ambion, Austin, TX, #1334).
2. Megascript T3 kit (Ambion #1338).

## 3. Methods

Whether to produce subtracted cDNA libraries or screen standard libraries with subtracted probes is the first consideration one must address. We advocate the construction of representative cDNA libraries that can later be screened with any desired probe. Indeed, such libraries can also be used to produce either synthetic driver or target mRNAs in large quantitites. This approach has the advantage that only one or two libraries must be constructed for each target/driver pair, and any type of probe can be used as may later be required. It has the disadvantage that more clones must be screened to ensure the representation of the rarest mRNAs.

The choice of subtraction protocol to be followed depends on the availability of mRNA from both target and driver cells or tissues. If the driver mRNA is not limiting (~50 µg available) then one can begin with **Subheading 3.7.**, photobiotinylation of driver mRNA, otherwise one should first construct a library which may then be used to generate driver RNA. Similarly, the abundance of target mRNA and the desire to produce a representative or subtracted library will dictate whether the target will be oligodT-primed first strand cDNA or random primed cDNA produced after in vitro transcription of the library (**Fig. 1**).

### 3.1. Preparation of Total RNA (Based on Ref. 10)

1. There are a number of quite good procedures for preparing high-quality RNA. In addition, several commercial kits are available for this purpose that work reasonably well, but are rather expensive. The method appropriate for your cells or tissue depends largely on the amount of endogenous RNases present. The following method is somewhat tedious but always gives high-quality RNA.
2. Homogenization—the tissue is homogenized in 5 vol of ice-cold GuSCN solution on ice with two 30-s bursts of the polytron at maximum speed. Be sure that the tissue is completely dispersed. The size of polytron generator used depends on the quantity of tissue, but for most cases, the 1-cm type is adequate. For hard tissues, such as bone, a generator with blades should be used. Cultured cells are first trypsinized, rinsed with PBS, and resuspended in a minimum volume of PBS. Each 1 mL of densely suspended cells is homogenized in ~10 mL of ice-cold GuSCN solution on ice as above. It is convenient to use sterile, disposable 50-mL polypropylene tubes for both homogenization and extractions.
3. Extract the homogenate with phenol:chloroform once and centrifuge to separate the layers for 5 min at 3000$g$. The upper, aqueous layer will appear somewhat milky, and is removed to a fresh tube. If you just use phenol instead of phenol chloroform, there will be no phase separation.
4. Extract the aqueous layer once with an equal volume of chloroform:isoamyl alcohol and centrifuge as above.
5. Centrifugation—the homogenate is layered onto 2.2-mL cushions of CsCl-EDTA solution prepared in SW-41 tubes. Centrifugation is for 18 h at 28,000 rpm at 20°C in an SW-41 rotor.

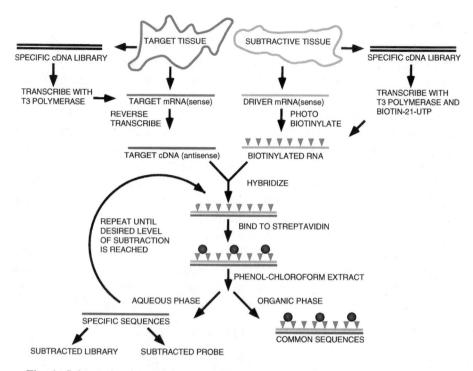

Fig. 1. Schematic view of the steps in constructing a subtracted cDNA library or subtracted probe. (*See* color plate 1 appearing after p. 368.)

If an SW-28 or SW-27 rotor is used, then the centrifugation speed should be decreased to 22,000 rpm. If other rotors are used, the volume of the cushion should be one-fifth of the total tube volume. Be sure to check the rotor manual to determine the maximum rotor speed with concentrated CsCl solutions.

6. Collection of the RNA—the tubes are removed from the centrifuge and placed in a rack.
7. The GuSCN layer is aspirated down to and including ~0.5 mL of the cushion. The tube is carefully filled with DEPC containing $H_2O$, allowed to stand 2–5 min, and then the water is aspirated. This is repeated twice.
8. After the final rinse, the tube is quickly inverted and allowed to drain. The bottom 2 cm of the tube are cut off with a fresh scalpel, and the tube placed upright in a rack. The RNA appears as a glistening button in the center of the tube. Small amounts of RNA may not be visible. The pellet is carefully rinsed with 0.5 mL of DEPC-treated $H_2O$, and the tube inverted to dry.
9. The pellet is then macerated with, and taken up into a tip containing ~100 μL of DEPC-treated $H_2O$ (or an appropriate amount for the RNA yield you end up with). The RNA is pipeted up and down and transferred to an Eppendorf tube. Be sure you transfer all the RNA. The RNA is then heated at 70°C for 2–30 min until dissolved. Large amounts of RNA need longer times and require more $H_2O$. The

insoluble debris is removed by brief centrifugation, and the supernatant removed to a fresh Eppendorf tube.

10. Add 0.5 vol of $NH_4$-acetate and 2 vol (i.e., 2X the volume of the RNA + acetate solution) of absolute ethanol. Precipitate for 15 min at –70°C or several hours to overnight at –20°C (preferred method). Centrifuge for 15–30 min at 4°C, and then drain the supernatant. Rinse the pellet three times with the ethanol-sodium acetate solution, centrifuging briefly between rinses.
11. Rinse the pellet once with 80% ethanol, once with 100% ethanol, and air-dry. Resuspend the RNA in an appropriate volume of $H_2O$, and store at –70°C.

## 3.2. Evaluation of RNA Quality

1. Quantitate the RNA by spectrophotometry. $OD_{260:280}$ ratios should be 1.8 or greater, although tissues containing significant quantities of proteoglycans may give lower ratios. If the ratio is too low, phenol-extract, chloroform-extract, and ethanol-precipitate the RNA one or more times.
2. Evaluate the integrity of the RNA by gel electrophoresis, either denaturing or nondenaturing.
3. Nondenaturing agarose gels: The simplest method is to run a 0.8% agarose gel containing 100 µg/mL ethidium bromide (EtBr), using TBE as the running buffer, just as for a normal DNA gel.
4. Denaturing gel—use a 0.8% formaldehyde-agarose gel run in MOPS-acetate buffer as described in ref. *(9)*. Stain with ethidium bromide for 30', and view under UV.
5. The 28S and 18S rRNA bands should be sharp and in a 2:1 ratio. Some organisms, e.g., *Drosophila melanogaster* and *Alligator mississipiensis*, have 28S rRNAs, which are nicked and migrate with the 18S bands in denaturing gels. Use nondenaturing gels with these organisms and if you do not see the 28S band in denaturing gels.
6. Storage of RNA—store at –70°C in $H_2O$ or as an ethanol precipitate at –20°C (preferred).

## 3.3. Preparation of Poly A⁺ RNA

1. Once the isolation of total RNA is complete, isolation of mRNA can be performed by affinity chromatography on oligo (dT) cellulose, since the vast majority of mRNAs of mammalian cells carry tracts of poly (A) at their 3'-termini. Several companies make kits for preparing poly A⁺ RNA. We have had success with those by Qiagen (75022 or 70042 [Los Angeles, CA]) and Invitrogen (K1520-02 [San Diego, CA]). We also describe here a reliable protocol that we have been routinely using before changing to the commercially available kit.
2. Equilibrate the oligo (dT) cellulose in sterile loading buffer. Binding affinity is batch-specific. However, a good rule of thumb is that 1 g of resin will bind 2–2.5 mg of poly (A)+ RNA.
3. Pour 1.0 mL packed volume of oligo (dT) cellulose in the Quik-Sep column.

4. Wash the column with, successively, three column volumes of sterile DEPC-treated $H_2O$, 0.1 $N$ NaOH, sterile DEPC-treated $H_2O$. Continue the last wash step until the pH of the column effluent is <8.0.

5. Wash the column with 5 vol of sterile loading buffer.

6. Dissolve the total RNA in sterile water. Heat to 65°C for 5 min. Add an equal amount of 2X loading buffer, and cool the sample to room temperature.

7. Apply the sample to the column, and allow to flow through by gravity. Collect the flow-through.

8. Heat the flow-through to 65°C, cool, and reapply to the column as above.

9. Wash the column with 5–10 column volumes of loading buffer, followed by four column volumes of low-salt loading buffer. The first RNA to elute off the column will be the poly (A)– fraction. The poly (A)+ fraction will elute with the no-salt elution buffer.

10. Elute the poly (A)+ RNA with two to three column volumes of sterile elution buffer. The eluted poly (A)+ RNA can be selected again on oligo (dT)-cellulose by adjusting the NaCl concentration of the eluted RNA to 0.5 $M$ and repeating **steps 6–10**. We generally elute into Falcon 2059 polypropylene tubes (17 × 100-mm).

11. Add sodium acetate (3 $M$, pH 5.2) to a final concentration of 0.3 $M$. Precipitate the RNA with 2.5 vol of 100% ethanol at –20°C overnight.

12. Spin the precipitated RNA at 10,000 rpm (12,000$g$) for 30 min. Carefully aspirate the supernatant, rinse the pellet with 70% EtOH, and air-dry. Resuspend the RNA pellet in an appropriate volume of DEPC-treated $H_2O$. Determine the RNA concentration spectrophotometrically. Reprecipitate and store as an ethanol precipitate at –20°C. RNA is most stable when stored under ethanol at –70°C.

## *3.4. Preparation of a Directional cDNA Library (Ref. 11)*

### *3.4.1. First-Strand Synthesis*

1. Use 20–400 µg of total cellular RNA (150 µg seems sufficient) or 1–5 µg of poly (A)+ RNA. Precipitate the RNA if necessary such that the final volume of the RNA will be 28 µL.

2. Denature the RNA by adding 5 µL of 0.1 $M$ methylmercuric hydroxide (*see* **Note 1**) and incubating for 5 min at room temperature. Quench the reaction with 0.5 µL of 5.8 $M$ 2-mercaptoethanol, and incubate for 5 min at room temperature.

3. The reaction mixture is then assembled as follows:

   33.5 µL RNA mixture (from **Subheading 3.4.1., step 2**);
   6.0 µL 10X first strand buffer;
   6.0 µL 20 m$M$ dNTPs containing 5-methyl-dCTP;
   3.0 µL 80 m$M$ Na-pyrophosphate;
   2.0 µL primer (5 mg/mL);
   2.0 µL $^{32}$P-dATP;
   3.0 µL placental RNAse inhibitor (30–100 U);
   5.0 µL AMV reverse transcriptase (50–100 U);

   Incubate at 42°C for 1 h. Add 3 µL reverse transcriptase and incubate for 1 h more.

4. Determine the percent incorporation by TCA precipitation. Remove 2 µL of the reaction mix. Add 8 µL of DEPC-treated $H_2O$, and spot 5 µL onto a Whatman GFC filter and reserve it. To the other 5 µL, add 25 µL of 2 mg/mL bovine serum albumin (BSA) and 100 µL of 20% trichloroacetic acid (TCA). Incubate on ice for 30 min, and then filter through GFC in a vacuum filtration device. Wash with 20 mL of 5% TCA, and then dry the filter for 20 min at 60°C or under a heat lamp. Transfer both filters to scintillation vials, add scintillation fluid, and count. Determine the percent incorporation of the trace label into cDNA. The yield in nanograms of cDNA is incorporation ×120 (nmol each nucleotide) ×4 (nucleotides) ×330 (g/mol of nucleotide).

5. Dilute the cDNA to 150 µL and load onto a 1 mL Sephadex G-50 spun column in TE. Spin for 3 min at 1000 rpm, and collect the flowthrough. Measure the volume by weighing the liquid and assuming 1 g/mL for $H_2O$. This step is necessary to remove 5-methyl-dCTP. Any remaining will be incorporated into the second strand and prevent cleavage at the 3' *Xho*I site.

### 3.4.2. Second-Strand Synthesis

1. The reaction mix is assembled as follows:
   - xx  µL first-strand reaction (~125–150 µL);
   - 30  µL 10X second-strand buffer;
   - 30  µL 20 m*M* dNTPs;
   - 3  µL RNase H (~2.5 U);
   - 1  µL 45 m*M* β-NAD;
   - 1  µL *E. coli* DNA ligase (~250 ng);
   - 12  µL DNA polymerase I (~80 U);
   - xxx  µL $H_2O$—to a final volume of 300 µL;
2. Incubate overnight at 14°C.
3. Heat at 80°C for 15 min, cool on ice, and spin briefly to collect the liquid.
4. Add ~20 U of T4 DNA polymerase, and incubate 1 h at 37°C (*see* **Note 2**).
5. Phenol extract, ethanol precipitate, rinse, and dry.

### 3.4.3. EcoRI Methylase Treatment

1. Resuspend the cDNA in 85 µL of $H_2O$. The reaction mixture is as follows:
   - 85 µL cDNA;
   - 11 µL 10X methylase buffer;
   - 10 µL *Eco*RI methylase (100–400 U).

   Incubate for 30 min at 37°C. Add 4 µL more methylase and incubate for 30 min longer.
2. Heat 80°C for 15 min, then cool on ice
3. Add 1 µL of 20 m*M* dNTP's and 1 U of T4 DNA polymerase. Incubate for 30 min at 37°C (*see* **Note 3**).
4. Phenol-extract, ethanol-precipitate, rinse, and dry the cDNA.

### 3.4.4. Kinase Linkers

1. 0.8 µg of [32]P-labeled linkers is needed for each cDNA preparation (*see* **Note 4**). An example reaction is the following:

8 μL linkers = 4 μg;
2 μL 10 kinase buffer;
1 μL α-$^{32}$P-ATP;
1 μL T4 polynucleotide kinase (~5 U);
8 μL H$_2$O.
Incubate for 15 min at 37°C.

2. Add:
1 μL 10X kinase buffer;
4 μL 10 m$M$ ATP;
1 μL T4 polynucleotide kinase;
4 μL ddH$_2$O.
Incubate for 45 min at 37°C.

3. Heat-kill the enzyme at 80°C for 15 min, and then cool on ice.

### 3.4.5. Linker Ligation (see **Note 5**)

1. Resuspend cDNA in 27 μL ddH$_2$O. An example reaction is the following:
27 μL cDNA;
5 μL 10X ligase buffer;
5 μL 10 m$M$ ATP;
6 μL $^{32}$P-labeled linkers—0.8 μg;
3 μL unlabeled linkers—3 μg;
2 μL T4 DNA ligase—10–20 Weiss U;
2 μL T4 RNA ligase—12–18 U (*see* **Note 6**).
Incubate for 1 h at room temperature then 14°C overnight.

2. Heat-kill the enzyme at 80°C for 15 min, and then cool on ice.

3. Remove 1 μL of the ligation mix to test for ligation (*see* **Note 7**).

### 3.4.6. EcoRI Digestion

1. An example reaction is the following:
49 μL ligation mix.
20 μL 10X buffer H.
126 μL ddH$_2$O.
5 μL *Eco*RI (~450 U).
Incubate for 1 h at 37°C.

2. Add 2 μL more enzyme and incubate 1 h more.

3. Remove 4 μL to check for digestion (*see* **Note 7**).

4. Optionally, save a portion for cloning into *Eco*RI cut vector.

### 3.4.7. XhoI Digestion (see **Note 8**)

1. Make the reaction mix 150 m$M$ in NaCl by adding NaCl and H$_2$O. *Xho*I cuts better in this increased salt concentration.

2. Add 5 μL of *Xho*I (~250–500 U), and incubate 2 h at 37°C.

3. Save a 5-μL aliquot to check the digestion.

4. Phenol-extract, ethanol-precipitate, rinse, and dry the cDNA.

## 3.4.8. Removal of Linkers and Small cDNAs

1. Pour a column in a 1-mL disposable pipet plugged with sterile polyester wool (*see* **Note 9**).
2. Make a reservoir with a 2-mL disposable Pasteur pipet cutting the bottom off at the appropriate level and slipping over the end of the column. Now cut the very top of the pipet off, fill, and rinse with 10 mL of column buffer.
3. Resuspend the cDNA in 50 µL of column buffer, and apply to the column (*see* **Note 10**).
4. Monitor the progress of the radioactivity into the column with a Geiger counter. When it gets one-third of the way (~250 µL), start collecting two-drop fractions into microfuge tubes. Continue collecting fractions until the major $^{32}$P peak (linkers) reaches the bottom.
5. Count the all of the fractions by Cerenkov counting, and plot the results. You should see two distinct peaks.
6. Run aliquots of alternate column fractions on a 1% agarose gel with labeled 1-kb ladder and unlabeled 1-kb ladder.
7. Dry the gel and expose to film. Pool the peaks containing cDNA from about 500 bp to the beginning of the column (*see* **Note 11**).

## 3.4.9. Ligation to Vector

1. Add an appropriate amount of vector to the pooled cDNA and ethanol-precipitate (overnight is best), rinse carefully, and dry (*see* **Note 12**). Try to use about 10X the weight in cDNA of λ arms. The use of less will result in a significant proportion of clones containing multiple inserts, although the total number of clones will increase.
2. Resuspend the cDNA/vector pellet in 5 µL of ligation cocktail, which contains the following:
   0.5 µL 10X ligation buffer;
   0.5 µL 10 m*M* ATP;
   2.0 µL T4 ligase (10–12 Weiss units);
   2.0 µL ddH$_2$O.
3. Be sure the pellet is completely dissolved, and centrifuge to put everything into the bottom of the tube.
4. Ligate at 14°C overnight.

## 3.4.10. Packaging and Titering the Library

1. Package the ligation mix as directed in the instructions accompanying the packaging mix (**Note 13**).
2. Dilute the packaged phage with 500 µL of SM, and add 20 µL of chloroform (**not** chloroform:isoamyl alcohol). Store in the dark at 4°C.
3. Plate 1, 10, and 100 µL of a $10^{-3}$ dilution on 300 µL of ER-1647 plating cells grown as described below. Adsorb the phage to the plating cells for 15–30 min at room temperature, and then transfer to 37°C for 5 min. Phage can adsorb to the

receptors, but cannot inject their DNA at room temperature. The transfer to 37°C produces a relatively synchronous infection. Incubate at 37°C for 8 h to overnight. It is essential that the strain used be deficient in methylcytosine restriction (mcrA⁻, mcrBC⁻).

4. Determining the fraction of recombinant clones in the library. Since ER-1647 is lac-, the library must be plated on an appropriate strain that allows α complementation and is McrABC⁻, like YS-1, or amplified first on ER-1647 and then plated on XL1-Blue. After amplification, the fraction of recombinants can be determined by plating on XL1-Blue with X-gal and IPTG as described above. Using UniZap-XR, a typical yield is 95+% recombinant phages.

### 3.4.11. Preparation of E. coli for Plating Libraries (see **Note 14**)

1. Maintain strains for growth on minimal plates, using maltose as the carbon source and supplemented with the appropriate auxotrophic requirements. ER-1647 requires histidine, methionine, and tryptophane.
2. Grow a single colony overnight in an appropriate volume of minimal medium with the required supplements.
3. Spin down the overnight cells and resuspend in 0.5 vol of 10 m$M$ MgSO₄. The cells are stable for about a month when resuspended in 10 m$M$ MgSO₄, but only a few days if not. Of course, fresh cells work better.
4. Use 200–300 µL of cells for a 100-mm plate or 600–800 for a 150-mm plate. In vitro packaging mixes contain a large excess of phage tails, so use the larger volume for them. Maltose-grown cells also help a lot.

### 3.4.12. Amplification of Libraries

1. It is beneficial to screen unamplified libraries, since one needs to screen fewer clones; however, it is better to amplify the library for permanent storage. A good practice is to reserve a portion of the library unamplified and amplify the remainder for permanent storage.
2. It is convenient to plate $100–150 \times 10^3$ phage/fresh 150-mm plate. We prefer to plate in the morning and observe phage growth during the day. When plaques are touching, overlay each plate with 15 mL of SM, and incubate at 4°C overnight with shaking if possible. Plaques grown this way will typically be only 1–2 mm in diameter.
3. Harvest the liquid and make it 5.0% in chloroform. Be sure to use fresh chloroform that has been stored in the dark, because photodegradation products of chloroform are reportedly toxic to phage.
4. Spin out the debris, and transfer the lysate to an appropriate container. For storage at 4°C make the lysate 5% chloroform and store in the dark. A foil-covered Erlenmeyer flask, or media bottle is a good choice. The chloroform inhibits the growth of molds, which cause the library titer to drop rapidly.
5. It is probably a good idea to store aliquots of the library at –70°C for permanent storage. Bring the lysate to 7% DMSO, and freeze conveniently sized aliquots.

### 3.5. λ *Phage Minipreps*

1. In order to prepare synthetic RNA from a λZAPII library, one must first prepare a sufficient quantity of phage DNA. With current RNA preparation technology, ~10 µg should be sufficient.
2. A high titer library ($10^{10}$ PFU/mL) will yield about 500 ng of phage DNA/mL, so 20–50 mL is about the right amount.
3. Add DNase I and RNase A to 50 µg/mL. Incubate at 37°C for 1 h with gentle shaking.
4. Transfer to a centrifuge tube, and spin at 12,000$g$ for 10 min. Carefully decant the supernatant to a fresh tube.
5. Add 1/4 vol of 20% PEG 8000, 2.5 $M$ NaCl, mix well, and incubate on ice for 1 h. Spin at 12,000$g$ for 10 min at 4°C. Carefully remove the supernatant taking care not to dislodge the phage pellet.
6. Resuspend the phage in a minimum volume of proteinase K reaction buffer (10 m$M$ Tris, pH 8.0, 5 m$M$ EDTA, 0.5% SDS). Take care to resuspend well at this point for maximum yields.
7. Transfer to a 1.5-mL microfuge tube, and add proteinase K to a final concentration of at least 200 µg/mL, and preferably 1 mg/mL. Incubate at 60°C for >30 min or 37°C overnight (*see* **Note 15**). Be sure that the phage do not clump up during the digestion. Pipet up and down if they do, or you will lose them in the phenol extraction.
8. Phenol:chloroform:isoamyl alcohol-extract and back-extract the organic phase with 20% of the original volume of TE. Pool the two aqueous phases.
9. Repeat the phenol extraction.
10. Transfer the aqueous phase to a fresh tube, and add 2 vol of EtOH. Now add 0.5 of the original volume of 7.5 $M$ NH$_4$-acetate and mix well (*see* **Note 16**). Mix well and store on ice for 30 min. Spin for 20 min at 4°C. Rinse the pellet well and dry.
11. Resuspend the pellet in a minimum volume of TE, quantitate, and check a small aliquot for purity by agarose gel electrophoresis.

### 3.6. *In Vitro Transcription of Phage DNA*

1. Digest 10 µg of phage DNA in a final volume of 100 µL with *Not*I or *Xho*I, depending on whether sense or antisense transcripts are desired.
2. Phenol-extract, ethanol precipitate, rinse, and dry the digested DNA. Resuspend at 1 mg/mL.
3. Prepare sense (T3 polymerase and *Xho*I digest) or antisense (T7 polymerase and *Not*I digest) RNA using the Ambion Megascript kit following the instructions with the kit. This kit reproducibly gives yields of 70–100 molecules of RNA/molecule of template. We typically use 2 µg of template and scale the reaction up 2X.
4. If desired, one can incorporate biotin-21-UTP during the transcription reaction for the production of driver RNA. Add biotin-21-UTP (Clontech) at 1/20 the final concentration of UTP in the transcription reaction.

5. Recover the RNA after transcription by adding an equal volume of 5 *M* LiCl and incubating at –20°C overnight.

6. Spin the precipitate for 20 min at 4°C, rinse well, and dry. Resuspend in DEPC-treated $H_2O$ and quantitate.

7. Add 0.5 volume 7.5 *M* NH4-acetate and 2.5 vol of ethanol, mix well, and store at –20°C. Calculate the new concentration of RNA, and determine the volume/µg of RNA.

### 3.7. Photobiotinylation of RNA (12,13)

1. How much RNA is needed? A maximum of about 60-fold molar excess of driver RNA is required for a proper subtraction. If necessary, this can be reduced by limiting the short hybridizations to about 4X molar excess and the long hybridizations to 10-fold excess. If one takes the trouble to calculate the amount of cDNA remaining after each subtractive hybridization, the total amount may be significantly reduced.

2. Mix poly (A)+ RNA or in vitro synthesized RNA (10–30 µg/reaction) with 50 µg of photoactivatable biotin acetate (Clontech). With the tube tops open and the lamp 6 in. from the sample, irradiate on ice for 15 min. Be sure to support the tube in a water bath rack, since the sunlamp rapidly melts the ice.

3. Add 1/10 vol 1 *M* Tris-HCl, pH 9.0, and extract repeatedly with TE-saturated 2-butanol to remove unreacted photobiotin (until the butanol phase is clear).

4. $CHCl_3$ extract the RNA, ethanol-precipitate, rinse, and dry. Resuspend in DEPC-treated $H_2O$ and repeat the photobiotinylation. At this stage, the pellet should be reddish if the biotinylation has been successful.

5. Pool identical RNAs, and store as ethanol precipitates at –20°C.

### 3.8. Subtractive Hybridization

1. The subtraction involves two different types of hybridization. We and others *(3,14)* have found that the typical long hybridizations are inefficient at removing abundant mRNAs, and yield libraries and probes containing significant quantities of housekeeping genes.

2. Prepare first-stranded target cDNA (*see* **Note 17**) from poly (A)+ mRNA as described above (preferred) or in vitro sense RNA derived from a target cDNA library. In the latter case one should use 5 µg of random hexamers as primers, since one cannot be sure that the poly A tail is present after *XhoI* digestion. Remove the RNA by adding 10 µL of 0.25 *M* EDTA and 30 µL 0.15 *N* NaOH, and incubating for 60 min at 65°C. Neutralize with 30 µL of 0.15 *N* HCl, ethanol-precipitate, rinse, and dry.

3. Sequence of hybridization—perform one short hybridization (**steps 4–10**) and then one long hybridization (**steps 11–17**) overnight. Repeat.

4. Short hybridization: spin an appropriate amount of biotinylated driver RNA precipitate (10-fold molar excess to the cDNA) rinse and dry it. Resuspend the RNA in 20 µL of hybridization buffer (HBS = 50 m*M* HEPES, pH 7.6, 0.2% SDS, 2 m*M* EDTA, 500 m*M* NaCl).

5. Transfer the resuspended RNA to the tube with the precipitated cDNA, and resuspend it as well.

6. Transfer the mixture to a 500 μL microfuge tube, and overlay with mineral oil. Boil for 3 min, and snap-cool on ice. Incubate for 30 min at 55°C.

7. Transfer the aqueous phase to 100 μL of hybridization buffer (HB = HBS minus SDS), and add 5 μg streptavidin. Mix well, and incubate at room temperature for 5 min.

8. Extract with 100 μL of phenol:CHCl$_3$ and centrifuge to separate the phases. Back-extract the organic phase with 20 μL of HB, and pool the aqueous phases.

9. Repeat the streptavidin and phenol extraction procedure twice.

10. CHCl$_3$-extract the pooled aqueous phases, ethanol-precipitate, rinse, and dry the cDNA.

11. Long hybridization: Spin an appropriate amount of biotinylated driver RNA precipitate (10-fold molar excess to the cDNA), rinse, and dry it. Resuspend the RNA in 20 μL of HBS.

12. Transfer the resuspended RNA to the tube with the precipitated cDNA, and resuspend it as well.

13. Carefully draw the mixture into the center of a baked, siliconized, 20-μL capillary tube, and seal the ends with a Bunsen burner. Transfer the sealed tube to a beaker of boiling water and heat for 5 min. Transfer the beaker containing the capillary to a 65°C water bath, and incubate for 24 h.

14. Remove the beaker from the water bath and slow cool to room temperature. Remove the capillary, and score both ends. Break one, and and transfer the capillary (cut end down) to an Eppendorf tube. While inverted, break the other end, and blow the mixture into the Eppendorf tube. Rinse the capillary with 20 μL of HB, and add another 80 μL of HB.

15. Add 5 μg streptavidin. Mix well and incubate at room temperature for 5 min. Extract with 100 μL of phenol:CHCl$_3$, and centrifuge to separate the phases. Back-extract the organic phase with 20 μL of HB, and pool the aqueous phases.

16. Repeat the streptavidin and phenol extraction procedure twice.

17. CHCl$_3$-extract the pooled aqueous phases, ethanol-precipitate, rinse, and dry the cDNA.

18. Calculate the yield of subtracted cDNA based on the amount of counts remaining compared with the starting material. One should expect to subtract >95% of the starting material.

### 3.9. Use of the Subtracted cDNA

1. Construction of a subtracted cDNA library. Begin with **Subheading 3.4.2., step 2**—second-strand synthesis.

2. Preparation of a subtracted cDNA probe. Divide the cDNA into three portions. Reagents for random priming are prepared as follows: 1 $M$ HEPES, pH 6.6, DTM (0.1 m$M$ each of dGTP, dTTP in 0.25 $M$ Tris-HCl, pH 8.0, 0.025 m$M$ MgCl$_2$, 0.05 m$M$ BME), OL (1 m$M$ Tris-HCl, pH 7.5, 1 m$M$ EDTA containing 90 OD U of random hexanucleotides [Pharmacia]/mL), LS (1 $M$ HEPES:DTM:OL [25:25:7] [v/v]) *(15)* (*see* **Note 18**).

3. The reaction mix contains 11.4 μL solution LS, 1 μL BSA (Sigma) (10 mg/mL), one-third of the subtracted cDNA fragment and H$_2$O to 37.5 μL.

4. Boil this reaction for 5 min, and then snap-cool on ice.
5. Add 5 µL each of α-[$^{32}$-P]dATP and dCTP (3000 Ci/m$M$) and 2.5 U of the Klenow fragment of DNA polymerase I (Pharmacia).
6. Incubate for 2 h to overnight at room temperature.
7. Remove the unincorporated nucleotides by Sephadex G-50 spun-column chromatography as above.

### 3.10. Library Screening

1. Plate the library to be screened at relatively low density (~5–10,000 phage/150-mm plate) to minimize the number of purification steps required. Concerning the amplification step described above, it is best to plate the phage in the morning, observe their growth during the day, and remove the plates to 4°C when plaques are relatively large (~2-mm), but still well isolated from each other.
2. Lift dupicate filters from each plate, the first for 3 min and the second for 6 min (*see* **Note 19**). Store the filters plaque side up on Whatman 3MM paper until all lifts are completed. Be sure to position registration marks carefully for accuracy in aligning the autoradiograms later.
3. Denature the phage by placing the filters (plaque side up) on pads of Whatman 3MM paper saturated with the following solutions for 3' each (10): Solution 1—0.5 $N$ NaOH, 1.5 $M$ NaCl; Solution—2 1.0 $M$ Tris-HCl, pH 7.5, 1.5 $M$ NaCl; Solution 3—2X SSC. Next, transfer the filters to 3MM paper, and allow to air-dry.
4. Interleave the filters between circles of filter paper, and secure the entire stack with tape. Bake at 80°C for 30 min.
5. After baking, transfer the filters to a dish of 50 m$M$ NaOH, and incubate for 15 min with shaking. This step reduces later background. Rinse with at least four changes of distilled H$_2$O for a total of 15 min.
6. Transfer the filters to a hybridization bag, and add 2 mL/filter of Church's buffer (7% SDS, 0.5 $M$ NaPO$_4$, pH 7.2) *(16)*. Prehybridize at 65°C for a convenient time, usually 15 min to several hours. Up to 20 filters (132 mm) can be processed/bag.
7. Remove the prehybridization solution and add fresh Church's buffer containing 5% w/v Dextran sulfate (Pharmacia) at 750 µL/filter. Denature the probe by adding 0.1 vol 2 $N$ NaOH and incubating at room temperature for 5 min. Add the denatured probe directly to the bag, seal, and mix well. Hybridize at 65°C overnight with shaking if possible.
8. Remove and save the hybridization solution at –20°C (add fresh probe to it for subsequent screening). Wash the filters 3 × 20' at 65°C in 0.5X SSC, 0.1% SDS. Expose to X-ray film with two intensifying screens overnight or longer.
9. Align the duplicate autoradiographs, and pick plaques with the wide end of a Pasteur pipet, which appear on both filters to 1 mL of SM buffer.
10. Plate several dilutions of each plaque stock, and incubate overnight at 37°C. Select a plate with ~50 plaques for the next round of screening. After this round, the individual plaques should be pure and can be picked with the narrow end of a Pasteur pipet.

11. Convert the phage to plasmids according to the Stratagene protocol provided with the ZAPII vector or library.

## 4. Notes

1. Be sure to use the appropriate laboratory technique when handling radioactive and toxic materials. Consult a laboratory safety manual if you are in doubt regarding what proper practices are. Methyl mercuric hydroxide is quite toxic and should be handled with extreme care in a fume hood. This toxicity is balanced by its extremely potent and reversible denaturing activity. Each new batch of methyl mercuric hydroxide should be tested for performance in denaturing gel electrophoresis as described in **ref. 9**. If sharp bands are not observed, purify the methyl mercuric hydroxide by stirring for 2 h at room temperature with a mixed-bed resin, such as Amberlite MB-1 or equivalent. Remove the resin and other debris by passing through a 0.2-μm syringe filter. Store in small aliquots at –70°C. Be sure to dispose of mercury waste appropriately.
2. This step is required to repair the cDNA and render it blunt-ended prior to linker ligation.
3. This additional blunting is required to repair any damage caused by *Eco*RI methylase. We have noticed variable amounts of nuclease activity in methylases and think this extra step is prudent.
4. We usually buy phosphorylated linkers as well as unphosphorylated ones, since it is more efficient to synthesize the linker with the phosphate on than to add it later enzymatically.
5. One should be careful that the linkers are in sufficient excess in this reaction. This, of course, depends on the yield of cDNA. Assume an average cDNA length of 1 kb in the first strand reaction and calculate the number of picomoles of ends. Be sure that the linkers are in 50- to 100-fold molar excess to ensure that cDNAs are not artifactually ligated to each other.
6. The addition of T4 RNA ligase stimulates blunt-end ligation up to 10-fold. Furthermore, RNA ligase is capable of ligating linker molecules to RNA remaining at the 5'-end of cDNA. It is quite probable that some of the longest cDNA molecules will have a few nucleotides of RNA left at the 5'-end, which RNase H can not remove. If RNA ligase is not added, then linkers can not be added to this end and these cDNAs will be lost from the library.
7. Check for ligation and digestion by running an 8% polyacrylamide gel in TBE before and after digestion samples. A typical protein gel apparatus is appropriate. Run the gel at 300–400 V, and stop it when the bromophenol blue goes half way down. Expose to film for several hours, or dry the gel down and expose for an appropriate time. Expect to see a ladder of linkers in the undigested sample, it should be gone in the *Eco*RI-digested sample, and there should be two bands near the bottom of the gel. If the *Xho*I digest was done separately, expect to see one additional band larger than the two in the *Eco*RI digest. If the ligation or either digestion did not work, then go back to the blunt-ending step, and repeat carefully checking each step individually.

8. Alternatively, the *Eco*RI and *Xho*I digests can be performed together using buffer H, but this does not permit checking that each step has worked. We usually do the digestions separately when testing new batches of enzymes and reagents, but otherwise do them together to save time.

9. Use a sterile, plugged, individually wrapped plastic pipet suitable for tissue culture. Score the pipet in the middle of the cotton. Break off and remove the cotton. Make a tiny ball of polyester wool, push into the column top with forceps, and blast into the bottom with compressed air or gas (available at most lab benches). Polyester wool used for aquarium filtration and is available at most pet shops.

10. Proper technique is critical here for good separation. Allow the liquid in the column to drain to the top of the bed. Apply 50 µL of cDNA carefully, and allow to run in. While this is going on, reattach the reservoir, and carefully add column buffer when the cDNA has fully entered the column. Of course, the column should not be allowed to run dry at any point or else resolution will be compromised.

11. Using Sepharose Cl-4B and the column system described above, the first fraction following the cDNA peak is about 500 bp. Pool the fractions from here to the beginning.

12. Take care not to overdry the DNA, or it will be quite difficult to resuspend. One to 2 min in the Speed-Vac are sufficient.

13. Stratagene recommends a 2-h incubation at room temperature. We have used up to 6 h with equally good results.

14. Although it is slightly more trouble to maintain cells on minimal plates and grow them in minimal media, the increased plating efficiency and reproducibility are worth it. Some strains, most notably C600 HflA[150] and Y1090, throw off resistant mutants at a high rate. Growth on media with maltose as the carbon source minimizes this phenomenon.

15. Proteinase K works well at elevated temperatures, but is denatured at temperatures above 65°C. This is a good step to let go overnight if desired.

16. When precipitating large amounts of DNA, you will get cleaner precipitates by adding the alcohol first, mixing well, and then adding the salt and mixing well again, Mix well, and store on ice, –20°C, or –70°C for 10+ min. For reasonable amounts of DNA, these three are about equivalent. The centrifugation time is much more critical. We typically use ice for 30 min followed by a 20 min spin at room temperature. For small quantities of DNA, overnight at –20°C gives superior recoveries.

17. If the production of a subtracted cDNA library is desired, then one is practically limited to using poly (A)+ RNA from the target tissue as the source of material. If a subtracted probe is being prepared, then either poly (A)+ RNA or RNA synthesized in vitro from an existing library is adequate. Whenever there is sufficient poly (A)+ RNA available, its use is preferred to minimize changes in complexity of the target RNA.

18. There are a number of high-quality kits available for labeling DNA by random priming. If you choose a commercial kit, be sure that it can be adapted to use two radionucleotides, since the probe must be of very high-specific activity to detect rare clones.

19. The choice of nitrocellulose or nylon membranes depends on the number of times the library will be screened. For one to three screenings we prefer to use sup-

ported nitrocellulose (e.g., BAS/NC, Schleicher and Schuell, Keene, NH) owing to its inherent high signal-to-noise ratio. If the library may be screened three or more times, them it makes sense to use a nylon membrane such as Nytram (Schleicher and Schuell). We have has intermittent difficulties with nylon membranes from other manufacturers and recommend each batch prior to using it.

## Acknowledgments

This work was supported by the American Cancer Society (Grant DB-36 to B. B.) and the G. Harold and Leila Y. Mathers Charitable Foundation (J. C. I. B.).

## References

1. St. John, T. P. and Davis, R. W. (1979) Isolation of galactose-inducible DNA sequences from *Saccharomyces cerevisiae* by differential plaque filter hybridization. *Cell* **16,** 443–452.
2. Sargent, T. D. and Dawid, I. B. (1983) Differential gene expression in the gastrula of *Xenopus laevis*. *Science* **222,** 135–139.
3. Wang, Z. and Brown, D. D. (1991) A gene expression screen. *Proc. Natl. Acad. Sci. USA* **88,** 11,505–11,509.
4. Aasheim, H. C., Deggerdal, A., Smeland, E. B., and Hornes, E. (1994) A simple subtraction method for the isolation of cell-specific genes using magnetic monodisperse polymer particles. *Biotechniques* **16,** 716–721.
5. Hara, E., Kato, T., Nakada, S., Sekiya, S., and Oda, K. (1991) Subtractive cDNA cloning using oligo(dT)30-latex and PCR: isolation of cDNA clones specific to undifferentiated human embryonal carcinoma cells. *Nucleic Acids Res.* **19,** 7097–7104.
6. Li, W. B., Gruber, C. E., Lin, J. J., Lim, R., D'Alessio, J. M., and Jessee, J. A. (1994) The isolation of differentially expressed genes in fibroblast growth factor stimulated BC3H1 cells by subtractive hybridization. *Biotechniques* **16,** 722–729.
7. Lopez-Fernandez, L. A. and del Mazo, J. (1993) Construction of subtractive cDNA libraries from limited amounts of mRNA and multiple cycles of subtraction. *Biotechniques* **15,** 654–659.
8. Sharma, P., Lonneborg, A., and Stougaard, P. (1993) PCR-based construction of subtractive cDNA library using magnetic beads. *Biotechniques* **15,** 610–612.
9. Maniatis, T., Fritsch, E. F., and Sambrook, J. (1989) *Molecular Cloning: A Laboratory Manual,* 2nd ed. Cold Spring Harbor Laboratory Press, Cold Spring Harbor, NY.
10. Han, J. H., Stratowa, C., and Rutter, W. J. (1987) Isolation of full-length putative rat lysophospholipase cDNA using improved methods for mRNA isolation and cDNA cloning. *Biochemistry* **26,** 1617–1625.
11. Blumberg, B., Wright, C. V. E., De Robertis, E. M., and Cho, K. W. Y. (1991) Organizer-specific homeobox genes in *Xenopus laevis* embryos. *Science* **253,** 194–196.
12. Forster, A. C., McInnes, J. L., Skingle, D. C., and Symons, R. H. (1985) Non-radioactive hybridization probes prepared by the chemical labelling of DNA and RNA with a novel reagent, photobiotin. *Nucleic Acids Res.* **13,** 745–761.

13. Sive, H. L. and St. John, T. (1988) A simple subtraction hybridization technique employing photoactivatable biotin and phenol extraction. *Nucleic Acids Res.* **16,** 10,937.

14. Sasaki, Y. F., Ayusawa, D., and Oishi, M. (1994) Construction of a normalized cDNA library by introduction of a semi-solid mRNA-cDNA hybridization system. *Nucleic Acids Res.* **22,** 987–992.

15. Feinberg, A. P. and Vogelstein, B. (1983) A technique for radiolabelling DNA restriction endonuclease fragments to high specific activity. *Anal. Biochem.* **132,** 6–13.

16. Church, G. M. and Gilbert, W. (1984) Genomic sequencing. *Proc. Natl. Acad. Sci. USA* **81,** 1991–1995.

# 38

# Differential Display of Eukaryotic mRNA

## Antonio Tugores and Juan Carlos Izpísua Belmonte

## 1. Introduction

Arbitrarily primed PCR is a method that was initially developed to analyze and compare genome complexity *(1,2)*. By using arbitrary primers, an array of stochastic sequences can be amplified using the polymerase chain reaction (PCR) and resolved by denaturing acrylamide gels, generating a characteristic pattern or fingerprint. This technique can be readily applied for the analysis of differentially expressed genes in two different cell types by simply converting the mRNA into cDNA *(3,4)*. A different fingerprint pattern will then reflect differential gene expression between the cell types analyzed. The number and intensity of the fingerprints depend largely on two parameters: the abundance of the original RNAs, and how well the primers match the primary sequence so that efficient amplification can take place. Taking this into consideration, the method appears to be more suitable for the comparison of cell types where a large number of differentially expressed genes are anticipated. A major advantage of this method when compared to other strategies used for the isolation of differentially expressed genes is that a very small amount of RNA is required. This makes the method very suitable for embryological studies, where sometimes the quantity of tissue that can be collected is limiting.

At least two major approaches can be taken to generate and amplify cDNA (**Fig. 1**). We will designate them as (1) arbitrarily primed PCR of RNA (RAP-PCR *[3]*), and (2) differential display of mRNA 5' ends *(4)*.

### 1.1. Arbitrarily Primed PCR of RNA

If a single primer is to be used as both upstream and downstream primer for amplification, cDNA can be synthesized with this single primer or with an oligodT primer to select for polyadenylated transcripts. Using the same primer

From: *Methods in Molecular Biology, Vol. 97: Molecular Embryology: Methods and Protocols*
Edited by: P. T. Sharpe and I. Mason © Humana Press Inc., Totowa, NJ

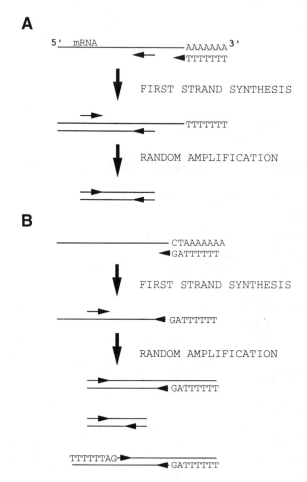

Fig. 1. The two major approaches that can be taken for the random amplification of RNA. In the first case (**A**), synthesis of the first strand can be driven with either oligodT or the same arbitrary primer that will be used for amplification. After random amplification, the product obtained should be identical in both cases. In the second case (**B**), the downstream (3') and upstream (5') primers are different. The first strand of cDNA is synthesized with a tailed oligodT that will only recognize a fraction of the mRNA population. Random amplification is then performed with a second arbitrary primer. Note that three different products can be obtained depending on the primer combination (*see* **ref. 16**).

for both the cDNA synthesis and the PCR is very convenient, since only one primer is required and good matches with the primer will be already selected during cDNA synthesis. On the other hand, priming with oligodT will select for polyadenylated transcripts, and thus, cDNA will not be made efficiently from ribosomal or heterogeneous nuclear RNAs, which are extremely abundant as compared to mRNA. A disadvantage of the latter is that nonpolyadenylated mRNAs will not be represented among the cDNA population. After oligodT-driven synthesis of cDNA, a single primer might then be used for the amplifications. This method also allows the use of nested primers that will select for population of PCR products that are in low abundance and thus underrepresented when a single primer is used *(5,6)*. The basis for this approach is discussed below.

## 1.2. Differential Display of mRNA 3' Ends

If different primers are to be used for the upstream and downstream positions, synthesis of cDNA should be driven with the downstrean primer. A strategy has been devised to generate different populations of cDNAs by using "tailed" oligodT primers for both the synthesis of cDNA and the PCR process. This method, developed by Liang and Pardee *(4)*, relies on the use of several downstream primers in combination with multiple upstream primers. This approach allows a myriad of combinations. Furthermore, additional products can form as a result of the priming of the individual primers with themselves. The cost of synthesizing and purifying the multiple primers used in this method can be considerable, and the authors have now commercialized a kit (GenHunter Corporation, 50 Boylston St., Brookline, MA 02146) that is definitively worth trying. This method has been proven successful in a variety of systems and could be a good choice. The major disadvantage is that this method will select preferentially for abundant mRNAs, and that the products will belong preferentially to the 3' noncoding regions that are the least conserved among related genes in different species. This can make data base searching less informative.

Both methods have been proven succesful in a variety of biological systems *(4,6–9)*. Unfortunateiy, it is impossible to predict what will be the result of choosing one strategy over the others until the experiments are actually done and the products analyzed. Once an authentic product has been isolated, isolation of a full-length cDNA will require the screening of a suitable cDNA library, although other PCR-based methods can also be used for this purpose *(10)*.

Here, we provide a protocol that can be followed among many others. This protocol is based on the RAP-PCR approach, as described originally by Welsh et al. *(3)*. A very detailed protocol of the second method has been described *(4)*, or can be obtained from GenHunter Corporation. Overall, what is really needed is not many protocols but a great deal of good luck.

## 2. Materials

All solutions listed should be made with the highest quality reagents available and with sterile double-distilled water.

## 2.1. RNA Extraction and Synthesis of cDNA

1. RNA STAT-60 (TEL-TEST"B", Friendswood, TX).
2. Chloroform.
3. Isopropanol.
4. 100% Ethanol.
5. Nuclease-free water.
6. Electrophoresis-grade agarose (Bio-Rad Laboratories, San Francisco, CA).
7. 10X TBE buffer, 0.89 $M$ Tris base, 0.89 $M$ boric acid, 0.02 $M$ ethylene diamino tetraacetate, disodium salt.
8. 10 mg/mL ethidium bromide (*see* **Note 1**).
9. DNAse I (RNase-free) (Boehringer Mannheim, Indianapolis, IN).
10. Human placental ribonuclease inhibitor: RNA guard (Pharmacia, Milwaukee, WI).
11. 2X Proteinase K (PK) buffer; 200 m$M$ Tris-HCl, pH 8.0, 350 m$M$ NaCl (5 $M$ stock), 25 m$M$ EDTA (0.5 $M$ stock), 1% SDS (20% stock), 10 mg/mL PK (Boehringer). Store at –20°C.
12. Phenol-SEVAG reagent: a mixture of 25 parts of ultrapure buffer-equilibrated recrystallized phenol, 24 parts of chloroform and 1 part of isoamylalcohol (*see* **Note 2**).
13. 10X First-strand buffer; 100 m$M$ Tris-HCl, pH 8.3 at 37°C, 40 m$M$ MgCl$_2$ 500 m$M$ KCl.
14. 100 m$M$ deoxynucleotide triphosphate (dNTP) stocks (Pharmacia).
15. Cloned MuTLV reverse trancriptase (Stratagene, San Diego, CA) (*see* **Note 3**).

## 2.2. Amplification of cDNA and Electrophoretic Analysis

1. Oligonucleotides. A various number of synthetic oligonucleotides will be required depending on the method of choice. These will be indicated in the methods sections where appropriate. The quality of the primers is of great importance, and purification is strongly recommended (*see* **Note 4**) for a brief protocol on how to purify synthetic oligonucleotides by using denaturing acrylamide gels).
2. 10X Second strand buffer;100 m$M$ Tris-HCl, pH 8.3 at 37°C, 20 m$M$ MgCl$_2$, 250 m$M$ KCl.
3. $\alpha^{32}$P-dCTP (3000 Ci/mmol, 10 mCi/mL) (*see* **Note 5**).
4. Recombinant Ampli*Taq* DNA polymerase (Perkin Elmer, Vaterstetten, Germany).
5. 5 $M$ Ammonium acetate (*see* **Note 6**).
6. 10 mg/mL Oyster glycogen.
7. Urea (Boehringer Mannheim).
8. A 5% (w/v) solution of acrylamide-methylen bis-acrylamide (29:1) (Boehringer Mannheim) in 1X TBE, 50% urea.
9. Formamide loading dye: 95% deionized formamide (Gibco-BRL, Gaithersburg, MD) (v/v), 20 m$M$ EDTA, 0.05% (w/v) bromophenol blue, 0.05% (w/v), xylene cyanol (w/v).
10. Talcum baby powder.
11. Single-sided emulsioned autoradiographic film (Kodak BioMax, Rochester, NY).
12. TE buffer: 10 m$M$ Tris-HCl, pH 7.6, 1 m$M$ EDTA.

## 2.3. Analysis of Differential Clones

1. Random oligonucleotide labeling solutions (*see* **Note 7**).
2. Deionized formamide (Gibco-BRL).
3. 37% Formaldehyde (Sigma-Aldrich Chemicals, St. Louis, MO).
4. 10X MOPS buffer; 0.4 $M$ MOPS pH 7.0, 0.1 $M$ sodium acetate,10 m$M$ EDTA.
5. 6X gel loading buffer: 25% (w/v) Ficoll 400, 0.25% (w/v) bromophenol blue, 0.25% (w/v) xylene cyanol, 1 m$M$ EDTA.
6. 0.5 $M$ $Na_2HPO_4$.
7. 0.5 $M$ $NaH_2PO_4$.
8. MAGNA NT nylon hybridization membranes (Micron Separations, Inc., Westboro, MA) or equivalent.
9. 20X SSC: 0.3 $M$ tri-sodium citrate, 3.0 $M$ NaCl.
10. UV crosslinker or 80°C oven.
11. 50X Denhardt's solution: 1% (w/v) bovine serum albumin, 1% (w/v) polyvinyl pirrilodone, 1% (w/v) ficoll 400. Filter-sterilize and store in aliquots at –20°C. Thaw at room temperature when needed. Can be temporarily stored at 4°C.
12. 10 mg/mL Salmon sperm DNA (*see* **Note 8**).
13. 20% Sodium dodecyl sulfate (SDS).
14. TA Cloning kit (Invitrogen) (*see* **Note 9**).
15. DNA sequencing kit (USB, Cleveland, OH).

## 2.4. Additional Equipment Required

1. Polytron (KINEMATICA AG, Lucerne, Switzerland).
2. Microcentrifuge.
3. UV spectrophotometer and quartz cuvet.
4. Thermal cycler (MJ Research or equivalent).
5. Thin-wall PCR tubes.
6. Sequencing gel setup.
7. Autoradiography equipment.

# 3. Methods

## 3.1. Preparation of RNA

For the preparation of embryo tissue RNA, we routinely use a commercial reagent called RNA STAT-60. This procedure is based on the method described by Chomczynski and Sacchi (*11*), but the RNA STAT-60 reagent contains the guanidine thiocyanate together with the phenol for easier handling. Other manufacturers supply similar reagents yielding identical results (*see* **Note 10**).

1. Embryonic tissue should be collected and washed in ice-cold PBS before addition of the extraction reagent. If the tissue of interest has to be collected over a long period of time, the tissue should be rinsed in PBS and flash-frozen into a tube inserted in a tank containing liquid nitrogen. Transferring of PBS to the freezing tube is not desirable. Samples can be stored frozen at –80°C.

2. Once a desired amount of tissue has been collected, the extraction solution should be added directly to the frozen tissue. Typically 4 vol of RNA STAT reagent are added/vol of sample in a 15-mL conical polypropylene graduated tube. Homogenization of the tissue with the chaotropic solution should be done as fast as possible. For this purpose, a Polytron is essential. The isolation of intact RNA greatly depends on the speed of this step. For small amounts of tissue, the homogeneization can be performed in a microcentrifuge tube: add the solution and shear the sample passing all the material through a 21-gage needle three or four times, until the material easily flows through the needle.

3. After homogeneization, 0.2 mL of chloroform are added/mL of RNA STAT solution used. The tubes are left at room temperature for 5 min.

4. Spin down the tubes at maximum speed in a tabletop centrifuge for 10 min. Alternatively, the mix can be transferred to 1.5-mL microfuge tubes that can be spun in a microcentrifuge.

5. Transfer the aqueous phase to 1.5-mL tubes (a maximum of 700 µL/tube), and add 700 µL of isopropanol. Precipitate for 1 h at room temperature. These tubes can also be stored at –20°C until needed.

6. Spin for 10 min, discard the supernatant, and rinse the RNA pellet with 70% ethanol to eliminate traces of salts and phenol. Once the ethanol has been completely removed (spin the tube shortly and remove the remaining liquid with a yellow tip), the pellet is resuspended in a minimal volume of nuclease-free water. Do not let the pellet dry, or resuspension will be very difficult.

7. Once resuspended, take a small volume of the sample, dilute it in a volume of water according to the size of the spectrophotometer quartz cuvet available, and measure the absorbance at 260 nm. Consider that 1 OD = 40 mg/mL of RNA. Absorbance at 280 nm should indicate contamination with proteins and/or phenol. A good-quality RNA should have a ratio of $OD_{260}/OD_{280} = 1.8–2.0$.

8. Load an amount equivalent to 1 µg of total RNA in a regular freshly made 0.8% agarose-1X TBE gel containing 0.2 µg/mL ethidium bromide to check the integrity of the RNA. Two major bands corresponding to the 28S and 18S ribosomal RNAs should appear clear and sharp.

9. Once the RNA has been checked, DNase treatment is required to remove DNA that will interfere with the PCR process, giving rise to bands that do not correspond to cDNA. Resuspend part of the RNA solution (5–50 µg) in 42 µL of nuclease-free water and add 5 µL of any clean 10X restriction enzyme buffer, 2 µL of DNase I RNase-free, and 2 µL of RNase guard. Incubate at 37°C for 30 min. Store the rest of the RNA at –80°C.

10. Add 50 mL of 2X PK buffer and 5 mL of 10 mg/mL PK. Incubate at 37°C for 30 min.

11. Extract twice with 100 mL of Phenol-SEVAG, recover the aqueous phase, and add 250 mL of 100% ethanol. Precipitate at least 30 min at room temperature. Do not precipitate at a cold temperature, since an excess of SDS and salt can also precipitate and interfere with subsequent steps.

12. Spin down the precipitate, rinse with 70% ethanol, remove all the remaining liquid after a 5-s spin, and resuspend the pellet in a small volume of nuclease-free

water. Measure the absorbance at 260 nm, and resuspend the RNA at 100 ng/mL. Integrity of the RNA should be checked at this point as mentioned above. To minimize repeated freezing and thawing, the RNA should be stored in small aliquots at –80°C. The integrity of the RNA is crucial for the success of the whole procedure.

## 3.2. Synthesis of cDNA and PCR

For cleaner results, polyadenylated RNA can be isolated as mentioned in the libraries chapter (Chapter 37). However, large amounts of RNA are required to obtain polyadenylated mRNA, so this choice is limited by the availability of the sample. Also, the time and effort employed in the extraction of more material do not necessarily correlate with an equal increase in performance. Therefore, total RNA will be the starting material used in these protocols.

The choice of the arbitrary primer to be used for amplification is of great importance. Primers should not form secondary structures with themselves. Therefore, internal hybridization sequences and palindromes should be avoided. The length of the primer for which the protocol below is designed should be 17–20 bases. For a trial experiment, either the M13 universal or reverse primers could be used.

General considerations must be taken when the first-strand synthesis is performed. First, cDNA should be made from three different dilutions of RNA from the same sample. This is especially important, since a particular band that does not appear with the same intensity in all three dilutions has a good chance of being nonspecific. Second, at least the first time a particular RNA is used with a particular primer, a similar reaction for cDNA synthesis should be performed in the absence of reverse transcriptase. This will be an indicator of the nonspecific background created by the PCR process on a particular sample.

1. Prepare dilutions of the RNA samples at concentrations of 100, 25, and 5 ng/mL in nuclease-free water. Mix in a sterile thin-walled PCR tube (amounts are in µL):

    | | |
    |---|---|
    | Nuclease-free water | 7.0 |
    | 10X first strand buffer | 2.0 |
    | 0.1 $M$ DTT | 4.0 |
    | 1 m$M$ dNTPs | 2.0 |
    | 10 pmol/mL primer* | 2.0 |
    | RNA | 2.0 |

2. Heat at 65°C for 5 min and chill on ice.
3. Add 5 U (1 µL) of MuTLV reverse transcriptase.
4. Incubate at 37°C for 1 h.
5. Heat at 94°C for 5 min to kill the room temperature. For convenience, these temperature steps can be programmed in a thermocycler. After first-strand synthesis, reactions can be stored at –20°C.

*Use arbitrary primer or oligodT.

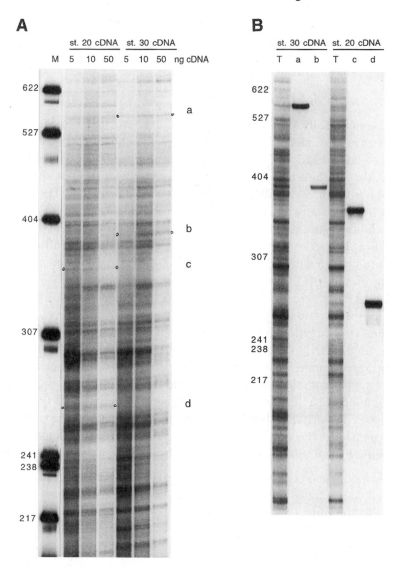

Fig. 2. Examples of a differential display analysis. Total RNA was isolated from chicken embryos at two different embryonic stages. Five, 10, and 50 ng of total RNA were subjected to reverse transcription, and afterward to amplification for 35 cycles using the M13 reverse primer, under the conditions described in the text. The amplification products were then separated on a denaturing polyacrylamide gel (**A**). Several bands were selected (observe the holes punched to recover the products from the gel), and after selection, the gel was re-exposed to verify recovery (not shown). The recovered material was subjected again to PCR under the conditions described in **Subheading 3.2.** As markers, 50 ng of cDNA were amplified independently as described in

6. Add to each tube 20 μL of a cocktail containing (in μL):

   | | |
   |---|---|
   | Nuclease-free water | 14.5 |
   | 10X second strand buffer | 2.0 |
   | 10 pmol/mL primer** | 2.0 |
   | $\alpha^{32}$P dCTP (*see* **Note 11**) | 1.0 |
   | *Taq* DNA polymerase | 0.5 |

   Subject the samples to a low-stringency amplification step: 94°C/5 min; 40°C/5 min; 72°C/5 min, followed by 30 high-stringency cycles: 94°C/1 min; 60°C/1 min; 72°C/2 min. At the end, add an additional step of 5 min at 72°C to allow the polymerase to elongate unfinished products. Transfer tubes to ice or store at –20°C if they are not going to be used immediately.

7. Transfer the reaction to a 1.5 μL microfuge tube containing 2 μL of 0.2 *M* EDTA, 1 μL of 10 mg/mL oyster glycogen, 43 μL of 5 *M* ammonium acetate, and add 260 μL of 100% ethanol. Precipitate at room temperature for at least 5 min. Spin down for 15 min at 12,000 rpm in a microcentrifuge, carefully discard the supernatant, and resuspend in 8 μL of formamide loading dye (*see* **Note 13**).

8. Together with the samples, a labeled size marker should be also loaded. Refer to **Note 14** for preparation of a size marker. Heat the samples at 100°C for 5 min, chill on ice, and load 3 μL on a 50% (w/v) urea, 1X TBE, 5% (w/v) acrylamide:bis-acrylamide (29:1), 50 cm long, 0.4-mm thick gel (*see* **Note 15**). Run at 55–60 W constant power until the xylene cyanol (the slowest migrating dye) band is about 5 cm from the bottom. In this gel percentage, xylene cyanol comigrates with DNA fragments of approx 160 bases.

9. Lift the gel with a piece of Whatman 3MM paper, and dry immediately without fixing. After drying, a little bit of baby powder might be applied to the surface of the gel to prevent sticking to the film during autoradiography. Expose the gel to a single-sided emulsioned autoradiography film (Kodak BioMax) at room temperature without intensifying screens. Be sure to place position markers, e.g., fluorescent dots, so that the gel can be perfectly aligned with the film after autoradiography.

10. A 5-h to overnight exposure should give enough signal to visualize the fingerprint pattern generated. Sometimes a difference in the footprint is not really obvious, and all lanes should be carefully checked. Bona fide products should appear in all three different dilutions (**Fig. 2**). If no differential bands are observed at this point, another oligonucleotide might be used (*see also* **Subheading 3.4.**).

11. Once convincing differentially expressed bands have been identified, punch holes at both sides of the band. Align the gel with the autoradiogram, and mark the position with a sharp pencil through the holes. Cut the region delimited by the

---

**Same arbitrary primer used for first-strand synthesis (*see* **Note 12**).

---

Fig. 2. *(continued)* (A), to check again for the appearance of the products selected. These products are now cloned, sequenced, and the major product will be considered as the "real" differentially expressed gene. Until Northern blot analysis is performed, there is no previous clue about the authenticity of the products.

pencil marks with a razor blade. and place the insert in a microfuge tube containing 100 μL of TE buffer. Close the tube firmly, heat the sample at 65°C for 1 h, and store at –20°C. After cutting the bands, the gel should be re-exposed to verify that the selected band has been accurately removed.

12. Once identity of the bands has been checked, thaw the sample, mix by vortexing, spin down the solid for 10 min, and place 5 μL of supernatant in a tube containing (in μL):

| | |
|---|---|
| Nuclease-free water | 18.5 |
| 10X Second-strand buffer | 4.0 |
| 10 pmol/mL same primer | 4.0 |
| 0.1 *M* DTT | 4.0 |
| 1 m*M* dNTP | 4.0 |
| *Taq* DNA polymerase | 0.5 |

Subject the sample to 35 cycles of amplification: 94°C/1 min; 50°C/1 min; 72°C/2 min. Take 5 μL at the end of the amplification, and repeat as above.

13. Load 20 μL of each sample on a 1.2% (w/v) agarose, 1X TBE gel containing 0.2 μg/mL ethidium bromide. Bands can be seen after the first round of amplification and should definitely be seen after the second one.

## 3.3. Analysis of the Products

The product of the amplification may contain several bands that migrate at the same position in the sequencing gel. The best way to assess the identity of the bands in the first place is to clone the PCR product into a suitable vector for cloning PCR products. Several clones should be sequenced to verify homogeneity of the initial product. Once a major clone has been identified, Northern blot hybridization should be performed using the cloned major PCR product as a probe. We do not recommend either dot-blot or Southern blot hybridization of the PCR products with a hot cDNA probe to verify positive clones, since repetitive sequences and labeled ribosomal RNA can give rise to nonspecific positive signals.

### 3.3.1. Cloning and Sequencing of the PCR Products

Because a single size band may contain several DNA species, several clones should be sequenced so that the real differentially displayed clones are not missed. If several sequences appear, individual clones should be tested as probes in Northern blot analysis, as described below. An alternative strategy has been described to clone specific products after Northern blot hybridization. This procedure involves the capturing and the reamplification of the hybridized cDNA from the membrane *(13)*.

Several kits are available to clone PCR products. The TA cloning kit from Invitrogen (San Diego, CA) allows direct cloning of homogeneous PCR products without purification and/or blunting of the DNA ends. If this kit is used, take 2 or 3 μL of the PCR mixture (after verification that a sufficient amount of the desired band is present) for the ligation reaction. Do not dry concentrate the product since dNTPs can inhibit the activity of T4 DNA ligase. An alternative kit form Stratagene requires blunt-ending of the cDNA prior to ligation. Both manufacturers provide excellent step-by-step protocols for the use of their products, so we will not provide a protocol for this procedure.

After cloning, clones should be sequenced for further analysis. This can be performed by using the dideoxy chain termination method using primers flanking the insert sequence (T3 and/or T7 primers) as initiators for the polymerase. All reagents required for sequencing can be purchased from USB (Cleveland, OH) as a kit, accompanied by instructions from the manufacturer.

### 3.3.2. Northern Hybridization

#### 3.3.2.1. LABELING OF THE PROBE

Approximately 25 ng of the PCR product should be labeled with random hexanucleotides according to the procedure of Feinberg and Vogelstein *(12)*. A detailed procedure including the solutions needed, can be found in Chapter 37 (**Subheading 3.9.**), and also **Note 7** in this chapter. Adding 10 pmol of the random oligonucleotide used for amplification to the labeling mix containing the random hexamers will help to reach a better specific activity.

### 3.3.3. Electrophoresis of RNA

1. Preparation of the gel: Add 1 g of agarose to 72.6 mL dH$_2$O. Melt in a microwave oven. Once the flask can be handled (55–60°C), add 10 mL of 10X MOPS buffer and 17.4 mL of 37% formaldehyde. Assemble a horizontal gel cast in a fume hood, because formaldehyde vapors are toxic. Allow the gel to solidify.
2. Preparation and running of samples: From the original RNA solutions made, resuspend 10–20 mg of RNA in 7 μL of nuclease free water. Then add 15 μL deionized formamide, 3 μL 10X MOPS, 5 μL 37% formaldehyde, and 1 μL of a 1 mg/mL ethidium bromide solution. Heat at 65°C for 5–10 min. Chill on ice, add 7 μL of 6X gel loading buffer, and load on the gel that should be in a gel apparatus in 1X MOPS running buffer. Run at 5.6 V/cm (distance between electrodes), until the bromophenol blue is at 3 cm from the bottom of the gel.

### 3.3.4. Blotting and Hybridization

Wash the gel twice for 30 min each time in 10 m$M$ phosphate buffer, pH 6.8. To make this solution, mix 51 mL of 0.5 $M$ NaH$_2$PO$_4$ with 49 mL of 0.5 $M$ Na$_2$HPO$_4$. Dilute the resulting 0.5 $M$ solution to 10 m$M$. Place the gel in a blotting apparatus, and transfer to a nylon membrane in 20X SSC. Once transfer is completed, check the position of the ribosomal markers with a handheld UV lamp (sometimes they are visible without it), and mark their position with a soft pencil. Also mark the orientation of the gel and the position of the wells. Wash the membrane in 2X SSC and fix the RNA in a UV crosslinker

(Stratalinker or similar) or in an 80°C oven for 1–2 h. Prehybridize the gel in a solution containing 50% formamide, 5X Denhardt's solution, 0.1% SDS, 20 m*M* phosphate buffer, pH 7.0, 5X SSC, and 250 µg/mL of denatured salmon sperm DNA *(16)* at 42°C for at least 1 h. To denature the DNA, heat it at 100°C for 10 min, and immediately chill on ice. After prehybridization, discard the solution, and add fresh solution containing the same ingredients plus 1–5 × 10$^6$ cpm/mL of probe. Remember that the probe has to be denatured prior to adding it to the hybridization solution. Leave hybridizing overnight at 42°C.

### 3.3.5. Washings and Autoradiography

Wash the membrane twice in the following solutions: 2X SSC/0.1% SDS at room temperature for at least 15 min each; 0.1X SSC/0.1% SDS at 42°C for at least 30 min each; same solution at 50°C for at least 30 min each. Check the extent of washing with a Geiger counter. If the background is high on the membrane (high counts where no RNA is expected), repeat washing with the same solution at 55°C for at least 30 min each. If the background is still high, wash briefly several times in the same solution at 60°C. Once washed, wrap the wet membrane in Saran Wrap, and expose to double-side emulsioned autoradiography film at –80°C using an intensifying screen. Sometimes bands require several days of exposure to be seen *(15)*. Once bands are detected, the size of the mRNAs can be roughly calculated as they migrate according to the logarithm of their molecular weight. The known sizes to draw a curve should be the 28 and 18S ribosomal RNAs whose position was marked with a pencil *(16)*.

### 3.4. Alternative Priming Methods

1. Nested priming. This method was developed to normalize the PCR relative to mRNA abundance. If two different messages have equally good matches with the first arbitrary primer, but their abundance differs by 10-fold, the differentially displayed bands will differ in intensity by 10-fold. The introduction of a nested primer with a new overhanging nucleotide at the 3'-end may give a chance to the least abundant message to be amplified if it happens that the new nucleotide matches the sequence. Therefore, by adding one nucleotide at the 3' end, 1/16 of the previously amplified molecules will be further amplified. If two nucleotides overhang, then only 1/256 parts will be amplified and so forth. A practical example of nested oligonucleotide design and their use are extensively discussed in **ref. 6**.

2. For nested primer amplification, proceed as in cDNA synthesis. Subject the sample to amplification as described earlier except that only 10 high stringency cycles are to be performed. After this, transfer 3.5 µL of each reaction to tubes containing 36.5 µL of: dH$_2$O, 22.0 µL, 10X second-strand buffer 4.0 µL, 10 pmol/mL nested primer 4.0 µL, 1 *M* DTT 4.0 µL, 1 m*M* dNTP 2.0 µL, *Taq* DNA polymerase 0.5 µL. Subject the sample to 40 high stringency cycles and procceed as above (**Subheading 3.2.1.**, step 7).

### 3.4.1. Use of Shorter Oligonucleotides

Alternatively, shorter oligonucleotides could be used for cDNA production and amplification. If, for example, 10 mers are used, the temperature of the high-stringency cycles should be dropped to as low as 37°C, and a slope period of 15 s should be allowed for the 37–72°C transition to prevent denaturation of the primer from the template. Also, shorter primers will have the tendency to generate less products than longer primers. The use of Stoffel fragment for amplification instead of Ampli*Taq* DNA polymerase will help to minimize this problem *(14)*. Except for these considerations, the protocol to follow is the same as that already described for longer primers.

## 4. Notes

1. Ethidium bromide is a suspected mutagen, and should be handled and disposed accordingly.
2. The preparation of Phenol-SEVAG is described in detail in **ref. 15**. For convenience, this reagent can also be purchased ready-to-use from USB, 0.1% 8-hydroxiquinolin (w/v) can be added to prevent oxidation of phenol and for easy visualization of the organic phase.
3. Owing to the large amounts of reverse transcriptase used, cloned MuLTV RT, is preferred over AMV RT because the former is considerably cheaper, although considered of lower performance when extending long products (i.e., when synthesizing cDNA to construct a library).
4. Purification of synthetic oligonucleotides: If the oligonucleotide has the trityl group on, detritylate it by resuspending the dry pellet in 200 µL 80% acetic acid. Leave at room temperature for 30 min. Dry in a Speed-Vac. Resuspend with 50 µL dH$_2$O, 50 µL, deionized formamide, and 25 µL of formamide loading dye. For a 15- to 25-mer, prepare a 2-mm thick 15% sequencing acrylamide, 1X TBE gel in a protein gel cast. Use a wide well (5 cm), or use several smaller wells. Heat the sample at 100°C for 2 min and load. Run at 250 V until bromophenol blue (the fastest migrating dye) is at 2 cm of the bottom (if in doubt, check **ref. 15**, for the relative position of fragments in denaturing acrylamide gels according to the migration of the dyes). Disassemble the cast, place the gel on Saran Wrap and on a silica gel TLC plate with fluorescence indicator (Sigma, T-6270), and illuminate the gel with a handheld UV lamp. Wear appropiate eye protection when using short wave UV light. The DNA will appear as a shade, since it absorbs the UV light that stimulates fluorescence on the TLC plate. Degradation and unfinished synthesis products will also appear, but the bona fide band should represent over 60% of the total. Cut the gel containing the DNA with a scalpel, and mince the gel piece thoroughly. Place the puree in a 15-mL conical tube, and add 2 mL of dH$_2$O and 1 mL of TE. Shake the tube overnight at 37°C. On the next day, recover the liquid by centrifugation, and concentrate with *n*-butanol. This is done by adding equal volumes of *n*-butanol to the DNA solution, mixing well, and spinning in a tabletop centrifuge. Discard the top butanol layer, and repeat

until the final volume is 1 mL. Make sure now that you are using a polypropylene tube. Add 5 mL of chloroform, mix, spin, and recover the top layer. Adjust the volume to 1 mL, and apply the sample to a Sephadex G25 NAP-10 column (Pharmacia) previously equilibrated with $H_2O$. Elute the sample with 1.5 mL of $dH_2O$, and dry in the Speed-Vac. Resuspend the pellet in 200 μL of $dH_2O$, and measure the absorbance of 2 μL in 1 mL of $H_2O$ at 260 nm (1 $OD_{260}$ = 30 μg/mL oligo). Calculate the concentration in mol/mL (the average molecular weight of a nucleotide is 330 g/mol).

5. Special care should be taken when handling radioisotopes. Refer to a laboratory safety manual for proper handling and disposal of $^{32}P$.

6. Never autoclave solutions containing ammonia. This solution should be filter-sterilized through a 0.2-μ filter.

7. (*See also* **Subheading 3.9.** and Chaper 37) This is also available as a commercial kit. Make sure the primers contained in the kit are random hexamers (i.e., Boehringer), since longer primers will label short DNA fragments less efficiently.

8. After preparation, sonicate extensively to shear the DNA and store at –20°C. DNA can also be sheared by passing multiple times through a 21-gage needle until the material flows easily through the needle.

9. This kit includes vector, T4 DNA ligase, 10X ligation buffer, and competent cells in order to clone the PCR clones obtained. Although expensive, its use greatly facilitates the cloning of PCR products.

10. RNA can be rapidly degraded by contaminating ribonucleases. For this reason, RNase-free material must be used, and gloves should be worn throughout the whole procedure.

11. $^{35}S$ dATP can also be used.

12. If oligodT was used for cDNA synthesis, then add the arbitrary primer at this point. The optimal primer concentration in the PCR reaction ranges from 0.3–10 μ$M$. This parameter will depend on the primer sequence and on the particular primer preparation used. To minimize effects of the latter, purification of oligonucle-otides through denaturing acrylamide gels is recommended. In a typical reaction, we use the primer at 1 μ$M$.

13. This precipitation step is added to remove most of the unincorporated nucleotide that would contaminate the lower buffer during electrophoresis. It also concentrates the product so that autoradiography can be performed in a shorter period of time.

14. Preparation of a sequence marker. Digest pBR322 with *Msp*I. After digestion, heat-inactivate the enzyme at 65°C for 20 min. Remove an aliquot containing approx 100 ng and label in 20 μL containing: 2 μL 10X First strand buffer, 2 μL 0.1 $M$ DTT, 3 μL of α$^{32}P$ dCTP, and 1 μL of MuTLV RT. Bring the volume to 20 μL with $dH_2O$. Incubate at 37°C for 1 h. Add 1 μL of 0.2 $M$ EDTA, 1 μL 10 mg/mL oyster glycogen, 22 μL of 5 $M$ ammonium acetate, and 132 μL of 100% ethanol. Precipitate 5 min at room temperature, spin for 15 min at 12,000 rpm, carefully discard the supernatant, and resuspend the pellet in 50 μL of formamide loading dye. Place 1 mL in a tube and count it without scintillation fluid in a scintillation counter. Load 3000–5000 cpm/gel. The bands generated are (in bp): 622, 527,

404, 307, 242, 238, 217, 201, 190, 180, 160 + 160, 147 + 147, 123, 110, 90, 76, and 67. Some double bands may appear if the marker has not been well dena- tured. This marker will be stable for a long time at –20°C, but the half-life of the isotope has to be considered (14.3 d).

15. Instructions for the preparation and casting of this type of gels are provided in **ref. *15***.

16. Several bands may appear if the original PCR product is not homogeneous. Also, a single gene may give rise to several mRNAs.

17. The size of ribosomal RNAs varies among species.

## Acknowledgments

A. T. and J. C. I. B. are supported by the Mathers Foundation. Laboratory work of J. C. I. B. is also supported by the National Science Foundation (grant IBN 9513859) and by a Basil O'Connor Research Award from the March of Dimes. A. T. is also supported by a fellowship from the Foundation, Dr. Manuel Morales (Tazacorte, Island of La Palma, Spain).

## References

1. Welsh, J. and McClelland, M (1990) Fingerprinting genomes using PCR with arbitrary primers. *Nucleic Acids Res.* **18,** 7213–7218.

2. Williams, J. G., Kubelik, A. R., Livak, K. J., Rafalski, J. A., and Tingey, S. V. (1990) DNA polymorphisms amplified by arbitrary primers are useful as genetic markers. *Nucleic Acids Res.* **18,** 6531–6535.

3. Welsh, J., Chada, K., Dalal, S. S., Ralph, D., Chang, R., and McClelland, M. (1992) Arbitrarily primed PCR fingerprinting of RNA. *Nucleic Acids Res.* **20,** 4965–4970.

4. Liang, P. and Pardee, A. B. (1992) Differential display of eukaryotic messenger RNA by means of the polymerase chain reaction. *Science* **257,** 967–971.

5. Welsh, J., Chada, K., Dalal, S. S., Cheng, R., Ralph, D., and McClelland, M. (1992) Arbitrarily primed PCR fingerprinting of RNA. *Nucleic Acids Res.* **20,** 4965–4970.

6. Ralph, D., McClelland, M., and Welsh, J. (1993) RNA fingerprinting using arbitrarily primed PCR identifies differentially regulated RNAs in mink lung (My1Lu) cells growth arrested by transforming growth factor b1. *Proc. Natl. Acad. Sci. USA* **90,** 10,710–10,714.

7. Liang, P., Averboukh, L., Keyomarsi, K., Sager, R., and Pardee, A. B. (1992) Differential display and cloning of messenger RNAs from human breast cancer versus mammary epithelial cells. *Cancer Res.* **52,** 6966–6968.

8. Liang, P., Zhu, W., Zhang, X., Guo, Z., O'Connell, R. P., Auerboukh, L., Wang, F., and Pardee, A. B. (1994) Differential display using one-base anchored oligo- dT. *Nucleic Acids Res.* **25,** 5763,5764.

9. Zimmermann, J. W. and Schultz, R. M. (1994) Analysis of gene expression in the preimplantation mouse embryo: use of mRNA differential display. *Proc. Natl. Acad. Sci. USA.* **91,** 5456–5460.

10. Loh, E. Y., Elliot, J. F., Cwirla, S., Lanier, L. L., and Davis, M. M. (1989) Poly- merase chain reaction with single-sided specificity: Analysis of T cell receptor δ chain. *Science* **243,** 217–220.

11. Chomczynski, P. and Sacchi, N. (1987) Single step method of RNA isolation by acid guanidinium thiocyanate-phenol-chloroform extraction. *Anal. Biochem* **162,** 156–159.
12. Feinberg, A. P. and Vogelstein, B. (1983) A technique for radiolabeling DNA restriction endonuclease fragments to high specific activity. *Anal. Biochem.* **132,** 6–13.
13. Li, F., Barnathan, E. S., and Karikó, K. (1994) Rapid method for screening and cloning cDNAs generated in differential mRNA display: application of Northern blot for affinity capturing of cDNAs. *Nucleic Acids Res.* **22,** 1764,1765.
14. Honeycutt, R., Sobral, B. W., and McClelland, M. (1997) Polymerase chain reaction (PCR) detection and quantification using a short PCR product and a synthetic internal positive control. *Anal. Biochem.* **248,** 303–306.
15. Sambrook, J., Fritsh, E. F., and Maniatis, T. (eds.) (1989) *Molecular Cloning. A Laboratory Manual.* Cold Spring Harbor Laboratory, Cold Spring Harbor, NY.
16. Guimaraes, M. J., Lee, F., Zlotnik, A., and McClanahan, T. (1995) Differential display by PCR: novel findings and applications. *Nucleic Acids Res.* **23,** 1832,1833.

# 39

## RT-PCR on Embryos Using Degenerate Oligonucleotide Primers

**Anthony Graham**

## 1. Introduction

The polymerase chain reaction (PCR) is an incredibly versatile technique that has made a profound impact on many areas of biology. PCR is based on the use of oligonucleotides, which flank the region of DNA of interest and which are complementary to sequences on either DNA strand, to prime the replication of each strand by *Taq* DNA polymerase. This enzyme is thermostable, and consequently, one can go through cycles of template denaturation and renaturation after each round of replication and exponentially amplify the sequence of interest. This procedure is very sensitive, and as such, it allows one to amplify, isolate, and utilize specific DNA sequences from vanishingly small quantities of starting material using either genomic or cDNA as a substrate.

PCR can also be used to facilitate the isolation of genes that are evolutionarily related, such as cognate genes from different species or multiple members of given gene families. In these cases, one is often trying to isolate DNA fragments whose sequence is not actually known. This can be done by taking advantage of the fact that the conservation of molecules is most evident at the protein level. In these cases, the sequence of the primers to be used is derived from conserved amino acid sequences that exist between the related genes. Since usually more than one codon encodes most amino acids (**Fig. 1**), and also since the amino acid sequence conservation between proteins may not be 100%, if the primers are to cover all of the possible combinations of DNA sequence that could encode a given amino acid motif, they will necessarily be degenerate. This chapter will focus on the use of degenerate oligonucleotide primers in PCR on cDNA synthesized from embryonic tissue. Using cDNA as

From: *Methods in Molecular Biology, Vol. 97: Molecular Embryology: Methods and Protocols*
Edited by: P. T. Sharpe and I. Mason © Humana Press Inc., Totowa, NJ

| Amino acid | | Codons |
|---|---|---|
| Alanine | Ala  A | GCA  GCC  GCG  GCU |
| Cysteine | Cys  C | UGC  UGU |
| Aspartic acid | Asp  D | GAC  GAU |
| Glutamic acid | Glu  E | GAA  GAG |
| Phenylalanine | Phe  F | UUC  UUU |
| Glycine | Gly  G | GGA  GGC  GGG  GGU |
| Histidine | His  H | CAC  CAU |
| Isoleucine | Isl  I | AUA  AUC  AUU |
| Lysine | Lys  K | AAA  AAG |
| Leucine | Leu  L | UUA  UUG  CUA  CUC  CUG  CUU |
| Methionine | Met  M | AUG |
| Asparagine | Asn  N | AAC  AAU |
| Proline | Pro  P | CCA  CCC  CCG  CCU |
| Glutamine | Glu  Q | CAA  CAG |
| Arginine | Arg  R | AGA  AGG  CGA  CGC  CGG  CGU |
| Serine | Ser  S | AGC  AGU  UCA  UCC  UCG  UCU |
| Threonine | Thr  T | ACA  ACC  ACG  ACU |
| Valine | Val  V | GUA  GUC  GUG  GUU |
| Tryptophan | Trp  W | UGG |
| Tyrosine | Tyr  Y | UAC  UAU |

Fig. 1. The genetic code. Each of the amino acids is shown along with its corresponding RNA codons.

a substrate with degenerate oligonucleotides rather than genomic DNA has the advantage that one does not need to worry about introns and also that if the appropriate fragment is amplified from a cDNA sample, one can therefore conclude that this gene is expressed in the tissue from which the RNA was isolated. This allows one to amplify cDNA fragments of genes of interest that have not already been isolated from that organism, and also to survey what members of a given gene family, be they previously identified or not, are expressed in a given tissue at any particular stage of interest and to isolate cDNA fragments for each gene.

## 2. Materials
### 2.1. RNA Preparation

1. All of the reagents should sterile and where possible treated with diethylpyrocarbonate to inactivate any RNases *(1)*. Solution D: 25 g guanidinium thiocyanate, 29.3 mL water, 1.76 mL 0.75 $M$ Na citrate, pH 7, 2.64 mL 10% sarkosyl, 38 mL β-mercaptoethanol. Store at $-70°C$.
2. Na acetate, pH 4.0.
3. Acid phenol: phenol that is water-saturated, but unbuffered.
4. Chloroform.
5. Glycogen.
6. Ethanol.

### 2.2. cDNA Synthesis

Use cDNA synthesis kit—Boehringer Mannheim (# 1 483 188), Promega (# A3500), or Pharmacia (# 27-9261-01).

### 2.3. Primers

There are a number of companies that will synthesize oligonucleotides to order.

### 2.4. PCR Reaction

1. *Taq* DNA polymerase, including 10X buffer (Mg-free or not) and $MgCl_2$ available from a number of suppliers, such as Promega.
2. Deoxynucleotide triphosphates (Boehringer Mannheim) or equivalent.
3. Paraffin oil.

## 3. Methods
### 3.1. RNA Preparation

This method is basically that of Chomczynski and Sacchi *(2)*—it works well from as little as $100–10^6$ cells and it can be scaled up if necessary.

1. Add 100 μL of solution D to $10^6$ cells or 10 mg tissue in a microcentrifuge tube, and vortex. Most tissues break up almost immediately, but tough tissues may need to be syringed.
2. Add 10 μL of 2 $M$ Na acetate, pH 4.0, and mix.
3. Add 100 μL of water saturated unbuffered phenol, and mix.
4. Add 20 mL of chloroform, and mix very well. The sample should become cloudy at this stage. If it does not, add extra chloroform.
5. Chill the sample on ice for 15 min, and then spin in a microfuge for 15 min.
6. Take the top phase into a new tube. If one is dealing with a small amount of tissue, one can add 1 μL of RNase-free glycogen at 20 mg/mL to aid precipitation.
7. Add 240 μL of ethanol, and put at $-20°C$ for 30 min.
8. Spin for 15 min in a microcentrifuge.

9   Wash the pellet with 70% ethanol, and drain.
10. Redissolve the RNA in 20 μL H$_2$O and store at –20 or –70°C with the latter being better for long-term storage.

## 3.2. cDNA Synthesis

For PCR, only the first cDNA strand needs to be synthesized, since the second strand is copied in the first cycle of the PCR reaction. Many companies sell first-strand synthesis kits, and these are recommended. This reaction uses the enzyme reverse transcriptase, which is an RNA-dependent DNA polymerase, i.e., it copies RNA into DNA. The two most commonly used enzymes are avian myeloblastosis virus (AMV) reverse transcriptase and Moloney murine leukemia virus (M-MuLV) reverse transcriptase. These enzymes have slightly different activities, but both function adequately. It should also be noted that AMV is used at 42°C, whereas M-MuLV works best at 37°C. The RNA extraction procedure outlined above yields a sample of total RNA and, therefore in addition to including mRNA, it also contains rRNA and tRNA. To copy mRNA preferentially in the reverse transcription reaction, the primer that is used is an oligo dT oligomer. This will bind specifically to the poly (A) tail of mRNA molecules and ensure that they are replicated in preference to the other RNAs in the sample. The other reagents include the appropriate buffer for the reverse transcriptase enzyme, deoxynucleotide triphosphates from which the DNA will be synthesized, RNase inhibitor, and water to make up the volume. The procedure is carried out according to the instructions of the manufacturer of the cDNA kit. In a first-strand synthesis reaction, I would use between 2 and 5 μL of the RNA extracted as above, depending on the amount of starting material.

## 3.3. Primer Design

There are a number of factors that should be taken into consideration when designing degenerate oligonucleotide primers. These will be illustrated using the isolation of chick activin type II receptors by degenerate RT-PCR as an example. The amino acid sequences of activin type II receptors from both mouse and *Drosophila* are shown in **Fig. 2** *(3)*. There are a number of regions that are well-conserved between these family members, and the amino acid sequences that were chosen to make primers against are highlighted, and their corresponding DNA sequences shown in **Fig. 2**. In this case, nested primers were employed, two external and one internal. The utilization of nested primers increases the specificity of the reaction owing to the fact that the substrate for the second and final round of PCR is considerably less complex than a total cDNA sample, since it has already been selectively amplified in the first round of PCR with the external primers. The two external primers were used in the first round of PCR, and then a small aliquot was taken, and in the second round this was subjected to further amplification using the internal primer and the 5'-external primer. The primers are spaced such that the final amplified sequence will be around 500 bp, which is a useful size, because it also allows one to use this fragment for other purposes, such as whole-mount *in situ* hybridization immediately after cloning it. The sequence of the primers to be used is dependent on the available amino acid sequences. The primers should be as nondegenerate as possible. This in effect means trying to avoid amino acid sequences that are comprised of amino acids encoded by multiple codons, such as arginine, serine, or leucine (**Fig. 1**). That having been said, an overriding consideration in primer design is that the 3' of the primers end be as close as possible to being unique. This in turn means looking for methionines or tryptophans in any peptide sequence, and since each of these amino acids is only encoded by one codon, making the 3'-end of the oligo correspond to these positions. If neither of these amino acids is conveniently located, then the next best option is to scan for amino acids that are only encoded by two codons, both of which will only vary at the third base. Therefore, in the case of the oligonucleotide which will prime the replication of the sense strand, to ensure that there is an exact match at the 3'-end the oligonucleotide should terminate at the second base of that codon. This is not a problem for the oligonucleotide that primes the replication of the antisense strand, since the two most 3'-bases of this oligo will be complementary to the first two positions of the codon, which are often invariant. As can be seen from **Fig. 2**, the activin receptor type II primers are all anchored on methionine and tryptophan codons and they cover sequence encoding six amino acids in all and should therefore be long enough to be stable at 72°C, which is the optimal temperature for *Taq* polymerase. These oligonucleotides also have restriction enzyme sites added at their 5'-ends—*Eco*RI in the case of the 5'-external oligo and *Xba*I for both the 3'-oligos—to aid the subcloning of the amplified fragment. To ensure that the restriction enzymes cut efficiently, two extra bases have been added to the 5'-end. The choice of the enzymes that are used and bases added at the 5'-end to enhance digestion are based on available information, such as that in the appendices of the New England Bioloabs catalog. It is of course possible that the amplified sequence may also contain an internal sites for one of the restriction enzymes that is chosen. If this is the case, then the PCR product can be cloned without prior digestion in the standard manner for blunt-end ligation, or it may be directionally cloned by only digesting with the enzyme that does not cleave internally and using the other end as a blunt end.

### 3.4. PCR Reaction

Because the oligonucleotides that are used are degenerate, there are no "standard" reaction conditions that will invariably work. The reaction parameters

**A**                                                                    1

```
Atr II

        IPTHEAEITNSSPLLSNRPIQLLEQKASGRFGDVWQAKLNNQDVAVKIFRMQEKESW

ActR-IIB2

        DIHEDPGPPPPSPLVGLKPLQLLEIKARGRFGCVWKAQLMNDFVAVKIFPLQDKQSW

ActR-II

        VPTQDPGPPPPSPLLGLKPLQLLEVKARGRFGCVWKAQLLNEYVAVKIFPIQDKQSW


Atr II        TTEHDIYKLPRMRHPNILEFLGVEKHMDKPEY--WLISTYQHNGSLCDYLKSHTISW

ActR-IIB2     QSEREIFSTPGMKHENLLQFIAAEKRGSNLEVELWLITAFHDKGSLTDYLKGNIITW

ActR-II       QNEYEVYSLPGMKHENILQFIGAEKRGTSVDVDLWLITAFHEKGSLSDFLKANVVSW


Atr II        PELCRIAESMANGLAHLHEEIPASKTDGLKPSIAHRDFKSKNVLLKSDLTACIADFG

ActR-IIB2     NELCHVAETMSRGLSYTHEDVPWCRGEGHKPSIAHRDFKSKNVLLKSDLTAVLADFG

ActR-II       NELCHIAETMARGLAYLHEDIPGLK-DGHKPAISHRDIKSKNVLLKNNLTACIADFG


                                         3

Atr II

        LAMIFQPGKPCGDTHGQVGTRRYMAPEVLEGAINFNRDAFLRIDVYACGLVLWEMVS

ActR-IIB2

        LAVRFEPGKPPGDTHGQVGTRRYMAPEVLEGAINFQRDAFLRIDMYAMGLVLWELVS

ActR-II

        LALKFEAGKSAGDTHGQVGTRRYMAPEVLEGAINFQRDAFLRIDMYAMGLVIWELAS


Atr II        RCDFA-GPVGEFQLPFEAELGLRPSLDEVQESVVMKKLRPRLLNSWRAHPGLNVFCD

ActR-IIB2     RCKAADGPVDEYMLPFEEEIGQHPSLEELQEVVVHKKMRPTIKDHWLKHPGLAQLCV

ActR-II       RCTAADGPVDEYMLPFEEEIGQHPSLEDMQEVVVHKKKRPVLRDYWQKHAGMAMLCE


                             2

Atr II        TMEECWDHDAEARLSSSCVMERFAQLNKYPSTQ

ActR-IIB2     TIEECWDHDAEARLSAGCVEERVSLIRRSVNGT

ActR-II       TIEECWDHDAEARLSAGCVGERITQMQRLTNII
```

Fig. 2. (**A**) portion of the amino acid sequences of the *Drosophila* Act-R gene and the mouse Type II Activin receptors *(3)*. The motifs shown in bold are those on which the PCR primers are based.

## B

5' external primer -1 -5' CGGAATTCAGATTCGGATGCGTATGG 3'

```
          C C   T   C   T   C
          G       G       G
          T       T       T
```

3' external primer - 2-5' GCTCTAGACTCAGCATCATGATCCCA 3'

```
          T   C   G   G   G
              G
              T
```

internal primer  - 3 -5' GCTCTAGAAAAAACCTCAGGAGCCAT 3'

```
        C GC   T   C   C
        G  G       G   G
        T  T       T   T
```

Fig. 2. *(continued)* **(B)** The sequences of the primers are shown underneath. Note that the 5'-primer also has an added *Eco*RI site and the two 3'-primers an *Xba*I site.

must be empirically determined for each primer pair. The two variables that are most frequently altered are the concentration of $MgCl_2$ in the reaction and the temperature at which the oligonucleotides anneal to the substrate. In a typical reaction, the author would initially try three final concentrations of $MgCl_2$, which span the normal range that is usually effective—1, 1.5, and 2 m$M$. If these concentrations are not effective, then concentrations from 0.5–5 m$M$ can also be tested. In the case of exact match primers, one tends to use an annealing temperature that is 10°C below the melting temperature of the oligo. This is obviously not applicable to degenerate oligos, which are a mix of oligos with a range of melting temperatures. Ideally, the higher the annealing temperature that can be used, the greater the specificity of the reaction but the temperature should also be low enough to allow the oligos to anneal to all of the potential target sites of interest. A useful annealing temperature to start with is 50°C. If at this temperature no sequence is amplified then the temperature can be dropped even to as low as 37°C, although reactions using these lower temperatures will invariably produce spurious amplification products at a much higher frequency. If annealing at 50°C produces multiple amplified products, then the annealing temperature should be raised. Some of the background may be owing to the primers annealing and *Taq* extending at lower temperatures as the reaction mix warms up. One way to reduce this sort of background amplification is to "hot start" the reaction. This is achieved by adding the *Taq* polymerase to reaction mix that has been denatured at 94°C and then cooled to 55°C or less effectively, but more simply by making up the complete PCR reaction mix, including Taq, on ice and then transferring the tubes straight to a block that has been prewarmed to 94°C.

The PCR reaction consists of the following components: *Taq* polymerase buffer, MgCl$_2$, deoxynucleotide triphosphates (dATP, dCTP, dGTP, dTTP) at a final concentration of 0.2 m$M$, oligonucleotide primers that should be in excess and are often used at concentrations of 0.5–1 m$M$, 1 U of *Taq* polymerase, target cDNA, and H$_2$O to make up the volume.

A typical PCR reaction of 100 μL using *Taq* polymerase purchased from Promega would be as follows:

1. Add the following to a microcentrifuge tube:
   10X Reaction buffer (Mg$^{2+}$ free) 10 μL.
   25 m$M$ MgCl$_2$ 4 μL (1 m$M$), or 6 μL (1.5 m$M$), or 8 μL (2 m$M$).
   dNTPs (20 m$M$) stock 1μLl.
   External primers (at a concentration of 0.5 μg/μL approx 50 μ$M$) 1 μL of each.
   *Taq* polymerase (5 U/mL) 0.2 μL.
   H$_2$O 78, 76, or 74 μL.
   cDNA 5 μL.
2. Overlay with 50 μL of paraffin oil to prevent evaporation. This step is not necessary if one is using a thermal cycler with a heated lid.
3. Place the sample in the thermal cycler and carry out the amplification. A typical amplification protocol is:
   First cycle (X1)—denature at 94°C for 2 min, ×1
   Subsequent cycles (×30)—denature, 94°C for 30 s.
            anneal 50°C for 2 min.
            extend 72°C for 2 min.
   Final cycle (×1)–extend 72°C for 10 min.
4. For the second round of amplification, use the internal primer and the appropriate external primer, and use 5 μL of the completed first-round PCR reaction mix as the DNA substrate. Mix the reagents and carry out the amplification as for the first round again with the range of Mg concentrations. One other factor that can be varied is the amount of first-round reaction that is added as a substrate for the second round. In some instance, the second round may be a lot cleaner if considerably < 5 μL is added.
5. At the end of the second round, run a 10-μL aliquot of all samples, i.e., all Mg concentrations from both the first and second rounds, on an agarose gel. **Figure 3** shows the results obtained with the activin type II receptor primers. As can be seen the first round of amplification with the external primers did not produce a band at any of the Mg concentrations used, but in the second round with the internal primer and the 5'-external primer, a fragment of the predicted size was

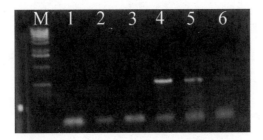

Fig. 3. Ethidium bromide-stained agaraose gel electrophoresis of the products PCR amplification. M is the marker track (1-kb ladder from Gibco-BRL), 1–3 are from the first round of amplification and 4–6 are from the second. The reactions run in lanes 1 and 4 were at 1 m$M$ MgCl$_2$, 2 and 5 at 1.5 m$M$, and 3 and 6 at 1.5 m$M$.

amplified under all Mg concentrations, although the 1-m$M$ concentration was clearly more efficient.

If the amplification has worked and an appropriate-sized fragment is present, the PCR reaction mix should then be cleaned up by phenol/chloroform extraction, especially if using oil, then ethanol-precipitated and redissolved in a small volume, and then digested with the appropriate restriction enzymes. The digest should then be run on a gel, the fragment isolated and subcloned into an appropriate vector, and the products sequenced. In some cases, all of the resulting clones will contain fragments of members of the gene family of interest, but in a number of cases, one will also find that other unrelated products have been amplified. This last point merely serves to stress the importance of sequencing the products of the amplification.

## References

1. Sambrook, J., Fritsch, E. F., and Maniatis, T. (eds.) (1989) *Molecular Cloning: A Laboratory Manual.* Cold Spring Harbor Laboratory, Cold Spring Harbor, NY.
2. Chomczynski, P. and Sacchi, N. (1987) Single step method of RNA isolation by guanindinium thiocyanate-phenol-chloroform extraction. *Anal. Biochem.* **162,** 156–159.
3. Childs, S. R., Wrana, J. L., Arora, K., Attisano, L., O'Connor, M. B., and Massague, J. (1993) Identification of a *Drosophila* activin receptor. *Proc. Natl. Acad. Sci. USA* **90,** 9475–9479.

# 40

# Single-Cell RT-PCR cDNA Subtraction

Damian L. Weaver, César Núñez, Clare Brunet, Victoria Bostock, and Gerard Brady

## 1. Introduction

A major problem in trying to understand complex developmental processes is one of heterogeneity at both the cellular and molecular levels. At the cellular level it is often difficult to identify cells that are undergoing developmental changes and to establish the stage of differentiation they have reached. At the molecular level, there is then a problem in establishing which of the many thousands of expressed genes are playing a role in development. Several approaches to identifying expressed candidate developmental genes are based on comparing the mRNA expression patterns in cells before and after transition points. Differential screening of cDNA libraries with labeled total cDNA probes from contrasting cell samples *(1)* provides a simple means of identifying genes that are expressed at high levels in one of the samples. cDNA subtraction protocols *(2,3)* have increased the sensitivity of this type of approach by removing sequences expressed in both samples. The limitation of these approaches lies in the need for large amounts of starting material necessary for cDNA preparation and subtraction. With the advent of the polymerase chain reaction (PCR) *(4)* and the development of techniques that allow amplification of target sequences expressed in single cells *(5)*, the limitations of applying cDNA subtraction and differential screening have been removed.

Here we describe the details of a cDNA subtraction approach previously applied to single cells *(6)*, which is based on the PolyAPCR method used to produce amplified cDNA representing all mRNAs present in samples as small as a single cell *(7,8)*. PolyAPCR has been applied successfully to a wide range of cell samples, including single micromanipulated cells *(6,7,9,10)*, antibody fractionated populations *(11)*, and fixed tissues *(12)*. The details of the method

From: *Methods in Molecular Biology, Vol. 97: Molecular Embryology: Methods and Protocols*
Edited by: P. T. Sharpe and I. Mason © Humana Press Inc., Totowa, NJ

are based on a genomic DNA subtraction protocol utilizing biotin addition and xvorganic extraction *(13)*, and the final enriched probes are used to screen full-length cDNA libraries. The advantages of this procedure over the related method of differential display are: the end products are full-length cDNAs, all differences between the starting samples are present after subtraction, and the flexibility of the initial *PolyAPCR* procedure enables subtraction to be applied to essentially any situation. The disadvantages of *PolyAcDNA* subtraction are that it can only be applied to relatively few samples, and any clone containing repetitive sequences is likely to be lost.

## 2. Materials

All reagents are of molecular biology grade and obtained from Sigma (St. Louis, MO) unless otherwise stated.

### 2.1. Preparation of PCR Material

1. 10X *Taq* buffer: 100 m$M$ Tris-HCl, 15 m$M$ MgCl$_2$, 500 m$M$ KCl, pH 8.3.
2. 25 m$M$ dNTP mix: 1:1:1:1 mix of 100 m$M$ dATP, dCTP, dGTP, and dTTP, (Boehringer Mannheim, Mannheim, Germany), store in 50-μL aliquots at –20°C.
3. Oligonucleotides (Pharmacia, Piscataway, NJ):
    *Not*IdT: cat ctc gag cgg cgg ctt ttt ttt ttt ttt ttt ttt ttt t: Primary PCR only.
    *Not*I24: gcg gcc gct ttt ttt ttt ttt ttt: Amplification of driver.
    *Not*IUnique: cat ctc gag cgg ccg ctt ttt ttt : Alternative tracer recovery oligo.
    T7dT: gat ctg cac gcg taa tac gac tcg agt ttt ttt ttt ttt ttt: Amplification of tracer.
    T7Unique: gat ctg cac gcg taa tac gac tc: Recovery of tracer after subtraction.
4. Water.
5. *Taq* polymerase (Boehringer Mannheim).
6. PCR reaction mix: 1X *Taq* polymerase reaction buffer, 0.25 m$M$ dNTP, 0.5 OD$_{260}$/mL oligonucleotide, 1.2 U/100 μL *Taq* polymerase.
7. 1.7% (w/v) agarose gel (SeaKem, FMC BioProducts, Rockland, ME).
8. 1XTBE: 0.9 $M$ Tris-borate, 0.002 $M$ EDTA, pH 8.0.
9. Ethidium bromide (10 mg/mL) (as a possible carcinogen, adequate safety precautions must be taken).
10. PCR product of known concentration.
11. Qiagen Tip-20.
12. QP buffer: 400 m$M$ NaCl, 40 m$M$ MOPS, pH 7.0.
13. QB-1 buffer: 850 m$M$ NaCl, 50 m$M$ MOPS, pH 7.0.
14. QF buffer: 1.2 $M$ NaCl, 50 m$M$ MOPS, pH 8.0, 15% ethanol.
15. QIAspin PCR purification kit (Qiagen, Chatsworth, CA).
16. HE buffer, pH 8.0: 10 m$M$ HEPES, 1 m$M$ EDTA.

### 2.2. Photobiotinylation of Purified Driver cDNA

1. Photobiotin: 1 mg/mL in water, aliquot and store at –20°C in lightproof tubes.
2. UV source: 300 W ultra vitalux bulb (Osram, Sylvania, Danvers, MA). (Do not allow exposure of skin to UV.)

3. 200 m*M* Tris-Cl, pH 9.0.
4. 7.5 *M* Ammonium acetate.
5. Ethanol: 100, 80, and 70%.
6. 1 *M* EPPS, pH 8.5 and 8.25, autoclave and filter-sterilize.
7. TE buffer: 10 m*M* Tris-HCl, 1 m*M* EDTA, pH 8.0.
8. Extraction buffer: 50 m*M* EPPS, pH 8.5, 0.5 *M* NaCl, 2 m*M* EDTA.
9. Streptavidin: 4 µg/µL in 50 m*M* EPPS, pH 8.25, store at –20°C in 5-µL aliquots to avoid freeze-thaw problems.
10. TE-saturated phenol: chloroform (1:1). (Prepare fresh immediately prior to use do not store). Add 700 µL phenol/chloroform (Gibco-BRL, Gaithersburg, MD) to 700 µL TE, pH 8.0, vortex mixture, centrifuge, decant aqueous phase, and repeat 3×.
11. TE-saturated 2-butanol: 25 mL of 2-butanol are mixed with 25 mL TE, pH 8.0. The mixture is centrifuged briefly and the lower aqueous phase is removed. This is repeated. The 2-butanol is in the upper phase.

## *2.3. Hybridization of Driver and Tracer*

1. 5 mg/mL tRNA: store at –20°C in small aliquots.
2. 3 *M* NaAc, pH 8.0
3. 10X Hybridization buffer: 100 m*M* EPPS, pH 8.25, 10 m*M* EDTA, pH 8.0, 1% SDS. Store at –20°C in aliquots.
4. 40% (w/v) PEG 8000, in water.
5. 5 *M* NaCl.
6. Subtractive hybridization buffer (prepare fresh): 12.5% PEG 8000, 1.5 *M* NaCl, 1X hybridization buffer.

## *2.4. Assessing the Enrichment of Probes Generated by Subtraction*

1. $^{32}$P dCTP 3000 Ci/mmol (Amersham, Amersham, UK).
2. dNTP labeling mix: 27 m*M* dATP, 27 m*M* dTTP, 27 m*M* dGTP, 16 m*M* dCTP.
3. Southern hybridization buffer: 50% deionized formamide (BDH), 1 *M* NaCl, 1%SDS, 10% dextran sulfate.

## 3. Methods

### *3.1. Preparation of PCR Material*

The amplification conditions described are those that have given reproducible results. However, optimum amplification conditions (e.g., $Mg^{2+}$, oligonucleotide, and dNTP concentrations) should be established empirically for each oligonucleotide. When designing the X-dT oligonucleotides (e.g., *Not*I dT), it was found that oligonucleotides with an X component of 15–36 bases gave good amplification, but it was found that shorter oligonucleotides generally gave a lower yield (*see* **Note 1**).

1. Amplification of driver. Add 1 ng of material previously amplified by *Not*I dT to a 400 µL PCR reaction using *Not*I 24 as the amplification oligonucleotide. Ther-

mal profile, 1 min 20 s at 94°C, 1 min 20 s at 42°C, 2 min 30 s at 72°C, × 25 cycles.

2. Amplification of tracer. Add 1 ng of material previously amplified by *Not*I dT to a 200 μL PCR reaction using T7dT as the amplification oligonucleotide. Thermal profile: 1 min at 94°C, 1 min at 42°C, 2 min at 72°C, × 25 cycles.

3. Purify driver-amplified material using a Qiagen Tip-20 following the manufacturer's protocol. However, to avoid losses in yield because of the small size of the PCR products, ensure that the wash buffer QB1 given above replaces the original QB buffer supplied by the manufacturer (*see* **Note 2**).

4. Concentrate the eluted material (final vol 0.8 mL) by precipitating with 80 μL 3 *M* NaAc and 650 μL isopropanol. Chill on wet ice for 20 min, and spin full speed in a refrigerated centrifuge for 15 min. Wash pellet with 70% ethanol, dry, and resuspend in 50 μL HE (*see* **Note 3**).

5. Purify tracer-amplified material by a QIAspin PCR purification column using the standard manufacturer's protocol.

6. Quantitate recovery of PCR material by gel electrophoresis, by comparing recovered material to a dilution series of a well-characterized PolyA PCR sample.

## 3.2. Photobiotinylation of Purified Driver cDNA

We have compared the efficiency of incorporating biotin in to the driver by the use of biotinylated oligonucleotides and nucleotides both separately and in conjunction with each other to that of the efficiency of photobiotinylation, and have repeatedly found that photobiotinylation is the most efficient and reliable method of biotin incorporation. However, despite the reliability of photobiotinylation, it is recommended that the efficiency of biotin incorporation into the driver is tested functionally as described prior to its use in the subtractive hybridization reaction.

1. Boil 40 μL (50–100 μg) driver cDNA for 2 min, then snap-cool on ice prior to adding 40 μL of photobiotin (1 mg/mL), and mix well. Place tube with the lid open 10 cm from the UV source, and irradiate for 5 min. Leaving the UV source on (to preserve the bulb), take out the sample, and mix by flicking the tube and replace under the UV for an additional 5 min.

2. Remove sample, add a further 40 μL of photobiotin, mix well, and replace under UV for a further 5 min. Then add an equal volume of 200 m*M* Tris-Cl, pH 9.0, to stop the reaction.

3. Free photobiotin may be removed by precipitating the DNA by adding 80 μL 7.5 *M* ammonium acetate and 560 μL 100% ethanol. Chill on ice 15 min, and spin full speed in a refrigerated centrifuge for 15 min. Discard supernatant, and wash pellet 1X with 80% ethanol, dry orange/brown pellet, and resuspend the biotinylated DNA pellet in 50 μL HE, and re-estimate concentration (*see* **Note 4**).

4. Efficient biotinylation may then be assessed by adding 50 μL of fresh extraction buffer and 2 μL of 4 μg/μL streptavidin to 2 μg biotinylated driver and incubating for 2 min at room temperature. At this stage, remove 10 μL for gel analysis, prior

to adding 50 μL TE-saturated phenol/chloroform. Vortex the sample for 30 s and then spin at full speed for 3 min. Remove the top three quarters of the aqueous phase (~25–30 μL) avoiding the interphase, and transfer to a fresh tube. Compare 10 μL of this extracted aqueous phase to the 10 μL removed before subtraction by agarose gel electrophoresis.

### 3.3. Subtractive Hybridization

Common tracer sequences are removed by solution hybridization with the biotinylated driver, followed by removal of all biotin-containing hybrids after addition of streptavidin and organic extraction.

1. To a 0.5-mL tube, add 200 ng tracer to 4 μg driver. Then add 5 μg tRNA to act as a carrier, adjust the volume to 80 μL with HE buffer, mix well and boil for 2 min. To precipitate, add 8 μL 3 *M* NaAC and 200 μL 100% ethanol. Vortex briefly, chill on ice for 20 min, and spin at full speed in a refrigerated centrifuge. Wash pellet with 70% ethanol and air-dry.
2. Carefully resuspend pellet in 9 μL 1X hybridization buffer, cover with 70 μL mineral oil, put on tube cap locks to prevent tubes from bursting open then boil for 5 min.
3. Place sample PCR machine preheated to 80°C. Run the following thermal profile: 5 min at 80°C, ramp 80⁻68°C over 15 min, 90 min at 68°C, 10 min at 80°C.
4. To the sample at 80°C on the PCR machine **immediately** add 91 μL of extraction buffer (preheated to 80°C). Centrifuge briefly to ensure all the aqueous layers have combined. Remove as much mineral oil from the top as possible.
5. Centrifuge again, and transfer the bottom 75 μL of hybridization/extraction mixture to a fresh tube containing 21 μL of fresh extraction buffer and 4 μL of 4 μg/μL streptavidin. Mix and incubate at room temperature for 2 min.
6. Add 100 μL TE-saturated phenol/chloroform, vortex for 30 s, and spin at full speed room temperature for 5 min. Remove top three-quarters of the aqueous phase (~75 μL) avoiding the interphase and transfer to a fresh tube.
7. Extract the transferred aqueous phase with 50 mL of chloroform vortex for 30 s, spin at full speed at room temperature for 5 min, and transfer as much of the aqueous layer as possible to a fresh tube. Use ~60 μL of this for the next round of subtraction. Save the remainder ~10 μL for later amplification. Use 5 μL of subtracted material for amplification with the T7Unique oligo.
8. The 60 μL of extracted tracer removed in **step 8** are then carried through further sequential rounds of subtraction, and the above steps are directly repeated with the extracted tracer substituting the 200 ng of starting tracer in **step 1**.
9. Repeat extraction cycles 3–4 times.
10. Amplification of subtracted material: Add 5 μL of subtracted material to a 100 μL PCR reaction using T7Unique as the amplification oligo. Thermal profile, 30 s at 94°C, 30 s at 60°C, 90 s at 72°C ×25 cycles.

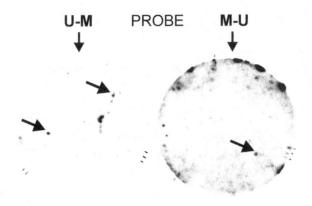

cDNA LIBRARY REPLICA FILTERS

Fig. 1. Southern hybridization of starting PolyAcDNA with enriched and nonenriched probes. The starting PolyAcDNA samples **U** and **M** are pools of PolyAcDNAs derived from 20 unilineage mouse hematopoietic precursors (**U**) and five multilineage mouse hematopoietic precursors (**M**) *(6)*. Replica filters of the starting samples were hybridized with unsubtracted (probes **U** and **M**) and with PolyAcDNA remaining after four sequential rounds of subtraction (probes **U–M** and **M–U**).

## 3.4. Hybridization with Subtracted Probes

Southern and colony hybridization can with labeled subtracted material be used to assess the efficiency of subtraction and to obtain full-length cDNA clones of differentially expressed genes (**Figs. 1** and **2**).

1. Prepare Southern filters of the starting samples and replica filters of λ cDNA libraries using standard protocols *(14)*.
2. Label PolyAcDNA probes either by random priming or by incorporating the radioactive nucleotide by PCR using the following conditions: 1X *Taq* buffer, 1:100 dNTP labeling mix, 0.5 $OD_{260}$/mL of the oligo used to generate the tracer, 1.2 U *Taq* polymerase, 5 μL $^{32}$P dCTP (final vol 100 μL). Thermal profile: 30 s at 94°C, 30 s at 60°C, 90 s at 72°C, ×20 cycles.

## 4. Notes

1. Hybridization conditions were optimized after a series of test subtractions in which tracer cDNAs were radioactively labeled and the removal of shared sequences monitored by measuring the amount of radioactivity remaining in the extracted aqueous phase. Over 80% of tracer sequences could be reproducibly removed in one round using the above hybridization conditions leading to 99% removal of common sequences after four sequential rounds of subtraction.
2. With some preparations of the oligonucleotide T7dT, we have noticed PCR artifacts which were generated in the initial conversion of the primary *Not*I dT-

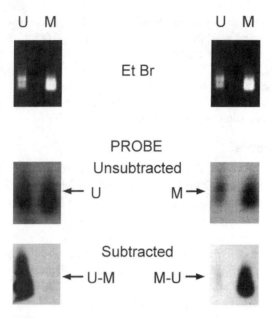

U  M                    U  M

Et Br

PROBE
Unsubtracted

←  U          M →

Subtracted

← U-M      M-U →

Fig. 2. Differential screening of a human λ cDNA library with subtracted PolyAcDNA probes. Two replica filters were produced from a single Petri dish containing approx 2000 λ plaques. One filter was hybridized with a probe enriched for sequences expressed in unilineage cells (**U–M**), and the other hybridized with a probe enriched for sequences expressed in multilineage cells. The arrows indicate clones that hybridized strongly with one probe and not the other.

amplified PolyAcDNA to T7 amplifiable tracer cDNA. Since these artifacts are not present in the *Not*I dT driver cDNA, they are enriched during subtraction and can form a significant part of the subtracted material if there are only minor differences between the subtraction partners. If this is a problem, to avoid this possibility, the tracer is maintained as with the may be avoided as follows.

**Subheading 3.1., step 1**: Substitute T7dT for *Not*I 24 in the PCR reaction to generate the driver.

**Subheading 3.1., step 2**: Substitute *Not*I dT for T7dT in the PCR reaction to generate the tracer.

**Subheading 3.3., step 6**: Substitute *Not*I Unique for T7Unique to amplify enriched tracer sequences after subtractive hybridization.

3. When washing Southern filters containing PolyAcDNAs hybridized with a PolyAcDNA probe, the probe will anneal to the polyA/T ends present in all PolyAcDNA molecules making it necessary to wash at high stringency (0.2X SSC) at temperatures up to 73°C.

4. Hybridization and washing conditions for library screening should be carried out at a lower stringency, particularly when probe and library are derived from dif-

ferent species. In the example shown in **Fig. 2**, a mouse was used to screen a human library, and washing stringency was kept low (45°C and 2X SSC).

5. The ultimate success of a screening procedure will often be judged by a functional test of the identified gene. For this reason, it is important to consider parameters, such as the average size of inserts in the library (reflecting the probability of having full-length cDNAs) and the presence of suitable selection markers and promoters in the cDNA vector.

## Acknowledgments

This work was supported by The Leukaemia Research Fund, The Cancer Research Campaign, and The Mark Richardson Memorial Trust. We would like to thank David Masters for contributing to the latter part of this work.

## References

1. St. John, T. P. and Davis, R. W. (1979) Isolation of galactose-inducible DNA sequences from *Sachromyces cerevisiae* by differential plaque filter hybridisation. *Cell* **16,** 443.

2. Zimmerman, C. R., Orr, W. C., Leclerc, R. F., Barnard, E. C., and Timberlake, W. E. (1980) Molecular cloning and selection of genes regulated in *Aspergillus* development. *Cell* **11,** 709.

3. Timberlake, W. E. (1980) Developmental gene regulation in *Aspergillus nidulans*. *Dev. Biol.* **78,** 497.

4. Saiki, R. K., Scharf, S., Faloona, F., Mullis, K. B., Horn, G. T., Erlich, H. A., and Arnheim, N. (1985) Enzymatic amplification of α-globin genomic sequences and restriction site analysis for diagnosis of sickle cell anaemia. *Science* **239,** 1350.

5. Rappolee, D. A., Wang, A., Mark, D., and Werb, Z. (1989) Novel method for studying mRNA phenotypes in single or small numbers of cells. *J. Cell Biochem.* **39,** 1–11.

6. Brady, G., Billia, F., Knox, J., Hoang, T., Kirsch, I. R., Voura, E. B., Hawley, R. G., Cumming, R., Buchwald, M., Siminovitch, K., Miyamoto, N., Boehmelt, G., and Iscove, N. (1995) Analysis of gene expression in a complex differentiation hierarchy by global amplification of cDNA from single cells. *Curr. Biol.* **5,** 909–922.

7. Brady, G., Barbara, M., and Iscove, N. N. (1990) Representative in vitro cDNA amplification from individual hemopoietic cells and colonies. *Meth. Mol. Cell. Biol.* **2,** 17–25.

8. Brady, G. and Iscove, N. N. (1993) Amplified representative cDNA libraries from single cells, in *Methods in Enzymology*, vol. 225 (Wassarman, P. M. and DePamphilis, M. L., eds.), Academic, San Diego, pp. 611–623.

9. Trumper, L. H., Brady, G., Bagg, A., Gray, D., Loke, S. L., Griesser, H., Wagman, R., Braziel, R., Gascoyne, R. D., and Vicini, S. (1993) Single-cell analysis of Hodgkin and Reed-Sternberg cells: molecular heterogeneity of gene expression and p53 mutations. *Blood* **81,** 3097–3115.

10. Dulac, C. and Axel, R. (1995) A novel family of genes encoding putative phero-mone receptors in mammals. *Cell* **83,** 195–206.
11. Cumano, A., Paige, C. J., Iscove, N. N., and Brady, G. (1992) Bipotential precur-sors of B cells and macrophages in murine fetal liver. *Nature* **356,** 612–615.
12. Cano-Gauci, D. F., Lualdi, J. C., Ouellette, A. J., Brady, G., Iscove, N. N., and Buick, R. N. (1993) *In vitro* cDNA amplification from individual intestinal crypts: a novel approach to the study of differential gene expression along the crypt-villus axis. *Exp. Cell Res.* **208,** 344–349.
13. Barr, F. G. and Emmanuel, B. S. (1990) Application of a subtraction hybridiza-tion technique involving photoactivatable biotin and organic extraction to solu-tion hybridization analysis of genomic DNA. *Anal. Biochem.* **186,** 369–373.
14. Sambrook, J., Fritsch, E. F., and Maniatis, T. (1989) *Molecular Cloning: A Labora-tory Manual.* Cold Spring Harbor Laboratory, Cold Spring Harbor, NY.

# 41

## *In Situ* Hybridization of Radioactive Riboprobes to RNA in Tissue Sections

**Radma Mahmood and Ivor Mason**

## 1. Introduction

The development of such techniques as transgenesis, saturation mutagenesis, and the polymerase chain reaction (PCR), all of which are described in detail elsewhere in this volume, has revolutionized experimental embryology. However, no single procedure has been more broadly applied across the field than that of *in situ* hybridization to RNA. This allowed researchers to determine rapidly the spatial and temporal expression of their gene of interest without having to resort to more tedious and less certain approaches requiring the production of antisera against its protein product. As such, *in situ* hybridization has become the standard first step toward the characterization of the developmental significance of a newly identified gene. Of course, *in situ* hybridization tells you nothing regarding translation of the mRNA into protein or about protein's subsequent localization, modification, or stability. However, *in situ* hybridization remains an extremely powerful tool in enabling the researcher to predict likely developmental functions for gene products rapidly.

Historically, this technique was pioneered by the Angerers working with sea urchin embryos *(1)*. It was refined by several groups, notably by McMahon and Wilkinson in studies of gene expression in sections of mouse embryos *(2,3)*. During the past five years or so, the use of radioactive *in situ* hybridization of tissue sections has been largely superseded by the application of non-radioactive approaches on whole embryos at early developmental stages and, to a lesser extent, on sections of older embryos and adults. The latter approaches offer considerable advantages in terms of their speed and, in the case of *in situ* hybridisations to whole embryos, further advantages in the ease of spatial interpretation of results. However, the radioactive approach remains more sensitive and is the only method

From: *Methods in Molecular Biology, Vol. 97: Molecular Embryology: Methods and Protocols*
Edited by: P. T. Sharpe and I. Mason © Humana Press Inc., Totowa, NJ

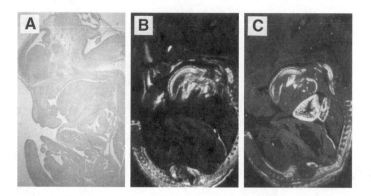

Fig. 1. Adjacent sagittal sections of an E14.5 mouse embryo. (**A**) Bright-field view of toluidine blue-stained section. (**B**) Same field viewed under dark ground optics to show hybridization of FGF-7 probe to a number of tissues, including a subset of muscles, nasal epithelium, cartilage capsules, and lung. Section was exposed for 14 d (**C**). Adjacent section hybridized with a cardiac α-actin probe showing hybridization to all skeletal and cardiac muscle. Section was exposed for 24 h.

by which certain "low-abundance" transcripts are detectable. Furthermore, for these reasons, it is likely that many studies of developmental gene expression undertaken by nonradioactive, whole embryo approaches are incomplete. As such, radioactive *in situ* hybridization remains a valid approach in embryological and other fields. The following protocol (*4*) is that used in our laboratory (*see* **Fig. 1**) and is a modification of an original method of Wilkinson and Green (*5*).

## 2. Materials
### 2.1. Preparation of Material for Sectioning

1. Diethylpyrocarbonate (DEPC) treated water. DEPC (Sigma, Poole, UK) is added to double-distilled water to a concentration of 0.1% (v/v). The solutions are shaken and then autoclaved. All solutions used for steps between sectioning tissue and the first post hybridization wash are prepared in DEPC water. All subsequent steps use solutions prepared in double-distilled water
2. 10X PBS: 1.3 $M$ NaCl, 70 m$M$ Na$_2$HPO$_4$, 30 m$M$ NaH$_2$PO$_4$. Check that pH is 7.0 (if required adjust slightly with HCl or NaOH as appropriate). Add DEPC to 0.1% and autoclave.
3. 4% (w/v) paraformaldehyde in PBS freshly prepared on the day of use for pretreatment of sections, but can be stored for fixation of embryos. Sixteen grams of paraformaldehyde (Fluka no. 76240, Gillingham, UK) are weighed out in a fume hood—wear mask and gloves. This is added to 400 mL of 1X PBS in distilled water and heated at 65°C in a water bath with occasional swirling until dissolved (about 90 min). It is cooled to 4°C prior to use or stored in 20-mL aliquots at –20°C.
4. 10X Saline: 8.3% (w/v) NaCl. Add DEPC to 0.1% and autoclave.

5. Howard's Ringer (0.12 $M$ NaCl, 0.0015 $M$ CaCl$_2$, 0.005 $M$ KCl. Per liter: 7.2 g NaCl, 0.17 g CaCl$_2$, 0.37 g KCl, pH 7.2, with very dilute HCl).
6. Absolute (100%) ethanol and graded ethanols 70, 80, and 95% (v/v) prepared with DEPC water.
7. Xylene (AnalaR, Merck BDH, Poole, UK).
8. Histoclear (National Diagnostics, Hull, UK).
9. Wax (Paraplast Plus, Sherwood, St. Louis, MO).
10. Plastic moulds for wax (e.g., R.A.Lamb, London, UK).

## 2.2. Preparation of Slides

1. Washed glass microscope slides and cover slips (Merck BDH).
2. 25-place metal slide racks (R.A. Lamb).
3. 400 mL capacity glass troughs with lids suitable to hold slide racks (R.A. Lamb).
4. Concentrated hydrochloric acid (Merck BDH).
5. Acetone (AnalaR, Merck BDH).
6. TESPA (3-aminopropyltriethoxysilane; Sigma) used in the fume hood.

## 2.3. Cutting and Mounting Sections

1. Suitable rocking microtome, disposable blades and holder (reduces chance of contamination with RNase from a generally used blade), bath for sections (well-cleaned with detergent, rinsed well with distilled water, and filled with DEPC water), and hot plate.
2. Plastic slide boxes suitable for storing slides with sections such that slides are well separated (e.g., R.A. Lamb).
3. Silica gel (Sigma).

## 2.4. Preparation of Template for Riboprobe Synthesis

1. Appropriate restriction endonucleases and manufacturer's buffers (e.g., Boehringer Mannheim, Lewes, UK).
2. Agarose (Life Technologies, Paisley, Scotland).
3. 10X TBE (0.89 $M$ Tris-borate, 0.89 $M$ boric acid, 0.02 $M$ EDTA; per liter 108 g Tris base, 55 g boric acid, 9.4 g EDTA).
4. Ethidium bromide (10 mg/mL). Stored at 4°C in the dark. **Caution:** ethidium bromide is a potent carcinogen.
5. Phenol (molecular biology-grade; Life Technologies). Melted at 65°C in a fume cupboard and equilibrated with 0.1 $M$ Tris. HCl, pH 7.0, followed by addition of a few crystals of 8-hydroxyquinoline, which will color the solution yellow. Stored in the dark at 4°C, and discarded if the solution begins to turn orange.
6. Chloroform (AnalaR; Merck BDH).
7. 3 $M$ sodium acetate, pH 7.0 (with acetic acid).
8. 100% (v/v) Ethanol and 70% (v/v) ethanol.

## 2.5. Synthesis of $^{35}$S-UTP-Labeled Riboprobes

1. RNA polymerases, manufacturer's buffers, and 200 m$M$ DTT (Promega, Southampton, UK and Boehringer Mannheim).

2. $^{35}$S-UTP (>1000 Ci/m$M$, Amersham International, Amersham, UK).
3. RNasin (Boehringer Mannheim).
4. RNase-free DNase 1 (Boehringer Mannheim).
5. Yeast total RNA (Sigma) 10 mg/mL dissolved in DEPC water, and stored as aliquots at –20°C.
6. 5 $M$ Ammonium acetate (Sigma) dissolved in DEPC water.
7. Aqueous scintillant, e.g., Aquasol (Dupont NEN, Hounslow, UK).

### *2.6. Hybridization to Tissue Sections*

1. Proteinase K (Boehringer Mannheim) 20 mg/mL in water stored in single-use (20 µL) aliquots at –20°C.
2. TE pH 8.0 (10 m$M$ Tris-HCl, pH 8.0, 1 m$M$ EDTA, pH 8.0) diluted in DEPC water and prepared from concentrated stocks dissolved in DEPC water.
3. Triethanolamine (Merck BDH).
4. Acetic anhydride (Merck BDH).
5. 20X SCC (3 $M$ NaCl, 0.3 $M$ sodium citrate).
6. Hybridization buffer: 50% (v/v) deionized formamide, 0.3 $M$ NaCl, 20 m$M$ Tris-HCl, pH 8.0, 5 m$M$ EDTA, pH 8.0, 10% (w/v) dextran sulfate, 1X Denhardt's solution, 0.5 µg/mL yeast tRNA. Store in aliquots at –70°C.

### *2.7. Posthybridization Washes*

1. TNE (0.5 $M$ NaCl, 10 m$M$ Tris-HCl, 5 m$M$ EDTA, pH 8.0).
2. High-stringency buffer: 50% v/v formamide, 2X SSC, 10 m$M$ DTT prepared fresh on day of use using solid DTT or a 100-m$M$ stock of DTT stored at –20°C.
3. RNase A (10 mg/mL) in water. Store at –20°C.
4. Graded (30, 50, 70, and 95% v/v) ethanols containing 0.3 $M$ (final) ammonium acetate. The presence of the ammonium acetate ensures that probe-target RNA duplexes remain intact during the dehydration process. Duplexes are stable in 100% (v/v) ethanol.

### *2.8. Autoradiography*

1. Emulsion (K-S no. 1355127, Ilford Photographic, Mobbesley, UK).
2. Glycerol (Merck, BDH).
3. Two-place plastic slide mailers (R.A. Lamb).

### *2.9. Developing and Mounting Slides*

1. D19 developer (Kodak, Paris, France): Dissolve 64 g D19 developer in 400 mL water. This can be used several times on the day of preparation, but discard after use.
2. Stop solution: 350 mL of 1% (v/v) acetic acid/1% (v/v) glycerol.
3. Fixing solution: 350 mL of 30% (w/v) sodium thiosulfate.
4. 0.25% Toluidine blue (Sigma) filtered through a Whatman general-purpose filter.
5. Graded ethanols (30, 50, 80, 95, 100% v/v).
6. Depex mounting medium (Merck BDH).

## 3. Methods

Between **Subheading 3.1., step 2** and **Subheading 3.6., step 20,** there must be no chance of contamination with RNases. Wear gloves, use disposable plastics instead of glass, and where possible, also use pipetmen designated for RNA use only, and DEPC-treat all solutions. If the presence of RNases is still suspected, then glassware can be treated also.

### 3.1. Preparation of Material for Sectioning

1. Dissect embryos in 1X PBS (mice) or Howard's Ringer (chicks). Fix in a large volume (at least five times the volume of embryonic tissue) of 4% (w/v) parafomaldehyde in PBS at 4°C overnight in a vessel that adequately contains the fumes (*see* **Subheading 2.3.**). Smaller embryos can be fixed for shorter periods of time (4 h minimum), and it is convenient to dechorionate zebrafish after fixation. Embryos can be fixed pinned out in Petri dishes, if required, in the designated refrigerator. Fixation times will vary with the size of the embryo from a few hours (mice up to embryonic [E] 9.5, chicks to stage 18) to 12 h (E18 mice) or even longer (newborn mice or chicks from stage 38; *see* **Note 1**). In our experience fixation for 2–3 d does not adversely affect results.
2. Incubate in 1X PBS on ice for 30 min (small embryos, i.e., chicks <E6.5, mice <E13.5) or 1 h (larger material). Use glass containers, since later treatments involve high temperatures and organic solvents.
3. Further dissect embryos if required, e.g., cut embryos away from surrounding membranes. Stage according to Theiler (*6*) or Hamburger and Hamilton (*7*) if required.
4. Remove PBS and replace with 1X saline. Incubate for 30 min (small embryos) or 1 h (larger material) on ice.
5. Replace saline with saline:ethanol (1:1), and incubate for 30 min (small embryos) or 1 h (larger material) on ice.
6. Replace saline:ethanol with 70% (v/v) ethanol (prepared with DEPC-treated water), and incubate for 30 min (small embryos) or 1 h (larger material) on ice.
7. Repeat **step 6**.
8. Replace 70% (v/v) ethanol with 80% (v/v) ethanol and incubate on ice 30 min.
9. Replace with 95% (v/v) ethanol, and incubate on ice for 30 min (small embryos) or 1 h (larger material) on ice.
10. Replace with 100% ethanol, and incubate on ice for 30 min (small embryos) or 1 h (larger material) on ice (*see* **Note 2**).
11. Place embryos in 100% ethanol for 30 min (small embryos) or 1 h (older embryos).
12. Replace with 100% ethanol for 30 min (small embryos) or 2 h (older embryos).
13. Replace 100% ethanol with xylene or (preferably) histoclear in glass bottles, and incubate on ice 30 min (small embryos) or 1 h (older embryos).
14. Repeat **step 13** with fresh xylene or histoclear.
15. Pour off and half-fill bottles with fresh molten wax (sections cut from blocks prepared with freshly melted wax have better elasticity than from wax that has remained molten for days), and place at 52°C for 45 min (small embryos) or 90 min (large embryos).

16. Change wax twice, and incubate each time for 60 min (small embryos) or 90 min (large embryos).
17. Embed tissue: fill molds with wax first (if possible, do this on a warmed plate to keep the wax molten), carefully transfer the embryo to the mold, orient with a hot mounted needle or warm forceps, and move to a cooled plate or onto the bench. Label one end of the block for reference after the wax is set and the embryo's orienation is no longer visible (e.g., by inserting a piece of card on which details of the specimen and its orientation are written) and allow the wax to set completely (1–2 h).
18. Blocks may be stored at 4°C for several years. Any degradation of RNA in such material is most likely owing to incomplete penetration of reagents during the above processing steps.

### 3.2. Preparation of Slides

Microscope slides need to be washed and silanized prior to collection of sections. Alternatively, pretreated slides (e.g., Superfrost, Merck BDH) are now commercially available.

1. Place metal racks of 25 slides for 10 s each in 400 mL of 70% (v/v) ethanol, 10% (v/v) concentrated HCl, then in distilled water, and finally in 95% (v/v) ethanol.
2. Wrap in baking foil to keep free from dust, and place the slides at 150°C for 30 min.
3. Cool on bench wrapped in foil.
4. Place the rack of slides for 10 min each first in acetone containing 2% (v/v) TESPA, then in two changes of acetone, and finally in distilled water (not DEPC-treated). Do this step in the fume cupboard.
5. Dry at 37°C overnight wrapped in foil.
6. Slides are stored individually in racks (to avoid scratches) at room temperature for up to 3 mo.

### 3.3. Cutting and Mounting Sections

1. Cut 7 to 10-μm sections as ribbons on a rotary microtome using a disposable blade.
2. Ribbons are placed on a pool of 20% (v/v) ethanol in DEPC-treated water on a slide, and cut into smaller strips with a clean razor blade if required.
3. The sections are floated out onto DEPC-treated water in a clean section bath at 45°C until creases in them disappear.
4. Sections are then collected onto TESPA-treated slides, and dried on a 48°C hot plate for 2 h or overnight at 37°C. It is useful to collect at least two adjacent sections per slide; this aids the subsequent identification of hybridizing tissues, since these should give identical results on both sections.
5. Store slides desiccated at 4°C such that they are free from moisture and sections still give acceptable results months after being cut. Before use, allow slides to warm to room temperature. If some slides are to be returned to storage, place them in the 37°C oven for 30 min prior to storage to reduce the moisture and store sealed with fresh desiccant.

## 3.4. Preparation of Template for Riboprobe Synthesis

1. The DNA fragment should have been previously cloned into a plasmid vector, such that prokaryotic RNA polymerase promoters flank it. Suitable vectors include the pGem series (Promega, Madison, WI) and the pKS or pBluescript series (Stratagene, La Jolla, CA). In our experience, T7 and T3 RNA polymerases give the best overall performance on a wide range of templates, whereas SP6 polymerase has problems transcribing certain templates.
2. Linearize plasmid (usually 10–20 µg) with a suitable restriction endonuclease using manufacturer's buffer and conditions (*see* **Note 3**).
3. Check that digestion is complete by analyzing 0.25 µg on a 1% (w/v) agarose gel in 1X TBE containing ethidium bromide (0.1 µg/mL) alongside some uncut plasmid. If the digestion is incomplete, add more enzyme, incubate for a further period, and reanalyze by gel electrophoresis.
4. Extract the remainder of the reaction mixture once with phenol, once with phenol:chloroform (1:1), and twice with chloroform.
5. Precipitate the plasmid DNA by the addition of 0.1 vol of 3 $M$ sodium acetate (pH 7.0) and 2.2 vol of ethanol followed by incubation on ice for 15 min.
6. Collect the precipitate by centrifugation at 10,000$g$ in a microfuge for 10 min, aspirate off the supernatant, and wash the pellet with 70% (v/v) ethanol.
7. After allowing the last traces of ethanol to evaporate, the pellet is resuspended in DEPC-treated water to give a final concentration of 0.5 µg/µL and stored at –20°C.

## 3.5. Synthesis of $^{35}$S-Labeled Riboprobes

1. Warm buffers, nucleotides, DTT, and templates to room temperature.
2. Assemble the following reaction mixture in this order: 10 µL $^{35}$S-UTP, 2 µL 10X transcription buffer, 1 µL 200 m$M$ DTT, 1 µL RNase inhibitor, 2 µL 2.5 m$M$ NTP, pH 8.O (a mix of CTP, GTP, and ATP to produce 2.5 m$M$ final of each), 1 µL DEPC water. Mix by gently pipeting up and down.
3. Add 1 µL of the appropriate RNA polymerase, and mix gently by pipeting (do not vortex).
4. Add 2 µL (1 µg) of DNA template, mix, and spin for 5 s in microfuge to collect all liquid in the bottom of the tube.
5. Incubate at 37°C for 1–2 h.
6. Add 10 µL DEPC water, 2 µL yeast RNA, and 2 µL RNase-free DNase, and incubate at 37°C for 15 min.
7. Add 50 µL 5 $M$ ammonium acetate (prepared with DEPC water) and 200 µL ethanol. Incubate on ice for 30 min, and spin in a microfuge for 15 min.
8. Remove supernatant, and immediately dissolve pellet in 30 µL 10m$M$ DTT.
9. Repeat **step 7** of **Subheading 3.5**.
10. Remove the supernatant, immediately redissolve probe in 20 µL 100 m$M$ DTT over 15–20 min with occasional vortexing, and count 1 µL in an aqueous scintillant. Calculate probe specific activity in dpm/µL/kb probe length: acceptable probes must be >$10^6$ dpm/µL/kb, i.e., specific activity of 1 µL × length of labeled RNA transcript in kb. Store probes at –70°C for up to 2 mo.

## 3.6. Hybridization to Tissue Sections

1. Early in the day, make up 400 mL fresh 4% paraformaldehyde in PBS, and store on ice until required. Also ensure that you have 3.5 L of DEPC-treated water for dilution of salines, etc. Use designated RNase-free glassware and racks for all of the following procedures, wear gloves throughout, and use DEPC-treated water and solutions.

2. Dewax slides in xylene twice each for 10 min in a fume cupboard. For this and all subsequent steps, use 350 mL of each solution for each incubation; sufficient to cover the slides completely. Slides are processed in metal racks holding up to 25 individual slides in appropriately sized staining troughs. Do not agitate the slides during incubations, since it may cause sections to detach.

3. Place in 100% ethanol for 2 min to remove most of the xylene. Meanwhile make up 1.6 L 1X PBS and 800 mL 1X saline.

4. Rehydrate sections by "dipping" through the following graded alcohol series 100% ethanol twice, 95% ethanol, 80% ethanol, 70% ethanol, 50% ethanol, and 30% ethanol. (Keep and reuse ethanols up to 10 times).

5. Place in 1X saline for 5 min and then in 1X PBS for 5 min. Discard used saline and PBS.

6. Place in the freshly made, cold 4% paraformaldehyde in PBS for 20 min in the fume hood. At the end of this, do not discard the fix, but keep for **step 11** of **Subheading 3.6.**

7. Meanwhile for 25 slides, prepare 15 mL of DEPC-treated TE, pH 8.0, mixed with 20 µL of proteinase K (20 mg/mL). Also wash sufficient cover slips of appropriate size in ethanol and leave to air-dry in a rack covered in foil.

8. Place slides in 1X PBS twice each for 5 min, and then lay flat on a sheet of foil on the bench with the sections uppermost.

9. Cover sections with 300 µL or more of proteinase K (sufficient to cover the sections), and incubate for exactly 7 min at room temperature.

10. Tip off the proteinase K, and place slides in a rack in fresh 1X PBS for 5 min.

11. Return the slides to the cold paraformaldehyde (**Subheading 3.6., step 6**) for 5 min in the fume hood.

12. Meanwhile, fill two troughs with DEPC water, and place in the fume hood. To the second (acetylation bath), add 6 mL triethanolamine and add 1 mL acetic anhydride, and mix with a small stir bar on a stirrer.

13. Dip slides briefly into the DEPC water, and then place in the acetylation bath for 10 min with stirring. (A small stir bar or "flea" will fit beneath standard slide racks and allow stirring in the presence of the latter. This step acetylates slides and sections, and helps prevent nonspecific binding of probe).

14. Place slides in fresh 1X PBS for 5 min and then fresh 1X saline for 5 min.

15. Dehydrate by "dipping" through the same ethanol series used in **Subheading 3.6., step 4** starting with 30% ethanol and increasing to the two 100% ethanols.

16. Place slides section side up to air-dry on foil. Add probe to them the same day.

17. Dilute probe in hybridization buffer to give $10^5$ dpm/µL/Kb.

18. Mix probe by vortexing and briefly spin in a centrifuge. Heat to 80°C for 3 min to remove secondary structures owing to internal hybridization of the probe. Allow to cool briefly on ice and then mix again.

19. Place 15–30 µL of probe solution on slide adjacent to sections, spread over sections with a piece of parafilm, and cover with a clean cover slip (**Subheading 3.6., step 7**) taking care not to introduce any bubbles.

20. Slides are collected on their sides in a rack that is then place inside a suitable box with a tightly sealed lid. Include a tissue soaked in 5X SSC, 50% formamide to prevent dehydration. Seal the box with suitable tape, and hybridize at 55°C overnight in an oven.

## 3.7. Posthybridization Washes

For all procedures after hybridization, we use separate glassware and graded ethanol solutions from those used prior to hybridization because RNase A is used during the washing procedures and will potentially contaminate all glassware and reused solutions.

1. Place slides in 4X SSC at room temperature until cover slips fall off (about 20 min).

2. Remove cover slips, and place slides in fresh 4X SSC for 1 h.

3. Meanwhile, make up 1.6 L of TNE buffer and 80 mL of high-stringency buffer/ 25 slides, prewarming the latter to 50°C or 65°C as appropriate (*see* **step 4**).

4. Place slides back to back in Coplin jars containing high-stringency buffer in a 65°C water bath (high stringency) or 50°C (low stringency) for 30 min (*see* **Note 4**).

5. Transfer slides to a rack, and place in 350 mL of TNE buffer for 10 min at 37°C. Then repeat this wash once more.

6. Place slides in a third trough containing 400 mL prewarmed TNE buffer and 800 µL of 10 mg/mL RNase A for 30 min at 37°C.

7. Wash slides for 10 min in fresh TNE buffer at 37°C.

8. Wash slides for 30 min in Coplin jars containing prewarmed high-stringency buffer at 65°C (or appropriate temperature).

9. Wash for 15 min in 2X SSC, and then for 15 min in 0.1X SSC both at room temperature.

10. Dehydrate through graded ethanols containing 2.5 $M$ ammonium acetate (the latter to prevent "melting" of the probe from target RNAs) from 30 to 100% by dipping as before.

11. Air-dry in a rack loosely wrapped in foil for 60 min minimum. The slides may be left overnight to dry.

## 3.8. Autoradiography

All steps are performed in a darkroom using a red safe light.

1. Put a foil-wrapped glass sheet onto a tray of ice to cool in the darkroom.

2. Place about 15 mL of emulsion ribbons into a 50-mL Falcon tube, and incubate in a 42°C waterbath to melt them.

3. Decant 8 mL of the melted emulsion into a slide mailer, and replace at 42°C.

4. Prepare 8.8 mL water containing 180 µL glycerol, and add to the melted emulsion. Mix gently with a glass pipet to expel bubbles, and leave in water bath for 10 min to warm to 42°C.

5. Dip slides briefly and with a uniform speed into the emulsion, wipe the back (nonsection side) of each, and lay section side up on cold tray for 10 min to set. If bubbles or streaks are visible, then redip the section.

6. Place slides in a box for emulsion to harden in complete darkness for 2 h, then transfer to a sealed, light-tight box containing silica gel packets, and store at 4°C for a few days to 2 wk depending on abundance of target transcript.

### *3.9. Developing and Mounting Slides*

Unless otherwise stated, all procedures are carried out in the darkroom using a red safe light.

1. Make up troughs containing 350 mL fresh developer, stop solution, and fixing solution. Allow or adjust the temperature of all solutions to reach 20°C, and place in the darkroom.

2. Remove slides from the refrigerator, and allow them to equilibrate to room temperature.

3. Place a rack of slides in the developer for exactly 2 min and 30 s, and then directly into the stop solution for 1 min and into the fixing solution for 5 min.

4. Wash in troughs of distilled water twice, each for 10 min. Slides can now be removed from the darkroom.

5. For counterstaining, place each slide in 50 mL of 0.25% (w/v) fresh filtered toluidine blue for exactly 1 min. Dehydrate by rapid dipping through a graded series of ethanols (30–100% ) washing twice in 100% ethanol.

6. Then clear with xylene (Histoclear cannot be used here, since it leaves an oily film on the slide) and mount with cover slips under Depex.

## 4. Notes

1. It is crucial during fixation, processing, and embedding in wax prior to sectioning, that all of the reagents completely penetrate and equilibrate within the embryonic tissue. Times vary according to the size of the embryo/tissue being prepared. The following illustrates modifications that have been used successfully on larger tissues (e.g., heads of E10 chick embryos and whole E18 mouse embryos).

    On d 1, fix in 4% paraformaldhyde in 1X PBS overnight at 4°C.

    On d 2, wash twice in PBS each for 30 min, wash twice in saline each for 30 min, and then incubate in 30% ethanol for 1 h, 50% ethanol for 1 h, 70% ethanol for 1 h, 100% ethanol for 3 h, and then in fresh 100% ethanol overnight.

    On d 3, incubate in fresh 100% ethanol for a further 5 h. Then incubate in three changes of Histoclear (or xylene) the first for 1 h, the second for 3 h, and the last overnight.

    On d 4, replace the Histoclear with freshly melted (52°C) wax, and incubate at 52°C in an oven for 3 h, replace the wax with fresh and incubate at 52°C for 5 h and then replace once more and incubate at 52°C overnight.

    On d 5, incubate with two more changes of wax each for 3 h, and then embed the tissue.

2. Embryos can be stored at –20°C for up to a week at this stage.
3. Restriction enzyme sites should be selected to produce a transcript of 300–1000 bp. If any larger, we find that nonspecific background signal increases. Also, if possible, select enzymes that produce a blunt end or a 5'-overhang, because 3'-overhanging ends facilitate "wraparound" transcripts resulting from the polymerase transcribing back along the complementary DNA strand. If there is no alternative to using an enzyme producing a 3'-overhanging end, the following procedure can be used after digestion to "end polish," generating a blunt end. After digestion, add Klenow polymerase to a final concentration of 5 U/μg DNA template, and incubate at 22°C for 15 min prior to proceeding with the rest of the transcription protocol.
4. Coplin jars containing 30–40 mL of solution are used at this stage and in **Subheading 3.7., step 8** to reduce the volume of high-stringency buffer used, the large amounts of DTT required would otherwise prove very expensive.

## Acknowledgments

Work in the authors' laboratory is supported by program grants from the MRC and Human Frontier Science Program and project grants form the MRC and the Wellcome Trust.

## References

1. Cox, K. H., Deleon, D. V., Angerer, L. M., and Angerer, R. C. (1984) Detection of mRNA in sea urchin embryos by in situ hybridization using asymmetric RNA probes. *Dev. Biol.* **101,** 485–502.
2. Wilkinson, D. G., Bailes, J. A., and McMahon, A. P. (1987) A molecular analysis of mouse development from 8 to 10 days post coitum detects changes only in embryonic globin expression. *Cell* **50,** 79–88.
3. Wilkinson, D. G., Bhatt, S., and McMahon, A. P. (1989) Expression of the FGF-related proto-oncogene int-2 suggests multiple roles in fetal development. *Development* **105,** 131–136.
4. Mason, I. J., Fuller-Pace, F., Smith, R., and Dickson, C. (1994) FGF-7 (keratinocyte growth factor) expression during mouse development suggests roles in myogenesis, forebrain regionalisation and epithelial-mesenchymal interactions. *Mech. Dev.* **45,** 15–30.
5. Wilkinson, D. and Green, J. (1990) *In situ* hybridization and the three-dimensional reconstruction of serial sections, in *Postimplantation Mammalian Embryos: A Practical Approach* (Copp, A.J. and Cockroft, D.L., ed.), IRL P, Oxford, pp. 155–171.
6. Theiler, K. (1989) *The House Mouse.* Springer-Verlag, Berlin.
7. Hamburger, V. and Hamilton, L. (1951) A series of normal stages in the development of the chick embryo. *J. Morphol.* **88,** 49–92.

# 42

## *In Situ* Hybridization to RNA in Whole Embryos

### Huma Shamim, Radma Mahmood, and Ivor Mason

## 1. Introduction

As indicated in the preceding chapter, few techniques have had such a major impact on progress in the field of developmental biology as *in situ* hybridization of labeled RNA or DNA probes to detect specific mRNAs in embryonic tissues. Initially, the technique was performed on sections of fixed material. However, this has largely been superseded by the development of techniques for hybridization to RNA in intact embryos. The latter approach involves less effort, and has the additional advantages that a complete temporal study in all tissues can readily be undertaken in one or two experiments and that spatial and temporal changes in gene expression are more readily appreciated (*see,* e.g., **Fig. 1A–C,F**). The following procedure is suitable for hybridization to embryos with RNA probes (riboprobes) derived from DNA sequences from the same species and, with slight modification, is frequently suitable for use with probes from other species. This protocol *(1–3)* is a modified version of that of Wilkinson *(4)*, and we have included procedures for "two-color" *in situ* hybridization, which allows the expression of two genes to be examined in the same embryo (*see* **Fig. 1F**). Protocols for the sectioning of material subsequent to *in situ* hybridization (*see* **Fig. 1D,E**) are also included.

## 2. Materials

### 2.1. General

1. Diethylpyrocarbonate- (DEPC) treated water. DEPC (Sigma, Poole, UK) is added to double-distilled water to a concentration of 0.1% (v/v), the solutions are shaken, allowed to stand for 1 h, and then autoclaved.
2. The following equipment is required: a dissection microscope offering up to at least 65× magnification (e.g., Nikon, Kingston upon Thames, UK or Zeiss,

From: *Methods in Molecular Biology, Vol. 97: Molecular Embryology: Methods and Protocols*
Edited by: P. T. Sharpe and I. Mason © Humana Press Inc., Totowa, NJ

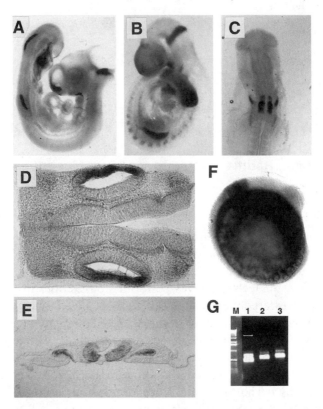

Fig. 1. **(A)** *In situ* hybridization of Fgf-8 to a stage 20 chicken embryo using a mouse cDNA sequence as probe. **(B)** The same probe as in (A) hybridized to a 9.5-d mouse embryo. **(C)** *In situ* hybridization of a chicken Fgf-3 probe to a stage 11 chicken embryo. **(D)** Vibroslice (thick, 50 μm) section of a chicken embryo hybridized with a ret probe showing transcripts in the lateral regions of the otic vesicles. **(E)** Thin (10 μm) section of the specimen in (C), showing Fgf-3 hybridization to the neural tube and branchial pouch endoderm. **(F)** "Two-color" *in situ* hybridization of Krox-20 (red) and Hox-B1 (blue) showing expression in adjacent segments (rhombomeres) of the embryonic zebrafish hindbrain. **(G)** A typical gel showing DIG-labeled riboprobes before (lane 1) and after (lanes 2 and 3) DNase treatment. Note that the intensities of the probe (lower) bands are much greater than that of the DNA template (upper) bands. Lane M is a DNA mol-wt standard ladder. Note that the RNA in lanes 1–3 appears as a "doublet" even though the enzymes used to linearize the template do not have 3'- overhangs. Whatever the source of the "doublet," we do not find that it affects the use or sensitivity of such probes. (*See* color plate 12 appearing after p. 368.)

Welwyn, Garden City, UK); transmitted (Nikon, Zeiss) and fiber optic (Schott, Zeiss, Leica, Milton Keynes, UK) light sources; heated water bath and microfuge (e.g., Eppendorf, Heraeus) for enzyme digests and riboprobe synthesis (e.g., Grant, Merck, Poole, UK); horizontal gel electrophoresis tank and power supply (e.g., Life Technologies, Paisley, Scotland); oven (e.g., Heraeus, Merck), water bath or Spin 'n' Stack incubator (Hybaid, Teddington, UK) for hybridizations and washes; rotating wheel or shaker at 4°C and at room temperature for incubations with antibodies and subsequent washes; suitable photomicroscopes and equipment for histological analysis. For sectioning a vibrotome and/or microtome are required.

## 2.2. Fixation, Processing, and Hybridization

All solutions are prepared in DEPC-treated water.

1. Phosphate-buffered saline (PBS; 10X concentrated): 1.3 $M$ NaCl, 70 m$M$ $Na_2HPO_4$, 30 m$M$ $NaH_2PO_4$. Check that pH is 7.0 (if required, adjust slightly with HCl or NaOH as appropriate). Add DEPC to 0.1% and autoclave.
2. Howard's Ringer: 0.12 $M$ NaCl, 0.0015 $M$ $CaCl_2$, 0.005 $M$ KCl. Per liter: 7.2 g NaCl, 0.17 g $CaCl_2$, 0.37 g KCl, pH 7.2, with very dilute HCl.
3. Paraformaldehyde: 16 g of paraformaldehyde (Fluka, Gillingham, UK, no. 76240) is weighed out in a fume hood wearing mask and gloves. It is added to 400 mL of 1X PBS in distilled water and heated at 65°C in a water bath in the fume hood with occasional swirling until dissolved (about 90 min). It is cooled to 4°C prior to use or stored in 20 mL aliquots at –20°C.
4. PBT is 1X PBS containing 0.1% (v/v) Tween 20 (Sigma).
5. Methanol (AnalaR; Merck BDH).
6. Hydrogen peroxide (30% w/w; Sigma, replace every 4 mo).
7. Proteinase K (Boehringer Mannheim, Lewes, UK) 20 mg/mL and stored in single-use (15–20 µL) aliquots at –20°C.
8. 25% (w/v) Glutaraldehyde (electron microscopy grade; BDH, Poole, UK). Glutaraldehyde is carcinogenic and should be used in a fume cupboard.
9. Prehybridization/hybridization solution: 50% (v/v) deionized formamide (Fisons, Loughborogh, UK), 5X SSC, pH 4.5 (20X SSC stock is 3 $M$ NaCl, 0.3 $M$ sodium citrate; use citric acid to pH), 50 mg/mL yeast RNA (Sigma), 1% (w/v) sodium dodecyl sulfate (SDS), 50 mg/mL heparin (Sigma). Heparin and yeast RNA are stored as concentrated 50 mg/mL solutions at –20°C.

## 2.3. Preparation of Template for Riboprobe Synthesis

1. Appropriate restriction endonucleases and manufacturer's buffers (e.g., Boehringer Mannheim).
2. Agarose (Life Technologies).
3. 10X TBE: 0.89 $M$ Tris-borate, 0.89 $M$ boric acid, 0.02 $M$ EDTA; per liter 108 g Tris base, 55 g boric acid, 9.4 g EDTA.
4. Ethidium bromide (10 mg/mL). Stored at 4°C in the dark. **Caution:** ethidium bromide is a potent carcinogen.
5. Phenol (molecular biology-grade; Life Technologies). Melted at 65°C in a fume cupboard and equilibrated with 0.1 $M$ Tris-HCl, pH 7.0 followed by addition of a few crystals of 8-hydroxyquinoline, which will color the solution yellow. Stored in the dark at 4°C and discarded if the solution begins to turn orange.
6. Chloroform (AnalaR; Merck BDH).

7. 3 $M$ sodium acetate, pH 7.0 (with acetic acid).
8. Absolute (100% v/v) ethanol and 70% (v/v) ethanol.

## 2.4. Synthesis of Digoxygenin (DIG) or Fluorescein-12 UTP (FITC) Labeled Riboprobes

1. RNA polymerases, manufacturer's buffers, and 200 m$M$ DTT (Promega, Southampton, UK, and Boehringer Mannheim). Aliquots of 200 m$M$ DTT are stored at –20°C.
2. 10X DIG and/or fluorescein RNA labeling nucleotide mixes (Boehringer Mannheim).
3. RNasin (Boehringer Mannheim).
4. RNase-free DNase-1 (Boehringer Mannheim).
5. TE, pH 7.0: 10 m$M$ Tris-HCl, pH 7.0, 1 m$M$ ethylenediamine tetraacetic acid, disodium salt (EDTA), pH 8.0, diluted from stocks dissolved in DEPC-treated water.
6. 4 $M$ Lithium chloride (prepared with DEPC-treated water).
7. 70% (v/v) Ethanol (prepared with DEPC-treated water).

## 2.5. Posthybridization Washes

1. Solution 1: 50% (v/v) formamide, 5X SSC pH 4.5, 1% (w/v) SDS.
2. Solution 2: 0.5 $M$ NaCl, 10 m$M$ Tris-HCl, pH 7.5, 0.1 % (v/v) Tween-20.
3. RNase A (Boehringer Mannheim) 20 mg/mL in double-distilled water and stored in aliquots at –20°C.
4. Solution 3: 50% (v/v) formamide, 2X SSC pH 4.5 (pH using citric acid).
5. 10X TBST: 1.4 $M$ NaCl, 27 m$M$ KCl, 0.25 $M$ Tris-HCl, pH 7.5, 1% (v/v) Tween-20. Dilute to 1X and add levamisole (Sigma) to 2 m$M$ (24 mg/50 mL) on day of use.
6. Goat serum (Sigma) heat-inactivated at 55°C for 30 min and stored in 1–2 mL aliquots at –20°C. Heat-inactivated sheep serum can also be used.
7. Embryo powder: Homogenize 10-d chick embryos or 14-d mouse embryos in minimum volume of PBS, add 4 vol of ice-cold acetone, mix, and place on ice for 30 min. Spin at 10,000$g$ for 10 min, and remove supernatant. Wash pellet with ice-cold acetone, and spin again. Spread the pellet out, and grind it into a fine powder on a sheet of filter paper. Allow it to air-dry. Store tightly capped at 4°C.
8. Anti-DIG or anti-FITC Fab fragments conjugated to alkaline phosphatase (Boehringer Mannheim).

## 2.6. Postantibody Washes and Histochemistry

1. NTMT, pH 9.5: 100 m$M$ NaCl, 100 m$M$ Tris-HCl, pH 9.5, 50 m$M$ MgCl$_2$, 0.1% Tween-20, 2 m$M$ levamisole (prepared freshly on day of use from concentrated stock solutions of the individual components).
2. Nitroblue tetrazolium (NBT) and 5-bromo-4-chloro-3-indolyl phosphate (BCIP/x-phosphate) both from Boehringer.
3. PBT containing 20 m$M$ EDTA.
4. NTMT, pH 8.0: 100 m$M$ NaCl, 100 m$M$ Tris-HCl, pH 9.5, 50 m$M$ MgCl$_2$, 0.1% Tween 20, 2 m$M$ levamisole (prepared freshly on day of use from concentrated stock solutions of the individual components).
5. Fast red TR/naphthol AS-MX (Sigma).

## 2.7. Sectioning After In Situ Hybridization

### 2.7.1. Thick Sections

1. Gel-albumin consists of 0.5% (w/v) gelatin (Sigma), 30% (w/v) egg albumin (Sigma), and 20% (w/v) sucrose (Sigma) in PBS. Heat the gelatin to dissolve in PBS (this takes quite a long time), cool and add egg albumin (powder), and allow to dissolve. Then dissolve the sucrose, filter through gauze, and store frozen in aliquots of 50 mL. This material can be repeatedly frozen and thawed.

### 2.7.2. Thin Sections

1. Tetrahydronapthalene (Sigma). Use in a fume cupboard, since it is very toxic.
2. Wax (Paraplast; Sherwood Medical Co., St. Louis, MO).
3. Depex (Merck BDH).
4. Plastic embedding molds (e.g., Merck BDH or R.A. Lamb).

## 3. Methods

### 3.1. General

1. All steps are performed at room temperature for 5 min with rocking or slow rotation unless otherwise stated (*see* **Note 1**). Times are increased to for older embryos (>embryonic day [E] 5.5 chicks or >E11 mice). The procedure as described below is suitable for avian embryos up about stage 26 (E5.5) chicks, E11.5 mice, and zebrafish at all stages. A more suitable protocol for *Xenopus* embryos is presented elsewhere in this volume.
2. To avoid damage, leave a small volume of liquid over embryos between each step.
3. Between fixation (**Subheading 3.2., step 2**) and washing after hybridization (**Subheading 3.5., step 1**), there must be no chance of contamination with RNases. Wear gloves and use disposable plastics instead of glassware where possible. Prepare all solutions with DEPC-treated double-distilled water.

### 3.2. Fixation, Processing, and Hybridization

1. Dissect out embryos (*see* **Note 2**) in 1X PBS (mice) or Howard's Ringer (chicks).
2. Fix in a large volume (at least five times the volume of embryonic tissue) of 4% w/v parafomaldehyde in PBS at 4°C overnight in a vessel that adequately contains the fumes (*see* **Note 3**). Smaller embryos can be fixed for shorter periods of time (4 h minimum), and it is convenient to dechorionate zebrafish after fixation.
3. Wash twice in PBT (for this and subsequent procedures up to hybridization **step 14** use an excess [e.g., 10-fold by volume] of solution over embryo tissue).
4. Embryos are then dehydrated in a graded methanol series: this dissolves membranes and helps to prevent gas bubbles from forming during the subsequent bleaching step. Wash with 25% methanol/75% PBT, and then 50% methanol/50% PBT, then 75% methanol/25% PBT each for 3–5 min with rocking. Finally, wash with 100% methanol twice each for 30 min (*see* **Note 4**).

5. Rehydrate embryos by washing with 75% methanol/25% PBT, then 50% methanol/50% PBT, and then 25% methanol/75% PBT each for 5–10 min with rocking. Finally, wash twice with PBT each for 5 min.

6. Bleach embryos with 6% (v/v) hydrogen peroxide in PBT for 1 h.

7. Wash three times with PBT.

8. Treat with 10 µg/mL proteinase K in PBT for an appropriate amount of time (*see* **Notes 5** and **6**) e.g., for chick embryos: 5–7 min for HH stage 10, 10 min for HH stage 17, and 20 min for HH stage 22 and above).

9. Rinse twice with PBT for 5–10 min (depending on age).

10. Postfix using 0.2% (w/v) gluteraldehyde, and 4% (w/v) paraformaldehyde in PBS for 20 min in the fume hood.

11. Wash three times with PBT.

12. Remove PBT, and rinse embryos once with prewarmed prehybridization solution.

13. Replace prehybridization solution with fresh prehybridization solution, and incubate at 70°C for at least 1 h (*see* **Note 7**).

14. Remove prehybridization solution and replace with prewarmed hybridization solution containing approx 1 µg/mL of riboprobe (usually 10 µL probe to 1 mL prehybridization solution; *see* **Subheading 3.4.**). For "two-color" *in situ*'s, add both DIG- and FITC-labeled probes simultaneously. Incubate at 70°C overnight. Use at least sufficient hybridization solution to cover the embryos completely.

## *3.3. Preparation of Template for Riboprobe Synthesis*

1. The DNA fragment should have been previously cloned into a plasmid vector such that prokaryotic RNA polymerase promoters flank it. Suitable vectors include the pGem series (Promega) and the pKS or pBluescript series (Stratagene). In our experience, T7 and T3 RNA polymerases give the best overall performance on a wide range of templates, whereas SP6 polymerase has problems transcribing certain templates.

2. Linearize plasmid (usually 10–20 µg) with a suitable restriction endonuclease using manufacturer's buffer and conditions (*see* **Note 8**).

3. Check that digestion is complete by analyzing 0.25 µg on a 1% (w/v) agarose gel in 1X TBE containing ethidium bromide (0.1 µg/mL) alongside some uncut plasmid. If the digestion is incomplete, add more enzyme, incubate for a further period, and reanalyze by gel electrophoresis.

4. Extract the remainder of the reaction mixture once with phenol, once with phenol:chloroform (1:1), and twice with chloroform.

5. Precipitate the plasmid DNA by the addition of 0.1 vol of 3 *M* sodium acetate (pH 7.0) and 2.2 vol of ethanol followed by incubation on ice for 15 min.

6. Collect the precipitate by centrifugation at 10,000$g$ in a microfuge for 10 min, aspirate off the supernatant, and wash the pellet with 70% (v/v) ethanol.

7. After allowing the last traces of ethanol to evaporate, the pellet is resuspended in DEPC-treated water to give a final concentration of 0.5 µg/µL and stored at −20°C.

### 3.4. Synthesis of DIG- or FITC-Labeled Riboprobes

1. Warm transcription buffer, nucleotides, DTT, and DNA templates to room temperature.
2. Prepare the following reaction mixture in this order: 11 μL DEPC-treated water, 2 μL 10X DIG, or FITC ribonucleotide mix, 2 μL 10X transcription buffer, and 2 μL 200 m*M* DTT. Mix by gently pipeting up and down. Next add 1 μL RNase inhibitor and 1 μL appropriate RNA polymerase, and mix in the same manner (*see* **Note 9**).
3. Add 2 μL (1 μg) of the linear DNA template, and mix gently by pipeting.
4. Incubate at 37°C for 1–2 h.
5. Check that the transcription reaction has been successful (**Fig. 1G**, lane 1). Analyze 1 μL of product on a 1% (w/v) agarose gel in 1X TBE containing ethidium bromide (0.1 μg/mL). If approx 10 μg of probe have been synthesized, then the DNA band and RNA band will have about the same intensity of fluorescence under UV transillumination (ethidium bromide intercalates poorly into RNA).
6. Meanwhile proceed with the purification of the riboprobe as follows. Add 2 μL RNase-free DNase-1, and incubate at 37°C for exactly 15 min.
7. Add 100 μL TE, 10 μL 4 *M* LiCl, 300 μL ethanol, and vortex.
8. Incubate at room temperature for 30 min, and spin in a microfuge at 10,000*g* for 15 min.
9. Remove the supernatant, immediately resuspend pellet in 100 μL TE and add 10 μL 4 *M* LiCl and 300 μL ethanol. Again, incubate at room temperature for 30 min, and spin in a microfuge at 10,000*g* for 15 min (*see* **Note 10**).
10. Carefully wash the pellet with 70% ethanol (made with DEPC-treated water).
11. Remove the supernatant and immediately dissolve the pellet in TE to give about 1 μg/10 μL as estimated from the gel (**Subheading 3.4., step 5**; usually 100 μL), add 2 μL RNasin, and store probe in 10 μL aliquots at –70°C.

### 3.5. Posthybridization Washes

1. Remove the hybridization solution, and wash embryos twice with prewarmed solution 1 each for 30–60 min at 70°C. For all washes, use an excess of wash solution (e.g., a 20-fold excess) over the approximate volume of the embryonic tissue.
2. Wash with solution 1:solution 2 (1:1) for 10 min at 70°C (*see* **Note 11**).
3. Wash three times with solution 2.
4. Wash twice with 100 μg/mL RNase A in solution 2 for 30 min at 37°C.
5. Wash with solution 2 for 10 min and then solution 3 for 5 min.
6. Wash twice with prewarmed solution 3 each for 30 min at 70°C.
7. Wash three times with 1X TBST.
8. Preblock embryos with 10% (v/v) goat serum in TBST for 60–90 min.
9. During preblocking, weigh out 3 mg embryo powder into a microfuge tube, add 0.5 mL TBST, and heat at 70°C for 30 min. Cool on ice, and add 5 μL goat serum and 1 μL anti-DIG-AP antibody (Boehringer-Mannheim). Shake gently at 4°C for at least 1 h to preabsorb the antibody. Spin in microfuge for 10 min. Dilute the supernatant to 2 mL with 1% goat serum in TBST (*see* **Notes 12** and **13**).

10. Remove the goat serum from the embryos, replace with preabsorbed anti-DIG-AP antibody (**step 9**), and rock overnight at 4°C.

### 3.6. Postantibody Washes and Histochemistry

1. Wash three times each for 15 min with TBST.
2. Change TBST hourly, continue to wash with TBST throughout the day, and then leave in a final wash overnight.
3. Wash three times for 15 min with freshly prepared NTMT (pH 9.5).
4. Incubate with NTMT containing 0.25 µg/mL NBT and 0.175 µg/mL BCIP. Wrap in foil and leave at room temperature until desired reaction product intensity is achieved (*see* **Note 14**). At this point, for *in situ* hybridizations using two probes, proceed directly to **step 6**. For single-color reactions, proceed to **step 5**.
5. For *in situ* hybridizations using a single probe, stop the reaction by washing thoroughly in PBT containing 20 m*M* EDTA or in 1X TBST. Refix in 4% paraformaldehyde prior to storing at 4°C or sectioning. Embryos can also be stored at 4°C in autoclaved 90% v/v glycerol, 1X PBS.
6. For *in situ* hybridizations using two probes, now rinse several times with TBST.
7. Inactivate alkaline phosphatase by incubating embryos in TBST at 70°C for 30–40 min.
8. Wash three times with TBST.
9. Preblock embryos again with 10% goat serum in TBST for 60–90 min.
10. Meanwhile preabsorb the anti-FITC-AP conjugate as in **Subheading 3.5., step 9.**
11. Remove the 10% serum from embryos, replace with preabsorbed anti-FITC-AP antibody, and rock overnight (or longer) at 4°C.
12. Repeat **steps 1** and **2** above.
13. Wash three times for 15 min with freshly prepared NTMT (pH 8.0).
14. Incubate with NTMT containing fast red TR/naphthol AS-MX. Wrap in foil and leave at room temperature until desired reaction product intensity is achieved (usually two to three times the amount of time required for DIG probes).
15. Stop the reaction as in **step 5**. Refix in 4% (w/v) paraformaldehyde prior to storing or sectioning.

### 3.7. Sectioning After In Situ Hybridization

Two procedures are described: the first is a rapid method for producing thick sections, and the second is a method for producing thin sections as permanent mounts.

#### 3.7.1. Thick Sections

1. Equilibrate embryos PBS; use several changes of PBS over 3–4 h.
2. Embryos are embedded in gel albumin. Thaw an aliquot of gel albumin, put 750 µL in a suitable mould, and mix with 75 µL glutaraldehyde (adjust volumes if required for larger embryos), and wait until a skin forms on top (a few minutes).
3. Put embryo (in a small volume of PBS) on top of setting gel albumin, suck off excess PBS, and orient the embryo as desired.

4. Mix separately 750 μL gel albumin with 75 μL glutaraldehyde, and quickly pour over the embryo.
5. Leave overnight at 4°C to set completely.
6. Sections are cut at 50 μm on a vibrotome, are mounted on slides in glycerol, coverslips are sealed with nail varnish and the preparations are stored at 4°C.

### 3.7.2. Thin Sections

The times indicated are for smaller embryos (e.g., stage 20 chicks or E10.5 mice) and should be increased for larger material.

1. Embryos are postfixed in 4% (w/v) paraformaldehyde in PBS for 4 h or overnight.
2. Place in glass vials, and dehydrate in 100% methanol for 5 min.
3. Remove the methanol, and equilibrate in propan-2-ol for 10 min.
4. Remove the alcohol, and clear in tetrahydronapthalene in a fume cupboard for 30 min.
5. Equilibrate in four changes of wax each for 1 h at 58°C.
6. Embryos are then embedded in wax in suitable moulds.
7. Sections are cut at a thickness of 10–20 μm and floated in a water bath.
8. Strips of 5–15 sections are collected onto gelatin- or polylysine-coated slides and allowed to dry. Then they are dewaxed in xylene (two changes each for 15 min) and mounted in DPX.

## 4. Notes

1. For larger embryos, some procedures, particularly washes after hybridization and after incubation with antibodies, are probably more efficiently performed in an incubator containing a slowly rotating wheel, e.g., "Spin 'n' Stack" (Hybaid).
2. If possible, it is worth opening the rostral neural tubes (e.g., the midbrain vesicle or the forebrain) of mouse and chick embryos older than E9.5 (mouse) and HH stage 14 chick to help reduce probe or antibody entrapment; any structure with a large lumen can pose this problem (e.g., also heart and gut) and may be worth dissecting for separate hybridization. As embryos get larger, the problems of high background and poor penetration increase, and it is advisable to dissect further embryos larger than E5.5 (chicks) and E11.5 (mice) into smaller pieces. If a particular tissue is of interest, results are considerably enhanced if it can be exposed or dissected free of surrounding tissue prior to fixation and subsequent processing. Such approaches have allowed the technique to be extended to late stages of development (*see,* e.g., **ref. 5**).
3. Fixation times will vary with the size of the embryo from a few hours (mice up to E10.5, chicks to HH stage 20) to 12 h or longer. In our experience, fixation for 2–3 d does not adversely affect results, but thereafter there is increasing reduction in sensitivity.
4. Some protocols suggest that embryos can be stored for a few months in methanol at this stage. However, it is our experience that this results in considerable loss of sensitivity, and we advocate long-term storage (>1 wk) at –20°C in prehybridization solution.

5. The optimum time for proteinase K treatment ideally should be determined for each new batch of protease. However, treatment for 7 min is a good starting point. For larger embryos, the position of the tissue of interest should be taken into consideration: peripheral tissues (e.g., epidermis, dermis) and small appendages (e.g., developing limbs) require shorter incubation times than deeper body organs. For very young zebrafish embryos (<10 somites), proteinase K treatment may not be required at all.

6. Embryos become very fragile after this step until after they are postfixed and must be treated very gently.

7. Embryos can be stored for months at –20°C in prehybridization solution before or after incubating at 70°C; precipitation of SDS from the buffer has no effect on the preservation of embryos or RNA under these conditions.

8. Restriction enzyme sites should be selected to produce a transcript of 300–1000 bp. At any larger size, nonspecific background signal occurs in embryos probably owing to trapping of probe. Also, if possible, select enzymes that produce a blunt end or a 5'- overhang, because 3'- overhanging ends facilitate "wraparound" transcripts, resulting from the polymerase transcribing back along the complementary DNA strand. If there is no alternative to using an enzyme producing a 3'-overhanging end, the following procedure can be used after digestion to "end polish," generating a blunt end. After digestion, add Klenow polymerase to a final concentration of 5 U/μg DNA template, and incubate at 22°C for 15 min prior to proceeding with the rest of the transcription protocol.

9. Either DIG- or FITC-labeled probes are suitable, and if expression of two genes is to be analyzed simultaneously, one probe is made with each. However, the signal produced from an FITC-labeled probe is weaker than that from the same probe labeled with DIG. Thus, when performing hybridizations with two probes, the probe for the most abundant of the mRNAs of interest generally should be FITC-labeled. In our experience, FITC-labeled probes should be coupled with the fast red substrate system (**Subheading 3.6.**) and DIG-labeled probes with the NBT/BCIP system to produce the most sensitive results. However, the red product of the fast red system cannot be visualized over the dark blue product of the NBT/BCIP system, and this must be taken into account when deciding which to label with DIG or FITC if the two genes being studied are expressed in the same tissue or organ.

10. The second precipitation can frequently be omitted but it helps to ensure that unincorporated nucleotides are efficiently removed, since their presence increases nonspecific "background" signal. This is especially true of FITC-UTP-labeled probes which sometimes benefit from a third precipitation.

11. The RNase treatment (**Subheading 3.5., steps 2–5**) should be omitted if probes are derived from DNA sequences of a different species. Frequently, it can be omitted anyway, but it should be borne in mind that its inclusion probably reduces overall background levels and the possibility of crosshybridization to transcripts derived from closely related members of a multigene family. For example, expression of transcripts of the vertebrate engrailed genes, En-1 and En-2, can only be distinguished if RNase treatment is included (C. Logan, personal communication).

12. In our experience, embryo powder prepared from a single vertebrate species can be used to prevent nonspecific binding of antibody to all vertebrate classes.

13. Blocking reagent (Boehringer Mannheim; BM 1096 176) can be used instead of embryo powder. Blocking reagent is made up as a 10% (w/v) stock in 1X MABT by heating to dissolve it, and then autoclaving, aliquoting, and freezing. 10X MABT is 500 m$M$ maleic acid, 750 m$M$ NaCl pH 7.5, and 0.5% (v/v) Tween-20, and 1 L is prepared by dissolving 58 g maleic acid, 44 g NaCl, and 35 g NaOH (the NaOH is required to dissolve the maleic acid). The pH is adjusted to 7.5 with 10 $N$ NaOH, and 5 mL Tween -20 is added. The dissolved blocking reagent is stored in aliquots at –20°C and used at 2% final concentration instead of embryo powder.

14. If embryos are to be sectioned, we have found that it is best to "overdevelop" the apparent level of signal on whole embryos in order to obtain sufficient intensity on sections.

## Acknowledgments

We thank our colleagues whose suggestions and discussion contributed to the development and revision of these protocols. Work in the authors' laboratory is supported by program grants from the MRC and Human Frontier Science Program and project grants form the MRC and the Wellcome Trust.

## References

1. Mahmood, R., Kiefer, P., Guthrie, S., Dickson, C., and Mason, I. J. (1995) Multiple roles for FGF-3 during cranial neural development of the chicken. *Development* **121,** 1399–1410.

2. Mahmood, R., Bresnick, J., Hornbruch, A., Mahony, C., Morton, N., Colquhoun, K., Martin, P., et al. (1995) FGF-8 in the mouse embryo: a role in the initiation and maintenance of limb bud outgrowth. *Curr. Biol.* **5,** 791–796.

3. Robertson, K. and Mason, I. J. (1995) Expression of the ret in the chicken embryo suggests roles in regionalisation in the vagal neural tube and somites and in development of multiple neural crest and placodal lineages. *Mech. Dev.* **53,** 329–344.

4. Wilkinson, D. G. (1992) Whole mount *in situ* hybridization of vertebrate embryos, in *In Situ Hybridization. A Practical Approach* (Wilkinson, D. G., ed.), IRL Pss, Oxford.

5. Iseki, S., Wilkie, A., Heath, J. K., Ishimaru, T., Eto, K., and Morriss-Kay, G. M. (1997) *Fgfr2* and *osteopontin* domains in the developing skull vault are mutually exclusive and can be altered by locally-applied FGF2. *Development* **124,** 3375–3384.

# 43

# Wholemount *In Situ* Hybridization to *Xenopus* Embryos

## C. Michael Jones and James C. Smith

## 1. Introduction

Whole-mount *in situ* hybridization was first used to detect gene expression in *Drosophila* embryos *(1)*. Various methods are now used to localize mRNAs in most species used for biological studies, and the methods have proven particularly useful when applied to vertebrate species. Here we provide a protocol commonly used for localizing transcripts in *Xenopus* embryos. The method is applicable to whole embryos and intact tissue explants. The methods are essentially as originally described by Hemmati-Brivanlou et al. *(2)* and then modified by Harland *(3)*.

Standard molecular biology techniques not described in detail here can be found in Sambrook et al. *(4)*.

## 2. Materials

| | |
|---|---|
| Digoxygenin-11-UTP (Boehringer-Mannheim 1209 256) | Torula RNA |
| Ribonucleotides | Heparin |
| 1 *M* DTT | Tween-20 |
| RNasin | CHAPS |
| RNA Polymerase (SP6,T3, T7) | EDTA |
| DNase I (RNase-free) | Maleic acid (optional) |
| MOPS | NaCl |
| EGTA | Lamb serum |
| MgSO$_4$ | MgCl |
| Formaldehyde | Tris, pH 9.5 |
| Methanol | Levamisol |

From: *Methods in Molecular Biology, Vol. 97: Molecular Embryology: Methods and Protocols*
Edited by: P. T. Sharpe and I. Mason © Humana Press Inc., Totowa, NJ

Formamide                                                    BSA
20X SSC                                                       Triton X-100
Nitro blue tetrazolium (NBT)
5-bromo-4-chloro-3-indolyl-phosphate (BCIP)
Dimethyl formamide

Other reagents and optional equipment are listed in the text where appropriate. Working concentrations and storage conditions for stock solutions are also detailed.

## 3. Methods

### 3.1. Probe Synthesis

A standard in vitro transcription reaction *(5)* incorporating digoxygenin-11-UTP is performed. SP6, T7, or T3 RNA polymerase may be used. Temperature sensitive reagents should be kept on ice, but the reaction should be mixed at room temperature to avoid precipitation of the DNA template. Reaction:

| | |
|---|---|
| 5X Transcription buffer | 10.0 µL |
| Linearized DNA template (1 µg/µL) | 2.5 µL |
| 1 $M$ DTT | 0.5 µL |
| 2.5 m$M$ NTP mix | 10.0 µL |
| RNasin | 1.0 µL |
| H$_2$O | 24.0 µL |
| RNA polymerase (20 U/µL) | 2.0 µL |

1. Incubate reaction at 37°C for 1–2 h.
2. Check synthesis by running 1 µL of the reaction on an agarose gel, and estimate the amount of RNA by comparing the intensities of the RNA and DNA template.
3. Add 2 µL of RNase-free DNase I, and incubate at 37°C for 10 min.
4. Purify the probe from unincorporated nucleotides by passing the reaction through a G50 spin column. These columns can be homemade. Alternatively, disposable cartridges can be purchased from IBI (R50 Spun Columns, cat. no. 06080).
5. Precipitate the flowthrough from the spin column.
6. Redissolve the purified pellet in H$_2$O. Probes can be reduced in size by limited alkaline hydrolysis if necessary (*see* **Note 1**). Store probe stock at –20°C. Probes are stable indefinitely.
7. Most probes work well when used at approx 1 µg/mL in hybridization buffer.

### 3.2. Fixation of Embryos

Albino embryos are preferred, because pigmentation in wild-type embryos can obscure the hybridization signal. If pigmented embryos or explants are to be used, they can be taken through the procedure and then bleached afterwards by incubating in 10% hydrogen peroxide in methanol. Satisfactory results can be obtained in this manner. The bleaching can take from several hours to several days depending on the pigmentation of the embryos.

Just before fixation, embryos should be removed from their vitelline membranes using sharpened watchmaker's forceps. For blastula and gastrula stage embryos, a hole should be placed in the blastocoel to prevent nonspecific trapping of the probe and/or antibody. For neurula stages, the archenteron should be pierced. Embryos are fixed in small glass vials or scintillation vials if a large number are to be processed simultaneously. The preferred fixative is MEMFA (0.1 $M$ MOPS, pH 7.4, 2 m$M$ EGTA, 1 m$M$ MgSO$_4$, 3.7% formaldehyde). A 10X stock solution of the salts can be kept at 4°C. To make fresh fixative, add 1/10 vol 37% formaldehyde and 1/10 vol 10X salts and water to make the volume needed. Embryos should be fixed for approx 1 h at room temperature with constant rotation. After fixation, remove the fixative, and add 100% ethanol or methanol to the embryos. Make two changes of the 100% alcohol, and store the embryos at –20°C. Embryos can be stored for several months or longer.

### 3.3. Hybridization

Embryos can be processed either in small glass vials or in mesh-bottomed baskets. Using glass vials is more labor-intensive, since each vial must be handled individually and solution changes require aspiration and refilling of each. Baskets can be processed in mass, and only involves lifting the baskets from solution to solution. Additionally, the risk of damaging or losing the tissues is reduced when baskets are used. Baskets can be made by attaching small mesh to an Eppendorf tube from which the bottom has been removed. Alternatively, baskets are available commercially from Costar (15-mm Netwell, 74-µm mesh, cat. no. 3477) and fit easily into 12-well plastic dishes (Costar 12-well cluster, cat. no. 3513), which makes transferring the baskets to different solutions easy. A large number of embryos, from different stages, can be processed simultaneously for each probe of interest. A sense control hybridization should be included to control for nonspecific signal and overall level of background staining. Embryos can be transferred by using a Pasteur pipet from which the end has been cut to increase the size of the opening. Each set of embryos is then processed as follows:

1. Rehydrate embryos by incubating for 2–5 min in 100% Methanol (MeOH), 75% MeOH/25% H$_2$O, 50% MeOH/50% H$_2$O, 25% MeOH/75% 1X PBS + 0.1% Tween-20 (PBST), and 100% PBST.
2. Wash three times for 5 min in PBST.
3. Incubate embryos in 10 µg/mL Proteinase K in PBST. The time must be determined for each batch of Proteinase K, but a good starting point is approx 30 min at room temperature. This treatment is optional, but it probably increases sensitivity to some degree.
4. Rinse twice for 5 min in 0.1 $M$ triethanolamine (pH 7.0–8.0).

5. Add 12.5 µL acetic anhydride/5 mL of triethanolamine and incubate for 5 min. Then, add another 12.5 µL of acetic anhydride and incubate for an additional 5 min. These steps block positively charged groups within the tissues *(6)*.
6. Wash twice for 5 min in PBST.
7. Refix embryos by incubating in 4% paraformaldehyde in PBST for 20 min.
8. Wash five times for 5 min in PBST.
9. Replace PBST with hybridization buffer (listed below). If 5-mL glass vials are used, 500 µL are a sufficient volume to cover a large number of embryos. If baskets are used, an appropriate volume to cover all embryos must be used. For the Costar netwells, use 2–2.5 mL. Hybridization buffer:
   50% Formamide, 5X SSC, 1 mg/mL Torula RNA, 100 µg/mL Heparin, 1X Denhardt's, 0.1% Tween-20, 0.1% CHAPS, and 5 m$M$ EDTA.
10. After embryos settle through the dense hybridization solution, replace the buffer with fresh hybridization buffer and incubate at 60°C. Prehybridize embryos 5–6 h.
11. After prehybridization, replace the solution with hybridization buffer containing the probe (1 µg/mL). Hybridize overnight at 60°C.

### 3.4. Washing

1. Keep the Hybridization Buffer containing the probe. It can be reused up to two times. Replace hybridization buffer with a solution of 50% formamide, 5X SSC (Mock-Hyb Buffer). Hybridization buffer can be used here, but it is more economical to use the Mock-Hyb Buffer. Wash at 60°C for 10 min.
2. Wash in 50% Mock-Hyb buffer/50% 2X SSC 10 min at 60°C.
3. Wash in 25% Mock-Hyb buffer/75% 2X SSC for 10 min at 60°C.
4. Wash twice for 20 min in 2X SSC/0.3% CHAPS at 60°C.
5. RNAse treat by incubating for 30 min at 37°C in 2X SSC/0.3% CHAPS containing 20 µg/mL RNase A and 10 U/mL RNase T$_1$.
6. Rinse in 2X SSC/0.3% CHAPS for 10 min at room temperature.
7. Wash twice for 30 min in 0.2X SSC/0.3% CHAPS at 60°C.
8. Wash twice for 10 min in PBST/0.3% CHAPS at 60°C.
9. Wash three times for 5 min in PBST at room temperature.

### 3.5. Antibody Incubation

Antibody incubations are performed in one of two buffer systems. One buffer consists of 1X PBS containing 2 mg/mL BSA and 0.1% Triton X-100 (PBT). The other is a maleic acid buffer developed by Tabitha Doniach (MAb = 100 m$M$ maleic acid, 150 m$M$ NaCl, pH 7.5). If background staining becomes a problem, the MAb is recommended. After the final wash in PBST, proceed as follows:

1. Wash three times for 5 min with PBT or MAb at room temperature.
2. Block tissues by incubating one hour at room temperature in PBT + 20% Heat-treated lamb serum. Lamb serum is heat-treated by placing it in a water bath at 55–60°C for 30 min. The serum is then stored frozen in aliquots. Alternatively, MAb containing 2% blocking reagent from Boehringer Mannheim (cat. no. 1096 176) and 20% serum can be used.

3. After blocking, replace the blocking buffer with the same solution containing a 1:2000 dilution of antidigoxygenin antibody (Fab fragments) coupled to alkaline phosphatase (BMB cat. no. 1093 274). Incubate overnight at 4°C, or at least 4 h at room temperature with constant shaking.

## 3.6. Antibody Washes

Wash tissues at least five times each for 1 h at room temperature in PBT or MAb. It is best if the tissues are moving gently within the wash solutions. Washes can be carried out overnight (or over the weekend) at 4°C if convenient.

## 3.7. Chromogenic Reaction

1. After antibody washes, rinse tissues twice for 5 min in alkaline phosphatase buffer (APB): 100 m$M$ Tris, pH 9.5, 50 m$M$ MgCl$_2$, 100 m$M$ NaCl, 0.1% Tween-20, and 5 m$M$ Levamisol. APB should be made fresh just before use, since the levamisol is not stable. Levamisol can be made as a 1 $M$ stock and stored at –20°C.
2. For color development, replace the last wash with APB containing 4.5 µL NBT (nitro blue tetrazolium, 75 mg/mL in 70% dimethyl formamide) and 3.5 µL BCIP (5-bromo-4-chloro-3-indolyl-phosphate, 50 mg/mL in 100% dimethyl formamide)/ mL. Alternatively, the color reaction can be carried out by placing the tissues directly into BM purple AP substrate, precipitating (BMB cat. no. 1442 074). The BM purple substrate tends to give less background staining than the NBT/ BCIP system, and is recommended if long reaction times are required. The appearance of staining is dependent on the level of gene expression and the probe that is used. Positive staining can become visible after several minutes, or can take up to 1 d.
3. When satisfied with the color development, stop the chromogenic reaction by placing the tissues in MEMFA. This postfixation stops the color reaction and stabilizes the stain. Postfix the tissues for at least 1 h, or they can be left overnight.

## 3.8. Alternative Protocol

An alternative to the processing steps listed in **Subheading 3.4.** can be used. For most markers that are routinely assayed, this shortened procedure works well and requires much less effort. After pretreatment and hybridization as described above, process the samples as follows:

1. Replace hybridization solution with 50% formamide/5X SSC for 10 min at 60°C.
2. Wash once in 25% formamide/2X SSC for 10 min at 60°C.
3. Wash twice for 30 min with 2X SSC/0.1% CHAPS at 60°C.
4. Wash twice for 30 min with 0.2X SSC/0.1% CHAPS at 60°C.
5. After the final wash at 60°C, wash twice at room temperature with PBT or MAb, and then proceed to the blocking step and antibody incubations as described above (**Subheading 3.5.**).

## 4. Notes

1. Limited alkaline hydrolysis of the probe to reduce its overall length to 200–300 bp is a consideration. Most probes work well when used as full length (2–3 kb probes have worked well), but others may function better if their length is reduced. To hydrolyze a probe, resuspend the precipitated probe in 50 μL of a solution containing 40 m$M$ sodium bicarbonate, and 60 m$M$ sodium carbonate. Heat at 60°C for 30 min to 1 h. After hydrolysis, increase the volume to 200 μL with water and precipitate. Redissolve the pellet in water, and use at a final concentration of 1 μg/mL for hybridization.

2. If persistent background is a problem, consider preincubating the anti-digoxygenin antibody with embryonic acetone powder prior to use. Acetone powder is easy to make and the protocol can be found in *Antibodies: A Laboratory Manual (7)*. Additionally, background is usually reduced when BM purple AP substrate is used for the color development instead of NBT/BCIP in APB.

3. Whole-mount specimens can be processed for histological sectioning, which allows more detailed analysis. If sectioning is desired, it is a good idea to develop the color as intensely as possible (overstain), monitoring to ensure that background does not become a problem. It is essential to postfix the tissue to stabilize the stain before processing for histology. To prepare the tissues for sectioning, dehydrate through a series of methanol (30, 50, 70, 90, and 100%), and clear with xylene or histoclear. Embed tissues in wax, and section at 10–20 μm. Thicker sections provide more intense signal. Tissues can be counterstained with eosin, which provides a nice contrast to the purple color reaction product.

## References

1. Tautz, D. and Pfeifle, C. (1989) A non-radioactive in situ hybridization method for the localization of specific RNAs in Drosophila embryos reveals translational control of the segmentation gene hunchback. *Chromosoma* **98,** 81–85.
2. Hemmati-Brivanlou, A., Frank, D., Bolce, M. B., Sive, H. L., and Harland, R. M. (1990) Localization of specific mRNAs in Xenopus embryos by whole mount in situ hybridization. *Development* **110,** 325–330.
3. Harland, R. M. (1991) In situ hybridization: an improved whole-mount method for Xenopus embryos, in: *Methods in Cell Biology*-Xenopus laevis: *Practical Uses in Cell and Molecular Biology* (Kay, B. K. and Peng, H. B., eds.), Academic, San Diego, CA, pp. 685–695.
4. Sambrook, J., Fritsch, E. F., and Maniatis, T. (1989) *Molecular Cloning: A Laboratory Manual.* Cold Spring Harbor Laboratory, Cold Spring Harbor, NY.
5. Melton, D. A., Krieg, P. A., Rebagliati, M. R., Maniatis, T., Zinn, K, and Green, M. R. (1985) Efficient in vitro synthesis of biologically active RNA and RNA hybridization probes from plasmids containing a bacteriophage SP6 promoter. *Nucleic Acids Res.* **12,** 7035–7056.
6. Hayashi, S., Gillam, I. C., Delaney, A., and Tener, G. M. (1978) Acetylation of chromosome squashes of Drosophila melanogaster decreases the background in autoradiographs from hybridization with [125]I-labelled RNA. *J. Histochem. Cytochem.* **36,** 677–679.
7. Harlow, E. and Lane, D. (1988) *Antibodies: A Laboratory Manual.* Cold Spring Harbor Laboratory, Cold Spring Harbor, NY.

# 44

# Wholemount *In Situ* Hybridization to Amphioxus Embryos

## Peter W. H. Holland

### 1. Introduction

Once a gene has been cloned the spatial distribution of its transcripts may be determined by *in situ* hybridization. This involves applying a labeled antisense RNA probe, with complimentary nucleotide sequence to the mRNA, either to tissue sections or to whole fixed specimens. The cephalochordate amphioxus has small embryos (100–300 μm in length) that are particularly will suited to whole mount *in situ* hybridization. The method has been successfully used to examine expression of many development expressed genes in amphioxus embryos, giving insight into the evolutionary conservation and divergence of gene expression between amphiouxus and vertebrates.

### 2. Materials

1. MOPS buffer: 0.5 $M$ NaCl, 1 m$M$ MgSO$_4$, 2 m$M$ EGTA, 0.1 $M$ morpholinopropanesulfonic acid buffer, pH 7.5. Sterile.
2. Paraformaldehyde: electron microscopy-grade (TAAB, Aldermastan, UK).
3. Nunclon four-well multiwell dishes (Nunc cat. no. 176740, Life Technologies, Paisley, Scotland).
4. PBS: 0.9 % (w/v) NaCl, 20 m$M$ sodium phosphate buffer, pH 7.4. Sterile.
5. PBT: PBS plus 0.1% Tween 20.
6. Proteinase K (10 mg/mL stock). Store frozen.
7. 0.1 $M$ Triethanolamine, pH 8.0. Sterile.
8. Acetic anhydride.
9. Hybridization buffer: 50% formamide (Fluka, Gillingham, UK), 1.3X SSC, pH 4.5, 5 m$M$ EDTA, pH 8.0, 500 μg/mL yeast RNA, 0.2% (v/v) Tween 20, 0.5% CHAPS, 100 μg/mL heparin, in DEPC-treated sterile water.
10. Oven with shaking platform, e.g., Hybaid Midi-Dual oven.

From: *Methods in Molecular Biology, Vol. 97: Molecular Embryology: Methods and Protocols*
Edited by: P. T. Sharpe and I. Mason © Humana Press Inc., Totowa, NJ

11. Wash solution A: 50% formamide (Fluka), 4X SSC, 0.1% Tween-20.
12. Wash solution B: 50% formamide (Fluka), 2X SSC, 0.1% Tween-20.
13. Wash solution C: 50% formamide (Fluka), 1X SSC, 0.1% Tween-20.
14. Wash solution D: 0.2X SSC, 0.1% Tween-20.
15. Blocking buffer: 5% (w/v) Boehringer Mannheim (Lewes, UK) digoxigenin blocking reagent, 2 mg/mL BSA in PBT. Takes over 1 h to dissolve at 70°C. Store frozen.
16. Amphioxus powder: Grind frozen adult amphioxus under liquid nitrogen, or homogenize in PBS. Then incubate in ice-cold acetone for 30 min. Recover powder by centrifugation, and grind further in a pestle and mortar. Store dry powder frozen.
17. Sheep serum: Heat to 55°C for 30 min. Store frozen.
18. Preabsorbed 1:3000 anti-DIG fab fragments: To 1.5 mg of amphioxus powder, add 400 µL PBT, and heat to 70°C for 10 min. Add 50 µL 20 mg/mL BSA, 50 µL pretreated sheep serum, 0.5 µL anti-DIG fab fragments (Boehringer Mannheim). Mix and incubate at 37°C for 1 h, and then overnight at 4°C. Centrifuge for 5 min to remove powder; to the supernatant, add 1 mL PBT containing 2 mg/mL BSA. Add 50 µL pretreated sheep serum. Store at –20°C in 200-µL aliquots.
19. APT buffer: 100 m$M$ NaCl, 100 m$M$ Tris-HCl, pH 9.5, 50 m$M$ MgCl$_2$, 1% Tween-20.
20. NBT/BCIP staining mix: Just before use, add 4.5 µL stock nitro blue tetrazolium salt (NBT) and 3.5 µL stock 5-bromo-4-chloro-3-indolyl phosphate (BCIP) to 1 mL APT buffer containing 2 m$M$ levamisole. Stock NBT is 75 mg/mL in 70% dimethyl formamide, and stock BCIP is 50 mg/mL in 100% dimethyl formamide; both stocks are supplied by Gibco-BRL as Immunoselect kit.
21. 37–40% Formaldehyde stock solution.

## 3. Methods

The protocol given is designed for amphioxus embryos and larvae, obtained as described in Chapter 33. The method is modified from **ref. *1*** and has been used to visualize RNA from several developmentally expressed genes of amphioxus (*see* **Note 1**). With modification, it has also been successfully applied to embryos of other marine invertebrates, notably ascidia (e.g., **ref. *2***).

1. Fix amphioxus embryos and larvae (up to 3 d postfertilization) in freshly made 4% paraformaldehyde in MOPS buffer, at 4°C for 12 h.
2. After fixation, transfer specimens through two changes of 70% ethanol, and store at –20°C. Do not allow to freeze.
3. Digoxigenin-(DIG) labeled riboprobes are made by standard protocols (Chapter 42); redissolve probes in 50% formamide at 100 ng/µL, and store at –20°C. (*see* **Notes 2** and **5**).
4. Transfer selected embryos into one or more wells of a disposable Nunclon four-well dish, and add 0.5 mL PBT. All subsequent incubation, wash, and hybridization steps are performed in the same Nunclon dish at room temperature, unless otherwise stated. Each change of solution is performed using a Gilson pipetman, while observing under a dissecting microscope to ensure that embryos are not lost.
5. Change solution for fresh PBT.

6. Change solution for 7.5 µg/mL proteinase K in PBT; incubate without agitation for 10 min.
7. Quickly wash with two changes of PBT. Leave second wash for 5 min.
8. Change solution for 4% paraformaldehyde in PBT; incubate without agitation for 1 h.
9. Wash with two changes of 0.1 $M$ triethanolamine, 2–5 min each.
10. Remove solution from embryos. Quickly add 2.5 µL acetic anhydride to 1 mL 0.1 $M$ triethanolamine in a microfuge tube, vortex, and add 300 µL onto embryos. Incubate without agitation for 5 min.
11. Repeat **step 10**.
12. Wash with two changes of PBT, 2–5 min each.
13. Change solution for a 1:1 mix of PBT and hybridization buffer; incubate for 5 min.
14. Change solution for hybridization buffer; incubate for 10 min.
15. Change solution for hybridization buffer prewarmed to 60°C; incubate at 60°C for at least 1 h, with horizontal rotation. A hybridization oven is useful.
16. Add 2 µL of DIG-labeled riboprobe from **step 3** to 200 µL hybridization buffer. Heat to 75°C for 2 min (*see* **Note 3**).
17. Remove solution on embryos, and replace with the heated and diluted probe. Seal sides of the multiwell dish with Parafilm, and place on foam microfuge rack floating in 70°C water bath for 2 min.
18. Move the dish to the shaking oven, and incubate overnight at 60°C with horizontal rotation.
19. Wash embryos in prewarmed solution A, 2 × 30 min at 60°C with agitation.
20. Wash embryos in prewarmed solution B, 2 × 30 min at 60°C with agitation.
21. Wash embryos in prewarmed solution C, 2 × 30 min at 60°C with agitation.
22. For more stringent washes, wash with solution D, 30 min at room temperature.
23. Wash with two changes of PBT, 5–10 min at room temperature (*see* **Note 4**).
24. Change solution for blocking buffer; incubate for 1 h at room temperature.
25. Change solution for preabsorbed 1:3000 anti-DIG fab fragments. Place on a rotating platform at room temperature for 10 min, and then incubate overnight at 4°C. Agitation is not necessary.
26. Carefully remove antibody solution from embryos. Store antibody frozen for reuse.
27. Wash embryos in PBT, 2× for 10 min at room temperature.
28. Wash embryos in PBT, 3× for 30 min at 37°C.
29. Wash embryos in APT buffer, 2× for 10 min at room temperature.
30. Wash embryos in APT buffer, 2× for 20 min at room temperature.
31. Replace solution with NBT/BCIP staining mix, and place in dark (e.g., cover in foil) until color develops (10 min to 48 h). If staining requires longer than overnight, change for fresh NBT/BCIP mix each day. Some recommend that this step be performed in wash glasses; we find no difference between glass and plastic dishes.
32. When all embryos have developed a consistent staining pattern, wash with two changes of PBT and fix in 4% formaldehyde in PBT for 1 h at room temperature.
33. Mount embryos on slides in 80% glycerol in PBT. For amphioxus embryos, support the coverslip with strips of autoclave tape stuck onto the slide.

## 4. Notes

1. Recent improvements to this protocol, incorporated above, include the high concentration of nonspecific RNA in the hybridization buffer (500 µg/mL), and long posthybridization and postantibody washing steps.
2. It is essential to use high-purity formamide in the hybridization buffer and wash solutions A, B, and C. Fluka formamide gives consistent results.
3. After prehybridization (**Subheading 3., step 15**), embryos can be stored at –20°C for later use.
4. In contrast to some other protocols, we omit an RNase step in the posthybridization washes; in amphioxus, it can reduce signal without enhancing specificity. Different results may be found in other species.
5. In our experience, the most critical variant is the quality of the probe. After synthesis, probes must be checked by agarose electrophoresis to assess concentration and degradation; storage in 50% formamide (**step 3**) greatly improves probe stability and dissolution.

## Acknowledgments

I thank Linda Holland, Nic Williams, Jordi Garcia-Fernàndez, and Steve Davis for their contributions to the development of this protocol.

## References

1. Holland, P. W. H., Holland, L. Z., Williams, N. A., and Holland, N. D. (1992) An amphioxus homeobox gene: sequence conservation, spatial expression during development and insights into vertebrate evolution. *Development* **116,** 653–661.
2. Yasuo, H. and Satoh, N. (1994) An ascidian homolog of the mouse *Brachyury (T)* gene is expressed exclusively in notochord cells at the fate restricted stage. *Dev. Growth Differ.* **36,** 9–18.

# 45

## *In Situ* Hybridization to Sections (Nonradioactive)

### Maria Rex and Paul J. Scotting

### 1. Introduction

In situ hybridization (ISH) takes advantage of the ability of mRNA within a cell to hybridize with exogenously applied complementary RNA (riboprobes) or DNA molecules. This interaction is visualized by labeling the applied nucleic acid probe with a detectable molecule (radioactive, such as [35]S, or nonradioactive, such as digoxygenin [DIG]). The technique allows patterns of gene expression to be visualized in many tissues or cell types simultaneously.

Nonradioactive ISH has several advantages over radioactive ISH. It lacks the biohazards associated with the use of radioisotopes, it takes days rather than weeks to get a visible signal, and it is cheaper. Using nonradioactive detection the degree of cellular resolution is significantly improved over that achieved with autoradiography. Another advantage of nonradioactive ISH is that it can be performed in combination with other assays with relative ease, in order to compare mRNA and protein distribution in the same tissues *(1)* as described in this chapter.

The procedures presented here are for ISH of DIG-labeled riboprobes to tissue sections from wax embedded tissues (methods for the use of frozen sections are generally similar but minor modifications may be required *[2,3]*). The use of tissue sections rather than whole mount embryos is preferable where the size of the embryo or the density of tissues leads to incomplete penetration of probe in whole mounts. The use of sections also allows a number of probes (for mRNA or proteins) to be used on the same or adjacent sections, an approach that is less straightforward in whole mounts. Both whole mount and section *in situ* hybridization protocols can therefore be used in a complementary way to give optimum results.

From: *Methods in Molecular Biology, Vol. 97: Molecular Embryology: Methods and Protocols*
Edited by: P. T. Sharpe and I. Mason © Humana Press Inc., Totowa, NJ

The procedures involved are:

## 1.1. Preparation of Probe

DIG-labeled riboprobes are synthesized by the transcription of sequences of interest which have been cloned into a vector such that they are flanked by two different RNA polymerase binding sites. The vector is first linearized with a restriction enzyme such that transcription produces "run-off" transcripts, that is, transcripts derived from the insert sequence alone. The restriction enzyme is chosen such that transcription yields an RNA probe which is complementary (antisense) to the target mRNA. The probe transcribed from the opposite strand (sense probe) can be use as a negative control.

The length of the probe is important. Probes of up to 1 kb are optimal. Longer probes penetrate the tissue less efficiently but can be partially degraded by alkaline hydrolysis to a more suitable size (*4*).

## 1.2. Preparation of Tissue

To preserve morphology, the biological material must be fixed. Crosslinking fixatives, such as paraformaldehyde, can give greater accessibility and retention of cellular RNA than precipitating fixatives.

Tissue sections can be cut either on a cryostat (frozen sections) or after embedding in a matrix, such as paraffin wax, on a microtome. Tissue morphology may be better preserved in embedded tissues. The methods presented here are for wax sections.

## 1.3. Prehybridization Treatment of Tissues

Before hybridization the wax is removed from the sections by dissolving in organic solvent. Proteinase treatment is used to render the target RNA more accessible to the probe by digestion of cellular protein. The optimal extent of Proteinase K treatment should be determined empirically since it varies between tissues. Sections must be re-fixed after proteinase treatment.

## 1.4. Hybridization of Probe to Tissue and Posthybridization Washes

The temperature for hybridization can be varied according to the specificity of interaction of probe and target required. For homologous probe and target as used in most cases we routinely use 70°C, use of lower temperatures may lead to nonspecific artefacts. Following hybridization, slides are washed at high stringency to remove unbound probe and probe that has bound to sequences similar to but distinct from the target sequence

## 1.5. Color Detection

Bound probe is visualized by incubation of sections with alkaline-phosphatase conjugated antibody against DIG and subsequent addition of substrates that yield an insoluble colored product. This reaction can be carried out for several days in order to get a stronger signal (though a strong signal can sometimes be obtained in a few hours). The limiting factor is the amount of nonspecific background staining obtained.

### 1.6. Immunohistochemistry

Slides are processed for immunohistochemistry to detect other markers immediately following color detection of the ISH signal. The immunohistochemical probe is visualized with Fast Red (**Fig. 1**). One potential problem of using this protocol is residual alkaline phosphatase activity from the ISH, which may produce a product with the Fast Red substrate. Controls in which no primary antibody was used during immunohistochemistry show that this is not generally the case. If residual alkaline phosphatase activity is a concern with other probes or reagents it can be inactivated by treatment with 0.1 $M$ glycine HCl, pH 2.2, prior to immunohistochemistry. Occasionally the proteinase K step used in the ISH protocol may destroy the antigen for subsequent immunohistochemical staining. In such cases a different method of antigen retrieval, such as pressure cooking or microwaving can be used instead *(5)*.

## 2. Materials

All solutions should be made to the standard required for molecular biology using molecular biology grade reagents and sterile distilled water.

### 2.1. Preparation of DIG-Labeled Single-Stranded RNA Probe (Riboprobe) by In Vitro Transcription

1. Riboprobe Gemini System II Buffers Kit (Promega, Southampton, UK) includes cold ribonucleotides, DTT, Transcription optimized buffer. Store at –20°C. NTPs go off, so kits to suit level of use should be bought.
2. T3 and T7 RNA polymerase (Promega, Southampton, UK). Store at –20°C.
3. DIG-UTP (Boehringer Mannheim, Lewes, UK). Store at –20°C.
4. RNasin Ribonuclease Inhibitor (Promega, Southampton, UK). Store at –20°C.
5. Restriction enzymes (single cut 5' and 3' of insert).
6. Phenol and chloroform.
7. 7.8 $M$ Ammonium acetate.
8. 4 $M$ Lithium chloride.

### 2.2. Preparation of Tissue (Fixation, Embedding, and Sectioning)

1. Phosphate-buffered saline (PBS).
2. 4% Paraformaldehyde is prepared on the day of use by adding 4 g paraformaldehyde (BDH/Merk Ltd, Poole, UK)/100 mL PBS, heating at 65°C until the paraformaldehyde has dissolved and cooling on ice (aliquots can frozen but only freeze-thaw once). Paraformaldehyde is toxic, handle solutions within a fume cupboard.

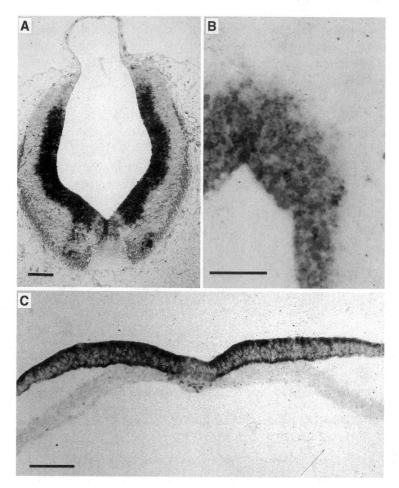

Fig. 1. Nonradioactive *in situ* hybridization on wax-embedded tissues in combination with immunohistochemical staining. Blue staining represents *in situ* hybridization signal and red is the reaction following immunohistochemical detection of other antigens. **(A)** *In situ* hybridization for *cSox21* (a transcription factor, *[8]*) plus immunohostochemical staining for neurofilament (antineurofilament monoclonal (Dako Ltd). **(B)** *In situ* hybridization for *cSox2 (9)* plus immunohistochemical detection of BrdU to identify proliferating cells (*see* **ref.** *10* for further details). **(C)** *In situ* hybridization for cSox3 (*8*) plus immunohistochemical detection of Brachyury (a nuclear transcription factor; *11*). (*See* color plate 5 appearing after p. 368.)

3. Saline (0.83% sodium chloride).
4. 100% Ethanol.
5. Histolene (Cellpath plc, Hemel Hempstead, UK). Less toxic alternative to xylene.
6. Paraffin wax, Tissue Tek pastilated.

## 2.3. Preparation of Subbed Slides (Silicon Coating to Aid Adherence of Tissue Sections)

1. 3-Aminopropyltriethoxysilane (TESPA) (Sigma, Poole, UK). Store at 4°C. TESPA is toxic.
2. Acetone.
3. 100% Ethanol.
4. Concentrated hydrochloric acid.

## 2.4. Prehybridization Treatment and Hybridization

1. Proteinase K is prepared as a 10 mg/mL stock and is stored at –20°C in small aliquots. On day of use dilute to 10 µg/mL with PBS.
2. PBT: 0.1% Tween-20 in PBS (pH 7.3).
3. 4% Paraformaldehyde prepared as in **Subheading 2.2.**
4. Methanol.
5. Hybridization solution: 50% deionised formamide (BDH/Merk Ltd, Poole, UK-molecular biology grade); 5X SSC (pH 4.5 with citric acid [BDH]); 50 µg/mL yeast tRNA; 1% SDS; 50 µg/mL heparin (Sigma). Store at room temperature for up to 1 mo.
6. Histolene.
7. 20X SSC: 20-fold stock solution of 0.15 $M$ NaCl and 0.015 $M$ sodium citrate.

## 2.5. Posthybridization Washes and Color Detection

1. Formamide-GPR (BDH/Merk, Poole, UK) Formamide is toxic. Avoid contact with the skin and use witin a fume cupboard.
2. 20X SSC (pH 4.4 with citric acid).
3. 10% SDS.
4. Sheep serum (Sigma): store in aliquots at –20°C.
5. Alkaline-phosphatase anti-digoxygenin Fab fragments (Boehringer Mannheim, Lewes, UK). Store at 4°C.
6. Embryo Powder: Homogenise suitable tissue, i.e., tissue being analyzed, in a minimum vol of PBS. Add 4 vol of ice-cold acetone, mix and place on ice for 30 min. Spin at 10,000$g$ for 10 min. Discard supernatant and wash pellet in ice-cold acetone and respin. Spread pellet on filter paper and grind to a fine powder. Allow to air dry. Store tightly capped at 4°C.
7. Blocking reagent for nucleic acid hybridization and detection (Boehringer Mannheim).
8. TBST: 10X TBS-135 m$M$ NaCl, 250 m$M$ Tris-HCl, pH 7.5. On day of use dilute to 1X and add 0.024 g levamisole and 0.5 mL Tween-20 to each 50 mL of 1X TBST.
9. Alkaline phosphatase buffer: 100 m$M$ NaCl, 50 m$M$ MgCl$_2$, 100 m$M$ Tris-HCl pH 9.5, 1% Tween-20. Make on the day from stock solutions.
10. Color solution is alkaline phosphatase buffer plus 45 µL/100 mL NBT (75 mg/mL, 4-nitro Blue Tetrazolium chloride, Boehringer Mannheim) and 35 µL/100 mL of

X-phosphate (50 mg/mL, 5-Bromo-4-chloro-3-indolyl-α-D-galacto-pyranose, Boehringer Mannheim), and 2 m*M* levamisole (Sigma). Levamisole is made fresh from solid every 2 wk and is stored at 4°C.

### 2.6. Immunohistochemistry

1. Sheep serum (Sigma): store in aliquots at –20°C.
2. Primary antibody/antiserum and biotinylated secondary antibody raised in the appropriate animal. The dilution of antibodies is determined empirically in each case. Store at 4°C.
3. PBS.
4. Alkaline phosphatase streptavidin (Vector Laboratories).
5. Fast Red (Sigma). Prepare according to suppliers instructions.
6. Mowiol 4-88 (Calbiochem-Novabiochem Corp). Mowiol is prepared by adding 2.4 g of Mowiol to 6 g of glycerol and 6 mL of water. The mixture is left at room temperature for several hours before the addition of 12 mL of 0.2 *M* Tris, pH 8.5. Heat to 50°C for 10 min. Store as aliquots at –20°C.

## 3. Methods

### 3.1. Preparation of DIG-Labeled Single-Stranded RNA Probe (Riboprobe) by In Vitro Transcription

1. Linearize DNA with an appropriate restriction enzyme (*see* **Note 2**).
2. Extract with phenol/chloroform and precipitate. Resuspend in sterile water at 1 mg/mL.
3. Warm buffers, nucleotides, DTT, and templates to room temperature (*see* **Note 3**).
4. Add the following, in the order shown, at room temperature:
   a. x μL DEPC-water (to final volume of 20 μL).
   b. 4 μL 5X transcription buffer.
   c. 2 μL 100 m*M* DTT.
   d. 2 μL rATP(10 m*M*).
   e. 2 μL rCTP (10 m*M*).
   f. 2 μL rGTP (10 m*M*).
   g. 1.3 μL rUTP (10 m*M*).
   h. 0.7 μL DIG-UTP (10 m*M*).
   i. 0.5 μL RNasin.
   j. 1 μL RNA polymerase (T3 or T7).
   k. x μL (~1 μg) template.
5. Mix gently, not vortex, pulse in microfuge.
6. Incubate at 37°C for 2 h.
7. Check 1 μL on a gel and in the meantime precipitate probe twice as follows: Add 100 μL water, 10 μL 4 *M* LiCl, 300 μL ethanol, 1 μL glycogen (4 mg/mL). Place at –80°C for at least 1 h. Spin and resuspend pellet in 30 μL of water; add 15 μL 7.8 *M* ammonium acetate, 1 μL glycogen and 100 μL ethanol. Place at –80°C for at least 1 h. Spin and resuspend pellet in 30 μL of water. Store at –80°C.

## 3.2. Preparation of Tissue (Fixation, Embedding, and Sectioning)

1. Fix tissue in 4% paraformaldehyde in PBS at 4°C overnight (or 4 h at room temperature).
2. Successively replace solution with the following (at least 30 min each with occasional agitation, at room temperature except where stated).
   a. Saline, twice at 4°C.
   b. 1:1 Saline:ethanol mix at 4°C.
   c. 70% Ethanol twice.
   d. 85% Ethanol.
   e. 95% Ethanol.
   f. 100% Ethanol twice.
   g. Histolene twice.
   h. 1:1 Histolene:melted paraffin wax for 20 min at 60°C.
   i. Wax, three times (20 min each at 60°C).
3. Transfer tissues to a mould at 60°C, orientate and allow wax to set.
4. Cut ribbons of 6 µm sections on a microtome. Float the ribbons of sections on a bath of distilled water at 40°C until free from creases. Collect on subbed slides. Dry slides at 37°C overnight.

## 3.3. Preparation of Subbed Slides

1. Dip slides in 10% HCl/70% ethanol, followed by distilled water, and 95% ethanol.
2. Dry in oven at 150°C for 5 min and allow to cool.
3. Dip slides in 2% TESPA (3-aminopropyltriethoxysilane) in acetone, 10 s.
4. Wash twice with acetone, and then once with distilled water.
5. Dry at 42°C.

## 3.4. Prehybridization Treatment and Hybridization

1. Dip slides in Histolene two times for 10 min to remove wax (*see* **Note 4**).
2. Rehydrate through graded alcohols (1 min in each): 100% MeOH (twice), 75% MeOH/25% PBT, 50% MeOH/50% PBT, 25% MeOH/75% PBT.
3. Immerse slides in PBT for 5 min.
4. Overlay the slides with Proteinase K, (10 µg/mL, prewarmed to 37°C) and leave for 15 min at room temperature.
5. Wash three times for 1 min with PBT.
6. Immerse the slides in ice-cold 4% cold paraformaldehyde in PBS and leave for 20 min at room temperature.
7. Wash the slides with PBT, twice for 1 min.
8. Add probe to hybridization solution (10–200 ng/slide—actual amount determined empirically). Heat to 80°C for 5 min to ensure uniform solution and quench in ice.
9. Apply the hybridization mix (approx 70 µL/slide) to the sections and cover with a coverslip. Incubate overnight at 40–70°C in a box humifidified with 50% formamide, 5X SSC (we routinely use 70°C for intact riboprobes 1–1.5 kb) (*see* **Note 5**).

## 3.5. Posthybridization Washes and Color Detection

1. Remove slides from box and submerge in 50% formamide, 5X SSC pH 4.5, 1% SDS at 65°C until the coverslips fall off (about 15 min).
2. Wash in 50% formamide 5X SSC pH 4.5, 1% SDS at 65°C, once for 15 min and once for 30 min.
3. Wash three times in 50% formamide, 2X SSC at 65°C; 30 min each.
4. Wash 3 times for 5 min in TBST at room temperature.
5. Block for 30 min at room temperature in blocking buffer.
6. Immerse slides in preabsorbed (*see* **Note 6**) alkaline-phosphatase anti-DIG Fab fragments (1/5000 dilution) and incubate at 4°C overnight or for 4 h at room temperature.
7. Wash three times for 20 min in PBST with a rotating stir bar.
8. Wash two times for 5 min in alkaline phosphatase buffer.
9. Incubate in color solution in the dark until desired reaction product intensity is achieved (between 4 h and 10 d).
10. Stop color reaction by incubating for 5 min in 10 m$M$ Tris-HCl pH 7.8, 10 m$M$ EDTA.

## 3.6. Immunohistochemistry

1. Block for 30 min in 20% sheep serum in PBS at room temperature.
2. Incubate with primary antibody/antiserum at the desired dilution in PBS plus 5% sheep serum for 1–2 h at room temperature.
3. Wash several times with PBS.
4. Incubate in biotinylated secondary antibody at appropriate dilution in PBS.
5. Wash in PBS.
6. Incubate in alkaline phosphatase streptavidin (1/100) in PBS for 30 min at room temperature.
7. Wash in PBS.
8. Incubate in Fast Red in the dark until desired reaction product intensity is achieved (15 min–2 h). Stop the reaction by washing in water.
9. Mount in Mowiol with a coverslip.

# 4. Notes

1. It is essential to avoid the degradation of the target of ISH, cellular RNA, prior to hybridization. Ribonuclease contamination should be avoided by autoclaving solutions; baking glassware and consumable supplies and reagents should be kept exclusively for ISH. Disposable gloves should be worn.
2. Extraneous transcripts have been reported to occur during the preparation of the riboprobe when the template contains 3' protruding ends, therefore do not use restriction enzymes that generate 3' overhangs. If there is no alternative restriction site the 3' overhang should be converted to a blunt end using Klenow DNA polymerase.
3. The mixture for the in vitro transcription protocol should be kept at room temperature during the addition of each successive component, since DNA can precipitate in the presence of spermidine (present in the transcription buffer) if kept at 4°C.

4. Histolene used for the initial dewaxing of sections can be kept for use in mounting slides in DePex.
5. The hybridization temperature may need to be altered. Oligonucleotide probes are hybridised at 30–37°C overnight *(2,6,7)*. cRNA probes are hybridized at 40–70°C *(4,7)*.
6. To pre-absorb antibody: Weigh out 45 mg chick embryo (or appropriate tissue) powder into a microfuge tube, add 7.5 mL TBST and heat at 70°C for 30 min. Cool on ice and add 75 µL sheep serum and 15 µL of anti-DIG Fab fragments. Shake at 4°C for 1 h to preabsorb antibody. Spin down solids from powder for 10 min. Remove supernatant and dilute it to 75 mL with 1% sheep serum in TBST. Antibody can be stored at 4°C with 0.02% sodium azide and reused at least once.

## References

1. Rex, M. and Scotting, P. J. (1994) Simultaneous detection of RNA and protein in tissue sections by nonradioactive *in situ* hybridization followed by immunohistochemistry. *Biochemica* (Boehringer Mannheim) **3**, 24–26.
2. Dickerson, D. S., Huerter, B. S., Morris, S. J., and Chronwall, B. M. (1994) POMC mRNA levels in individual melanotropes and GFAP in glial-like cells in rat pituitary. *Peptides* **15**, 247–256.
3. Wang, D. and Cutz, E. (1994) Simultaneous detection of mRNA for bombesin/gastrin-releasing peptide and its receptor in rat brain by nonradiolabeled double ISH. *Laboratory Investigation* **70**, 775–780.
4. Angerer, L and Angerer, R. (1991) Localisation of mRNAs by in situ hybridization. *Meth. Cell Biol.* **35**, 37–71.
5. Norton, A.J., Jordan, S., and Yeomans, P. (1994) Brief, high temperature heat denaturation (pressure cooking): a simple and effective method of antigen retrieval for routinely processed tissues. *J. Pathol.* **173,** 371–379.
6. Brouwer, N., Van Dijken, H., Ruiters, M. H. J., Van Willigen, J.-D., and Horst, G. J. (1992) Localisation of dopamine D2 receptor mRNA with non-radioactive in situ hybridization histochemistry. *Neurosci. Lett.* **142**, 223–227.
7. Komminoth, P., Merk, F. B., Leav, I., Wolfe, H. J., and Roth, J. (1992) Comparison of [35]S- and digoxigenin-labeled RNA and oligonucleotide probes for ISH. *Histochemistry* **98**, 217–228.
8. Rex, M. Uwanogho, D. A., Orme, A., Scotting, P. J., and Sharpe, P. T. (1997) *cSox21* exhibits a complex and dynamic pattern of transcription during embryonic development of the chick central nervous system. *Med. Dev.* **66**, 39–53.
9. Uwanogho, D., Rex, M., Cartwright, E., Pearl, G., Scotting, P. J., and Sharpe, P. T. (1995) Embryonic expression of the chicken *Sox*2, *Sox*3 and *Sox*11 genes suggests an interactive role in neuronal development. *Mech. Dev.* **49,** 23–36.
10. Rex, M., Church, R., Tointon, K., and Scotting, P. J. (1997) Combination of non-isotopic in situ hybridization with detection of enzyme activity, bromodeoxyuridine incorporation and immunohistochemical markers. *Histochem. Cell Biol.* **107,** 519–523.
11. Rex, M., Orme, A., Uwanogho, D., Tointon, K., Wigmore, P. M., Sharpe, P. T., and Scotting, P. J. (1997) Dynamic expression of chicken Sox2 and Sox3 genes in ectoderm induced to form neural tissue. *Dev. Dynamics* **209**, 323–332.

# 46

# Immunohistochemistry Using Polyester Wax

## Andrew Kent

## 1. Introduction

For the preparation of sections for immunohistochemistry at the level of the light microscope, the choice of sectioning medium usually lies between frozen (fresh or fixed-cryoprotected) or paraffin wax. If the antigen survives fixation and processing at 60°C, paraffin wax is used. If the antigen survives fixation, but not heating to 60°C, frozen sections from fixed cryoprotected tissue are prepared, but if the antigen does not survive chemical fixation, sections are prepared from fresh, rapid-frozen tissue, and a trade-off in morphological preservation is accepted. However, if the antigen survives chemical fixation, but is sensitive to temperatures above 40°C (as many are to some degree) or is soluble in the clearing agent, polyester wax offers an opportunity for immunostaining without having to resort to frozen sections.

Steedman *(1,2)* introduced polyester wax as an alternative sectioning medium to paraffin wax for histology, claiming improved morphological preservation. Polyester wax is alcohol-soluble (i.e., negates the need for an additional clearing reagent, e.g., xylene or chloroform) and has a melting point of 37°C. These two properties contribute to milder solvent conditions, causing less extraction and thereby improve morphological preservation *(3)*. By analogy, polyester wax should be inherently more suited to antigen preservation and therefore immunohistochemistry than paraffin wax.

The use of this wax for immunohistochemistry has previously been reported in combination with both formalin fixation *(4)* and acid-ethanol fixation *(5)*. Its routine use for immunohistochemistry was described in 1991 *(6)*, and it has been used in my laboratory as the method of preference for immunohistochemistry for 11 yr (e.g., *7,8*). It can be used for any immunostaining procedure,

From: *Methods in Molecular Biology, Vol. 97: Molecular Embryology: Methods and Protocols*
Edited by: P. T. Sharpe and I. Mason © Humana Press Inc., Totowa, NJ

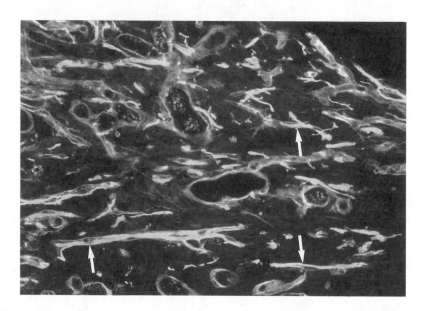

Fig. 1. Regrowing axons, sprouting from a transected peripheral nerve, are always seen in association with Schwann cells (→). Migrating Schwann cells are labeled with antilaminin and detected with antirabbit IgG-FITC. Axons are labeled with anti-(-tubulin isoform III and detected using biotinylated antimouse IgG and Extravidin-TRITC (×220). (*See* color plate 7 appearing after p. 368.)

e.g., immunofluorescence or immunoperoxidase, with or without counterstaining, or combined with *in situ* hybridization (ISH) *(9)*.

Many antigens benefit from processing in polyester wax. In general, immunostaining is much improved in polyester wax than paraffin wax and a strong signal can usually be obtained by an indirect method rather than a more lengthy amplification method. In some cases, immunostaining in polyester wax sections can be achieved where a result is only possible in paraffin wax sections after antigen retrieval (trypsinization or pressure cooking), e.g., labeling of Schwann cells with antibody to S100, clone SH-B1 (Sigma-Aldrich, Poole, Dorset, UK) (*see* **Figs. 1** and **2**).

## 2. Materials

### 2.1. Processing and Sectioning

1. Fixative, either: (a) 4% paraformaldehyde (PF) in phosphate-buffered saline (PBS), pH 7.4. For 1 L dissolve 40 g of PF (Taab laboratories Ltd., Aldermaston, Berks., UK) in 900 mL of distilled water at 60°C with two to three drops of 1 $M$ NaOH. Add NaCl (8 g), KCl (0.2 g), anhydrous $Na_2HPO_4$ (1.15 g) and $KH_2PO_4$ (0.2 g). Make up to 1 L with distilled water. (b) Ethanol-acetic acid: For 100 mL,

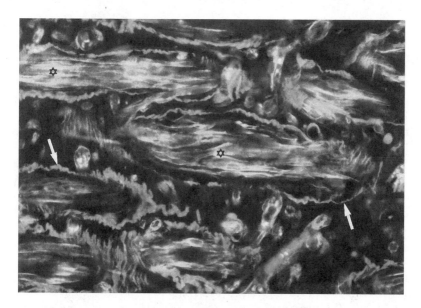

Fig. 2. Schwann cells from the proximal segment of a transected peripheral nerve invade the myotubes of an acellular muscle graft. The sarcolemmal basement membrane (→) is labeled with antilaminin and detected with antirabbit IgG TRITC. The Schwann cells (*) are labeled with anti-S100β and detected using biotinylated antimouse IgG and Extravidin-FITC (×450). Both micrographs were taken with dual excitation on an Olympus Provis photomicroscope using Fujichrome Provia (1600 ASA). (*See* color plate 8 appearing after p. 368.)

      add 5 mL of glacial acetic acid to 95 mL of pure ethanol. **NB**. Other aldehyde or alcohol-based fixatives can be used, e.g., Bouin's fixative.

2. Industrial methylated spirit (IMS) for dehydration of tissue and dewaxing sections.
3. Polyester wax. For 500 g stock, melt 480 g of 400-polyethylene glycol distearate (PEGD) (Aldrich Chemical Co.) at 40°C overnight. Melt 20 g cetyl alcohol (Merck, Poole, UK) at 60°C. Warm up the PEGD to 60°C, add the cetyl alcohol, mix well, pour into convenient size trays, and allow to set. Store at 4°C.
4. Rotary microtome.
5. Pelcool unit or equivalent (Bright Instrument Co. Ltd., Huntington, Cambs., UK) with associated cooled chuck.
6. Glass Ralph knifemaker (e.g., Taab Histoknifemaker).
7. Glass strips (16 in. × 1.5 in. × 6 mm, Taab Superglass) or disposable, low-profile Feather blades (Raymond Lamb, London, UK).
8. Ralph knife holder or disposable blade holder (Raymond Lamb).
9. PTFE aerosol spray, Sprayflon Plus 3 (Filcris, Royston, Herts, UK).
10. Section adhesive, 1% aqueous gelatin or either Biobond (British BioCell International, Cardiff, Wales, UK) or Vectabond (Vector Labs Ltd., Bretton, Peterborough, UK).

## 2.2. Immunostaining

1. Humidified staining box either an adapted "sandwich" box or purpose-made black perspex box with transparent perspex lid (Genex Ltd., Coolsdon, Surrey, UK).
2. Blocking solution, either bovine serum albumin (BSA) (0.1–1%), or normal serum (1–5%) from the species in which the secondary antibody was raised, diluted in PBS.
3. Primary antibody, diluted appropriately with blocking solution.
4. Secondary antibody diluted with blocking solution (usual range 1 in 100 to 1 in 200).
5. Tertiary antibody, e.g., Extravidin-FITC (Sigma) or ABC-peroxidase (Vector Labs. Ltd.) at the recommended dilution in PBS.
6. For immunoperoxidase:
   a. Suitable substrate, e.g., diaminobenzidine (DAB). For 10 mL, dissolve 6 mg DAB in 10 mL PBS and, immediately before use, add 3 μL of 30% $H_2O_2$, or a DAB alternative, e.g., Vector VIP.
   b. Suitable counterstain, e.g., for DAB, 0.1% toluidine blue.
   c. A setting mountant, e.g., DPX (Merck).
7. For immunofluorescence:
   a. An anti quenching mountant. For 100 mL, dissolve 2.5 g 1,4-diazabicyclo [2.2.2] octane (DABCO) (Sigma) in 90 mL of glycerol at 37°C, and 0.1 g sodium azide in 10 mL of PBS. Then combine, mix well, and store in a dark bottle at 4°C.
   b. Nail varnish for sealing cover slips onto slides.

## 3. Method

## 3.1. Tissue Processing

1. Tissue is excised and immersed in fixative at 4°C as quickly as possible. If tissues have a minimum dimension > 5 mm, then slices of this width are prepared. This ensures that over the fixation period of 4–24 h (time dependent on the size of the tissue and on the antigen) all regions of the tissue are adequately fixed, and the antigen is retained uniformly.
2. Following fixation with paraformaldehyde, the tissue is washed in PBS prior to dehydration with IMS using a series of 50, 70, and 90%, and three absolute IMS changes (30 min each).
3. If the tissue is fixed with acid-ethanol, it can be immediately transferred to absolute alcohol.
4. From the final alcohol change, tissues are transferred to a mixture of polyester wax and alcohol (50:50) at 40°C and subsequently to a 75:25 mixture before changing to 100% wax. Infiltration times are dependent on the size of the tissue, but for the size recommended above, 2 h in each of the mixtures and 4 h (2 changes) in wax should suffice. For larger pieces of tissue, the infiltration times should be extended accordingly. However, it should be remembered that the longer the tissue remains at 40°C, the greater is the risk of antigen loss. Extended infiltration (>24 h) also encourages hardening of tissue, which makes sectioning more difficult.
5. Following infiltration, tissue is blocked out and allowed to set at room temperature.
6. Blocks are best stored at 4°C.

## 3.2. Sectioning

1. For sectioning, blocks are mounted on a cooled chuck connected to a rheostat (e.g., Pelcool Unit) for temperature control. Sections can be cut in the range of 2 *(5)* to 30 µm *(3)* thick, with 7 µm being suitable for most applications. The temperature of the block face is maintained at 10–15°C and room temperature below 22°C. Blocks can be cut on low-profile disposable metal knives, e.g., Feather blades, but glass Ralph knives, thinly coated with PTFE, are preferable. Ribbons can be stored at 4°C.

2. Sections intended for routine immunohistochemistry are mounted on clean glass slides, and sufficient adherence is obtained by allowing sections to decompress on a pool of 1% (w/v) aqueous gelatin with careful warming. (Placing the slide on a tissue-culture flat flask filled with warm water [30°C] is ideal).

3. Excess gelatin is removed, and sections are dried on overnight at room temperature. If trypsinization is necessary to expose the antigen, then sections should be mounted on Biobond- or Vectabond-coated slides. For ISH, sections are also mounted on Biobond or Vectabond coated slides, but additionally, following removal of wax in 100% IMS, the slides are removed and dried for 30 min at 37°C *(9)* prior to equilibration in PBS (*see* **Notes 1–6**).

## 3.3. Immunostaining

### 3.3.1. Quick Method for Abundant Antigens

Following dewaxing in a descending series of alcohols (100 to 50% v/v) and equilibration in PBS, the following routine protocols are recommended. All incubations are carried out in an humidified chamber to prevent drying out at any stage during immunostaining.

1. Incubate for 2 h in primary antibody, diluted in 0.1% (w/v) BSA/PBS, at room temperature.
2. Wash in PBS (twice for 15 min).
3. Incubate for 1 h in species-specific anti-IgG antibody conjugated to FITC, and diluted in 0.1% (w/v) BSA/PBS (1 in 100 recommended).
4. Wash in PBS (twice for 15 min).
5. Mount in glycerol-based antiquench mountant, and seal cover slip with nail varnish.

### 3.3.2. Amplification Method for Less Abundant Antigens

1. Preblock in 1% (v/v) normal serum (from the species in which the intended secondary antibody is raised), 0.1% (w/v) BSA in PBS for 30 min.
2. Incubate overnight at 4°C in primary antibody, diluted in the blocking solution.
3. Wash in PBS (twice for 15 min).
4. Incubate for 1–2 h at room temperature in biotinylated secondary antibody diluted in the blocking solution (1 in 200 recommended).
5. Wash in PBS (twice for 15 min).
6. Incubate for 1 h at room temperature in Extravidin-FITC diluted in PBS (1 in 100 recommended).

7. Wash in PBS (twice for 15 min).
8. Mount in glycerol-based antiquench mountant, and seal cover slip with nail varnish (*see* **Notes 7–10**).

### 3.3.3. Immunoperoxidase Method for Less Abundant Antigen

1. Follow **steps 1–5** in **Subheading 3.3.2.**
2. Incubate for 1 h at room temperature in Vectastain Elite ABC reagent (prepared 30 min before use).
3. Wash in PBS (2 × 15 min).
4. Incubate with appropriate substrate, either DAB (brown) or Vector SG (black) or Vector VIP (deep purple).
5. Wash in distilled water (5 min).
6. Counterstain, dehydrate in IMS, clear in xylene, and mount in DPX.

## 4. Notes
### 4.1. Problems with Processing and Sectioning

1. Sections fragment during decompression. Inadequate fixation, dehydration, or infiltration—reassess size of block or processing times. Section left to decompress for too long or at too high a temperature—take more care when sections decompress.
2. Sections compress too much, and do not regain original block dimensions. Knife blunt—shift to new area of knife or change knife. Block temperature too cold or too warm—alter Pelcool setting.
3. Section exhibits chatter. Knife angle wrong—check angle is set to 5/6°. Loose component—check security of knife holder, knife, and block.
4. Examination of stained sections reveals wrinkles. Decompression inadequate—increase time or temperature to allow sections to decompress optimally.
5. Sections fail to ribbon. Room temperature too high—turn on air conditioning or seek alternative venue. Block poorly trimmed—use a sharp blade to retrim and align sides of the block.
6. Sections do not adhere to slide. Sections dried onto slide too briefly—dry sections overnight. Sections not mounted on either an adhesive solution or a coated slide—check mounting solution is 1% gelatin or that slides are coated as intended. Proteinase is contaminating the diluent—use fresh diluent (bacteria grow readily in warm diluent!).

### 4.2. Problems with Immunostaining

7. No positive immunostaining obtained. Antigen does not survive processing—check antigen's sensitivity to fixation and dehydration using fresh frozen, and fixed, cryoprotected, frozen sections. Antigen is not present in tissue—use a positive control. Primary antibody is at fault—use a positive control and a new aliquot or batch. Secondary antibody or detection system is at fault—check potency of secondary antibody using an alternative primary antibody. If using an indirect method of detection, this may not be sensitive enough—try an amplification system.

8. Patchy immunostaining obtained. Gelatin solution crept over surface of section during decompression—remount new sections with additional care. Nonuniform fixation—reassess fixation protocol. Sections allowed to dry out during staining—restain using an adequate volume of immunoreagent and ensure humidity of chamber.

9. Background staining too high. Concentration of primary antibody too high—reduce titer to minimum. Secondary antibody is interacting with the tissue–e.g., for mouse monoclonals used on rat tissue, use rat adsorbed antimouse IgG, or use control lacking primary antibody. Tissue has (a) autofluorescence or (b) endogenous enzyme, e.g., peroxidase—(a) View section without antibodies or (b) react section with enzyme substrate alone. Antibodies binding via ionic interaction—increase pH from 7.4–8.2., increase salt concentration, use detergent, e.g., 0.1% Tween-20.

10. Immunofluorescent stain is rapidly lost from the section during microscopy. Antibody (probably primary) is labile in mountant—refix sections in 4% PF for 5 min after immunostaining and prior to mounting in glycerol-based mountant.

## References

1. Steedman, H. F. (1957) Polyester wax. A new ribboning embedding medium for histology. *Nature* **179,** 1345.
2. Steedman, H. F. (1960) Polyester wax, in: *Section Cutting for Microscopy.* Blackwell Scientific Publications, Oxford, UK, pp. 52–55.
3. Sidman, R. L., Mottla, P. A., and Feder, N. (1961) Improved polyester wax embedding for histology. *Stain Tech.* **36,** 279–289.
4. Kusakabe, M., Sakakura, T., Nishizuka, Y., Sano, M., and Matsukage, A. (1984) Polyester wax embedding and sectioning technique for immunohistochemistry. *Stain Tech.* **59,** 127–133.
5. Roholl, P. J. M., Dullens, H. F. J., Kleijne, J. Dubbink, E. J., and Den Otter, W. (1991) Acid ethanol fixation and polyester waxembedding combines preservation of antigenic determinants with good morphology and enables simultaneous bromodeoxyuridine (BRDU) labeling. *Biotech. Histochem.* **1,** 55–62.
6. Kent, A. P. (1991) The routine use of polyester wax for immunohistochemistry. *Microsc. Anal.* **24,** 37.
7. Mathewson, A. J. and Berry, M. (1985) Observations on the astrocyte response to a cerebral stab wound in adult rats. *Brain Res.* **327,** 61–69.
8. Berry, M., Hall, S., Rees, L., Carlile, J., and Wyse, J. P. H. (1992) Regeneration of axons in the optic nerve of the adult Browman-Wyse (BW) mutant rat. *J. Neurocytol.* **21,** 426–448.
9. Richardson, L. L. and Dym, M. (1994) Improved adhesiveness of polyester wax sections for immunohistochemistry. *Biotechniques* **17,** 846–848.

# 47

# Immunohistochemistry on Whole Embryos

## Ivor Mason

## 1. Introduction

Immunohistochemical studies on intact embryonic material offers the same advantages as similar *in situ* hybridization techniques, namely that spatial relationships of tissues expressing the antigen of interest are more readily appreciated (*see* **Fig. 1**). In addition, the lack of prior processing and sectioning of material considerably reduces the time required for such analyses, and material can still be sectioned subsequently if required. However, it should be noted that relatively few antibodies or antisera are suitable for this technique owing to crossreactivity with other antigens, low affinity, or low titer. Crossreactivity, even with low affinity produces unacceptable levels of nonspecific or "background" staining, and is most commonly a feature of polyclonal antisera. Thus, although relatively few monoclonal antibodies (MAbs) are high affinity and high titer, it is this type of reagent that is most frequently suitable for use on intact embryos. The procedure described below is slightly modified from that in **refs.** *1* and *2*.

## 2. Materials

1. Phosphate-buffered saline (PBS; 10X concentrated, pH 7.2) /L: 2.1 g $KH_2PO_4$, 90 g NaCl, 7.26 g $Na_2HPO_4 \cdot 7H_2O$. Autoclaved.
2. Paraformaldehyde (Fluka, Gillingham, UK).
3. Hydrogen peroxide (30% w/w; Sigma, replace every 4 mo).
4. Triton X-100 (Sigma, Poole, UK).
5. Sodium azide (Sigma).
6. Basal medium (Eagle) containing 25 m$M$ HEPES (Life Technologies, Paisley, Scotland).
7. Calf serum (Life Technologies).
8. Normal goat or donkey sera (Sigma).

From: *Methods in Molecular Biology, Vol. 97: Molecular Embryology: Methods and Protocols*
Edited by: P. T. Sharpe and I. Mason © Humana Press Inc., Totowa, NJ

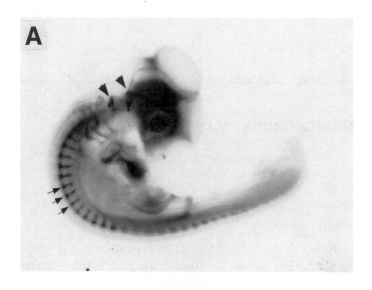

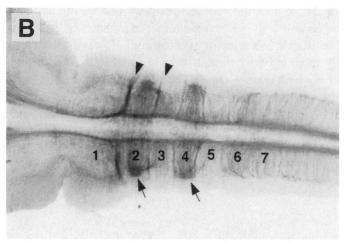

Fig. 1. **(A)** Whole-stage 19 chick embryo stained with an MAb against the cell-adhesion molecule L1/Ng-CAM, which is expressed on a subset of axons in the nervous system. Cranial (arrowheads) and dorsal root (arrows) ganglia of the peripheral nervous system are visible. **(B)** The hindbrain dissected from the same embryo and mounted flat to visualize axons within hindbrain segments (rhombomeres) and accumulating of axons at exit points (arrows) in rhombomeres 2 and 4 and in interrhombomeric boundaries (arrowheads) are indicated.

9. Peroxidase-conjugated secondary antibodies (Jackson Laboratories).
10. 0.1 *M* Tris-HCl, pH 7.2.
11. Diaminobenzidine (DAB; 10 mg tablets from Sigma).

## 3. Method

1. Fix embryos overnight in 4% w/v paraformaldehyde in PBS at 4°C (*see* **Notes 1** and **2**) using a large (20- to 50-fold by volume) excess of fixative.
2. Wash in PBS three times each for 30 min.
3. Block endogenous peroxidase enzymes with PBS containing 0.05% (v/v) hydrogen peroxide, 1% (v/v) Triton X-100 at 4°C overnight on a rocker or slowly rotating wheel (*see* **Note 3**).
4. Wash in PBS containing 1% (v/v) Triton X-100 three times each for 1 h.
5. Dilute antibody (*see* **Note 4**) in basal Eagle's medium containing 10% calf serum, 1% (v/v) Triton X-100 and 0.02% (w/v) sodium azide (*see* **Note 5**). Incubate for 2–4 d at 4°C with rocking.
6. Wash in PBS containing 1% v/v normal serum of same species as peroxidase-conjugated secondary antibody (e.g., normal goat serum for goat antimouse secondary antibodies) and 1% (v/v) Triton X-100, three times each for 1 h at 4°C.
7. Incubate with peroxidase-conjugated secondary antibody appropriate for the primary antibody used in **step 5** (e.g., peroxidase-conjugated goat antimouse IgG or IgM for mouse MAbs). The secondary antibody is diluted in PBS containing 1% (v/v) Triton X-100 and 1% (v/v) normal serum as in **step 6** at 4°C overnight. The precise dilution of secondary antibody should be determined empirically, but we generally find that dilutions of 1:100 are about right.
8. Wash as **step 6**.
9. Wash in 0.1 *M* Tris-HCl, pH 7.2, twice for 30 min.
10. Incubate in 0.5 mg/mL DAB in 0.1 *M* Tris-HCl, pH 7.2, for 3 h at 4°C in the dark with rocking (*see* **Note 6**).
11. Replace with 0.5 mg/mL DAB in 0.1 *M* Tris-HCl, pH 7.2, containing 1 µL/mL hydrogen peroxide, and allow color to develop (usually 5–15 min; longer may be required if embryo is to be sectioned subsequently).
12. Reaction is stopped by several quick changes of tap water followed by several changes of PBS over a few hours.
13. Embryos can be further dissected at this point, and then cleared in 90% (v/v) glycerol, 1% PBS, 0.02% (w/v) sodium azide for storage or mounting for photography.
14. For sectioning, embryos are dehydrated through graded ethanols, equilibrated in xylene, and then embedded in wax as described in Chapter 41 (radioactive *in situ* hybridization to tissue sections, I.J.Mason) followed by sectioning at thicknesses of 7–15 µm.

## 4. Notes

1. This protocol works well on intact embryos up to at least mouse embryonic d 12, chick Hamburger and Hamilton *(3)* stage 24, and zebrafish to 72 h. For older embryos, it is likely that some block dissection will be required prior to fixation to keep background levels low. Alternatively, some tissues, such as the neural tube, can be dissected after performing the procedure on intact older embryos.
2. It is worth trying a range of fixatives before ruling out the suitability of an antibody or antiserum for this procedure. In addition to standard paraformaldehyde

fixation (**Subheading 3., step 1**), it is worth trying 0.5% (w/v) paraformaldehyde, 0.5% (v/v) glutaraldehyde; 4% (w/v) paraformaldehyde, 0.1% (v/v) glutaraldehyde; 5% (v/v) acetone, 95% (v/v) ethanol; perfix (Fisher Scientific, Loughborough, UK); dry AnalaR acetone. The latter three fixatives are precipitating (rather than crosslinking) fixatives and, thus, are probably not worth trying if the antigen is a small soluble protein (e.g., $M_r < 20,000$). For a detailed consideration of antibody–antigen interactions including fixation, *see* **ref. *4*.

3. Some antibodies may bind more strongly at 37°C, and for these the use of a combined rotating wheel and incubator, such as a "Spin 'n' Stack" (Hybaid, Teddington, UK), is recommended.
4. Required dilution must be determined empirically and will depend on titer and affinity. We have used dilutions varying between 1:2 and 1:500 for MAbs raised against different antigens.
5. The diluted antibody can be frequently reused one or more times. It should be retrieved and stored at 4°C between uses.
6. DAB is carcinogenic. All materials in contact with it and all waste solutions should be treated overnight with excess bleach to break it down.

## Acknowledgments

Thanks are owed to Andrew Lumsden for introducing me to this technique. Work in the author's laboratory is supported by the MRC, Human Frontier Science Programme and The Wellcome Trust.

## References

1. Lumsden, A. G. S. and Keynes, R. J. (1989) Segmental patterns of neuronal development in the chick hindbrain. *Nature* **337,** 424–428.
2. Guthrie, S. and Lumsden, A. (1992) Motor neuron pathfinding following rhombomere reversals in the chick embryo hindbrain. *Development* **114,** 663–673.
3. Hamburger, V. and Hamilton, L. (1951) A series of normal stages in the development of the chick embryo. *J. Morphol.* **88,** 49–92.
4. Morris, R. J. (1995) Antigen-antibody interactions, in *Monoclonal Antibodies: Production, Engineering and Clinical Application* (Ritter, M. A. and Ladyman, H. M., eds.), Cambridge University Press, Cambridge, UK.

# 48

## Whole Embryo Assays for Programmed Cell Death

**Anthony Graham**

## 1. Introduction

Cell death occurs by two processes that are cytologically and biochemically distinct. Pathological cell death, or necrosis, is characterized by early cell swelling and the lysis of intracellular organelles with nuclear breakdown following at a later stage. In contrast, nonpathological cell death or apoptosis is associated with early nuclear fragmentation, cytoplasmic shrinkage, and preservation of intracellular organelles *(1)*. Necrosis is usually associated with cell injury; ATP levels fall early, and consequently, protein synthesis is inhibited. On the other hand, in apoptosis, ATP levels remain near normal, and new protein synthesis is required to initiate the process. One result that stems from the early nuclear fragmentation that is associated with apoptosis is that DNA degradation, resulting in the formation of a DNA ladder, is frequently observed.

During development, programmed cell death plays a vital role in eliminating unwanted cells, and it is invariably apoptotic in nature with the dying cells displaying the hallmarks of this process *(2)*. Programmed cell death will act to prune cells that have been produced in excess, such as neurons. It is also involved in the eradication of cells of structures that are no longer necessary, such as the tail of tadpoles or the Müllerian duct in males *(2)*. The apoptotic elimination of cells is also important in morphogenesis, and in both the limb and the cranial neural crest, it functions to ablate populations of cells locally and, as such, spatially separates the surviving cell populations from each other *(2,3)*.

In developing systems, one can often not obtain enough material to analyze programmed cell death by biochemical means, such as assaying for DNA

From: *Methods in Molecular Biology, Vol. 97: Molecular Embryology: Methods and Protocols*
Edited by: P. T. Sharpe and I. Mason © Humana Press Inc., Totowa, NJ

laddering, and therefore one must use alternative methods. The most definitive method of determining whether or not apoptosis, as opposed to necrosis, is occurring in a group of cells is to analyze the cells ultrastructurally using electron microscopy (EM). This is a powerful tool because it will allow one to identify unequivocally a cell as being apoptotic or necrotic. The problem with using EM to analyze cell death in developing embryos is that it is a cumbersome technique that cannot be used practically to detect apoptosis or to gain an impression of a spatiotemporal profile of apoptosis. This chapter will describe two methods for detecting apoptosis in whole embryos, both of which will give a complete picture of apoptosis in any system. Both of these techniques are based on detection of the DNA fragmentation that occurs early in the apoptotic program. One point that should be noted is that although both of these techniques will pick out apoptotic sites in the embryo, the fact that the cells are actually dying by apoptosis should also then be confirmed independently by EM analysis.

### 1.1. Visualization of Programmed Cell Death in Whole Embryos Using Acridine Orange

Vital staining with the dye, acridine orange, has been previously used to analyze patterns of apoptosis in both vertebrate and invertebrate embryos *(3,4)*. During apoptosis, cells fragment into a number of apoptotic bodies that are chromatin-rich and are intensely stained by acridine orange. In contrast, healthy cells exclude this dye. Thus, when used vitally, i.e., on living specimens, it can only pick out dead cells. The method described below will work, with slight modifications, on any type of embryo.

### 1.2. Visualization of Programmed Cell Death in Whole Embryos Using TUNEL

The terminal transferase-mediated dUTP-biotin Nick End Labeling (TUNEL) method to detect apoptotic cells is based on the fact that terminal transferase will catalyze a template-independent addition of dNTPs to the 3'-OH ends of DNA strands and that one of the early events during apoptosis is the cleavage of a cell's DNA *(5,6)*. Thus, cells undergoing programmed cell death are detected by incubating the embryo, which has been fixed, with terminal transferase and a labeled nucleotide, dUTP-biotin conjugated, with the latter being added to the nicked ends of the DNA fragments and subsequently detected with a streptavidin conjugate. Again, this method will work with slight modification for any type of embryo (*see* **Note 2**).

## 2. Materials

### 2.1. Acridine Orange Staining

1. Phosphate-buffered saline (PBS): 0.13 $M$ NaCl, 7 m$M$ Na$_2$HPO$_4$, 3 m$M$ NaH$_2$PO$_4$. Adjust pH to pH 7.0 and autoclave.
2. Howard's Ringer: 125 m$M$ NaCl, 1.5 m$M$ CaCl$_2$·2H$_2$O, 5 m$M$ KCl adjust to pH 7.2 with 1 $M$ NaOH and autoclave.
3. Acridine orange: (Sigma, Poole, UK) or equivalent—Make a 5 mg/mL stock in H$_2$O, and dilute from that, stepwise, into the physiological salt solution.

### 2.2. TUNEL

1. 4% Paraformaldehyde, (w/v), in PBS: *see* **Subheading 2.1., item 1** for PBS recipe.
2. Triton X-100.
3. Terminal transferase (Boehringer Mannheim, Mannheim, Germany), which comes with the terminal transferase buffer as a 5X stock and 25 m$M$ CoCl$_2$.
4. Deoxy uridine triphosphate (dUTP) and biotin-16-deoxy uridine triphosphate (Biotin-16-dUTP) (Boehringer Mannheim).
5. Streptavidin–fluorescein reagent (Amersham, Little Chalfont, UK).
6. Streptavidin–horseradish peroxidase conjugate reagent (Amersham).
7. 1,4-Diazabicyclo [2,2,2] octane (DABCO) (Sigma).
8. 3',3'-Diaminobenzidine (DAB) (Sigma).
9. Hydrogen peroxide (30% w/v) (Sigma).

## 3. Methods

### 3.1. Acridine Orange Staining

1. Dissect out the embryos into an appropriate physiological salt solution, such as PBS, for mouse embryos or Howard's Ringer for chick embryos, and process directly.
2. Incubate embryos at 37°C in PBS, or Ringer's, containing acridine orange at a concentration of 5 µg/mL for 15 min.
3. Wash the embryos twice, with shaking, for 1 min each in PBS.
4. Mount the specimens under a glass coverslip in either PBS or Ringer's and analyze the specimens immediately under rhodamine epifluoresence.
5. Photograph the results.

Results obtained with acridine orange staining of whole chick embryos are shown in **Fig. 1**. The images shown here were taken using a Bio-Rad MRC 600 (Bio-Rad, Hemel Hempstead, UK) confocal microscope, and the rhodamine and phase images have been merged to give a clear picture of the relationship between areas of acridine orange staining and embryonic anatomy. During the early phase of neural crest production in the chick hindbrain, a focus of acridine orange staining cells is observed over rhombomere 3 (**Fig. 1A**), whereas at slightly later stages, the focus over this segment is enlarged, and there is also prominent staining over rhombomere 5 (**Fig. 1B**). The importance of apoptosis

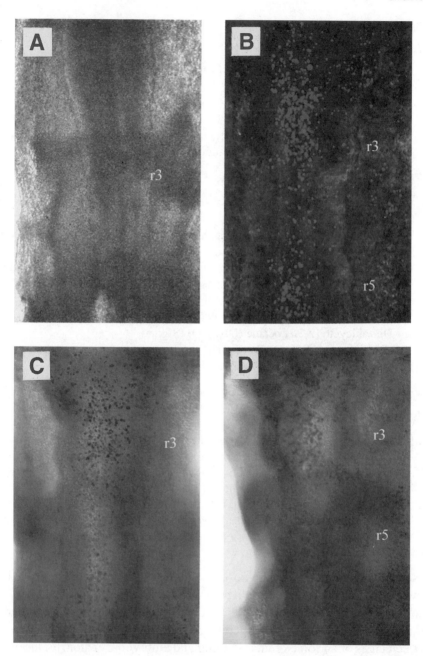

Fig. 1. Apoptosis in the rhombencephalic neural crest of the chick embryo as revealed by acridine orange staining (**A,B**) and TUNEL (**C,D**). The embryos shown in (A) and (C) are at stage 10, but those in (B) and (D) are older, stage 11 (*see* **Note 1** and color plate 3 appearing after p. 368).

in this system is that by clearing these two territories (rhombomeres 3 and 5) of neural crest cells, the neural crests that emerge from the adjacent regions and that are spatially preprogrammed are kept separate from each other *(3)*.

Acridine orange staining has the advantages that it is extremely rapid, cheap, and easy to perform. On the other hand, it has a number of disadvantages. Firstly, because it is used vitally, and since the dye is not fixable and consequently the stain is transient, the embryos cannot be processed for other detection procedures, such as immunohistochemistry. Acridine orange staining also has the drawback that the embryos must be processed swiftly and viewed almost immediately, at least within 1 h of staining, since the tissue dies and the intense staining of the apoptotic bodies is lost. It should also be pointed out that although the signal may be more intense under flourescein epifluoresecne, necrotic cells will also be picked up under this illumination. This method also has the disadvantage that since there is no permeabilization step, apoptotic cells that lie deep within tissues may not be detected. One way of improving this is to cut the embryo prior to staining so that there is greater access to the tissue of interest. For example, with the central nervous system, one can slit the neural tube dorsally.

## 3.2. TUNEL

1. Fix the embryos in 4% paraformaldehyde in PBS either for 2 h at room temperature or overnight at 4°C.
2. Wash the embryos in PBS containing 1% Triton X-100 three times for 30 min each.
3. Wash the embryos once in 1X terminal transferase buffer, 2.5 m$M$ CoCl$_2$, 1% (v/v) Triton X-100.
4. Incubate the embryo for 3 h at 37°C in 1X terminal transferase buffer, 2.5 m$M$ CoCl$_2$, 0.5 µ/mL terminal transferase, 10 m$M$ dUTP (2:1 dUTP:dUTP-biotin), 1% (v/v) Triton X-100.
5. Wash the embryo three times in PBS, 1% (v/v) Triton X-100 for 30 min.
6. Incubate overnight at 4°C with rocking in an appropriate streptavidin conjugate, such as strepevidin-fluorescein or streptevidin–horseradish peroxidase.
7. Wash the embryos three times in PBS, 1% (v/v) Triton X-100 for 1 h each. Then proceed either according to **step 8** or **steps 9–12** as appropriate.
8. In the case of streptavidin–fluorescein detection, mount the specimens under 90% (v/v) glycerol, 1X PBS containing the antiquenching agent DABCO at 2.5% (v/v), and view under fluorescent optics.
9. Wash the embryos twice in PBS for 30 min each.
10. Incubate the embryos in DAB in PBS (5 mg/10 mL PBS) for between 1 and 3 h, depending on the stage of the embryo at 4°C with rocking in the dark.
11. Exchange the above solution for the same, but this time add hydrogen peroxide (60% or 200 vol), 10 µL/10 mL DAB in PBS, and develop in the dark for 5–15 min.
12. When the background staining starts to appear, stop the reaction by washing several times with PBS.

13. Wash in PBS twice each for 30 min, and then mount the specimens under a glass coverslip in 90% glycerol/10% PBS and view under bright-field optics.

## 4. Notes

1. **Fig. 1C,D** shows the results obtained using TUNEL, and streptavidin–horseradish peroxidase detection, on whole chick embryos during the period of hindbrain neural crest apoptosis. As was found with the acridine orange staining, during the early phase of crest production, there is staining over rhombomere 3 (**Fig. 1C**), and at a slightly later stage, this staining is more pronounced and extensive, and also apoptotic cells are also found over rhombomere 5 (**Fig. 1D**).

2. The disadvantages of using TUNEL are that it is relatively expensive and takes somewhat longer than the acridine orange staining, but it has many advantages that together make it the method of choice. One big advantage is that this method uses embryos that are fixed. This means that one can accumulate specimens at 4°C and that they can be processed at leisure. The fixation also allows one to permeabilize the embryos using detergent, Triton X-100, and therefore even apoptotic cells deep within the embryo can be detected. The fact that this method can use fluorescein or horseradish peroxidase detection also means that this technique can be coupled with other procedures, such as immunohistochemistry or axonal tracing, and at least when using horseradish peroxidase detection, the preparations are permanent. The horseradish peroxidase-reacted embryos can also be sectioned afterward to allow a more complete analysis of the distribution and the cell type of the apoptotic cells.

## References

1. Wylie, A. H., Kerr, J. F. R., and Currie, A. R. (1980) Cell death: The significance of apoptosis. *Int. Rev. Cytol.* **68,** 251–305.
2. Raff, M. C. (1992) Social controls on cell survival and cell death. *Nature* **356,** 397–400.
3. Graham, A., Heyman, I., and Lumsden, A. (1993) Even-numbered rhombomeres control the apoptotic elimination of neural crest cells from odd-numbered rhombomeres in the chick hindbrain. *Development* **119,** 233–245.
4. Wolf, T. and Ready, D. F. (1991) Cell death in normal and rough eye mutants of *Drosophila. Development* **113,** 825–839.
5. Gaverilli, Y., Sherman, Y., and Ben-Sasson, S. A. (1992) Identification of programmed cell death *in situ* via specific Labeling of nuclear DNA fragmentation. *J. Cell Biol.* **119,** 493–501.
6. White, K., Grether, M. E., Abrams, J. M., Young, L., Farell, K., and Steller, H. (1994) Genetic control of programmed cell death in *Drosophila. Science* **264,** 677–683.

# 49

## Gene Interference Using Antisense Oligodeoxynucleotides on Whole Chick Embryos

*Optimal Ring and Roller-Bottle Culture Technique*

### Jonathan Cooke and Alison Isaac

## 1. Introduction

Antisense oligodeoxynucleotides are being widely used to interfere with specific gene activities in cell-culture systems *(1,2)*, and there are possible analytical advantages to partial and timed interference in the whole embryo, for certain genes, as compared with targeted null mutations ("knockouts"—*see* e.g., **ref. 3**). Despite this, and the logical possibility of controlling rigorously for the sequence specificity of effects observed, use of antisense oligos will continue to attract a healthy level of controversy, because of occasional evidence for nongenetically based sequence specificity of particular cellular effects *(4)*, and also the possibility of inadvertently targeting other, unknown genes containing matching base sequences. These matters are further discussed below, and in **Subheading 4.** The chick blastoderm consists of two or at most three-cell layers at the time many genes with vital roles in development of the basic body plan are being activated, and only one layer, the epiblast, resembles a true epithelium enough to constitute a potential barrier for oligo access to the other two. In fact, there is evidence that epiblast itself is good at uptake of oligos, and it should be remembered that there is evidence for extensive bulk uptake by "cell drinking" (pinocytosis) by ectoderm and neurepithelium, the equivalent cell layer of the young mammalian embryo. This blastoderm can be kept, with or without incubation, in a protein-free nutrient medium for up to 3 h, and then cultured onward in either of two ways. It proceeds, within 15 h, from gastrula/neurula stages to those where the normality or otherwise of body plan formation can readily be assessed. Even relatively transient and incomplete

From: *Methods in Molecular Biology, Vol. 97: Molecular Embryology: Methods and Protocols*
Edited by: P. T. Sharpe and I. Mason © Humana Press Inc., Totowa, NJ

degradation of gene product levels by antisense will thus greatly help in analysis of a some dynamically expressed genes' early roles, and supplement "knockout" analysis in mice. Thus, control over the timing and possibly degree of interference might enable analysis of genes where the mouse null mutant develops insufficiently for much to be said about what went wrong, or halts development at a stage precluding analysis of separate and slightly later developmental functions of the same gene. Furthermore, mammalian genes are documented for which acute antisense interference in cell culture produces a functional deficit that cannot be occurring in the corresponding cells of the null mutant whole embryo, whose phenotype is much less severe *(5,6)*. This suggests functional compensation mechanisms during the prolonged sequence of development, whereby a gene's normal role is masked.

Antisense oligos exert their functional interference by binding to their target sequence on nascent or working messenger RNA, inhibiting its translation, causing degradation through RNase H, or having both these effects. There is consensus that the half-life of normal phosphodiester DNA oligos (P-oligos), both in culture media and within the cell, tends to be minutes long and is thus inadequate to perturb levels of even the most rapidly turned over gene products enough for developmental effects. A variety of chemical modifications to the DNA structure have therefore been developed, which cause resistance to the exonucleases—widespread in serum and within cells—that degrade short DNA stretches (preferentially from the 3'-end). Of these, the currently most widely used are phosphorothioated DNA oligos (S-oligos), in which sulfur is substituted for one of the oxygen atoms in the phosphate (sugar) backbone *(7,8)*. It is believed that their half-life in biological situations is lengthened by more than an order of magnitude, that they enter cells relatively readily though maybe by different mechanisms from natural DNA, and that their stability is traded off against somewhat less binding stability for a given DNA–RNA hybrid sequence, and possibly diminished ability of the hybrid to act as substrate for RNase H. If the last is the case, such oligos will act only by allowing decay of protein levels through nonreplacement by translation, rather than also acting as a catalytic (recycling) degrader of message within the cell. The literature reports successful cases of sequence-specific gene interference with and without the specific RNA degradation following their use *(9,10)*. In our laboratory, most successful work has utilized this modification, which is by far the least expensive currently on the market, in 15-mer antisense sequences. Thus, in **Subheading 3.**, it will be understood that S-oligos are used, but our experience with alternatives that may be better in the future is briefly discussed in **Subheading 4.**

Our recent work has been with two closely related genes encoding zinc-finger-containing putative transcription factors, with different early developmental expression patterns and apparent functions. We have achieved the following evidence for the genetically meaningful sequence-specificity of the "phenotypes" of developmental disturbance caused by the antisense treatments (*see* **Subheading 4.** for further comments):

1. A large number (>30 in all) of other oligo sequences of comparable length and overall base composition, either of random sequence, sense to the same genes, antisense to other genes, or sequence parallel to the active ones but with 6-base internal 5'- to 3'- reversed sectors, are without effect in the procedure at up to twice the normal concentrations used.
2. The "phenotypes" are different and nonoverlapping for the two genes sequence-targeted, and are comprehensible in each case in terms of the normal sites and times of dynamic expression of that gene's mRNA. Thus in one case, a particular cell-adhesive-behavioral transition normal to each expression site is inhibited *(3)*. In the other, somite segmentation and left–right anatomical asymmetry of the body are disturbed, the gene being dynamically expressed in nascent somites and, at a separate site, with striking left–right asymmetry *(11)*.
3. Antisense to sequences at two separate, comparable positions within the coding sequence give the same distinctive "phenotype" in the case of each gene. At least in one of the cases, there is a strong interactive effect such that a given micromolarity of antisense oligo gives much stronger interference if composed of equal amounts of the two sequences than if given as either sequence alone. This would theoretically be expected under conditions where each oligo species was independently compromising a high proportion of all mRNA molecules at a particular time.
4. Direct evidence for gene-specific mRNA degradation in the hours following treatment has been obtained in the case of one but not both of the genes, whereas none of the oligos involved appear to lessen mRNA amounts generally in the cultured embryos. The literature suggests that specific RNA degradation in the case of antisense-active phosphorothioate oligos is not necessarily to be expected *(9)*.

Taken together, particularly with inclusion of **item 3**, we regard these as rather powerful evidence for the biological meaningfulness of the effects. Demonstration of specific degradation of gene product at either RNA or protein level is always valuable if the necessary reagents are available. However, insistence on it before work is accepted would seem unnecessarily restrictive, in that only genes on which much resource-intensive work has already been done (quantitative PCR or raising of an appropriate antibody) could then be considered for the approach. In itself, the approach can be used as a relatively low-cost screen for interest, applied to any new chick gene with a dynamic "early developmental" expression pattern ($600 or £400 oligo-related costs/ sequenced gene, for 10 experiments involving 60 experimental and 60 control blastoderms, in labs equipped for chick culture). The main overall evidence for validity of antisense-specific effects is that the great majority of oligo sequences, including those that are sense to known genes, fail to affect development. It is known that a statistical majority of sequences within any one gene will be ineffective targets for antisense (e.g., *9*), and the question of design of effective sequences is alluded to in **Subheading 4.**

**Subheading 3.** describes:

1. The preparation of ring-culture setups for the chick blastoderm by a method modified from the original owing to New *(12)* and the removal of blastoderms into protein-free medium with optimal physiological treatment of the embryos.
2. The administration of phosphorothioated oligodeoxynucleotides with adjunct use of a lipofection procedure, which we believe to enhance intracellular availability (*see* **Subheading 4.** for possible alternative improvements to availability).
3. Replacement of blastoderms into ring culture for development of up to 20 h longer, leading to potentially normal development of the 10–15 somite stage.
4. Replacement of blastoderms from **step 2.** into an alternative form of culture *(13)*, allowing good development over 36 h to the 30+ somite stage (early limb buds and considerable craniofacial morphogenesis), as well as the possibility of continuing presence of oligos, thus maintaining gene interference, in the medium.

## 2. Materials

1. Pannett and Compton (PC) saline for bird embryos (original ref. given in *12*) (PC saline). If filter-sterilized or autoclaved in the two parts when making up, this can be stored at room temperature for months. PC saline is needed in liter quantities per experiment dealing with 10–20 chick eggs, 500-mL bottles are thus convenient.
2. Hank's Balanced Salt Solution (BSS), made up with 0.1X the originally specified concentration of divalent cations $CaMg^{2+}$. Use of the tissue-culture pH indicator dye facilitates important monitoring for freedom from the highly alkaline ovalbumin before administration of P-oligos. Made up sterile as for PC saline, this is best stored in 20-mL universals of which 1/typical experiment will be required.
3. Liebovitz L-15 air-buffered tissue-culture medium without antibiotics, but with glutamine (TCM): Gibco-BRL (Paisley, Scotland). Best aliquoted into 10-mL tubes and stored at 4°C (shelf life 3 mo+ if true sterility maintained). One to two tubes needed per experiment.
4. Gentamycin as antibiotic (Sigma, Poole, UK): Stable for months at 4°C as a stock at 10 mg/mL (×200 working concentration) in sterile distilled water.
5. New ring-culture medium: Eight parts of the thin albumen from fresh eggs (avoid any nonfluid albumen—*see* **Subheading 3.** for opening of eggs): 1 part TCM: 1 part HBSS: 50 μg/mL Gentamycin. With good lab practice (i.e., washing all implements, containers, and surfaces in copious plain water between sessions), use of antibiotics should ideally be superfluous for chick embryo culture on the open bench because of the bacteriostatic properties of lysozyme in albumen. Their addition saves frustration in overcrowded multipurpose lab spaces, however. Such newly prepared medium can be kept at 4°C for 2–3 d in glass or tissue-culture plastic, but is normally prepared fresh for each experiment, 3 mL/planned ring culture should be available.
6. S-Oligos: Phosphorothioated, HPLC-purified and sequence-certified DNA-oligos, normally 15-mer. The most consistent, quickest, and best-value commer-

cial supplier of these is currently the U.S. Company, Oligos Etc. Inc., whose product takes up in sterile distilled water immediately with almost no residue, for use with no further purification. The "1-μm" synthesis scale normally yields 2+ mg of a 15-mer S-oligo, enough for 10 single-oligo experiments as described in the next section. Store according to manufacturer''s instructions. We store 12.5-μL aliquots in 0.5-mL Eppendorf tubes at 2 mmol, and thaw from –70°C just before use. Such aliquots often become appreciably less effective after 5+ mo. Lyophilized storage should improve this.

7. Components of New ring cultures (*see* **Fig. 1**):
    a. Glass watchglasses, approx 50 mm diameter, with as "deep" a curvature as possible. BDH (Poole, UK) has been a good supplier.
    b. Standard-size plastic tissue-culture (Petri) dishes, ideally 60-mm diameter, but in any case large enough for watchglasses to be placed inside with forceps.
    c. Disk filter papers cut to fit within dishes, and with central disk cutout for location and through-lighting of the watchglass.
    d. Rings, at least 22 mm inner diameter and 3 mm depth, but with not more than 26 mm outer diameter. One face especially (the lower when set up) should form a plane and be smooth-surfaced. They are traditionally made of glass, but the skill to make them adequately is in short supply. We find rings cut from T.C.-grade stainless-steel tubing of 1.5 mm wall thickness, outer diameter 25 mm, and of 3 mm depth are ideal. One face and its edges should be polished, and the other with advantage left somewhat rougher to grip vitelline membranes.

8. Shelf air-incubator set to 38.5°C for ring culture and transient dish culture with oligos. We have not found $CO_2$-buffered chick cultures advantageous, but some labs believe them to be. In that case, the TCM component for the culture itself should be different, although it should be pointed out that chick egg albumen itself is extremely alkaline and presumably matched to the embryo's requirements in this way.

9. Straight-sided, screw-capped plastic bottles of 5 mL for roller culture *(13)* (Sterilin, Elkay, Shrewsbury MA).

10. A roller-tube incubator, set so that the 5-mL bottles themselves rotate at 45 rpm. A cabinet fan-heated to a controlled 38.5°C houses a set of rollers with long axes inclined 10–15° off horizontal, humidification being unnecessary as tube caps are sealed. B.T.C. Engineering, Milton, Cambridge, UK, makes a compact bench-top model, the standard setting of whose roller apparatus at 30 rpm produces the required rotation of these tubes for chick culture.

11. Instruments for culture and manipulation of chick embryos:
    a. Heavy blunt smooth forceps with ridged tips, for cracking eggshells, pulling albumen off yolk without tearing vitelline membranes and lifting watchglasses and rings (**Subheading 3.1., Part 1**).
    b. Fine forceps (in UK no. 5 jeweler's) with close-fitting tips ground only slightly blunter than as from manufacturer, for gripping by their peripheries and moving individual blastoderms while off their vitelline membranes.
    c. Two pairs of forceps as in **item b**, but with tips ground to be accurately fitting but relatively blunt and smooth, for manipulating vitelline membrane (**Subheading 3.1., Part 2** and **Subheading 3.4., steps 2** and **3**).

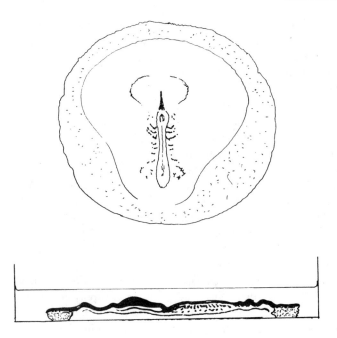

Fig. 1. Ring-culture and off-membrane "oligo incubation": A young head-process-stage blastoderm (a typical stage for treatment) is seen in plan view at top. The middle diagram shows how such blastoderms are incubated, most often epiblast layer upward, in shallow layers of medium in plastic dishes (not to scale with blastoderm). The bottom sectional view is of such a blastoderm in the membrane-ring-watchglass assembly of New ring culture *(12)* over albumen medium, showing the germ-layer-inverted configuation and the adhesion of the epiblast periphery (normally, and ultimately again after successful replacement) to the membrane.

    d. Ideally, small curved-tipped forceps, with finely ridged tips, for lifting culture rings slightly after vitelline membranes have been wrapped and stretched

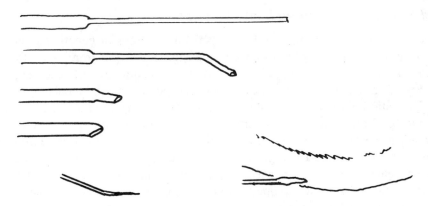

Fig. 2. Instruments for ring culture: In order from top to bottom, approximately real size, are shown a normal Pasteur pipet (the starting form), the special curved pipet for underfilling and draining cultures and settling blastoderms onto membranes, and the medium- and the wide-mouthed pipets for handling blastoderms. An appropriate shape for the smooth tungsten needle tip is shown in side view (left) and from above, to scale with the periphery of a blastoderm (right).

on them (**Subheading 3.1., Part 2** and **Subheading 3.3.**). Failing these, the forceps of **item c** are used for this.

   e. Medium straight-bladed scissors with pointed tips for cutting vitelline membrane rings off yolks (**Subheading 3.1., Part 1**).

   f. Glass Pasteur pipets with carefully flame-smoothed mouth openings and sizes as shown in **Fig. 2**. These should have bulbs (teats) of the firmest rubber variety available. Within reason, the "tougher" the squeeze required on the bulb, the more delicate and controlled can be the sucking and blowing movements during set up of cultures and washing of blastoderms. The soft plastic "pastettes" (Alpha Labs. Ltd.), can be used as throw-away alternatives for all except the special curved pipet shown in the figure, which is essential for ring-culturing. In this the plane of the mouth should be sloped at some 45°, in order to be near-horizontal to bench surface when the pipet is held relaxed, with wrist on the bench.

   g. An electrolytically smoothed tungsten needle with something like the tip shape and size, in relation to a blastoderm indicated in **Fig. 2**. A needle-mounter that is fat like a writing pen helps the hand stay relaxed for small controlled movements (**Subheading 3.1., Part 2**).

12. Last but strangely, not necessarily easiest to find are smooth-surfaced, circular or oval, flat-bottom glass or plastic dishes, of such a depth that a layer of saline just covering egg yolks comes within a centimeter or two of the rim (**Subheading 3.1., Part 1**). Black color at the dish bottom is advantageous, but a cleaned black microscope tile or something similar can be inserted for visualizing membranes and blastoderms.

## 2.1. Egg Supply, Storage, and Incubation

It is best to use, within each lab, eggs from one flock or genetic strain kept by a particular supplier of free-range fertile eggs. *See* **Subheading 4.** for further remarks. *See* **ref. *14*** for typical incubation times leading to particular stages at 38.5°C. Eggs should be stored at 12–14°C between laying and incubation.

Population genetics and evolution theory dictate that the incidence of even single-base mismatches between any oligo-length sector of a given gene sequence as present in different flocks or even strains of chicken is extremely low for the great majority of genes. Even so, one feels comfortable if antisense experiments are directed against embryos of the same chicken flock from which the investigated gene was cloned, and if more than one independent isolate of the relevant gene sector was sequenced with complete concordance.

## 2.2. Microscope Requirements

Dissecting microscopes for setting up these procedures should use a flat stage close to bench height with heat-filtered incident (top) illumination, though later access to transillumination is useful in examining photographing results at the whole-embryo level. Importantly, the lowest power objective of the microscope should allow viewing of the entire culture ring (*see* **Subheading 2., item 7d**) within its central field of good focus.

## 3. Method

The descriptions of procedure that follow will seem overdetailed to many, and of course, as with all procedural skills, different adept practitioners come to use quite individual personal techniques, including the apparent ability to perform what is described with knitting needles and gardening tools while discussing the next experiment! However, we aim below at enabling a novice to all biological micromanipulation techniques to become skilled at the procedures with little personal tuition. They are not extreme, so for the great majority of younger (or even not so young) biologists, this will take remarkably little time.

## 3.1. Ring Culture Procedure (Modified After ref. [12] with Preincubation in Oligos)

Part 1 is carried out in the dish (*see* **Subheading 2., item 12**) with a black area on the bottom, in a layer of PC saline covering the yolks, at between 25 and 30°C (where they rapidly cool to 20°C; *see* **Notes 3** and **4**). Initially, process 6 yolks maximum/25-cm diameter dish, though with skill (speed in this case), 15 or more can be set up from such a dish. This is a naked-eye operation with good overhead lighting, but many workers could also use an engineers "anglepoise" lamp, which is in a ring around a large low-power magnifying lens that can be pulled over saline surface or bench.

1. Wipe eggs, straight from 38.5°C incubator, with 70% ethanol. Leave with blunt end upward for 5 min before taking first one. Crack carefully around with taps of butt end of big forceps, then insert tips parallel with the shell, removing a "cap" of shell from the blunt end of one-fourth to one-third the depth of the egg. Using the edge of the shell and the blunt-ridged forceps, separate the yolk from its hammock of thick chalazal albumen and then discard the latter by tipping out of shell. The "thin" albumen for the culture medium can then be removed with the blunt-mouthed pipet and pooled, from several eggs, into a small flask. Then tilt the yolk carefully into the saline, and turn it to float blastoderm uppermost. When all yolks are in the dish, go around them clearing off as much as possible of remaining jelly-like albumen coating their upper hemispheres using scissors and curved blunt-toothed forceps. The more of this removed at this stage (preferably right out of the dish), the better.

2. Line up yolks round edge of dish, and work by bringing each yolk opposite you in succession by smoothly turning the dish. Using the (washed or different!) big forceps again, place a watch glass and ring assembly under the saline on the black background in middle of dish, and remove the ring to one side. Using one of the specially blunt-tipped no. 5 forceps and the straight scissors, cut around yolk **slightly** above the "equator" to give a disk of vitelline membrane with blastoderm very near its center and still toward the saline surface. Cut toward forceps as they rotate yolk toward you. With a firm and steady pulling motion by both pairs of small blunt forceps, peel the membrane disk by its edges away from you, and off the yolk onto the watchglass, inner side uppermost. Centralize disk on watchglass, and remove major folds. With skill, this is done with movements of the ring itself while held in the forceps, just before it is placed polished face downward to pin the membrane, with blastoderm near its center, in the center of watchglass. This movement should be done within a few seconds, since a thin albumen layer on the back of the membrane otherwise contracts it to near ring size.

   Develop skill in using the forceps to ensure that at the beginning of the peeling movement, yolk separates from membrane to leave a rather "clean" membrane, but with attached blastoderm and a peripheral skirt of yolk in the middle. Also, younger blastoderms are more likely to remain adherent to membrane if the use of a two-handed pulling technique ensures that folding and tension lines across them are minimal. Do not grip membrane too hard or very near its edges, or it will tear there. Temperature of the saline, age of eggs, stage of egg incubation, and length of time in saline before performing this step all affect the likelihood of blastoderm staying on membrane and yolk separating cleanly. Only experience with each stage of blastoderm can help optimize this. If much yolk remains on membranes, it is best to blow it off with pipets at his stage (hence the 6 eggs/dish recommended for starting!). *See* **steps 5** or **6** if blastoderm has stayed on yolk or floated off into saline, but also proceed to **steps 3** and **4** with each membrane/ring assembly as is needed for each blastoderm to be cultured. Speed makes for success of the whole culture experiment, since the more briefly each yolk/blastoderm is in saline only, the more likely it is to stay on membrane for transfer and the better its subsequent resumption of development.

3. Preferably (for quickness) without moving to microscope at this stage, even out membrane wrinkles or folds, clear away spare albumen strands and yolk, and stretch the blastoderm slightly, but evenly on membrane within the ring, maximizing the amount of membrane disk outside the ring. This is done with the blunt fine forceps and the special curved fine Pasteur pipet (but remember, no membrane holes!). With small movements, points on the ring can be lifted slightly while membrane is pulled beneath it, and with skill, many small, uneven membrane disks can be "rescued" at this point. Start by cutting overlarge disks, and with practice they can be made smaller and neater, since having just the necessary membrane to overwrap the ring (*see* **Part 2**) is advantageous. If preferred, this step can be done after "lifting out" of each watchglass/ring assembly into its culture dish as follows, and under the microscope.

4. Lift out the entire assembly with the large forceps, and keeping it almost horizontal but draining some saline from one edge, place in the Petri dish on the filter paper ring. By pipeting saline off from around the ring, then gently into the ring, and then lifting the ring gently at one point, it can be assured that the blastoderm itself sinks well under the saline meniscus and is not bulging up to touch this, which may disrupt its structure. Then, partially replace the PC saline with 1:1 Liebovitz TCM:Hank's BSS mixture inside the ring. Bring each blastoderm from one dish of yolks or, for each experiment, to this stage before proceeding. The provision of TCM:BSS at this point, an addition to the original New culture technique, ensures in our experience that streak stage 4 and even younger streak and prestreak stage blastoderms can wait at room temperature (around 20°C rather than warmer) for up to 3 h or longer without loss of capacity to develop onward. Even so, in experiments involving the younger embryos, waiting times before **Part 2** below should be shorter if possible. Proceed to **Part 2**.

5. As a parallel step at this point, if blastoderms that came off membranes at **step 2** are to be used in the experiment (virtually obligatory in work on very young stages and, on some days, with most streak stages), these should be removed individually at **step 2** by a blunt-mouth pipet from the PC saline dish, and spread out in a 3-mm deep layer of TCM:BSS in a Petri dish at room temperature. These seem to stay in best condition if manipulated to be epiblast (former membrane-facing) side up in the medium. They can then join blastoderms removed manually from membranes (**Part 2, step 2**), and should be in equivalent physiological condition.

6. In addition, the very youngest blastoderms (prestreak to **stage 2**) sometimes stay on the yolk surface itself at **step 2**, and can be removed and stored in a way that optimally preserves their developmental potential. This is again done for each individual after bringing its membrane setup to the end of **step 4**, so that blastoderms that have stayed behind on the yolk are not left in the large PC saline dish for more than a few minutes each. A low-power engineer's magnifier lamp or equivalent over the dish is very desirable for this. The blastoderm is cut out, together with a shallow yolk floor underlying the subgerminal space, by excavating movements with two pairs of the no. 5 forceps. One fine-tipped pair and one

as used in holding vitelline membrane are optimal, and care is necessary to cut just outside the true cellular border of blastoderms, usually visible as a white line on pale yellow yolk. This whole is then transferred to a plastic Petri dish in a layer of TCM:BSS using the blunt-mouthed pipet, and stored blastoderm epiblast uppermost as in **step 5**. This and **step 5** represent the second-best storage condition out of the egg at room temperature, after the condition described in **step 4** above or, best of all, the end of **Part 2, step 1**. About 20 such blastoderms can be stored/60-mm plastic dish in 5 mL of medium, but these very young ones should only be separated from their yolky masses close to the time of their incubation in oligos, and so forth, or their direct transfer back into completed ring cultures. This separation can be done with one pair of fine no. 5s and the flattened mounted needle, by inverting, cutting into the upturned subgerminal cavity through the yolk floor with cross-shaped cuts until the white hypoblast cellular surface is clearly seen, and peeling off the yolky quadrants. Cooling from room temperature in a 4°C refrigerator for a few minutes only will bring a dish of these blastoderms to a condition where the yolk separates from the cells with least likelihood of damage.

Part 2 is completed with each culture under the lowest power of the dissecting microscope so that the entire ring can be viewed at once. Because depth perception is much better, incident lighting rather than transillumination is best for this and all subsequent operations (except for examining development itself without terminating the cultures).

1. The pairs of small blunt forceps are used to complete the process of centralizing the blastoderm, evenly stretching the membrane and then folding it up to wrap over the (rougher) upper face of the ring so as to form a potential seal all around the ring bottom. To do this, saline must first be pipeted from inside and outside the ring until the top face is free of it. Forceps tips are best kept near the same point, opposite your body on the ring periphery, and the whole watchglass or Petri dish rotated as you work, rather than the hands following the changing angle of the ring edge. Initially, the ring alone is lifted slightly while the other forceps wraps a region of membrane completely over the top face from the outside. Subsequently, one forceps grasps membrane-ring-membrane at the point where wrapping over has just been done, and lifts it very slightly off the watchglass, while the other grasps the edge of the membrane and slides it up the outside of the ring, stretching it slightly along the ring while pulling it over. By adjusting the degree of this stretch at each location, the blastoderm can with practice be centralized within the ring. The procedure is quickly learned, with fewest punctured membranes, when the wrists are supported and relaxed, and the ring turned rather than "following" its curvature with your hands. Once a hole has been made however small, in the membrane inside the ring or along the lower half of the outer ring wall (further up will be okay with luck), then abandon and start another assembly, since the ring will inevitably flood with albumen medium later on. Hence the importance of blunt smoothed forceps and totally smoothed pipet mouths. Remember that a

certain number of spare membrane-ring setups (from casualty blastoderms) act as backups for the transfer of good blastoderms from holed membranes!

If the timing of an experiment allows for it, or if two operators are working together, the optimal condition for holding blastoderms while accumulating numbers for oligo treatment is to then half-replace the fluid under the lightly stretched vitelline membrane with albumen culture medium (*see* **Subheading 3.3., step 3** *below* for use of special-angled pipet), keeping a shallow layer of TCM:BSS above the blastoderm, but accumulation at **step 4** of **Part 1** is normally more practical and adequate.

2. Removal of blastoderms from vitelline membrane: The inside of the ring should now be half-full of TCM + saline, and sufficient albumen medium should lie beneath the slightly stretched membrane to form a cushion. Attempts to remove well-adherent blastoderms without first stretching membranes over rings are more likely to result in membrane puncture, so that blastoderm needs a new membrane setup. Overlying saline should be clear enough to allow good visibility without reflection from the incident light. Now is the time to remove pieces of the hypoblast/endoderm if this is to be done. We think this offers no advantage in cases of genes initially activated in epiblast, even if the cells concerned are presumptive mesoderm, but it may improve access of oligos to the already mesodermal cells of gastrulating stages. Hypoblast is a papery thin epithelium, closely adherent to underlying cells near the streak midline, but less so more peripherally. Good oblique incident light and the smooth, flat-tipped mounted needle are necessary to get underneath it, and loosen and remove a piece in the axial (usually the most relevant) region. This does not in itself affect subsequent development if the removed window is not too big and the streak well established (st. 3+ onward). Otherwise, it does cause developmental delay and a tendency to deficiency of anterior axial structures.

   To take off tightly adherent blastoderms, the smooth flat needle is first used, working on the area opaca periphery nearest your body and bringing new regions to this location by rotating the whole assembly toward the needle point. Experience is required to locate the exact edge where the cellular epiblast is migrating on the membrane, in examples of different ages removed from yolks under different temperature conditions (affecting the amount of yolky material that is left overlying this junction). A very small wound must initially be made to insert the needle tip, almost flat, into the space between blastoderm and membrane without puncturing the latter. Then, using a sideways movement toward the periphery, unseal the first bit of this from the membrane. It is always worth quickly looking for a place to begin, however small, where the periphery has come unstuck by itself. On "good" days, most blastoderms up to headfold stages have at least one such place, and life is quicker because from this point on, the fine smoothed jeweler's forceps can be inserted in the gap. The whole periphery can then be unsealed with a quick succession of small sideways movements of the closed forceps tips, with much less chance of damage to blastoderm or membrane. Blastoderms freed in this way should be folded in two inside the ring and removed to be stored flat in dishes of TCM:BSS as in **Subheading 3.1.**, **Part 1**, **steps 5** and **6**. Use the medium-mouthed pipet. Using the wide-mouth pipet in the restricted fluid volume inside the ring is hard to control without disruption of the blastoderm and/or punching a hole in the membrane.

## 3.2. Treatment of Blastoderms with Phosphorothioate Deoxyoligonucleotides

Blastoderms/embryos younger than stage 4, the full-length streak, tend to show poor subsequent development, such as small axial structures or abortive posterior segmentation, if incubated in the way to be described for more than 1–1.5 h. Older ones up to stage 9 (6–7 somites—beyond which blastoderms are difficult to replace stably into ring culture) are amenable to up to 3+ h of treatment. In any case, the conditions of the static off-membrane culture itself, rather than toxicity of oligos, appear to be the physiological limitation unless oligo concentrations are above 80 $\mu M$.

1. Remove subgerminal yolk from any stored blastoderms that left the membrane by themselves. Then gently wash all blastoderms by passing in and out of a medium-mouth pipet or pulling them around by the edges with smooth forceps, in a new dish of room temperature TCM:BSS, before careful transfer to the final 35-mm plastic Petri culture dish containing a measured quantity, such as 3 mL of the same medium. Such a dish can appropriately contain 12 streak-stage blastoderms, or fewer older ones, without overlap of the spread cell layers. The point of knowing the volume of the bath, from here on, is that with precise removals and additions of further components, the final concentrations of oligo DNA/lipofection agents present can be known. With practice ("tough" pipet bulbs and relaxed supported wrists), blastoderms can be passed singly or in groups from the end of the blunt or medium pipet without contact with the surface, and with addition of only negligible volumes of medium to this "oligo bath". Concentration of the original egg albumen in the oligo bath should be <0.5% to avoid danger of phosphorothioate oligo precipitation. This is easy to achieve, but as a beginning check, mix 1:200 of the albumen medium into a dish of TCM:BSS to see the pronouncedly more purple color this produces. As a precaution against premature oligo degradation by exonucleases, we advise transition during this step to use of glass and metal instruments that are reserved solely for contact with oligo incubations, and can be water-washed, alcohol-treated, and flamed between uses.
2. The simple addition of oligos: Using automatic pipeters, such as Gilson's, the TCM:BSS in the dish is reduced to 0.8 mL, a layer just covering the spread blastoderms, meanwhile setting 0.7 mL aside by using it to take up the oligo from two to three freshly thawed aliquots in 0.5-mL Eppendorf tubes. Convenient aliquots of 12.5 $\mu L$ of oligos at 2 m$M$ in water will contribute 25 $\mu M$ oligo/mL each to the final mixture. Thus, in the present case with final volume of 1.5 mL, two aliquots will give 34 $\mu M$ and 3 aliquots, 50 $\mu M$ treatment, with the option of studying the combined effects of different oligo sequences. The 0.7 mL is simply

dripped on from successive fillings of a fine Gilson tip, taking care to cover all the embryos. With practice, the dripping can be of a size and height that visibly stretches and deforms the blastoderms' cellular surface without tearing it, which may help in penetration between cells, and so on, where a microlayer not subject to good bulk mixing may otherwise exist. This detail is probably only important for liposomes, however (**step 3**), which are supramolecular particles and not diffusing. Finally, blastoderms are pulled through the medium by their edges, or subirrigated by the fine pipet with medium already in the dish (no bubbles!), ensuring access to the side not covered by the dripping on. Incubate lidded dish for 1–3 h at 38.5°C, depending on blastoderm stage, swirling gently and respreading blastoderms at half-time. For crowded blastoderms, a small supplement of fresh medium at this time-point relieves acidification.

3. Addition of oligos with a lipofection procedure: In our hands, a procedure using Lipofectamine (Gibco-BRL), a recently available polycationic liposome-forming reagent (*see,* e.g., *15–17* for lipofection), gives appreciable if not dramatic potentiation of the sequence-specific effects per micromole of antisense DNA. Because the lipofection procedure involves the settling out of a fine suspension rather than presence of all the oligo in true solution, blastoderms should probably be placed with the surface uppermost that is believed most relevant for interference with the gene concerned. Otherwise, epiblast-up culture seems somewhat better physiologically, and we have in fact seen only modest evidence of the relevance of germ layer presentation. Lipofectamine is used at 0.5 μL (1.0 μg)/mL per micromole of oligo in the final lipofection medium, but this involves much less than the molar ratio of lipid to DNA believed optimal for liposome formation (*see* manufacturer's recommendations for transfections with "genetic" length DNA molecules). Chick blastoderms can be incubated in up to 25 μL/mL Lipofectamine for over an hour with normal long-term development, but absence of accompanying DNA makes the cells extremely sticky for plastic. We thus typically premix 1/15 of the oligo DNA complement for each incubation with the whole of the Lipofectamine for 20 min at room temperature, giving about the optimal DNA/lipid ratios (manufacturer's instructions) before addition to the blastoderms. This is immediately followed by the balance of the naked oligo DNA to give the planned total oligo micromolarity. There is evidence from cell-culture systems that access of the additional naked DNA into the relevant cell compartments (nucleus and cypoplasm) may be potentiated by the original lipofection (*9*). Incubation with such a liposome preparation alone, thus, including 2–3 μM S-oligo DNA with acceptable Lipofectamine concentrations, does not give effects on whole chick embryos. We thus give below a typical treatment, avoiding DNA precipitation and giving optimal sequence-specific effects from 50 μM phosphorothioate oligos (15-mer), though with many S-oligo samples the simple addition of all the DNA to the lipid in a total volume of at least 0.5 mL, 20 min before presenting to the blastoderms, works as well;

a. Take up the oligo or oligos from aliquots as in **step 2** *above* (e.g., 3 aliquots of 12.5 μL at 2 m*M* for a final 1.5 mL of 50 μ*M*), in 0.3 mL total TCM:BSS, and take to one tube. Then place a further 0.2 mL of such medium into each of two small plastic tubes of materials recommended in the technical data sheet for Lipofectamine. Original Eppendorf tubes as well as some others appear to be suitable. To one tube add 25 µL (50 µg) Lipofectamine; to the other, 1/15 (i.e., 22 µL in total) of the solution of naked DNA oligo or mix of oligos. Mix each tube rapidly, then mix together in one of the tubes, but without foaming, and leave for 20 min at room temperature. This mixture is left at room temperature for 20 min for liposome formation, so the step is suitably done before the final assembly, washing, and sorting of off-membrane blastoderms into the treatment dishes (experimental and matched control group, and so on). The result should be an opalescent appearance in the tube, but not a turbid precipitate. Precipitation follows either contamination of the TCM:BSS with egg albumen, or allowing the local concentration of oligo to rise above 150 µ$M$ on mixing with the lipid.

b. Remove medium from each "oligo dish" containing its blastoderms, to leave the shallow layer of 0.8 mL, then drip the lipofection mixture (0.4 mL) onto the chosen surface of the blastoderms in the dish as for simple oligo treament (**step 2**). Immediately follow this with the (0.3 mL) balance of the naked oligo DNA in TCM:BSS medium, bringing the volume up to approx 1.5 mL.

c. Using flamed smooth fine forceps, "pull" the embryos about through the medium layer to mix the liposome-containing medium intimately into the cellular surface, and subirrigate with a fine pipet avoiding air bubbles. Leave them spread out epiblast- or hypoblast side uppermost (depending on desired target site for oligos). Incubate the lidded dish at 38.5°C for 1–3 h (2 h are our most frequent period with headfold stage embryos of 1–3 somites), possibly remixing gently and "feeding" (*see* **step 2** *above*), once during the incubation period.

## 3.3. Replacement of Blastoderms into the Ring Cultures and Completion of Setting Up

The antisense effect in this work is relatively transient within the time scale of development (one of its potential analytical advantages), so that it works best if little developmental pause in real time occurs immediately following administration of oligos. Chick blastoderms at streak headfold stages are very susceptible to a delay in resumption of development even at normal incubation temperature, following experience of a rapid downward temperature shift. Ideally, therefore, the return to ring culture (and maybe even the whole procedure from the egg) would be at incubation temperature throughout. This is impracticable, but the local bench temperature, for the dish in which blastoderms have been incubated with oligos should be as high as possible during the period required for transfer of a set of them back to the culture setups. In particular, use of a larger dish as a warm water jacket to the "oligo dish" so that there is no downward temperature shock to blastoderms while awaiting transfer has led to the most rapid and normal resumption of development and the strongest relative antisense effects. Certainly, chilling of blastoderms at this stage may even prejudice the final result, and not merely the time schedule of subsequent "control" development.

1. The depth of TCM:BSS must be increased at this stage to facilitate pipet transfer (while preserving some of the oligo concentration around embryos). This added medium should be prewarmed to near-incubation temperature (*see above*). By gently oscillating the dish, blowing with a fine medium-filled pipet, or the use of forceps, it should be ensured that each blastoderm is individually prepared for the rapid pipet transfer of the following step by being free-floating, and not stuck to the plastic floor of the dish. This is much easier for those incubated "epiblast-up."

2. Lift the ring/membrane assembly at one point (with the small curved-toothed forceps) and withdrawing the remaining fluid from beneath it using the special bent, fine smooth-mouthed pipet. Rolling the ring around helps remove more of this underlying saline, and having the overlying space about one-third filled with TCM:BSS also helps by "pressing" saline out more thoroughly from beneath. The optimal pipet is so angled that its oblique smooth mouth lies horizontal and is easily placed beneath or (*see later*) into the space around the inner edge of the ring.

3. Place two successive Pasteur pipet bulbs-full of the albumen ring-culture medium beneath the membrane using the same pipet (no bubbles!) to total about 1.75 mL for a 25-mm diameter ring. In between the bulbs-full, blastoderms are replaced into the ring space in a minimal depth of protein-free medium, epiblast side down to the membrane as originally. Medium is added from the oligo dish to that within the ring space, so that when the blastoderm is delivered from the blunt or medium-mouth pipet, it can be spread over the central clear space on the slightly domed (convex) membrane using fine smooth forceps. Spreading is helped by the shallowness of the fluid layer at this position.

4. After the second bulb-full of albumen is beneath the membrane and the ring released, the meniscus overlying the membrane will, ideally, pin the blastoderm to its central position and allow it to spread, as medium is progressively withdrawn from around it with the same special pipet. In draining this space, concentrate on the gutter that develops next to the inner ring face as the membrane becomes convex from pressure of underlying albumen medium, and of course avoid damaging the blastoderm or sucking a hole in membrane. Cultures that are well drained before the start of incubation produce the best development. The above is an idealized, or even a personal description of what works best. As a procedural skill, this is the trickiest step to become adept at, but the learning process is nevertheless surprisingly rapid! Since any strands of nonfluid albumen can interfere with settling blastoderms on membranes, some prefer to use two, marked separate special pipets of the same design, one for albumen-pipeting steps and one for blastoderm-settling ones (i.e., under-membrane and above-membrane, respectively). Another way of optimizing the spread of the replaced blastoderm is by gentle stroking outward with the flat of the needle (**Fig. 2**).

For blastoderm transfer, the medium-mouthed pipet gives much greater control than the wide-mouthed, with the blastoderm folded or rolled up hypoblast side inward. It can then be unfolded, centralized, and if necessary turned hypoblast/endoderm side up with the fine-pointed forceps, using the area opaca edge, which can take local wounding. The saline depth is adjusted at this stage so that as soon as more is drained from the periphery, the blastoderm is abruptly pinned centrally by the meniscus onto the slightly convex membrane, rather than sliding to one side or the other. Underfilling with medium and draining to give full convexity is then completed. With a little more practice, as a final step, clean protein-free medium can be washed in and out under the blastoderm edges on a very small scale using fluttering movements on the bulb of the special pipet. With this technique, one can flatten out folds in the area opaca and centralize the flattened blastoderm on the dome of membrane, even removing any free lumps of yolk that sometimes are trapped beneath blastoderms. These particles, remaining in what will be the chick's equivalent of the mammal pro-amnion cavity, can interfere mechanically with closure of neural folds.

As in previous steps, most people find it best to learn to work at the part of the ring and blastoderm edge furthest from the body, and when necessary, rotating the whole assembly to bring new parts toward the hand holding the pipet or other instrument. The bend in the shaft and angling of the mouth of the special pipet is to facilitate all the steps of ring culturing, repositioning blastoderms, and draining the space within the ring.

5. The filter paper should always be saturated with water (or the saline) to preserve 100% humidity. Development can be observed from the ventral aspect in reasonably clean ring cultures. After about 18 h, if further culture is needed, they are redrained above and "fed" with a small amount of more medium beneath to preserve the membrane convexity and air/fluid interface, and the filter paper resaturated for humidity. Blastoderms set up in such cultures at streak stages will develop for up to 30 h, though with increasing mechanical compression of the head and heart region, and deficiency in the normal blood supply. In reality, around the 18 somite stage is the outer monitoring point for effects of oligos within this culturing system, and the most suitable genes for study with it are those that are required for normal expression of the basic body plan (outline of the CNS anatomy, heart, and first 10 somites).

For full and final examination (fixation for histology, *in situ*, and so forth), the ring should be flooded with warm PC saline, starting by dripping directly onto the embryo so that it remains intact below the meniscus rather than being lifted onto the fluid surface. The entire blastoderm is then peeled off the membrane using needle and/or fine forceps. Transfer by pipet to a saline wash removes spare protein and solid particles from the structure, greatly improving quality of specimens for whole-mount *in situ* or structural histology.

## 3.4. Roller-Bottle Culture After Oligo Treatment

Our laboratory has recently reported successful culture of chick embryos in a version of the roller-tube method used for mammal development *(13)*. This permits more extended normal development (30 +somites), probably because the extraembryonic blood-vascular supply becomes as profuse as that *in ovo*, whereas this does not occur in ring culture. We now have evidence that phosphorothioate oligos simply added to these cultures can maintain interference with the same genes that have been successfully treated by the modified New ring-culture method described in **Subheadings 3.1.–3.3.** above. However, probably for reasons connected with the oligo uptake lipofection procedure, the strongest sequence-specific effects are seen in roller-tube culture after oligo treatment as in **Subheading 3.2.**, the static incubation in a shallow layer of protein-free medium. We therefore include a description of the current roller-bottle method. Such embryos develop resembling fish larvae in that they are linearly extended along the top of a "yolk-sac" derived from the extraembryonic area pellucida and area vasculosa. Their two abnormalities are that normal torsion of the embryo does not develop and the heart tube displays dorso-ventral looping only, and that local failures of dorsal neural closure are liable to remain, even though CNS regional structure, neural crest emigration, and so forth, are normal. The latter feature, which could even offer an advantage for continued oligo treatment of the generative layer of the CNS, is seen just in individuals that were set up during the most intense folding movements of neural plate formation, rather than before or afterward.

Our current roller-culture medium is 100% Liebovitz TCM as above, plus 5 (when oligos are in medium) to 20% fetal calf serum. We already have evidence for necrosis, however, particularly in the notochord/floorplate region of the cross-section, in advanced embryos following growth in this medium. It is very likely that serum-free media, such as those developed for advanced chick organ culture (e.g., *18*), will give even better and more extended development, since these and other authors have evidence for toxicity of serum to chick embryo structures. This offers the exciting possibility that half-life of phosphorothioate oligos in such cultures could be prolonged, since serum frequently contains exonucleases. Since we have not yet done the necessary testing ourselves, we draw attention to this, but do not detail the alternative media.

1. Remove the oligo incubation dish from incubator, and place on a warm bench or in a larger dish with a layer of warm water, thus avoiding cooling shock to blastoderms. Increase the depth of TCM:BSS to 5–6 mm with warmed medium, such that blastoderms of the ages in the dish can be folded in two without interference from the liquid surface.

2. Fold each blastoderm in two along the embryo's midline (*see* **Fig. 3**), endoderm side inward, so that the midline is slightly stretched out and folds in the area opaca are eliminated, and crimp together the inner faces of the area opaca (the yolky celled, peripheral ring of the blastoderm) at two or three pairs of opposing positions around the perimeter to maintain this folded configuration. This is best done with two pairs of the smoothed, relatively blunt, but well-fitting no. 5 forceps.

3. Holding the folded blastoderm by its curved periphery with forceps near one end of the axis, cut around the entire periphery with iridectomy scissors, just inside or outside the area opaca/ area pellucida junction, but avoiding damage to the poste-

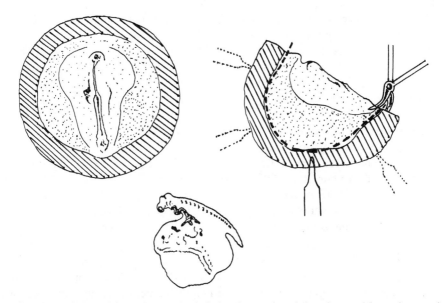

Fig. 3. Roller-bottle culture: At left is shown in plain view a blastoderm with somewhat older head process than that of **Fig. 1**. This is a good stage to commence with, although blastoderms down to stage 3 respond to the technique. At right is indicated the folding along the midline with epiblast outward, crimping with forceps at a few positions to stabilize the folding, and then cutting near the pellucida/opaca junction with iridectomy scissors. Below is sketched an embryo after approx 24 h of culture from such stages, showing how the extraembryonic disk, including its blood vascular system, develops as a "yolk-sac" with the embryo proper extended along the top.

rior end of the streak or future tailbud. This creates a sealed, somewhat flattened lemon-shaped pocket that has the embryonic axis along one edge and the seal along the other. **Step 2** and and this step are best done in succession on each blastoderm. The size of iridectomy scissors is obviously influential, as is the fact that younger blastoderms are relatively thicker and less stably foldable than older ones, but full-streak and even younger-staged blastoderms can be prepared and roller cultured in this way, as well as embryos of up to 10 somites at the start of culture. Cutting just outside the area opaca junction is probably optimal for prolonged culture, since the annular blood vessel normal to this region then develops fully. Folding hypoblast/endoderm out before cutting, giving a layer configuration equivalent to that in a mammal egg cylinder, does not result in normal development in the bird embryo.

4. Using the medium-mouth pipet, transfer such preparations in groups of up to three to the small plastic roller bottles described in **Subheading 2.**, each containing 1 mL of the culture medium prewarmed (*see* introduction to this subsection). Lightly screw-seal caps and place on the inclined roller to rotate immediately. Present experience suggests that lower concentrations of oligos or oligo-

lipofection mixtures than those typically in static preincubation dishes are sufficient to maintain a level of antisense interference in the roller culture.

It appears that serum is hardly required for the first 15 h of culture from streak/headfold stages, but is required for longer culture periods. Medium should be half-replaced every 12–15 h, using 20% serum for older embryos beyond 20 somites, for the longest culture. It is not yet known what the most advanced processes are in chick development whose genetic control can be investigated with antisense treatment of the whole embryo in this way.

Some aspects of final anatomy in these embryos are best seen after saline washing and under incident light, but before cutting off the extraembryonic sac, which leads to contraction of the unfixed tissue. Considerable fixation before this is cut away will prevent contraction of ventro-lateral parts of embryos and axial curling, which would otherwise make them less interpretable in whole-mount *in situ* and sectional series than embryos fixed from ring culture.

## 4. Notes

1. Suitable genes, the choice ("design") of target sites, and the appropriate controls: Many genes are rapidly deployed just before the period of utilization of their products in distinct developmental events, and then have dynamically changing expression levels and sites over the hs-long time scale of early development. Such dynamic RNA expression patterns are themselves evidence for relatively rapid message turnover (short half-life), but could only be part of developmental control mechanisms if accompanied by correspondingly rapid turnover of the protein product. These are the criteria recognized as favorable for antisense interference with gene function in cell systems, viz. protooncogenes and other cell-cycle- or cell-state-change-related genes (*1,2,19*), so that many genes distinctively expressed in early development are suitable candidates for the whole chick embryo approach. Low or modest transcription levels (genes whose signals take relatively long to "come up" in developing *in situ* specimens) are also favorable, because a relatively lower level of interference with their transcription might limit gene function. Many transcription factor and intercellular signal-encoding developmental genes are suitable for this criterion as well.

   Not all such genes are suitable, however. For instance, antisense oligos to several sites in the chick Brachyury gene have been tried in our lab, at relevant stages in view of the known, vertebrate-conserved mutant phenotype of Brachyury. These are without gene-related or other effects, but on consideration, this would not be surprising even if they were acting significantly on the mRNA stability or translation. Although vital for one or a few specific functions in development, this gene's RNA (at least in chick) is transcribed very abundantly within the lineages of cells that will require it (as well as in many that never will), and well ahead of the time of the protein's necessary function in these cells as deduced from the mouse and fish mutants. This suggests, though it does not prove, long half-life RNA and/or protein and thus an expected "perdurance" of gene function in the face of antisense action. Even if antisense effects are exerted in a particular case at molecular level, the gene's early expression pattern does not firmly predict that early developmental steps will be compromised. This is a fascinating scientific matter that relates as much to targeted mutagenesis (knockout) studies of such genes in mice as it does to antisense studies. Thus, oligos targeted to several sites in both the follistatin and the noggin genes of chick have also failed to give effects within the early axial phase of development that the current technique covers. The enticing early expression sites, and corresponding overexpression effects obtained experimentally in Xenopus with these gene products, have meanwhile prompted mouse gene knockouts. From these, it appears that neither gene is required (subject to confirmation) until unexpectedly later steps in development, after induction of the basic axial body plan (*20* and unpublished work).

Only well-controlled positive results of antisense studies are informative about a gene's role. In the nature of the case, unlike that of targeted mutagenesis, absence of effects is currently noninformative. Apart from the considerations just made, the main reason for this is the imperfect criteria for choice of sites, within the transcribed sequence of any gene, that are suitable targets for antisense oligos. There is an extensive literature on such "design criteria" that it is inappropriate to review here. A broad consensus is:

a. That 15-mer to 18- or 20-mer offer the best compromise between likelihood of binding to sequences unique within the transcribed genome, and ease of entry to relevant sites in cells.

b. That short oligo sequences with obvious self-pairing stretches leading to hairpin structures or stable homodimers should be avoided.

c. That the GC/AT content should not be too imbalanced in relation to that of the DNA generally.

d. That computer software can scan the relevant species database (very fragmentary in the case of chick) for possible matches within other genes, and also predict the melting temperature of the DNA/RNA hybrid for each sequence sector. These melting temperatures should obviously be somewhat above the incubation temperature.

An overriding consideration might be sequence-determined secondary structure features of RNA molecules under physiological conditions, determining accessibility as target, but there is little agreement about software that will adequately predict this. Few papers describe comparative results of reasonable numbers of targeted sequences within particular genes (e.g., *9*), but perhaps somewhat less than one-third of all sequence stretches at random would be good for antisense interference. Oligo sequences to be used as mixtures should obviously not pair stably with each other. Striking interference has been reported for target sectors at all relative positions within mRNA, including untranslated 3'-stretches. In the absence of counter indications in the sequence, however, the translation start site and its neighborhood is a first choice.

The choice of the corresponding sense sequence as control oligo for a candidate antisense-interfering one is as illogical as the practice of specifically creating sense strand riboprobes for *in situ* hybridization controls. Apart from length, both display only random degrees of similarity or matching of possibly relevant properties, as sequences go, to the antisense one. In the case of gene interference, there is even evidence for a significant incidence of sense-specific effects (dependent on oligo chemistry, e.g., *21*), exerted by binding and interference with transcription itself. With both the RNA detection and gene-interference effects of antisense, the cumulative evidence for genetic specificity overall must be the distinctive, gene-expression- related effects of sequences antisense to various genes, coupled with the general absence of effects of other sequences in the same procedure. For oligo (as opposed to translation length) antisense, this includes other sequences that happen to be antisense to genes. A complete strategy must take account of the occasional report of sequence-specific, but apparently gene-unrelated effects of S-oligos on cellular properties (e.g., *4*). With time, therefore, the strategy becomes to design antisense oligos to each newly addressed gene, and when and if repeatable specific effects are identified, to think about more stringent individual control strategies for the gene-relatedness of these effects. We choose to use oligo sequences that are parallel with the experimental ones, but have short (6-base) 5'-3' inversions, and are screened for being antisense to known chick genes. They thus match for overall base composition, in case this should be relevant to chemical and not biological specificities of effect. Since most sequences do nothing, we additionally regard the finding of the same "phe-notype" with more than one sequence antisense to the same gene, and especially a more than linear cooperative effect between such sequences, as very strong evidence for gene specificity of effects.

2. Alternatives for nuclease resistance, and access into relevant cellular compart-ments, of oligo DNA: Chemical modifications to oligos to increase their half-life while preserving antisense properties fall into two categories: modifications to the bases themselves and modifications of the sugar-phosphate backbone. In the former category, the substitution of C5-propyne derivatives of pyrimidine bases in DNA may be an exciting breakthrough, affording greatly increased sequence-specific binding while preserving the ability of the hybrid sequence to act as substrate for RNase H so that oligo catalytically degrades message or premessage *(9)*, but these are not yet commercially available or widely stud-ied. In the latter category, we have experience of chimeric oligos in which the end stretches are made with p-ethoxy derivatization of the backbone, leaving a central 7- to 8- base stretch of normal DNA. It is believed that RNase H substrate properties of these oligos may be preserved, and that binding constants are not greatly lowered per sequence. Most importantly, we find that they are not pre-cipitated by the albumen in the medium, so that longer-term persistence of gene interference in ring culture might be expected. In the case of the slug gene (*see* **ref. 3**), the effects of a particular antisense 15-mer sequence in its phosphoro-thioated form, and their degree of persistence during development after the intial off-membrane treatment, were well known to us for S-oligos. We found that these could be replicated when blastoderms were treated only briefly in protein-free medium with the corresponding p-ethoxy chimeric oligo sequences, and then returned to ring-culture with a much lower oligo concentration. Persistent effects

were seen even in blastoderms treated once by subinjection through the periphery into the space between epiblast and vitelline membrane, at the start of culture. Unfortunately, these oligos, and also further new modifications available through our normal commercial source (*see* **Subheading 2.**), are currently too expensive to allow exploration of many sequences in new genes.

Particularly S-oligos are almost certainly taken up into cells mainly by a pinocytotic process leading into membrane-bound vesicles, from which passage into the truly intracellular locations (cytoplasm, nucleus) of their RNA targets is vastly inefficient owing to enzymic degradation. Lipofection procedures potentiate uptake, possibly by destabilizing these intracellular vesicles as well as increasing incorporation from outside the cell. It is even reported that lipofection releases previously taken up, fluorescently tagged oligo from vesicular sites to allow immediate concentration in the nuclear space *(9)*. Attention has recently been paid to an alternative potentiating vehicle for passage of molecules from outside cells into truly intracellular sites; the so-called fusogenic peptides derived from the sequence of the flu virus envelope protein that is influential in its entry to the cell. Such peptides have the property of pH-dependent conformational change that destabilizes lysosomal membranes, greatly increasing their turnover of intact contents into the cell *(22)*. They can apparently act significantly as vehicles for oligonucleotides in this way merely by copresence in the medium *(23,24)*, and not necessarily by being covalently linked to them. In preliminary work (with S. Wharton, this institute), we have indications that a 20 amino acid fusogenic peptide of this kind moderately potentiates sequence-specific effects of particular concentrations of oligos. It is currently unclear whether the optimal effects follow from preparing oligo-bearing lipofection mixture with copresence of the peptide (on the order of 10 $\mu M$), adding peptide independently after lipofection, or adding to medium with the naked S-oligos. These effects have followed the ring-culture version of oligo treatment with brief off-membrane preincubation. Since there is no evidence for toxicity of the required concentrations of peptide, it now remains to combine this with the recent finding that continuous presence of oligo is effective in roller-bottle culture as described above.

3. Problems with embryo quality: Eggs from appropriate suppliers (*see* **Subheading 2.**) are usually 90%+ fertile, and develop normal embryos with high reliability. From late autumn to January (under UK conditions), there is a decline in fertility %, and during hot midsummer weather, one in quality of embryos, coupled with an increase in spontaneous abnormality. The latter may be owing to extensive periods spent by unincubated eggs at temperatures too close to incubation (25°C+). These times of year are therefore not recommended for starting this work, although with experience they need not stop continuation of studies.

4. Preparing to start work: The setting up of ring cultures, removal of embryos from membranes, and their incubation in dishes and replacement as described should all be practiced until good normal development is frequent after these procedures from the stages it is desired to work on. Development to heart formation and anterior somite segmentation can be achieved after following the whole proce-

dure with prestreak blastoderms, including brief (<1 h) off-membrane treatment, but this requires much more attention to detail than success with those from stage 4 (full streak) onward. It is hard to practice the lipofection without actual oligo, but all other aspects benefit from such practice before actual experiments are attempted. All the relevant sections of **Subheading 3.** should be read before starting out. The quicker the embryos pass through the stages of the method apart from the incubation itself, the better will be the biological results, though a careful experiment involving oligo treatment of say 20 blastoderms, to be ring-cultured as age-matched control and experimental samples of 10, involves most of a day. A good general guide is that embryos should be either incubating in oligos, or resting at low 20°Cs in a shallow layer of TCM:BSS, in their ring setups, or in a dish. They should not experience a sharp drop from incubation to resting temperature while being replaced on membranes after oligo treatment.

# References

1. Wagner, R. W. (1994) Gene inhibition using antisense oligodeoxynucleotides. *Nature* **372,** 332–335.
2. Brysch, W. and Schlingensiepen, K.-H. (1992) Design and application of antisense oligonucleotides in cell culture, in vivo and as therapeutic agents. *Cell. Mol. Neurobiol.* **14,** 555–567.
3. Nieto, A. M., Sargent, M. G., Wilkinson, D. G., and Cooke, J. (1994) Control of cell behavior during vertebrate development by *Slug*, a zinc finger gene. *Science* **264,** 835–839.
4. Watson, P. H., Pon, R. T., and Shiu, R. P. C. (1992) Inhibition of cell adhesion to plastic substratum by phosphorothioate oligonucleotide. *Exp. Cell Res.* **202,** 391–397.
5. Caceres, A. and Kosik, K. (1990) Inhibition of neurite polarity by tau antisense oligonucleotides in primary cerebellar neurons. *Nature* **343,** 461–463.
6. Harada, A., Oquchi, K., Okabe, S., Kimo, J., Terada, S., Ohshimer, T., et al. (1994) Altered microtubule organisation in small-calibre axons of mice lacking *tau* protein. *Nature* **369,** 488–492.
7. Loke, S. L., Stein, C., Zhang, H., Avigan, M., Cohen, J., and Neckers, L. M. (1988) Delivery of c-*myc* antisense phosphorothioate oligodeoxynucleotides to hematopoietic stem cells in culture by liposome fusion: specific reduction in c-*myc* protein expression correlates with inhibition of cell growth and DNA synthesis. *Curr. Top. Microbiol. Immunol.* **141,** 282–289.
8. Matsukura, M., Shinozuka, K., Zon, G., Mitsuya, M., Reitz, M., Cohen, J. S., et al. (1987) Phosphorothioate oligodeoxynucleotides: Inhibitors of replication and cytopathis effects of human immunodeficiency virus. *Proc. Natl. Acad. Sci. USA* **84,** 7706–7710.
9. Wagner, R. W., Matteuci, M. D., Lewis, J. G., Gutierrez, A. J. Moulds, C., and Froehler, B.C. (1993) Antisense gene inhibition by oligonucleotides containing C-5 propyne derivatives. *Science* **260,** 1510–1513.
10. Schlingensiepen, K.-H. and Brysch, W. (1992) Phosphothioate oligomers: inhibitors of oncogene expression in tumor cells and tools for gene function analysis, in

*Gene Regulation: Biology of Antisense RNA and DNA* (Erikson, R., and Islant, J., eds.), Raven, New York, pp. 317–328.

11. Isaac, A., Sargent, M. G., and Cooke, J. (1997) Control of vertebrate left-right asymmetry by a snail-related zinc finger gene. *Science* **275,** 1301–1304.

12. New, D. A. T. (1955) A new technique for the cultivation of the chick embryo *in vitro. J. Embryol. Exp. Morphol.* **3,** 326–331.

13. Connolly, D. J., McNaughton, L. A., Krumlauf, R., and Cooke, J. (1995) Improved *in vitro* development of the chick embryo using roller-tube culture. *Trends Genet.* **11,** 259–262.

14. Hamburger, V. and Hamilton, H. L. (1951) A series of normal stages in the development of the chick. *J. Morphol.* **88,** 49–92.

15. Hawley-Nelson, P., Ciccarone, V., Gebeyehu, G., Jessee, J., and Felgner, P. L. (1993) Lipofectamine™ reagent: a new, higher efficiency polycationic liposome transfection reagent. *Focus* **15(3),** 73–79.

16. Felgner, P. L., Gadek, T. R., Holm, M., Roman, R., Chan, H. W., Wenz, M., et al. (1987) Lipofection: a highly efficient, lipid-mediated DNA transfection procedure. *Proc. Natl. Acad. Sci. USA* **84,** 7413–7417.

17. Behr, J. P., Demeneix, B., Loeffler, J. R., and Perez-Mutul, J. (1989) Efficient gene transfer into mammalian primary endocrine cells with lipopolyamine-coated DNA. *Proc. Natl. Acad. Sci. USA* **86,** 6982–6986.

18. Stern, H. M. and Hauschka, S. D. (1995) Neural tube and notochord promote in vitro myogenesis in single somite explants. *Dev. Biol.* **167,** 87–103.

19. Cohen, J. S. (Ed) (1989) Oligodeoxynucleotides, Anitsense Inhibitors of Gene Expression. CRC, Boca Raton, FL.

20. Matzuk, M. M., Lu, N., Vogel, H., Sellheyer, K., Roop, R., and Bradley, A. (1995) Multiple defects and perinatal death in mice deficient in follistatin. *Nature* **374,** 360–363.

21. Augustine, K., Liu, E. T., and Sadler, T. W. (1993) Antisense attenuation of Wnt-1˙ and Wnt-3a expression in whole embryo culture reveals roles for these genes in craniofacial, spinal cord and cardiac morphogenesis. *Dev. Genet.* **14,** 500–520.

22. Wharton, S. A., Martin, S. R., Ruigrok, R. W. H., Skehel, J. J., and Wiley, D. C. (1988). Membrane fusion by peptide analogues of influenza virus haemagglutinin. *J. Gen. Virol.* **69,** 1847–1857.

23. Akhtar, S., Fisher, M., Lentz, B., and Juliano, R. L. (1992) Enhancing intracellular bioavailability of antisense oligonucleotides. *J. Pharm. Pharmacol.* **44 (Suppl.),** 1046.

24. Hideaki, H., Yagisawa, H., Takahashi, S., and Hirata, H. (1994) Amphiphilic peptides enhance the efficiency of liposome-mediated DNA transfection. *Nucleic Acids Res.* **22,** 536–537.

# 50

## Protein Techniques

*Immunoprecipitation, In Vitro Kinase Assays, and Western Blotting*

### David I. Jackson and Clive Dickson

### 1. Introduction

The regulation of complex developmental pathways involves the spatial and temporal interaction of specific gene products. Several powerful genetic approaches, including the tagging of gene products or cells, have allowed their distribution in space and time to be determined. The advent of procedures to create transgenic animals with acquired or abrogated gene functions has allowed the phenotypic consequences of genetic alteration to be analyzed by design rather than chance mutation. For example, the introduction of a portion of the Y chromosome into a female mouse can change its sex, demonstrating the presence of a sex-determining gene *(1)*. In combination with a range of molecular biology techniques, such as RNA protection, PCR amplification, *in situ* hybridization, immunohistochemistry, and the use of reporter genes, to indicate how, when, and where a gene is expressed. This avalanche of information has started to provide a real insight into the molecular mechanisms that regulate developmental processes. However, studies on the gene products are still crucial for an understanding at the cellular level processes, such as pattern formation and organogenesis.

Proteins may be modified by glycosylation, phosphorylation, proteolysis, or myristylation, and each of these modifications affects the properties of the protein. For example, specific changes in the level of phosphorylation of certain proteins have been observed during the compaction of the mouse embryo, and it is a central paradigm for cell:cell signaling by tyrosine kinase receptors *(2–5)*. To determine the presence of a particular protein or to investigate possible

From: *Methods in Molecular Biology, Vol. 97: Molecular Embryology: Methods and Protocols*
Edited by: P. T. Sharpe and I. Mason © Humana Press Inc., Totowa, NJ

posttranslational modifications, such as glycosylation, phosphorylation, proteolytic processing, and protein:protein interactions, there are several basic protein separation techniques that can be applied, and some general recipes for these are described (*6–9*).

## 2. Materials

All solutions should be made of Analar-quality reagents and distilled water. Stock solutions of 20% (w/v) SDS, phosphate-buffered saline (PBS), 5 $M$ NaCl, 0.5 $M$ EDTA pH 8.0, 1 $M$ dithithrietol, 1 $M$ Tris HCl at pH 8.8, 8, 7.5, and 6.8 are required.

1. Lysis buffers:
   a. RIPA lysis buffer: 150 m$M$ NaCl, 1% (v/v) NP-40, 0.5% (w/v) deoxycholate, 0.1% (w/v) SDS, 50 m$M$ Tris-HCl, pH 8.0, and protease inhibitors. NP-40 lysis buffer: 150 m$M$ NaCl, 1% (v/v) NP-40, 50 m$M$ Tris, pH 7.5, and protease inhibitors.
   b. Protease inhibitors: A mixture of 1 m$M$ phenylmethylsulfonyl fluoride (PMSF), (the 1 $M$ concentrated stock dissolved in acetone), and 1 mg/mL each of antipapain, chymostatin, leupeptin, and pepstatin (*see* **Note 1**).
   c. 2X dissociation buffer: 4% (w/v) SDS, 120 m$M$ Tris-HCl, pH 6.8, 0.002% (w/v).
   d. Bromophenol blue, 20% (v/v) glycerol 0.1 $M$ dithiothreitol (*see* **Note 2**).
2. Antibody coupling solutions: 0.1 $M$ borate buffer pH 8.2, 20 m$M$ ethanolamine pH 8.2, 0.2 $M$ triethylamine, pH 8.2, dimethyl pimelimidate, and protein A Sepharose beads (Amersham Pharmacia Biotech, Little Chalfont, UK).
3. Kinase buffer: 100 m$M$ NaCl, 25 m$M$ HEPES pH 7.5, 5 m$M$ MnCl$_2$, 10 m$M$ MgCl$_2$, 1 m$M$ Vanadate, 10 μ$M$ Mg-ATP (Sigma), $\gamma^{32}$P-ATP (3000 Ci/mmol).
4. SDS polyacrylamide gel electrophoresis (PAGE) solutions.
   a. Resolving gel: 12.5 mL 30% (29:1) acylamide/*bis*-acrylamide mix, 7.5 mL 1.5 $M$ Tris-HCl, pH 8.8, 0.3 mL 10% (w/v) SDS, 9.9 mL H$_2$O, 0.3 mL 10% (w/v), ammonium persulfate (made fresh), 12 μL TEMED.
   b. Stacking gel: 0.83 mL of 30% (29:1) acylamide/*bis*-acrylamide mix, 0.63 mL 1 $M$ Tris-HCl, pH 6.8, 50 μL 10% (w/v) SDS, 3.4 mL H$_2$O, 50 μL 10% (w/v) ammonium persulfate, 5 μL TEMED.
   Running buffer: 25 m$M$ Tris base, 192 m$M$ glycine, 0.1% (w/v) SDS.
5. Blotting solutions:
   a. Semidry blotting buffer: 48 m$M$ Tris base, 39 m$M$ glycine, 0.037% (w/v) SDS, 20% (v/v) methanol.
   b. Western blocking buffer: PBS 5%(w/v), BSA 0.05% (v/v), Tween-20, or PBS, 5% (w/v) nonfat milk powder.
   c. Western washing buffer: PBS 0.05% Tween-20.
   d. ECL reagent commercially available from Amersham Pharmacia Biotech.
6. Antibody-labeling solutions:
   25 m$M$ KI, Iodogen (Pierce, Warriner, UK), CHCl$_3$, Sephadex G-25 (Amersham Pharmacia Biotech), and $^{125}$I (100 mCi/mL).
7. Gel-staining solutions:
   a. Gel-fixing solution: 45% (v/v) methanol, 9% (v/v) acetic acid.

b. Coomassie stain: 45% (v/v) methanol, 9%(v/v) acetic acid containing 0.25% (w/v) Coomassie blue R250.

# 3. Methods

The embryonic material can be used in several ways as indicated in **Fig. 1**; the various techniques are described in order below.

## 3.1. Radiolabeling Cell/Tissue Cultures

1. The cultures or tissue explants are rinsed in culture medium missing the appropriate substrate. Thus, the medium must be methionine-free or phosphate-free if these are to be the radiolabeled compounds.
2. The cultures are then incubated in the same medium containing 1 mCi/mL of the radiolabeled compound, at 37°C, in tissue-culture conditions for 2–3 h.
3. The radiolabeled medium is removed, and the embryonic tissue is washed once in "cold" medium before being dissociated in either 2X dissociation buffer or RIPA/NP40 lysis buffer.

## 3.2. Preparation of Extracts

To immunoprecipitate proteins from embryonic tissue or cell culture, the material has to be disrupted under conditions that solubilize the proteins (*see* **Notes 3** and **4**).

1. Embryos or cell cultures are collected and washed in PBS prior to homogenization using a dounce homogenizer, or other commercial disruption device, with ice-cold RIPA or NP40 lysis buffer.
2. Centrifuge the lysate at 12,000g for 10 min at 4°C to remove the insoluble material; in the case of NP-40 lysates, this will also cause the unlysed nuclei to sediment.

## 3.3. Coupling of Antibodies to Protein A Beads

In order to immunoprecipate antigens, the revelant antibody is preferably coupled to the solid support (protein A-sepharose beads), but alternatives are mentioned in **Note 5**.

1. Wash the protein A-sepharose beads with PBS, and then add the monoclonal antibody (MAb) (*see* **Note 6**) to 2 mg/mL of packed protein A-sepharose beads in PBS for 2 h at 4°C. Fifty millimeters of ascites or 1 mL of polyclonal antiserum or tissue-culture supernatants/mL of packed beads can also be used.
2. Wash in 0.1 *M* borate buffer, pH 8.2, three times, and then once with 0.2 *M* triethylamine, pH 8.2.
3. Add 5 vol of 0.2 *M* triethylamine, pH 8.2, containing 20 m*M* dimethyl pimelimidate for 45 min to crosslink the protein to the beads.
4. Spin down the beads, and add 20 m*M* ethanolamine, pH 8.2, to stop the reaction. After 5 min, wash three times with 0.1 *M* borate buffer, and finally resuspend the coupled beads in RIPA or NP40 lysis buffer (**Note 7**).

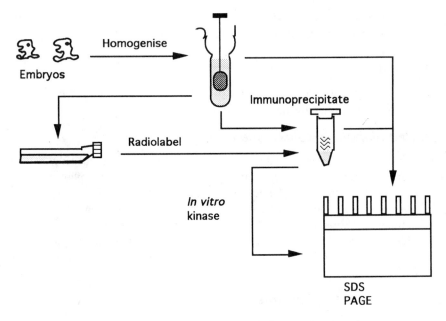

Fig. 1. Schematic diagram of the relationship of the protocols described in this chapter.

## 3.4. Immunoprecipiation

Immunoprecipitation is a technique that allows the biochemical characterization of protein from relatively few cells, and is an important technique where material may be a limiting factor. It can also be used to study various types of posttranslational modification (*see* **Note 8**). The amount of material required will depend on the abundance of the antigen (*see* **Note 9**).

1. The lysates are precleared twice for 1 h at 4°C with 25 µL of protein A-Sepharose beads (*see* **Note 10**).
2. The precleared lysates are then immunoprecipitated with 10 µL of 2 mg/mL antibody protein A-Sepharose beads for 4 h to overnight at 4°C.
3. The beads are washed six to eight times in lysis buffer for 1 min with as much of the solution as possible being removed at each wash (**Note 11**).

At this stage, the immunoprecipitate can be analyzed by standard fractionation techniques, such as SDS-PAGE, or assessed for enzyme activity (e.g., kinase activity).

## 3.5. In Vitro Kinase Assay

1. The beads are resuspended in 30 µL of kinase buffer.
2. 10 µCi of $\gamma^{32}$P-ATP are added to the immune precipitates and incubated at ambient temperature for 15 min.

3. 0.7 mL of lysis buffer containing 30 m$M$ EDTA, pH 8.0, is added to stop the reaction, and the beads are then washed three times in kinase buffer. The samples are then run on an SDS polyacrylamide gel as described below.

## 3.6. SDS Polyacrylamide Gel Electrophoresis (PAGE)

The basic SDS-PAGE method is based on that described by Laemmli *(10, 11)* and as an aid the 12.5% polyacylamide gel, is described in **Subheading 2.4.** (*see also* **Note 12**). SDS polyacrylamide gels are electrophoresed in running buffer (**Note 13**). Gels containing SDS separate on the basis of molecular weight, and it is possible to calibrate the gel using a sample of appropriate marker proteins; these are commercially available as prestained, or [14]C-labeled markers, which allow them to be visualized by autoradiography or on immunoblots.

1. The resolving gel is poured (adding the TEMED and ammonium persulfate last), and the surface is flattened by overlaying with a small amount of butan-1-ol.
2. After polymerization the butan-1-ol is drained off, and the stacking gel poured on top of the resolving gel.
3. To the sample, an equal volume of 2X dissociation buffer is added and boiled for 3 min prior to loading onto the gel (*see* **Fig. 1**).

After electrophoresis, the gels can be fixed and stained with Coomassie blue to visualize the proteins if the gel is not to be immunoblotted (*see* **Note 14**). The gels are fixed for 30 min, and then stained with Coomassie blue for 4 h, followed by destaining in "fixing solution," which can be accelerated at 50°C.

## 3.7. Immunoblotting

1. After electophoresis, the proteins within the gel are transferred onto a nitrocellullose membrane using a semidry electroblotter in semidry blotting buffer at 0.3 A for 40 min (*see* **Subheading 2.5.** and **Note 15**).
2. The remaining protein bindings sites on the membrane are blocked for 1 h in Western blocking buffer.
3. Rinse once in PBS and twice in Western washing buffer.
4. Incubate in 0.1–2 µg/mL of specific antibody in Western washing buffer containing 0.5% (w/v) milk powder for 1 h at room temperature. The sensitivity may be increased by incubating overnight at 4°C. However the background may also increase.
5. Wash three times for 5 min in Western washing buffer.
6. Incubate with the second antibody (e.g., antimouse HRP-labeled secondary antibody for primary MAbs) in Western washing buffer containing 0.5% (w/v) milk powder for 1 h at room temperature.
7. Wash six times for 5 min in Western washing buffer.
8. Develop in ECL reagent for 1 min, and expose to X-ray film (*see* **Note 16** for alternative methods).

## 3.8. Antibody-Labeling Method

An alternative to ECL is the use of radiolabeled probes, such as [125]I labeled protein A or second antibody raised against the detecting antibody, which are also commercially available (*see* **Note 17**).

1. Two holes are made in the bottom of a 1.5-mL Eppendorf tube with a 19-gage needle, and the bottom is plugged with a small piece of nylon wool. The Eppendorf is then filled with 1 mL of swollen G25 Sepharose resin.
2. The "column" is washed with 10 mL of PBS containing 10 mg/mL BSA. The column is finally washed with 2 mL of PBS containing 25 m$M$ KI, which has been freshly prepared. Dissolve 1 mg Iodogen in 1 mL of $CHCl_3$ and pipet 4 µL of the solution into a 1.5-mL Eppendorf tube, and evaporate to dryness.
3. Add about 30 µg of antibody in 200 µL PBS to the Iodogen-coated tube with 400 µCi of [125]I at room temperature for 10 min.
4. 200 µL of the labeled antibody are then loaded onto the G25 Eppendorf "column" and centrifuged at 170$g$ for 1 min in a Falcon 2063 test tube containing 20 µL of 10 mg/mL BSA. The labeled antibody can be stored at 4°C for up to 3 wk, and is used at $5 \times 10^5$ cpm/mL in Western blocking buffer containing 1 mg/mL of unlabeled antibody.

Prior to autoradiography, the gel must be dried onto 3MM paper under vacuum at 60°C for 2 h, after fixing for 30 min. The gel is then exposed to X-ray film with an intensifying screen at –70°C (**Notes 18** and **19**).

## 4. Notes

1. A 20-µmol aqueous solution of PMSF at pH 8.0 has a half-life of 35 min, and therefore, this inhibitor must be added just prior to use *(12)*. The inclusion of 10 m$M$ iodoacetamide is also useful to prevent crosslinking of proteins owing to the presence of free sulfhydryl groups during immunoprecipitation.
2. The dissociation buffer is usually made reducing by the addition of 0.1 $M$ dithiothreitol or 1.4 $M$ β-mercaptoethanol prior to boiling, under which conditions disulfide-linked chains will become dissociated. Such reagents maybe omitted, and this may be useful in immunoprecipitation techniques where the immunoreactive Ig chains in their reduced form (25 and 50 kDa) obscure the antigen in question.
3. The way cells are lysed has a critical effect on the composition of the extract *(13,14)*. Nonionic detergents, such as Nonident P-40 (NP-40), solubilize antigens in their native state and preserve the nuclei in an intact state, so that they may be removed by centrifugation. Mild detergents such as CHAPS, digitonin, and Brij 96, enable macromolecular complexes to be immunoprecipitated allowing protein associations to be investigated. However, the most widely used and versatile immunoprecipitation buffer is RIPA. This lysis buffer is often the best to use when little is known about the solubilization behavior of the antigen in question.
4. Strong ionic detergents, such as SDS, denature the antigen, and so can be used with antibodies that are considered to be "nonimmunoprecipitating" antibodies.

Such antibodies often include antipeptide antibodies, since they are usually generated against linear epitopes, which may be masked by the tertiary structure of the native protein. Thus, SDS denaturation of the protein allows the masked epitope to be exposed and accessible to the antibody: to do this, lyse the cells in 10 vol of 2% SDS, 50 m$M$ Tris, pH 7.5, and place in a boiling water bath for 10 min. The DNA is then sheared by passing through a 21-gage needle several times, or by using a probe sonicator with several short bursts. Pellet the insoluble material by centrifugation at 10,000$g$ for 10 min, and then dilute the supernatant 20-fold in PBS containing 2% (w/v) BSA.

5. Antibodies or antiserum can be added directly to the cell lysate at 4°C for 1 h or overnight. It is often worth trying both procedures, since the former usually leads to cleaner immunoprecipitates, but if the antibody is of low affinity, the latter may be necessary to achieve optimum binding to antigen. The antibody, including the antibody/antigen complexes; is then recovered using a second antibody against the first, although it is far more usual to use protein A or protein G linked to Sepharose beads. These reagents provide high-affinity binding to the Fc component of the antibody, and make washing and recovery of the antigen very easy (*see* **Note 11**).

6. Most MAbs or polyclonal antibodies used in immunodetection procedures are either mouse or rabbit antibodies. With the exception of mouse IgG subclass 1, all these antibodies can be coupled to protein A-sepharose, allowing maximal correct orientation of the antibody for immunoprecipitation. The use of the crosslinker dimethyl pimelimidate *(15)* also means that the antibody remains attached to the protein A beads during electrophoresis producing "cleaner" Western blots.

7. The percentage attachment of the antibody can be determined by measuring the optical density at 280 nm before and after coupling. If the coupling is low, then either protein G-sepharose or high-salt coupling can be used; for high-salt coupling, the first step of the low-salt coupling procedure is replaced by washing the protein A-sepharose beads three times in PBS, 3 $M$ NaCl, pH 9.0, and then add the antibody for 2 h at 2 mg/mL in the same buffer.

8. Immunoprecipitation can also be used in combination with radiolabeled compounds in order to study protein modifications, such as phosphorylation, or the study of newly synthesized proteins (by $^{35}$S-methionine incorporation). The addition of radioisotopes to cell-culture media to follow protein modification of embryonic cell lines is routinely performed, and could be adapted to incubating, over short time spans, early embryos or dissected tissues from fetuses. Experiments have been performed on early embryos for $^{32}$P *(16)*, and $^{35}$S-methionine labeling of proteins *(17)*. Microinjection of radiolabeled compounds into embryos has also been tried, for example, the labeling of proteoglycans with [$^{35}$S] sulfate *(6)*.

9. From early stage embryos, moderately abundant proteins can be detcted by immunoprecipitation, but not by simple staining procedures, such as with Coomassie blue or the more sensitive silver-staining technique. As a guideline, immunoprecipitation has been used to detect C-cadherin using two 48-cell embryos and subsequent separation by SDS-PAGE and immunoblotting using an

ECL detection system (*18*, and *see* **Subheadings 3.6.** and **3.7.** for details of techniques). However, between 230 and 310 eight-cell mouse embryos were needed to detect E-cadherin by the same method (*16*). In general, the amount of material needed for this type of experiment is really a matter of trial and error. The affinity of the antibody is also a major factor with affinities of $> 10^7/\text{mol}$ required to immunoprecipitate antigens efficiently.

10. A preclearing step is not essential, but usually leads to more specific immunoprecipitation. For clean immunoprecipitates, which is especially important if radiolabeled antigens are being immunoprecipitated, an extra preclearing step is required. This is usually with nonspecific antibody coupled to protein A-sepharose for 1 h at 4°C, i.e., mouse immunoglobulins in the case of MAbs.

11. The use of a 100 mL Hamilton syringe with a very narrow bore and blunt end (point style 3) allows the liquid to be removed without any beads.

12. This can only be considered to be a very brief description, and more comprehensive reviews are available elsewhere (*19*). A 12.5% polyacrylamide gel is suitable for proteins between 12 and 45 kDa, whereas a 7.5% gel is suitable for 50–150 kDa proteins. Gradient gels allow the separation of most proteins of interest, and can be poured by mixing equal volumes of 5 and 15% acrylamide solutions in a gradient maker.

13. As guide to running conditions that will depend on the apparatus used: 4 h at 140 V, or 16 h at 35 V. These conditions will also vary depending on the size and percentage of the gel.

14. A more sensitive staining method is to use silver, and there are commercial kits available based on the method of Sammons et al. (*20*). However, staining is not usually appropriate for immunoprecipitates that invariably are associated with radiolabeling or immunoblotting detection procedures.

15. The semidry blotting technique gives a higher yield of transfer and is quicker. However, the small buffer volumes used can sometimes lead to asymmetric transfer of proteins across the gel, and it is also possible to "overblot" where the protein passes through the blotting membrane. For semidry blotting, the membrane can either be nitrocellulose or specially developed membranes, such as Immobilon P (PVDF, Millipore). In the case of PVDF membranes, they should be "activated" for a few seconds in methanol and then washed in transfer buffer, whereas nitrocellulose membranes should just be washed in transfer buffer. Wet blotting is often more effective when a PVDF membrane is used with a transfer buffer composed of 10 m*M* CAPS (Sigma), pH 11.0, and the transfer is carried out at 0.3 A, 60 V at 4°C overnight.

16. Colorimetric development of immunoblots probed with phosphatase and peroxidase-labeled antibodies is possible. However, these procedures are relatively insensitive, the color can fade, and they are generally less versatile than other techniques. Electrochemi-illuminescence (ECL) is just as easy to perform and gives a permanent record in the form of an autoradiograph, the exposure time is relatively short, and the length of exposure can be varied in order to produce the correct intensity of band. ECL reagents are readily available in kits, and therefore, the technique will not be described here.

17. Immunoprecipitation is a sensitive technique, especially when it is combined with detection by [125]I-labeled antibodies. Labeling the first antibody with [125]I produces a more specific reagent and, therefore, cleaner immunoblot, although it may be less sensitive, since it does not have the amplification of signal (and background), which occurs with methods incorporating another layer in the detection system.

18. The autoradiograph produced may be intensified by the use of fluorography ([3]H by 1000-fold, and [35]S by 5- to 10-fold), and can be performed using commercially available products, such as Amplify (Amersham), or by the following method: after fixing, the gel is rinsed twice in DMSO for 45 min, and then in DMSO plus 22% (w/v) diphenyloxazole (PPO) for 45 min. The gel is finally rinsed under a stream of water for 1 h and dried before autoradiography.

19. In recent years, the general availability of phosphoimagers has replaced autoradiography and fluorography as a detection system for radiolabeled proteins. It has several advantages, including a linear response over a wide range of activities, and in conjunction with the appropriate software, can be used for easy quantitation for comparative studies.

## References

1. Koopman, P., Gubbay, J., Vivian, N., Goodfellow, P. N., and Lovell-Badge, R. (1991) Male development of chromosomally female mice transgenic for Sry. *Nature* **351,** 117–127.

2. Bloom, T. and McConnell, J. (1990) Changes in protein phosphorylation associated with compaction of the mouse preimplantation embryo. *Mol. Reprod. Dev.* **26,** 199–210.

3. Fantl, W. J., Johnson, D. E., and Williams, L. T. (1993) Signalling by receptor tyrosine kinases. *Ann. Rev. Biochem.* **62,** 453–481.

4. Wassarman, D. A., Therrien, M., and Rubin, G. M. (1995) The ras signalling pathway in Drosophila. *Curr. Opin. Genet. Dev.* **5,** 44–50.

5. Dickson, B. (1995) Nuclear factors in sevenless signalling. *Trends Genet.* **11,** 106–111.

6. Itoh, K. and Sokol, S. Y. (1994) Heparan sulfate proteoglycans are required for mesoderm formation in Xenopus embryos. *Development* **120,** 2703–2711.

7. Brown, T. L., Fisher, W. C., Collins, M. D., De, B. K., and Scott, W. J., Jr. (1994) Identification of a 100-kDa Phosphoprotein in developing murine embryos as Elongation Factor 2. *Arch. Biochem. Biophys.* **309,** 105–110.

8. Schneider, D. S., Jin, Y., Morisato, D., and Anderson, K. V. (1994) A processed form of the Spätzle protein defines dorsal-ventral polarity in the Drosophila embryo. *Development* **120,** 1243–1250.

9. Mathieu, M., Kiefer, P., Mason, I., and Dickson, C. (1995) Fibroblast Growth Factor (FGF) 3 from Xenopus Laevis (XFGF3) binds with high affinity to FGF receptor 2. *J. Biol. Chem.* **270,** 6779–6787.

10. Laemmli, U. K. (1970) Cleavage of the structural protein during the assembly of the head of the bacteriophage T4. *Nature* **227,** 680–685.

11. Sambrook, J., Fritsch, E. F., and Maniatis, T. (eds.) (1989) *Molecular Cloning. A Laboratory Manual,* 2nd ed. Cold Spring Harbor Laboratory, Cold Spring Harbor, NY.
12. James, G. T. (1978) Inactivation of the protease inhibitor phenylmethylsulphonyl flouride in buffers. *Anal. Biochem.* **86,** 574–579.
13. Lazarovitz, A. I., Osman, N., Le Feuvre, C., Ley, S. C., and Crumpton, M. J. (1994) CD7 is associated with CD3 and CD45 on Human T cells. *J. Immunol.* 3957–3966.
14. Jackson, D. I., Verbi, W., Lazarovitz, A. I., and Crumpton, M. J. (1995) T lymphocyte activation: The role of tyrosine phosphorylation. *Cancer Surveys* **22,** 97–110.
15. Schneider, C., Newman, R. A., Sutherland, D. R., Asser, U., and Greaves, M. F. (1982) A one-step purification of membrane proteins using a high efficiency immunomatrix. *J. Biol. Chem.* **257,** 10,766–10,769.
16. Aghion, J., Gueth-Hallonet, C., Antony, C., Gros, D., and Maro, B. (1994) Cell adhesion and gap junction formation in the early mouse embryo are induced prematurely by 6-DMAP in the absence of E-cadherin phosphorylation. *J. Cell. Sci.* **107,** 1369–1379.
17. Christians, E., Campion, E., Thompson, E. M., and Renard, J-P. (1995) Expression of the HSP 70.1 gene, a landmark of early zygotic activity in the mouse embryo, is restricted to the first burst of transcription. *Development* **121,** 113–122.
18. Fagotto, F. and Gumbiner, B. M. (1994) β-Catenin localisation during Xenopus embryogenesis: accumulation at tissue and somite boundaries. *Development* **120,** 3667–3679.
19. Hames, B. D., and Rickwood, D., eds. (1981) *Gel Electrophoresis of Proteins—A Practical Approach.* IRL Press, Oxford University Press, Oxford, UK.
20. Sammons, D. W., Adams, L. D., and Nishizawa, E. E. (1981) Ultrasensitive silver-based color staining of polypeptides in polyacylamide gels. *Electrophoresis* **2,** 135.

# VIII

## Microscopy and Photography

# 51

## Microscopy and Photomicrography Techniques

### Richard J. T. Wingate

## 1. Introduction

By its nature, embryology is highly dependent on microscopy. Many experimental procedures are carried out under either dissection or compound microscopes, and photomicrographs are often the principle data in embryological studies. Most researchers are therefore familiar with achieving an adequate illumination, resolution, and contrast in observing or manipulating specimens. However, when faced with a choice of objectives, contrast enhancement optics, film types and image processing options, high image quality cannot be realized by trial and error. Achieving the highest optical performance for photomicrography requires that a variety of options within the microscope are optimized with respect to the preparation of the specimen. Photomicrographic reliability and quality also rely on a knowledge of how film behaves with respect to illumination and how subsequent image processing can enhance information within a specimen.

It is important to note that although this chapter outlines how image quality is limited by optical and photographic constraints, it can only supplement the essential technical information that is contained in every microscope instruction manual.

### 1.1. Dissection Microscopes

Dissection microscopes have low-magnification lenses with long working distances, which give a good depth of field, but a poor resolution. Nevertheless many specimens are photographed unmounted on dissection microscopes equipped with a 35-mm camera attachment. For general observation, there are no rigid principles. Lighting is a matter of personal preference and depends on

From: *Methods in Molecular Biology, Vol. 97: Molecular Embryology: Methods and Protocols*
Edited by: P. T. Sharpe and I. Mason © Humana Press Inc., Totowa, NJ

the specimen. However, where live specimens are being observed, "cold" light sources (where the lamp is remote and the light transmitted to the specimen by fiber optics) are preferred to lamps integrated into the microscope stand (which generate heat). Most dissection scopes give the opportunity for transillumination (light shining through the specimen) and transparent Sylgard- (*see* **Subheading 2.1.**) coated dishes have therefore generally replaced wax as a substrate for pinning out specimens. For photomicrography, translucent agarose-filled dishes provide a less rigid support for the specimen, but give a more diffuse background light (*see* **Subheading 2.2.**). In terms of camera focusing and choice of film, the same principles apply as for compound microscopes (*see* **Subheading 4.**).

## *1.2. Compound Microscopes*

Apart from routine activities, such as observation and photomicrography, inverted and fixed-stage compound microscopes are used to monitor cell injection and patch clamping. For photomicrography, standard format compound microscopes may have an integral camera, or a connecting tube mounted on a trinocular head with fittings for video or CCD cameras, or a camera body for 35mm film.

Various factors affect the quality of image that can be achieved. The way in which the specimen is prepared and mounted will influence the type of objectives that can be used, and which type of contrast enhancement filters are appropriate.

## 2. Materials

1. Sylgard (Dow Corning, Barry, South Glamorgan, UK): This is made up 9 parts to 1 with a hardener. Sylgard dissolves in solvents, such as Histoclear and xylene.
2. Agarose-filled dishes: Use 1% agarose in distilled water. Heat to >60°C to dissolve and cool to set.
3. Cleaned coverslips: Wash in 70% ethanol/1% conc. HCl and then 2% (TESPA) 3-aminopropyl- triethoxysilane/3-(triethoxysilyde)-propylamine for 10 min. Rinse in acetone for 10 min and then distilled water before drying.
4. Gelvatol (Monsanto Chemicals, Knowsley, UK)/ Mowiol (Hoechst, AG, Frankfurt, Germany): Use 2.4 g Mowiol to 6 g of glycerol. Stir to mix. Add 6 mL water and leave for several h at room temperature. Add 12 mL of 0.2 *M* Tris buffer (pH 8.5), and heat to 50°C for 10 min to dissolve. For fluorescence, add 1,4-diazobicyclo-(2.2.2)-octane (DABCO).

## 3. Methods
## *3.1. Specimen Preparation*

Different techniques require that tissue be fixed and processed in a variety of different ways. This in turn influences which mounting media may be used. Different kinds of mounting media require different optical considerations. In particular, the refractive index of the mountant will affect the spatial resolution that can be achieved for a given objective.

### 3.1.1. Whole-Mount Specimens

Particular care must be taken in mounting relatively large, intact or partially dissected embryos under a coverslip. Although such material may be photographed through a dissection microscope, mounting a specimen under coverglass and in a mountant with a high refractive index will give a better spatial resolution. It may be desirable to clean slides and cover-glass prior to use (*see* **Subheading 2.3.**) and to ensure that the thickness of cover-glass is exactly 0.17 mm where optimal resolution or transmission of, in particular, fluorescent light is required (*see* **Subheading 3.2.6.**). Whole-mount specimens are mounted under raised coverslips, which use small drops of silicon grease at each corner of the coverslip as supports. The coverslip is thus held above the specimen, and its final height can be adjusted by carefully applying downward pressure. Without silicon grease supports, delicate embryonic tissue is easily flattened. Excess mountant can be removed using absorbent Postlip paper (Hollingsworth & Vose Co., Winchombe Glos., UK). If a nonsetting mountant (e.g., glycerol) is being used, then the edges of the coverslip can be sealed by painting a heavy layer of nail varnish around the edges of the glass.

Even with the support of a silicon grease plinth, applying too much downward pressure to the specimen is potentially disastrous. Therefore, cleaning the cover-glass above a whole-mount preparation requires care. Dust and grease can be removed by dragging a piece of lens tissue, which has been moistened with a drop of 70–100% alcohol, across its surface.

### 3.1.2. Mountants

Choice of mountant is influenced by whether the specimen is dehydrated, and affects not only the permanence of the preparation, but also the clarity of the image. Embryos that are small and virtually transparent can be fixed whole or sectioned and cleared with glycerol/PBS, which itself can be used as a mountant. Glycerol remains liquid and cannot be regarded as a reliable permanent mounting medium. By contrast, mountants that set to form a solid matrix are not only more stable over time, but also allow the use of oil immersion objective lenses which give optimal visual resolution. Oil immersion lenses should not be used on anything other than permanent preparations (*see* **Subheading 3.2.4.**).

1. Nonpermanent preparations—a convenient mountant, which also clears delicate whole mounts is a mixture of 90% glycerol/0.02% azide in PBS (*see* **Note 1**). For fluorescent specimens 2.5% DABCO (1,4-diazobicyclo-[2.2.2]-octane) can be added to this mixture to prevent fading. Stains that use FITC as a fluorochrome

should be mounted in a medium with a pH of at least 8.5. For a more durable preparation, the cover-glass can be sealed with nail varnish.

2. Permanent preparations—water-soluble permanent mountants (i.e., which set) usually suffer from shrinkage, which can ruin a specimen. The more viscous they are, the less of a problem this is likely to be. However, viscosity makes the mountant hard to handle. Probably the best aqueous, setting mountant is Gelvatol (Monsanto)/Mowiol 4-88 (Hoechst), which hardens overnight and which can be stored as frozen aliquots (*see* **Subheading 2.4.**)

Some tissues can only be cleared adequately by xylene or its less toxic equivalent, Histoclear (National Diagnostics, Hessle Hull, UK), having first been dehydrated through an alcohol series. Some counterstains, such as cresyl violet (Nissl), require that the tissue is dehydrated to xylene/Histoclear. Tissue which to be cleared in this way can be mounted in DePeX or, if fluorescent, in Fluoromount (Sigma), both of which set solid over time.

## 3.2. Choice of Objectives

A typical compound microscope will be set up with a range of objectives of increasing magnification, which will all be approximately parfocal (*see* **Note 2**) (a standard range would be ×2.5, ×10, ×25, ×40, ×60 or ×100). Parfocality means that it is possible to switch objectives without adjustment to the focus. Where eyepieces have an adjustable focus collar (diopter adjustment), parfocality will also depend on whether the diopter of the eyepieces has been correctly set (*see* **Note 3**). Low-power objectives usually have a longer working distance between specimen stage and objective lens (an exception being long-working distance [LWD or ELWD] high-power objectives, which are used to monitor intracellular dye injection visually). Therefore, it is always advisable to bring the specimen into focus with a low-power objective first, moving subsequently to higher magnifications and shorter working distances. Apart from magnification, lenses vary in a number of optical parameters, which are engraved on the side of the objective (**Fig. 1**). It is important to ensure that an appropriate objective is being used with respect to the microscope and the specimen preparation. The following features are usually described.

### 3.2.1. Lens Type

Lenses vary in the degree to which they are corrected for spherical aberration and wavelength transmission. With lower degrees of correction and, therefore, a lower quality, color fringes may appear at the edges of the image in certain lighting conditions (reflecting residual spherical aberration), and there may be some anomaly in the perceived depth of objects of different colors within the sample.

1. Achromatic—spherical aberration corrected for one color only, whereas chromatic aberration is corrected for two colors. Green or yellow/green illumination

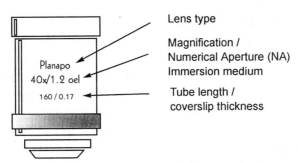

Fig. 1. Objective lenses have information engraved on their casing that gives details of their correct use. The various features that might be described are explained in **Subheadings 3.2.1.–3.2.6**.

reduces this residual aberration. Because of the incomplete color correction, photomicrographic images may not match the visually perceived image.

2. FL (semiapochromatic/fluorite)—well corrected for chromatic aberration, but may have a slight curvature of field.

3. Apo (apochromatic)—corrected for red, green, and blue spectra. This allows higher numerical apertures (NA) to be achieved, optimizing resolution and also color transmission for photomicrography.

4. Plan (plan-achromatic)—corrected for spherical aberration and, therefore, producing a flat visual field, but with minimal color correction.

5. Plan Apo (plan apochromatic)—corrected for spherical aberration and spectral continuity. These objectives may contain up to 15 separate lenses, as reflected in their price. For black-and-white photography and in the correct lighting conditions, a Plan lens may equal a Plan Apo lens in resolution. However, for color photomicrographs, the latter is preferred.

6. Fluor/Neofluar—optimized for transmission of UV epifluorescence. Unless lenses are corrected within the UV range of the spectrum, the fluorescent image and light image will be out of focal register. Fluorescent samples will usually look significantly worse with any other kind of objective.

7. Other markings, for example: LWD, ELWD—long and extra-long working distance. Ph—indicates an objective with a phase plate, which is used for phase-contrast microscopy (*see* **Subheading 3.5.1.**). The code will match a particular phase setting on the substage condenser.

### 3.2.2. Magnification

The magnification engraved on the side of the lens is an approximation. The discrepancy between the nominal magnification and the actual power of the lens may be up to $\pm \times 0.3$. This variability may be important in the calibration of some computer-based imaging systems (such as confocal microscopes) where images are usually scaled according to the nominal magnification factor, unless otherwise corrected.

### 3.2.3. Numerical Apertures (NA)

The NA of the lens is engraved beside or beneath the magnification, and reflects the optical quality of the lens and its working distance. NA is dependent on the refractive index of the medium between the lens and the specimen (*see* **Note 4**) and, therefore, is greater for immersion lenses (*see* **Subheading 3.2.4.**). Higher values also entail shorter working distances, greater light-gathering capacity, and give a higher spatial resolution (*see* **Note 5**).

Depth of field is inversely proportional to the square of the NA of the objective. Therefore, for a given magnification, if a greater depth of field is required, for example, when scanning a specimen, high-power eyepiece lenses can be combined with a lower-power objective, which will have a smaller NA. Greater depth of field can also be produced by closing (stopping down) the aperture diaphragm of the condenser lens (*see* **Subheading 3.4.1.**).

### 3.2.4. Immersion Medium

Unless otherwise marked, objective lenses do not require any immersion medium. Immersion media improve the NA of the lens, specifically allowing values >1 to be achieved. Immersion lenses are usually optimized for one particular immersion medium (immersion oil [oil, oel, H.I.], glycerol/glycerin, or water [W.I., WAS]), a drop of which is place between the objective front lens and the coverslip. Some objectives have a rotating collar, which compensates for the refractive index of different media (air = 1.00:water = 1.33:glycerol = 1.47:oil = 1.52).

The density of the immersion fluid should be as closely matched as possible to the density of the mountant. For example, tissue mounted in a permanent mountant, such as DePeX, should be viewed using oil immersion lenses (*see* **Note 6**). Because the working distance and depth of field of such lenses are small, considerable care must be taken in advancing the objective toward the specimen. It may not be possible to resolve through the depth of some particularly thick specimens. Many lenses are spring-mounted to cushion any impact, but contact with the coverslip should be avoided. It is always advisable to focus on the specimen with a parfocal low-power objective before changing to a high-power objective.

### 3.2.5. Tube Length

Most objectives are designed for a microscope whose overall tube length is 160 mm (very occasionally 170 mm). An exception is, for example, the Zeiss Axiophot system, which has infinity optics (designated by ∞).

### 3.2.6. Coverslip Thickness

Cover-glass thickness has been standardized at 0.17 mm, and many objectives have this figure engraved on their casing. Some objectives (such as those used for metallurgy) may be marked with a 0 or NCG, designating no cover-glass. For high-quality dry objectives, a correction collar may be included to compensate for variations in glass thickness (as much as ± 0.3 mm). This collar can be adjusted to maximize the contrast in the specimen. For fluorescence, coverslip thickness should fall between 0.15 and 0.19 mm.

### 3.3. Light Source and Field Diaphragm

The lamps for both bright-field (tungsten or halogen light) and fluorescent light sources must be correctly centered and focused to achieve optimal imaging. For older microscopes with a reflecting mirror under the substage condenser, the light source is focused directly onto the specimen ("critical illumination"). For the majority of research microscopes, which are designed for Köhler illumination (*see* **Subheading 3.4.2.**), the filament image is focused onto the aperture diaphragm of the substage condenser by a collector lens system which produces a homogeneous illuminated field. Apart from a variable control on the voltage passing to the light source, the field of view illuminated by the lamp is regulated by the field diaphragm, which is situated before the substage condenser in the light path (**Fig. 2**).

For epifluorescence, the image of the filament is focused directly onto the specimen (critical illumination). On some fluorescence microscopes, the images of the filament and a second reflected, virtual image can be viewed and aligned on a focusing window. On others, an image of the lamp can be focused by placing a piece of paper where the slide would be and observing the lamp directly. As with tungsten/halogen light sources, the field of illumination can be regulated via a field diaphragm. Since fluorescence may fade rapidly, it may be desirable to use the field diaphragm to reduce the scatter of light outside the area of interest. Although lamps vary, mercury vapor bulbs, once lit, should burn for at least 30 min and should not be reignited for 30 min after they are switched off. Each time the lamp is switched on and off, approx 2 h are lost from the life-span of the lamp.

### 3.4. Substage Condenser

The substage condenser contains a lens complex, an aperture diaphragm, and may also house prisms and filters used in contrast enhancement. The condenser can be focused to optimize the illumination of the specimen. The aperture diaphragm works independently of the field diaphragm to optimize contrast, depth of field, and resolution by altering the NA of the condenser lens (**Fig. 2**). The maximal NA of the condenser lens is usually engraved on its side. On some condensers, a swung-in lens can be used to alter the power of the condenser lens according to the choice of objective lens. Stopping down the aperture diaphragm will reduce the amount of light reaching the specimen, but unlike the field diaphragm, will not produce vignetting when the specimen is viewed through the microscope eyepieces.

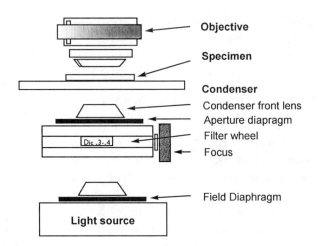

Fig. 2. The standard arrangement of light source, substage condenser, and specimen stage on a microscope designed for Köhler illumination. Two separate diaphragms, at the light source (field diaphragm) and on the condenser (aperture diaphragm), regulate the passage of light to the specimen.

As with objectives, the lenses in the condenser can be optically corrected in various ways, although this is not of primary importance for the resolution of the final image. Similarly, although it is possible to use immersion oil on virtually any high N.A. condenser lens, gains in image quality may be negligible.

### 3.4.1. Focusing the Condenser—"Köhler Illumination"

Bringing the condenser into focus is a simply a matter of bringing the edges of the field diaphragm into sharp focus **(Fig. 3A)**.

1. Focus the objective lens so that the image of the specimen is sharp through the eyepiece.
2. Where the condenser has a range of front lenses and, particularly, if it has a range differential interference contrast (DIC) prisms, check that the NA marked on the condenser lens and on the DIC adjustment collar matches that of the objective.
3. Close the field diaphragm so that its edges are in view, and center the circle of illumination with adjuster screws on the substage condenser.
4. Adjust the focus of the condenser such that the edges of the field diaphragm are sharp.
5. Open the field diaphragm to illuminate the field of interest, but to minimize the amount of stray light. The edges of the field diaphragm should remain just inside the field of view.
6. Repeat this procedure every time an objective is changed.

### 3.4.2. Aperture Diaphragm Adjustment

This adjustment matches the NA of the condenser to that of the objective, optimizing the cone of light that passes through the specimen **(Fig. 3B)**.

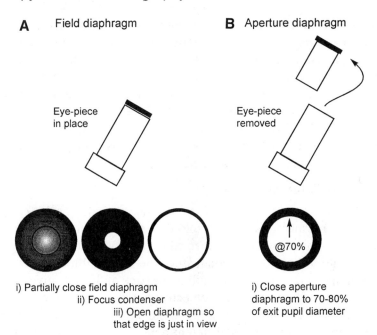

**A** Field diaphragm

Eye-piece
in place

i) Partially close field diaphragm
ii) Focus condenser
iii) Open diaphragm so
that edge is just in view

**B** Aperture diaphragm

Eye-piece
removed

@70%

i) Close aperture
diaphragm to 70-80%
of exit pupil diameter

Fig. 3. The steps involved in focusing the condenser, optimizing the field diaphragm (**A**) and the condenser aperture diaphragm (**B**) (*see* **Subheadings 3.4.1.** and **3.4.2.**). Schematic views through the eyepiece for each step are shown below.

1. Focus the condenser as above (*see* **Subheading 3.4.1.**).
2. Remove the eyepiece, and ensure that the field diaphragm is sufficiently open to fully illuminate the field of view.
3. Close the aperture diaphragm on the substage condenser so that its radius is 70–80% of the illuminated field (the "exit pupil" of the objective). Screening off the outer 20–30% of the exit pupil gives an approximate match between the NAs of condenser lens and objective lens.
4. Replace the eyepiece.
5. Repeat this procedure every time an objective is changed.

The 20–30% screening of the exit pupil is a "rule of thumb," and the final adjustment of the aperture diaphragm alters image quality in various ways. As the aperture diaphragm is closed, resolution decreases, but both contrast and the depth of field increase (*see* **Note 7**). As stated above, there should be no vignetting when the aperture diaphragm is adjusted, only a uniform variation in lighting intensity.

### 3.5. Contrast Enhancement: Phase-Contrast/DIC (Nomarski)/ Dark Field

The principle of all contrast enhancement systems is to turn otherwise invisible variations in material (density, polarizing ability, and so forth) into differences in perceived light intensity. For embryological specimens, some kind of contrast enhancement is usually extremely useful. For example, with fluorescent preparations, standard counterstains may either quench fluorescence or be autofluorescent themselves. There are occasions, however, when a particular contrast enhancement may reduce the information within an image; for example, Nomarski optics may make small labeled objects (such as cell nuclei) less easy to see. It is always worth viewing a specimen both with and without contrast enhancement before photographing. The use of three common image enhancement systems is described below.

### 3.5.1. Phase Contrast

Phase contrast converts differences in refractive index or thickness within an otherwise transparent specimen into differences in light intensity. This is achieved by illuminating the specimen with a ring of light by means of a phase "annulus" within the substage condenser. The ring of illumination is aligned with a complementary phase plate within the back of the objective. Since these elements must be in register, only objectives of the same geometrical characteristics can be used with a given phase annulus. The phase condenser may therefore contain a number of different annuli for different objectives. The phase annulus (in the condenser) and phase plate (in the objective) are aligned by means of either a focusing telescope (which replaces the eyepiece) or an Amici-Bertrand lens that can be swung into the light path. Phase-contrast optics should be used with thin specimens.

### 3.5.2. DIC or Nomarski

Differential interference contrast can be used with both thin and thicker specimens to generate a "pseudorelief" image of the preparation. DIC filters enhance local gradients in optical density by prismatically splitting then recombining polarized light. From the lamp, light passes first through a polarizing filter, and then a "Wollaston" prism within or beneath the condenser. Having passed through the specimen, light is refracted by a second Wollaston prism (usually mounted just behind the objective) and filtered by a second polarizing filter, the "analyzer" filter situated closest to the eyepieces. The analyzer is easily swung in or out, and should always be removed for fluorescence imaging, since it reduces the amount of light passing to and from the slide. The polarizer can be rotated to optimize the DIC effect. For good DIC, the field and aperture diaphragms should be optimized for Köhler illumination (*see* **Subheadings 3.4.1.** and **3.4.2.**, **Fig. 3A,B**).

Since this system utilizes polarized light to generate changes in light amplitude, naturally polarizing media, such as mica and Plexiglas, may generate optical artifacts when placed in the light path. In addition, some biological materials, for example, structures consisting of parallel oriented fibers, are also inherently polarizing or "birefringent." With only the polarizer and analyzer filters of the DIC system present, birefringent structures will therefore appear as either very bright or very dark if oriented at 90° to their inherent optical axis. Birefringence may be useful in identifying particular kinds of structures within a specimen.

### 3.5.3. Dark Field

Dark-field or oblique illumination is used to observe dark, particulate staining or for example, darkly labeled axons against a homogenous bright-field background. By producing what is effectively a negative image of the specimen, such features become clearer as bright objects against a dark background. The condenser directs an oblique illumination through the specimen, and the objective lens only receives light if it is reflected or refracted by opaque objects in the illuminated field. Where parts of the specimen are transparent, light passes uninterrupted through object space and does not enter the objective. Darkly stained particles or fibers within the specimen interrupt the light path and appear as bright objects.

Certain stains are considerably clearer under dark-field illumination, despite there being no real gain in spatial resolution. However, specimens have to be carefully prepared: the image will be impaired if tissue is not well cleared (the background will not appear black), or if the slide is contaminated with grease or dust (which will be visible as very bright points of light).

## 3.6. Fluorescence Microscopy

Fluorescent labeling of specimens offers numerous advantages, not the least of which is the use of multiple labels linked to different fluorochromes on the same preparation. Fluorescent specimens can be viewed using a compound microscope fitted (usually) with an epifluorescent lamp (for epi-illumnation, the light is both projected and detected through the objective lens), filter blocks which contain dichroic beam-splitting mirrors, and objectives designed to transmit fluorescence (*see* **Subheading 3.2.1.**). The basic principle is to excite a fluorochrome within a specimen with wavelengths of UV light that are optimized with respect to the absorption spectrum of that particular molecule. The resulting fluorescence emission is then filtered according to the appropriate maxima of the emission spectrum for that fluorochrome. Where two fluorochromes are being used, filtering can be optimized to ensure that the emission spectra of each label do not overlap.

### 3.6.1. Choice of Fluorochromes

In most experimental situations, a choice of fluorochromes is available. Where two fluorochromes are being used in the same specimen, it is important that their fluorescence signals can be clearly distinguished (for example, different populations of dye-labeled axons or different immunolabels). It is important to compare the specifications of the excitation and emission filters with the absorption and emission spectra of the fluorochromes chosen. Other factors that should be taken into account are the brightness and photostability of a particular fluorochrome. Some fluorescent dyes survive fixation better than others. For example, for intracellular dye injection, the highly fluorescent 5,6-carboxyfluorescein cannot be fixed whereas the relatively less bright lucifer yellow can be cross-linked by paraformaldehyde fixation.

### 3.6.2. Sample Preparation

Some special conditions should be noted for fluorescent samples. Glutaraldehyde is auto-fluorescent and so its use should be avoided. If there is no alternative to this fixative, its autofluorescence can be countered by treatment with $NaBH_4$. Dyes may be pH sensitive; for example FITC fluorescence yield is significantly reduced if the pH of the mountant is <8.5–9.0. Some counterstains are autofluorescent (e.g., Feulgen and Nissl stains). Fluorescent latex beads, which can be used as axonal tracers, or as implants for the slow release of a variety of soluble factors, are dissolved by alcohol and xylene (*see* **Note 8**). In general, background fluorescence increases with time stored. Background can be reduced by keeping tissue at 4°C, but it is advisable to view and, in particular, photograph fluorescent material as soon as possible. Permanent, nonaqueous mountants tend to transmit more light than glycerol-based mountants decreasing photographic exposure times and hence the risk of photobleaching.

### 3.7. Confocal Microscopy

Confocal microscopes use a scanning laser epi-illumination to excite fluorochromes within the specimen and detect the resulting fluorescence by means of a photomultiplier tube (PMT). The performance of the microscope is ultimately limited by the quality of its conventional optical components, and the same considerations for optimization exist as for conventional fluorescence. Confocal microscopy requires objective lenses of high NA. The range of possible excitation wavelengths depends on the sophistication (and hence cost) of the laser light source. As for conventional fluorescence imaging, filtering both the excitation and emission spectra relies on dichroic beam-splitting mirrors.

The advantage of confocal microscopy is in the ability to focus on a defined optical plane at a given depth within the tissue. This is achieved by collecting light through a small pinhole in front of the PMT. Reflected light from outside the focal plane diverges and falls outside the pinhole aperture. The smaller the pinhole, the narrower the thickness of the optical section and the greater the spatial resolution. The optical section gives a very clean image without interference from scattered light outside the plane of focus. However, narrow optical sectioning also results in an image that is much dimmer than a conventional fluorescence image. Full fluorescence intensity is restored when thin optical sections are taken in series through a specimen and then digitally recombined to produce a projected image through the entire depth of the tissue. Since each element of the resulting projection is in sharp focus, there will be less background than with a conventional fluorescence image.

PMTs can also be used to collect bright-field pictures by collecting transmitted light gathered through the condenser lens. Such images lack any color, contrast information and their resolution depends on the optical quality and correct focusing of the condenser.

### 3.7.1. Practicalities of Digital Image Acquisition

The aim of confocal microscopy is the production of an optimized black-and-white digital image. Subsequent image manipulation can enhance the image by addition of pseudocolors, but the amount of information must be maximized at the point of image acquisition. The quality of the PMT determines the resolution (number of pixels) and number of gray shades in this final image (for example, an 8-bit image has 256 gray levels). However, regardless of the type of PMT, the aim is always to utilize the available gray levels fully. The points of interest in the image must be spread over the maximum tonal range. In some cases, this may involve setting a nonlinear collection ramp such that background is either suppressed or the intensity of faint objects within the image enhanced. Performing nonlinear filtering after the image is collected will only result in a loss or distortion of the original tonal information. Optimizing the image always involves setting an arbitrary "black-level" at an intensity of 0, and ensuring that the brightest elements of the image fall just below or at the maximum available pixel brightness (saturation point) by setting a gain control. Too much gain produces a saturated image that contains very little tonal information and appears very "flat." Where the fluorescence emission is low, increasing gain also leads to increased noise levels in the PMT. Noise can be compensated for by collecting a final image that is an average of successive scans of the same view. The number of scans to be averaged depends on the gain of the PMT and can be set empirically by determining how many averaged passes are required before the image stabilizes. These principles also apply to other digital image capture devices.

### 3.8. Photomicrography

Cameras used for photomicrography generally consist of either a standard "single-lens-reflex" (SLR) 35-mm camera body mounted via a tube to the microscope (which uses its own integral exposure meter) or consist of a specially designed unit that is integrated into a dedicated "photomicroscope" (which will generally have more sophisticated exposure controls and which may use either 35-mm or larger format film). Although film is still widely used, digital image acquisition is increasingly common. However, regardless of the type of camera, the central principle of photomicrography is always to optimize the image (in terms of contrast, resolution and color) at the point of collection, be it on a photographic negative, CCD camera, video camera, or PMT. This must be achieved by having the correct focus and lighting conditions for optimal contrast and color faithfulness.

Since microscope cameras have a fixed aperture, exposure times depend only on lighting levels and the sensitivity of the digital collection device or photographic media. Long exposure times, particularly in fluorescence photomicrography, are usual. For digital cameras, long exposure times result in problems related to thermal noise. Consequently, images are usually averaged (*see* **Subheading 3.7.1.**), and highly sensitive CCD chips are used at low temperatures. For film, extended exposures (beyond as little as perhaps 2 s) result in particular problems for color photomicrography relating to reciprocity failure (*see* **Subheading 3.9.1.**).

### 3.8.1. Focusing the Camera

Light from the specimen must be directed to the camera and focused onto the surface of the film. This usually involves redirecting or splitting the light path from the eyepieces to the camera tube and ensuring that what is in focus for the eyepiece is also in focus on the camera. Such parfocality cannot always be assumed for a microscope with a 35-mm camera attachment via a camera tube, and it is advisable to check that the image is correctly focused for the camera through the eyepiece of the SLR camera body. Photomicroscopes are designed so that camera and eyepieces are parfocal when the diopter of each eyepiece is correctly adjusted for the observer (*see* **Note 3**). This is usually achieved by focusing the crosshairs of a reticule placed or projected onto the eyepiece diaphragm. The diopter ring is adjusted such that the reticule image is sharp when the eyes of the observer are relaxed. The simplest way to relax eye muscles is to glance briefly at a distant object before focusing the crosshairs. When the specimen is focused, crosshairs should also remain sharply in focus. Particular care must be taken with objectives of a low magnification and especially with dissection microscopes, where the increased depth of field allows the observer's eyes to accommodate across a greater depth within the specimen. In such situations, a "focus telescope" may be useful to focus accurately on a specific optical plane.

### 3.8.2. Black and White Photography

Black and white is still widely used for publication of photomicrographs and offers certain advantages over color photography. Monochrome films are less sensitive to variations in exposure times and, in particular, are unlikely to suffer reciprocity law failure where exposure times are under 1 min (*see* **Subheading 3.9.1.**). Films vary in their speed (measured by ASA or DIN) reflecting their sensitivity to light: generally, faster film speeds result in a coarser-grain image. Films may also vary in their gradation or contrast, which can be characterized by the "$\gamma$" of their emulsion. Within a reasonable exposure time, the response of an emulsion to increasing exposure is linear. Gamma ($\gamma$) is defined as the tan of the angle of the linear portion of this response curve ($\alpha$) and is a useful guide to the contrast of the film. If $\gamma > 1$ ($\alpha > 45°$), then the emulsion has an enhanced reaction to changes in intensity and, therefore, a "hard" contrast. If $\gamma < 1$ ($\alpha < 45°$), then the film has a "soft" contrast. Emulsion may also react slightly differently to different colors of light. A panchromatic film will have an approximately even sensitivity across all wavelengths, but may still be less responsive to deep-reds. A selection of films used in black-and-white photomicrography is as follows.

> T-MAX 100 ASA (soft contrast)
> ILFORD FP4 (intermediate contrast)
> KODAK TECHNICAL PAN 125 ASA (high contrast)
> T-MAX 400 ASA (for fluorescence)
> AGFA SCALA 200 ASA (reversal film for B&W transparencies)

#### 3.8.2.1. FILTERS

Black-and-white photomicrography allows the use of color filters to optimize the contrast of objects within the specimen. Color filtering can be used to convert differences in color contrast into differences in perceived intensity. An absorption filter of one color will remove wavelengths of the complementary color deepening the contrast between such objects and the background. For example, a blue-green or cyan filter will darken the reddish-brown (DAB) product used in histological peroxidase reactions. The effects of different colors of contrast filters are shown in the following table (adapted from **ref. 1**).

### 3.9. Color Photography

The most frequent problem encountered in color photomicrography is matching the colors recorded on film to those perceived by the observer through the eyepiece. In particular the color of the photographic image will depend on the color temperature of the light source. Tungsten filaments give off a reddish light, which will not be obvious to the observer, but will be clear in photomicrographs taken with a standard "daylight"

**An object stained:**

| Filter used | Blue | Green | Yellow | Red |
|---|---|---|---|---|
| | | **becomes:** | | |
| Red | Very dark | Dark | Light | Very light |
| Green | Dark | Very light | Light | Very dark |
| Blue | Very light | Dark | Dark | Very dark |
| Cyan | Light | Light | Dark | Very dark |
| Magenta | Light | Very dark | Light | Light |
| Yellow | Very dark | Light | Very light | Light |

film. Films designed for tungsten light sources (e.g., Fujichrome 64T) are therefore adjusted for a color temperature of or 3200 K. Some microscope lamps have a "PHOTO" or "3200 K" setting, which produces a standardized bright illumination for photography. Manufacturers may alternatively supply a color temperature vs voltage curve, which enables lamp intensity to be adjusted accordingly. The high color temperatures required for color constancy (usually equivalent to a lamp burning at its brightest) may demand impossibly rapid shutter speeds. Since altering the brightness of the lamp would also change the color temperature of the illumination, light intensity can only be reduced by placing neutral density filters in the light path. These absorb energy equally across the spectrum and can be used additively to achieve a reasonable photographic exposure time without changing color temperature.

Daylight films are adjusted for a color temperature of 5500–6000 K and when used with tungsten illumination, a blue filter (Kodak Wratten 80A) can compensate for the orange cast of artificial lighting conditions **(Fig. 4A)**. Daylight films should be used for color fluorescence photomicrographs (UV lamps have a higher color temperature), although, unless two different fluorochromes are being photographed, black-and-white film generally gives better resolution and exposure reliability than color film (*see* **Subheading 3.9.1.**). Typical films are listed below. All are color-reversal films (producing transparencies).

> FUJI 64 T (for standard tungsten illumination)
> EKTACHROME 400 ASA (fluorescence)
> KODAK P1600X (fluorescence)

---

Fig. 4. *(opposite page)* **(A)** Color temperature: The effect of a blue filter when a daylight film is used to photograph a specimen under tungsten illumination. The filter compensates for the pinkness of the illumination and increases the contrast. **(B)** Pseudocolor: Three examples of the use of pseudocolor with black-and-white digital images. (Top) A pseudocolor LUT is used to convert the gray scales within a "sharpened" confocal microscope image of a DiI-labeled embryonic brain into changes in color complements. (Middle) Black-and-white

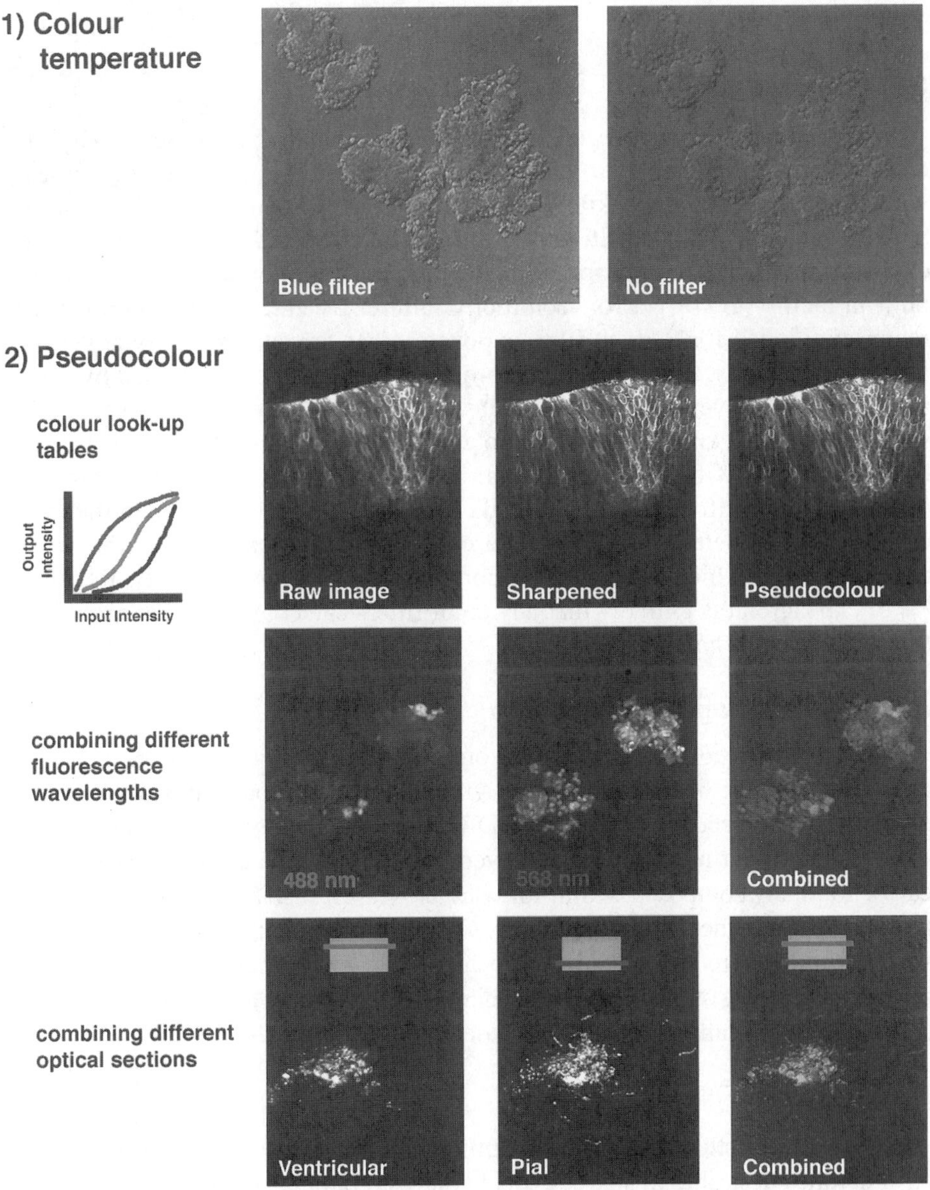

**1) Colour temperature**

Blue filter — No filter

**2) Pseudocolour**

colour look-up tables

Output Intensity / Input Intensity

Raw image — Sharpened — Pseudocolour

combining different fluorescence wavelengths

488 nm — 568 nm — Combined

combining different optical sections

Ventricular — Pial — Combined

Fig. 4. *(continued)* images of the same specimen illuminated for different fluorescent dyes are color-coded and merged to produce an image equivalent to a photographic double exposure. This image is then merged with a DIC picture of the same specimen. (Bottom) Confocal microscope optical sections are color-coded and merged to show the differences in cell dispersal at different depths in the tissue. Pseudocolor pictures were prepared within Adobe Photoshop from 8-bit $768 \times 512$ pixel images produced on a Bio-Rad MRC-600 confocal system mounted on a Nikon Diaphot inverted compound microscope. (*See* color plate 4 appearing after p. 368.)

### 3.9.1. Reciprocity Failure

When exposure times exceed much more than maybe 2 s, the sensitivity of color-reversal films begins to change. Effectively the film's speed decreases with time, a phenomenon known as failure of reciprocity (the relationship between exposure and sensitivity). This can be corrected for by multiplying exposure times by the appropriate factor and most photomicroscopes have a built-in facility to correct for reciprocity failure. By entering the appropriate reciprocity index for a given film, exposure times are automatically adjusted for long exposures. If necessary, reciprocity failure can be determined by trial and error for a given film type (increasing exposure 2× on successive exposures). Between 2 and 8 s, there is also a shift in the sensitivity of the different color pigments within the emulsion. For example, as the green pigment becomes less sensitive, the film will take on a purplish tinge. Film manufacturers may supply details of appropriate color compensation filters within this exposure range, which can be used to correct shifts in color sensitivity. Beyond 8 s of exposure, it is unlikely that any color filters can compensate entirely for the shifts in color balance *(2)*.

## 3.10. Digital Image Manipulation

Having produced the optimal photomicrograph, computer software packages allow a range of further image manipulation. The superimposition of, for example, a fluorescence image and a DIC illuminated view of the same specimen, which might normally be achieved by a photographic double exposure, can be routinely composed within software programs, such as Adobe Photoshop (Adobe Systems Inc., Mountain View, CA). Further manipulation can involve spatial filtering to reduce noise or sharpen contrast, and perhaps most frequently, the pseudocoloring of black-and-white digital images. These techniques also raise the possibility of fairly sophisticated "touching up" of data.

### 3.10.1. Spatial Filtering

Even with a picture that has been optimized for contrast and illumination, it is sometimes desirable to sharpen local contrast gradients or smooth an image to remove random noise. A common filtering strategy, called a kernel operation, is to compare the intensity value of a pixel with that of its immediate neighbors, and use any differences to increase or reduce the intensity of this central pixel by a computed factor. The degree of filtering depends both on the algorithm used by the kernel operation and the area over which pixel values are sampled.

1. Sharpening filters (*see* **Fig. 4**) produce a crisper image by accentuating differences in intensity between a given pixel and its neighbors. For example, a

Laplacian filter equalizes the intensity over areas of low contrast and accentuates the intensity changes where gradients in pixel intensity are sharp. This produces an image composed largely of edge information (effectively a first-order derivative of the original), which when subtracted from the original pixel values, produces an image in which boundaries and the contrast of fine structures are enhanced.

2. Noise reduction filters remove the random pixel values that may be generated when, for example, a digital camera is working at maximum gain. Such noise is usually reduced during digital image collection by averaging successive images of the same field. If such real-time filtering is not practicable, or still leaves a noisy image, random intensity fluctuations can be filtered post hoc. Noise reduction filters take a block of pixels of a predetermined size, and replace the intensity value of the central pixel with a value based on the average or median pixel intensity within the block. This has the effect of eliminating rogue, high or low single pixel values. Algorithms based on averages may result in an unacceptable loss of contrast or even the appearance of pseudoresolution artifacts *(3)*. However, such artifacts can be avoided using filters that "rank" rather than average pixel values, such as the "median" filter included in many software packages.

### 3.10.2. Pseudocolor

Pseudocolor is used to accentuate the information in black-and-white images by translating differences in intensity into differences in color. Three examples of its use in confocal microscopy are given in **Fig. 4**. In the first, a confocal image of cells within a chick hindbrain stained with the fluorescent dye DiI is first sharpened and then intensity values converted into color differences. The color conversion is achieved by differentially adjusting the relationship between the original intensity of the black-and-white image (input) and the intensity of red, green, and blue brightness (output) for a given input value. This relationship is shown on the graph to the left where "input" is on the $x$-axis and "output" is on the $y$-axis. This kind of plot is known as a color or output look up table (LUT) and can usually be user-defined within a given software package.

In the second example, an aggregate of heterogeneously labeled chick hindbrain cells was scanned with two different excitation wavelengths (488 nm for green/yellow and 568 nm for red/yellow fluorescence). This reveals two populations of cells. The black-and-white images have been pseudocolored and then recombined. In the 488 nm image, red and blue were completely removed from the image leaving a green intensity spectrum. In the 568 nm image, all green and blue were removed from the image, leaving red. The combined green and red fluorescence images were then superimposed onto a DIC filtered view of the same preparation.

In the third example, pseudocolor coding is used to identify populations of similarly labeled cells at different depths through a preparation. A cluster of cells, all labeled with DiI (which emits a red/orange fluorescence), were scanned at two different depths through the thickness of a chick hindbrain using a confocal microscope. The two black-and-white optical sections were then color-coded in red and green according to depth and recombined. This allows the relative dispersal of labeled cells at different layers of the developing brain to be contrasted. This kind of image could not be produced by conventional double-exposure film photomicrography.

### 3.10.3. Excessive Digital Manipulation

Information in a digitized image can be altered, almost seamlessly, to produce "cleaner" results. A digital paintbox can remove dust and scratches from a scanned 35-mm transparency, compensate for uneven illumination, or produce a uniform background color. Although this may appear similar to the traditional "dodging and burning" used during black-and-white printing, the ease and extent to which a digital image can be altered are considerably greater. Similarly, since in many laboratories images are routinely generated on computer for publication, the opportunities for excessive image manipulation have substantially increased. Ultimately, in such an environment, producing reliable data is a matter of personal responsibility. Some self-evident principles are:

1. Use sharpening, edge detection, or noise reduction filters globally across the image.
2. Preserve the data intact. Never digitally paint over areas that show data in an attempt to "clean up," for example, staining patterns. Never move or "clone" areas of the image where data are represented.

### 3.11. Cleaning Optical Components

Simple precautions will reduce the need for cleaning optical components within the microscope. Most obviously, dust covers should be used for all microscopes. Glass components should be handled with great care and should never be in contact with bare skin. When cleaning objective lenses, immersion oil can be drawn off by applying lens tissue to the edge of the front lens. It should only be necessary to apply very light, if any, pressure to the lens itself. If the lens is to be cleaned directly with lens tissue, then movements should always be outward from the center of the lens. It is important to use fresh lens tissue for every action. If a cleaning solvent is needed, then it is important to refer to the recommendations of the lens manufacturer: some solvents will dissolve the glue, which holds the front lens in place. Dry objectives should never be used with immersion fluid. Other coated glass components, such as fluorescence filters, should never be handled and should not be cleaned with lens tissue or solvents. Pressurized air in containers contains a solvent that may condense on surfaces that are held too close to the jet of air.

### 3.12. Troubleshooting

Related section numbers are given in parentheses.

1. Cannot see an image, or image is vignetted.
    a. Check that light path to specimen is clear. Is the specimen evenly illuminated?
        i. Is a condenser lens of the correct maximum NA in place? For lower magnifications, it might be necessary to remove "swung-in" lenses—objectives of ×1 magnification are normally used without a substage condenser.
        ii. Is the dark-field condenser prism out of the light path?
        iii. Are neutral density or color filters fully out of the light path?
        iv. Is field diaphragm sufficiently open?
        v. Is aperture diaphragm sufficiently open?
        vi. Is DIC analyzer filter swung out (particularly for fluorescence illumination)?
    b. Check that light path to eyepieces is clear, and in particular, that the beam is being directed to the eyepieces and not to the camera.
    c. For fluorescent specimens, check that the appropriate filter combination is in place, and that the fluorescent lamp is correctly aligned and focused.
2. Cannot bring the specimen into focus at high power.
    a. Is the thickness of the specimen too great for NA of the objective lens (**Subheading 3.2.3.**)? Specifically, is the objective touching the slide?
    b. Is a dry objective being used incorrectly with immersion oil or vice versa (**Subheading 3.2.4.**)?
    c. Is the immersion medium matched to the refractive index of the mountant (**Subheading 3.2.4.**)?
    d. Inverted compound microscopes only—is slide placed correctly with cover-glass facing downward?
3. Insufficient contrast:
    a. Focus condenser to achieve "Köhler illumination" (**Subheading 3.4.1.**).
    b. Adjust aperture diaphragm of substage condenser to correct NA (**Subheading 3.4.2.**).
4. DIC not producing a "pseudorelief" image (**Subheading 3.5.2.**)
    a. Is the analyzer filter in place?
    b. Are DIC prisms in the substage condenser matched to the NA of the objective (in more sophisticated microscope systems)?
    c. Are aperture and field diaphragms correctly adjusted (*see* **Note 3**)?
    d. Is polarizer rotated to achieve optimal illumination?
5. Photomicrograph exposure is seemingly infinite:
    a. Is all light (particularly for fluorescence) being directed to the camera (i.e., no image should be seen through the eyepieces)?
    b. Is film speed set correctly?

c. Are camera batteries exhausted (where a separate 35-mm camera body is attached via a tube)?
6. Color photomicrographs have variable color balance and brightness.
   a. Was camera voltage supply set to PHOTO, 3200 K or a constant voltage?
   b. Was an artificial (tungsten) light source film used?
   c. If using "daylight" film (*see* **Subheading 3.9.**), was a compensating blue filter (Kodak Wratten 80A) in place?
   d. Were exposure times longer than 2 s? If so:
      i. Was the correct reciprocity index programmed into camera exposure mechanism with respect to film type?
      ii. Were the correct color compensation filters used?

## 4. Notes

1. Specimens that are to be further processed for *in situ* hybridization with RNA probes or that are labeled with a volatile dye, such as DiI, can be mounted in fixative in the same way.
2. A general exception is ×1 objectives, which are highly unlikely to be parfocal with any other objective lens.
3. Each eyepiece has a diopter adjustment ring, which can be aligned with engraved markings on the eyepiece tube. Observers who would normally wear glasses should adjust the diopter with respect to each eye. With the ring aligned to the zero position, the specimen should be brought into focus at high power. To achieve parfocality, the same field should be viewed at low magnification and the specimen focused again using the diopter adjustment ring on the eyepiece. This should be repeated for each of the lower-power objective lenses and separately for each eyepiece.
4. NA is calculated as the product of the refractive index of the medium between the objective lens and the specimen multiplied by the sine of the half-angle of the cone of light entering the objective. This cone becomes narrower (and therefore the NA smaller) as the working distance increases *(4)*.
5. Spatial resolution is proportional to the wavelength of transmitted light divided by the NA.
6. For an oil immersion lens, using a mounting medium with a refractive index lower than glass or oil will result in a corresponding loss in numerical aperture (effectively that achievable with the less dense medium). Any area of low refractive index in the object space, e.g., a water or air bubble, will have this effect.
7. Stopping the aperture diaphragm right down can produce "false resolution," a diffraction image that gives the impression of a fine ultrastructure within the specimen.
8. If dehydrating through an alcohol series, reduce the overall exposure to approx 2 min.

## References

1. Langford, M. J. (1979) *Basic Photography*, 2nd ed. Focal Press, London, p. 376.

2. Langford, M. J. (1980) *Advanced Photography,* 4th ed. Focal Press, London, p. 355.
3. Russ, J. C. (1995) *The Image Processing Handbook,* 2nd ed. CRC Press, London, UK, p. 674.
4. James, J. (1976) *Light Microscope Techniques in Biology and Medicine,* Martinus Nijhoff Medical Division, The Hague, The Netherlands, p. 336.

# Index